Power Systems

For further volumes:
http://www.springer.com/series/4622

Hartmut Spliethoff

Power Generation from Solid Fuels

Springer

Dr. Hartmut Spliethoff
TU München
Institut für Energiewirtschaft
und Anwendungstechnik
Arcisstrasse 21
80333 München
Germany
spliethoff@tum.de

ISSN 1612-1287
ISBN 978-3-642-02855-7 e-ISBN 978-3-642-02856-4
DOI 10.1007/978-3-642-02856-4
Springer Heidelberg Dordrecht London New York

Library of Congress Control Number: 2009942919

Cover design: deblik, Berlin

Printed on acid-free paper

Springer is part of Springer Science+Business Media (www.springer.com)

Preface

Today, fossil fuels dominate worldwide primary energy consumption. In 2000, about 40% of total primary energy was used for electricity generation, and of this, coal was the fuel for 40%, making it the most important primary energy carrier for power production. Forecasts of future energy consumption predict a further increase of worldwide coal utilisation in the coming 20 years. In comparison to natural gas and oil, coal has the advantage of being the most abundant fossil energy carrier.

Fossil fuels are the major source of CO_2 emissions and cause global warming with all its negative impacts. It is generally accepted today that huge efforts have to be undertaken to limit the emissions of CO_2 and to reduce the impact of global warming. Mitigation scenarios indicate that this can only be achieved if all options for CO_2 reduction are followed. The principle possibilities for reducing CO_2 emissions are more efficient energy utilisation, the substitution of fossil fuels by renewable energies or nuclear energy and carbon capture.

It is the intention of the author to explain the technical possibilities for reducing CO_2 emissions from solid fuels. The strategies which will be treated in this book are more efficient power and heat generation technologies, processes for the utilisation of renewable solid fuels, such as biomass and waste, and technologies for carbon capture and storage.

The book introduces the different technologies to produce heat and power from solid fossil (hard coal, brown coal) and renewable (biomass, waste) fuels, such as combustion and gasification, steam power plants and combined cycles. The technologies are discussed with regard to their efficiency, emissions, operational behaviour, residues and costs. Besides proven state of the art processes, the focus will be on the potential of new technologies currently under development or demonstration.

Chapter 1 gives an overview of current worldwide primary energy consumption and its future development. The impact of CO_2 emissions on global warming is summarised and the strategies for CO_2 reduction are identified.

Chapter 2 deals with the origin and classification of solid fuels. Reserves of solid fossil fuels are indicated and the energy potential of biomass and waste is estimated. The fuel properties are characterised with regard to thermal conversion processes.

Chapter 3 provides the thermodynamic fundamentals of the thermal cycles which are required to convert the chemically bound energy of the fuels into power.

The focus of Chapter 4 is the technology of the steam power plant, which is the dominant process for power plants. The fundamentals of steam generation are introduced and the design principles of a conventional state-of-the-art steam power plant are explained. In comparison to this reference plant, the different possibilities for efficiency increase and the impact of advanced steam conditions on the steam generator is discussed. A summary of the design data of the most advanced operated power plants in the world is included in the outlook for the further development of steam power plants.

Chapter 5 deals with combustion, which is the dominant technology of fuel conversion. Starting from the principles of solid fuel combustion and the fundamentals of pollutant formation, the different combustion technologies of fixed bed, fluidised bed and pulverised fuel combustion are compared. Emission reduction technologies, either primary measures within the combustion process or secondary flue gas cleaning, are examined. Operational problems such as slagging, fouling and corrosion, which have to be related to ash properties and process conditions and which are of great importance for solid fuel combustion, are discussed. The production of mineral residues is inevitable in solid fuel combustion; the options to use the residues are described.

Although the technologies for biomass and waste conversion follow the same principles as for coal, substantial differences arise due to the differing fuel quality and the smaller capacity of such power plants. Therefore, biomass and wastes are treated separately in Chapter 6. Besides biomass combustion, biomass gasification, waste combustion and co-combustion technologies are the focus of this chapter. It explains how ash-related problems in biomass and waste conversion are even more pronounced than for coal and will effect the operation of biomass/waste plants and limit the electrical efficiency. Co-utilisation of biomass in coal-fired power stations is a further process option, and the impact on emissions and operational problems is discussed.

Gas turbine-based combined cycles for natural gas offer the highest efficiencies in power generation, of up to about 60%. The focus of Chapter 7 is to show the state of development of combined cycle processes for solid fuels. After describing the technology of natural gas-based combined cycles, the processes, the potentials and the development stages of the integrated gasification combined cycle (IGCC), the combined cycle with pressurised fluidised bed combustion (PFBC), the combined cycle with pressurised pulverised coal combustion (PPCC) and the externally fired combined cycle (EFCC) will be explored.

Along with the efficiency increases and the use of renewable energy sources, CO_2 capture and storage methods offer a possible means of CO_2 reduction in fossil fuel-fired power plants. Chapter 8 gives an overview of the options for CO_2 separation, transport and storage for power plants.

This book developed over the years of my activities at the University of Stuttgart, the Technical University of Delft and now the Technical University of Munich. Results from various research projects are included in the book. The basis of this book was my habilitation "Combustion of solid fuels", which was published in 2000 in German. Since that time, a lot of new developments have emerged, while other areas within the field have progressed only slightly. This is reflected in the book.

I would like to thank all those who provided materials, contributions and comments to the different chapters of this book: Dr. Oliver Gohlke, Dr. Michael Müller, Dr. Arnim Wauschkuhn, Mr. Sven Kjaer, Mr. Helmuth Brüggemann, Mr. Kendel, co-workers from my chair Energy Systems at the Technical University of Munich and my colleagues from my former employers the Technical University of Delft and the University of Stuttgart. Furthermore, I would like to thank Herbert Rausch for translations and Patrick Lavery for proofreading. Special thanks go to Mrs. Brigitte Demmel for requesting copyrights and Mrs. Korinna Riechert for drawing figures.

München
August 2009

Contents

List of Figures

List of Tables

List of Symbols

Symbol	Unit	Meaning
a	%	part load
A	m^2	cross section, surface
b	J/kg	specific anergy
b	m	width
U	flh/a	utilisation factor (full-load operating hours per year)
c	kJ/(kg K)	specific heat capacity
c_p	kJ/(kg K)	specific heat capacity at constant pressure
C	$kJ/(kg \cdot K)$	specific heat capacity
C_f	e/GJ	fuel costs
$C_0 = 5,77 \times 10^{-8}$	$W/(m^2\ K^4)$	coefficient of radiation of the black body
CoC	1/a	cost of capital
d	m	diameter
e	J/kg	specific exergie
h	m	height
h	J/kg	specific enthalpy
H	J	enthalpy
HR	kJ/kWh	heat rate
HHV	kJ/kg	higher heating value
I	e	investment costs
K_o	–	Konakow number
k	kg/s	reaction velocity
LHV	kJ/kg	lower heating value
m	kg	mass
$\underline{m}$	kg/s	mass flow
n	–	number
P	W	power
P	m	perimeter
p	bar	pressure

Q	J	heat
$\underline{Q}$	W	heat flux
q	J/kg	specific heat
$\underline{q}$	W/m^2	specific heat flux
$\bar{R}$	J/(mol K)	general gas constant
$R1$	–	efficiency criteria for waste
S	J/K	entropy
s	J/(kg K)	specific entropy
s	m	length
T	K	thermodynamic temperature
t	s	time
t	°C	temperature
t_{P}	m	tube pitch
u, v, w	m/s	velocity components
V	m^3	volume
W	J	work
w	J/kg	mass-related work
x	–	steam mass fraction
β	Grad	helix angle
ε	–	emissivity
ζ	–	exergetic efficiency
η	–	efficiency
κ	–	loss
κ	–	adiabatic coefficient
λ	–	air ratio, stoichiometry
υ	–	stoichiometric coefficient
Φ	kg/(m^2 s)	mass flow density

Indices

1,2, j	states
12	state change 1–2
0	base case, without losses
aux	auxiliary
a	ambient
A	air
Ad	adiabatic
b	boundary
B	boiler
Chem.	chemical
diff	diffusion
el	electrical

F	fuel
F	furnace
FE	furnace exit
FG	flue gas
FL	flame
FW	feed water
FW	furnace wall
Gen	Generator
GT	gas turbine
i	inner
i	isentropic
l	lower
LS	live steam
m	mechanical
m	mean
ne	net
p	particle
P	pipe
RC	radiation convection
S	steam
S	slag
ST	steam turbine
tot	total
th	thermal cycle
T	turbine
u	upper
U	unburnt
W	wall
W	water

Chapter 1
Motivation

1.1 Primary Energy Consumption and CO_2 Emissions

1.1.1 Development of Primary Energy Consumption in the Past 40 Years

The global consumption of primary energy has been marked by a strong increase in the past 40 years. Figure 1.1 presents the development of primary energy consumption, broken down into groupings, namely industrial countries of the OECD; former Soviet Union; and emerging economies (i.e. developing countries). In 1965, the worldwide consumption of primary energy amounted to only 3,860 MTOE (million tonnes of oil equivalent); by 2005, it had increased to 10,224 MTOE. This corresponds to an increase of 172% or an annual rate of increase of 2.5% (BP 2008). In industrial countries, the increase was around 107% for 40 years, corresponding to an annual rate of increase of almost 2%. In the emerging economies, which started from a lower absolute level, the increase was 640%, which corresponds to an annual rate of increase of more than 5%.

Figure 1.2 shows the share of primary energy consumption of the different countries and regions for the year 2005. A conspicuous fact here is the high share of North America, where in the USA alone almost a quarter of the entire primary energy of the world is consumed.

In 2005, the fossil energy sources, i.e. crude oil, natural gas and coal, comprised 87% of primary energy consumption (see Fig. 1.3).

1.1.2 Developments Until 2030

Predictions of the development of primary energy consumption are based on various assumptions about the total population, the gross national product and the energy efficiency of different countries and are highly dependent on general political conditions. The following shall present predictions of the development of the energy demand up until 2030, which predominantly rely on an extrapolation of the current developments and general conditions. The effect of political measures introduced

H. Spliethoff, *Power Generation from Solid Fuels*, Power Systems,
DOI 10.1007/978-3-642-02856-4_1,

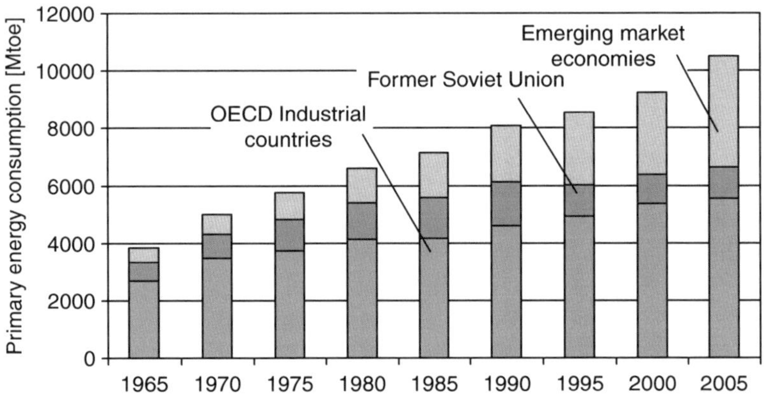

Fig. 1.1 Global primary energy consumption 1965–2005 by country groupings (BP 2008)

until now is reflected; future possible and also probable measures are not taken into consideration. The reference scenario of the International Energy Agency (IEA) in 2006, for instance, assumes a growth of the world population to 8.1 thousand million up to the year 2030 (2004: 6.4 thousand million), an increase of the gross national product of 4% at the beginning, levelling off at 2.9% per year, and natural oil prices of somewhat more than $60 per barrel (real price 2005). Technologies of power supply and energy utilisation (end-use technologies) become ever more efficient.

The predictions illustrated in Figs. 1.4, 1.5, 1.6 and 1.7 of global primary energy demand, electric power generation, installed power plant capacities and CO_2 emissions rely on data of the IEA and the US Department of Energy (DoE) (IEA 2002,

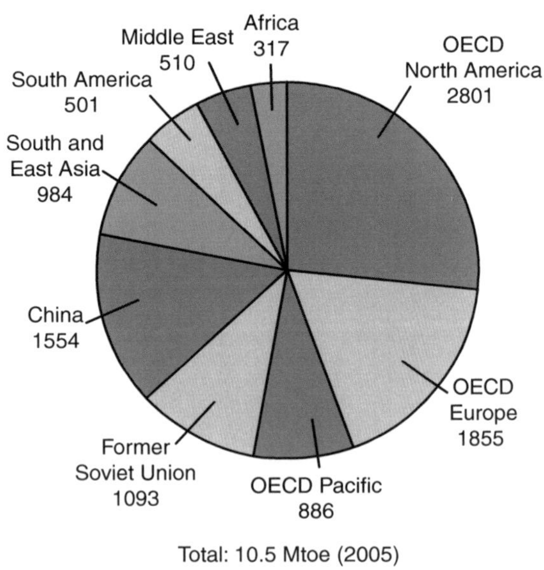

Fig. 1.2 Primary energy consumption in 2005 by regions and countries (BP 2008)

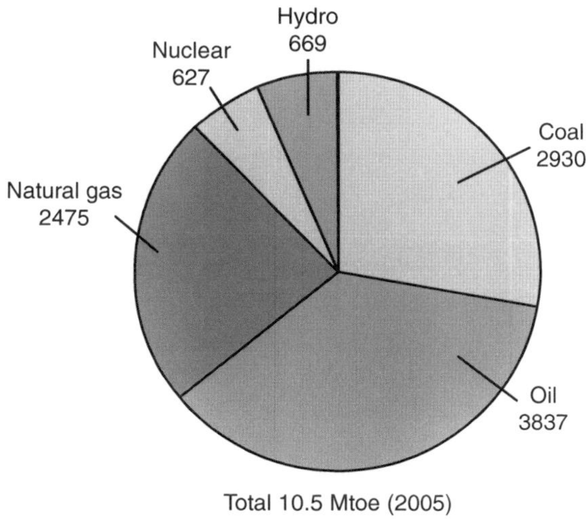

Fig. 1.3 Primary energy consumption in 2005 by primary energy sources (BP 2008)

2006b, a; DoE 2007). They describe probable development if no dramatic measures are taken to reduce energy consumption and CO_2 emissions. Possible measures shall be discussed in Sect. 1.3.

According to Fig. 1.4, global primary energy consumption will increase by more than 60% (in comparison to the year 2000) by 2030, which corresponds to an annual rate of increase of about 1.6%. This increase can be explained on the one hand by the growth of the world population and on the other by the accumulated needs of the emerging economies, such as China and India. Predictions, for example for China, say that the energy consumption will double in such countries. Fossil energy sources will continue to cover more than 80% of the primary energy consumption in 2030, with crude oil still making up the most important energy source, with a rough fraction of about 35%. Natural gas among all the energy sources shows the highest rates of increase – the consumption of it will double with respect to the year 2000 and its relative fraction will rise to 26%. The fraction of coal will decrease slightly,

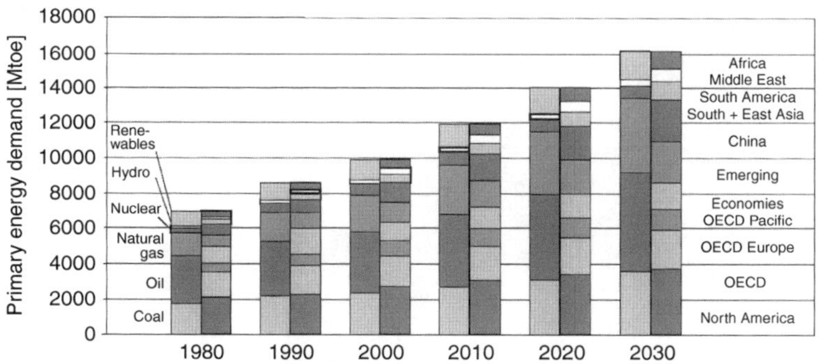

Fig. 1.4 Primary energy demand 1980–2030 of countries and regions with respect to primary energy sources (IEA 2002, 2006b; BP 2008)

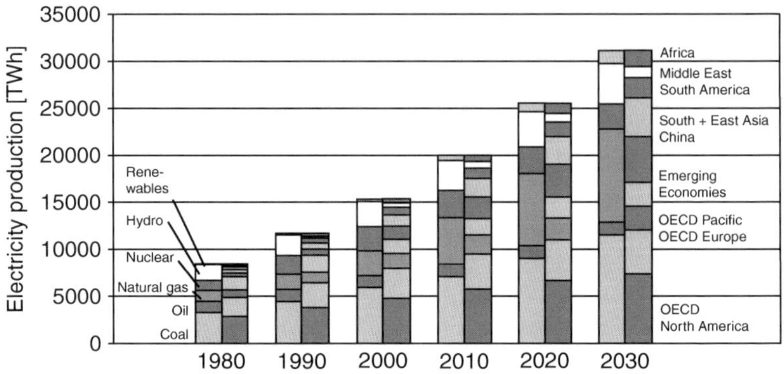

Fig. 1.5 Electric power production 1980–2030 of countries and regions with respect to primary energy sources (IEA 2002, 2006b)

arriving at about 22% by 2030. In the absolute, though, the consumption rises by 50% from the year 2000.

Electric power (see Fig. 1.5) will still further consolidate its great importance as an end-use energy source. The consumption of electric power will about double between 2000 and 2030, the rates of increase of 2.4% per year ranging clearly above the growth rates of primary energy consumption. Coal, with about 37%, will be the most important primary energy source in electric power generation; natural gas will increase its share to more than 30%.

The predicted rise of electric power consumption requires the installation of new power plants on a considerable scale (see Fig. 1.6). The power plant capacity installed worldwide amounted to about 3,400 GW in 2000 and is supposed to rise to 7,060 in 2030. Taking into consideration that old plants have to be removed

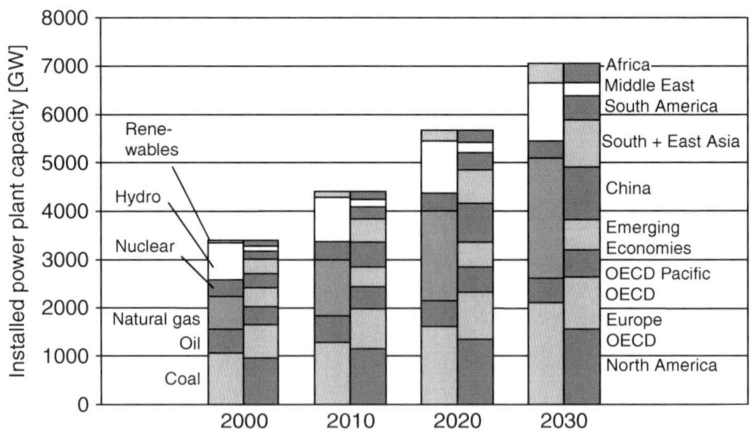

Fig. 1.6 Installed power generation capacity 2000–2030 (IEA 2002)

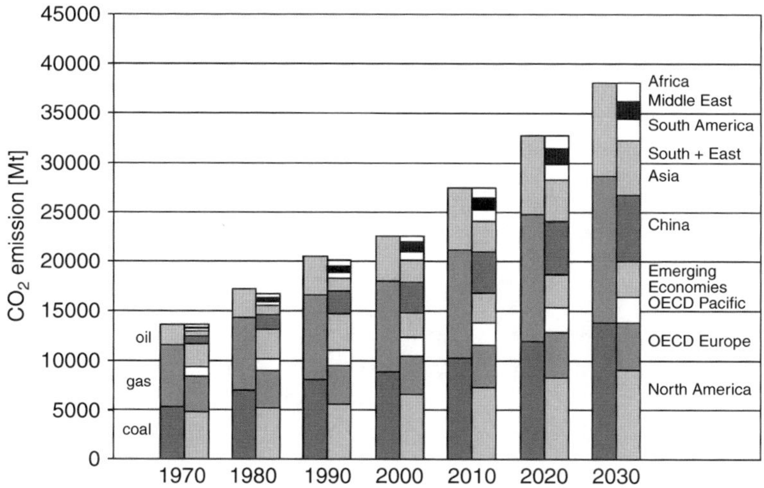

Fig. 1.7 CO_2 emissions 1970–2030 (IEA 2002, 2006b)

from service and replaced, it follows that, by 2030, electricity-generating plants with a total capacity of 4,800 GW will have to be erected throughout the world. This corresponds to 9,600 power plants with an electrical power output of 500 MW. One has to assume in this respect that new power plants will be built predominantly for primary energy sources such as natural gas (about 2,000 GW) and coal (about 1,500 GW). In China alone, thermal power plants, for example, with a total power of 720 GW shall have to be installed by 2020; per year, between 30 and 40 new coal-fired power plants with a capacity of 600 MW are currently being built. While in the emerging economies and developing countries, new power plants cover the added demand, new power plants in Europe are planned mainly as substitutes for existing old plants. By the year 2020, about 200 GW of power station capacity shall be newly installed in Europe.

The CO_2 emissions illustrated in Fig. 1.7 suggest a likely rise to about 38 thousand million tonnes of carbon dioxide per year until 2030. Referring to the year 2000, this corresponds to a rise of about 68%.

1.2 Greenhouse Effect and Impacts on the Climate

The climate of the Earth is vital for the living conditions of the entire living world. The discussion about possible future climatic changes has reached all strata of our society and has in many fields an influence on political and economic action, both on a national and on the international scale. The standard of knowledge of international climate research is compiled in the assessment reports of the Intergovernmental Panel on Climate Change (IPCC) (www.ipcc.ch).

1.2.1 Greenhouse Effect

Some gases contained in the atmosphere have a filtering effect: they let the majority of short-wavelength solar radiation pass through, while partly absorbing infra-red radiation emitted from the Earth, leading to a heating-up of the lower layers of the atmosphere. These gases, accounting for this so-called greenhouse effect, are hence termed greenhouse gases. They bring about a natural net warming of about 33°C. Without the present composition of the Earth's atmosphere, a temperature of −18°C would predominate on Earth. The atmosphere and the oceans balance the heat budget and provide for heat exchange between day and night, summer and winter, polar and equatorial zones. Without an Earth-like atmosphere, temperature differences of 250°C between day and night occur, for example on the Moon's surface, to draw a comparison (Borsch 1992).

A distinction is made between the natural greenhouse gases and those produced by man, the so-called anthropogenic greenhouse gases. Some of the greenhouse gases are both of natural and anthropogenic origin. Table 1.1 shows the contribution of the various greenhouse gas types to the natural and anthropogenic greenhouse effect. The most significant greenhouse gas is carbon dioxide (CO_2). It is produced through energy consumption in the combustion of carbonaceous fossil fuels such as coal, natural gas and crude oil. In the process, dead organic substance becomes oxidised to carbon dioxide, which is given off to the atmosphere. The quantities discharged this way to the atmosphere amount to about 26 thousand million tonnes of carbon dioxide[1] per year (2005). Added to this, there are further, inexactly quantifiable, amounts of emitted carbon dioxide from forest clearing and through soil degradation. The contribution of these emissions is estimated at about 3–7 thousand million tonnes of carbon dioxide per year.

Table 1.1 Present concentrations of greenhouse gases and their contribution to the natural and anthropogenic greenhouse effect (data from IPCC (2007b) and Beising (2006))

Greenhouse gas	Carbon dioxide CO_2	Methane CH_4	Chloro-fluorocarbons CFCs	Nitrous oxide N_2O	Ozone O_3	Water vapour
Concentration: pre-industrial time (about 1800)	280 ppm	0.7 ppm	0	270 ppb	–2.6%	
Today (2005)	379 ppm	1.8 ppm	0.5 ppb	319 ppb	25 ppb	2.6%
Increase rate (2005)	+1.9 ppm/a	+2 ppb/a		0.8 ppb/a		
Emissions (2005)	26 Gt/a	400 Mt/a	0.4 Mt/a	15 Mt/a	0.5 Gt/a	
Contribution to natural greenhouse effect = temperature rise	26%	2%	–	4%	8%	60%
Contribution to anthropogenic greenhouse effect	61%	15%	11%	4%	9%	–

[1] One tonne of carbon corresponds to 3.67 tonnes of carbon dioxide.

The CO_2 emissions of anthropogenic origin may be low compared with those of natural origin, but then the natural CO_2 emissions are counteracted by reactions of decomposition in the same order of magnitude. CO_2 emissions of 120 thousand million tonnes of carbon per year released through respiration and decay are in turn extracted from the atmosphere by photosynthesis (Heinloth 2003). The atmospheric CO_2 reservoir, which is an essential part of the global carbon cycle, being the base material for the carbon in the biosphere, amounts to about 750 thousand million tonnes of carbon at present. Referring to this reservoir, annual anthropogenic CO_2 emissions constitute about 1%, half of which remain in the atmosphere, the rest mainly dissolving into the oceans. On the whole, CO_2 emissions have led to a rise in CO_2 concentrations in the atmosphere through the years and hence to an increase of the atmospheric CO_2 reservoir. At the moment, the annual increase amounts to about 1.9 ppm. The CO_2 concentration reached in 2005 was at about 379 ppm. The CO_2 concentration before the industrial revolution (about 1750–1800) has been reconstructed through ice cores sampled in Antarctica and was determined at about 280 ppm (IPCC 2001b, 2007b; Borsch 1992; IPCC 2001a, 2007a).

In addition to CO_2, other greenhouse gases are discharged into the atmosphere through human activities. This group of gases includes methane (CH_4), nitrous oxide (N_2O) and chlorofluorocarbons. The impact of the various greenhouse gases in causing the greenhouse effect arises, besides from the emitted quantity, from the residence time of the gases in the atmosphere and their molecular structure which determines the heat absorption capacity. The concentrations of all greenhouse gases are evaluated corresponding to their climatic effect and indicated as CO_2 equivalent. In 2005, the sum of all long-lived greenhouse gases was 455 ppm, with CO_2 making the greatest contribution. About 50% of the anthropogenic greenhouse effect has to be attributed to the energy sector (inclusive of the entire transportation sector; 80% of this fraction is caused by CO_2).

In order to determine the effect of natural or anthropogenic factors on the radiative balance in the atmosphere, the current assessment reports of the IPCC apply the concept of radiative forcing. It indicates the change of the net irradiance out of solar irradiance and terrestrial radiation. Figure 1.8 shows the change of radiative forcing due to anthropogenic greenhouse gases and aerosols and the changes in solar irradiance and in land use for the period from 1750 to 2005. It can be noticed that the long-lived greenhouse gases involve a marked increase of the radiation flux, with the impact of CO_2 of more than $1.5\,W/m^2$ dominating. The contributions of the other factors to radiative forcing are significantly smaller, with both negative and positive impacts being implied.

It should be taken into consideration, though, that the scientific state of knowledge about radiative forcing is very heterogeneous in regard to the individual fields. Only in the case of the greenhouse gases is the level of knowledge high; concerning the effect of the aerosols and other substances, the level is low or very low.

The greenhouse effect induced by human activity through the intensified emission of climate-relevant trace gases is held, for the predominant part, responsible for the rise of the temperature by 0.74°C in the past 100 years (IPCC 2007b).

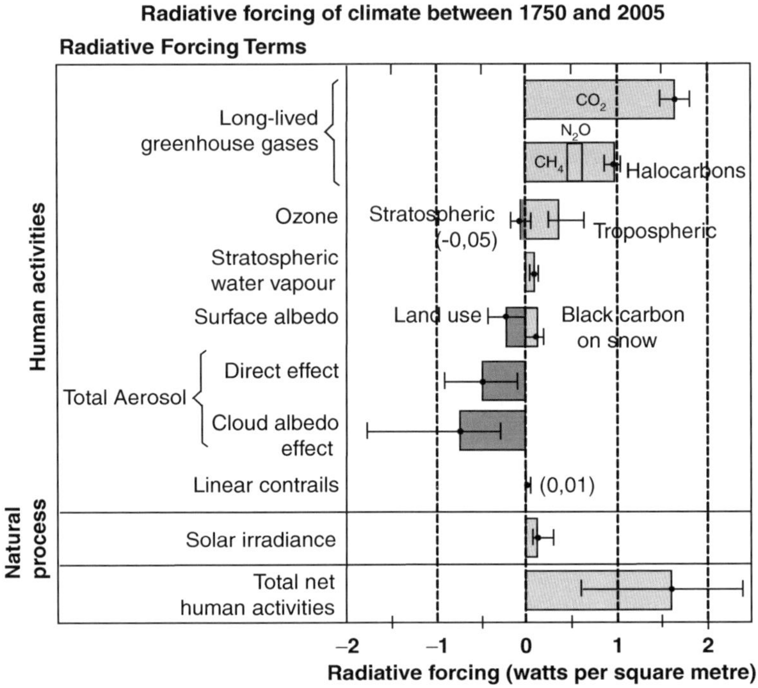

Fig. 1.8 Change in radiative forcing in the period 1750–2005 (IPCC 2007b)

1.2.2 Impacts

A small temperature increase of even few degrees can lead to a far-reaching change of the global climate. A warming process will shift the climatic zones. The subtropical dry zones, for example, will expand poleward into the currently fertile regions in southern Europe, the USA, China, South America and Australia. On top of this, climatic variations and climate extremes like storms, hurricanes, storm tides, periods of drought and heavy rains will become more frequent and stronger. The sea level will rise because of melting ice masses on land and through the expansion of water, thus threatening coastal regions. In what way and to what extent plants and animals are capable of adapting to the climate change depend on the rate at which the climate alters (Heinloth 2003).

1.2.3 Scenarios of the World Climate

The IPCC's assessment reports provide a comprehensive presentation of the current standard of knowledge in climate modelling (IPCC 2001b, a, 2007b, a). The task of climate modelling is to determine the climate system's reactions to natural or

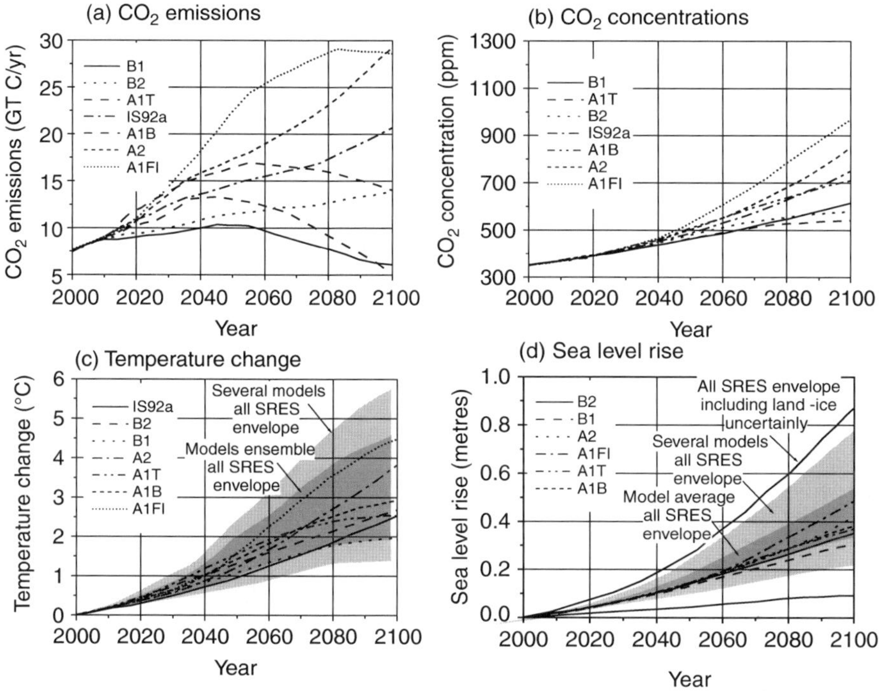

Fig. 1.9 Scenarios of the global CO_2 emissions (**a**), CO_2 concentration (**b**), temperature rise (**c**) and sea level (**d**) (IPCC 2001b)

anthropogenic changes, such as the increase of the CO_2 concentration, and thus the resilience of the system. A summary of the calculations is presented in Fig. 1.9.

Scenarios of the global energy consumption and the associated emissions up to the year 2100 (Special Report on Emission Scenarios (IPCC 2001c) (SRES 2001)) are intended to cover a wide range of possible developments, and they form the basis for the calculation of the world's climate in the long term. Figure 1.9a shows the CO_2 emissions for different scenarios which are used for numerical climate simulations. Complex climate models are based on the conservation of mass, impulse and energy in a three-dimensional grid encompassing the globe and have to take into account atmosphere, oceans, continental surfaces, the cryosphere, the biosphere and their interactions as individual components. The further development of the partly very simple models is in progress.

The different scenarios of the CO_2 emissions assume a rise of the CO_2 concentration in the atmosphere to values between 540 and 970 ppm up to the year 2100 (see Fig. 1.9b) (IPCC 2001c; SRES 2001). According to the assessment report of 2007, temperature increases of the global mean surface temperature between 2.5 and 4.1°C by the end of this century in comparison to the mean value between 1961 and 1990 were determined for selected scenarios (see Fig. 1.9c). The source of uncertainty on the one hand lies in uncertainties of the climate model calculations and, on the other, in the wide range of emission scenarios investigated. According to Fig. 1.9d, the average sea level will rise by 21–51 cm; in higher latitudes, though, up to 1 m; in the North Sea, it will rise by 50 cm (IPCC 2007b).

Even if the CO_2 concentrations were frozen at today's level (which is tantamount to an almost complete reduction of the CO_2 emissions worldwide), both the temperature and the sea level would continue to rise. This can be put down to the interaction between troposphere and ocean. While the troposphere responds to changes of concentrations and the associated radiative forcing on a timescale of less than 1 month, the timescales in the case of near-surface sea water range between years to decades, and even centuries in the case of the deep ocean and ice caps. So, even with freezing today's CO_2 concentrations, the temperature would still rise by about 0.5–0.6°C on the whole, with the biggest part of the increase happening within the next 100 years.

These relationships underline the need for a quick and drastic reduction of CO_2 emissions, precisely because our climate reacts with great inertia to the increase of greenhouse gases. It also becomes clear, though, that global warming can only be limited, not negated, even by intensive abatement efforts. In the so-called stabilisation scenarios, CO_2 emission is reduced to achieve a stable equilibrium concentration in the atmosphere.

1.3 Strategies of CO_2 Reduction

For reduction of CO_2 emissions from the energy sector there are principally three different strategies, as shown in Fig. 1.10:

- Energy saving
- Substitution (C-lean/free for C-rich energy sources)
- CO_2 capture and storage (Carbon capture and storage, CCS)

1.3.1 Substitution

The primary energy sources produce CO_2 emissions to various extents. Fossil fuels, for instance, depending on the fuel composition, more or less involve high CO_2 emissions. Figure 1.11 shows the specific CO_2 emissions of fossil fuels with respect to their calorific values. Fuels like natural gas, with a lower carbon fraction, produce in consequence lower and fuels like bituminous coal or lignite, higher specific CO_2 emissions.

By substituting natural gas as the lower carbon fuel for lignite, bituminous coal or crude oil as the higher carbon fuels, it is possible to correspondingly reduce the emissions of carbon dioxide. What stands in the way of using natural gas, however, are the smaller reserves of this energy source.

Renewable energy sources or nuclear energy involve only small CO_2 emissions in the power generation process. So if fossil energy sources are replaced by them, CO_2 emissions are almost completely avoided.

1.3.2 Carbon Capture and Storage (CCS)

Pollutants from combustion processes of fossil energy sources, such as sulphur dioxide, nitrogen oxides and particulates, are to a great extent removed nowadays. For separating (and thereby removing) these pollutants, which are even before removal in low concentrations, an amount of energy is needed such that the efficiency of the plant is diminished by 1–2%. Carbon dioxide, in contrast to these pollutants, is the main product of combustion and arises in great amounts. Its capture is possible from the technical point of view. Various concepts in this respect are being pursued at present, and projects are in progress for constructing coal-fired power plants with CO_2 capture. Carbon dioxide capture and transport to a storage location involve a marked diminution of the efficiency by 8–10%.

The different possibilities of CO_2 capture and storage from coal-fuelled power generation processes are discussed in Chap. 8.

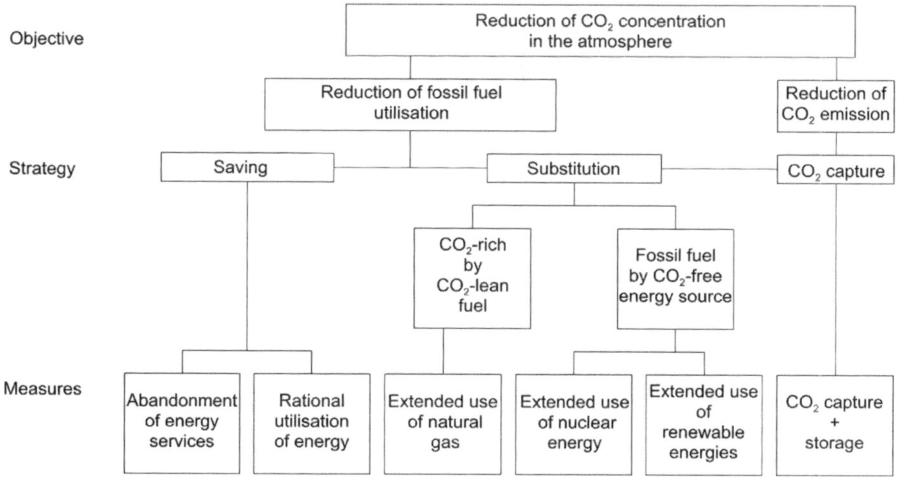

Fig. 1.10 Strategies to reduce the CO_2 emissions to the atmosphere from the energy sector

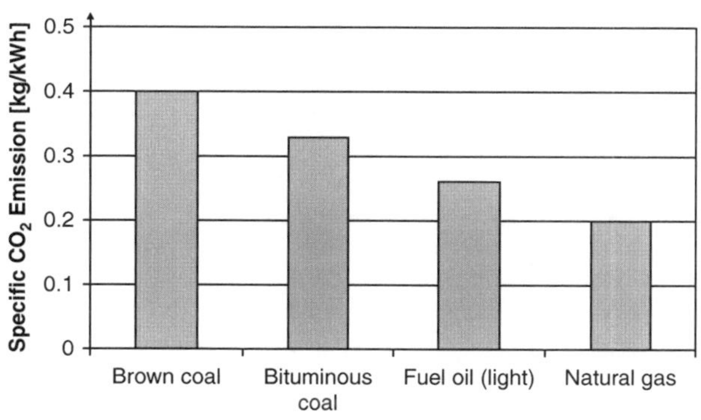

Fig. 1.11 CO_2 emissions of fossil fuels in respect to their calorific value

1.3.3 Energy Saving

Primary energy serves to provide useful energy or power services in the form of process heat, room heat, drive force or light. Cutting down on primary energy consumption and hence reducing CO_2 emissions can be achieved, for example, by doing without power services or by producing the same useful energy from less primary energy (more efficient energy utilisation). More efficient ways of utilising energy can substantially contribute to the abatement of CO_2 emissions. Efficient energy utilisation comprises on the one hand avoiding conversion losses on the part of the end-user, for instance through building insulation, and, on the other, reducing conversion losses in energy conversion processes. Modern power plant technologies aiming at boosting the efficiency of electric power generation belong to the more efficient ways of energy utilisation.

1.3.4 Mitigation Scenarios

Mitigation scenarios serve to define the reductions necessary to limit the impacts of the greenhouse effect to certain extents and to point out required measures. There are a great number of calculations for this purpose, which determine the allowable CO_2

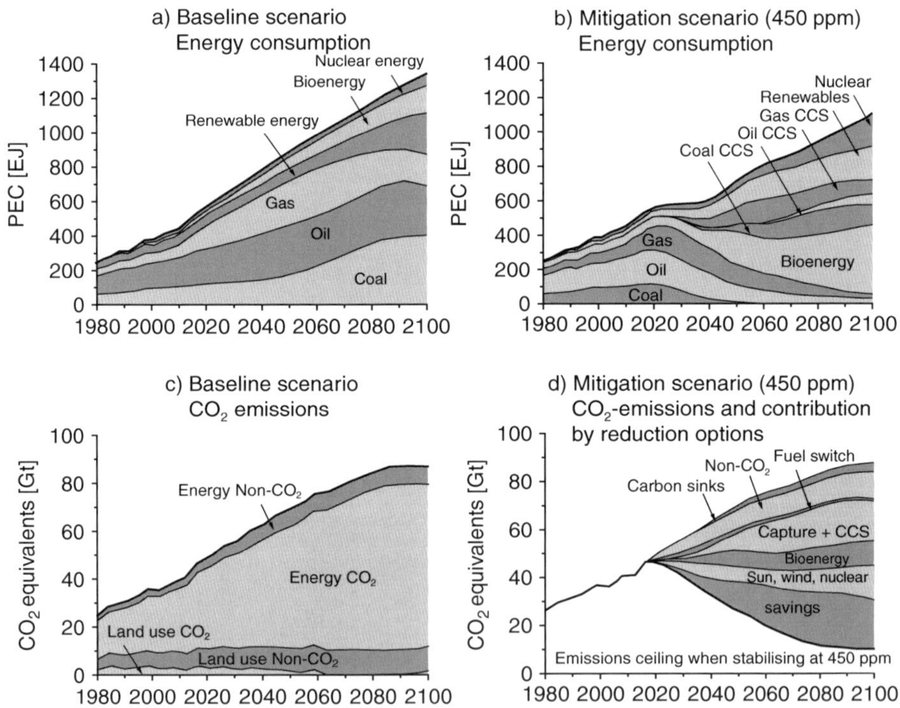

Fig. 1.12 Primary energy use for the baseline scenario (**a**) and for the mitigation scenario (**b**) and CO_2 emissions of the baseline scenario (**c**) and the mitigation scenario (**d**) (van Vuuren 2006)

emissions or, according to the different scenarios shown in Fig. 1.9, the necessary reduction to maintain a stable, defined CO_2 concentration in the atmosphere (IPCC 2001b, 2007b). The following shall present the example of a calculation, without, however, stating a plan for translation into practice (van Vuuren 2006).

The starting point of the calculation is the target to limit global warming to a rise of 2°C. With a stabilised CO_2 concentration in the Earth's atmosphere at 450 ppm CO_2 equivalents or less, it can be assumed with a probability greater than 50% that this aim will be achieved. The results of the calculations are compiled in Fig. 1.12. In order to achieve a stable CO_2 concentration at 450 ppm, the CO_2 emissions worldwide have to be reduced by 40% up to the year 2050 and by 70% up to 2100 in comparison to 1990 values.

The primary energy consumption is plotted in Fig. 1.12a for the baseline scenario and in Fig. 1.12b for the mitigation scenario. Figure 1.12d shows a CO_2 emission reduction scenario and the contribution of the different measures taken to achieve those reductions in comparison to the baseline scenario (Fig. 1.12c). Without entering a discussion of the individual measures, it becomes clear that, for achieving that aim, all possible options have to be taken into consideration. Increasing the energy efficiency ought to always be the first action.

References

Beising, R. (2006). Klimawandel und Energiewirtschaft – Eine Literaturrecherche, Stand Oktober 2006. Essen, VGB PowerTech.

Borsch, P. (1992). Was wird aus unserem Klima? Fakten, Analysen & Perspektiven. München [u.a.], Bonn Aktuell.

BP (2008). Statistical review of world energy 2008, from www.bp.com.

DoE (2007). International Energy Outlook, Energy Information Administration, Department of Energy, from www.eia.doe.gov/oiaf/ieo/index.html.

Heinloth, K. (2003). Energie und Umwelt – Klimaverträgliche Nutzung von Energie. Stuttgart, Teubner.

IEA (2002). World energy outlook 2002. Paris, IEA.

IEA (2006a). Energy Technology Perspectives, Scenarios and Strategies to 2050. Paris, OECD/IEA.

IEA (2006b). World energy outlook 2006. Paris, IEA.

IPCC (2001a). Climate change 2001: mitigation. (Third Assessment report WG3). Cambridge, Cambridge University Press.

IPCC (2001b). Climate change 2001: the scientific basis. (Third Assessment report WG1). Cambridge, Cambridge University Press.

IPCC (2001c). Special report on emission scenarios (SRES). Cambridge, Cambridge University Press.

IPCC (2007a). Climate change 2007 – mitigation of climate change, working group III contribution to the fourth assessment report of the IPCC intergovernmental panel on climate change. Cambridge, Cambridge University Press.

IPCC (2007b). Climate change 2007 – the physical science basis, working group I Contribution to the fourth assessment report of the IPCC intergovernmental panel on climate change. Cambridge, Cambridge University Press.

van Vuuren, D., Berk, M., Farla, J. and de Vos, R. (2006). From climate objectives to emissions reduction. Netherlands Environmental Assessment Agency, Publication 500114003/2006, from www.mnp.nl/en.

Chapter 2
Solid Fuels

2.1 Fossil Fuels

2.1.1 Origin and Classification of Coal Types

Coal, oil and natural gas are called fossil fuels because they are the remains of plant and animal life preserved in sedimentary rocks. It is generally believed that coal was formed from plant matter and oil formed from marine organisms (Drbal 1996).

Brown and hard coal developed through a process of partial decomposition under air-deficient conditions of plant matter that had accumulated on land and in swamps during previous geological periods. By continued deposition of sediments and plant debris, the older sediments gradually sank to greater depths and, with growing pressure and a resulting dewatering process, became compacted. Under anaerobic conditions, the organic substance underwent, by pressure and heat, a metamorphic process called coalification. Peat formation and the formation of soft brown coal are the first steps of the coalification process. With greater depths, higher pressures and rising temperatures, coalification proceeds (thermal metamorphosis), hard brown coal develops from soft brown coal and, eventually, hard coal is formed.

The coalification process involves an increase in the fraction of solid carbon and a decrease in the volatile matter content of the material. In the early stages of decomposition, the formation of H_2O, CO_2 and N_2 predominates; in advanced stages, CH_4 is mainly formed. The increasing pressure pushes the water content down further and further. The moisture content decreases from about 70% (in peat) to about 15% (in anthracite). Volatiles diminish from a fraction of 75 to 10%. As a consequence of the release of CH_4 and CO_2, the C content increases from about 50 to more than 90%.

Coal types are commonly differentiated from one another according to their content of volatile components (for definitions, see Sect. 2.1.2) on a dry and ash-free matter basis and according to the characteristics of the coke. The USA, Great Britain and Germany each use their own classification systems, which are all based on the volatile content (see Fig. 2.1) (Skorupska 1993).

An international classification system is in place that assigns a three digit number to each bituminous coal. For an assessment of the combustion characteristics,

H. Spliethoff, *Power Generation from Solid Fuels,* Power Systems,
DOI 10.1007/978-3-642-02856-4_2, © Springer-Verlag Berlin Heidelberg 2010

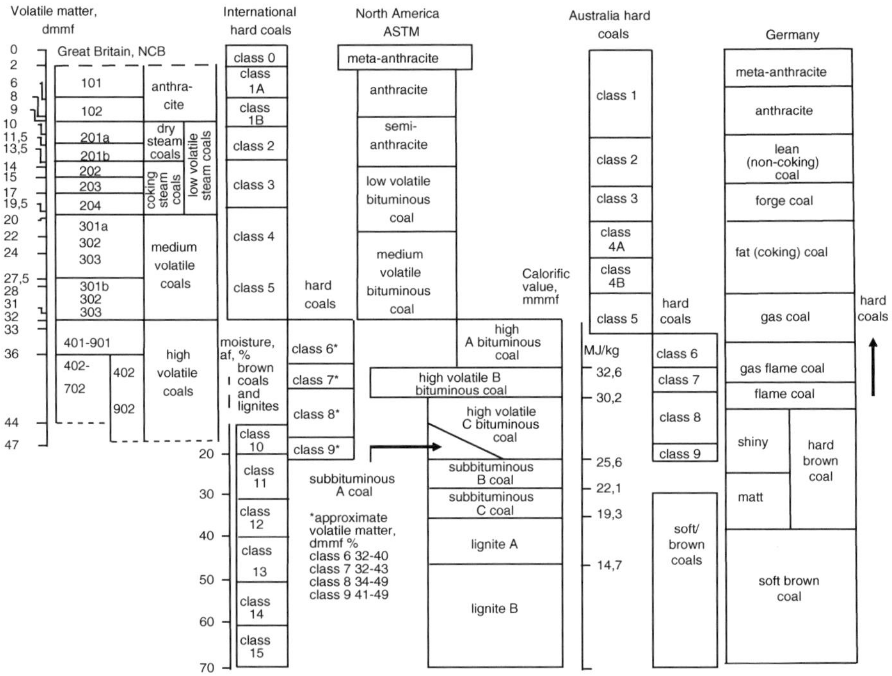

Fig. 2.1 Comparison of different coal classification systems (Skorupska 1993)

however, this system – apart from the classification according to the volatiles content – is of minor importance because it focusses on carbonisation- and gasification-engineering characteristics related to coking (JBDT 1985). Besides the volatiles content, it takes into account the caking and the coking ability.

Table 2.1 compiles the characteristics of different coal types Alstom Power as source (Effenberger 2000).

2.1.2 Composition and Properties of Solid Fuels

Coal is a mixture of organic material and mineral matter. The organic matter is responsible for the energy content of the fuel, while it is the mineral matter that presents significant challenges in the design and operation of a power plant. Several types of analysis are performed to evaluate the coal properties that affect the design and operation of power plant components and systems. These analyses are the determination of the heating value, the proximate analysis, the ultimate analysis, the mineral analysis of the ash, the determination of the ash fusion temperature, the analysis of the grindability and the determination of the swelling index. In addition, other physical characteristics of the coal may be determined, such as the bulk density and the particle size distribution. The methods for performing the various tests on fossil fuels have been developed by various standards organisations such

Table 2.1 Composition of hard and brown coals (Effenberger 2000) and Alstom Power

Coal type		Dry ash-free matter basis							Raw coal		
Origin	Site of deposit	Volatiles (%)	C (%)	H (%)	O (%)	N (%)	S (%)	Higher heating value HHV (MJ/kg)	Ash (%)	Water (%)	Lower heating value: LHV (MJ/kg)
	Peat										
IRL	Derrygreenagh	69.6	58.0	5.6	34.9	1.2	0.3	23.86	1.5	55.0	7.7–7.9
GR	Philippi	68.5	57.5	5.4	33.5	2.8	0.8	23.0	20–22	40–45	7.3–7.9
	Soft brown coal										
D	Rhineland	55.0	68.3	5.0	27.5	0.5	0.5	26.38	5–20	50–62	6.3–9.6
D	Helmstedt	59.4	72.6	5.8	16.7	0.4	4.4	29.75	12–22	42–46	9.2–10.5
D	Schwandorf	55.0	63.6	5.0	26.1	1.3	4.0	25.33	6–20	50–58	6.3–7.5
D	Lausitz	55.0	67.5	5.2	25.5	1.0	0.8	25.37	2–5	55–60	8.2–8.5
D	Leipzig	63.0	71.6	6.1	19.5	0.7	2.1	28.35	5–7	50–55	9.0–11
D	Halle-Bitterfeld	57.5	72.0	5.5	18.3	0.8	3.4	29.81	5–7	52–56	9.6–10.0
GR	Ptolemais	57.0	65.3	5.3	26.5	1.6	0.5	25.25	6–22	52–60	3.6–6.7
GR	Megalopolis	62.0	60.5	6.2	30.6	1.3	1.4	24.45	13–17	60–64	2.8–4.0
AUS	Yallourn	51.5	67.5	4.8	26.7	0.7	0.3	25.54	1–2	63–72	5.0–7.5
PL, D	Patnow, Lusatia	58.4	73.6	5.1	19.7	0.5	1.1	28.56	6–15	52–58	8.0–8.8
H	Gyöngyös/ Visonta	63.0	63.8	4.8	26.8	1.1	3.5	24.83	15–30	46–54	5.0–6.7
TR	Elbistan	67.0	61.4	5.1	29.6	0.8	5.1	23.69	8–24	48–62	3.3–6.2
	Hard brown coal										
D	Peissenberg	52.0	74.0	5.5	14.5	1.4	4.6	29.23	12–20	8–12	19.7–23.0
A	Fohnsdorf	47.0	72.5	5.4	16.3	1.2	4.6	30.35	8–16	8–14	20.0–22.6
A	Köflach	56.0	67.7	5.7	25.0	1.2	0.3	27.21	6–10	30–35	13.0–14.7
SLO	Trbovlje	53.0	72.5	5.6	17.2	1.2	3.5	28.47	30–35	20–24	10.0–11.7
HR	Raša	50.4	75.2	5.4	6.9	1.1	11.5	34.12	6–20	2–4	27.6–30.1
CZ	Most (Brüx)	48.0	77.5	5.8	14.6	1.0	1.2	32.32	5–15	15–25	18.8–22.2
CZ	Falknov	54.5	73.5	6.0	17.9	1.1	1.5	30.9	4–14	25–35	15.1–18.4
H	Tatabanya	52.0	73.0	5.8	17.7	0.9	2.6	31.4	6–12	12–14	23.0–24.3
TR	Tuncbilek	44.5	76.4	5.8	13.8	2.5	1.5	32.19	14.22	14–24	15.0–18.1

Table 2.1 (continued)

Coal type		Dry ash-free matter basis							Raw coal		
Origin	Site of deposit	Volatiles (%)	C (%)	H (%)	O (%)	N (%)	S (%)	Higher heating value HHV (MJ/kg)	Ash (%)	Water (%)	Lower heating value: LHV (MJ/kg)
	High-volatile bituminous coal										
D	Ruhr Basin, Aachen	33.7	85.9	5.5	6.2	1.6	0.8	33.9	6–9	7–10	27.6–28.5
D	Saar Basin	38.2	82.7	5.2	9.4	1.2	1.5	33.0	5–8	3–5	28.0–28.9
GB	Scotland	41.5	81.4	5.4	10.3	2.1	0.8	33.9	4.6	13.8	26.3
	Medium-volatile coal										
D	Ruhr Basin	33.7	85.9	5.5	6.2	1.6	0.8	35.2	6–7	8–10	27.6–28.0
CZ	Ostrava-Karviná	38.0	83.4	5.0	9.5	1.4	0.7	33.7	12	4.5	27.1
PL	Upper Silesia	33.2	84.5	5.2	7.2	1.6	1.2	32.4	8–13	4–10	26.2–27.0
GB	Yorkshire	34.4	84.3	5.2	8.0	1.7	0.8	34.8	6.8	2	31.7
AUS	Queensland	35.9	84.7	5.4	7.7	1.7	0.5	34.2	6.8	5	29.0
	Fat coal										
D	Ruhr Basin	24.4	88.7	5.0	4.1	1.6	6.7	36.2	6–9	7–10	28.5–29.3
D	Saar Basin	32.5	86.9	5.2	5.4	1.3	1.1	35.6	7–9	8–10	28.0–28.4
USA	Pennsylvania	24.6	83.2	5.1	3.4	1.4	1.9	36.3	8	3	31.2
ZA		28.2	82.5	4.5	9.9	1.7	1.3	34.2	15	8	25.4
	Forge coal										
D	Ruhr Basin	12.4	90.7	4.1	2.1	1.7	1.4	36.2	6–9	7–10	28.5–29.3
D	Aachen	13.8	89.8	4.8	2.8	1.5	0.5	35.8	6–9	8–10	28.5–29.3
	Lean Coal										
D	Ruhr Basin	10.5	90.8	3.8	2.7	1.7	0.8	36.0	6–9	7–10	28.7–29.3
F	Nord-Pas de Calais	12.0	89.8	3.8	4.8	1.0	0.6	36.4	6	3	32.3
	Anthracite										
D	Ruhr Basin	7.7	91.8	3.6	2.6	1.4	0.7	35.9	4–7	3–5	31.0–31.4
UA	Donets	4.0	94.4	1.8	1.4	0.7	1.7	34.1	5	5.7	30.0

Source: Alstom Power and Effenberger (2000).

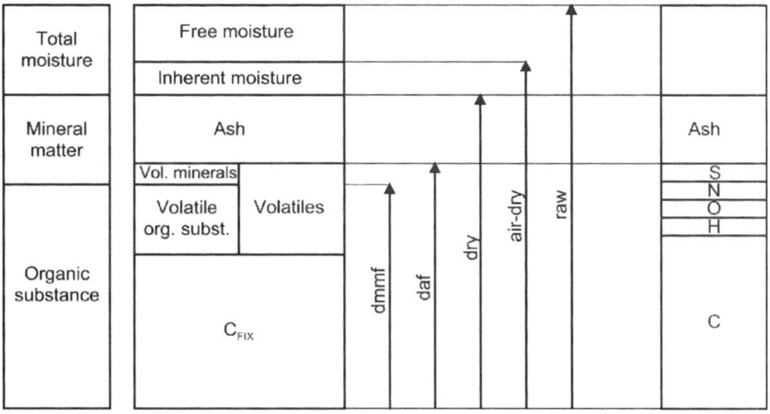

Fig. 2.2 Coal composition

as the American Society for Testing Materials (ASTM), the Deutsches Institut für Normung (DIN), the British Standards Institution, Australian Standards (AS) and the International Standards Organisation (ISO).

Figure 2.2 shows the general composition of a coal. The raw coal, besides the combustible organic substance, contains inert material, which is made up of mineral matter and water. Since the determination of the mineral content requires rather sophisticated methods, the common practice is to use the ash content instead (JBDT 1976; Gumz 1962; Adrian et al. 1986; Ruhrkohle 1987).

The proximate analysis includes the determination of the total moisture, the air-dried moisture, the volatile matter, the fixed carbon and the ash. It involves heating the sample to various temperatures for different periods of time and noting the weight loss in the sample.

A proximate analysis reports moisture in only two categories: total and air-dried. Air-dried moisture is also referred to as inherent moisture. The total moisture content is composed of free or surface moisture and inherent moisture. While free moisture adheres to the outside surface of the fuel, inherent moisture is bound in the capillaries inside the grain. Drying at room temperature makes the free moisture evaporate; the air-dried sample remains. Further heating to 105 °C makes the remaining, inherent moisture evaporate, and the dry, "moisture-free" coal remains. Chemically bound water, in the form of hydrates of the mineral matter, such as clay minerals, remains in the coal. These hydrates are not taken into consideration in the conventional moisture content determination at 105 °C (Ruhrkohle 1987).

Heating the dry, moisture-free sample to 900 °C in an inert atmosphere releases the volatile components. In this process, a multitude of vapours and gases escape. The remaining matter is called char. From the weight loss in this process, the volatile matter content is calculated. It should be taken into consideration when assessing this value that, because of the dissociation and release of carbonates, the volatile matter content may appear higher than it actually is. The combustible fraction of the char is described as fixed carbon (fixed C); the incombustible fraction is termed ash.

The content of fixed C is not the same as the C content of the fuel which, besides the fixed carbon, also includes the carbon in the volatile matter.

The volatile matter content determined according to the standards does not correspond to the volatile matter released in a real combustion process, because the temperature, heating rate and residence time in an industrial furnace differ from the respective values under laboratory conditions. In industrial firing plants the amount of released volatile matter may be considerably higher.

The ash content of a coal is determined by means of the residue left over after burning a sample with air at 815 °C (German standard DIN 51719). This content is not identical to the mineral matter content, because the ash is only the mineral matter residue from combustion. In combustion engineering it is common, though, to give the ash content as a measure of the mineral substances in the fuel. The procedure for the determination of the mineral matter content is more sophisticated than that for the determination of the ash content. The procedure consists of chemical processes in which the sample becomes demineralised by hydrochloric and hydrofluoric acids (Ruhrkohle 1987). The mineral matter content can include inherent mineral matter spread throughout the coal seam as well as extraneous mineral matter from the roof or floor of the seam. Some of the inherent mineral matter in coal is derived from inorganic compounds associated with plant life. This mineral matter is generally responsible for about 1–2% of the ash in the coal. The extraneous mineral matter comprises the bulk of the ash in the coal (Drbal 1996).

The mineral matter undergoes a chemical conversion in the combustion process. For hard coals, the conversion and release of the volatile products of decomposition has a weight-reducing effect on the ash. The weight of the ash (the residual matter from combustion) is lower than the weight of the original mineral matter content.

In the process of combustion of hard coal, hydrates and carbonates bound to mineral components are released, while alkalis volatilise, and pyritic sulphur decomposes. Mineral components are partly transformed into an oxidic form during combustion. However, describing the ash composition only in terms of oxides of the elements found in the ash analysis is inaccurate.

A part of the decomposition products of combustion is taken into account in the determination of the volatile matter content. For example, the mineral matter content of coals from the Ruhr basin, on average, is 9% higher than the ash content (Ruhrkohle 1987). For coals which contain alkaline earths as part of the mineral matter, e.g. brown coals, there may also be an increase in the weight of the ash during incineration as a consequence of the absorption of sulphur oxides.

Table 2.2 shows a compilation of the mineral elements occurring in coals, while Table 2.3 gives the main components of hard and brown coal ashes.

Ultimate analysis determines the contents of carbon, moisture, nitrogen, sulphur and chlorine. The difference in the balance between the sum of the contents determined by the ultimate analysis and the total dry ash-free (d.a.f) weight is commonly assumed to be oxygen. The elemental composition is the basis for the combustion calculations of the stoichiometric oxygen demand, the flue gas quantity and the flue gas composition.

Table 2.2 Coal minerals (Adrian et al. 1986)

Mineral	Formula	Fraction (percentage) by weight
Clay minerals		Up to 50
Kaolinite	Al_2O_3*$2SiO_2$*H_2O	
Illite	K_2O*3(Al, Fe)$_2O_3$*16SiO*$4H_2O$	
Carbonates		Up to 20
Calcite	$CaCO_3$	
Dolomite	$CaMg(CO_3)_2$	
Siderite	$FeCO_3$	
SiO_2 group		1–15
Quartz	SiO_2	
Chalcedony	Si*O_2	
Sulphides		Up to 20
Pyrite	FeS_2	
Marcasite	FeS_2	
Accessory minerals		
Feldspar	(K, Na)$AlSi_3O_3$	
Apatite	$Ca_5F(PO_4)_3$	
Hematite	Fe_2O_3	
Rock salt	NaCl	
Rutile	TiO_2	

Table 2.3 Main components of coal ash (Adrian et al. 1986)

Ash component		Hard coal (%)	Brown coal/lignite (%)
Silica oxide	SiO_2	30–50	1–50 (mostly 10)
Aluminium oxide	Al_2O_3	15–30	1–35 (mostly 8)
Iron oxide	Fe_2O_3	2–22	4–25
Calcium oxide	CaO	1.5–15	15–60
Magnesium oxide	MgO	1–8	1.5–12
Sulphur trioxide	SO_3	1–5	4–40
Phosphoric acid	P_2O_5	0.2–1.5	0.1–1.8
Potassium and sodium oxides	$K_2O + Na_2O$	1–5	0.5–2

The calorific or heating values are a measure of the thermal energy released in complete combustion. The reference temperature is 25 °C in accordance with German standard DIN 51900.

Water is contained in the fuel before combustion (the moisture of the fuel) and is formed during the combustion of the hydrous compounds. The higher heating value (HHV) or gross calorific value (GCV) assumes water to be present in a liquid state after combustion. In contrast, the lower heating value (LHV) or net calorific value (NCV) counts both water fractions as being in a vapour state. The higher heating value is higher than the lower heating value by the heat of evaporation of the fuel moisture and the water formed at 25 °C (2,443.5 kJ/kg). Since the heat of evaporation is normally not used in industrial processes, it is common to apply the lower heating value. The higher heating value is determined by a bomb calorimeter (German standard DIN 51900); the lower heating value is calculated from the HHV minus the latent heat of the water vapour.

Higher and lower heating values can also be determined by correlations between the heating value and analysis values from statistical studies. The values calculated this way, however, are only approximate.

The ash fusion behaviour allows some conclusions about the behaviour of the mineral components and the fouling and slagging behaviour during combustion to be drawn. For investigation purposes, a sample body of ash is heated. Changes of shape occur at specific temperatures, giving information as to the characteristics of the sample. The ambient atmosphere is either air (oxidising) or a mixture of CO and CO_2 (reducing).

In different countries, the methods to determine the ash fusion behaviour are similar but different shapes of sample bodies are used. According to the American ASTM Standard D 1857, the ash is pressed in a triangular pyramid of 19 mm in height and a 6.35 mm triangular base (Stultz and Kitto 1992). The test sample according to German standard DIN 51730 has a cylindrical or cubic shape of 3 mm height and 3 mm diameter/width (see Fig. 2.3). Photographs are taken of the shape of the compacted sample body as it changes, and the temperature at each photograph is recorded. The specific temperatures characterising the fusion behaviour are as follows:

- Initial deformation temperature (ID): when the first signs of a change in form are visible.
- Spherical or softening temperature (ST): when the sample has deformed to a spherical shape where the height of the sample is equal to the width at the base ($H = W$).

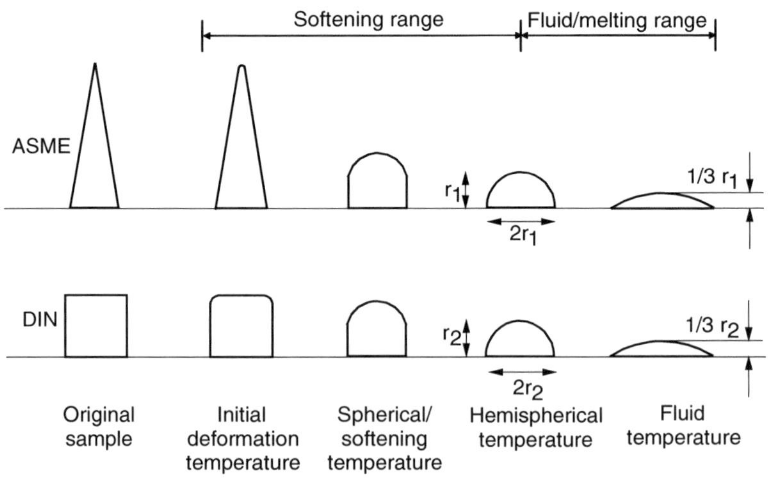

Fig. 2.3 Characteristic ash fusion temperatures according to DIN and ASME

- Hemispherical temperature (HT): when the sample body has changed to a hemispherical shape. Its height equals one half the width of the base ($H = 1/2\,W$).
- Fluid temperature (FT): when the sample body has melted down to a flat layer with a maximum height of about one third of its height at the hemispherical temperature.

The temperature range between the initial deformation and hemispherical temperature is defined as the softening range, the range between hemispherical and fluid temperature as the fluid temperature range. When the difference between the hemispherical temperature and ash fluid temperature is small, then the slag is referred to as "short"; a large difference occurs when the slag is "long".

The results of the above-described investigations are transferable to an industrial scale only to a limited extent, because the laboratory conditions do not correspond to the conditions in industrial firing systems, either in the way the samples are prepared, or in the procedure of the method.

2.1.2.1 Petrographic Analysis

Petrographic analysis classifies the coal according to its structural constituents – the macerals (Chiche 1970, 1973). This information is used to gain an insight into the process of the coal formation, so as to relate the decayed organic matter to the coal.

Maceral is the term for the smallest structural constituent recognisable by an optical microscope. The macerals can be distinguished from one another by their reflectance. In the analysis of maceral groups of hard coal, three maceral types – vitrinite, exinite and inertinite – are distinguished. Vitrinite comes from wood matter, while exinite mainly consists of products of digested sludge. The third maceral group, inertinite, which requires further analysis before being confirmed as originating from the vegetable matter, is relatively unreactive (Ruhrkohle 1987; Adrian et al. 1986). With brown coal, the maceral groups distinguished are huminite, liptinite and inertinite, where huminite and liptinite, as far as their origin is concerned, correspond to the hard coal maceral groups of vitrinite and exinite, but with a lower degree of decomposition (Zelkowski 2004). Table 2.4 gives a general compilation of the maceral groups and macerals of hard and brown coals.

The various maceral groups are distinguished by their contents of volatile matter and their reflectance. In the case of hard coal, exinite has the highest volatile matter content and the lowest level of reflectance, while inertinite has the lowest content of volatile components and the highest reflectance of the maceral groups. With higher coalification degrees, the volatile matter contents of all maceral groups decrease while converging towards each other (see Fig. 2.4) (Ruhrkohle 1987).

Hard coals of the northern and the southern hemispheres differ markedly as to their maceral composition. Coals of the northern hemisphere show a dominance of vitrinite, the content being about 60–80%, with the contents of both exinite and inertinite varying, with a maximum of 30% each. Coals of the southern hemisphere have a significantly higher inertinite content of more than 50%. There is a direct correlation between the volatile matter in a coal and the reflectance of vitrinite

Table 2.4 Macerals of brown and hard coals (Zelkowski 2004)

Brown coal		Hard coal	
Maceral group	Maceral	Maceral group	Maceral
Huminite	Textinite, ulminite attrinite, densinite gelinite, corpohuminite	Vitrinite	Telinite, collinite vitrodetrinite
Liptinite	Sporinite, cutinite, resinite, suberinite, alginate, liptodetrinite chlorophyllinite	Exinite	Sporinite, cutinite resinite, alginite liptodetrinite
Inertinite	Fusinite, semifusinite, macrinite, sclerotinite inertodetrinite	Inertinite	Micrinite, macrinite semifusinite, fusinite inertodetrinite

(see Fig. 2.5). This correlation is used to determine the distribution of the contents of volatile matter. Results serve to infer whether the fuel in question is a pure coal or a blended type. For example, despite having the same volatile matter content, the coal types in Fig. 2.6 exhibit clear differences in the distribution of macerals (Ruhrkohle 1987).

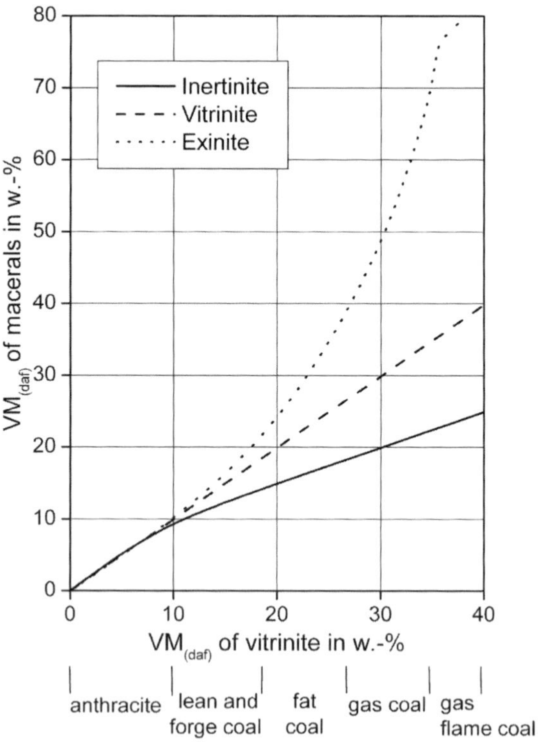

Fig. 2.4 Volatile matter of macerals as a function of the coal type (Ruhrkohle 1987)

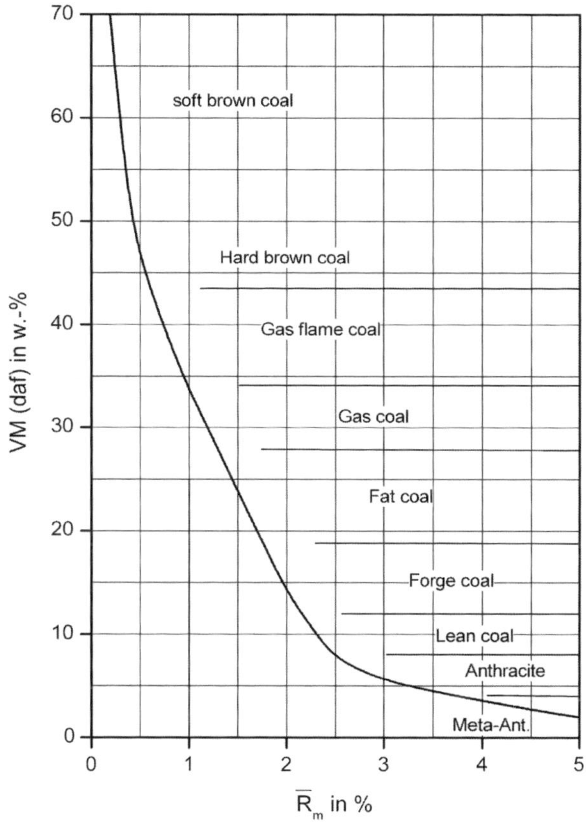

Fig. 2.5 Correlation of the volatile matter content to the reflectance R_m of vitrinite (Ruhrkohle 1987)

2.1.3 Reserves of Solid Fuels

Amongst the group of fossil energy carriers, coal has the highest reserves and resources. The geographic distribution of coal deposits is considered to be well known, being located mainly by exploration. The large coal basins are concentrated in the northern rather than the southern hemisphere of the Earth:

North America: Appalachians, central continental area and western states
Europe: From England across northern France, Germany and Poland
Russia/Ukraine: Very large coal basins with hard and brown coals
China: Large deposits, predominantly in the north
Australia: Large basins of hard coal in the eastern part of the continent (New South Wales and Queensland)
South Africa: Thick coal-bearing seams

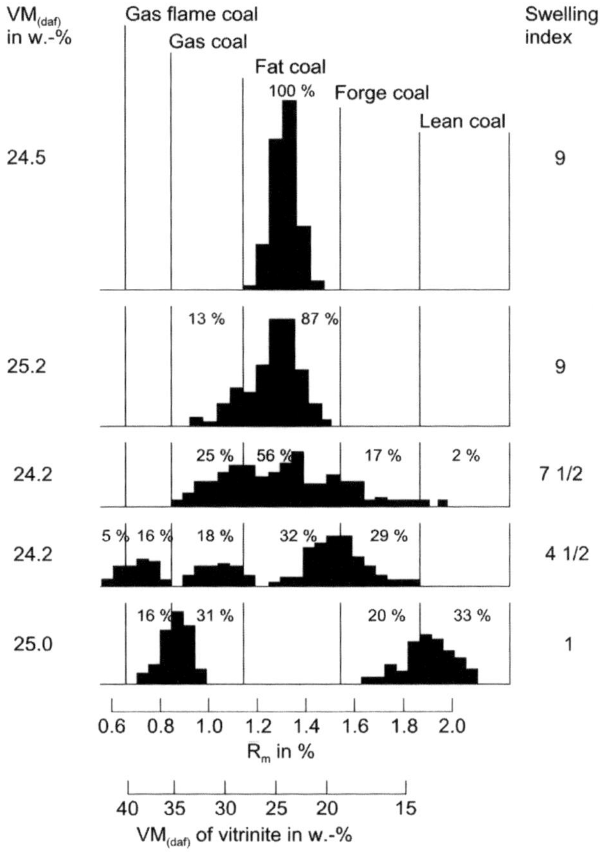

Fig. 2.6 Reflectance analysis for coals with a similar volatile matter content (Ruhrkohle 1987)

On the global scale, the proven reserves were 726 thousand million tonnes of coal equivalent (TCE) in 2006 (BMWi 2008). Proven reserves are present if geological and engineering information indicates with reasonable certainty that exploitation is possible under existing economic and operating conditions. As a comparison, in 2006 the proven reserves of natural gas were 162 thousand million TCE and of crude oil, 201 thousand million TCE. From this data, on the basis of the actual global consumption, the reserves to production (R/P) ratio of coal (indicating the time that coal will last) is 168 years; for natural gas, the result is 61 years and for oil, there are 41 years (BMWi 2008; BGR 2008; BP 2008).

The regional distribution of hard and brown coal reserves and resources is shown in Fig. 2.7. The highest share of the total reserves can be found in the USA (27%), followed by China (19.8%) and Russia (13.7%). The amount of resources is about one order of magnitude higher than the reserves. About 40% of the global resources can be found in China. Resources differ from reserves by being the amount physically present or expected to be present with a certain probability, where reserves are those currently accessible and economic.

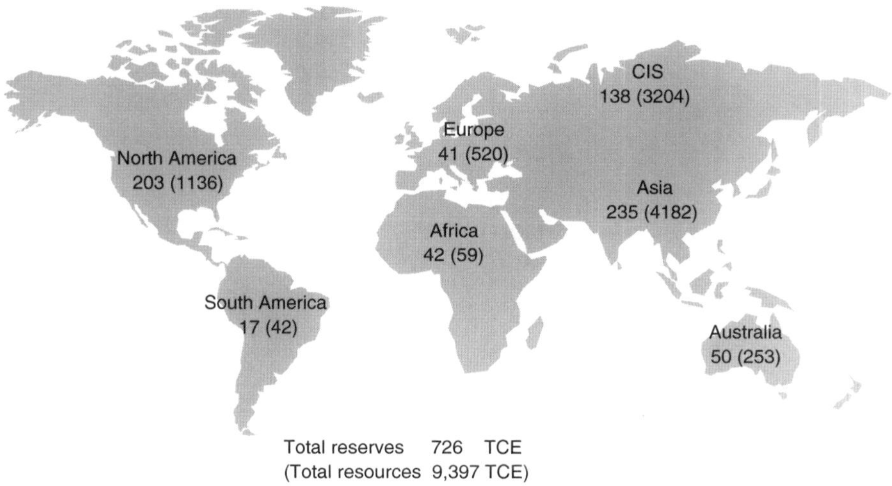

Fig. 2.7 Distribution of coal reserves and resources (data from BMWi 2008)

The production of hard and brown coal in the world as a whole reached 6.2 thousand million tonnes of coal in 2006, corresponding to 4.3 thousand million tonnes of coal equivalent (TCE) (BP 2008). Of this, hard coal comprised 93% and brown coal 7% (BGR 2008).

The relative fractions of the solid fossil fuels in primary energy consumption will remain relatively constant in the near future, as explained in Sect. 1.1. At the same time, absolute coal production and consumption will increase. This is illustrated in Table 2.5, which shows coal production for selected years in the past and predictions for the future until 2030, divided into OECD and other countries. One may notice the steep rise in Asian countries, which account for 80% of the rise in coal production up to 2015.

Table 2.5 World coal production and exports (in million tonnes) (IEA 2006)

Production	1980	2004	2015	2030
OECD North America	687	1,080	1,248	1,376
OECD Europe	1,163	834	855	905
OECD Pacific	183	399	450	453
Eastern Europe	842	736	809	707
Africa	93	193	211	248
China	626	1,881	3,006	3,867
India	114	441	636	1,020
Asia, other countries	64	202	295	419
Latin America	18	34	44	63
Total	**3,822**	**5,558**	**7,328**	**8,858**
Export	172	619	819	975

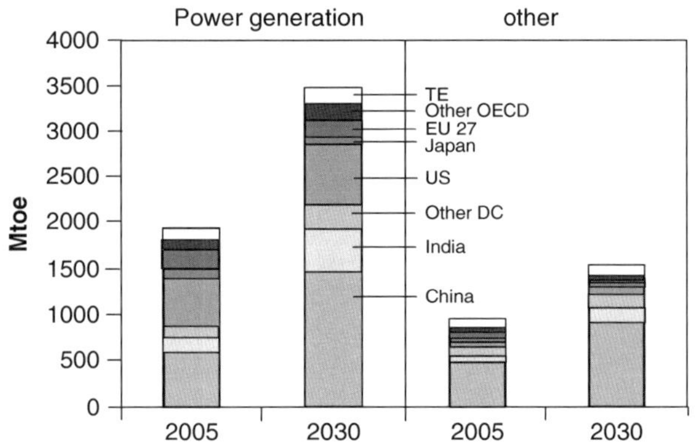

Fig. 2.8 Coal consumption in the power generation sector and other sectors (data from IEA 2007)

Coal is predominantly used for power production. Figure 2.8 shows coal consumption in the power generation sector and in other sectors (mainly from coke utilisation in the steel industry). Figure 2.8 clearly shows that China and India account for 78% of the growth of coal use in the power generation and 90% of the growth in other sectors (IEA 2007).

In this context, coal is for the most part used in the proximity of the coal mining site. It is estimated that 60% of the coal used in power production is sourced within a radius of 50 km of the plant. Compared to other fossil fuels, the trade in hard coal is less developed. In 2004, about 11% of the total production was exported. The hard coal trade, however, is expected to grow strongly because the consumption in countries with small deposits of their own, such as Japan and other South and East Asian countries, will rise while subsidised coal mining in Europe will further decrease. In Asian regions in particular a strong trade will develop.

The price trend for imported steam coal in Germany is plotted in Fig. 2.9 and compared to the costs of natural gas and crude oil, using the basis of 1 TCE. It is obvious that the increase in coal prices is smaller in comparison to oil and natural gas. Prices of all fossil fuels will rise further in the future, but due to its high flexibility of supply, it is assumed that a bottleneck will not occur for coal.

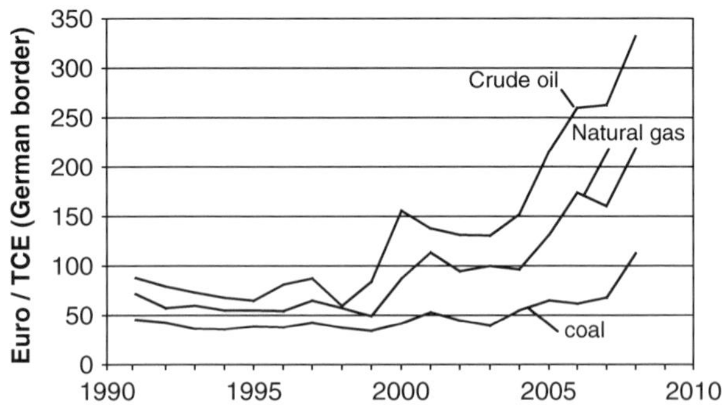

Fig. 2.9 Price trend of hard coal in comparison to oil and natural gas (data from BMWi 2008)

2.2 Renewable Solid Fuels

Definition of biomass: The term biomass describes material of organic origin, be it living or dead. Biomass therefore includes plant and animal life, their respective waste or residual material and in the broader sense all conversion products such as paper or cellulose, organic residuals from the food industry and organic waste from households, trade and industry. The distinction between biomass and fossil fuels begins with peat, which is not defined as biomass (Kaltschmitt 2001; Kleemann and Meliß 1993; CMA 1995).

The line between biomass and waste is drawn differently from country to country. In some countries the term biomass is used for any plant-derived organic matter available on a renewable basis, thus including dedicated energy crops and trees, agricultural food and feed crops, agricultural crop waste and residues, wood wastes and residues, aquatic plants, animal wastes, municipal wastes and other waste materials. In other countries the term biomass is defined more strictly and takes biomass to mean only fuels arising from agricultural and forestry sources, using a separate category, waste fuels, for the waste products of human, urban and industrial processes. In the context of this book the latter definition will be used.

2.2.1 Potential and Current Utilisation

The stock of biomass on the land mass of Earth is currently estimated on an energy basis at 1,000,000 million TCE. Biomass as a whole grows at a rate of about 100,000 million TCE per year, i.e. about 10% of the biomass stock on Earth (Kleemann and Meliß 1993). With a fraction of 90%, forests are the biggest biomass source (CMA 1995).

Looking at the figures, the energy contained within the biomass that grows each year is 6–7 times as much as the total world primary energy consumption. However, about 50% of the biomass, such as roots and leaves, is not exploitable for energy recovery. About 2% of the world biomass production is used as food and forage, 2% is used in combustion and 1% is industrially processed to make wood products and paper and fibre materials. The fraction used as fuel, more than 1,700 million TCE per year, covers about 10% of the world primary energy consumption (IEA 2006).

In developing countries, biomass use mostly takes the form of wood combustion. Sustainable forestry strategies are generally not practiced. Worldwide, only about 10% of woodland area is used for forestry; wood that grows in nature remains unused to a large extent. Even in industrial countries where forests are cultivated, wood is predominately used by the wood-processing industry. The use of wood for energy recovery, i.e. as fuel, has minor importance in these countries. Considering the world as a whole, only a fraction of wood is used as fuel (CMA 1995).

Biomass data distinguishes between three different potentials:

- the total or theoretical potential, which describes the total accumulated biomass quantity,
- the technical potential, which is the quantity that could actually be used, and
- the economic potential, which indicates the yield that can today, or within several years, economically compete with other fuels (i.e. fossil fuels).

The technical potential is smaller than the total, and the economical lower than the technically usable potential (Kaltschmitt et al. 2006).

Estimating the biomass potential for energy consumption, a distinction has to be drawn between residual or waste biomass on the one hand and biomass from sites used exclusively for energy purposes on the other. Residual biomass that can be used for energy purposes is produced

- in farming, in the form of cereal straw, residuals from foliage plants and animal waste,
- in forestry, in large quantities in the form of residual wood, and
- in waste management, in the form of household organic residual matter and industrial waste (see Sect 2.2.1.2).

The potential of energy crops is given by the available arable land which could be used for the plantation of cereals, reed and grass plants, or fast-growing trees.

2.2.1.1 Biomass from Farming and Forestry

By-Products of Farming

Residuals and by-products from farming can be used as fuels for power production. Straw is obtained as a by-product in the production of cereals. In sugar production from sugar cane, bagasse is a by-product which is widely used as a fuel, as are pressing residues arising in the production of vegetable oils, if they do not have a use as food supplement in the feeding of livestock.

In Germany, in terms of farming residuals as fuels, straw is essentially the only one. The straw yield can be estimated from the data on the area under cultivation and the straw obtained from the respective cereal type. The amount is about 46 million tonnes/year (Schneider et al. 2007). Of the gross yield of straw, however, only a fraction can be exploited for energy purposes – the fraction that remains after farming uses has been exhausted. These uses include the ploughing of the straw into the soil to improve the soil structure and/or for the formation of humus and using it as litter or forage for livestock (Kaltschmitt et al. 2006). Based on the assumption that only about a fifth of the straw is usable as an energy carrier, the result is an energy potential of about 130 PJ/year or 4.4 million TCE/year, corresponding to a fraction of the primary energy consumption of 0.9%.

By compiling worldwide data on the fractions of herbaceous residual matter and by-products which can be used as fuels, and taking into account the relevant restrictions, it can be estimated that there is a global technical potential of about 17,000 PJ/year (580 million TCE). The biggest energy potentials in this context are found in Asia. In Europe, straws from cereals, rape and maize arise in farming. Cereals, with a cultivated area of about 33 million ha, are the most significant of these. Assuming a 20% utilisation of the straw produced, the technical potential of straw amounts to about 485 PJ/year in the EU 15 and to 721 PJ/year for the EU 30 (EU 27, Norway, Switzerland, Turkey) (Schneider et al. 2007). Including other herbaceous biomass fractions such as grass, the potential amounts to 1,000 PJ/year in the EU 15 and 1,500 PJ/year in the EU 30.

Residual Wood

In Germany, completely naturally grown forests hardly exist nowadays, apart from a few exceptions such as the Bavarian Forest National Park. Instead, forests are cultivated to obtain wood for industrial use. Besides trunk wood as the main product, the processes of thinning out and trunk wood harvesting produce residual material which today remains in the forest, to a large extent unused. This material consists of trunk wood sections and thick branches which are not suitable for industrial purposes but can be used as fuel. The additional biomass in the forest, such as withered branches and twigs, bark and leaves cannot be utilised as fuel in an industrially reasonable way and should remain in the forest to conserve the humus and nutrient cycle.

For the regional distribution of the yield, the points of reference are the woodland areas. In Germany, the well-wooded southern federal states are characterised by higher and the sparsely wooded northern states by lower potentials (Kaltschmitt et al. 2006).

In trunk wood processing, residual matter is produced in particular in sawmills and in the processing of the timber. These residues, however, are for the most part utilised as feedstocks for the paper industry and in chipboard manufacturing or as a fuel in the wood-processing industry. Wood biomass is also sourced from waste wood, i.e. wood no longer used for its original purpose (Kaltschmitt 2001; Frühwald 1990; Wegener and Frühwald 1994).

The technical potential of residual wood in Germany amounts to about 424 PJ/year of forestry residues (waste timber, bark, etc.), 57 PJ/year from the wood-processing industry and 78 PJ/year of waste wood. The total potential is 570 PJ/year, corresponding to a fraction of the primary energy consumption of 4%.

The worldwide potential can be calculated on the basis of existing wooded areas and the average of the different wood yields. The result of such a calculation is a potential of approximately 42,000 PJ/year or 1,400 million TCE. Broken down, this amount is composed half by the wood yield theoretically exploitable as a fuel, 13 and 17% by the production residuals from timber cutting and further industrial processing, respectively, 7% by the waste wood produced each year and 8% by other kinds of wood residues. The biggest potential for the exploitation of wood as an energy source is found in North America due to the currently unused large yield of wood. In the countries of the European Union, the potential yield of woody biomass, including waste wood, amounts to some 3,200 PJ/year in the EU 15 and to almost 5,000 PJ/year in the EU 30.

Energy Crops

For areas of arable land no longer needed for food production, one potential use under discussion is the plantation of energy crops. The biomass types in question are the following (Kaltschmitt et al. 2006; Kaltschmitt 2001; Lewandowski 1996):

- *Conventional cereals (barley, rye, triticale, maize).* Cereals, besides producing grain for food and forage, can also be grown for use in power production. In this process, the above-ground parts of the cereal plant (the straw and the grain) are used as a solid fuel. The advantage of the plantation of cereals to produce a solid energy carrier is the known, mature technology for cultivation and harvest. Depending on the local conditions, the resulting average annual yields of dry matter (straw and grain) for cereal crops such as triticale, winter wheat, winter barley and rye range between 9 and 13 tonnes of dry matter/ha. Arguments against the combustion of these crops for power production, which could also serve as food, are the ethical and moral concerns which arise from the context of the continued, widespread hunger around the world.
- *Fast-growing reed and grass plants.* Fast-growing reed and grass plants are C_4 plants, which in the process of photosynthesis, consuming carbon dioxide from the atmosphere, build up a compound with four carbon atoms as a first building block. The group of these plants includes maize, millet and sugar cane. In contrast, most of the plants on Earth, and almost all European plants, are C_3 plants (Borsch 1992). Due to their more efficient photosynthetic mechanism, C_4 grasses consume less water per kilogram of produced dry matter while also providing a higher yield per acre (Lewandowski 1996). The plant, dry after the growth period, can be used as a solid fuel.

The advantage of C_4 plants is their high yield of biomass; the drawback is the scant experience of large-scale cultivation and harvesting. Among the plants suitable for cultivation for energy purposes in Germany, those most suitable are those characterised not only by high yields but also by relatively low requirements for soil, climate and care.

Due to its high yields, *Miscanthus sinensis* in particular has become known as a potential energy carrier. *Miscanthus*, also called the Chinese reed, is a C_4 plant native to East Asia, which belongs to the Poaceae family. In contrast to annual grass plants such as cereals or maize, *Miscanthus* is a perennial plant which has subterranean perennial organs (rhizomes) from which new shoots develop in spring (Lewandowski 1996). *Miscanthus* is grown for a period of about 10 years, producing full yields from the third year or so. The anticipated high yields, of up to 40–50 tonnes of dry matter/ha, have in practice, in Europe, not met expectations. Instead, yearly maximum yields of 20–25 tonnes dry matter/ha from the third year seem to be realistic from fields in central Europe (Hartmann and Strehler 1995). Depending on local and climatic conditions, the yield may also be a lot lower (Kaltschmitt 1993). In central Europe, frost in winter may damage the rhizomes and hence diminish the yield.

Other C_4 plants are the reed and the giant reed, types of millet also belonging to the Poaceae family. Compared to *Miscanthus*, however, they are expected to produce lower yields under central European conditions. The millet types which can be cultivated in Germany are C_4 plants of tropical origin too. In conditions of heavy precipitation and mild climate, the achievable dry matter yields range between 15 and 22 tonnes/ha yearly.

- *Fast-growing trees (willow, poplar)*. Biomass can also be produced through fast-growing tree types, such as willow or poplar, which are grown as short rotation crops. After a breeding phase, the above-ground biomass is mechanically harvested after 4–20 years. In the form of woodchips, it can be used as a solid fuel. The tree stumps sprout again. The biomass can be harvested again after 2–12 years, respectively, depending on the site, climate and the tree type. In Germany, the respective yearly yields are 12–15 tonnes of dry matter/ha (Kaltschmitt 1993; Hartmann and Strehler 1995).

The fundamental parameter for the technical potential of energy crops is the area available for cultivation. Worldwide, this area is estimated to be between 350 and 950 million ha. In industrial nations, the area of the existing arable land which can be assumed to be available for the cultivation of energy crop averages around 7%. In developing countries, the area theoretically available and suitable for energy crop growing is on average considerably higher. Supposing that a mix of plants suited to the given location was cultivated on these areas, a technical energy potential can be calculated. The calculated potentials vary between 37 and 82 EJ/year. The highest potential in this respect is in Africa. The potential in Europe is limited. The countries of the European Union offer a potential in the range of 1.8–3.8 EJ/year. In Germany, in the medium term, a maximum area of 2 million ha will probably be useable for energy crop cultivation, so a potential of about 365 PJ/year has to be assumed.

Summary of Potentials and Current Utilisation

Table 2.6 compiles the above-discussed potentials for Germany and shows the extent of current use. At present, almost all residual wood from forestry and industry, as well as all waste wood, is exploited. Other sorts of wood and straw remain unused, so there is a potential to increase the share of biomass in primary energy consumption from the current 2% or so up to about 8%. Other authors state a potential use of solid biomass of between 2 and 15% of the primary energy consumption.

The dominant renewable energy source in Europe is biomass, with a share of 4.5% of the primary energy consumption in 2005 and 68% of total renewables. Biomass provides 30% of the PEC in Latvia and nearly 20% in Finland. Most of this is wood. Sweden is not far behind with 17.5% (Eurostat 2007). The specific differences between the countries result from differing boundary conditions, such

Table 2.6 Biomass potential and utilisation in Germany (Schneider 2007)

	Potential	Utilisation	Potential/PEC	Utilisation/PEC
	in PJ/yr		Share in %	
Residual forest wood	169	147–165	3.0	1.0–1.1
Small wood	123			
Additional forestry wood	132			
Wood industry residuals	57	51	0.4	0.4
Waste wood	78	62	0.5	0.4
Other woody biomass	10	1	0.1	0
Straw	130	3	0.9	0
Grass, other	48–77	0	0.4–0.6	0
Energy crops	365	0	2.6	0
Total	**1,112–1,141**	**261–279**	**7.8–8.0**	**1.8–2.0**

PEC: Primary energy consumption

as the fraction of forest area, the fraction of agriculturally productive land, the climatic conditions or national policies. Furthermore, in countries which, compared to Germany, have a higher use of biomass, the potentials are higher than the current utilisation.

Worldwide, though, the share of biomass in primary energy consumption is significantly higher than in Europe. Table 2.7 shows the worldwide potentials of wood

Table 2.7 Biomass potential, current utilisation and share of PEC in different regions of the world (Schneider 2007; Van Loo 2008; Kaltschmitt et al. 2009)

	North America	Latin America	Asia	Africa	Europe	Middle East	Former SU	Total
Potential [EJ/a]								
Wood	12.8	5.9	7.7	5.4	4.0	0.4	5.4	**41.6**
Herbaceous biomass	2.2	1.7	9.9	0.9	1.6	0.2	0.7	**17.2**
Dung	0.8	1.8	2.7	1.2	0.7	0.1	0.3	**7.6**
Biogas	(0.3)	(0.6)	(0.9)	(0.4)	(0.3)	(0.0)	(0.1)	**(2.6)**
Energy crops	4.1	12.1	1.1	13.9	2.6	0.0	3.6	**37.4**
Total	**19.9**	**21.5**	**21.4**	**21.4**	**8.9**	**0.7**	**10.0**	**103.8**
Current utilisation [EJ/a]								
Trad. biomass		1.2	22.5	9.7				**33.4**
Modern biomass	4.1	2.4	3.6	2.3	3.4		0.7	**16.8**
Total	**4.1**	**3.6**	**26.1**	**12**	**3.4**	**0**	**0.7**	**50.2**
PEC [EJ/a]	120.4	21.8	154.8	25	78.9	19.5	46.5	**473**
Utilisation/ PEC [%]	3	17	17	48	4	0	2	**10.6**
Potential/ PEC [%]	17	98	14	86	11	1	22	**22**

and herbaceous residual matter and energy crops differentiated by region and related to the primary energy consumption. Globally, a technical potential of biomass of about 100 EJ/year can be surmised, which corresponds to a share of 22% of the total primary energy consumption in 2006. The current utilisation of biomass, as a percentage of the primary energy consumption, is 10.6%. This high share comes about from traditional biomass use in fast-developing and developing countries, for example as firewood. Table 2.7 distinguishes between modern and traditional biomass utilisation. Modern refers to modern technologies, such as biomass combustion for combined heat and power production. The share of modern biomass in PEC is around 3.5% worldwide (Van Loo and Koppejan 2008; Schneider et al. 2007).

2.2.1.2 Wastes

Waste is an unwanted or undesired material or substance. The European Union, under the Waste Framework Directive (EU 2008), more precisely defines waste as an object the holder discards, intends to discard or is required to discard.

The waste management ambition in Germany and Europe is to avoid the production of waste, for instance by using low-waste production processes. If waste is produced it should be used as a material (recycling) or thermally to convert the energy content of the waste to useful thermal or electrical energy (recovery). The disposal of wastes is the option used as the last resort. Disposal includes on the one side dumping (to landfill) but also thermal conversion processes, where disposal is the primary objective.

This means that the thermal treatment of waste can be classified either as disposal or as recovery. The distinction between recovery (or utilisation) and disposal is based on the energy efficiency of the process. This is laid down in the European Waste Framework Directive (EU 2008), where an energy efficiency criterion, $R1$, is defined. The utilisation of waste in plants having an efficiency above a certain value is considered recovery, and below this value it is considered disposal. The $R1$ criterion is defined in a footnote to Annex II of the Waste Framework Directive (EU 2008) and is discussed in Sect. 6.4.

Political specifications and laws have affected a change in the total wastes produced and their division. The total waste volume in Germany is going down, and at the same time the proportion of recovered matter is increasing. Usable materials such as paper, glass, metal and plastics are collected separately or get separated from other municipal wastes once collected. Only particular waste types have a calorific value.

Of the total waste volume in Germany of 331 million tonnes in 2005 (Becker et al. 2007), only a minor part was of organic origin. Given that a definite distinction between organic wastes and wastes of fossil origin is not possible in most cases, Table 2.8 presents an overview of the entire waste volume in Germany. It distinguishes between waste volumes from manufacturing industries and the wastes collected by public waste disposal services (Bilitewski et al. 2000).

A major fraction of waste arises in the building and mining industries, mainly as building rubble (180 million tonnes per year) and overburden from mining (52

Table 2.8 Amount of wastes in Germany (Becker et al. 2007)

	2002	2003	2004	2005
Waste volume	1,000 t			
Total	381,262	366,412	339,368	331,876
Building rubble and demolition waste (incl. roadway rubble)	240,812	223,389	187,478	184,919
Mining spoil (non-hazardous waste)	45,461	46,689	50,452	52,308
Wastes from production processes and industry	42,218	46,712	53,005	48,094
Municipal wastes	52,772	49,622	48,434	46,555

All values in thousands of tonnes

million tonnes per year). However, these wastes have little or no exploitable energy (Becker et al. 2007). The 48 million tonnes (approximately) of waste per year in the producing industries is distributed over a multitude of material groups, each of which can be partly utilised for energy purposes. Examples are residual matter from paper production, wood treatment, petroleum finishing and coal beneficiation and plastic and textile waste.

Municipal solid waste (MSW) had a share of 14% of the total waste volume, corresponding to 46 million tonnes, in 2005. Figure 2.10 shows the amount, the utilisation and disposal of MSW in Germany. MSW predominantly refers to household waste (domestic waste), and sometimes also to commercial wastes collected by a municipality. Due to the increasing proportion of separated fractions, such as paper, plastics, glass in Germany, the amount of mixed household waste decreased

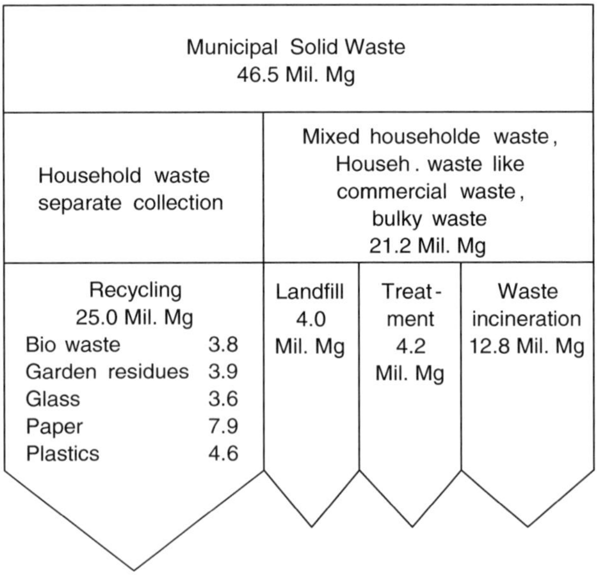

Fig. 2.10 Amount, utilisation and disposal of MSW in Germany in 2005 (data from BMU 2007a)

continuously in the last years to about 14 million tonnes per year in 2005 (Becker et al. 2007). Mixed household waste is also termed residual waste (or household refuse). The average calorific value of residual waste ranges between 6 and 10 MJ/kg (Thomé-Kozmiensky 1994). With its heterogeneous composition and its diverse types of hazardous matter, these waste types are disposed of in specially designed waste incineration plants, mostly stoker-fired furnaces. In 2007, 18 million tonnes was thermally treated in 72 waste incineration plants (BMU 2007b). In accord with national law (*TA Siedlungsabfall*, German Technical Specifications for the Disposal of Municipal Waste) dumping of wastes with an organic fraction of greater than 5% has been forbidden in Germany since 2005.

2.2.1.3 Refuse-Derived Fuels

In Germany and in other European countries, municipal, industrial and residual wastes are partially pretreated and then prepared into fuels (refuse-derived fuel (RDF) or secondary recovered fuel (SRF)). The aim of the preparation is to improve the quality of the waste stream in a way that the substitute fuel produced can be burned in plants without operational problems and without pollution loading. The use as a fuel in an RDF power plant or as a supplementary fuel in coal-fired power plants or cement kilns in this respect imposes various requirements on the fuel. The production of RDF as a rule uses the following waste streams as feedstocks:

- Mixed household waste (residual waste)
- Bulky waste
- Household waste like commercial waste
- Homogeneous, mostly industrial, waste streams such as plastics, paper, wood and textiles

The purpose of the treatment is to produce a homogeneous, highly calorific, chemically and biologically stable and low-pollution fuel. There are a great number of methods available which treat the feed material in a mechanical, thermal or biological way. Typical process steps of the preparation are

- drying by thermal or biological methods,
- mechanical separation of partial streams by selection and classification (using an air classifier or rotary drum screens),
- separation of iron and non-ferrous metals,
- separation or reduction of impurities, for example chlorine, and
- size reduction and homogenisation.

The processing usually consists of the sieving out of the fraction of fines, crushing, metal separation and drying and usually increases the heating value. The separated metals are sold. It is also possible to obtain reduced chlorine contents by carefully selecting specific waste streams, especially in the case of commercial waste. There are various RDF utilisation schemes around the world. In most countries the RDF is processed on the site of the energy from waste (EfW) plant.

In Germany, many decentralised plants with a combination of mechanical and biological treatment of waste have been built in recent years. The purpose is to produce a fuel which can be utilised in EfW plants elsewhere. Two process variants are distinguished:

- Conventional mechanical – biological treatment (MBT) first separates metals and highly calorific components from the feed waste. The highly calorific components are used as a substitute fuel/RDF for co-firing in coal-fired power plants or as the only fuel in RDF power plants. The remaining fraction goes to landfill after biological treatment (aerobic digestion).
- The aim of mechanical – biological stabilisation (MBS) is to dump no or only small amounts of mineral wastes and to use most of the feed waste for the production of substitute fuels (stabilate). The feed waste is first dried in the biological process by the reaction heat that is produced. The dried wastes are then sorted into recyclable fractions (substitute fuels, ferrous and non-ferrous metals, etc.). The substitute fuel/RDF is then used for co-combustion in coal-fired power plants or RDF power stations.

The energy balance of a mechanical – biological process is very much dependent on the process configuration. A typical ratio of the energy output of the RDF fuel to the energy input of the feed waste is about 60–70% for MBT and 80–90% for MBS.

Both variants can only be used when there are sufficient capacities in industrial fuel-burning plants capable of handling substitute fuels (produced from highly calorific fractions of the wastes) or stabilates.

Mechanical – biological waste treatment – as opposed to thermal waste treatment – is not an independent disposal process but divides the waste into various groups and prepares these for disposal or recycling. MBT processes therefore require integration into other waste management processes for the further disposal of the waste fractions produced.

The total capacity of the mechanical – biological waste treatment plants in Germany currently ranges around 5–6 million tonnes per year. After completion of all the plants planned in 2006, 66 mechanical – biological waste treatment plants with a capacity of about 7.1 million tonnes/year will be available (UBA 2008).

2.2.1.4 Sewage Sludge

Sewage sludge shall be discussed here as an example of a homogeneous waste type obtained in great quantities. Sewage sludge is the residual matter from treatment processes of household and industrial wastewaters. The quantity of it depends on the number of households in the treatment plant catchment, the industrial wastewater load and the efficiency of the sewage treatment plants (Bilitewski et al. 2000). A distinction is made between raw sludge and digested sludge.

In 2003, in Germany, about 2.2 million tonnes of dry solid matter of sewage sludge was obtained from municipal wastewater treatment (Schmelz 2006). This quantity, however, does not correspond to the actual loading of wastewater treatment plants, because sewage sludge has a moisture content of 92–98%. Common practice in this respect is to reduce this content by mechanical dewatering to obtain a moisture content of 30–45% in the dry matter; in a few cases, the sewage sludge is afterwards thermally dried to a moisture content of 5–10% in the dry matter. Due to the high water content, the energy content of sewage sludge is low. At a moisture content of 30% of dry matter, it ranges around 1–2 MJ/kg.

The purpose of the dewatering and thermal treatment of sewage sludge at sewage treatment plants is disposal and volume reduction only. Energy is generally not produced for more than in-plant use. The weight reduction obtained by sludge treatment is shown in Fig. 2.11 (Gerhardt et al. 1996). The greatest volume reduction, greater by a factor of 5–10, is achieved by the mechanical sewage sludge dewatering process on the premises of the sewage plant. Subsequent thermal drying at the sewage plant or in combination with a power plant again reduces the volume down to between half and a quarter of the volume after mechanical dewatering. The combustion of the organic components reduces the volume only by a factor of 2. Combustion is necessary to produce waste which is dumpable according to *TA Siedlungsabfall* (the German Technical Specifications for the Disposal of Municipal Waste).

In 2004, 56% of sewage sludge produced in Germany was used for agriculture or recultivation and 38% was burned. It is expected that the use in agriculture will decrease due to more stringent limits on trace metals and the falling public acceptance of such use, thus promoting thermal sewage sludge utilisation (Schmelz 2006).

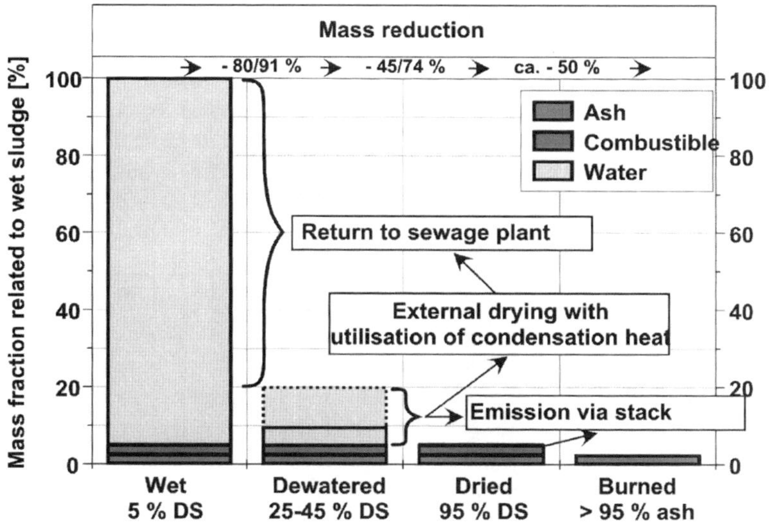

Fig. 2.11 Effect of treatment on the volume reduction of sewage sludge (Gerhardt et al. 1996)

2.2.2 Considerations of the CO_2 Neutrality of Regenerative Fuels

Carbon dioxide is produced from the combustion of biomass as well as from fossil fuels. However, an equivalent quantity of CO_2 is taken up from the atmosphere by the plant during its growth. Thus in agricultural systems, which follow regulated cultivation methods, the growth period has an effect of balancing out the CO_2 emissions from the utilisation of biomass as a fuel. When biomass is overexploited, such as in the case of the slash-and-burn of tropical forests, the growth period following utilisation does not adequately compensate for the CO_2 produced during combustion. The bound carbon is released, and so slash-and-burn has to be seen as the same as fossil fuel utilisation (Schmidt 1992).

When residual matter such as straw or forest wood residue is used for energy purposes, most of the emitted CO_2 is extracted from the atmosphere again during the growth period of the cycle. However, because there is no strict interdependence between the production of the biomass and its use as a fuel, the CO_2 capture of the growth period cannot be set against the CO_2 release during its utilisation for energy; no reduction of CO_2 emissions follows from it at first. If, on the other hand, this biomass is not utilised for energy, the carbon is released to the atmosphere again in the form of CO_2 or as methane (which is much worse) during natural decomposition. The same is true for the organic fractions of household refuse or sewage sludge (Kaltschmitt et al. 2006).

2.2.2.1 Comparison of *Miscanthus* and Hard Coal on a Greenhouse Gas Emissions Basis

A comparative study was made between the use of coal and the use of cultivated *Miscanthus* as a fuel (Kicherer 1996). *Miscanthus* was grown on permanent fallow land and, once harvested, co-combusted in an existing pulverised hard coal fired power plant. As a basis, it was assumed that the biomass was transported 50 km on average and that it substituted coal directly.

When comparing the CO_2 emissions, the CO_2 generated during the production processes were taken into account, for example the operation and maintenance of machines and buildings. Additionally, the CO_2 emissions involved in the production of goods such as fertilisers were considered. However, the CO_2 emissions from the construction of machines and buildings were ignored. CO_2 emissions were also produced during the transport and preparation of *Miscanthus*, and this was accounted for. Furthermore, the estimated additional N_2O emissions from the soil as a result of the cultivating of *Miscanthus* were included and converted into CO_2 equivalent emissions using N_2O's greenhouse-CO_2 equivalency factor. Figure 2.12 shows the percentage contributions to the total greenhouse gas emissions of the various steps in *Miscanthus* processing. It is conspicuous that nitrogen fertilisation contributes almost 50% of the greenhouse gas emissions. On one hand, this has to be put down to the energy which has to be expended for the production of the fertiliser and, on the other, to the N_2O emissions released by the nitrogen fertiliser when spread. The contribution of the transport of the biomass over distances of 50 km, in con-

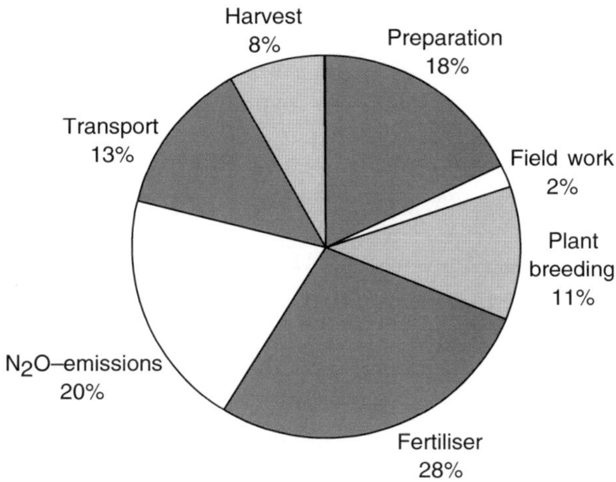

Fig. 2.12 Breakdown of the CO_2 emissions in *Miscanthus* processing (Kicherer 1996)

trast, is only a small fraction. The assumed preparation method of the biomass was pulverisation.

Considering the emissions released in the combustion of *Miscanthus*, it can be observed that more CO_2 per MJ is released than in the combustion of coal, i.e. 103 kg CO_2/GJ (Fig. 2.13). This amount, though, is to a great extent compensated by a negative contribution from the uptake of CO_2 during the growth period. In total, the resulting CO_2 emissions of production and thermal utilisation of *Miscanthus* amount to 6.2 kg/GJ when factoring in the growth period. In the combustion of hard coal, in comparison, 93.2 kg/GJ is released directly, while during the mining and preparation processes, additional CO_2 emissions of 3.4 kg/GJ are made. Utilising *Miscanthus* reduces CO_2 emissions by 93% compared to the combustion of hard coal.

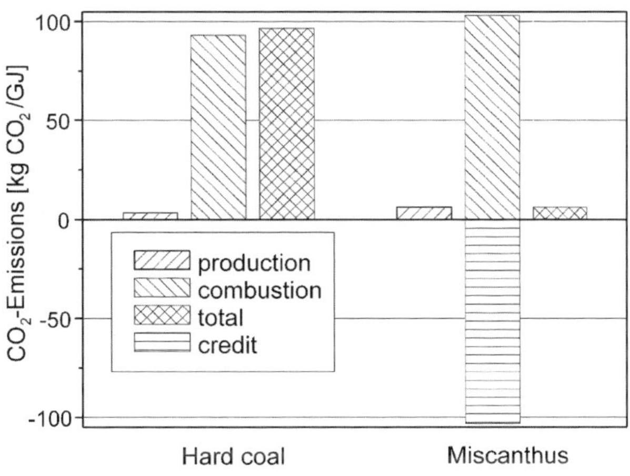

Fig. 2.13 CO_2 emissions from the combustion of *Miscanthus* and hard coal

2.2.2.2 Harvest Ratios

The result of the evaluation of the regenerative energy utilisation of a fuel is its energy balance. It is given as an output/input ratio by means of so-called harvest ratios, where the useful energy of an energy medium is set in proportion to the expenditure of energy necessary for its production (Hartmann and Strehler 1995; Born 1992). If the harvest ratio is above 1, this means that, using the technology, and for the fuel considered, energy is released and CO_2 abated. Harvest ratios below 1 often occur but such crops are not realistic candidates for energy production because in those cases more energy is expended during growth and preparation than is gained through utilisation.

Biogenous solid fuels yield harvest ratios between 10 and 20 or so. In the study mentioned above, a harvest ratio of 14 was calculated for *Miscanthus* (cultivation and utilisation). According to others (Hartmann and Strehler 1995), the harvest ratio for *Miscanthus* is over 19 (see Fig. 2.14). Liquid energy media such as rapeseed oil or ethanol from sugar beet or sweet sorghum have lower harvest ratios.

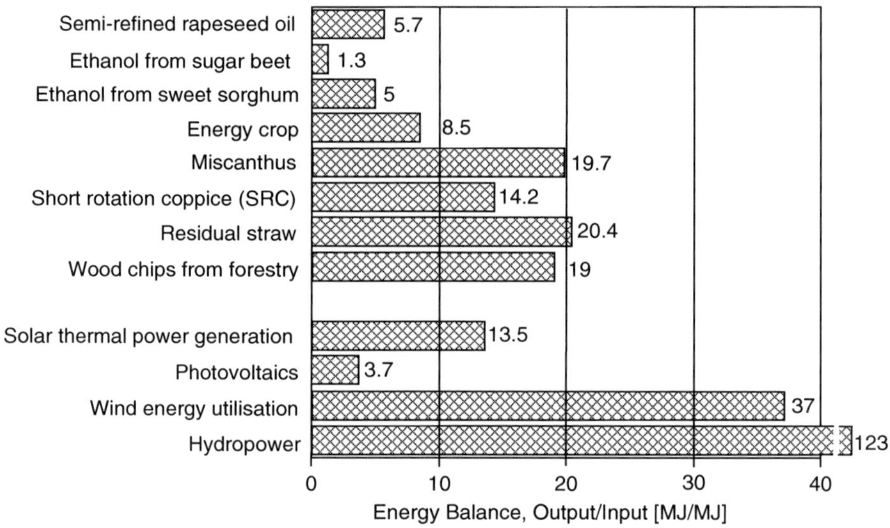

Fig. 2.14 Harvest ratios of various biomass types (Hartmann and Strehler 1995)

2.2.3 Fuel Characteristics of Biomass

2.2.3.1 Biomass from Farming and Forestry

Molecular Structure

Biomass essentially consists of macromolecular organic polymers – lignin, cellulose and hemicellulose. Cellulose is by far the most common organic substance. It is a polysaccharide consisting solely of glucose chains which are held together by

Table 2.9 Components of biomass (% by wt) (Kicherer 1996)

	Lignin	Cellulose	Hemicellulose	Ash	Other
Hardwood	26–31	40–48	19–25	1	3
Softwood (coniferous wood)	22–25	35–43	21–30	1	3
Wheat straw	18	32	37	8	5
Miscanthus	18	40	34	3	7

hydrogen bonds in crystalline clusters, forming the framework of the cell walls. Cellulose is an important raw material for the chemical industry (cellulose production). Hemicellulose or polyoses are structurally similarly to cellulose, but also contain other sugar types as basic building blocks, not only glucose chains. Lignin, one of the lignocellulose substances, is a three-dimensional aromatic branched-chain macromolecule; it acts as a binder for the cellulosic tissue. Lignin is responsible for the lignification of the cell walls. Table 2.9 shows the molecular composition of the various biomass types. It is observable that woods have higher lignin contents than herbaceous plants (Kleemann and Meliß 1993; Kicherer 1996; CMA 1995).

Moisture Content

The moisture content of fuel derived from biomass is generally higher than the respective moisture content of hard coal. Straw and whole cereal plants immediately after the harvest may have moisture contents up to 40%, but they can be reduced to below 20% within 2–3 days by field drying, provided the weather is favourable (Hartmann and Strehler 1995; Clausen and Schmidt 1996). With energy-grass crops like *Miscanthus*, moisture contents below 20% can also be achieved by choosing to harvest in spring, after the leaves and petioles have dried (Lewandowski 1996). Values below 20% are required for herbaceous biomass so that it can be stored while avoiding the formation of moulds and spores (Wieck-Hansen 1996; Clausen and Schmidt 1996).

Wood in a fresh state contains between 40 and 60% moisture. This content can be reduced by partially drying the unchopped, uncut wood in the forest or, in the case of woodchips, by a subsequent drying process in a storage area. With coarse woodchips, the dry state is achieved by natural air circulation, while for fine woodchips, forced ventilation is necessary. Given sufficient drying time (several months) and ventilation, the moisture content can also be reduced to less than 20% (Hartmann and Strehler 1995; Kaltschmitt 2001).

Calorific Value

The lower heating value (LHV) of dry ash-free ligneous and herbaceous biomass ranges between 17 and 21 MJ/kg; the calorific value is between 16 and 20 MJ/kg. Ligneous biomass has a somewhat higher calorific value than herbaceous biomass. Basically, however, the calorific value of biomass is determined by its moisture

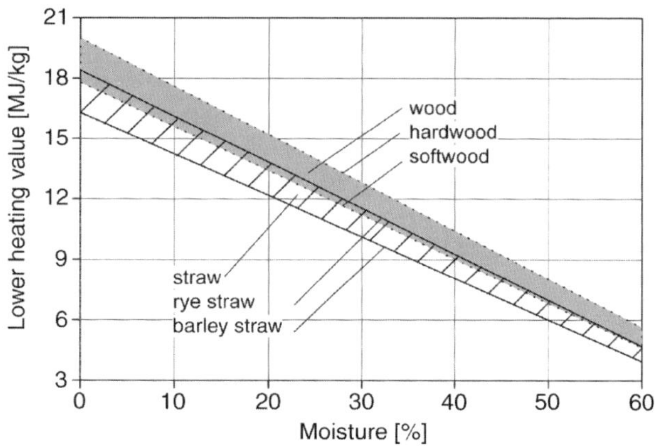

Fig. 2.15 Calorific value as a function of the moisture content

content; starting out from the dry matter, it diminishes with an increasing moisture content (see Fig. 2.15). Up to 60% moisture, the calorific value of wood may be between 6 and 18 MJ/kg. Air-dried wood with 15–20% moisture has a calorific value between 14 and 15.2 MJ/kg.

Volatile Matter, Residual Char, Ash

Figure 2.16 compares the contents of volatile matter, fixed carbon and ash of straw, wood, hard coal and brown coal. Biomass has a markedly higher volatile matter content than hard coal. As the fuel is heated in the furnace, the volatile matter is released

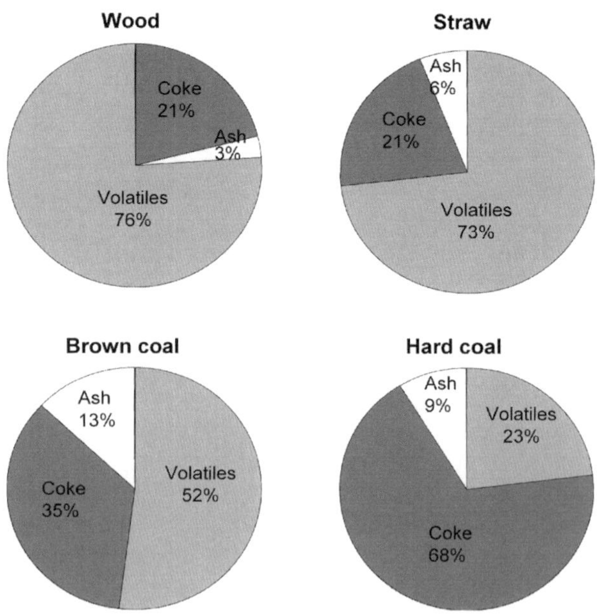

Fig. 2.16 Volatile matter, residual char and ash contents of various biomasses and coals

and homogeneously burned. This way, a small residual char fraction remains, which has a high porosity and hence is very reactive. Ligneous biomass, as a rule, has a low ash content. Herbaceous biomass types have ash contents similar to hard coal if the ash content is referred to the calorific value.

Elemental Composition

Table 2.10 shows the composition of different biomass types, including typical values for the constituents as well as their ranges. Biomasses have significantly lower fractions of carbon, while their oxygen contents exceed that of coal many times over. The hydrogen fractions are somewhat higher than that of coal. The high oxygen fractions and the associated partial oxidation of fuel molecules mean a lower calorific value of dry ash-free matter in comparison to coal.

Relevant to pollutant formation are the trace elements nitrogen, sulphur and chlorine. Figure 2.17 displays the contents of these compounds in various solid fuels (with respect to their calorific values).

Compared to hard coal, all biomass types are distinguishable by significantly lower sulphur contents (again, with respect to the calorific value). On top of this, SO_2 that is formed during the combustion of biomasses may be bound by the ash, so that the SO_2 emission limits can be met without sophisticated desulphurisation engineering.

The content of nitrogen in the fuel depends on the biomass type and the way it is cultivated. While wood contains very little nitrogen, straw as fuel can mean nitrogen inputs to firing in the same order of magnitude as, or higher than, hard coal. Nitrogen contained in the grain of whole cereal plants is significantly higher in concentration. For perennial grass plants like *Miscanthus*, a transfer of the nutrients (nitrogen, potassium, phosphorus) from the sprouts to the rhizome occurs in late summer, so that the nitrogen content in the plant matter above ground decreases (Lewandowski 1996). Biomass in general is an excellent fuel in regard to applying primary combustion-engineering measures, given that most of the nitrogen is released into the gas phase during the combustion of volatile matter.

A much more problematic constituent than nitrogen and sulphur in the fuel is chlorine, which is the cause of operational problems as well as pollutant emissions problems. Chlorine contents in herbaceous plants are in some cases far higher than that of coal. Cereal straw, in this respect, has the highest values. Wood, in contrast, has low chlorine contents. Chloride is taken up from the soil by the roots of energy crops. Chloride is found naturally in soils but is also part of fertilisers, in the form of potassium chloride (KCl). In coastal areas, the chlorine content of plants is higher, due to the higher salt concentration in the air. Tests are being carried out to reduce the chlorine content of biomass by replacing the potassium component of the fertiliser. Results of such tests are that the chlorine content could be reduced to a third. In the case of open-air storage of straw, most of the chloride is leached by rain (Wieck-Hansen 1996).

Table 2.10 Fuel composition of biomass types (Kaltschmitt 2001; Lewandowski 1996; Hartmann and Strehler 1995; Clausen and Schmidt 1996; Obernberger 1997; Spliethoff et al. 1996)

	Straw		Wood		*Miscanthus*		Whole cereal plants		Hard coal (comparison)	Brown coal (comparison)
	Typical value	Range	Typical value	Range	Typical value	Range	Typical value	Range	Göttelborn	Fortuna
Moisture content [%]	15	10–20	45	20–60	20	10–30	15	10–20	7	55
LHV, raw [MJ/kg]	14.8	12.5–16.4	9.6	5.7–15.5	14.0	11.2–16.6	14.9	12.5–16.6	27.9	8.7
LHV, dry ash-free [MJ/kg]	18.7	17.5–19.0	19.5	18.5–20.0	18.5	18–19	18.7	17.5–19	30.2	22.2
Ash % dry	4.5	3–7	0.5	0.3–4	2.5	1.5–5.0	4.0	3–7	8	9
Volatile matter % dry	78	75–81	80	70–85	80	78–84	78.0	75–81	35.1	53
C	47.0	46–48	50	49–52	48	47–50	47.0	46–48	74.3	62.8
H	6.0	5.4–6.4	5.8	5.2–6.1	6.0	5.2–6.5	6.0	5.3–6.8	5	4
N	0.5	0.3–1.5	0.2	0.1–0.7	0.3	0.1–0.4	1.4	0.4–1.7	1.5	0.5
S	0.15	0.10–0.2	0.05	< 0.1	0.1	0.02–0.13	0.1	0.07–0.11	1	0.5
Cl	0.4	0.1–1.1	0.02	< 0.1	0.3	0.1–0.4	0.3	0.25–0.5	0.2	
O (difference)	41.5		43.4		42.8		41.2		9.5	23.2

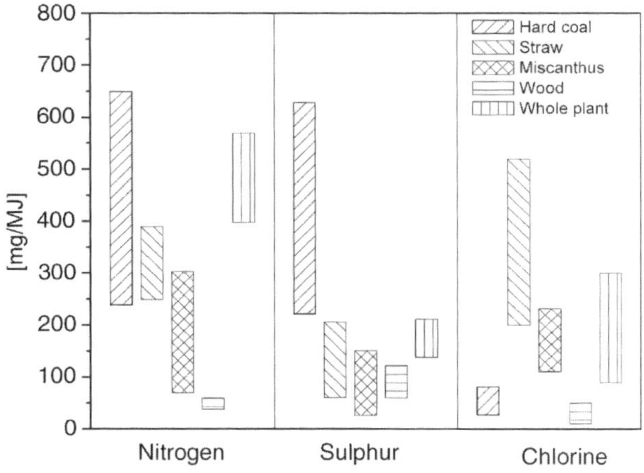

Fig. 2.17 Ranges of nitrogen, sulphur and chlorine contents in biomass compared to hard coal

Ash Fusion Characteristics

Wood has ash fusion temperatures like hard coal, in the range of 1,200–1,400 °C. Straw has significantly lower initial ash deformation temperatures (ca. 900 °C), so more severe fouling and slagging problems have to be expected. Figure 2.18 draws a comparison between the ash fusion characteristics of various types of biomass and fossil fuels. The comparison also reveals the great scattering of values within the same biomass type.

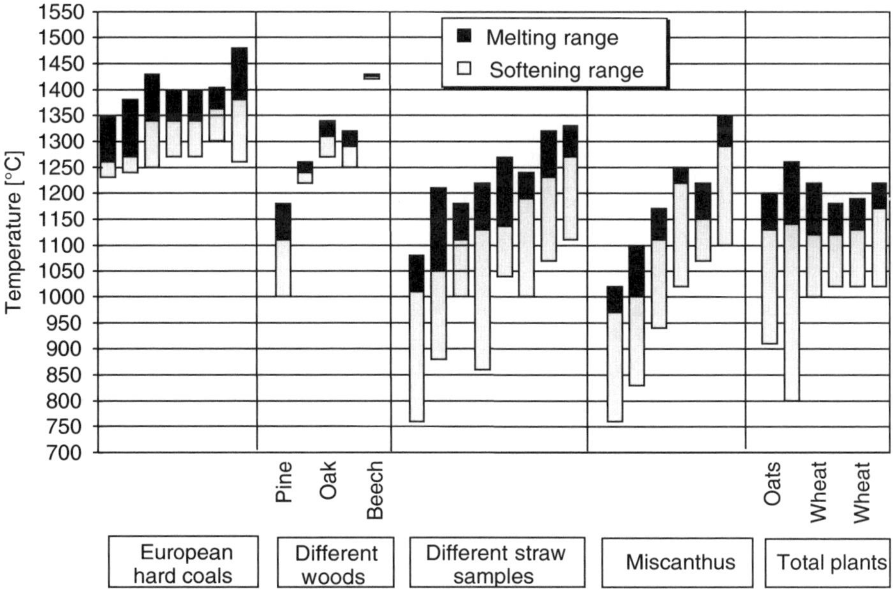

Fig. 2.18 Ash fusion temperatures of various biomass types

Table 2.11 Ash composition (%) of a wood (spruce) and a straw compared with one hard and one brown coal type

	Straw	Spruce	Hard coal	Brown coal
SiO_2	65.43	29.61	43.46	11.07
Al_2O_3	0.59	2.59	27.83	8.05
Fe_2O_3	1.17	6.73	9.93	5.03
CaO	9.47	37.06	5.21	31.19
MgO	1.76	5.38	2.75	4.02
K_2O	18.07	9.52	3.54	0.10
Na_2O	0.20	1.97	1.18	0.10
SO_3	0.98	3.21	4.42	40.24
TiO_2	0.10	0.31	1.08	0.20
ZnO	0.00	0.21	0.10	0.00
P_2O_5	2.25	3.42	0.49	0.00

The low fusion temperatures of herbaceous biomass can be put down to the composition of the inorganic ash components. Comparing the components, it can be seen that Si, Al and Fe dominate in the ash of hard coal, while Si, K and Ca dominate in biomass ash. For ash of herbaceous biomass in particular, the melting point is lowered by its high potassium content, which, with respect to the calorific value, is about 4–20 times as much as the content in hard coal. Table 2.11 shows the ash compositions for a wood type (spruce) and a straw type compared to one hard and one brown coal.

Densities of Biomass Types

The density of a fuel type has an influence on the transport method and the associated costs, the necessary storage space and the required fuel preparation and feeding. For biomass, this density is significantly lower than for fossil fuels and depends not only on the fuel type (straw, wood, cereals, C_4 grass plants), but also on what form the fuel is in (i.e. bales, chaff, chips, pellets, powder, shavings).

Table 2.12 shows the density of various types of biomass, including variations for different forms of particular biomasses.

Table 2.12 Densities (at a moisture content of 15%) of various biomasses (kg/m^3) (Kicherer 1996; Hartmann and Strehler 1995)

Biomass	Density		Bulk density	
Herbaceous biomass:	Large-size cubic bales	Round bales	Chaff	Pellets
Straw	150	120	70	520
Miscanthus	130		120	
Whole cereal plants	220	190	130	560
Grain			Grain 700	
Wood	Cordwood 300–500		Chips 200–300	Pellets 650

Table 2.13 Energy densities of various biomasses

Fuel	Density ρ [kg/m^3]	Lower heating value (LHV) [MJ/kg]	Energy density [GJ/m^3]
Straw, large-size cubic bales	150	14.4	2.2
Straw, chaff	70	14.4	1.0
Straw, pellets	520	14.4	7.5
Whole plant, large-size cubic bales	220	14.4	3.2
Miscanthus, large-size cubic bales	130	14.4	1.9
Wood chips	250	15.3	3.8
Hard coal	870	28	24.4
Brown coal	740	10	7.4

The form of preparation that has become generally accepted for ligneous biomass is that of woodchips; for herbaceous biomass, according to experience in Denmark, big bale systems seem to be most suitable for straw. Further compaction in the field or in the forest is not beneficial for transport, but means additional costs and energy expenditures.

Due to the low densities of biomasses and their low calorific value, the resulting energy densities lie about one order of magnitude below the density of hard coal and significantly below the density of brown coal (see Table 2.13).

2.2.3.2 Waste

The fuel properties of residual wastes differ a lot from region to region depending on the relative fractions of the material groups (such as plastics, paper, cardboard, wood and organics) in the waste. Table 2.14 shows the distribution of the material groups for one type of residual waste in Germany. Based on the moisture contents and the calorific values of each group, it is possible to determine the average values of a residual waste as a whole. In the given case the result is a mean moisture content of 33% by weight and a mean calorific value of about 8.5 MJ/kg (Hoffmann et al. 2008). The upper and lower limits of the fuel properties of residual waste are given in Table 2.15 (Reimann and Hämmerli 1995).

In the past few decades, the lower heating values (LHVs) of municipal wastes have risen substantially in industrial countries. This is partly due to an increased consumption of paper and plastic materials. The widespread introduction of the separate collection of organic waste, with its relatively low heating value, has also contributed. Whereas in the 1980s, the average LHV was in the range of around 6 MJ/kg, the value increased in Germany to 8.7 MJ/kg in 1992. Today, for the design of a municipal waste incinerator, a design heating value of 9.5–10 MJ/kg is chosen (Bilitewski et al. 2000). Figure 2.19 shows the variations of the lower heating values for different countries.

Table 2.14 Composition of residual MSW (example) (Hoffmann 2008)

Fraction	Fraction of waste [wt%]	Moisture [wt%]	LHV [kJ/kg]
Organics	35.0	65.0	7,000
Paper, cardboard	8.0	25.0	11,000
Wood	3.0	31.0	15,000
Fine fraction (< 10 mm)	19.0	23.0	3,500
Combined materials	6.0	12.0	12,000
Other	5.0	5.0	6,000
Textiles	4.0	28.0	14,000
Plastics	10.5	6.0	22,500
Fe metal	2.0	0	0
NF metal	0.5	0	0
Glass	3.0	0	0
Minerals	3.0	0	0
Pollutants	1.0	0	5,000
Average		33.0	8,438

Table 2.15 Variations of fuel characteristics and the composition of residual MSW in Germany (Effenberger 2000)

Ultimate analysis (%)	Fusion behaviour (fly ash) (°C)	Heavy metals (g/kg raw)
H = 4–5 (water free)	Deformation temp. 1,100	Pb = 0.6–2
S = 0.2–0.7 (water free)	Fluid temp. 1,260	Cu = 0.12–0.78
O = 17–30 (water free)		Fe = 10–100
N = 0.3–0.45 (water free)		Zn = 0.44–2.3
Cl = 0.5–1.5 (water free)		Sn = 0.05–0.32
		Cr = 0.02–0.88
Ash ≅ 25	Bulk density in kg/m^3	Cd = 0.003–0.012
Moisture ≅ 30		Ba = 0.084–1.225
Combustable = 45	Bulk 90–120	
	Collection vehicle 350–550	
LHV=8,300–10,500 kJ/kg	Receiving bunker 200–300	

2.2.3.3 Refuse-Derived Fuel (RDF)

Table 2.16 shows the composition of various refuse-derived fuels produced from different input materials (and different mechanical – biological treatment methods for MSW). As described in Sect. 2.2.1.3, the preparation methods serve to produce a homogeneous, highly calorific fuel with reduced levels of pollutants which can be burned in an RDF power plant or co-fired in a coal-fired power plant. It is noticeable that the calorific values are significantly higher (up to 25 MJ/kg) than the basic waste. Utilisation problems can be posed in particular by the contents of chlorine and heavy metals.

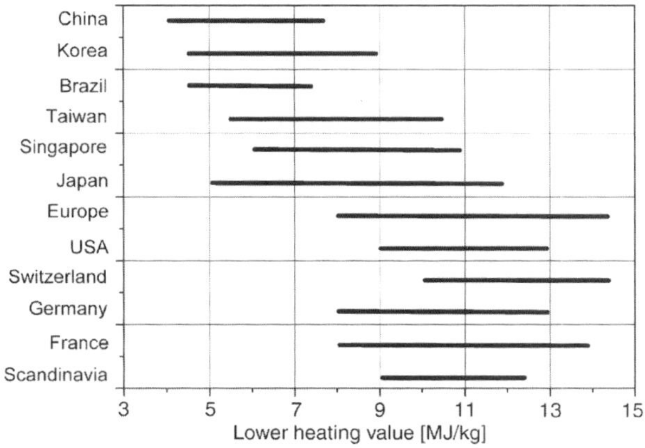

Fig. 2.19 Lower heating value of waste in different countries (*Source*: Martin)

2.2.3.4 Sewage Sludge

Moisture and Ash Content, Calorific Value

In municipal sewage treatment plants, raw sludge or, more commonly, digested sludge is produced. For raw sludge, a moisture content of about 96% is typical. The dry solid matter, on average, consists of 65% organic and combustible components and 35% ash. Digested sludges have a higher ash content because part of the organic matter of the sewage sludge is converted either into CH_4 (in anaerobic conditions) or CO_2 (in aerobic conditions) in the digestion process. In addition, the moisture content may be diminished during the longer period of storage. The dry solid matter of digested sewage sludge is composed half of organic matter and half of ash (Gerhardt 1998; Spliethoff et al. 1996).

The calorific value of sewage sludge is determined by the moisture and the ash contents. Figure 2.20 explains these correlations. For purely organic matter, a calorific value of about 21 MJ/kg can be taken as a basis. The variation in calorific value of sewage sludge from different sewage treatment plants, and from the same plant at different times, ranges around ±1 MJ/kg (Gerhardt et al. 1997). For sewage sludge with an ash content of 35%, the calorific value of the dry solids is about 14 MJ/kg, while digested sewage sludge with 50% ash has a dry solids calorific value of about 10.5 MJ/kg.

Due to the high moisture content, sewage sludge produced in a sewage treatment plant has no or a negative calorific value because heat has to be used to vaporise the water. The common and energy-saving method is mechanical dewatering at the sewage treatment plant. The resulting dewatering degree depends on the sewage sludge, the dewatering method and the addition of conditioning agents. Incompletely digested sewage sludge cannot be stored for a long time after dewatering because of the development of odours and build-up of flammable gases. As Figure 2.20 shows, the calorific value of an undigested sewage sludge (type C) with a dry solid matter content of around 20% lies between 0.5 and 1.2 MJ/kg.

Table 2.16 Composition of various RDFs, showing the influence of the input material (Fehrenbach et al. 2006)

		Municipal solid waste				Household-like commercial waste					
			hcf MBT	DS MBS	DS MPT	Rich in paper and cardboard		Rich in plastics		Bulky waste	
Input material		Input	RDF			Input	RDF	Input	RDF	Input	RDF
Moisture	[%]	33.8	10.7	14.8	14.7	21.1	7.8	20.2	7.4	19.5	13.8
Carbon, fossil	[%]	10.1	27.6	17.1	16.9	11.2	23.9	15.1	31.4	11.4	17.3
Carbon, organic	[%]	12.8	19.4	21.5	21.7	14.9	23.3	12.9	19.3	22.3	28.5
Chlorine	[%]	0.48	0.62	0.78	0.77	0.85	0.99	1.43	1.6	0.17	0.13
Sulphur	[%]	0.2	0.17	0.25	0.25	0.27	0.15	0.27	0.15	0.14	0.08
Cadmium	[mg/kg]	6.7	7.03	6.7	6.6	11.6	9.8	19.4	18.8	2.9	1.42
Mercury	[mg/kg]	0.24	0.24	0.27	0.26	0.27	0.27	0.5	0.51	0.1	0.074
Antimony	[mg/kg]	11.7	12.6	8.25	8.2	11.9	12.9	15.1	18.1	20.7	7.45
Arsenic	[mg/kg]	3.2	2.1	2.24	2.2	2.8	1.7	2.9	1.65	5.3	1.68
Lead	[mg/kg]	204	168	228	127	356	189	436	284	112	76.7
Chromium	[mg/kg]	256	290	332	329	267	342	274	344	120	43.9
Fe metal	[%]	3.41	0.01	0.01	0.01	2.7	0.008	2.7	0.001	8	0.01
Non Fe metal	[%]	0.39	0.02	0.001	0.002	0.4	0.0012	0.4	0.001	0	0.001
LHV	[MJ/kg]	9.6	21.6	17.4	17.4	11.2	21.2	11.8	23.3	13.7	19.4

MBT: mechanical – biological treatment, MBS: mechanical – biological stabilisation, MPT: mechanical – physical treatment (drying), hcf: high calorific fraction from MBT, RDF: refuse-derived fuel

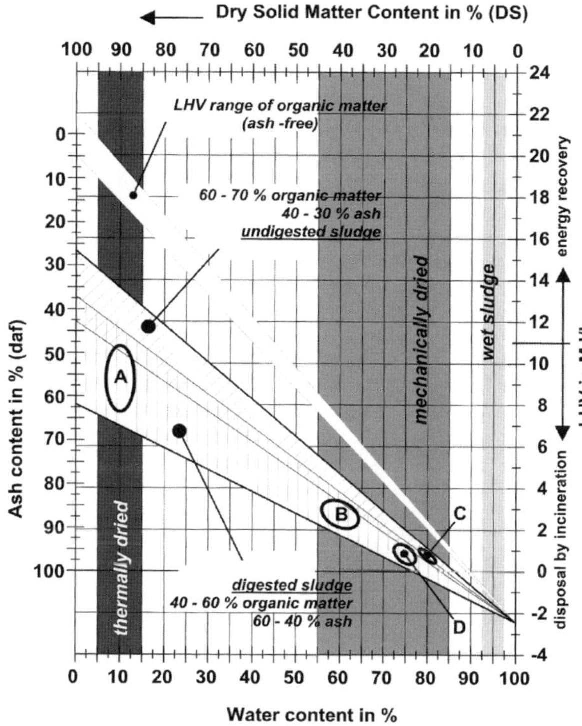

Fig. 2.20 Calorific values of municipal sewage sludge (Gerhardt 1998)

Digested sludge is more effectively dewatered by mechanical means than by other means. Figure 2.20 shows the range of values of a badly dewatered (D) and a well-dewatered type of sewage sludge (B). Digested sludge, at a moisture content of 60%, has a net calorific value of 2–3 MJ/kg. By thermal drying, the calorific value can be markedly increased, but this requires energy to vaporise the water. It can be noticed that the calorific value of the thermally dried digested sludge (range A) generally lies below 11 MJ/kg.

Composition

Table 2.17 shows the analytical data for the dry state of different sewage sludge types in comparison to hard and brown coal. Sewage sludge has a higher ash content because of the input of sand and other inert material. The volatile matter corresponds mainly to the organic substances in the sludge. A conspicuous result of the ultimate analysis is the low carbon content and the high oxygen content. The nitrogen content of sewage sludge is significantly higher than that of coal.

The mineral fraction of sewage sludge consists of about 40% acidic oxides, such as silicon oxide (SiO_2) and aluminium oxide (Al_2O_3), and 40% basic oxides such as CaO, Fe_2O_3, K_2O, MgO and Na_2O. The remaining 20% is composed of phosphates, sulphates and carbonates. In contrast, the fraction of the acidic oxides in hard coal almost reaches double this value (ca. 80%) whereas the fraction of the basic

Table 2.17 Fuel composition of sewage sludge (Gerhardt et al. 1997; Gerhardt 1998)

	Dewatered sewage sludge			
	Typical value	Range	Hard coal	Brown coal
Moisture content [%]	55 (dewatered) 5 (thermally dried)		7	55
Lower heating value (LHV) raw [MJ/kg]	3.6 (dewatered) 10.2 (thermally dried		27.9	8.7
Lower heating value (LHV) dry [MJ/kg]	10.9	8.8–14.4	30.2	22.2
Ash % dry	46.9	39–53	8.3	
Volatile matter % dry	51	28–55	34.7	50
Fixed C dry	2.5	1–24	57	38
C	25.5	20–40	72.5	63
H	5	2–5	5	4
N	3.3	2–5	1.3	0.5
S	1.1	0.6–7	0.9	0.5
Cl	0.1	0.02–0.6	0.2	0.1

oxides is around 20%. The nature of the ash composition of sewage sludge means a lower ash softening point in comparison to hard coal ash. The initial ash deformation temperature lies, depending on the sewage sludge type, around 1,100–1,200°C.

References

Adrian, F., Quittek, C. and Wittchow, E. (1986). *Fossil beheizte Dampfkraftwerke. Handbuchreihe Energie, Band 6*, Herausgeber T. Bohn. Technischer Verlag Resch, Verlag TÜV Rheinland.

Becker, B., Knichel, H., Thomas, J. and Hauschild, W. (2007). Nachhaltige Abfallwirtschaft in Deutschland, Ausgabe 2007. Statistisches Bundesamt.

BGR (2008). Reserven, Ressourcen und Verfügbarkeit von Energierohstoffen 2006, Jahresbericht 2006, Bundesanstalt für Geowissenschaften und Rohstoffe.

Bilitewski, B., Härdtle, G. W. and Marek, K. A. (2000). Abfallwirtschaft: Handbuch für Praxis und Lehre. Berlin [u.a.], Springer.

BMU (2007a). Aufkommen, Beseitigung und Verwertung von Abfällen im Jahr 2005.

BMU (2007b). Bericht zur Siedlungsabfallentsorgung 2006, Stand 1.9.2006, from www.bmu.de.

BMWi (2008). Zahlen und Fakten: Energiedaten, BMWi.

Born, P. (1992). CO2-neutrale Energieträger aus Biomasse? *BWK* **44**(6): 271–274.

Borsch, P. (1992). Was wird aus unserem Klima?: Fakten, Analysen & Perspektiven. München [u.a.], Bonn Aktuell.

BP (2008). Statistical review of world energy 2008, from www.bp.com.

Chiche, P. (1970). Grundlagenforschung über Chemie und Physik von Kohle und Koks III. Forschungshefte Kohle. Luxemburg, Europäische Gemeinschaft für Kohle und Stahl.

Chiche, P. (1973). Grundlagenforschung über Chemie und Physik von Kohle und Koks IV. Forschungshefte Kohle. Luxemburg, Europäische Gemeinschaft für Kohle und Stahl.

Clausen, J. C. and Schmidt, E.R. (1996). Specifications for solid biofuels in Denmark. Tagung "Biomasse als Festbrennstoff–Anforderungen, Einflussmöglichkeiten, Normung" Schriftenreihe "Nachwachsende Rohstoffe", Band 6, Landwirtschaftsverlag, Münster.

CMA (1995). Nachwachsende Energie aus Land- und Forstwirtschaft. *Broschüre*.

Drbal, L. F. (1996). Power plant engineering. New York [u.a.], Chapman & Hall.

Effenberger, H. (2000). Dampferzeugung. Berlin, Heidelberg, Springer.

EU (2008). EU Commission: Directive 2008/98/EC of the European Parliament and of the Council of 19 November 2008 on waste and repealing certain directives (Waste Framework Directive).

Eurostat (2007). Measuring progress towards a more sustainable Europe, 2007 monitoring report of the EU sustainable development strategy. E. Communities.

Fehrenbach, H., Giegrich, J. and Möhler, S. (2006). Behandlungsalternativen für klimarelevante Stoffströme. Heidelberg, ifeu.

Frühwald, A. (1990). Holzbe- und -verarbeitung. VDI-Berichte Nr. 794, 1990, pp. 9–21

Gerhardt, T. (1998). Thermische Behandlung von kommunalen Klärschlämmen in Kohlenstaubfeuerungen. Essen, VGB-Kraftwerkstechnik, Verl. Techn.-Wiss. Schriften.

Gerhardt, T., Spliethoff, H. and Hein, K. R. G. (1996). Thermische Nutzung von Klärschlämmen in Kraftwerksfeuerungen. Untersuchungen an einer Staubfeuerung im Pilotmaßstab. *Entsorgungspraxis* (3).

Gerhardt, T., Spliethoff, H. and Hein, K.R.G. (1997). Bedarf von thermischen Behandlungsverfahren für kommunale Klärschlämme. *Entsorgungspraxis* (3): 40–47.

Gumz, W. (1962). Kurzes Handbuch der Brennstoff- und Feuerungstechnik. Berlin, Göttingen, Heidelberg, Springer.

Hartmann, H. and Strehler, A. (1995). Die Stellung der Biomasse im Vergleich zu anderen erneuerbaren Energieträgern aus ökologischer, ökonomischer und technischer Sicht, Schriftenreihe Nachwachsende Rohstoffe, Band 3.

Hoffmann, G., Wünsch, C. and Biletewski, B. (2008). *Ersatzbrennstoffe aus Siedlungsabfall – Eine Energiebilanz*. 4. Fachtagung "Verfahren und Werkstoffe für die Energietechnik": Biomasse & Abfall – Regionale Brennstoffe richtig nutzen, Sulzbach-Rosenberg, Dorner.

IEA (2006). World energy outlook 2006. Paris, IEA.

IEA (2007). World energy outlook 2007. Paris, IEA.

JBDT (1976). Jahrbuch der Dampferzeugertechnik, 3. Ausgabe, Vulkan, Essen.

JBDT (1985). Jahrbuch der Dampferzeugertechnik, 5. Ausgabe, Vulkan, Essen.

Kaltschmitt, M. (1993). Energieträgerproduktion auf pflanzlicher Basis. *Landtechnik* **48**(8/9): 400–406.

Kaltschmitt, M. (2001). Energie aus Biomasse: Grundlagen, Techniken und Verfahren. Berlin [u.a.], Springer.

Kaltschmitt, M., Hartmann, H. and Hofbauer, H. (2009). Energie aus Biomasse: Grundlagen, Techniken und Verfahren. Dordrecht, Heidelberg, London, New York, Springer.

Kaltschmitt, M., Streicher, W. and Wiese, A. (2006). Erneuerbare Energien, Systemtechnik, Wirtschaftlichkeit, Umweltaspekte. Berlin, Heidelberg, Springer.

Kicherer, A. (1996). Biomasseverbrennung in Staubfeuerungen – technische Möglichkeiten und Schadstoffemissionen. Düsseldorf, VDI.

Kleemann, M. and Meliß, M. (1993). Regenerative Energiequellen. Berlin [u.a.], Springer.

Lewandowski, I. (1996). *Einflußmöglichkeiten der Pflanzenproduktion auf die Brennstoffeigenschaften am Beispiel von Gräsern*. Tagung Biomasse als Festbrennstoff – Anforderungen, Einflussmöglichkeiten, Normung, Schriftenreihe Nachwachsende Rohstoffe, Band 6, Landwirtschaftsverlag, Münster.

Obernberger, I. (1997). Nutzung fester Biomasse in Verbrennungsanlagen unter besonderer Berücksichtigung des Verhaltens aschebildender Elemente. Graz, dbv.

Reimann, D. O. and Hämmerli, H. (1995). Verbrennungstechnik für Abfälle in Theorie und Praxis. Bamberg, Reimann.

Ruhrkohle (1987). Ruhrkohlenhandbuch. Essen, Glückauf.

Schmelz, K.-G. (2006). Klärschlammmengen und Entsorgungskosten im Vergleich zum europäischen Ausland. Perspektiven der Klärschlammverwertung, Bonn.

Schmidt, A. (1992). Bioenergie – Ein Vergleich der biologischen mit den technischen Möglichkeiten der Nutzung von Solarenergie. *BWK* **44**(5): 227–231.

Schneider, S., Deimling, S. and Kaltschmitt, M. (2007). Leitfaden Bioenergie : Planung, Betrieb und Wirtschaftlichkeit von Bioenergieanlagen. Biogene Brennstoffe als nachwachsende Energieträger. Gülzow b. Güstrow, Fachagentur Nachwachsende Rohstoffe.

Skorupska, N. M. (1993). Coal specifications – impact on power station performance. London, IEA Coal Research.

Spliethoff, H., Siegle, V. and Hein, K.R.G. (1996). *Erforderliche Eigenschaften holz- und halmgutartiger Biobrennstoffe bei der Zufeuerung in existierenden Kohlekraftwerken.* Tagung: Biomasse als Festbrennstoff - Anforderungen, Einflussmöglichkeiten, Normung. Landwirtschaftsverlag, Münster, Schriftenreihe "Nachwachsende Rohstoffe", Band 6.

Stultz, S. C. and Kitto, J. B. (1992). Steam, its generation and use. Barberton, OH, The Babcock & Wilcox Company.

Thomé-Kozmiensky, K. J. (1994). Thermische Abfallbehandlung. Berlin, EF für Energie- und Umwelttechnik.

UBA. (2008). Retrieved 17.8.2008, from http://www.umweltbundesamt.de/abfallwirtschaft/entsorgung/index.htm.

Van Loo, S. and Koppejan, J. (2008). The handbook of biomass combustion and co-firing. London, Earthscan.

Wegener, G. and Frühwald, A. (1994). Das CO2-Minderungspotential durch Holznutzung, Holz als Energieträger. *Energiewirtschaftliche Tagesfragen* **44**(7): 421–425.

Wieck-Hansen, K. (1996). Parameters influencing Straw Quality. Tagung "Biomasse als Festbrennstoff–Anforderungen, Einflussmöglichkeiten, Normung". Schriftenreihe "Nachwachsende Rohstoffe", Band 6, Landwirtschaftsverlag, Münster.

Zelkowski, J. (2004). Kohlecharakterisierung und Kohleverbrennung. Essen, VGB PowerTech.

Chapter 3
Thermodynamics Fundamentals

3.1 Cycles

3.1.1 Carnot Cycle

Named after the French scientist Nicolas Carnot, the ideal Carnot cycle converts a maximum fraction of heat input into work. In this process, work is delivered without heat exchange and without losses, and heat is added and taken out without any change in temperature. As a reference process, the Carnot cycle illustrates fundamental knowledge about the thermodynamics of energy conversion (Hahne 2004; Meyer and Schiffner 1989; Strauß 2006). The Carnot cycle combines two process steps with isentropic changes of state and two process steps with isothermal changes of state to form a closed reversible cycle. These steps are shown in Fig. 3.1:

1–2: isentropic compression with work input w_{12},
2–3: isothermal expansion at a constant upper process temperature T_u with heat input $q_{23} = q_{in}$,
3–4: isentropic expansion with work output w_{34},
4–1: isothermal compression at a constant lower process temperature T_l with heat output $q_{41} = q_{out}$.

The energy added to the cycle in the form of heat is only partially converted into useful work; the other portion is released to the environment. The lines of state of the four process steps of the Carnot cycle form a rectangle in the $T-s$ diagram. The area beneath the isotherm T_u gives the heat input:

$$q_{in} = T_u (s_3 - s_2) \tag{3.1}$$

and the area beneath the isotherm T_l gives the heat output:

$$|q_{out}| = T_l (s_4 - s_1) = T_l (s_3 - s_2) \tag{3.2}$$

H. Spliethoff, *Power Generation from Solid Fuels,* Power Systems,
DOI 10.1007/978-3-642-02856-4_3,

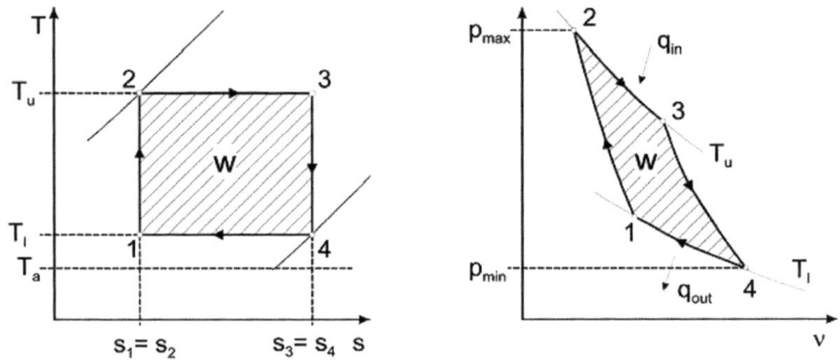

Fig. 3.1 Carnot cycle $T - s$ and $p - V$ diagrams

The useful work of the Carnot cycle is described as follows:

$$|w| = q_{\text{in}} - |q_{\text{out}}| \tag{3.3}$$

which, in the $T - s$ diagram, corresponds to the rectangular area enclosed by the lines of state.

The thermal efficiency (the ratio of useful work to input heat) is calculated for the Carnot cycle as follows:

$$\eta_{\text{th}} = \frac{|w|}{q_{\text{in}}} = \frac{q_{\text{in}} - |q_{\text{out}}|}{q_{\text{in}}} = 1 - \frac{|q_{\text{out}}|}{q_{\text{in}}} \tag{3.4}$$

Consequently, for the Carnot cycle, this becomes

$$\eta_{\text{th}} = \frac{T_{\text{u}}\,(s_3 - s_2) - T_{\text{l}}\,(s_3 - s_2)}{T_{\text{u}}\,(s_3 - s_2)} = \frac{T_{\text{u}} - T_{\text{l}}}{T_{\text{u}}} = 1 - \frac{T_{\text{l}}}{T_{\text{u}}} \tag{3.5}$$

hence the thermal efficiency of the reversible Carnot cycle, also called the Carnot factor, only depends on the constant temperatures of heat input and output. The Carnot factor is greater the higher the temperature T_{u} of the heat input and the lower the temperature T_{l} of the heat output.

The Carnot factor is always less than 1 because the heat release temperature always lies above the ambient temperature of about 280–300 K. There is no cycle which has a better efficiency for a given temperature gradient $T_{\text{max}} - T_{\text{min}}$. To achieve high efficiencies, one tries to bring real processes closer to the Carnot cycle.

3.1.2 Joule–Thomson Process

The Joule–Thomson process is the idealised reference process for gas turbines. A simple, open gas turbine process, shown in Fig. 3.2, consists of a compressor, a combustion chamber and a gas turbine. The air sucked in from the environment at p_1 and T_1 becomes compressed by the compressor to pressure p_2. The compressed air in the combustion chamber oxidises the fuel, turning it into a hot flue gas with temperature

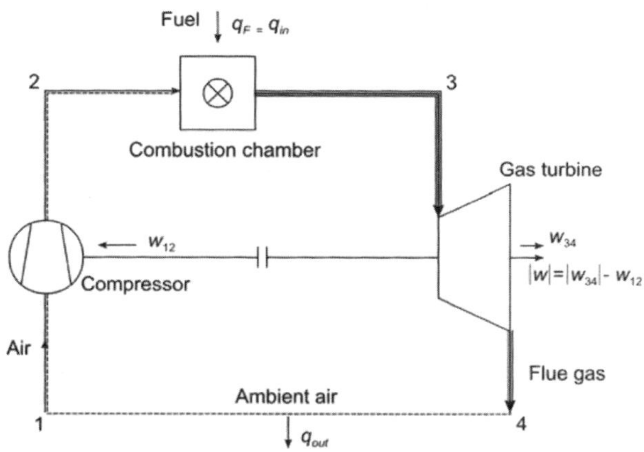

Fig. 3.2 Schematic diagram of an open gas turbine process

T_3, which afterwards does work in the turbine, expands and is cooled down to the gas turbine exit temperature. The waste gas is released to the environment.

For the ideal Joule–Thomson process, the assumption is that both the compression and the expansion processes are isentropic, i.e. reversible. The Joule–Thomson process therefore consists of two isentropes and two isobars. If the discharge of the cooled-down but still hot gases to the environment is conceived as an isobaric heat dissipation, the course of the process can be represented as a cycle. The corresponding $p - V$ and $T - s$ diagrams are shown in Fig. 3.3.

For the heat input and output quantities, assuming an ideal gas[1] with a constant c_P, the following holds true:

$$q_{\text{in}} = h_3 - h_2 = c_p (T_3 - T_2) \tag{3.6}$$

and

$$|q_{\text{out}}| = h_4 - h_1 = c_p (T_4 - T_1) \tag{3.7}$$

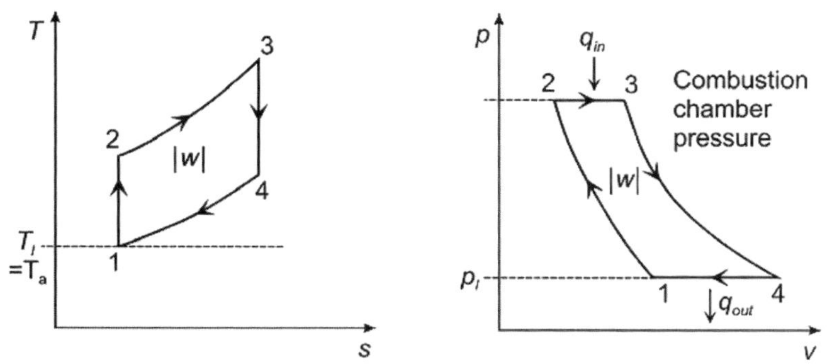

Fig. 3.3 $p - V$ and $T - s$ diagrams for the ideal Joule – Thomson process

[1] For real gases c_p is a function of the temperature. In this case the medium specific heat capacity $\overline{c}_p$ between the corresponding temperatures has to be used for the calculations.

It holds true that for the work w_{12} to be done by the compressor:

$$w_{12} = h_2 - h_1 \tag{3.8}$$

and that for the work w_{34} produced by the turbine:

$$w_{34} = h_4 - h_3 \tag{3.9}$$

and that for the gain in work w:

$$|w| = |w_{34}| - w_{12} \tag{3.10}$$

Hence, the efficiency of the Joule–Thomson process is

$$\eta_{th} = \frac{|w|}{q_{in}} = 1 - \frac{|q_{out}|}{q_{in}} = 1 - \frac{c_p\,(T_4 - T_1)}{c_p\,(T_3 - T_2)} = 1 - \frac{T_4 - T_1}{T_3 - T_2} \tag{3.11}$$

From the equations of state for the isentropes of the process

$$\frac{T_2}{T_1} = \left(\frac{p_2}{p_1}\right)^{\frac{\kappa-1}{\kappa}} \tag{3.12}$$

and

$$\frac{T_3}{T_4} = \left(\frac{p_3}{p_4}\right)^{\frac{\kappa-1}{\kappa}} \tag{3.13}$$

the relation

$$\frac{T_2}{T_1} = \frac{T_3}{T_4} \tag{3.14}$$

is given by $p_2 = p_3$ and $p_4 = p_1$.

Thus, by putting Eq. (3.14) into Eq. (3.11) and by transforming it, one gets the following expression for the thermal efficiency of the ideal Joule–Thomson process:

$$\eta_{th} = 1 - \frac{T_1}{T_2} = 1 - \left(\frac{p_1}{p_2}\right)^{\frac{\kappa-1}{\kappa}} \tag{3.15}$$

Thus the efficiency of the ideal Joule–Thomson process is only dependent on the pressure ratio. An increase of the pressure ratio, though, also results in an increase of the temperature T_2. Given that the heat in the combustion chamber:

$$q_{in} = \frac{|w|}{\eta} \tag{3.16}$$

has to be supplied, turbine inlet temperatures T_3 can arise which are not feasible because of the physical constraints of the materials that are currently available.

In a real gas turbine process, however, irreversibilities occur in all components which sum to produce a deviation from the ideal Joule–Thomson process. In a gas turbine process, in contrast to a steam process, not only the irreversibilities in the turbine but also those in the compressor are important factors. The pressure losses in the combustion chamber also result in deviations from the ideal process.

A real Joule–Thomson process is shown in Fig. 3.4. Irreversibilities in the compressor and the turbine are defined by means of the isentropic compressor efficiency

$$\eta_{i,c} = \frac{h_{2,\mathrm{id}} - h_1}{h_2 - h_1} \tag{3.17}$$

and the turbine efficiency

$$\eta_{i,T} = \frac{h_3 - h_4}{h_3 - h_{4,\mathrm{id}}} \tag{3.18}$$

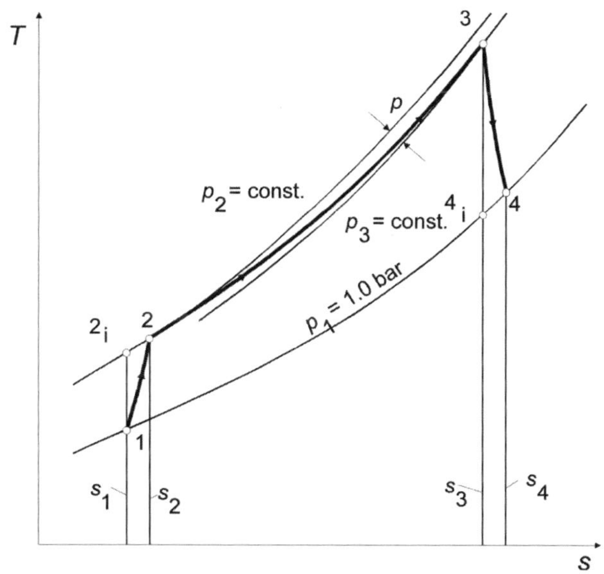

Fig. 3.4 $T - s$ diagram of the real Joule – Thomson process

3.1.3 Clausius–Rankine Cycle

In contrast to the Joule–Thomson process, the Clausius–Rankine cycle is based on water and steam as the working media. Its principle, a simple steam power cycle, is shown in Fig. 3.5. The phase change of liquid into gas occurs in the steam generator, while the phase change of gas into liquid happens in the condenser. The feed pump

Fig. 3.5 Schematic diagram of a simple steam-electric power plant

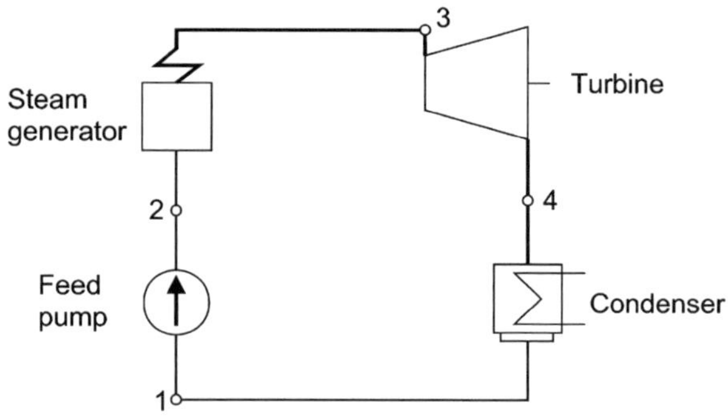

transports the water into the steam generator, where it gets preheated, evaporated and superheated. In the turbine, the steam is expanded at constant entropy, imparting mechanical work in the process. In the condenser, heat is extracted and the waste steam condensed. The condensate in turn is fed to the feed pump.

Figure 3.6 shows the reversible Clausius–Rankine cycle in $h - s$ and $T - s$ diagrams. This cycle serves as a thermodynamic reference process for steam generation processes. The steps are as follows:

1–2: isentropic compression in the feed pump by work input w_{12}
2–3: isobaric heat supply $q_{23} = q_{\text{in}}$ in the steam generator (preheating, evaporation, superheating)
3–4: isentropic expansion in the turbine with work output w_{34}
4–1: isobaric heat dissipation $q_{41} = q_{\text{out}}$ in the condenser (Hahne 2004).

In the turbine, the steam imparts mechanical work, whereas for raising the pressure of the feed water, work has to be supplied. Hence the gain in work in the cycle process is

$$|w| = |w_{34}| - w_{12} = (h_3 - h_4) - (h_2 - h_1) \tag{3.19}$$

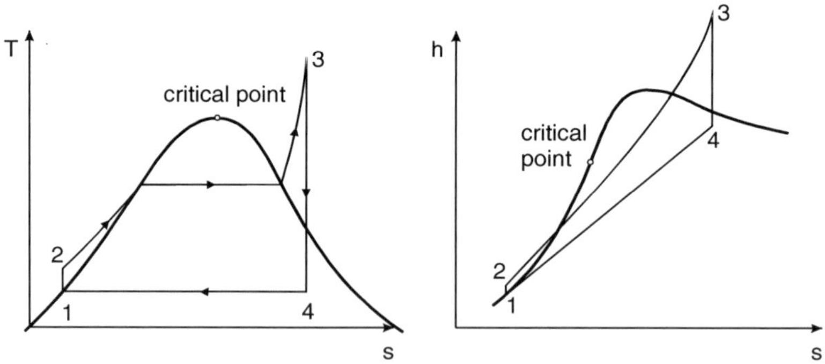

Fig. 3.6 Ideal Clausius–Rankine cycle $T - s$ and $h - s$ diagrams

The thermal efficiency of the Clausius–Rankine cycle is then

$$\eta_{\text{th}} = \frac{|w|}{q_{\text{in}}} = \frac{h_3 - h_4 - (h_2 - h_1)}{h_3 - h_2} \tag{3.20}$$

For comparison with the Carnot cycle, the thermodynamic mean temperature of the heat supply

$$T_{\text{m,in}} = \frac{q_{\text{in}}}{s_3 - s_2} = \frac{h_3 - h_2}{s_3 - s_2} \tag{3.21}$$

and that of the heat extraction

$$T_{\text{m,out}} = \frac{|q_{\text{out}}|}{s_4 - s_1} = \frac{h_4 - h_1}{s_4 - s_1} \tag{3.22}$$

are defined so that, for the reversible Rankine cycle, the Carnot factor can also be calculated:

$$\eta_{\text{th}} = \frac{T_{\text{m,in}} - T_{\text{m,out}}}{T_{\text{m,in}}} = 1 - \frac{T_{\text{out}}}{T_{\text{in}}} \tag{3.23}$$

The thermal efficiency of the Clausius–Rankine cycle thus becomes greater the higher the mean thermodynamic temperature of the heat supply and the lower the mean thermodynamic temperature of the heat extraction. Another consequence is that, at a given maximal heat supply temperature and a minimal heat dissipation temperature, the result of the idealised isothermal heat exchange processes of the Carnot cycle is in each case the highest efficiency. In the Clausius–Rankine cycle, feed water preheating, evaporation and superheating inevitably result in a lower average temperature of the heat input, so the efficiency of the Rankine cycle is lower than the Carnot factor. Therefore, measures to raise the thermal efficiency of the steam power cycle can be assessed with reasonable adequacy by means of the average thermodynamic heat input temperature.

Like in every technical plant, there are losses in steam-electric power plant processes, making a reversible course of the Clausius–Rankine cycle impossible. Irreversibilities develop in the form of pressure losses in the cycle–by friction, turbulence and mixture losses in the turbo-machinery and in other components–and also in the form of heat transfer losses in processes with finite temperature differences (all heat exchangers). The irreversibilities result in an increase in entropy. This entropy increase in the turbine is included in the isentropic efficiency:

$$\eta_{i,T} = \frac{h_3 - h_4}{h_3 - h_{4,\text{id}}} \tag{3.24}$$

3.2 Steam Power Cycle: Energy and Exergy Considerations

The energy efficiency, η, is the ratio of the power delivered or produced by a process to the power which is supplied to it (Adrian et al. 1986). In condensation power plants, fuel power is used exclusively for electrical power production. The electrical capacity of a power plant is described by both the gross installed capacity and also the net output capacity. The gross installed capacity P_{gr} is the capacity measured at the generator, whereas the net output capacity P_{ne} is the power output delivered to the network. The difference between the gross and net output capacities is given by the so-called electrical auxiliary power $P_{aux,\,el}$ which is needed to supply all electrical auxiliary devices, e.g. for coal milling, for driving the feed pump (only when there is an electrical feed pump), the combustion air and flue gas fans and to cover the loss of the station service transformer:

$$P_{ne} = P_{gr} - P_{aux,el} \tag{3.25}$$

The total or net efficiency of a power plant producing only electrical power is the quotient of the electrical power output and the supplied fuel power, the latter of which is the product of the fuel flow $\dot{m}_F$ and the lower heating value, LHV, of the fuel:

$$\eta_{ne} = \frac{P_{ne}}{\dot{m}_F \cdot \text{LHV}} \tag{3.26}$$

The efficiency of a power plant is made up of various single efficiencies which, multiplied with each other, add up to the total efficiency:

$$\eta_{ne} = \eta_B \cdot \eta_{th} \cdot \eta_m \cdot \eta_{Gen} \cdot \eta_{aux} \cdot \eta_P \tag{3.27}$$

where η_B is the steam generator efficiency and η_{th} is the thermal efficiency. Efficiency η_m reflects the mechanical losses of the turbine; the generator efficiency, η_{Gen}, covers electrical and mechanical losses of the generator.

The auxiliary power efficiency η_{aux} takes into account the electrical and the mechanical power demand (if not included in η_{th} already), while efficiency η_R represents the heat losses of the live steam and reheater pipes which connect the steam generator and the turbine.

For the boiler or steam generator the efficiency becomes:

$$\eta_B = \frac{\sum \dot{m}_{S,j} \cdot \Delta h_j}{\dot{m}_F \cdot \text{LHV}} \tag{3.28}$$

where $\dot{m}_{s,j}$ are the individual mass flows of the working medium (water/steam) supplied with heat from combustion in the steam generator. Δh_j are the increases of enthalpy attained in each mass flow. For the simple steam cycle shown in Fig. 3.5, this becomes

$$\eta_{\mathrm{B}} = \frac{\dot{m}_{\mathrm{S}}(h_3 - h_2)}{\dot{m}_{\mathrm{F}} \cdot \mathrm{LHV}} \tag{3.29}$$

The efficiency of the steam generator is, however, determined mostly indirectly – by the losses of the steam generator. The steam generator losses with respect to the fuel power are (Doležal 1990)

- loss through unburned combustibles (κ_{U}),
- loss through sensible heat of the slag (κ_{S}),
- flue gas loss (κ_{FG}) and
- loss through radiation and convection of the steam generator (κ_{RC}).

Accordingly, the steam generator efficiency is

$$\eta_B = 1 - \kappa_{\mathrm{U}} - \kappa_{\mathrm{S}} - \kappa_{\mathrm{FG}} - \kappa_{\mathrm{RC}} \tag{3.30}$$

For the thermal efficiency of the real cycle, which represents the ratio of the inner power output of the turbine P_{i} (the power of the turbine without mechanical losses) to the steam energy supplied, this becomes

$$\eta_{\mathrm{th}} = \frac{P_{\mathrm{i}}}{\sum \dot{m}_{\mathrm{S},j} \bullet \Delta h_j} \tag{3.31}$$

where $\dot{m}_{\mathrm{s},j}$ are the individual mass flows of water/steam and Δh_j stands for the respective increases of enthalpy attained in the steam generator. For the simple steam process shown in Fig. 3.5, analogous to Eq. (3.20), this is

$$\eta_{\mathrm{th}} = \frac{P_{\mathrm{i}}}{\dot{m}_{\mathrm{S}}(h_3 - h_2)} \tag{3.32}$$

The efficiency of the cycle η_{th}, in contrast to the efficiency of the loss-free process $\eta_{\mathrm{th},0}$, is decreased by friction losses during expansion in the turbine. These losses are taken into account by the isentropic turbine efficiency $\eta_{\mathrm{i,T}}$:

$$\eta_{\mathrm{i,T}} = \frac{\eta_{\mathrm{th}}}{\eta_{\mathrm{th},0}} = \frac{h_3 - h_4}{h_3 - h_{4,\mathrm{id}}} \tag{3.24}$$

With the inner power of the turbine P_{i} and the mechanical output of the turbine shaft P_{m}, the relevant equation for the mechanical efficiency η_{m} is

$$\eta_{\mathrm{m}} = \frac{P_{\mathrm{m}}}{P_{\mathrm{i}}} \tag{3.33}$$

for the generator efficiency

$$\eta_{\mathrm{Gen}} = \frac{P_{\mathrm{Gen}}}{P_{\mathrm{m}}} \tag{3.34}$$

and for the auxiliary power efficiency

$$\eta_{\mathrm{aux}} = \frac{P_{\mathrm{ne}}}{P_{\mathrm{Gen}}} \tag{3.35}$$

If the feed pump is driven electrically, and also if driven by a steam turbine, the driving power of the feed pump is commonly added to the auxiliary power. In the case of a turbine-driven feed water pump, the power of the feed pump turbine is taken into account in calculating the thermal efficiency of the real cycle and added to the power output of the main turbine in Eq. (3.31).

Often, the turbine or turbine generator efficiency η_{T} is used, which represents the ratio of the gross electrical output and, if necessary, the mechanical power output (in the case of feed pumps with a steam turbine drive) to the steam energy input:

$$\eta_{\mathrm{T}} = \frac{P^{*}{}_{\mathrm{Gen}}}{\sum \dot{m}_{\mathrm{s},j} \cdot \Delta h_j} = \eta_{\mathrm{th}} \cdot \eta_{\mathrm{m}} \cdot \eta_{\mathrm{Gen}} \tag{3.36}$$

with

$$P^{*}_{\mathrm{Gen}} = P_{\mathrm{Gen}} + P_{\mathrm{aux,m}} \tag{3.37}$$

If the feed pump is driven by a steam turbine, the power output of the turbine generator $P^{*}{}_{\mathrm{Gen}}$ increases, surpassing the gross output P_{Gen} by the amount of the mechanical output of the turbine drive $P_{\mathrm{aux,\ m}}$. Where the feed pump is driven electrically, the power output $P^{*}{}_{\mathrm{Gen}}$ equals the generator output P_{Gen}. The turbine generator efficiency, in contrast to the thermal efficiency of the cycle, also takes into account the losses occurring in the turbine and the generator.

Therefore the auxiliary power efficiency becomes

$$\eta_{\mathrm{aux}} = \frac{P_{\mathrm{ne}}}{P^{*}{}_{\mathrm{Gen}}} = \frac{P^{*}{}_{\mathrm{Gen}} - P_{\mathrm{aux,el}} - P_{\mathrm{aux,m}}}{P^{*}{}_{\mathrm{Gen}}} \tag{3.38}$$

Besides an energy efficiency, it is also possible to develop an expression for the total and single exergy efficiencies:

$$\zeta_{\mathrm{ne}} = \zeta_{\mathrm{B}} \cdot \zeta_{\mathrm{th}} \cdot \zeta_{\mathrm{Gen}} \cdot \zeta_{\mathrm{aux}} \cdot \zeta_{\mathrm{m}} \cdot \zeta_{P} \tag{3.39}$$

Given that the fuel energy and exergy differ only very slightly, the total energy and exergy efficiencies are almost equal. Significant differences, however, arise for the single efficiencies, in particular in the process of energy conversion in the steam generator and in the energy conversion process of the real cycle. The mechanical efficiency and the generator efficiency have the same values if the friction heat is not utilised (Herbrik 1993).

3.2.1 Steam Generator Energy and Exergy Efficiencies

Analogous to the energy efficiency η_B of the steam generator, and in accordance with Eq. (3.29), the exergy efficiency ζ_B of the boiler can be defined as

$$\zeta_B = \frac{\dot{m}_S(e_3 - e_2)}{\dot{m}_F \cdot e_F} \tag{3.40}$$

where $\dot{m}_S$ is the steam mass flow, $\dot{m}_F$ is the fuel mass flow and e_F stands for the fuel's, e_2 for the water's and e_3 for the superheated steam's exergy.

From Eq. (3.29), it follows by transformation that

$$\frac{\dot{m}_S}{\dot{m}_F} = \eta_B \frac{\text{LHV}}{h_3 - h_2} \tag{3.41}$$

For the input of exergy, the ambient temperature T_a is incorporated:

$$e_3 - e_2 = h_3 - h_2 - T_a(s_3 - s_2) \tag{3.42}$$

If Eqs. (3.41) and (3.42) are inserted into Eq. (3.40), then the following expression is derived:

$$\zeta_B = \frac{\text{LHV}}{e_F} \eta_B \left(1 - T_a \frac{s_3 - s_2}{h_3 - h_2}\right) \tag{3.43}$$

or

$$\zeta_B = \frac{\text{LHV}}{e_F} \eta_B \frac{e_3 - e_2}{h_3 - h_2} \tag{3.44}$$

for the boiler exergy efficiency.

The boiler exergy efficiency indicates which part of the supplied fuel exergy e_F is maintained as exergy of the steam. This efficiency, in contrast to η_B, assesses the energy conversion in the steam generator.

The exergy efficiency thus essentially depends on two factors. The first factor, η_B LHV/ e_F, represents the losses through flue gas and radiation. The second factor can be calculated from the feed water inlet and exiting live steam state quantities. This factor implicitly includes the considerable exergy losses through the irreversibilities of combustion and heat transfer.

The changes of state of the water are shown in Fig. 3.7. The water entering at temperature T_2 first gets preheated, vaporised and superheated. The area below

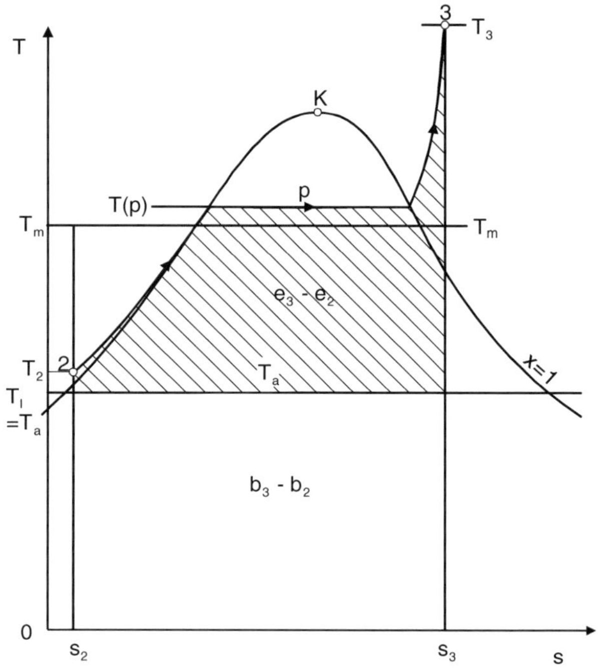

Fig. 3.7 Isobaric state changes in the evaporator (Baehr and Kabelac 2006)

the isobar of the boiler pressure indicates the increase of the water's enthalpy, expressed as

$$h_3 - h_2 = q_{23} \tag{3.45}$$

This increase corresponds to the heat that the water absorbs in the steam generator. The area between the isobar of the boiler pressure and the isotherm T_a of the ambient temperature corresponds to the increase of the water's exergy, $e_3 - e_2$.

If the mean temperature of the heat addition

$$T_{m,in} = \frac{h_3 - h_2}{s_3 - s_2} \tag{3.46}$$

is put into Eq. (3.43), the result is

$$\zeta_B = \frac{\text{LHV}}{e_F} \eta_B \left(1 - \frac{T_a}{T_{m,in}}\right) \tag{3.47}$$

While the energy efficiency of a steam generator typically lies above 0.9, the corresponding value for the exergy efficiency ranges around 0.5.

This low value is caused by

- exergy losses via flue gas and irradiation – about 6%,
- the exergy loss of combustion – about 15% and
- the exergy loss of the heat transfer – about 30%.

Losses through the sensible heat of the flue gas and through irradiation are taken into account in both the energy and exergy efficiency. Losses through the irreversibilities of combustion and heat transfer are only included in the exergy efficiency. Irreversible combustion and heat transfer convert about half of the fuel exergy input into anergy, while exergy cannot be made use of in the following energy conversion steps, having to be discharged as waste heat.

3.2.2 Energy and Exergy Cycle Efficiencies

Analogous to the energy efficiency of the Clausius–Rankine Cycle:

$$\eta_{\mathrm{th}} = \frac{P_{\mathrm{i}}}{\dot{m}_{\mathrm{S}}(h_3 - h_2)} = \frac{w}{h_3 - h_2} \tag{3.48}$$

it is possible to define an exergy efficiency:

$$\zeta_{\mathrm{th}} = \frac{P_{\mathrm{i}}}{\dot{m}_{\mathrm{S}}(e_3 - e_2)} = \frac{w}{e_3 - e_2} \tag{3.49}$$

This efficiency specifies what part of the exergy taken up in the steam generator is converted into useful work. If the cycle is reversible, the thermal efficiency ζ_{th} becomes 1; divergences from this ideal value represent thermodynamic losses. To break these down, the useful work is calculated as

$$\begin{aligned} w = h_3 - h_4 - (h_2 - h_1) &= e_3 - e_4 - (e_2 - e_1) - T_{\mathrm{a}}[(s_3 - s_4) - (s_2 - s_1)] \\ &= e_3 - e_2 - (e_4 - e_1) - e_{L34} - e_{L12} \end{aligned} \tag{3.50}$$

So the useful work obtained is the exergy taken up in the steam generator $(e_3 - e_2)$ minus the exergy losses – the exergy delivered in the condenser $(e_4 - e_1)$ and the exergy losses caused by irreversibilities in the feed pump $(e_{\mathrm{L}12})$ and in the turbine $(e_{\mathrm{L}34})$.

Hence, for the exergy cycle efficiency, the expression becomes

$$\zeta_{\mathrm{th}} = 1 - \frac{e_4 - e_1}{e_3 - e_2} - \frac{e_{L12} + e_{L34}}{e_3 - e_2} \tag{3.51}$$

The losses of exergy are pictured in Fig. 3.8. The exergy loss of the feed pump, $e_{\mathrm{L}12}$, is small in contrast to the exergy loss of the steam turbine, $e_{\mathrm{L}34}$. The exergy loss of the steam turbine depends on the isentropic efficiency of the turbine.

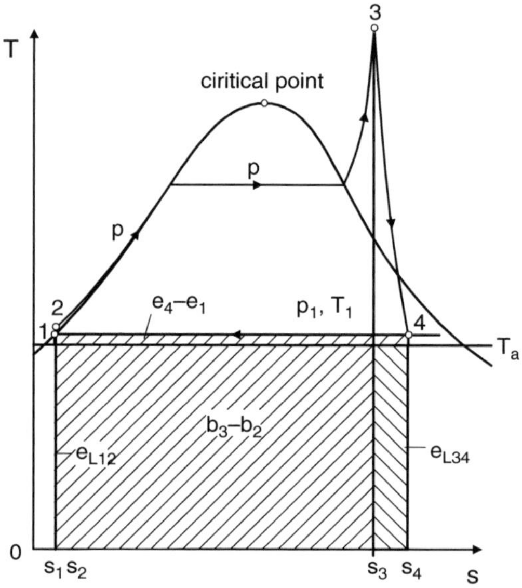

Fig. 3.8 Exergy losses of a simple steam cycle (Baehr and Kabelac 2006)

Given that in the condenser, the exergy $e_4 - e_1$ is transferred to and then discharged to the environment with the cooling water, it has to be regarded as an exergy loss. A reduction of the exergy losses can be achieved by bringing the condensation temperature as close as possible to the ambient temperature by using a large heat transfer surface and a large cooling water mass flow.

In the condenser, the heat q_{41} (which can be represented by the rectangular area below isobar $4-1$ in the $T-s$ diagram) is given off to the cooling water flow. It can be expressed as

$$q_{41} = b_3 - b_2 + e_L = b_3 - b_2 + (e_4 - e_1) + e_{L12} + e_{L34} \tag{3.52}$$

Besides the exergy losses of the cycle, which arise through irreversibilities and convert exergy into anergy, the heat q_{41} also comprises the anergy $b_3 - b_2$ taken up in the steam generator with the heat q_{23}. From the condenser, therefore, the entire anergy load is discharged to the environment.

Typical exergy efficiencies of the cycle, which are around 0.9, are significantly above the typical energy efficiencies of about 0.45.

3.2.3 Energy and Exergy Efficiency of the Total Cycle

There is no influence on the overall efficiency by this differentiated – i.e. energetic or exergetic – approach. There are, however, clear differences when considering the steam generator efficiency and the thermal efficiency of the cycle. The exergy efficiency defines the place where the thermodynamic losses originate and hence better indicates the potential for efficiency increases (Baehr and Kabelac 2006).

The greatest exergy losses and thus the greatest potential for improving the efficiency are found in the steam generator section of the process. The losses in the turbine are significantly smaller.

References

Adrian, F., Quittek, C. and Wittchow, E. (1986). Fossil beheizte Dampfkraftwerke. Handbuchreihe Energie, Band 6, Herausgeber T. Bohn. Technischer Verlag Resch, Verlag TÜV Rheinland.

Baehr, H. D. and Kabelac, S. (2006). Thermodynamik: Grundlagen und technische Anwendungen. Berlin, Heidelberg, Springer.

Doležal, R. (1990). Dampferzeugung: Verbrennung, Feuerung, Dampferzeuger. Berlin, Heidelberg, New York, Springer.

Hahne, E. (2004). Technische Thermodynamik: Einführung und Anwendung. München [u.a.], Oldenbourg.

Herbrik, R. (1993). Energie- und Wärmetechnik. Stuttgart, Teubner.

Meyer, G. and Schiffner, E. (1989). Technische Thermodynamik. Leipzig, Fachbuch.

Strauß, K. (2006). Kraftwerkstechnik: zur Nutzung fossiler, nuklearer und regenerativer Energiequellen. Berlin [u.a.], Springer.

Chapter 4
Steam Power Stations for Electricity and Heat Generation

4.1 Pulverised Hard Coal Fired Steam Power Plants

4.1.1 Energy Conversion and System Components

Power plants produce electricity, process heat or district heating, according to their task (Stultz and Kitto 1992). Electric power is the only product of a condensation power plant and the main product of a power plant with extraction – condensation turbines, where extraction steam is a by-product. Power plants for process heat generation or combined heat and power (CHP) stations generate electrical power, steam and district heat as their main products. Simultaneous heat and/or steam utilisation, along with power generation, is an effective method to diminish waste heat losses at the cold end of the turbine.

Figure 4.1 shows the main components of a modern coal-fired power plant. They can be divided into the following plant sections:

- Fuel supply and preparation
- Steam generator with furnace
- Turbine and generator
- Heat rejection unit, condenser, cooling tower
- Units for emissions reduction and disposal

The generation of electrical power in a steam power plant involves the multiple conversion of the primary energy contained within the fuel (Schröder 1968). In combustion, the fuel is oxidised by the oxygen of the combustion air, thus generating hot flue gas. In this process, the fuel's latent chemical energy is released. The hot flue gas transfers its thermal energy by radiation and convection to the working media (water/steam) via the boiler heat exchanger surfaces.

In the turbine, the thermal energy of the steam is converted into mechanical energy which, in turn, is transformed into electrical energy in the generator. These conversions involve certain losses of the fuel energy input, as seen in Fig. 4.2. The major loss, of 50% of the fuel heat input, occurs during the energy conversion in the turbine. This loss can be explained by the cycle efficiency. The waste heat is dissipated to the environment through the condenser. Further significant losses occur

H. Spliethoff, *Power Generation from Solid Fuels,* Power Systems,
DOI 10.1007/978-3-642-02856-4_4, © Springer-Verlag Berlin Heidelberg 2010

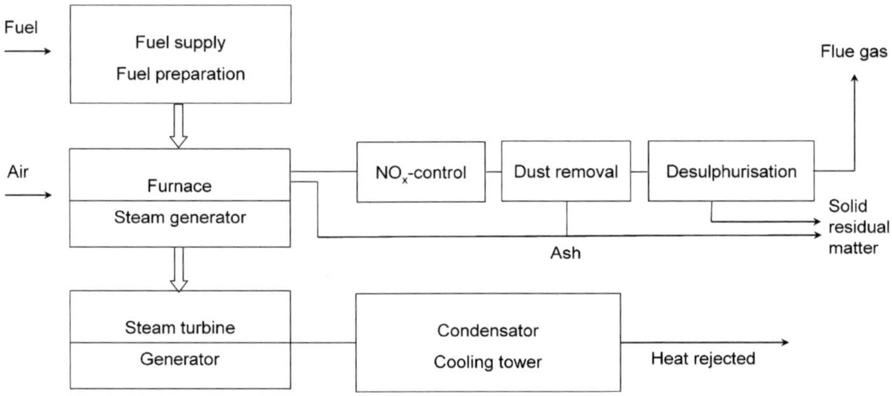

Fig. 4.1 Components of a steam power plant

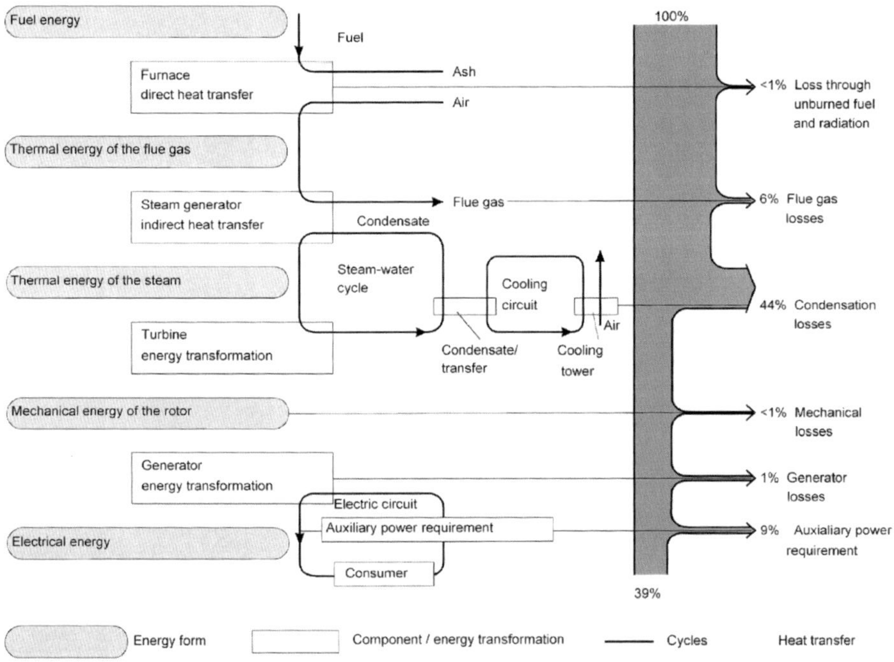

Fig. 4.2 Energy transformation or conversion, circulation of energy-carrying media and efficiency in a condensation power plant

in the steam generator, mainly as flue gas losses of about 6%. The auxiliary power requirements of about 9% of the fuel energy input add to these losses.

4.1.2 Design of a Condensation Power Plant

Figure 4.3 shows the simplified schematic design of a modern pulverised coal fired power station unit.

The fuel, coal, is transported from the coal storage area of the power station to the coal bunkers, which are arranged inside the boiler house and have a storage capacity of up to 1 day. Feeders transport coal from the bunkers to the mills for drying and pulverising. The milling fineness of the pulverised coal is adjusted according to the requirements of the firing. The combined drying and pulverising process of hard coal fired furnaces uses hot air that is heated up to 350–400°C in an air preheater. The high moisture content of brown coals requires hot flue gas for drying.

The pulverised coal is transported to the burners by the transport air flow, which is also used for the drying process. The transport air is further used in the combustion process as primary air. Complete combustion of the fuel is achieved by injecting secondary air, heated in the preheater to 300–400°C, into the furnace.

In the furnace, the pulverised coal burns almost completely, radiating heat to the furnace walls, producing flame temperatures between 1,400 and 1,600°C. The volumetric flow increases about 10-fold in the process, while it decreases again to nearly the input volume in the flue gas cooling path. The furnace wall, made of

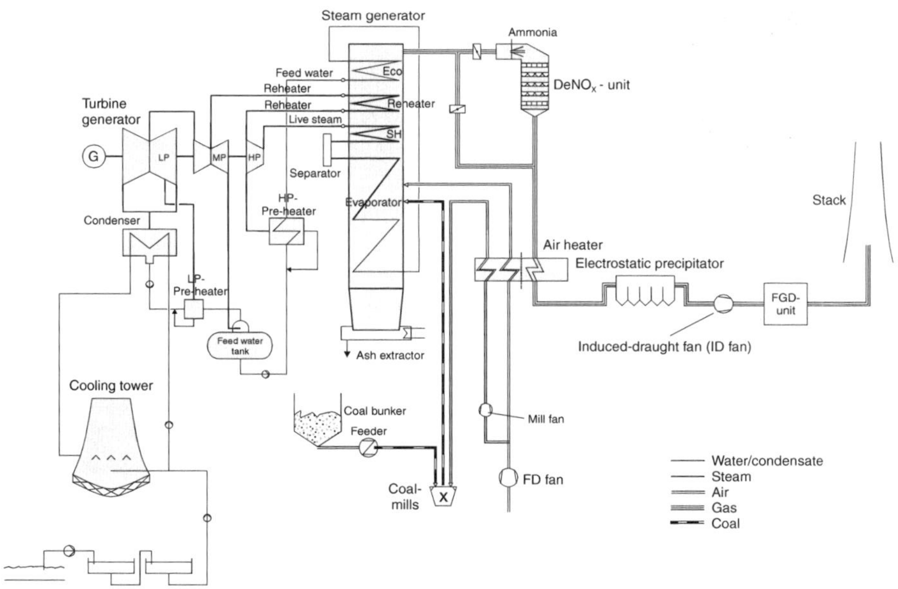

Fig. 4.3 Schematic diagram of a hard coal fired thermal power station

tightly welded membranes, forms the evaporative heating surface, which vaporises the feed water. After the flue gases are cooled to about 1,200–1,300°C at the end of the furnace, they are further cooled down by the convective heating surfaces of the superheater (SH), the reheater (RH) and the feed water preheater, also called the economiser. Then nitrogen is removed from the flue gas in a $DeNO_x$ unit at a temperature range of 300–400°C. In the air heaters the flue gases transfer their residual heat to the combustion air, during which they are cooled to the exit flue gas temperature of the steam generator.

For further cleaning, the flue gas is conducted through an electrostatic precipitator (ESP) to remove dust and, through a flue gas desulphurisation unit, to meet the allowed sulphur dioxide emission standards. The gases are discharged to the environment via a stack or natural-draught cooling tower. One or more induced-draught fans transport the flue gas from the furnace to the outlet. In the course of retrofitting measures in various power plants, further series-connected induced-draught fans have been added to the existing equipment to transport the flue gas through the desulphurisation and $DeNO_x$ units. In new power stations, equipped with flue gas desulphurisation and $DeNO_x$ units from the outset, one or more induced-draught fans are connected in parallel to overcome the pressure loss of all installations and components in the flue gas train.

In the steam generator, the energy released in combustion is transferred to the steam – water cycle, and the enthalpy of the steam is converted into mechanical work by the turbine. The turbine exhaust steam is turned to water in the condenser.

The steam – water cycle is a substantial parameter in the overall design of the power plant. The thermodynamic data of the water – steam cycle is the basis for the steam generator and turbine configurations and determine the power plant's efficiency.

Condensate pumps transport the condensate to the feed water tanks via low-pressure preheaters (LP preheaters), which are heated by steam from the lower pressure-staged turbine extraction. In the feed water tanks, the condensate is further preheated and degassed by steam from the mid-section turbine extraction in a direct-contact heater. The high-pressure feed water pump sets the operating pressure in the water – steam section of the boiler and transports the feed water to the boiler inlet via the high-pressure preheaters, which are heated by steam from the upper pressure turbine extraction stages. The feed water is preheated to the entry temperature of the boiler in 6–9 stages. In the preheater, the extraction steam is cooled, condensed and possibly supercooled and drained back into the condensate or feed water flow before the preheater. The higher the feed water temperature of the respective preheating stage is, the higher the boiling temperatures have to be, and hence the extraction pressure of the associated extraction steam flow. The last preheating stage before the boiler is fed with steam taken from the cold reheater in a conventional design or from the HP turbine extraction in an advanced design.

In the boiler, the preheated feed water is further heated in the economiser, the last convective heating surface in the flue gas path, and then conducted to the evaporator heat exchanger surface. The superheater heats the steam coming from the evaporator up to exit temperature of the superheater, i.e. to the level of the so-called live steam temperature. The level of the turbine entry temperature is slightly lower, by the amount of the temperature drop in the connecting high-pressure steam piping. After partial expansion in the HP turbine, most power plants heat the steam up to levels such as the live steam temperature or higher in a so-called reheater (exchanging heat with the flue gas). Higher temperatures in the reheater are possible due to the lower pressure.

In the condenser, the turbine exhaust steam condenses, with the waste heat being transferred to the cooling water circuit. Closed cooling water circuits are mostly equipped with natural-draught cooling towers for the re-cooling of the water. The buoyancy in the cooling tower makes the heated cooling air flow upwards after it has taken up heat from the cooling water in a trickle cooler. The heated cooling air exits to the environment via the cooling tower mouth at the top.

4.1.3 Development History of Power Plants – Correlation Between Unit Size, Availability and Efficiency

The block power station was born out of the need for higher power plant capacities (due to increasing energy demands), changing expectations with respect to lower investment costs and the desire for a higher reliability in power supply. Besides other parameters, it is, in particular, the

- unit output,
- efficiency and
- availability

that describe the development of the block power station unit.

Given the high availability of each of the large plant parts, modern hard coal fired power plants are generally designed as block units, meaning all the process units are contained together in one "block". The direct physical interactions of steam generators, turbines and auxiliary installations involve less investment because of shorter connecting pipes. In addition, the pressure and heat losses are lower than the range-type power stations that were common earlier in the 1900s. In range-type power stations, several boilers feed one steam range which can supply several turbines.

From the early 1950s, condensation power plants were built as block units with simple reheating for base and for intermediate loads. At the beginning, the unit capacities were some 60 MW or more; live steam and reheater temperatures were at 525°C, while the live steam pressure was at about 125 bar.

The maximum block capacity rose with the maximum capacities of single plant components. Step by step, the power station unit has been supplemented by additional components and plants. Today, the largest unit capacities are 1,010 MW in Europe, which will increase to 1,100 MW by 2010, and 1,300 MW in the USA (see Fig. 4.4) (Eitz 1996; Smith 1996). Conventional live steam conditions proven in operation are 180–250 bar and 540°C, with reheater temperatures at 540°C as

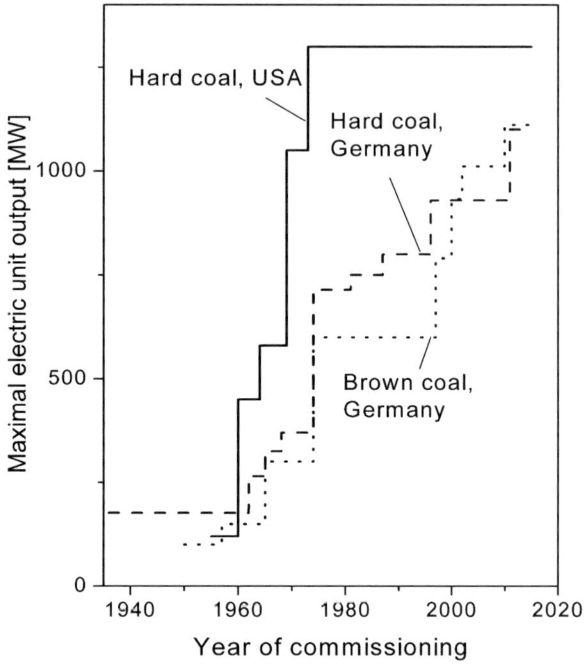

Fig. 4.4 Maximum unit capacity

well. All over the world, one can see a trend towards higher live steam conditions. Figure 4.5 shows the development of live steam conditions in Germany.

With the unit capacities and the live steam conditions increasing, the efficiency levels rose as well (see Fig. 4.6). The power station costs decreased, depending on

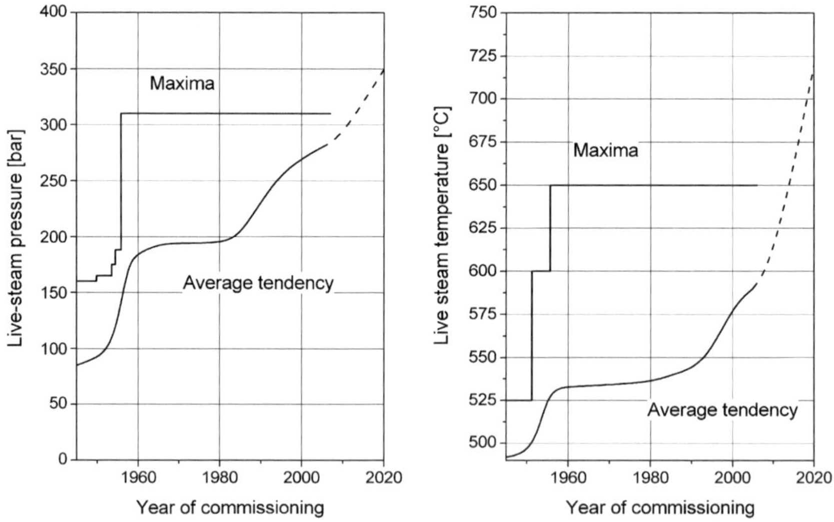

Fig. 4.5 Evolution of live steam conditions of German plants

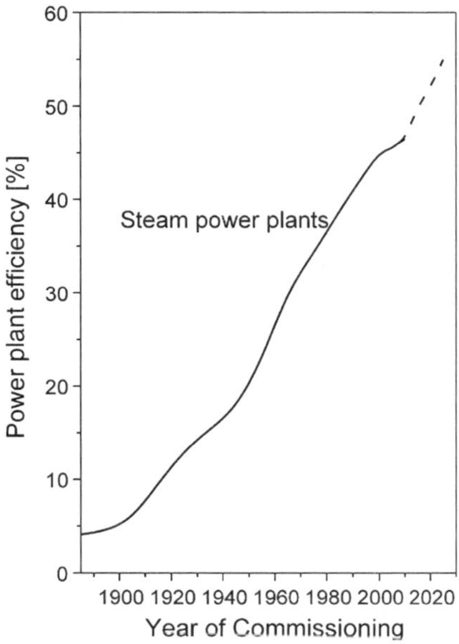

Fig. 4.6 Evolution of the efficiency level of German plants

the capacity, making efficiency-enhancing measures become more cost-effective. Higher efficiencies of large units can also be explained physically: specific surface heat losses of boilers and losses of rotating machinery due to leakiness diminish with higher capacities.

Availability of technology becomes important with increasing capacities, the need for more pollution control equipment and the desire for technical developments towards higher efficiency levels. High availability is desirable for reliable electricity production and is a necessary comparative criterion of technical developments. Further development of steam power plants should therefore be based on the comparable availability of proven power plant concepts.

Besides being economically significant, availability also has an environmental impact. The lowest CO_2 emission level is achieved by a generation system when the power plants with the highest efficiency are of comparably high availability. Lower availability rates, in consequence, deteriorate the gain in efficiency. Until the second half of the 1960s, the development of the power plant unit efficiency was sustained by development of the plant's thermal efficiency. There are numerous scientific studies on this subject (Knizia 1966). In the 1970s, the efficiency was further enhanced along with increasing unit sizes from 150 via 300 to more than 600 MW. At the same time, the availability rate was increased and thus the effect of the efficiency enhancement improved.

While the flue gas particulate collector was a fixed component in the plant design from the very beginning, the plants were augmented by flue gas desulphurisation units only from the mid-1970s and by nitrogen oxide control devices from the mid-1980s on. The availabilities of these components were at first low but then increased as they developed. For example, in Germany gas cleaning devices for SO_2 and NO_x

Table 4.1 Data for the reference power plant (Spliethoff and Abröll 1985)

Power plant unit	
Gross rated power	740 MW
Net rated power	690 MW
Efficiency	39%
Mechanical capacity of the feed pump	21 MW
Auxiliary power requirement	50 MW
Mode of service	Intermediate load range (170 starts p.a.)
Steam generator	
Capacity	2250 t/h (625 kg/s)
Construction	Once-through boiler
Live steam condition	209 bar, 535°C
Steam condition after reheater	39.6 bar, 535°C
Entry temperature of feed water	250°C
Firing	
Air ratio	1.3
Flue gas temperature	130°C
Coal mills	4 × 74 t/h
Forced-draught fan (FD fan)	1 × 100%
Induced-draught fan (ID fan)	1 × 100%
Range of control	40–100%
Steam generator efficiency	94%
Boiler feed pump	1 × 100% duty turbine-driven pump 1 × 50% duty motor-driven pump
Steam turbine generator	
Construction	Condensation turbine with single reheating
Operational mode	modified sliding-pressure operation with throttling of the intake valves (5%)
Turbine pressure sections/number of extractions	4 (1 × HP, 1 × MP, 2 × LP)/6
Live steam condition	190 bar/530°C
Exhaust steam pressure	0.0549 bar
Back-cooling system	
Cooling tower construction	Natural-draught wet-type cooling tower
Heat rejection	894 MW
Air temperature	10°C, max. 35°C
Cold water temperature	16.6°C, max. 29°C
Flue gas cleaning unit	
Particulate collector	Electrostatic precipitator (ESP)
Nitrogen oxide control device	High-dust catalyst before air preheater
Desulphurisation unit	Wet desulphurisation with limestone
Flue gas off-take	Stack, reheat after FGD unit

became required by law in 1988 with the inception of ordinances of the German *Bundesimmissionsschutzgesetz (BImSchG)*, or Federal Pollution Control Act. Any power plant with emission levels exceeding the prescribed standards concerning dust, SO_2 and NO_x may be operated only at limited duty or not at all.

Different national standards in some countries have in consequence differing nitrogen oxide control methods. Higher limits make it possible to develop and apply less complex emission control techniques as well as the more advanced technologies. In such situations, more lenient emission standards may mean higher energy conversion efficiencies and lower losses as compared to power stations with stricter emission standards.

The environmental stipulations that have an impact on the efficiency of inland power plants also limit the use of cooling water and once-through cooling cycles. Comparisons of efficiency and availability across national borders should take these differences into account.

4.1.4 Reference Power Plant

Operating experiences and technological developments are introduced into the planning of new power plants and thus form the basis of the respective technical state of the art. Developments build upon this state of the art. For this reason, a reference power plant, the data for which is compiled in Table 4.1, shall be the basis for further discussion in this chapter. This power plant corresponds to the state of the art from the 1980s in Germany. The reference power plant will be used in the following sections to explain fundamentals and design by way of comparison with the further development of steam power plants.

4.2 Steam Generators

In a steam generator fired with fossil fuels, the chemically bound energy in the fuel is released through combustion and transferred to the generator's steam – water heating surface system. The high-pressure water is evaporated and superheated. The capacity range of steam generators lies between 0.4 t/h for process steam generators, up to 4,500 t/h for large power plant boilers for electricity production (Stultz and Kitto 1992). The steam parameters are determined by the requirements of the process. Large steam generators for electricity production are operated at steam temperatures of 540°C and steam pressures of 180–240 bar (STEAG 1988). New plants are designed for live steam pressures up to 300 bar and live steam and reheat temperatures up to 600°C/620°C.

The fossil fuels used are coal, oil or gas. In electricity production, steam generation from nuclear energy plays an important part too. Other energy sources used today for steam production are fuels of biogenetic origin and residual matter from industrial processes, e.g. peat, wood, wood residues, biogases, straw, waste liquors and gases from chemical processes as well as blast furnace gas from pig iron production.

The schematic design of a simple steam generator, a shell boiler, is shown in Fig. 4.7 (Doležal 1990). Tubes that conduct flue gas, immersed in a boiling water

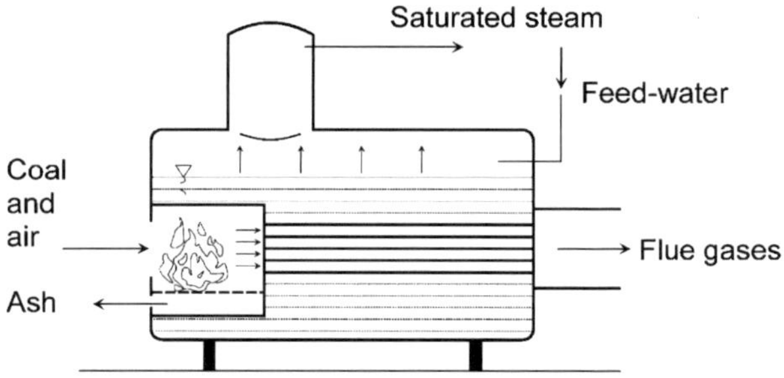

Fig. 4.7 Schematic graphic of a shell boiler

bath, transfer heat for steam generation. In order to maintain a continuous process, steam extraction from the steam space and water supply to the water bath are synchronised according to the water level. Shell boilers are suited to low steam pressures only and so are utilised for low capacities up to 54 t/h of steam output and steam pressures up to 35 bar (Sobbe 2004).

In steam generators of higher capacity, the heat exchange surfaces consist of complex parallel tube systems. A great number of small water and steam flows, conducted through tubes with a small inner diameter, take up heat along the heated stretches of the tubes. Both technically and economically, this is the most effective method to generate steam at high pressures. This method is also utilised to preheat, evaporate and superheat the working medium, water, up to saturation temperature. Accordingly, a steam generator consists of various heat exchange surfaces, such as the feed water heater or economiser, evaporator, superheater and reheater, which operate with different heat flux densities depending on the firing and the hot flue gases. The increases in the volumetric flows are provided for by branching of the heated single tubes, introducing more flow capacity. The relative heat absorptions of the economiser, evaporator and superheater are dependent on the pressure, as the evaporation enthalpies decrease with higher pressures. The heat absorptions of the economiser and the superheater increase with higher pressures.

The various steam generator systems differ in the configuration of the evaporator, while there is no difference in the superheater and economiser units. A distinction is made between circulation and once-through systems. The course of evaporation in the tubes is shown in Fig. 4.8 for partial (circulation) and for complete evaporation (once through) (Stultz and Kitto 1992). In circulation steam generators, water is heated to boiling temperature in the heated vertical evaporator tubes, forming steam bubbles. In the drum mounted above the heated tubes, the rising water – steam mixture is divided, with the steam flow being fed to the superheater and the water flowing back through downcomer pipes to re-enter the heated evaporator tubes. In this case, the process is a mere partial evaporation in the evaporator tube. Complete evaporation is achieved only after several recirculations. Circulation systems have a fixed liquid – vapour phase transition point in the drum.

In contrast, in once-through steam generators, the water in the evaporator tube is in one stage preheated, evaporated and partially superheated. Because steam leaves

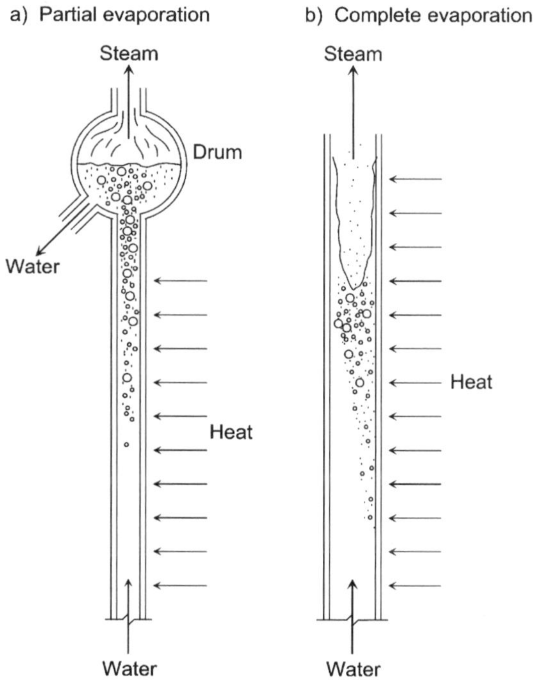

Fig. 4.8 Evaporation process in vertical evaporation tubes

the evaporator, this system does not need a water – steam division drum. In once-through systems, the liquid – vapour phase transition point is not fixed.

The required heat for steam generation is transferred to the heat exchange surfaces by radiation and convection. The heat exchange conditions in the evaporator – water-wetted tube walls and high mass flow densities – make it possible to achieve high heat transfer rates in the evaporator. The furnace walls, which have the highest heat flux density in a steam generator, due to the flame radiation, are therefore mostly designed as evaporative heating surfaces.

4.2.1 Flow and Heat Transfer Inside a Tube

Parameters of great importance for the design of steam generators are the heat transfer and the flow in the evaporation area. Figure 4.9 presents the processes during evaporation in a long, vertical-flow tube with homogeneous heating (Adrian et al. 1986). The water first enters the tube as an under-cooled liquid, cooling the tube by convective heat transfer. In this area, where there is only water flow, the heat transfer between the tube wall and the medium is good and only depends on the velocity.

The first steam bubbles form when the water touching the inside wall reaches the boiling point, thus developing a bubble flow. This is termed sub-cooled boiling, because the liquid in the centre of the tube flow has not yet reached boiling temperature. The temperature continues to rise until the entire medium reaches the

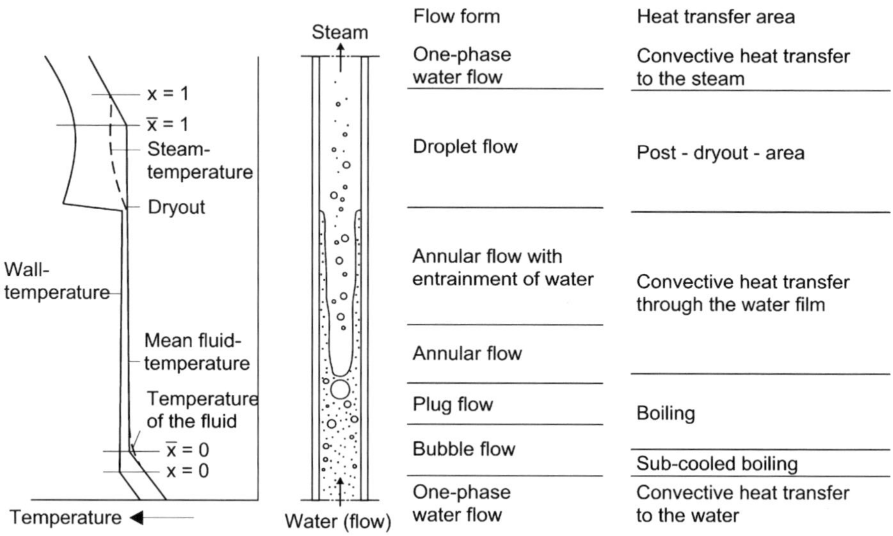

Fig. 4.9 Schematic diagram of the evaporation processes in a vertical tube (Adrian et al. 1986)

boiling temperature. The steam bubbles forming in the boiling process merge, the flow changes from a bubble flow to a plug flow and subsequently to an annular flow, where the wall is eventually wetted by just a water film. This water film becomes so thin that evaporation in it is suppressed. The heat in this state is transferred by convection and thermal conduction through the water film, and vaporisation takes place at the interface between the water and the steam. Boiling and bubble movement result in a high heat transfer coefficient, the highest in a steam generator.

When the water film becomes sparse, wetting the wall only incompletely, the heat transfer diminishes and the wall temperature rises considerably. A boiling crisis occurs, also called dryout, with the tube wall drying more and more. The heat transfer becomes small, because the wall is wetted only in parts and steam cooling has not yet become effective due to partial evaporation. The location of the boiling crisis and the level the wall temperature rises to depend on numerous factors, such as the heat flux density, the mass flow density, the tube design and the steam quality.

Figure 4.10 shows the influence of the heat flux density on the wall temperature (Stultz and Kitto 1992).

In the region immediately following the dryout region, some water droplets are still present, although at this stage the steam is already slightly superheated. With the remaining water droplets evaporating, the steam quality and its velocity rise, so that cooling improves and the wall temperature falls slightly. After evaporation is complete, the flow becomes a steam flow with convective heat transfer. Both the temperature of the steam flow and the temperature of the tube increase thereafter.

For the design of steam generators, boiling crises are of great importance, because they can lead to excess temperatures in the tube walls, which have to be taken into account in the design stage. There is a distinction between a "first-

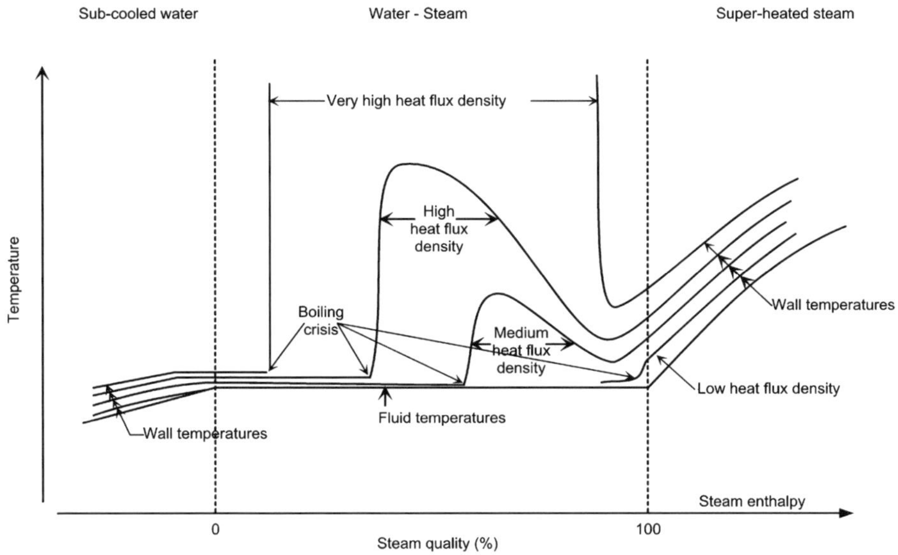

Fig. 4.10 Tube wall temperatures at different heat flux densities (Stultz and Kitto 1992)

kind" and a "second-kind" boiling crisis. The "first kind" of boiling crisis, called DNB, from "departure from nucleate boiling", is caused by excessively stressed heat exchange surfaces. This crisis can occur anywhere in the evaporation area, from the sub-cooled boiling region to the annular flow region, when a so-called critical heat flux density is reached and then exceeded. The higher the steam quality and the higher the pressure the lower the critical heat flux density. A steam film forms on the wall, which impedes the heat transfer. During design of a steam generator, the DNB crisis has to be designed out. By improved cooling of the tubes, e.g. by using smaller tube diameters or internally finned tubes, the critical heat flux can be raised.

The "second kind" of boiling crisis occurs during the transition from annular to droplet flow, through a drying out of the water film. The effects of this boiling crisis, though, are of minor consequence compared to the DNB crisis. They are a systematic phenomenon with once-through steam generators. In circulation steam generators, due to the partial evaporation, the liquid – vapour phase transition point is not reached (Strauß 2006).

Figure 4.11 shows the qualitative impact of internally finned tubes on the location and temperature of the boiling crisis. In vertically mounted plain tubes, the water in the evaporation area flows partially as a film on the wall and partially in the form of dispersed droplets in the steam centre. In this condition, the boiling crisis, i.e. the drying out, occurs at a steam quality considerably less than 1. Insufficiently high flow velocities in partial evaporation can then result in high tube wall temperatures. In tubes with internal helicoid fins, the flow is set into a twisting movement by the helical guidance of the fins. Centrifugal force makes the dispersed water droplets settle on the wall, which keep the wall covered with a wet coating up to high steam qualities of $x > 0.9$. This way, the flow velocities are already high when dryout

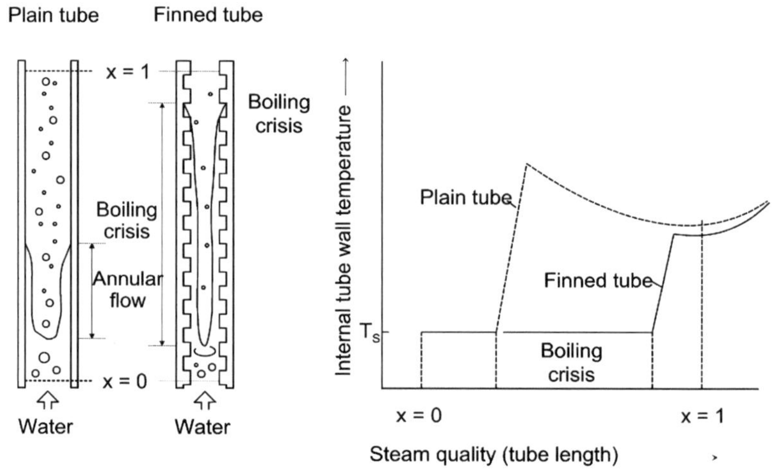

Fig. 4.11 Flow patterns and wall temperatures in plain and internally finned vertical evaporator tubes (Kefer et al. 1990)

occurs. The effect is a good heat transfer and thus low tube wall temperatures (Kefer et al. 1990).

Figure 4.11 shows vertically mounted and evenly heated tubes. However, for once-through steam generators with helically wound tubing in the furnace, the tubes are inclined and heated on one side only (see Fig. 4.12). In the evaporation area, this configuration may result in the formation of a transition zone where only part of the

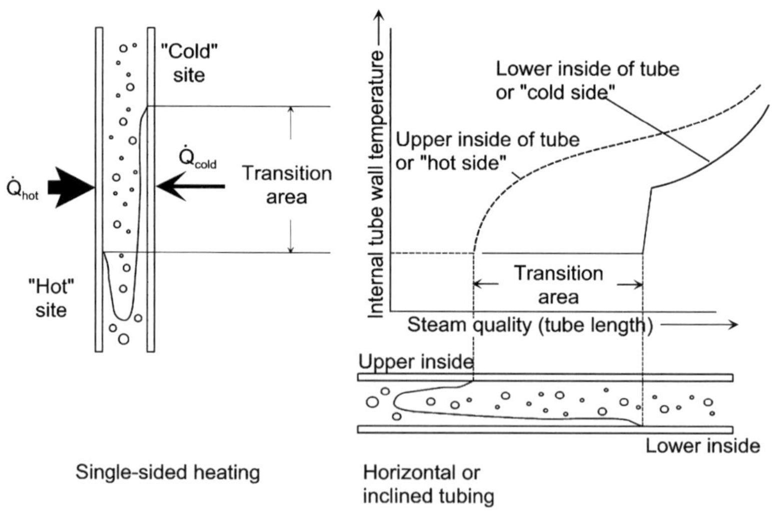

Fig. 4.12 Flow patterns and wall temperatures in a single-sided heated, horizontal or inclined evaporator tube (Kefer et al. 1990)

inner perimeter of the evaporator tube is wetted. With the heating only on one side, the water film dries on the heated side faster than on the cold side.

If the tube is inclined or mounted horizontally, gravity causes a segregation of water and steam. The water flows – mainly as a film or in droplets – in the lower part of the tube cross-section, while the more light-weight steam flows in the upper part. This way, the wetting is maintained at differing lengths in the upper and lower parts of the cross-section, possibly resulting in differing inner wall temperatures between the upper and lower sides. In the extreme case, the upper inside of the tube is dry before evaporation begins, while the wet coating on the lower inside only dries when there is total vaporisation.

Due to the variable wetting, transient thermal stress may occur, causing damage to the tube. A comparison of the temperature conditions in a horizontal tube with a segregated water/steam flow to a vertical tube without segregated flow revealed that the temperature maxima of the horizontal tube was lower. This can be explained by the thermal conduction between the cold lower and hot upper parts of the tube and the eddies that form during vaporisation (Kefer et al. 1990).

4.2.2 Evaporator Configurations

As already described in Sect. 4.2.1, steam generator systems are divided into circulation and once-through systems. The various state-of-the-art designs are shown in Fig. 4.13.

4.2.2.1 Natural Circulation

Natural-circulation steam generators typically consist of economisers and an evaporator with risers that form the heated furnace wall, a drum for the separation of water from steam and unheated down pipes and superheaters (see Fig. 4.14).

Water with a temperature just below the boiling point flows down through the downcomer pipes. In the vertically mounted evaporator risers with upward flow, a water – steam mixture forms through heating and is again divided into water and steam in the drum.

In a natural-circulation steam generator, a circulating flow forms because of the density difference between the falling water in the unheated downcomer and the water – steam mixture in the heated riser (see Fig. 4.15). Besides accelerating the rising water, the density and pressure differences also compensate for the flow resistance in the evaporator and in the drum.

With the heating increasing, the flow speeds up in stable circulation until a maximum is reached. With a further increase in the heating, the pressure loss in the evaporator tubes also increases, while the change of the density difference, the driving force, is only insignificant, meaning that the flow slows down (an unstable area). In designing natural-circulation boilers, the fact that a rise in the thermal load always leads to higher circulation velocities (in stable circulation) must be taken into

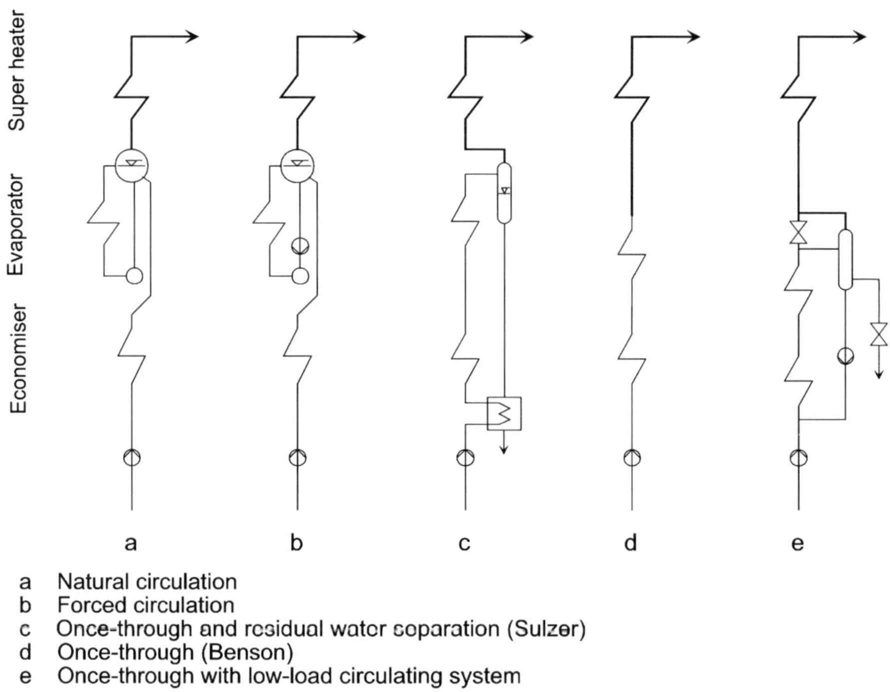

Fig. 4.13 Evaporator configurations

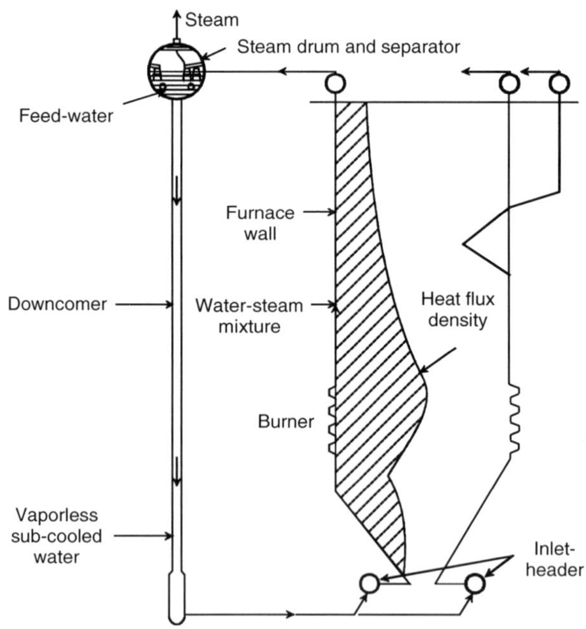

Fig. 4.14 Schematic diagram of a natural-circulation steam generator (Stultz and Kitto 1992)

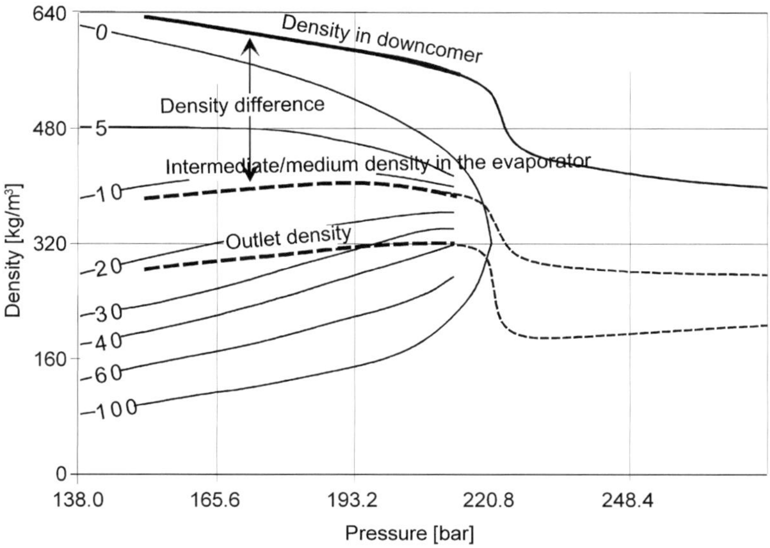

Fig. 4.15 Density differences in a natural-circulation steam generator (Stultz and Kitto 1992)

account. In this case, the flow necessary for cooling is determined by the heating (Stultz and Kitto 1992).

Natural-circulation steam generators work effectively where the density differences between water and water – steam mixture are high. Natural circulation is limited by the circulation ratio, which decreases with a higher design pressure and a higher steam quality, meaning cooling is no longer ensured at pressures above 185 bar in the drum – corresponding to 170–180 bar before the turbine. The absolute upper limit for a drum boiler is the critical pressure, because at this point and above, a two-phase water – steam mixture no longer exists.

The advantages of natural circulation are its simple construction and the low power demand of the feeding pump. In addition, the requirements for the feed water quality are lower than in once-through systems because impurities accumulate in the drum, meaning they can be blown down. Disadvantages are due to the necessarily thick wall of the drum, which may restrict the allowable rate of load change, and the restriction on the system pressure (Doležal 1990). Thick-walled high-pressure parts are more sophisticated with regard to manufacturing and quality approval.

4.2.2.2 Forced Circulation

In forced-circulation systems, the buoyancy of the steam, as the only driver of the water or the water – steam mixture circulation, is supported by a circulating pump, preferably installed at the bottommost point of the downcomer, where there is the highest water pressure. Forced circulation is limited to a range of about 200 bar to ensure sufficient water – steam separation in the drum (Adrian et al. 1986).

Since the circulating pump can balance out the pressure losses in the riser and downcomer parts of the evaporator, it is possible for the design to include components with higher pressure losses. It allows the choice of narrower tubes for better cooling; forced distribution of the water at the inlet of the evaporator tube; and drum inserts that are more effective for water separation but have higher pressure losses.

The applications of forced-circulation steam generators, like natural-circulation systems, are low-pressure and intermediate-pressure plants with capacities up to 500 t/h and also heat recovery steam generators, whereas for high-pressure steam generators, in Germany, it is preferred to use once-through forced circulation. In several countries, though, and in the USA in particular, the forced-circulation system is the preferred system even for large plants, with capacities up to 2,000 t/h and pressures up to 170 bar.

The drawback of a higher power demand for forced circulation, compared to natural circulation, may be balanced out by the financial advantages of material savings. Since forced-circulation steam generators feature lower circulation ratios (3–5) at a higher steam quality, they can be built in considerably smaller dimensions than natural-circulation systems. The result is more cost-effective construction types, especially with higher pressure configurations (Strauß 2006).

4.2.2.3 Once-Through Systems

In once-through systems, evaporation and a slight superheating take place in one stage in the evaporator tube. In contrast to circulation systems, the liquid – vapour phase transition point in the evaporator tube changes its position depending on the load or, for control processes, along with the change of the fuel-to-feed water flow ratio.

Well-known once-through steam generators are the Benson and the Sulzer boilers, or in Russia the Ramsin boiler. However, they are only rarely utilised now in their original design. The Benson boiler shown in Fig. 4.16 had an evaporator consisting of several vertical tubes with upward flow, mounted in series-connected banks, which at the same time defined the furnace perimeter. The liquid – vapour phase transition point was in the so-called final evaporative bank which, for salt deposit considerations, had been installed after the furnace in the convective heat exchanger range, with low heat transfer rates. In the Sulzer boiler, several parallel evaporator tubes meandering through the furnace formed the evaporator (see Fig. 4.17) (Doležal 1990; Wauschkuhn 2001).

The difference to the Benson boiler was that this way each tube ran the entire length of the evaporator. Typical features of the Sulzer boiler were the wet operating regime of the evaporator and the following downstream water separator, which was designed to separate a residual water content of 5%. The mineral-containing residual water was disposed of as boiler blowdown. The differences between the Sulzer and Benson boilers have vanished as the development of the boiler systems has advanced – modern once-through boilers are largely identical.

Once-through steam generators usually operate with circulating devices, which in the lower load range ensure flow stability in the evaporator and sufficient cooling

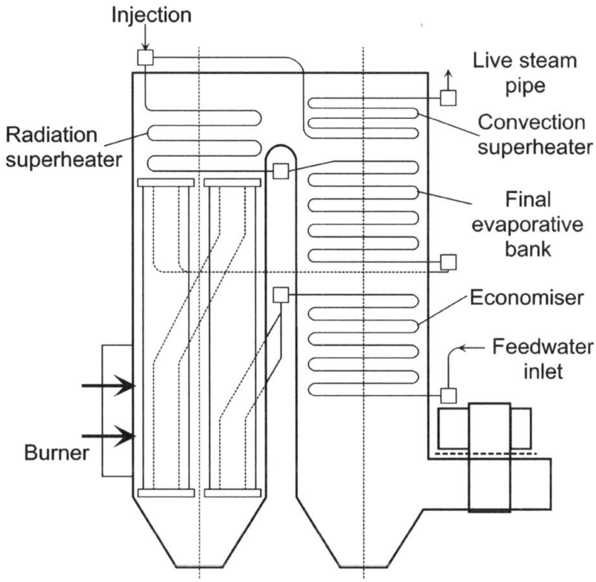

Fig. 4.16 Benson boiler (Doležal 1990)

of the evaporator tubes (see Fig. 4.13). Water at the end of the evaporator that has not vaporised is separated as residual water, collected and recirculated. This circulation fixes the liquid – vapour phase transition point, in particular in the low load range. In the upper load range, the evaporator is operated in once-through mode without the circulating pump and without water separation. In this case, the liquid – vapour phase transition point migrates, occurring after, or near the end of, the furnace, i.e.

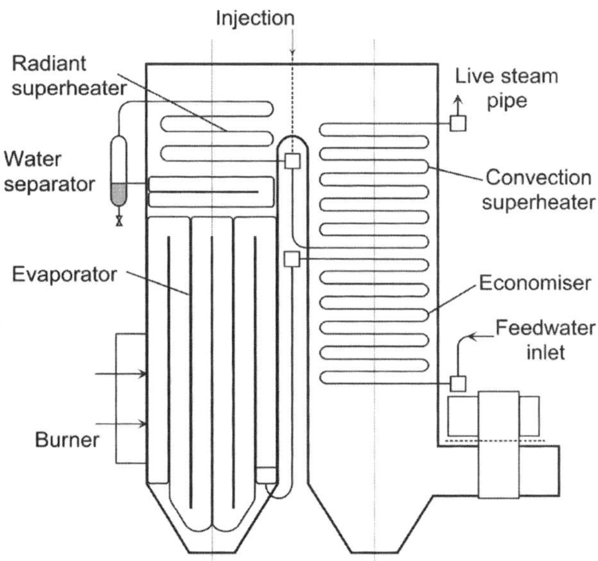

Fig. 4.17 Sulzer boiler (Doležal 1990)

in areas of low heat flux density. This is usually in the area where convective heat transfer has started (Adrian et al. 1986; Baehr 1985).

Evaporators of once-through steam generators today are made of tightly welded membrane tube walls. The mass flow density in the evaporator tubes has to be set such that excessive tube wall temperatures are avoided even with low heat transfer coefficients of the inside tube walls. Helically mounted evaporator tubes are a measure to ensure that the mass flow densities required for cooling are also suitable for large radiant heat fluxes in the furnace. Heating differences due to high heat fluxes in the wall centre and relatively low fluxes in the furnace corners, as well as unbalanced combustion, are compensated by each of the tubes running through all the walls. Another possible measure is internally finned evaporator tubes.

Membrane tube walls in a helically wound pattern, however, are not able to carry, without additional support, the weight of the furnace, the structural bracings, the insulation and the water contained within it, as well as the possible fouling and slagging deposits. Nor can they sustain tubular offsets, caused by changes in the furnace pressure. The load is borne by vertical sling straps, which are welded on. Tubular offsets of the membrane walls caused by (furnace) pressure forces working vertically against the tube walls are limited by structural sling straps mounted horizontally around the perimeter at different heights. In the upper section of the furnace, it is possible to change to vertical tubing under lower heat flux densities (see Fig. 4.18) (Franke et al. 1993, 1995; Wittchow 1995).

In Germany, all large-scale steam generators since 1960 have been designed as once-through systems. Disadvantages of such systems are the more complex construction of the furnace walls, due to the wound pattern of the tubes; the higher auxiliary energy demand for the feed water pressure increase; and the higher control requirements. Despite lesser material requirements, the more complicated manufacturing and assembly result in higher costs compared to drum boilers with vertical tubing. The fact, however, that once-through systems have no thick-walled components, such as the drum in circulation systems, offers advantages such as the ability to operate with a sliding pressure, faster start-up and a greater flexibility to make fast load changes. This is important, especially for large coal-fired power stations

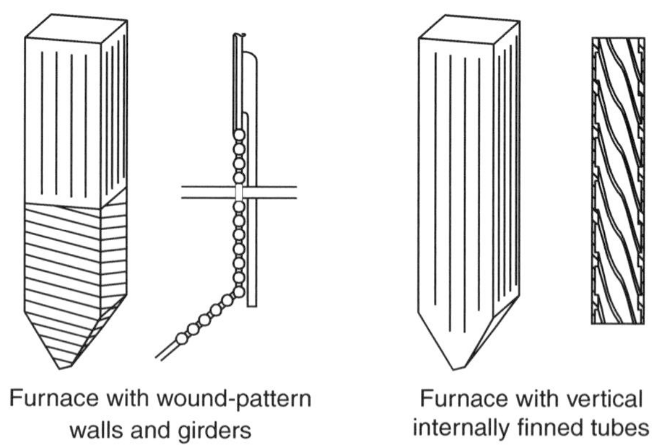

Fig. 4.18 Evaporators with wound-pattern furnace walls and with vertical tubing for once-through steam generators (Wittchow 1995)

which are used for medium load operation, where some stations are started up and shut down daily. In contrast to circulation systems, once-through systems can also be applied in advanced steam generators with higher steam parameters, because it is the only system suited to supercritical pressures. Higher steam temperatures, though, can also be used in circulation systems (Stultz and Kitto 1992; Doležal 1990; Strauß 2006; Wittchow 1982).

A further development of once-through steam generator technology is the use of internally finned tubes. The more intensive cooling of these tubes allows lower mass flow densities in the evaporator, thus making it possible to use vertical evaporator tubes. In contrast to the state-of-the-art wound-pattern walls, this new construction involves both lower costs and a number of additional operational advantages, which will be discussed in Sect. 4.3.5.3 in the context of evaporator design. For future power stations with advanced steam parameters, the forced once-through circulation concept has many advantages (Wittchow 1995; Lehmann et al. 1996).

4.2.3 Steam Generator Construction Types

The introduction of membrane walls that were fully joined by welding changed the design, manufacture and assembly of steam generators.

Until the 1960s, the evaporator tubes were individually mounted in the refractory lining of the furnace wall, without fins connecting adjoining tubes. The evaporator tubes only had the function of heat absorption. Fireside sealing-off against the boiler house was ensured by the wall construction of firebricks, insulation and metal casing.

Membrane walls fulfil both functions. The wall construction consists only of gas-tight evaporator tube banks and the insulation behind them. Fireproof lining is unnecessary. In addition to the lengthwise expansion of the tubes, the membrane walls also expand in the traverse direction, so that the expansion forces have to be countered by adequate reinforcement.

The membrane wall being welded from bottom to top limits the free design of the steam generator. Whereas previously, the evaporator could be adapted to the heat absorption (for instance by the final evaporative bank), today the furnace dimensions pre-set the size of the evaporator.

4.2.3.1 Single-Pass Boilers and Two-Pass Boilers

Steam generator designs (or construction types) are divided into two groups: single-pass and two-pass boilers. In single-pass or tower boilers, the convective heating surfaces (the superheater, reheater and economiser) are mounted above the furnace, so that the flue gases only have to be redirected after the last water vapour/steam heating surfaces. This helps to minimise erosion, in particular with high-ash coal types. Only after being cooled down to 400°C are the flue gases conducted to the air preheater through an uncooled blank pass.

The free space of the blank pass, having a temperature equal to that of the flue gas, is very often utilised for catalytic NO_x control. Further advantages of the single-pass boiler are

- simple mounting and assembly of supports and heating surfaces,
- little heat stress,
- straight flow paths with few bends,
- thermal expansion of the boiler body in only one direction (downwards in the case of a fixed support point on the furnace roof) and
- tube lanes of the convective heating surfaces that widen from top downwards against the current of the flue gas flow, so that deposits can fall through.

The disadvantage of the single-pass boiler is its height. Compared to two-pass boilers, the boiler of, for example, a 700 MW_{el} hard coal fired furnace is about 20–30 m higher (Fig. 4.19). The required base is roughly the same for both construction types (Strauß 2006).

Two-pass boilers offer more favourable conditions for heat transfer by introducing a second pass and adapting its cross-section to the volumetric flow through it, which decreases with falling flue gas temperatures. Two-pass boilers can be built with hanging superheater surfaces – the so-called plate heating surfaces – with wide spacings of about 1 m, hanging from the ceiling of the first pass. These heating surfaces are suited to high temperatures of more than 1,230°C. Incorporated into the design, they create a rather compact boiler construction, meaning 5–10% lower investment costs in comparison to single-pass boilers. Hanging heating surfaces, though, are not suitable for a frequent start-up/shutdown operation mode, because they cannot be drained. If horizontal superheater surfaces are also used in a two-pass

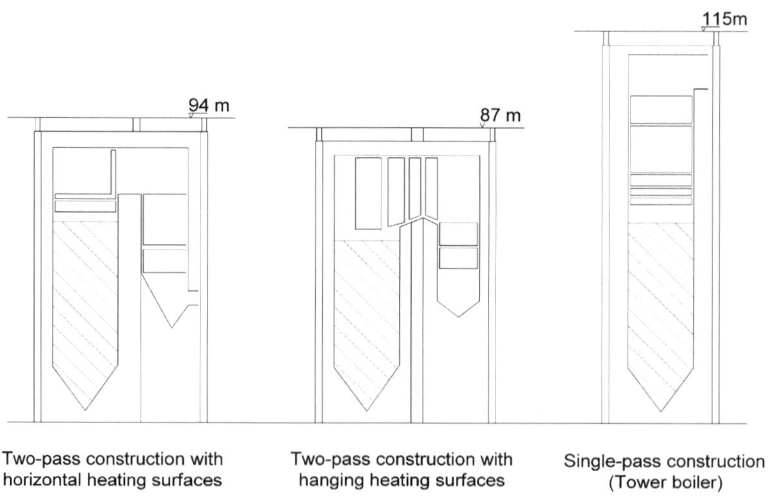

Fig. 4.19 Comparison of single- and two-pass boilers (Strauß 2006)

boiler, the investment costs are roughly the same. Two-pass boilers can be erected faster, because both passes can be assembled at the same time.

The choice of the boiler construction type depends on factors such as the ash content, the ash composition and the service mode of the steam generator. In Europe, single-pass boilers are preferred, because coal-fired furnaces are used for intermediate load ranges. Outside Europe, two-pass boilers with hanging superheater surfaces are usually constructed (Strauß 2006; Adrian et al. 1986).

4.2.4 Operating Regimes and Control Modes

4.2.4.1 Operating Regimes

Power stations can be categorised according to the duty they operate under: peak, intermediate and base loads.

A peak load power station is operated for only a small number of hours per day and only a fraction of the days in the year. Its annual output corresponds to about 2,000 annual full-load hours (equivalent hours of full-load operation per year), where there are 8,760 h in each year. (The actual amount will be greater because of start-up, shut down and partial-load operation). Such a station should reach its rated power within a short time (i.e. start-up time), and it should be possible to shut it down very quickly. The energy losses during start-up and shutdown should be small. Power stations such as pumped storage power stations or gas turbine power plants are used for peak load.

The base load power station, in contrast, is designed for inexpensive fuels, high efficiency levels and a small number of start-up and shutdowns. The output per year corresponds to about 6,000–8,000 annual full-load hours. It features a relatively small load control range between about 70 and 100% of its rated power, where the load change capability is not a very important criterion because the plant is mostly operated at its rated power. With few outages per year, start-up and shutdown times are of minor importance.

The yearly output of mid-range power stations lies between 2,000 and 6,000 annual full-load hours. Such a plant should be capable of dealing with peak load operation, with daily start-ups and shutdowns, as well as base load operation, with long operating periods and part-load conditions. It features a wide control range of about 30 or 40–100% of its rated power, and a good dynamic transient response and an efficiency as high as possible are sought.

Based on the operating regime of the power plant, the number of start-ups has to be specified in the design phase. Start-ups are classified into cold, warm and hot start-ups:

- Hot start-up: after an outage of maximum 8 h. Such an outage typically occurs overnight. For a hard coal fired power station operated in the mid-range such as the reference power plant, about 3,000–4,500 hot start-ups are scheduled for the lifetime of 40 years.

- Warm start-up: after an outage of 8–72 h. The outage is typically over the weekend. For a medium-range power plant the number of warm start-ups is about 1,000 over the station lifetime.
- Cold start-up: after an outage of more than 72 h. This start-up is quite rare; the total number for the medium-range power plant is about 200 (Zehtner 2009).

4.2.4.2 Primary, Secondary and Tertiary Control

The generation of power within a network such as the UCTE (Union for the Coordination of Transmission of Electricity) network in central Europe needs to be controlled and monitored for a secure and high-quality supply of electricity. The goal of the control is to maintain a balance between generation and consumption (demand) of electricity. The key control variable is the frequency of the net, which should be kept stable at 50 Hz, or 60 Hz in the USA or parts of Japan. In case of a drop in the frequency, caused by a higher consumption in comparison to the generation, power plants have to increase their load in order to stabilise the frequency. The rules of load-frequency control and requirements of power station performance are given in the Transmission Code for the West European UCTE net (UCTE 2004). The guidelines distinguish between primary, secondary and tertiary control.

Primary Control

The objective of primary control is to rapidly re-establish the balance between generation and consumption within the synchronous area by using turbine speed or turbine governors. By the joint action of all interconnected units, primary control stabilises the system frequency at a stationary value after a disturbance in a timeframe of seconds, but without restoring the reference values of system frequency and power exchanges. Outside periods of correction, the set-point frequency or scheduled frequency value is 50 Hz. Primary control is activated if the frequency deviation exceeds ±20 mHz.

All power stations have to be capable of delivering a maximum primary control reserve of 2% of the rated power within 30 s. The maximum reserve has to be activated at a frequency deviation of 200 mHz and has to be maintained over a period of 15 min. At lower frequency deviations, the required increase is correspondingly smaller, though the load change speed of 1% load per 15 s remains the same.

Primary control is supported by the self-regulation of consumption and generation in the network. The self-regulation is assumed to be 1%/Hz, meaning a load decrease of 0.2% occurs in case of a frequency drop of 200 mHz.

Secondary Control

Secondary control restores primary control reserves and maintains a balance between generation and consumption of electricity within each control area in a timeframe of seconds to, typically, 15 min. Accordingly, load variations of differing magnitudes must be corrected in the control area within this timeframe. Secondary control is based on secondary control reserves which are under automatic control by the operator of the network area. Secondary control is accomplished by increasing the fuel input of a power plant and thus puts requirements on the dynamic behaviour of power plants. The recommendations of the *Deutsche Verbundgesellschaft* (the association of German transmission system operators), located in Heidelberg, in 1991 fixed a required load change rate of coal-fired power plants of between 4 and 8%/min, referring to the rated load, within a load range of 40–100% (Verbundgesellschaft 1991). With the expanded European interconnected network system, these strict limits are no longer valid. It is now the responsibility of the operator of the network area to cater for a sufficient secondary control reserve (Verbundgesellschaft 1996; VDN 2007).

Tertiary Control

Tertiary control reserve is required to restore the secondary control reserves. Tertiary control reserve is usually activated manually after activation of secondary control and frees secondary reserve. Tertiary control is achieved by re-scheduling power generation of operating plants or start-up of additional plants. Tertiary control thus corresponds to the operation planning of all power plants within a network area.

4.2.4.3 Constant-Pressure and Sliding-Pressure Operation

The output of a condensation power station is set by means of the live steam mass flow $\dot{m}_{LS}$ (Doležal 1990). The mechanical power, P_m, of the turbine shaft depends on the live steam pressure p_{LS}, the cross-section of the opening A, or the lifting of the turbine intake valves, and the live steam temperature, T_{LS}, according to the following relation:

$$P_m \approx \dot{m}_{LS} \approx A \frac{p_{LS}}{\sqrt{T_{LS}}} \tag{4.1}$$

The live steam temperature should remain constant throughout the whole load control range, so that a high efficiency rate is also achieved during part load and to avoid stress on the turbine caused by temperature changes. The turbine output and the live steam mass flow to the turbine are set during steady-state conditions, either when the live steam pressure is at a constant cross-section of the turbine intake valves (sliding or variable pressure) or when the intake cross-section is at a constant steam pressure (constant or fixed pressure).

Constant-Pressure Operation

The control in constant-pressure operation is subdivided into throttle control and governing control. In constant-pressure governing control, the first turbine stage is designed as a control wheel and is preceded by sets of nozzle valves (see Fig. 4.20). As the load increases, the nozzle valves are sequentially opened.

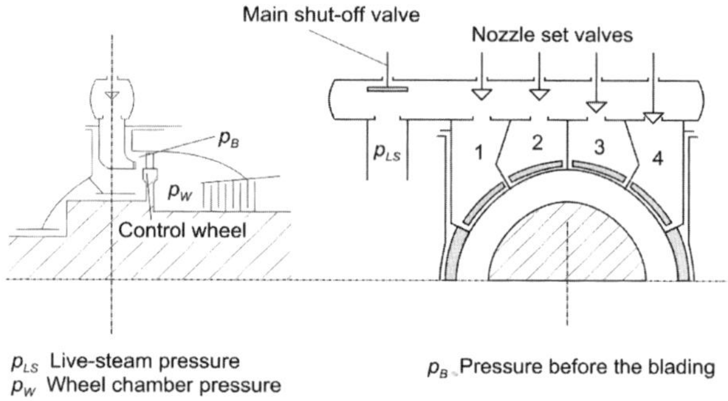

Fig. 4.20 Turbine with nozzle set and control wheel (Traupel 2001)

Under any load condition, only one of the valves is partially open, so loss through throttling only occurs there. The other valves are already fully opened or still closed. Because only a partial flow is subject to throttling losses, the part-load efficiency of the turbine is high. In such a case, the first stage of the turbine, the control stage, is charged by a high pressure p_B only for part of the circumference, where this pressure is slightly lower than the live steam pressure (i.e. constant pressure). The control stage cuts the pressure back to the wheel chamber pressure p_W and homogenises the steam distribution over the blading of the circumference of the following turbine stage (Traupel 1982).

In constant-pressure operation with throttle control, the total live steam mass flow is controlled by throttling the steam pressure through all the live steam valves at the same time. The turbine does not need a control stage, since the first turbine stage is charged uniformly and with lower pressure than the nozzle set governed stage. The pressure losses in throttling have a disadvantageous effect on the heat rate in part-load operation. In full-load operation, the heat rate may be somewhat better than in constant-pressure operation with the nozzle set governing, because there is no efficiency-decreasing impact of the control stage.

In the balanced steady-state conditions of a power station unit, the steam produced and the steam consumed by the turbine are equal. Fuel flow and steam generation correspond. The steam production is controlled by the fuel mass flow, the changes of which, however, have a delayed effect due to the thermal inertia of the steam generator. In constant-pressure operation, the steam energy stored in the boiler is used to control rapid load changes. By further opening the intake cross-section of the turbine control valve, additional steam is extracted from the steam generator and used to bridge the interval until the conditions are balanced out by the fuel supply.

Sliding-Pressure Control

In sliding-pressure operation, the turbine output and the steam flow are adjusted by the pressure at the outlet of the boiler. In natural sliding-pressure operation, the live steam valves of the turbine are completely opened, and the cross-section of the turbine intake is constant throughout the whole load range.

An output change using this control type can only be carried out by changing the fuel flow, a consequence of which is a long delay control characteristic of a change in the steam generator. Given that, in sliding-pressure operation, the pressure rises with increasing output, it is necessary that an increased steam flow is produced by the boiler before the output of the turbine increases.

In industrial practice, in order to diminish the disadvantages of the delayed control characteristic of natural sliding-pressure operation, modified sliding-pressure control is used. The live steam valves in this operation are opened about 95% during steady-state conditions, so that in the case of a power demand similar to that of constant-pressure operation, the valves open and thus increase the steam flow to the turbine. By this slight throttling of the turbine intake valves, a limited loss is chosen in favour of better control dynamics (Baehr 1985).

Advantages of sliding-pressure control are a load-independent temperature distribution in the turbine, a lower pressure stress on the steam generator and a lower power demand of the boiler feed water pump in part-load operation. Disadvantages are the changes of the boiling temperature in the evaporator, due to the pressure changes. The advantage of the decreasing power requirement for boiler feed pumping is stronger when the live steam pressure becomes higher. The general outcome in applying natural sliding pressure is a heat rate (including that of the boiler feed pumping power) which is slightly better than with nozzle-governed constant-pressure operation; with modified sliding pressure the heat rate is higher (Adrian et al. 1986; Baehr 1985). See also Sect. 4.4.4.

4.2.4.4 Impacts on the Turbine by Sliding-Pressure or Constant-Pressure Operation

The comparison of the different control modes in Fig. 4.21 shows that, in constant-pressure operation with the nozzle set controlling, the pressure p_B after the turbine inlet valves and before the blading remains almost constant over the load range. In sliding-pressure control, in contrast, and also in constant-pressure operation with throttle control, the pressure shows a linear rise with the output.

Both in sliding-pressure and in constant-pressure operation with throttle control, the stage pressures change to the same degree depending on the output, so that the stage temperatures are constant. In constant pressure operating with the nozzle set controlling, the pressure drop over the control stage $p_B - p_W$ becomes steeper with a decreasing output, so that the stage temperatures of the stages drop as well (Strauß 2006).

The influence of the control mode on the temperature in the high-pressure section is shown in Fig. 4.22. Load changes in constant-pressure operation cause considerable changes in temperature in the area of the first stage of the high-pressure turbine. Thermal stress arising in the process therefore limits the load change rate, in particular in the case of high-capacity turbines. In sliding-pressure operation, the temperature conditions in the turbine remain almost constant, so load changes are possible even with large turbines, even abruptly. This means that in sliding-pressure operation, the steam generator determines the dynamic performance of the power

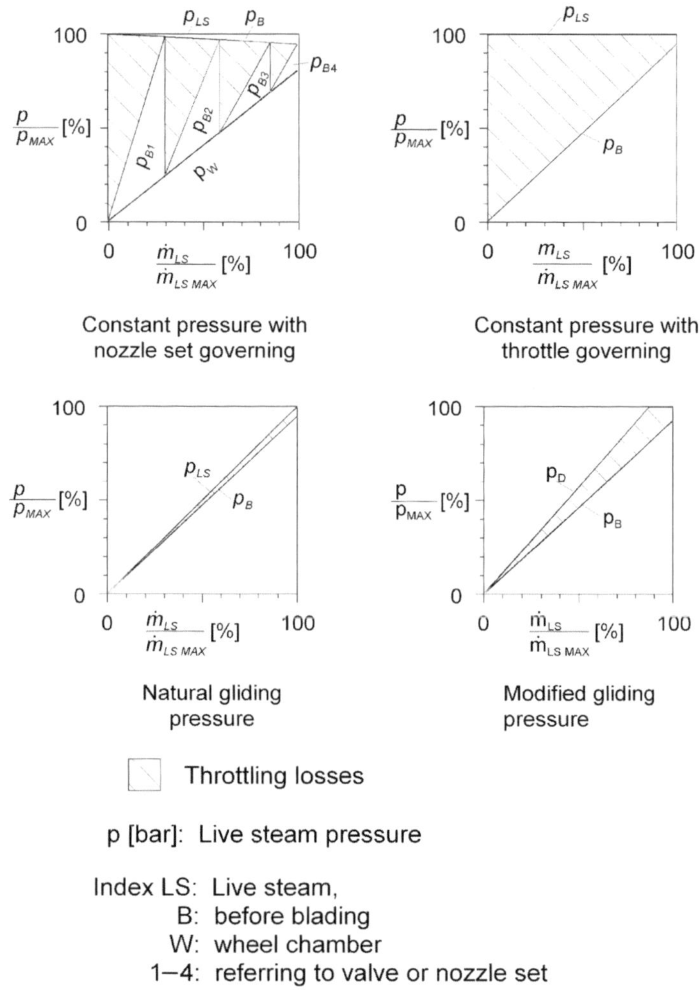

Fig. 4.21 Influence of the control mode on the pressure pattern at the turbine intake (not to scale) (Baehr 1985)

station unit, with the turbine having much higher allowable load change rates than in constant-pressure operation.

4.2.4.5 Impacts on Circulation or Once-Through Steam Generators by Sliding-Pressure or Constant-Pressure Operation

Circulation or once-through steam generators can in principle be operated with sliding or constant pressure. Circulation steam generators, however, are not operated with sliding pressure because it would involve considerable restrictions on load changes. Circulation systems are almost exclusively operated with constant-pressure

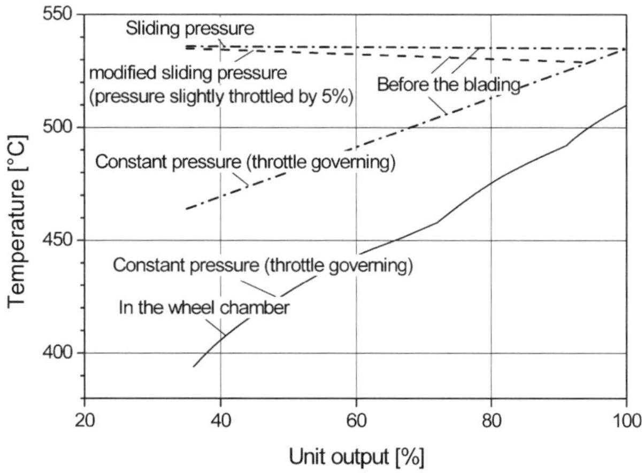

Fig. 4.22 Temperatures in the high-pressure section of the turbine with different control modes (Wittchow 1982)

control, while once-through systems mostly use sliding pressure, though in some cases constant pressure as well. For steam generators operated with constant pressure only in the evaporator, sliding-pressure operation does have the well-known operating advantage, though not the economic advantage, of the disproportionately decreasing power consumption of the boiler feed pump in part-load operation.

Different system characteristics determine different degrees of suitability of drum boilers and once-through boilers for rapid load changes. While the thick-walled drums of circulation steam generators limit the allowable rate of load change, the stress of a once-through boiler is lower at the same pressure rating due to the thinner walls of the separators. However, with higher pressures and temperatures involved, thick-walled construction parts of once-through steam generators, such as separators, do limit the allowable load change rates.

In the case of a short-term increased power output demand of about 5%, the output can be increased by opening the turbine valves, which is possible both using modified sliding pressure and at constant-pressure control. Steam released in the first 20 s comes essentially from the live steam pipe and the superheater. Only afterwards does the evaporator add to the extra steam supply. The greater storage capacity of the drum boiler is an advantage in this case compared to once-through boilers. Delays in steam production if a step load change occurs can be bridged for a longer period until the compensation by the firing rate takes effect (Wittchow 1982).

In both boiler systems, greater output changes are always initiated by increasing the firing rate. Drum and once-through boilers differ in controlling the feed water. In drum boilers, the feed water is designed to be controlled by the drum water level. The feed water control is coupled with the fuel control via the evaporator and the circulation system. Changes in the feed water flow do not immediately influence the flow through the superheater. When the firing rate is increased, delayed steam generation in the circulation system, due to the large storage capacity of the evaporator, may result in insufficient superheater cooling. With rapid load changes, the spray attemperators often do not suffice to control the live steam temperatures, so this circumstance places another limit on the load change rate in drum boilers.

The once-through boiler, compared to the drum boiler, has less steam storage capacity. In addition, in sliding-pressure operation, a large load change involves the boiler being more highly pressurised. For the once-through boiler, the enthalpy after the evaporator is used as the controlling variable for the feed water control. By means of a short increase in the feed water flow, the pressurising can be accelerated and the cooling of the superheater ensured. The limits of the once-through boiler thus result from delays in steam production in consequence to fuel flow changes.

So it can be said that different control modes and operation of once-through steam generators determine both the dynamic behaviour of the unit and the load-dependent heat rate. Once-through boilers are capable of coping with load change rates of 5–8% per minute, which is higher than the rates of 2–3% per minute that drum boilers can deal with (Wittchow 1982). The influence of the different control modes on the heat rate is described in Sect. 4.4.4.

4.2.4.6 Start-Up

The operation of a power station unit in the lower intermediate load range and peak load range also involves frequent start-ups and shutdowns. Start-up losses should be kept at a minimum in order not to impair the economic efficiency of power generation.

These losses are smaller with shorter start-up times, and the earlier the electrical unit output reaches the minimum output that allows the shutting of steam bypasses to the turbine.

After ignition, fuel flow and electric power consumption rise very quickly, but they cannot be used for power generation until the turbine generator is connected to the electrical grid. After connection to the grid, the start-up losses decrease as turbine bypasses are closed.

Once-through and circulation steam generators today are usually started up with water – steam separation behind a filled evaporator, which ensures that only steam is fed to the superheater. In all steam-generating systems, sufficient cooling of all heating surfaces must be guaranteed in the start-up process. Additional restrictions may arise due to thick-walled parts (Adrian et al. 1986; Wittchow 1982).

During start-up, a natural-circulation steam generator can only slowly increase its firing rate, because sufficient cooling of the heated risers becomes effective only when the circulating flow starts, that is, after evaporation has set in. It is also because steam must be available for the cooling of the superheater. In once-through or circulation systems, the evaporator and each tube already have a defined flow before ignition of the burners, both in the initial water phase and in the following water – steam phase. Due to the small storage capacity of water/steam in the system, steam generation can quickly be increased.

The reliable cooling of all superheater surfaces is a prerequisite for a rapid increase in the firing rate. It is ensured by an adequate turbine bypass system (see Fig. 4.23) (Adrian et al. 1986). Separated bypass systems for the high-pressure section (HPS), and the intermediate- and the low-pressure sections (IPS, LPS) of the turbine allow independent charging of the turbine parts while maintaining the

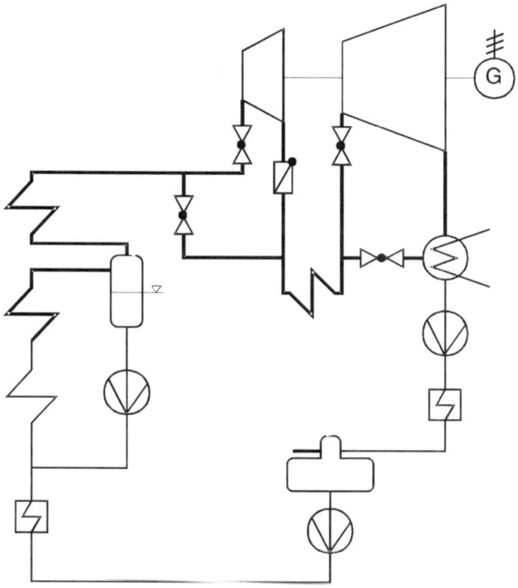

Fig. 4.23 Startup system of a power plant unit (Wittchow 1982)

cooling of the reheater. The pressure systems of the boiler and the turbine parts are decoupled. This makes it possible to operate them in independent regimes, such as during the start-up and the shutdown processes, and in accidents. In short-term failures of the turbine generator, caused by network disturbances for instance, it is possible to keep the boiler operating at any output rate, so that after the fault is cleared, the turbine generator can be charged again within a very short time. During start-up, the bypass heats those plant components which are unheated to wall temperature, so they can be charged with steam in duty operation and interconnected rapidly.

In Germany, the above-described bypass system is utilised in all power plants in order to make use of the favourable start-up behaviour of the once-through boilers which are used there almost exclusively.

This system has advantages for other boiler systems as well. The usual long start-up times for units with drum boilers in other countries can be put down to the start-up systems used, which often lack turbine bypasses with sufficiently large dimensions.

A further criterion for assessment of the start-up process for each plant is the allowable temperature gradient across the thick-walled construction parts. Figure 4.24 shows the corresponding values for the drum of a 660 MW boiler and for the separators of a 720 MW once-through boiler. Correlated with the pressure-dependent boiling temperature, it is possible to calculate the warm-up times for these parts. Owing to the thermal flexibility of its construction, the once-through boiler, compared to drum boilers, has advantages when starting up from a cold state and after weekend shutdowns. In contrast, there are no remarkable differences for warm or hot start-ups, provided that the pressure in the drum boiler has not dropped too low before start-up (Wittchow 1982).

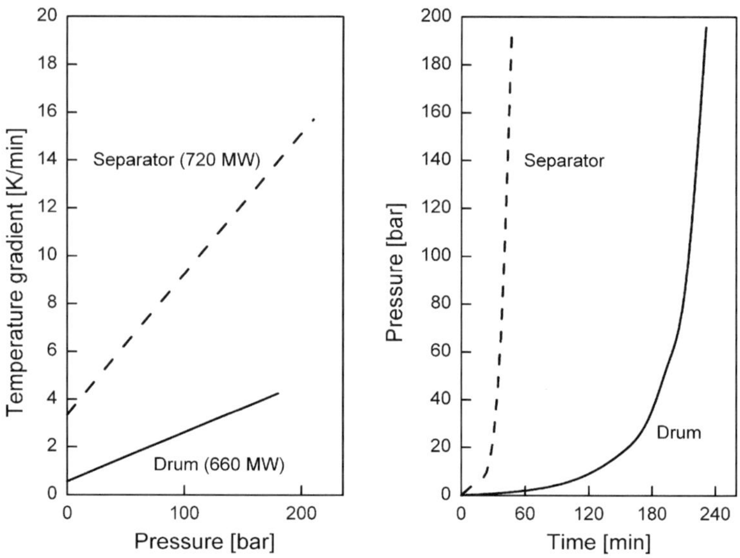

Fig. 4.24 Allowable temperature gradients and warm-up times of thick-walled construction parts of drum and once-through boilers (Wittchow 1982)

4.3 Design of a Condensation Power Plant

This chapter presents the design procedure for a condensation steam power plant firing pulverised hard coal, with particular focus on the thermal and fluid design of the steam generator. The general design fundamentals shall be explained using the example of a pulverised coal fired mid-range load power plant with conventional steam conditions. The parameters of this power plant, designated in the following as the reference power plant, are given in Table 4.1.

4.3.1 Requirements and Boundary Conditions

The design of a condensation power plant and, in particular, the steam generator, is subject to a range of requirements with respect to the

- Plant capacity
- Fuel
- Operating regime
- Boundary conditions and official directives
- Efficiency
- Availability
- Investment and operating costs
- Serviceability
- Service life, maintenance and repair (STEAG 1988; Baehr 1985)

Because the requirements are partly contradictory, the design in each case is a compromise between the different requirements. The task of a plant design is the optimisation for the given case. From the beginning, the plant capacity, fuel, operating regime and location are usually fixed design parameters (Stultz and Kitto 1992; Adrian et al. 1986; Baehr 1985).

4.3.1.1 Fuel

The planned fuel is a key factor for the design of the plant. Compared to a gas-fired power plant, a coal-fired power plant is much more complex and requires additional, sophisticated components such as installations for the unloading, transport, storage and mixing of solid fuels, as well as machinery for fuel preparation, equipment for the cleaning of heating surfaces, devices for ash transportation and disposal and additional flue gas cleaning units. The design of the furnace, the steam generator and other components is dependent on the fuel. For this reason, designing a power plant includes the specification of a design fuel and the range of fuels fired.

4.3.1.2 Operating Regime

The plant design has to take into account the planned operating regime – base load, mid-range load or peak load (see Sect. 4.2.4). The number of expected start-ups per year, classified into cold, warm or hot starts, and the necessary load control ranges and daily load changes between the minimum and the rated power have to be determined prior to the design.

Both the fuel costs and the utilisation factor (the number of maximum-equivalent hours per year the plant is operated) of the plant determine the economic optimum of the investment costs. For a base load power plant, the higher investment costs of the desired higher efficiency rates are more economic than for a mid-range load plant. If a plant is almost only full-load operated, thick-walled components and the resulting limits to the load change rate can be tolerated. It is sufficient to design such a plant for operating regimes with small load changes and a small number of start-ups.

The design of mid-range load plants, however, involves more compromise and therefore requires a more considered design with regard to the behaviour during load changes, start-ups and shutdowns, the minimum power and the efficiency over the load range. Modern hard coal fired plants can usually be operated in a load range from about 35 to 100% of the rated power. Loads below 35% are in general only possible with oil or gas as backup firing.

4.3.1.3 General Conditions and Official Directives

The conditions specific to the location have to be exactly determined prior to designing a power plant. An important part of these conditions, which have to be incorporated into the power plant design, is the legislative directives. The legislator stipulates allowable emission levels which have to be complied with by installing flue gas cleaning and noise insulation. Water withdrawal for process cooling and the discharge of wastewater have to be planned and carried out in compliance with the ordinances referring to water rights. In Germany, to give an example, the thermal stress that it would impart upon rivers may no longer allow the operation of once-through cooling in the summer. This restriction can be avoided by back-cooling processes, which are mostly used for inland locations. The height of natural-draught towers can also be limited by directives. Locations near the seaside allow once-through cooling with seawater. Aspects of the design that impact upon waterways, railways and highways have, as a rule, to comply with directives of local authorities as well.

Further location-specific factors are climatic conditions such as the temperature and humidity of the air and the air pressure. The surrounding infrastructure, residential areas and use of the environment, the geographical and geological conditions and, in particular, the available surface area have an important influence on the type of construction.

4.3.1.4 Efficiency

High overall efficiencies of conventional steam power plants can be achieved by the following features:

- High temperatures and pressures of the generated live steam before it enters the turbine
- High temperatures of the single or multiple reheat cycle in intermediate pressure stages
- Regenerative air heating and fuel drying
- Regenerative feed water heating
- Low exhaust steam pressures of the turbine before condensation
- Low losses of all plant components
- A low electric auxiliary power demand

The different methods to raise the efficiency will be dealt with in Sect. 4.4. These methods, however, inevitably result in higher construction and maintenance costs. The strength of the metallic material exposed to high temperatures deteriorates with time. Plant components with higher efficiency rates require parts with thicker walls to withstand higher temperatures and pressures. When fast temperature changes occur, stronger thermal stresses evolve in these parts, leading to levels that can exceed the allowable design strength and consequently to a shorter service life of the components. Therefore, advanced power plants necessarily involve longer start-up times and thus greater start-up losses and lower load change rates.

4.3.1.5 Availability

A high availability of technology implies a high-quality standard of plant components, standby components and care in operating, control and maintenance. For financial reasons and because of an achievable high level of availability, large single components such as boilers, forced-draught fans (FD fans), induced-draught fans (ID fans), turbines, cooling towers, generators and transformers are designed as mono-devices (i.e. one unit operating at full load instead of two or more at partial load).

As regards plant equipment designs, for example of FGD units and catalytic NO_x control units, there is a tendency towards single-line design. In the case of other plant components, standby options have to be discussed on the basis of their individual availability and the extra costs. In the case of an interconnected network system, the considerations about unit availability can include the existing reserve capacity of the network.

4.3.1.6 Costs

Costs are classified as variable costs, which depend on the operating period of the plant, and fixed costs. Variable costs are basically the fuel costs and the operating and maintenance costs. Fixed costs are the capital and personnel costs. The costs for the personnel depend on the serviceability of the plant.

The costs of power production are largely influenced by the plant unit size. Both the specific investment costs per unit of the capacity and the maintenance and the personnel costs decrease when the size of the unit is greater. Large hard coal fired power plants have the cost of capital as their greatest part of the fixed costs. Figure 4.25 shows how the unit investment costs of the entire plant and of its main components decrease as the capacity increases. The cost decrease lessens with high unit capacities, so a rise of the capacity will yield less financial advantages (STEAG 1988; Kotschenreuther and Klebes 1996).

Figure 4.26 shows the breakdown of investment costs for a large hard coal fired power plant. In Germany, the specific investment costs of large power station units amounted to about 1,000 Euro/kW$_{el}$ around 2005. For power plants planned and built in Asia, the costs are about 30–40% lower due to lower manufacturing costs and less demanding directives/regulations. Competition in the past induced a decrease of the specific investment costs; however, recently the huge worldwide metal demand has caused an increase in investment costs by 50% (2008).

The economic optimum for a specific power plant configuration is determined by balancing the cost reductions achieved through higher efficiencies against the additional costs of the efficiency increase.

The correlation between the economically feasible investment ΔI and an advantage of consumption $\Delta HR/HR_0$ results from the following formula:

$$\frac{\Delta I}{\Delta HR/HR_0} = \frac{HR_0 \cdot P_{el} \cdot U \cdot C_F \cdot 10^{-5}}{CoC} \tag{4.2}$$

with

$\frac{\Delta I}{\Delta \mathrm{HR}/\mathrm{HR}_0}$ = economically feasible additional investments referring to the heat rate improvement [Euro/%]

HR_0 = basic heat rate (net) [kJ/kWh]

U = utilisation factor (full-load operating hours per year) [flh/a]

P_{el} = electric net power output at full load [MW]

C_{F} = fuel costs [€/GJ]

CoC = cost of capital/debt service factor as a function of financing and operation period (STEAG 1988) [1/a]

The level of economically feasible investment costs per percent of heat rate improvement and installed electrical kW net power are shown in Fig. 4.27 for the reference 750 MW hard coal fired power plant. The reference power plant, equipped

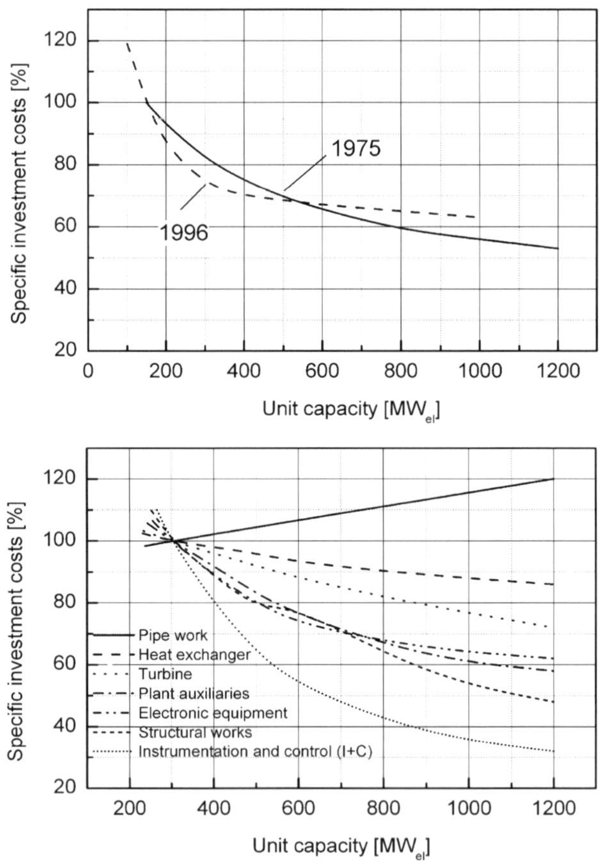

Fig. 4.25 Decrease of specific costs for the plant entity and for the plant components with increasing unit capacity (STEAG 1988; Kotschenreuther and Klebes 1996)

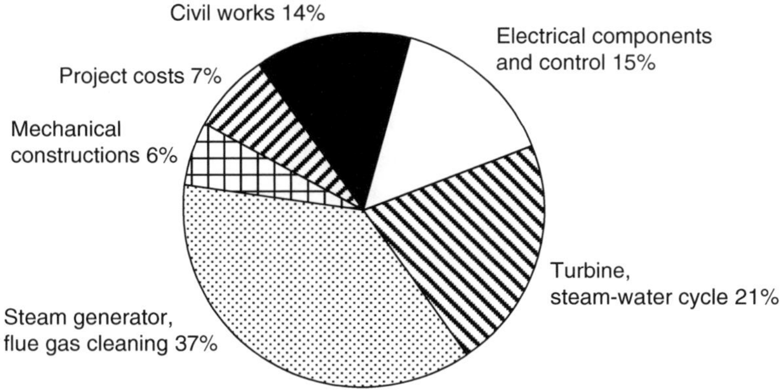

Fig. 4.26 Breakdown of investment costs of a large pulverised coal fired power plant

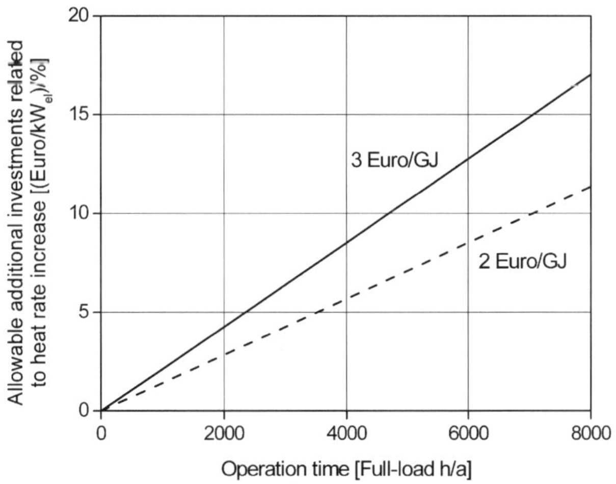

Fig. 4.27 Economically feasible additional investments per percentage of heat rate increase as a function of fuel price and operation time

with a wet-type cooling tower, was designed for mid-range load and has a net heat rate of 9,230 kJ/kWh$_{net}$ ($\eta_{net} = 39\%$). The economically feasible investment costs are calculated with a debt service factor of 0.13 per annum. The level of economically feasible investments depends on fuel price and degree of utilisation of the power plant.

4.3.1.7 Serviceability

The serviceability of the individual plant parts and the power plant installation as a whole are based on the applicability of the instrumentation and control (I + C) equipment. A more sophisticated power plant control system has to be balanced against the reduction of personnel costs.

4.3.1.8 Design Life

An important parameter in the power plant design is the planned lifetime. Conventional and advanced designs are planned for a lifetime of 200,000 h of operation. Together with the planned operating regime, the design life is mainly determined by the design of the main components, i.e. the steam turbine and, in particular, the high-pressure and superheated steam pipework and the respective steam generator components and vessels which are subject to regular inspection according to law. Base load power stations are mainly subject to creep rupture stresses, while mid-range load power stations are usually subject to alternating stress. Both types of stress result in consumption of the design life or fatigue of construction parts. The inspection of the components and the determination by calculation of the expended lifetime are laid down in technical rules such as the European Standard (or Norm) EN 12952 or formerly the "German Technical Rules on Steam Generators" (*Technische Regeln Dampferzeuger* (*TRD*)). Apart from that, the design will take into account regular scheduled outages for replacing worn parts or for improvement or retrofitting purposes, without factoring in the availability of such improved technology. The recording of operational conditions, for instance to identify the actual and the allowable temperature transients, may be reasonable in order to detect and avoid an undue reduction of service life. Based on the knowledge of the required plant service life, the design should provide that the individual components have accordingly a design life and should include specified parameters for the operating regime.

4.3.2 Thermodynamic Design of the Power Plant Cycle

The thermodynamic design of the cycle comprises the determination of the

- Process flow configuration
- Steam parameters
- Preheater configuration
- Heat dissipation (Baehr 1985)

The thermodynamic design determines the conditions of the closed steam – water cycle, yielding a power plant cycle diagram such as plotted in Fig. 4.28 for the reference power station (Spliethoff and Abröll 1985).

The choice of the preheater configuration defines the number of stages and the design of the individual heater stages, thus determining the final feed water temperature. In the steam generator, the feed water is heated to boiling temperature, evaporated and superheated and reheated after a partial expansion step.

In designing the process flow, only the live steam conditions and the conditions of the reheated steam are defined at first, without specifying the heating surfaces for the heat transfer from the flue gas to the water – steam system. The pressure losses occurring in the steam generator are calculated by approximation. The process flow design comprises the conversion of the thermal energy of the steam into the mechanical energy in the turbine. This includes the definition of the exhaust steam condition in the condenser, i.e. the exhaust steam temperature, as well as the type of drive of the feed water pump.

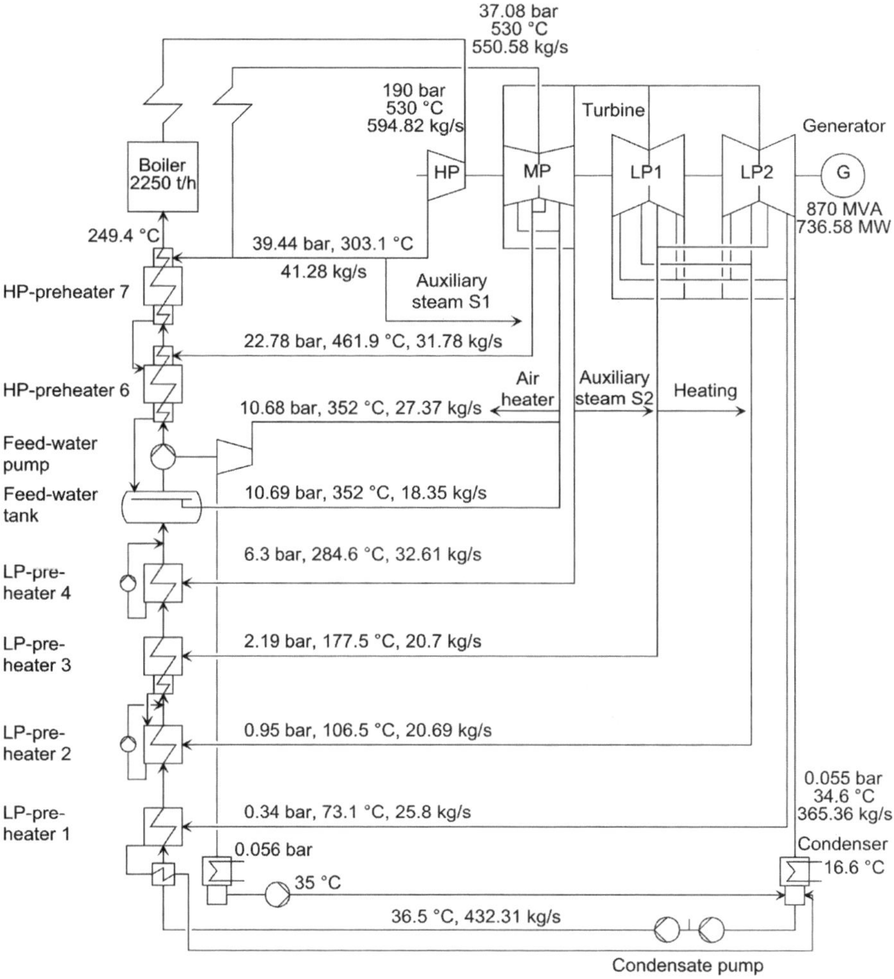

Fig. 4.28 Cycle of a conventional steam power plant with hard coal firing (reference power plant) (Spliethoff and Abröll 1985)

The essential factor for the choice of the parameters is the order of magnitude of the capacity of the power station unit (Kotschenreuther and Klebes 1996). Figure 4.29 shows the guideline and empirical values for the definition of the design parameters as a function of the generator capacity, established on the basis of installed condensation power plants (Baehr 1985). The guideline values are determined largely by economic factors, as well as by process-engineering factors. The higher the capacity of the plant, the higher the economically feasible investments. Note that power plants with advanced steam conditions, to be discussed in Sects. 4.4 and 4.5, are not taken into consideration in this part.

The capacity of the planned unit fixes the live steam pressure of the steam generator and also defines other process parameters as well as the cycle of the process.

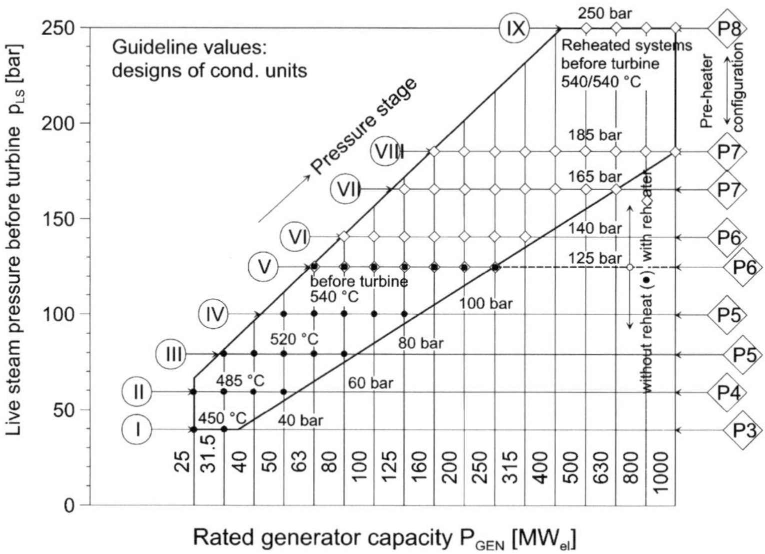

Fig. 4.29 Guideline values for the design of steam power plants (Baehr 1985)

Higher pressure stages also justify more complex technology. Reheating is provided for pressures of more than 125 bar. The recommended live steam temperatures are also defined by the pressure stage.

Figure 4.29 presents several recommended pressure stages for a given generator capacity. The choice of higher pressure stages is reasonable for high fuel costs and full-load plants, the low-pressure stages for mid-range or peak load and low fuel costs.

The exhaust steam temperature of the turbine is determined by the temperature of the cooling medium, which takes the waste heat, and the temperature gradient of the waste heat transfer defined in the design. The location also determines the choice of the cooling medium, e.g. seawater or ambient air, and their respective seasonal average temperature.

Lower exhaust steam temperatures bring about higher efficiency contributions to the production of electric power by the "cold end". On the other hand, the investments rise with decreasing temperature differences between the condensate and the cooling medium. The justifiable expenditures have to be estimated by means of a cost-effectiveness optimisation.

The "thermal cornerstones" – the live steam conditions, the reheater steam conditions, the regenerative feed water preheating by turbine extraction and the cold end of the turbine – determine the thermal efficiency and the heat rate of the condensation turbine.

Figure 4.30 shows the turbine heat rate for the configurations shown in Fig. 4.29. Today, cycle simulation software is commonly utilised for designing and optimising the thermodynamic cycle. Section 4.4 discusses measures to increase the thermal efficiency in the heat flow design of a power plant.

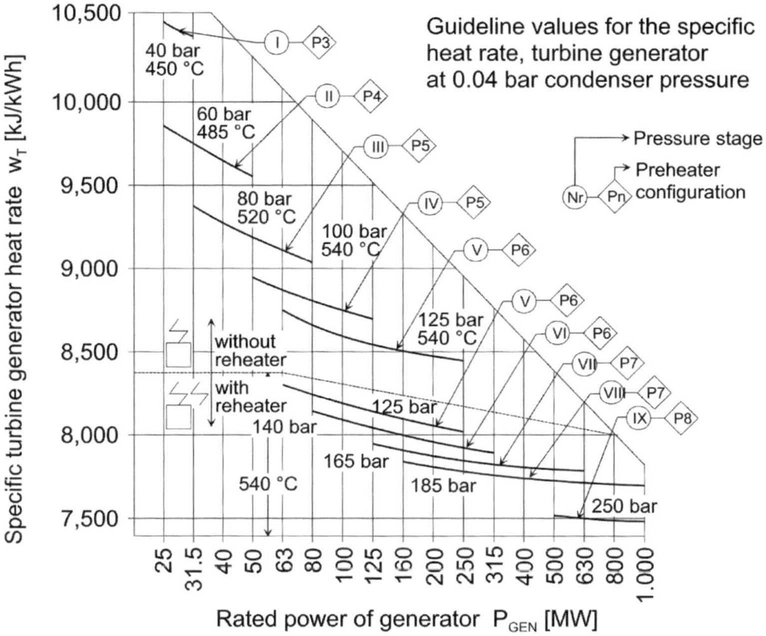

Fig. 4.30 Specific heat rate of the turbine generator (Baehr 1985)

The net efficiency of a power plant is calculated by the various individual efficiencies:

$$\eta = \eta_B \cdot \eta_T \cdot \eta_{aux} \cdot \eta_P \tag{4.3}$$

where η_B is the boiler or steam generator efficiency and η_T is the efficiency of the steam turbine unit (see also Sect. 3.2). The auxiliary power efficiency η_{aux} takes the electrical and mechanical power requirements into account; the efficiency η_P comprises the heat losses of the live steam and the reheater pipework that connects the steam generator and the turbine. For the turbine efficiency η_T, which represents the ratio of the electrical and mechanic power output to the steam energy input, the following equation applies:

$$\eta_T = \eta_{th,0} \cdot \eta_{i,T} \cdot \eta_m \cdot \eta_{Gen} \tag{4.4}$$

where $\eta_{th,0}$ is the thermal cycle efficiency at loss-free (isotropic) expansion, $\eta_{i,T}$ is the inner turbine efficiency and η_{Gen} is the generator efficiency. The mechanic losses of the turbine shaft are taken into account by η_m.

The calculation of net efficiencies requires the knowledge of individual efficiency rates of the plant components. If pertaining data is not yet sufficiently exact, nor other data in the planning stage available, the values can at first be estimated based on guideline values of previously constructed plants. Having designed the components, these estimations have to be corrected later and the calculations have to be repeated.

4.3.3 Heat Balance of the Boiler and Boiler Efficiency

In the boiler or steam generator, the chemically bound energy of the fuel is converted into thermal energy of the flue gas and then transferred to the steam – water cycle. For a steam generator with a single reheating heat exchange stage, the heat balance can be calculated according to Fig. 4.31:

$$\dot{Q}_F + \dot{Q}_A = \dot{m}_{LS}\,(h_{LS} - h_{FW}) + \dot{m}_{RS}\,(h_{RS2} - h_{RS1}) + \dot{Q}_{LOSS} \tag{4.5}$$

The boiler efficiency can be calculated directly when the steam conditions and flows and the heat addition into the furnace are known:

$$\eta_B = \frac{\dot{m}_{LS}\,(h_{LS} - h_{FW}) + \dot{m}_{RS}\,(h_{LS2} - h_{LS1})}{\dot{m}_F \cdot \mathrm{LHV} + \dot{m}_A \bar{c}_{PA}\,(t_A - t_o)} \tag{4.6}$$

For the indirect calculation of the boiler efficiency, only the losses of the boiler have to be known. Initially, they will be based on experience values. With the loss through unburned matter (K_U), the loss through sensible heat of the slag (K_S), the flue gas loss (K_{FG}) and the loss through radiation and convection of the external surfaces of the boiler (K_{RC}), the boiler efficiency can be calculated:

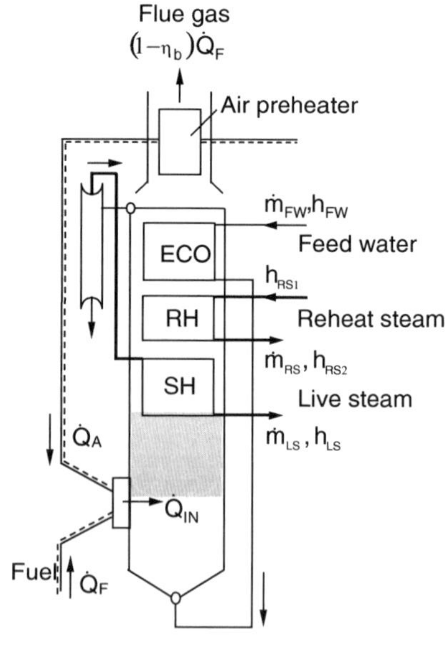

Fig. 4.31 Heat balance of a steam generator

$$\eta_B = 1 - K_U - K_S - K_{FG} - K_{RC} \tag{4.7}$$

4.3.4 Design of the Furnace

The definitions in the previous section help to determine the required fuel mass flows. Mass flows of air and flue gas are determined by combustion calculations. The results of these calculations are the mass flow data necessary for the design of the furnace and the steam generator (Stultz and Kitto 1992).

The furnace and the combustion system (fuel preparation, combustion and air guidance) have to be designed for complete combustion at low emissions. The design of the furnace determines in part the construction type and size of the steam generator.

The essential parameters for the furnace design are the fuel mass flow, the primary fuel and the planned range of mixtures with secondary fuels. For solid fuels, other important design parameters besides the calorific value and the moisture and ash contents are the grindability of the coal, the fraction of volatile matter, the elemental and the ash composition and the ash melting behaviour.

The choice of the firing system configuration (frontal firing, opposed firing, tangential firing, down-firing, bottom firing) is followed by the determination of the number and arrangement of the burners, including the mills. An important consideration in this process is the requirements for part-load performance.

In setting the dimensions of the furnace, the following performance aims have to be taken into consideration:

- Stable ignition
- Complete burnout
- Prevention of slagging and corrosion inside the furnace
- Prevention of fouling and corrosion on the convective heating surfaces

The depth and breadth of the furnace have to be adapted to the flame form in a way that the flame can expand as freely as possible, thereby ensuring that the walls will not be touched. Contact of the wall by the flame would lead to soot formation and also to corrosion damage of the walls. The furnace height of the firing plant is chosen such that the fuel can burn out completely (Baehr 1985).

The cross-section and height of the furnace have to be chosen according to the fuel type such that slagging and fouling inside the furnace, as well as on the subsequent heating surfaces, are within acceptable limits.

For coal types with a slagging tendency, a much larger cross-section will be chosen. The ash deformation temperature of the fuel defines the necessary furnace outlet temperature at the furnace end before the convective heating surfaces, in order to avoid sticky deposits on the convective heating surfaces. Hard coal combustion systems have a furnace outlet temperature of about 1,250°C and brown coal combustion systems about 1,050°C. According to the composition of the ash, they can have higher or lower values. While in small steam-generating units, the key variables

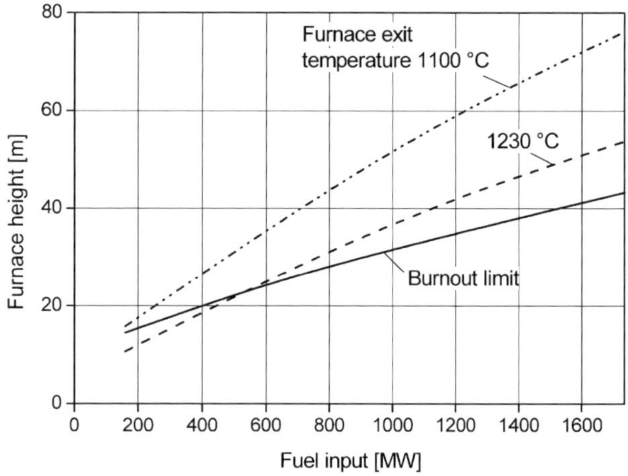

Fig. 4.32 Burnout limits and furnace exit temperatures in hard coal fired tangential combustion systems (Strauß 2006)

for the dimensioning are the necessary burnout rate and thus the residence time needed for the combustion of the remaining char, higher capacity units base their dimension of the construction height on the furnace exit temperature (see Fig. 4.32) (Strauß 2006).

In practice, characteristic values are used for the design of the furnace cross-section, the burner configuration and the furnace height, as shown in Fig. 4.33 (Baehr 1985).

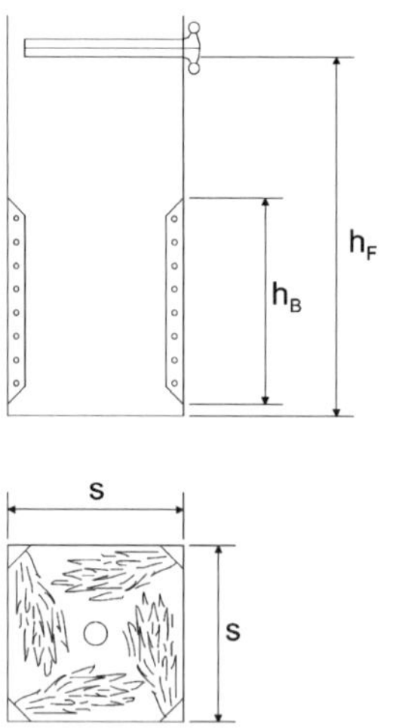

Volumetric heat release $q_V = \frac{\dot{Q}_{in}}{V_F} = \frac{\dot{Q}_{in}}{s^2 h_F}$ (MW/m³)

Cross-sectional area heat release $q_C = \frac{\dot{Q}_{in}}{A_C} = \frac{\dot{Q}_{in}}{s^2}$ (MW/m²)

Surface heat release $q_W = \frac{\dot{Q}_{in}}{A_F} = \frac{\dot{Q}_{in}}{4\,s\,h_F}$ (MW/m²)

Burner-belt heat release $q_B = \frac{\dot{Q}_{in}}{A_B} = \frac{\dot{Q}_{in}}{4\,s\,h_B}$ (MW/m²)

$\dot{Q}_{in}$ = Heat input
V_F = Furnace volume
A_C = Furnace cross-section
A_F = Furnace surface
A_B = Burner belt surface

Fig. 4.33 Reference values for steam generators

4.3.4.1 Volumetric Heat Release Rate

The furnace volumetric heat release is a measure of the residence time in the furnace and thus makes it possible to evaluate the burnout. It is defined by the given cross-section and the furnace outlet temperature.

4.3.4.2 Cross-Sectional Area Heat Release Rate

The cross-sectional area heat release rate is one of the key variables in the design of the furnace and is a measure of the flue gas velocity. It depends on the fuel and on the steam generator capacity (see Fig. 4.34) and sets the cross-section of the furnace for the design.

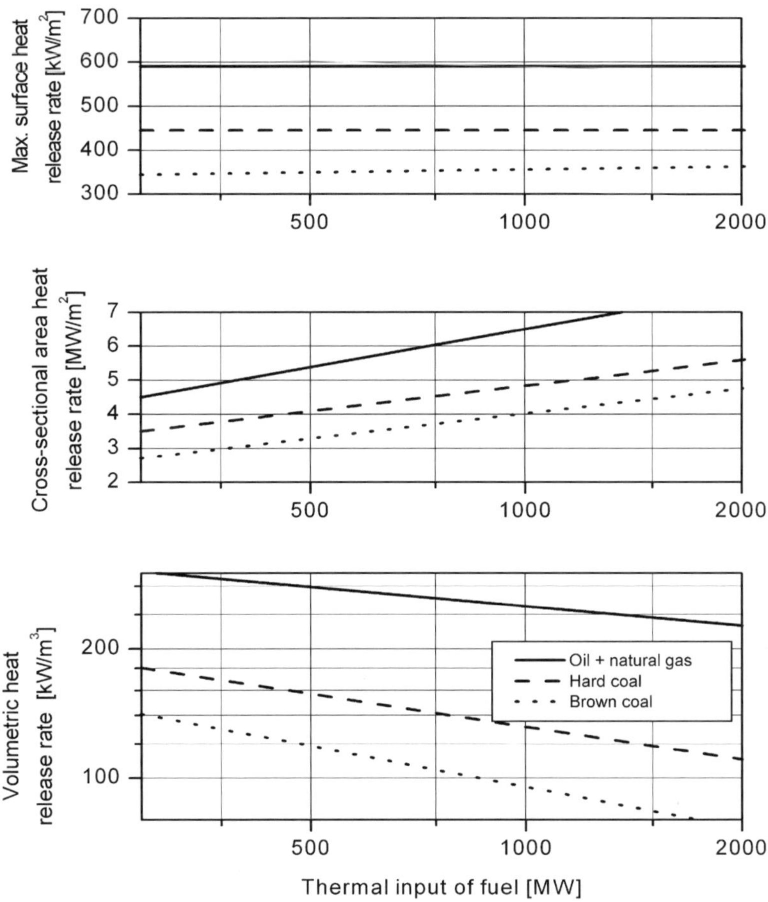

Fig. 4.34 Allowable heat release rates in furnaces (Adrian et al. 1986; Strauß 2006; Baehr 1985)

4.3.4.3 Surface Heat Release Rate

The mean surface heat release rate is a measure of the average temperature decrease. It is determined by the furnace exit temperature at a given cross-section. The local allowable surface heat release maximum depends on the fuel.

4.3.4.4 Burner-Belt Heat Release Rate

The burner-belt heat release rate is an indication of the flame temperature in the burner area, as it represents the ratio of the thermal input to the cooling surface in the burner area. Its order of magnitude depends on the fouling rate of the fuel, among other parameters. For normal hard coal not prone to slagging, the value is about $1\,\mathrm{MW/m^2}$.

Figure 4.34 gives guideline values for the volumetric, cross-sectional and maximum surface heat release rates. The comparative values mentioned above give reference values for the design of a furnace, but are not a substitute for the calculation of the heat transfer processes (Baehr 1985).

4.3.4.5 Calculation of the Flue Gas Cooling

Whereas the cross-section of the furnace is defined by the chosen firing system and the allowable cross-sectional heat release, the furnace height or (wall) heating surface area of large steam generators is determined by the necessary flue gas cooling to the furnace exit temperature. The height defines the threshold between radiative and convective heating surfaces.

For assessing the heat exchange between the flue gases in the furnace and the enclosing walls, one starts from a mean flue gas temperature in the furnace T_{FG} and a mean wall temperature T_{W} (Doležal 1990; Strauß 2006).

The flue gases in the furnace transfer the heat flux $\dot{Q}_{\mathrm{F}}$ to the furnace walls (evaporator) by radiation:

$$\dot{Q}_{\mathrm{F}} = \varepsilon_{\mathrm{FW}} \cdot C_0 \cdot A_{\mathrm{FL}} \left(T_{\mathrm{FG}}^4 - T_{\mathrm{W}}^4\right) \tag{4.8}$$

with the variables

$\varepsilon_{\mathrm{FW}}$ = emissivity between flame and wall
C_0 = coefficient of radiation of the black body ($5.67 \times 10^{-8}\,\mathrm{W/m^2\,K^4}$)
T_{W} = the mean wall temperature
T_{FG} = the mean flue gas temperature in the furnace
A_{FL} = the flame surface
A_{W} = the wall surface

If a flame fills the furnace completely, the surface of the flame A_{FL} equals the surface of the furnace A_{W}. In other cases, ratios are given between the two surfaces (Ledinegg 1966).

The emissivity between the flame and the wall depends on the emissivities of the surface wall and the flame and can be calculated:

$$\varepsilon_{FW} = \left(\frac{1}{\varepsilon_F} + \frac{1}{\varepsilon_W} - 1\right)^{-1} \tag{4.9}$$

The surface emissivity of an oxidised steel surface is between 0.6 and 0.8. Furnace ash deposits affect the heat transfer. The emissivity of deposits depends on the chemical composition, structure and porosity of the layer. The apparent emissivity, which describes the combined deposit and substrate emissivity, lies between 0.5 and 0.8 for most deposits (Stultz and Kitto 1992).

The flame emissivity can be calculated by

$$\varepsilon_F = \varepsilon_\infty(1 - \exp(-ks)) \tag{4.10}$$

where ε_∞ is the emissivity for a very thick flame. The parameter s is the thickness of the flame or beam length and k depends on the character of the flame. The parameter k varies between 0.75 for luminous flames and 0.5 for blue flames.

Typical values for the emissivity ε_∞ are as follows:

Hard coal, brown coal	0.55–0.8
Oil	0.6–0.85
Natural gas	0.4–0.6

The resulting emissivity is, for a hard coal fired furnace, in the range of 0.4–0.7, mainly depending on fouling and slagging.

The mean furnace temperature of the dry bottom furnaces is calculated as the geometric mean of the adiabatic combustion temperature T_{ad} and the furnace outlet temperature T_{FE}:

$$T_{FG} = \sqrt{T_{ad} \cdot T_{FE}} \tag{4.11}$$

The heat flux in the furnace $\dot{Q}_F$ is transferred from the flue gas mass flow $\dot{m}_{FG}$, having a specific heat $\bar{c}_{p_{F6}}$, which cools from the adiabatic flame temperature T_{ad} down to furnace exit temperature T_{FE}:

$$\dot{Q}_F = \dot{m}_{FG} \cdot \bar{c}_{p_{F6}} (T_{ad} - T_{FE}) \tag{4.12}$$

The resulting heat balance is

$$\varepsilon_{FW} \cdot C_0 \cdot A_W \left(T_{ad}^2 \cdot T_{FE}^2 - T_W^4\right) = \dot{m}_{FG} \cdot \bar{c}_{p_{F6}} (T_{ad} - T_{FE}) \tag{4.13}$$

and can be expressed as

$$\left(\frac{T_{FE}}{T_{ad}}\right)^2 + Ko \cdot \frac{T_{FE}}{T_{ad}} = \left(\frac{T_W^4}{T_{ad}}\right)^2 + Ko \qquad (4.14)$$

where

$$Ko = \frac{\dot{m}_{FG} \cdot \bar{c}_{p_{F6}}}{\varepsilon_{FW} \cdot C_0 \cdot A_W \cdot T_{ad}^3} \qquad (4.15)$$

Ko is an undimensional similarity coefficient, called the Konakow number.

The relation above serves to calculate the exit temperature of a given furnace or, in case of a given outlet temperature, the surface necessary for the cooling of the flue gases. In the calculation of modern steam generators with water-cooled tubes and vaporisation temperatures below 400°C, ${T_W}^4$ can be neglected. Fouling and slagging of furnace walls make the temperatures rise considerably.

The calculation of furnace wall heating surfaces and the preselected form (design) and dimensions of the cross-section together define the furnace height. By means of additional internal heating surfaces, such as a division wall that divides the furnace vertically, it is possible to reduce the furnace height (Doležal 1990).

The prediction of the radiant heat transferred to the walls of the furnace is one of the most important steps in designing a steam generator and has to be more exact than the calculation method described above, which only allows a rough estimation of the furnace exit temperature. The objective of such a calculation is to determine the local heat fluxes towards the furnace walls and to determine the distribution of the temperature and heat flux densities inside the furnace and at the furnace end (Baehr 1985).

In most cases, simpler, partially empirical models are employed. The results of a one-dimensional plug flow model based upon a mean cross-sectional temperature are shown in Fig. 4.35. The maximum heat flow density in the upper burner area ranges around 0.27 MW/m^2 during standard operation.

Firing conditions deviating from standard operation, such as those during fuel changes, when changing burner combinations, while there are unbalanced fuel and air distributions, during load change, or furnace wall fouling, can lead to locally higher heat flow densities. In the design and calculations of firing and heat transfer conditions, these cases are usually taken into account using empirical values (Stultz and Kitto 1992).

The calculation of the combustion course, in particular for new firing and burner concepts, employs three-dimensional numerical models which consider flow, reaction and heat transfer and determine the distribution of heat flow densities at the furnace walls. This way it is possible to determine and describe the impacts of deviations from standard firing conditions.

4.3.5 Design of the Steam Generator and of the Heating Surfaces

In designing the steam generator, it is necessary to dimension the heating surfaces such that the temperatures and mass flows defined in the cycle design can be met while taking the allowable material temperatures into consideration. Designing the thermal configuration and the steam generator is an iterative procedure. Given that at the beginning not all data is available and that guideline values have to be relied on, the design has to be repeated until the required mass flows and temperatures are met over the entire load range. This iteration is first carried out for the steam generator and then for the total cycle.

After designing the thermal configuration and the steam generator, it is possible to design the pressure parts and to begin to develop details. In Europe the "Pressurised Equipment Directive PED 97/23/EG" defines the boundary conditions for pressurised equipment. The design and construction regulations for steam generators are specified in the European Standard (or Norm) EN 12952, which replaced the design rules "*Technische Regeln für Dampfkessel (TRD)*" or "Technical Rules for Boilers" in Germany. Alternative to the European regulations, the "ASME Boiler and Pressure Vessel Code" by the American Society of Mechanical Engineers (ASME) can be used (even in Europe).

The steam generator heating surfaces are the membrane furnace walls and the flue gas pass, as well as the tube banks across the flue gas cross-section.

The usual construction for furnace and flue gas pass walls are tube-fin bar-tube wall constructions which are connected and welded together to make gas-tight

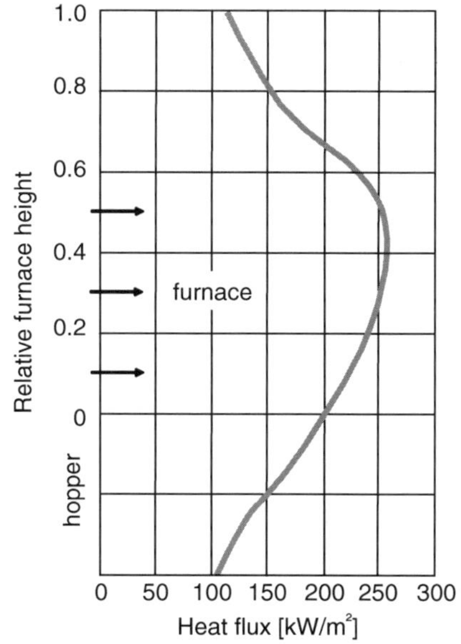

Fig. 4.35 Calculated heat flux distribution across the height of the furnace (Effenberger 2000)

membrane walls. These walls are only exposed to thermal radiation on one side; the other side is insulated against the outside in order to avoid heat loss.

The components in the flue gas path that follow after the furnace are the convective heating surfaces of the superheater, the reheater and the economiser. They consist of a great number of parallel tubes which are mounted crossways to the flow. Heat is mostly transferred by convection.

Evaporator and superheater surfaces are exposed to much higher temperatures on the side facing the fire and the flue gas than on the water/steam-cooled side. Heat transfer conditions on the inside and the outside surfaces of the heated tubes characterise the tube wall temperatures. At high inner pressures and defined flow velocities, they range only a little above the steam temperatures and are thus much lower than the furnace and the flue gas temperatures.

The allowable tube wall temperatures can be above the temperature of the working medium by a maximum of 50 K for radiant heating surfaces and of 20 K for convective heating surfaces. The temperatures must not exceed the tube wall temperature limits, which are dependent on the materials and the design pressure. Unheated tube, header and vessel walls will take, approximately, the temperatures of the steam flows.

Figure 4.36 shows a schematic drawing of the heating surface configuration of a single-pass or tower boiler, with the same furnace and flue gas duct cross-section over the height of the tower – a widespread construction type in Europe and Japan.

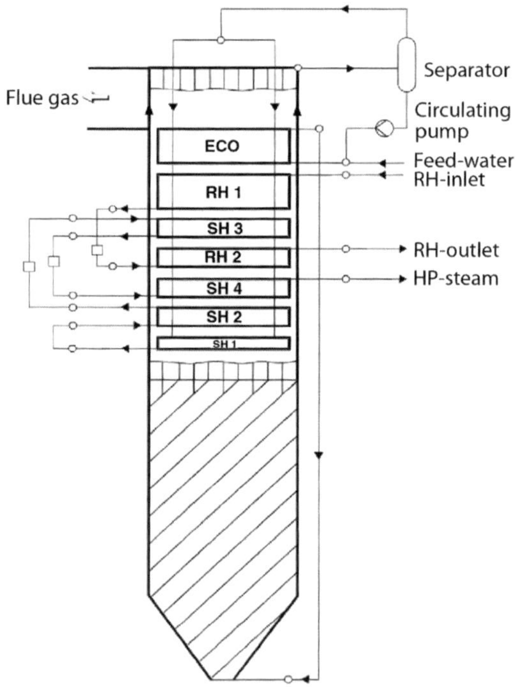

Fig. 4.36 Heating surface configuration of a single-pass boiler ("tower boiler")

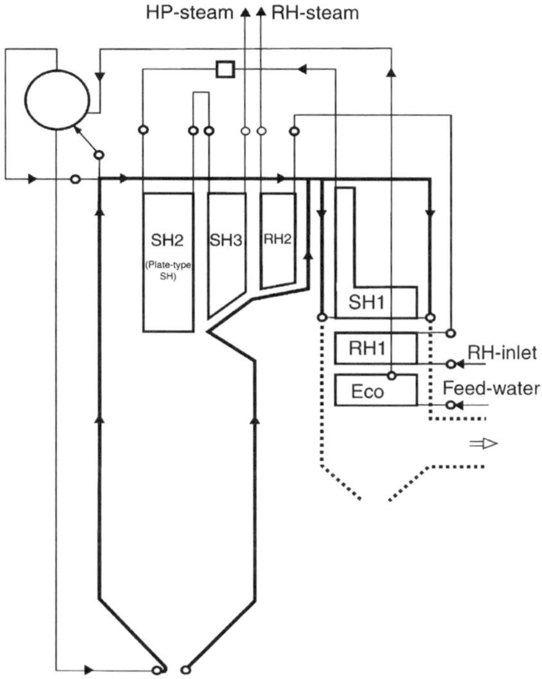

Fig. 4.37 Heating surface configuration of a two-pass boiler

Figure 4.37 shows the configuration of a steam generator in two-pass construction – the common type in the US and other countries.

For economic reasons, the steam generator should be designed such that the total heating surface area is a minimum. This minimum is met by a configuration that leads the hot flue gases in a counter-current flow to the working medium (i.e. water/steam) – a design that is only partly feasible.

The furnace walls are the heating surfaces with the highest temperatures (occurring on the flue gas side) and highest heat flux densities. The water/steam has to guarantee sufficient cooling in order keep the tube-furnace wall temperatures below the allowable material temperatures. However, this requirement cannot be met by the low heat transfer of steam. So in practice these heating surfaces are used to vaporise the medium, because this process with its two-phase mixture provides a good heat transfer.

The construction of the area of the convective heating surfaces uses a counterflow configuration. An exception to this is made only for some superheaters, in order to ensure constant temperatures of the live steam as a function of the load. Figure 4.38 charts the flue gas temperatures and the material temperatures along the flue gas path of the reference power plant. It can be observed that the heat flux density declines along the flue gas path. The heat transfer coefficient, too, shows a decrease towards the end of the steam generator, with the exception of the economiser stage.

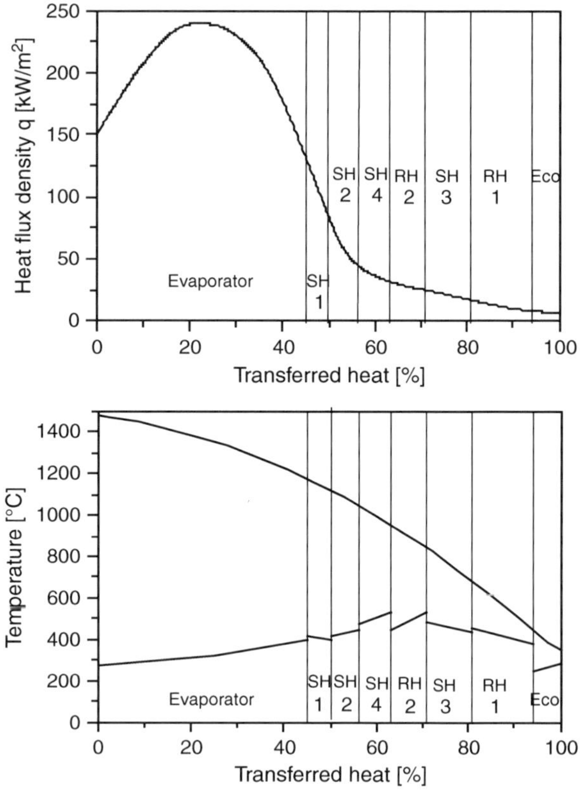

Fig. 4.38 Flue gas, temperature of the working medium and heat flux density of the reference power plant

4.3.5.1 Impact of the Live Steam Pressure

The steam conditions defined in the thermal design pre-set the conceptual design of the evaporator (natural circulation, forced circulation or once through) and the heat levels necessary for preheating, vaporisation, superheating and reheating. The furnace exit temperature defines the heat to be transferred in the furnace and by the downstream convective heating surfaces.

The next step is to integrate the live steam pressure defined by the steam conditions into the design of the heating surfaces. Figure 4.39 shows $h - p$ diagrams for a low-pressure and for a high-pressure boiler. At low pressures, the heat of evaporation predominates, whereas at higher pressures, the vaporisation enthalpy decreases and the heat demand for superheating increases (Doležal 1990).

In designing steam generation systems with a fixed liquid – vapour phase transition point, it is possible for over-determination to occur. This is because the vaporisation heat decreases with higher design pressures, while the flue gas cooling requirements and the evaporator capacity are fixed. As Fig. 4.40 shows, the entire furnace is required to act as an evaporative heating surface at low pressures, as the feed water is preheated and steam superheated only on convective heating surfaces. Given the lower vaporisation enthalpy at high pressures, the flue gases are not suf-

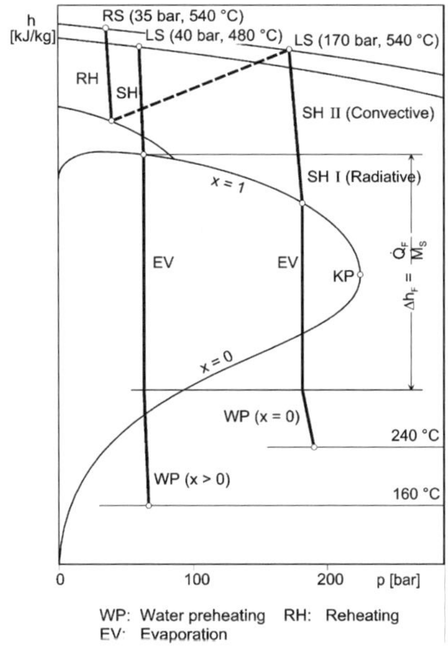

Fig. 4.39 $h - p$ diagram for LP and HP boilers (Doležal 1990)

ficiently cooled as they flow towards the furnace end if no additional measures are taken. The resulting furnace outlet temperature at a live steam pressure of 170 bar thus amounts to 1,300°C. But because a great number of coal types have lower ash deformation temperatures than this, the flue gas has to be further cooled by additional measures (Wittchow 1982).

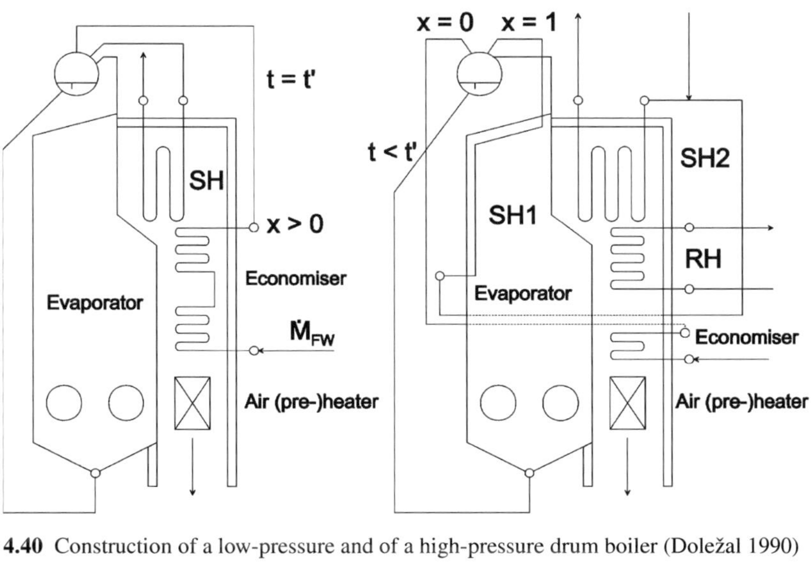

Fig. 4.40 Construction of a low-pressure and of a high-pressure drum boiler (Doležal 1990)

One of the options, pictured in Fig. 4.40, is to mount additional heating surfaces for superheating in the furnace. Such additional surfaces, however, are undesirable in practice as their incorporation into the evaporator wall is difficult due to the differing steam temperatures.

Another possibility is flue gas recirculation, which shifts the heat absorption further into the convective area. The disadvantage here is the higher auxiliary power demand. Plants in the USA utilise hanging plate heating surfaces as the first superheater surface in the flue gas flow. These can be located anywhere where flue gas temperatures are up to 1,400°C. They are also relatively insensitive to slag deposits because of their construction. When considering the furnace dimensions, it is useful to know that such heating surfaces are very compact.

The correlation of flue gas cooling and evaporator design only holds for steam generators with a fixed liquid – vapour phase transition point for natural or forced circulation. Limits resulting from the water and steam do not exist in dimensioning the furnace of once-through steam generators. For once-through steam generators with variable start and end points of vaporisation, the furnace and flue gas duct walls form a single heating surface where the last stage of preheating, the vaporisation and the first superheating stage take place. The transition from vaporisation to superheating migrates within the evaporator tubes, occurring mostly in the upper furnace section. With an increasing pressure, the liquid – vapour phase transition point shifts further down, and the furnace wall is used for superheating to a greater extent. Because the heat flux density in the upper evaporator and furnace walls is already below the mean heat flux density of furnace walls, levels that exceed the allowable tube wall temperatures are not expected in the range of the boiling crisis of the "second kind" (see Sect. 4.2.1).

4.3.5.2 Design of the Evaporator

In the furnace, the radiant heat transferred to the evaporator wall determines, via the heat flux distribution, the mass flow density necessary for cooling the evaporator tubes. At the furnace wall, at about the height of the burners, the highest heat flux densities occur. They decrease towards furnace end, falling further afterwards, in the area of the convective heating surfaces (see Fig. 4.38). Increasingly, the heat is transferred by convection, which also occurs in the flue gas duct walls.

For the design of a steam generator, it is necessary to know beforehand the maximum tube wall temperature, which is a function of the gas-side heat flux density (about 300–350 kW for hard coal firing systems), and the mass flow density of the steam – water mix. This is in order to avoid the allowable material temperatures being exceeded where the boiling crisis occurs. In once-through steam generators, the water/steam mass flow used for cooling decreases with the load, whereas the heat flux densities in the burner area decrease only to a minor extent, so it is the partial-load condition that determines the design. In general, the mass flow density of an evaporator with plain tubes lies between 700 and 800 kg/m^2 s at a minimum output of 30–40% (Franke et al. 1993). The mass flow density at the rated power lies between 2,000 and 2,500 kg/m^2 s.

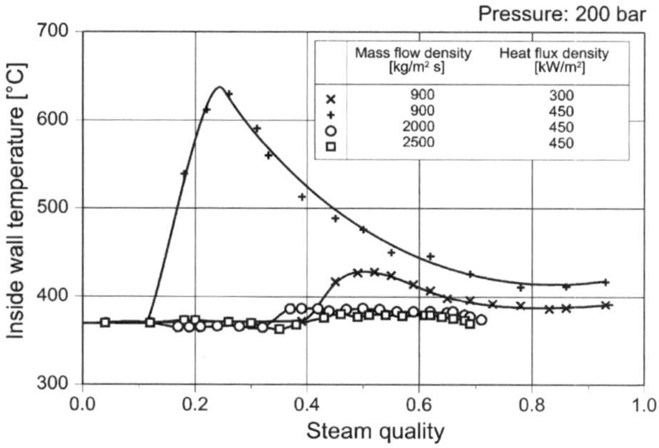

Fig. 4.41 Inside wall temperatures of a heated plain tube (Franke et al. 1993)

Figure 4.41 shows the inside wall temperatures as a function of the steam quality for plain tubes at different heat and mass flux densities. At a strong heating density of 450 kW/m^2, a too-low mass flow density of 900 kg/m^2 s causes a strong rising of the tube wall temperatures. With reduced heat flux densities, such as occur in partial-load conditions, the temperature rise is less dramatic (Franke et al. 1993).

For a forced once-through steam generator with plain tubes, the high heat flux densities inside the furnace require a helically wound furnace wall (see also Sect. 4.2.2.3). The number n of the welded parallel tubes depends on the mass flow density Φ required for cooling at the partial load a, the inner tube diameter d_i and the steam flow $\dot{m}_\mathrm{s}$ (Strauß 2006):

$$a\dot{m}_\mathrm{s} = n \cdot \frac{\pi d_\mathrm{i}^2}{4} \Phi \tag{4.16}$$

or expressed in terms of n:

$$n = \frac{a\dot{m}_\mathrm{s}}{\frac{\pi}{4} d_\mathrm{i}^2 \Phi} \tag{4.17}$$

If the tubes, with the tube pitch t_P, are welded together in parallel into a band, the equation for the bandwidth b of the helically wound wall applies, shown in Fig. 4.42:

$$b = nt_\mathrm{p} = t_\mathrm{p} \cdot \frac{a\dot{m}_\mathrm{s}}{\frac{\pi}{4} d_\mathrm{i}^2 \Phi} \tag{4.18}$$

Both the tube diameter and the tube pitch cannot be chosen freely. The tube diameter is confined at the upper limit by the heat transfer and the tube weight and at the lower limit by the pressure loss. Large steam generators have tube diameters usually between 30 and 50 mm. The tube pitch is influenced by the fin-bar width as well as the tube diameter. The allowable fin-bar width is between about 12 and

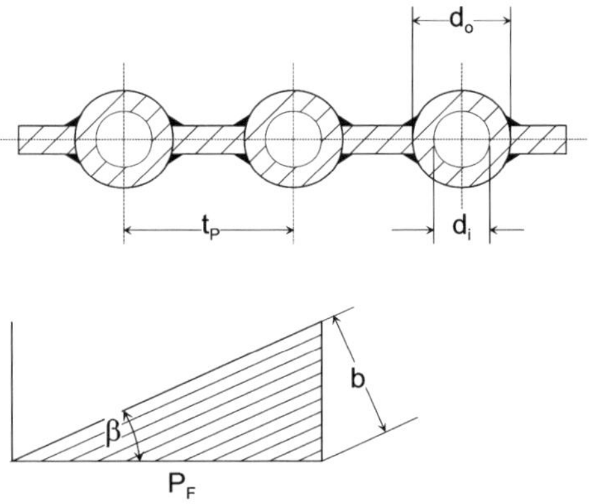

Fig. 4.42 Schematic drawing of the helical winding (Doležal 1990)

15 mm. The upper limit of the fin-bar width is given by the hazard of scaling of the fin bar, the lower one is given by constructive and economic points of view.

The bandwidth b is smaller than the furnace perimeter P_F, so the band has to be wound helically around the furnace to completely line the furnace wall. The helix angle β can be calculated as

$$\beta = \arcsin(b/P_F) \tag{4.19}$$

The helix angle increases with the boiler size.

In the upper area of the steam generator, before the convective area begins, the helical winding transforms into vertical tubing. Because vertical tubing is more economical than helical winding, it should be designed to begin at the lowest possible furnace height. The helical winding and vertical tubing are joined by clevises (see Fig. 4.43). In the vertical tubing, the mass flow density of the working fluid, then in a vapour state, is diminished by increasing the number of tubes by a factor of 3–4. If the transition to vertical tubing is carried out at a furnace height which is too low, it is possible that, with high gas-side heat flux densities and low mass flow densities of the cooling fluid, excessive tube wall temperatures arise in the vertical tubes. In contrast, when the helical winding is too high, non-uniform heating can have stronger effects due to the longer tubes of the helix, thus also causing the tube wall temperatures to exceed the allowable limit (Kefer et al. 1990).

Commonly, the helical tube winding finishes with the furnace, while the vertical tubing begins in the convective section. Where heat flux densities decrease, for instance when flue gas temperatures fall to between 750 and 800°C, the number of tubes is diminished. It becomes possible to double the tube pitch, because the fin-bar temperatures are below the scaling temperature of the material. The larger tube pitch of the membrane walls facilitates the insertion of the superheaters and reheaters, which have narrowing tube pitches. The tube pitch in the upper section of the vertical pass – commonly 100–120 mm – defines the tube pitch in all wall areas. Figure 4.43 shows the tube pitches of the reference power plant.

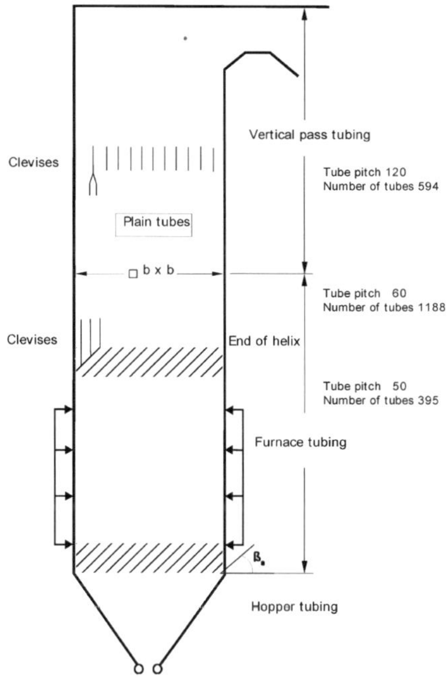

Fig. 4.43 Wall tubing of a single-pass boiler with helical winding in the furnace section (*Source*: Alstom Power)

4.3.5.3 Evaporators with Vertical Internally Rifled Tubes

The helical winding of tubes, as opposed to vertical tubes, requires a more complex construction, because the tubes are not self-supporting. Vertical mounting of the evaporator tubing in the furnace, using plain tubes, would require a great number of parallel tubes at adequately low mass flow densities which, even then, would not ensure sufficient cooling (Franke et al. 1993; Wittchow 1995).

Internally rifled evaporator tubes allow lower densities of the water/steam mass flow at the same heat flux density, owing to the more intensive heat transfer from the inner tube wall to the working fluid, so that the evaporator tubes can also be mounted vertically in the furnace (see Fig. 4.44). The helically wound tubing of the reference power plant (740 MW_{el}, 1,900 MW thermal input of fuel), with about 400 tubes sized 38×5.6 mm, is designed for a mass flow density at full load of $2,100\,kg/m^2$ s. The angle of the helix is 16°. A steam generator of the same size, to draw a comparison, needs about 1,500 internally finned, vertically mounted evaporator tubes, sized 34×6 mm. The mass flow density in full load operation is only $1,000\,kg/m^2$ s. For a sufficient cooling at the heat flow densities at full load, mass flow densities of $250–350\,kg/m^2$ s are satisfactory (Franke et al. 1993).

The low mass flow densities of internally rifled tubes allow sufficient cooling at low minimum capacities, without causing high pressure losses due to high mass flow densities at full load. In comparison to inclined tubes, vertical mounting avoids segregation processes. The minimum load of the steam generator can be lowered from 35–40 to 20%. A lower minimum load could decrease the number of start-ups

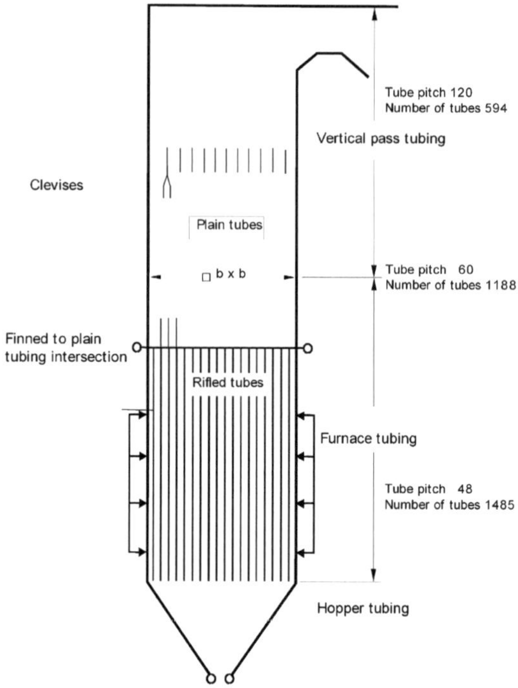

Fig. 4.44 Wall tubing of a single-pass boiler with vertical tubes in the furnace section (*Source*: Alstom Power)

and shutdowns, which would have a positive effect as regards both the fatigue of elements and the fuel consumption, because start-up and shutdown losses are avoided. In addition, the plant could do without a circulating system for low load.

Lower allowable mass flow densities also entail operational advantages. Internally rifled tubes have higher pressure losses per metre tube length than plain tubes at the same mass flow and the same dimensions. The pressure losses of a steam-generating system equipped with internally rifled tubes decrease considerably though, because of the low mass flow density and the shorter tube length. Whereas conventional evaporators involve pressure losses between 25 and 10 bar, it is possible to achieve levels of pressure loss as low as 5 bar by using vertical internally rifled evaporator tubes (Franke et al. 1995).

Lower mass flow densities and vertically mounted tubes improve the buoyancy conditions in a once-through steam-generating system. The outcome is a so-called natural-circulation characteristic, where extra heating typically results in a better cooling of the tube, similar to a drum boiler.

The pressure gradient along the tube evolves through fluid friction and geodetic or hydrostatic pressure of the steam column. If the fluid friction, or friction loss, along the pressure gradient predominates at high mass flow densities, the additional heating leads to an increased steam fraction in the boiling water flow, to a higher flow velocity and to a rise of the pressure loss. Yet since the pressure difference is the same in all parallel tubes, the throughput of the more strongly heated tubes decreases.

Fig. 4.45 Throughput characteristic of a tube with 25% extra heating (Wittchow 1995)

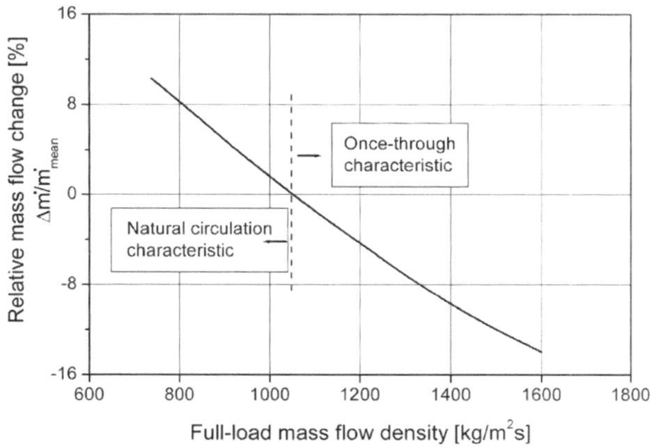

If the geodetic pressure loss predominates, the additional heating leads to higher mass flow densities. Due to the increased steam formation, the geodetic pressure loss of a tube with constant mass flow diminishes, because the steam column becomes lighter. The decrease of the geodetic pressure drop is higher than the rise of the friction loss. The pressure loss being given, however, the mass flow through the additionally heated tube rises (natural-circulation characteristic, see Fig. 4.45). The impact of the extra heating on the steam temperatures at the evaporator outlet is minimised by the self-regulating effect. This can be an advantage for the application of higher steam conditions, since the difference between the fluid temperature in the evaporator and the allowable material temperature may be smaller (Franke et al. 1993, 1995; Wittchow 1995). On the other hand, the counterbalance of the heating by the helically wound tubes does not apply.

Though one might expect a higher price for the tubes, financial benefits of the steam generator of up to 10% have been found, because the evaporator can be designed as a self-supporting construction (Wittchow 1995). Also, manufacturing and mounting are simpler than for helically wound tubing, which may be an advantage if the manufacturing is to be done in newly industrialised countries. Investigations in large-scale industrial plants with a test configuration of several vertical internally finned tubes mounted in parallel with helically wound tubes confirm the advantages of this concept (Franke et al. 1995; Kral et al. 1993).

In circulation steam generator construction, the more economical vertical tubes are used. The maximum heat flow density of about $0.4\,\mathrm{MW/m^2}$ common in coal-fired furnaces requires mass flow densities in the evaporator of around $600\,\mathrm{kg/m^2\,s}$, which have to be controlled by the natural circulation. Since the circulation ratio decreases with rising pressure in natural circulation, limits of approximately 185 bar arise for the maximum pressure in the evaporator, which corresponds to a live steam pressure before the turbine of about 175 bar. In a forced-circulation system, the circulation mass flow of $1{,}000$–$2{,}000\,\mathrm{kg/m^2\,s}$ is controlled by the circulating pump (Strauß 2006).

4.3.5.4 Evaporator Stability

Different operating modes of and uneven fuel flows to the burners of a burner group cause asymmetric firing conditions and non-uniform heat fluxes to the furnace walls. Given their great lengths and temperature rises, evaporator tubes of forced once-through steam generators react to heating differences with differing temperatures in the tube wall and at the evaporator outlet. The helical winding still ensures a good heating balance because each of the parallel tubes runs along all four walls of the furnace (Franke et al. 1993).

The design of a steam generator has to ensure an even flow through all the parallel tubes of the evaporator as well. Impacts of additional heating on the flow conditions in the evaporator tubes depend on the characteristic response of the evaporator. If the extra heating of a tube causes the flow through it to diminish (once-through characteristic), the possible consequence is that the temperatures exceed the allowable limit for the material. For the previously described natural-circulation characteristic, a temperature rise through extra heating is counterbalanced by the self-regulating rise of the boiling water vapour flow in the tube in question. This characteristic depends on the mass flow density and the fluid friction of the fluid involved. Low mass flow densities (below 1,000 kg/m^2 s) favour the operator-preferred natural-circulation characteristic response (see Fig. 4.45) (Wittchow 1995).

One option for checking whether a stable and even flow in the evaporator has been achieved is to consult the characteristic curves of the evaporator (Baehr 1985). Figure 4.46 shows the correlation between pressure loss and steam mass flow with the heating as a parameter. While the characteristics of tubes filled with a water flow

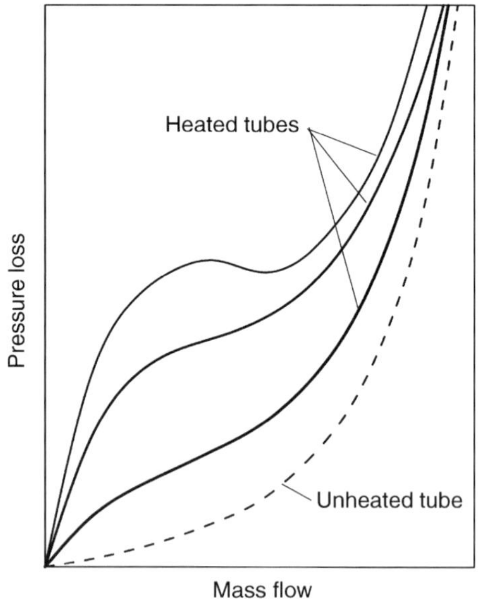

Fig. 4.46 Characteristic curves of the evaporator (Baehr 1985)

correspond to a second-order parabola, tubes which are filled by a flowing two-phase mixture, i.e. boiling water and steam, give a third-order curve.

An unstable flow occurs if the curve has a saddle-like behaviour, the consequence of which can be that three different mass flows evolve for the same pressure gradient. If a mass flow has a lower rate than needed for cooling the tubes, the effect can be damage to the tubes. The stability of steam generators and measures to raise the stability are dealt with in detail in Doležal (1990).

4.3.5.5 Design of the Convective Heating Surfaces

The units in the flue gas path following the furnace are the convective heating surfaces of the superheater, the reheater and the economiser. While the superheater and the reheater heat the steam up to the required turbine inlet temperatures, the economiser cools the flue gases down prior to the air heater and preheats the feeding water to a level close to the boiling point.

The convective heating surfaces consist of a great number of parallel tubes in a cross-flow arrangement to the flue gas flow. In contrast to the heat transfer to the evaporator surfaces by radiation, transfer by convection applies heat to the whole tube circumference, which is why the heating surface banks are smaller for the same temperature difference (Strauß 2006). The heating surface dimensions being decided previously, the heat transfer depends on the flue gas velocity and the driving temperature difference. The tubes should be mounted in the flue gas duct as close to each other as possible in order to achieve a high heat transfer level. The distance of the tube, however, is limited by the increasing pressure loss on the flue gas side and by possible fouling due to fly ash deposits (Stultz and Kitto 1992).

Superheater and Reheater

Heating surfaces used for superheaters and reheaters can be hanging or horizontal tube bundles. In Germany, where the single-pass construction is commonly built, only horizontal, drainable heating surfaces are used. In boilers in two-pass construction, hanging heating surfaces are often used in the cross-pass for super- or reheating. The distance between the tubes, the so-called tube pitch, depends on the flue gas temperature and the flue gas dust concentration (ash content of the coal). With the decreasing temperature, the tube pitch narrows in the direction of the flue gas flow. With low ash contents of the coal, it is possible to use smaller tube pitches and hence to build a more compact steam generator.

Two-pass boilers of the US type often have a hanging plate-type superheater which can be used in areas of high temperatures of around 1,400°C. With these heating surfaces, the predominant method of heat transfer is radiation. The tubes, wound closely to each other in a plane, form a plate, with large distances, of more than 1 m, between the plates. Such plate superheaters are insensitive to ash deposits. Figure 4.47 shows the tube pitches as a function of the flue gas temperature for US-type steam generators. In the case of single-pass or tower boilers, the upward-diminishing transverse pitch enables the dropping through of ash deposits that have come off.

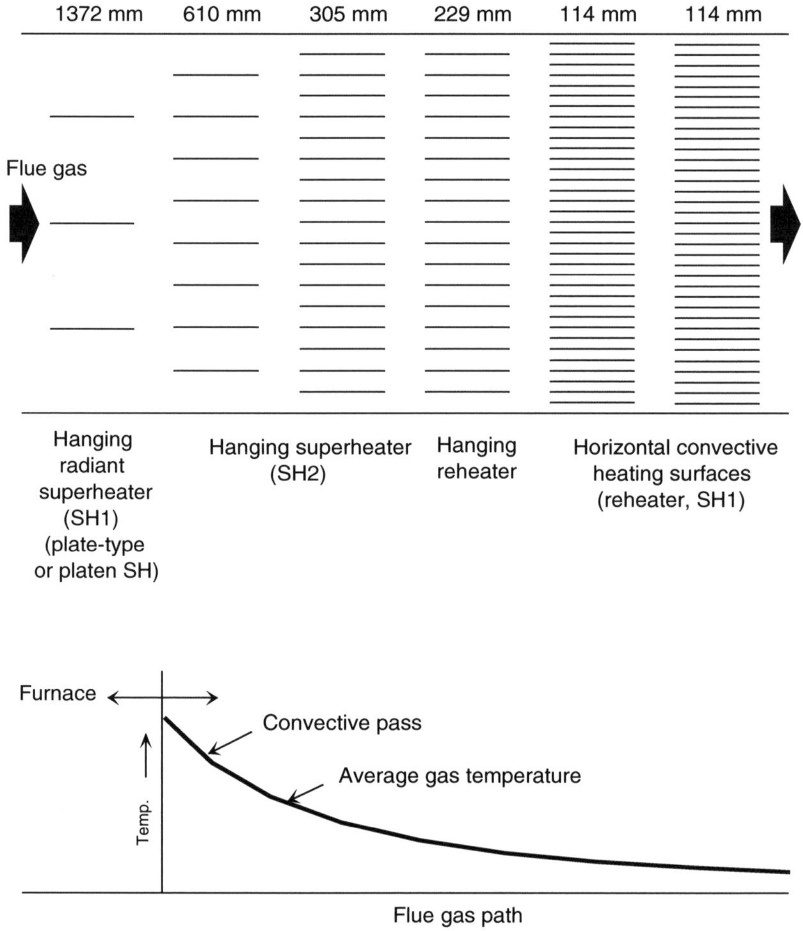

Fig. 4.47 Heating surface divisions in US constructions (Stultz and Kitto 1992)

The superheater and the reheater are designed for high steam temperatures, which only allow low temperature differences between the different tubes of a heating surface. Material temperatures in excess of the allowable limits may arise via an uneven flow through the tubes or by an imbalanced heating of some of the tubes. Asymmetric fireside temperatures have more of an effect when the temperature rise of a heating surface is higher.

Design and construction therefore have to guarantee an even flow and to counteract the impacts of an imbalanced heating. Temperature discrepancies between individual tubes are balanced out by dividing the superheater system into several stages, combining and mixing all single-tube steam flows in one stage and then re-establishing the division in the following stage into single-tube steam flows. Disadvantages of multistage superheaters are the higher costs and higher pressure losses due to headers and manifolds. Large steam generators usually have superheaters divided up into several piping runs connected in parallel. By crossing the piping

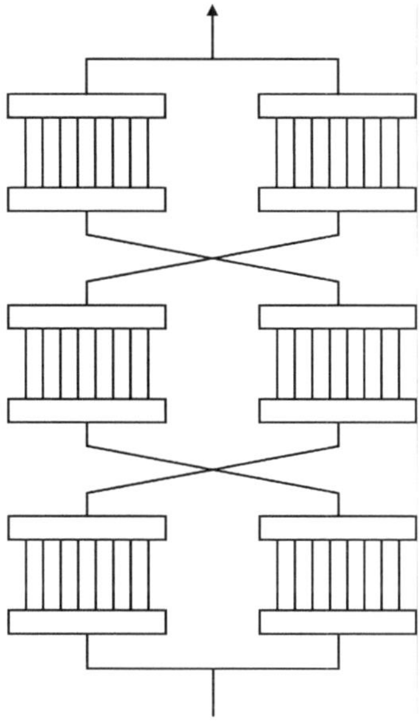

Fig. 4.48 Crossing of multistage superheaters

runs between the superheater stages, it is possible to counteract uneven heating (see Fig. 4.48).

With this construction the steam flows in the runs change their position in the flue gas pass from one outside to the other or, in the case of four piping runs, from outside to inside and vice versa (Strauß 2006; Baehr 1985).

For the control and limitation of live steam and reheater steam temperatures, attemperation is commonly applied. High-pressure feed water (HP feed water) is injected before or after the last superheater or reheater stage in attemperators.

Maintaining Constant Live Steam Temperatures

Falling high pressures and reheater steam temperatures in partial-load conditions diminish the mean temperature of the heat addition and hence the thermal efficiency. The live steam and reheater steam temperature should therefore be constant throughout the whole load range.

The design and location of the heating surfaces determine the temperature characteristics as a function of the load for each superheater and reheater. Heating surfaces in areas of high temperatures, above about 1,200°C, take up heat predominantly by radiation, and heating surfaces in areas of low temperatures, mainly by convection (Strauß 2006). With the output diminishing, the radiant heating surfaces in the furnace take up relatively more heat (radiation characteristic) whereas the heat share of the convective heating surfaces decreases (convection characteristic). Care should be taken that, for superheating and reheating convection surfaces, both convection and radiation characteristics are incorporated into the design, in order to

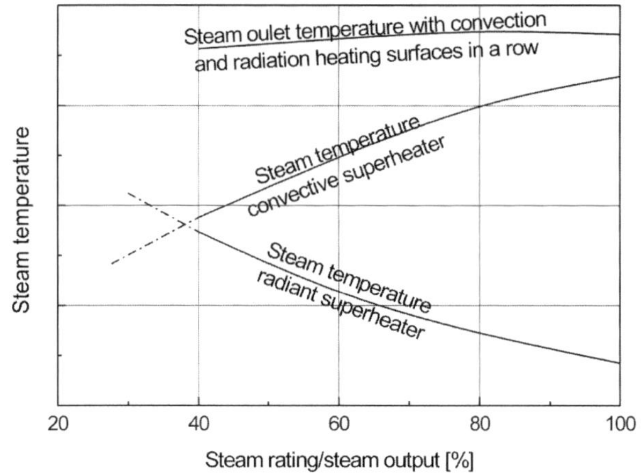

Fig. 4.49 Characteristics of radiation and convection heating surfaces

achieve a constant steam temperature throughout the output range (see Fig. 4.49) (Stultz and Kitto 1992; Doležal 1990; Adrian et al. 1986).

A heat flux shift between the evaporator and the convective heating surfaces has different impacts on the superheater and the live steam temperatures, which also depend on the evaporator system (Wittchow 1982).

During part-load operation of once-through steam generators with a variable liquid – vapour phase transition point, the vaporisation area migrates to the beginning of the evaporator, with the heat absorption in the furnace increasing and the superheating area in the membrane wall becoming larger. The lower heat uptake in the convective heating surfaces has therefore only a minor effect on the live steam temperature. In once-through boilers, the live steam temperature is kept constant up to about 35% of the load capacity by the setting of the ratio of the fuel flow to the feed water flow. The ratio of attemperator water to feed water flow remains unchanged at about 5% throughout the whole output range. It is not necessary to take measurements on the flue gas side.

The injection of the spray water in the high-pressure section of the steam generator causes a reduction of the efficiency only if the temperature of the spray water differs from the entrance temperature of the feed water. Generally though, water for injection is extracted after the high-pressure preheaters, so no efficiency loss is involved.

In circulation steam generators with a fixed liquid – vapour phase transition point, a higher heat absorption in the vaporisation area results in a greater steam flow which, with the consequently smaller convective heat flux, becomes less superheated. Therefore, below outputs of about 50–60%, the live steam temperature of such generators decreases.

In steam generators of the US-type circulation configuration, hanging plate superheaters are used, with a radiation characteristic at high temperatures, for superheating and reheating at the furnace end. By means of measures on the flue gas side, such as shifting the firing to the upper burner levels, or by additional plant

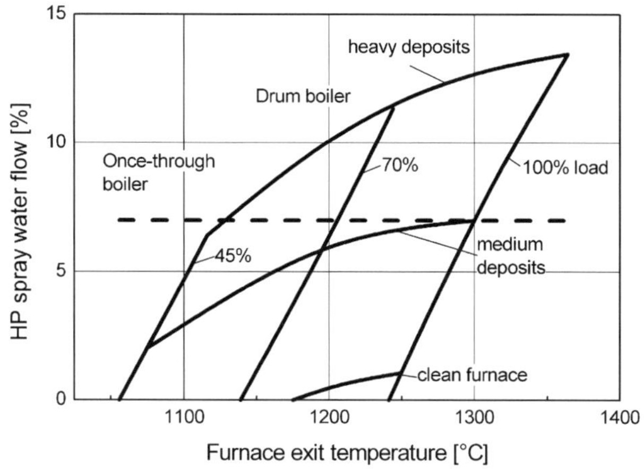

Fig. 4.50 Dependence of the HP spray water flow on the unit output and on the fouling state of the furnace (Wittchow 1982)

components, such as flue gas recirculation, it is possible to extend the control range for constant steam temperatures towards lower outputs.

Firing conditions and heat flux distributions deviating from the normal state evolve through the fouling of the heating surfaces. Ash and slag deposits on the furnace evaporator walls move heat to the convective heating surfaces. The radiation heat fraction taken up by the furnace walls and the evaporator decreases. Figure 4.50 shows, for once-through and circulation steam generators, the impacts of fouling on the spray water mass flow in relation to the output-dependent flue gas temperature at the furnace exit (Wittchow 1982).

Once-through steam generators adapt to these changes by shifting the liquid – vapour phase transition point in the direction of the evaporator end, so that the superheater area becomes smaller. The greater convective heat flux fraction is balanced out by the altogether smaller effective superheating surface, while the steam temperatures and the spray water flows remain nearly constant, independent of the fouling state in the steady-state condition.

Circulation steam generators with a fixed liquid – vapour phase transition point in the drum react to fouling by producing less steam. At the same output, the steam flow leaving the drum and entering the superheater section decreases, although the convective heat flux for superheating has increased due to the fouling in the furnace. In order to avoid excessive tube wall temperatures, it is necessary to provide for and correspondingly include in the design sufficiently large additional cooling flows, i.e. spray water flows.

The spray water flowrate needed for heavy fouling is affected by the requirement that, when the furnace wall is clean, the flow is still sufficient to control the steam temperature. The shift of heat absorption into the convective area has to be manageable by spraying under all operating conditions. If, for the fouling state, the maximum spray water flow would be reduced by choosing a smaller superheater surface, the steam temperatures, inversely, can no longer be achieved for the high-load conditions in clean state.

Figure 4.50 shows the impacts of fouling on the total spray water flow for a plant in service. The change of the spray water flow by fouling with a factor of 2 or more is considerably stronger than the output-dependent change of the spray water flow.

Maintaining Constant Reheater Temperatures

As in the case of superheating in high-pressure zones, heating surfaces with convection and radiation characteristics should be utilised in order to keep constant reheater steam temperatures. Reheating does not involve the balancing influence on the live steam temperature by migrating vaporisation and superheating zones in the furnace wall of a once-through steam generator.

The operating regime of a steam generator – fixed or sliding pressure – can have an influence, however, on the necessary temperature rise. In fixed-pressure operation, the reheater must be supplied with relatively more heat because the reheater inlet steam temperature drops as the output decreases. But in sliding-pressure operation, the reheater inlet temperature is nearly independent of the output.

A relatively simple method to control and limit the reheater outlet temperature is to spray feed water between two subsequent reheat surfaces at a pressure similar to the exit steam pressure of the high-pressure turbine. In this case, the reheater is designed to be larger for full load and its steam exit temperature is limited to the allowable temperature by spray water admixing. When output diminishes, the necessary spray water flow decreases as well.

Reheater spraying for temperature control, however, has the consequence of a loss in efficiency, because the high-pressure zone of the steam generator is bypassed, and only steam at the reheater pressure is produced and exploited. The heating of the spray water by mixing at a low reheater pressure results in a lower temperature of heat addition.

Other methods of temperature control avoid the disadvantageous effect of reheater spraying on the steam generator efficiency, for instance by transferring heat between the live steam and the reheater steam system or by shifting the heat flux through flue gas recirculation or tilting burner, to set constant reheater steam outlet temperatures. In order to control the heat transfer between the high-pressure superheater and the reheat surfaces, the reheat surfaces are designed to be either larger or smaller than without this control. Designed larger, they transfer heat to the live steam system in the upper output range. Designed smaller, they take heat from the high-pressure superheaters in lower output range. The heat is exchanged either in heat exchangers outside the flue gas duct or in live steam and reheat tubes mounted coaxially and heated by flue gas from the outside (the "Triflux system"). Excess heat of the reheater can also be used to preheat the feed water. This kind of temperature control only involves pressure losses (Adrian et al. 1986).

In Germany, the measure usually adopted is reheater temperature control by spraying, with sufficient excess air in part-load operation. Usually, the reheater spray water flow of a forced once-through steam generator at full load operated with sliding pressure comprises 1% of the feed water flow.

Economiser

The economiser (sometimes shortened to "eco"), or feed water preheater, is a steam generator's penultimate fireside heating surface and at the same time its first heating surface on the steam side. The entrance temperature of the feed water is 250°C for the reference power plant, while the flue gases are cooled from about 450 to about 350°C. In once-through steam generators, the last part of preheating before boiling starts occurs in the evaporator, avoiding premature vaporisation in the economiser. In circulation steam generators, the preheated feed water, for the same reason, is fed into the evaporator drum before the boiling stage.

As a consequence of the small temperature difference between the two working media, the economiser needs a very large heat exchange surface. The raw material utilised for the economiser is usually unalloyed steel. Plain tubes are used as a rule. External fins improve the fireside heat transfer if they are kept free from ash deposits (Stultz and Kitto 1992).

The risk of corrosion from flue gases cooling below their dew point must be avoided. Cold feed water must therefore not be fed to the economiser. The regenerative feed water preheating, which heats the feed water up to the above-mentioned temperature of 200–300°C before it enters the economiser and which lets the flue gas cool down to temperatures of, at most, above that level, determines the fireside outlet temperature, depending on the terminal temperature difference (TTD) of the economiser.

If nitrogen oxide control is necessary, further requirements for the flue gas temperature between the economiser and the air preheater may arise (Reuter and Hönig 1988). In so-called high-dust configurations, the catalyst is mounted between the economiser and the air preheater. A catalytic flue gas $DeNO_x$ reactor needs a reaction temperature of about 350°C, which is provided in this location.

4.3.5.6 Air Preheater

The air preheater transfers flue gas heat from the lower flue gas temperature region to the combustion air. This low-temperature heat transfer diminishes the necessary fuel energy on one hand and, on the other, influences the ignition and the combustion course of the firing by higher temperatures of the combustion air.

The use of regenerative feed water preheating to raise the cycle efficiency requires combustion air preheating, as the medium, water, cannot be used to make use of the flue gas waste heat, because of the higher temperatures. Air preheating raises the combustion temperature (in the furnace) and, owing to the higher temperature drop between flue gas and steam, makes it possible to use smaller heating surfaces.

In the air preheater of the reference power plant, the flue gas cools down from the temperature after the economiser of 350°C to a temperature of 130°C, which lies above the acid dew point of the flue gas. In the counterflow, the combustion air of about 45°C, after being preheated by a steam air heater, is heated up to the combustion temperature of 310°C. Low outlet flue gas temperatures minimise the flue gas energy losses of the steam generator. The acid dew point of the flue gas sets the low-temperature limit, as temperatures below this point would result in corrosion and fouling. Measures to increase the efficiency by limiting the flue gas energy losses make use of existing design reserves, but may apply restrictions on the coal feedstock (see also Sect. 4.4.2.2).

Since the combustion air flow is about 10% smaller than the flue gas flow, a combustion air temperature correspondingly below the flue gas temperature develops. The terminal temperature difference (TTD) of the heat exchanger should not lie below 20 K, in order to limit the construction size (Strauß 2006).

The combustion air temperature (the air preheat temperature) depends on the requirements of the furnace. For coal-fired furnaces, the temperature ranges from 350 to 400°C, where higher temperatures are chosen for slag-tap furnaces and lower ones for dry-bottom furnaces. For fluidised bed furnaces, with their low combustion temperatures determined by the system, the level usually reaches up to 250°C. For stoker-fired furnaces, the design air preheat temperatures may only go up to 150°C in order to prevent caking on the stoker. In all combustion systems, low calorific coal types are generally combusted with hotter combustion air and correspondingly higher preheat temperatures than high calorific coals. Stable combustion is another reason why it is necessary to preheat the combustion air when using low calorific coals.

For financial reasons, regenerative heat exchangers are utilised almost exclusively for combustion air preheating in steam generators. At first, the flue gas heat is transferred to a heat accumulator, which transfers it, with a delay, to the air to be preheated. The heat accumulator consists of plate packages where air and flue gas flow alternatively through.

The construction is such that either the heat accumulator is the rotating component and the two inlet and outlet hoods are the stator (Ljungstroem construction) or the two hoods are the rotating component and the heat accumulator is the stator (Rothemuehle construction). Despite gaskets between the movable and the fixed parts, it is not possible to completely avoid leakage from the airside under pressure. The leakage flow of combustion air typical for regenerative heat exchangers lies in the order of magnitude of 5% and causes an increase of the flue gas mass flow (Stultz and Kitto 1992).

If a regenerative heat exchanger is charged with cold combustion air, the plate temperatures at the air inlet fall below the dew point temperature. For this reason, the cold side is fitted with corrosion-resistant material, e.g. enamelled plates, and with cleaning equipment to remove deposits. Deposits left in place increase both airside and fireside pressure losses. The temperature drop below the dew point can be prevented if the regenerative air preheater is preceded by a steam air heater, which raises the temperature of the supplied air above the dew point temperature of the flue gas. Such a steam air heater is also used in part-load operation because during such operation the medium plate temperature can also be below the dew-point temperature.

The air preheater of pulverised hard coal furnaces is designed dual flow – for primary air and secondary air. This way, the different air temperature requirements of the furnace and for combined drying and pulverising can be met. The secondary air temperature corresponds to the necessary combustion air temperature. The primary air temperature is determined by the necessary heat for coal drying. The primary air system is operated above atmospheric pressure in order to balance out the flow resistance of the primary air ducts, the mill and the fuel powder transport from the mill.

4.3.6 Design of the Flue Gas Cleaning Units and the Auxiliaries

4.3.6.1 Design of the Flue Gas Cleaning Units

The allowable emission standards require installations for dust removal, nitrogen control and desulphurisation. The boundary conditions are the pre-determined flue gas mass flow and the necessary removal efficiencies, which are determined by the dust content, the sulphur dioxide and nitrogen oxide concentrations in the raw gas and the respective emission standards. The techniques of emission reduction are dealt with in Chap. 5 in the context of firing technology.

4.3.6.2 Design of the Auxiliaries

The forced-draught fan supplies the burners with the air mass flow required for combustion (determined during the design). The necessary overpressure of the secondary air is determined by the resistance of the air inlet, the air preheater, the air ducts and the burners. Booster fans produce the rise in pressure of the primary air necessary to surmount the additional resistance in the mills, classifiers, pulverised coal supply pipes and the burners. The pressure losses of the secondary air range around 70 mbar; those of the primary air are about 160 mbar.

The power demand of the induced-draught (ID) fans for transporting the flue gases depends on the flue gas mass flow and on the pressure drop along the flue gas path. The furnace is operated with some mbar of underpressure. The pressure drop along the flue gas path before the ID fan, which transports the total flue gas, amounts to 40–50 mbar at the rated power of the plant, depending on the fireside flow resistance.

4.4 Possibilities for Efficiency Increases in the Development of a Steam Power Plant

One solution for the reduction of CO_2 emissions from power plants fired with fossil fuels is to increase their efficiency. All fossil fuels have a content of carbon, either higher or lower. Coal, among them, is a fuel with a comparatively high carbon content and at the same time the fuel with the highest percentage use worldwide in power production.

Research and development is currently being conducted, aimed at reducing CO_2 emissions by increasing the efficiencies of all the units in a power plant discussed in this book. The possibilities in this respect are distinguished for stationary operation as follows:

- Increases in the thermal efficiency of the cycle
- Measures to minimise the losses
- Measures to reduce the auxiliary power requirements

The stated efficiency rates usually refer to the rated power. However, the efficiency of the plant in part-load operation and the losses at start-up and shutdown should be taken into account as well.

4.4.1 Increases in Thermal Efficiencies

Improvements of thermal cycles aim at attaining a high mean temperature of the heat addition and a low mean temperature of the heat extraction.

High mean temperatures of the heat addition and therefore high thermal efficiencies are achieved by

- increasing the live steam conditions (temperature and pressure),
- single or double reheating,
- regenerative feed water preheating,
- reducing reheater spraying and lowering mean temperatures of the heat dissipation and
- low exhaust steam temperatures in the condenser.

The conversion processes associated with losses are presented in Fig. 4.2.

4.4.1.1 Increasing the Live Steam and Reheater Steam Conditions, Single or Double Reheating and Reheater Spraying

High mean temperatures of the heat addition contribute to a high thermal efficiency. They can be achieved by a high pressure in the high-pressure steam generator (HP steam generator), by a high live steam temperature, by regenerative feed water preheating and by reheating to high reheater temperatures.

A higher live steam pressure entails correspondingly high boiling water temperatures, which raise the heat input temperatures to a higher mean level, with the outlet temperature remaining the same, thus increasing the thermal efficiency. Lower live steam pressures and hence lower boiling water temperatures decrease the mean temperature of heat addition and the efficiency. However, higher pressures require more power for the feed water pump. Further pressure increases give diminishingly greater thermal efficiencies, which are eventually cancelled out, and then exceeded by, the efficiency losses due to the increased feed water pump power requirements.

The pressure level at which the pressure impact on the efficiency becomes inverted lies considerably higher, though, than the live steam pressure levels common

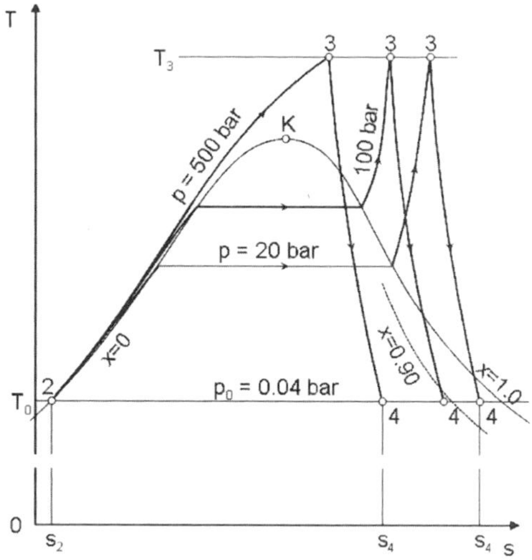

Fig. 4.51 Pressure influence on the exhaust steam conditions (Baehr 2006)

today, which are limited by the strength of the available construction materials (Adrian et al. 1986).

With increasing pressure, the live steam conditions shift towards smaller entropies. Correspondingly, the exhaust steam conditions also shift to lower steam and higher water contents (see Fig. 4.51). However, for technical reasons, the so-called exhaust moisture $(1 - x_4)$ must not exceed values of about 0.1. With an excessively high exhaust moisture, droplet impact occurs in the last stages of the turbine, which leads to erosion of the final-stage blades. The prescribed exhaust moisture limits the choice of the live steam pressure for a simple steam-generating plant without reheating or makes it necessary to install reheating (Baehr and Kabelac 2006). Since the reheated steam after expansion has a higher entropy with a higher steam quality, damage of turbine blades through droplet impact is less likely.

Higher live steam and reheater outlet steam temperatures also result in higher mean temperatures of the heat input, and thus in a higher thermal efficiency. Figure 4.52 shows the influence of pressure and temperature on the efficiency of the cycle, given as the relative heat rate gain. For the temperature range of up to 600°C, a rule of thumb is an increase of the net efficiency of 0.02% (absolute) per degree of temperature increase (with the live steam temperature equalling the reheater temperature). In the range from 600 to 700°C, the increase is about 0.016% per degree of temperature increase. The influence of the temperature increase of the live steam is in this process somewhat higher than that of the reheater temperature. In the pressure range up to 250 bar, a rise of the live steam pressure results in an efficiency improvement of 0.01%/bar; higher pressures of up to 300 bar result in an improvement of about 0.008 bar. With yet higher live steam pressures, the gain in net efficiency diminishes again (Klebes 2007; Adrian et al. 1986; Billotet and Johänntgen 1995; Kotschenreuther et al. 1993).

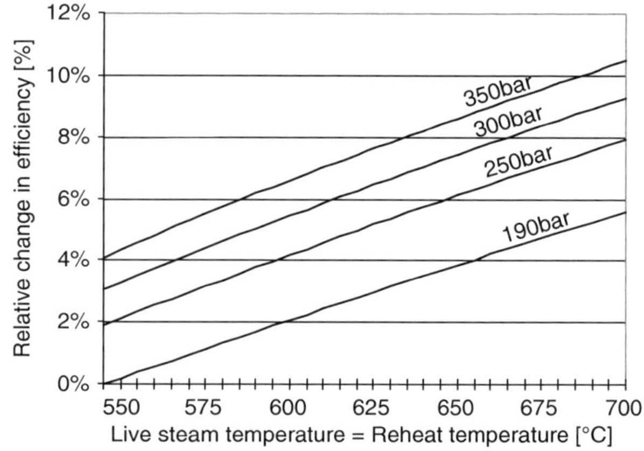

Fig. 4.52 Influence of live steam pressure and temperature on heat rate

High live steam pressures and temperatures are limited by the available construction material. Particularly with new materials, the operating behaviour of the plant has to be taken into account in design. The influences of advanced live steam conditions on the steam generator design is discussed in Sect. 4.5.

Reheating raises the mean temperature of the heat input (see Fig. 4.53) since the mean reheating temperature is higher than that of the simple steam process. For the reference power plant with conventional steam conditions (190 bar, 530°C, 530°C

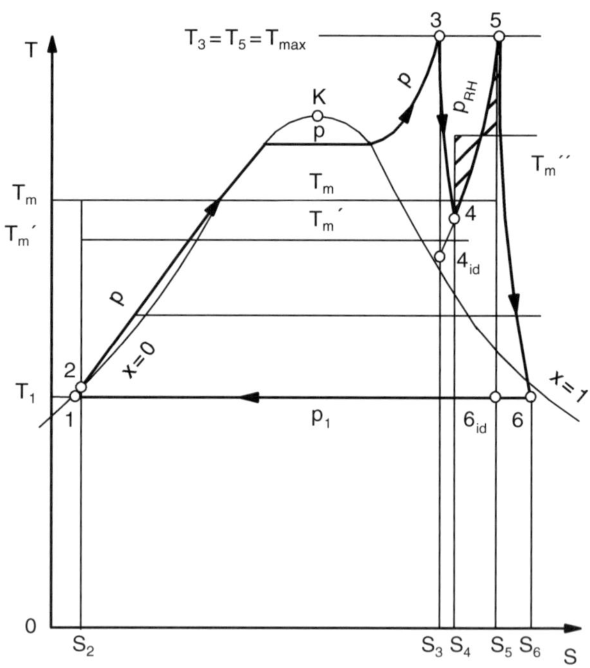

Fig. 4.53 Changes of state in the process with reheating (Baehr and Kabelac 2006)

(turbine inlet)) the mean heat input temperature in the high-pressure part of the steam generator lies at $t_{m'} = 364°C$, while the medium temperature of heat addition in the reheater is $t_{m''} = 430°C$, resulting in an overall mean temperature of heat addition $t_m = 376°C$. For a power plant with advanced steam conditions (285 bar, 600°C, 620°C) the medium temperature of heat addition lies at $t_m = 415°C$, while in the high-pressure part of the steam generator the mean temperature is $t_{m'} = 400°C$ and in the reheater $t_{m''} = 470°C$.

In designing a power plant, optimum pressure ranges arise both in single and in double reheating. The optimum pressure depends on the live steam pressure. As a rule of thumb, the ratio between live steam and reheater pressure in modern power plants is between 5 and 6. The optimum can be explained with Fig. 4.53. Reheating results in a maximum increase of efficiency, if the cold reheat temperature T_4, which is a function of the reheat pressure, is at the level of the medium temperature of heat addition $t_{m'}$ in the high-pressure part of the steam generator. In this case, reheating increases the medium temperature of heat addition of the steam generator. If the cold reheat temperature is lower, at least part of the heat addition in the reheater results in lower efficiencies. Additionally, a reheat pressure that is too low can result in superheated steam at the turbine exit and thereby increase the medium temperature of heat dissipation.

Figure 4.54 shows the optimum of a double reheating regime in the form of equidistant efficiency curves. Deviations from the optimum pressures entail a deterioration of the efficiency. The optimum depends on the chosen live steam pressure (Rukes et al. 1994).

Assuming conventional steam conditions – such as those of the reference power plant for instance – introducing double reheating raises the net efficiency rate by up to 2%. Higher live steam pressures increase, while higher live steam temperatures decrease the gain in efficiency (Adrian et al. 1986). For a power plant with steam conditions of 280 bar, 585°C (live steam), 600°C (reheat steam), double reheating raises the net efficiency by 0.7% (Kotschenreuther et al. 1993). A drawback is the increased pressure loss. Double reheating can have a disadvantageous effect on the operating regime. In an investigation into the use of a double reheating process for a base-load power plant, an allowable load change rate between 2 and 4% per minute was reported. In comparison, load change rates of 4–8% per minute are required in Germany for a mid-range load power plant. Double reheating, as opposed to single reheating, makes additional investments necessary for the steam generator, the tubes and the turbine of between 2.5% (Kjaer 1993) and 3% (STEAG 1988) of the total investment costs. In Germany, double reheating was installed 11 times in total between 1950 and 1983 (VGB 1995). Because of the higher costs and the disadvantages for mid-range load operation, this technology was not employed after that time and is not taken into consideration in current projects in Germany.

Double reheating allows low condenser pressures, because the exhaust steam moisture is reduced, avoiding droplet erosion of the last-stage blades (Kjaer 1993). With the low condenser pressures associated with seawater cooling, double reheating allows the use of higher live steam pressures and lower condenser pressures.

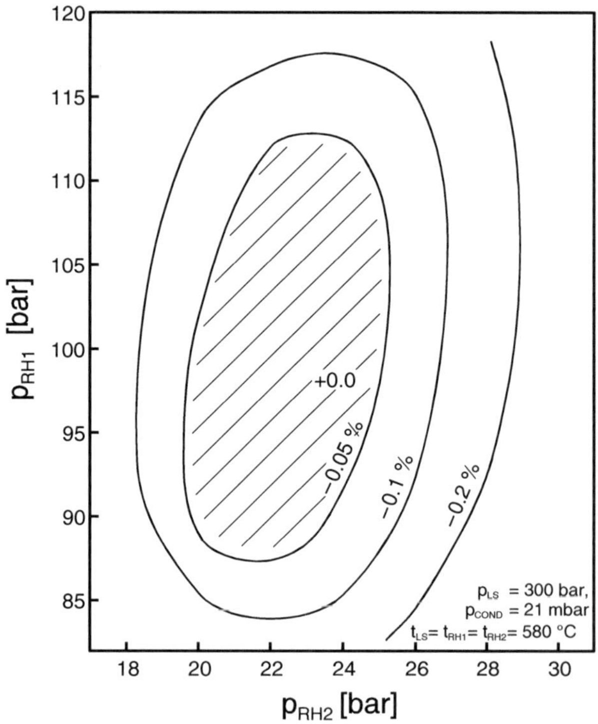

Fig. 4.54 Equidistant efficiency curves with the deviation from the optimum net efficiency as a function of the reheater pressures with double reheating (Kjaer 1990)

The application of double reheating with live steam conditions of 302 bar, 592°C, 605°C, 605°C and a condenser pressure of 23 mbar results in a net efficiency improvement of 1.4% in comparison to a power plant with single reheating and live steam conditions of 288 bar, 597°C, 605°C and a condenser pressure of 23 mbar. The drawback of double reheating is the above-mentioned additional cost.

A new double reheating concept design currently under development is promising costs comparable to single reheating. The idea of the so-called Master Cycle is to reduce the exergy loss of the heat transfer of superheated bleed steam from the IP turbine to the feed water. Superheated bleed steam from the IP turbine is used for feed water preheating at the level of its condensation temperature, resulting in high exergy losses. The exergy losses are higher for double reheating than for single reheating. The exergy loss by extracted steam can be reduced by shifting the IP extraction stages to a separate turbine fed by steam from the first cold reheat steam line. The new turbine expands the cold reheat steam to temperatures and pressures of the extraction stages, delivering about 4% of the net power. The steam flow through both reheating stages is reduced, resulting in lower capital costs of the cycle. Calculations give a Master Cycle efficiency improvement of 1.45% (326 bar/592°C/605°C/605°C, 23 mbar) over single reheating (Kjaer and Drinhaus 2008).

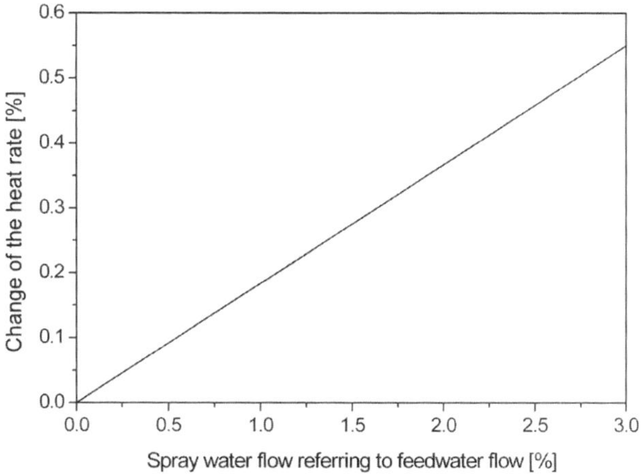

Fig. 4.55 Influence on the efficiency of reheater spraying (Baehr 1985)

As reported in Sect. 4.3.5.5, controlling the reheater temperature by a spray attemperator diminishes the efficiency, because the high-pressure range of the steam generator is bypassed by doing so, and steam is produced at a low pressure and temperature. Figure 4.55 shows the influence on the efficiency of the reheater attemperator mass flow (Baehr 1985). In the case of the reference power plant, the spraying mass flow at full load is about 0.9% of the feed water mass flow. New power plant designs limit the temperature-controlling spraying mass flow to 0.2% of the feed water mass flow (Breuer et al. 1995).

The measures described above have an effect only on the thermal and on the turbine efficiency, but not on the energetic steam generator efficiency. They are included in the exergetic steam generator efficiency rate, though (see Sect. 3.2).

4.4.1.2 Influence of Feed Water Preheating

Based on the condensate temperature, regenerative preheating of the feed water raises the feed water temperature, via several stages, up to the temperature at which it enters the steam generator. During this process, steam is extracted that has already been through the turbine. The extraction steam flows reduce the exhaust steam flow and thus the loss of exhaust steam heat.

With the live steam pressure and the superheater outlet temperature staying constant, preheating the feed water raises the mean temperature of the heat input.

The water flowrate and the extraction pressure define the preheating of a regenerative feed water heater:

- The steam pressure of the turbine extraction and the flow of the water to be heated (i.e., in LP heaters the condensate flow and in HP heaters the feed water flow) define the single-extraction steam flows.
- The extraction steam transfers its heat inclusive of the condensation heat. The water, before leaving the heater, heats up almost to the boiling temperature of the extraction steam pressure.

Nowadays, six to nine feed water heaters with feed water outlet temperatures between 250 and 300°C are commonly used for large thermal power plants. Higher feed water outlet temperatures are chosen as live steam pressures increase. The heater configuration of the reference power plant without raised live steam conditions is shown in Fig. 4.28. Four LP heaters and two HP heaters preheat the feed water up to 250°C, with the feed water tank and pumps necessary parts of the process.

The last, upper-most HP heater is usually heated by extracted steam from the cold reheat line behind the HP turbine. The reheat pressure, derived from optimisation calculations for the entire cycle, thus defines the feed water outlet temperature (see Fig. 4.56) (Rukes et al. 1994).

The feed water heating temperature can be further raised by inserting an additional preheater, heated by extraction steam from the high-pressure section of the turbine. Such additional extraction from the HP turbine section uncouples the reheater pressure and feed water outlet temperature. Figure 4.57 shows a heat flow diagram, with stages, where the feed water is preheated to 300°C.

A feed water heating temperature increase from 250 to 290°C, by additional extraction of steam from the HP turbine section, results in an efficiency increase of 0.7% (Billotet and Johänntgen 1995); the result of an increase from 270 to 300°C is an absolute improvement of 0.75% (Kotschenreuther et al. 1993). Figure 4.58 shows the impact of an increase in the feed water temperature – a relative decrease of the heat rate, which is dependent on the pressure level (Klebes 2007).

The rise of the feed water outlet temperature comes up against limiting factors with regard to the steam generator design. It is imperative to prevent boiling in the economiser in order to avoid flow instabilities and to ensure a steady charge of the evaporator tubes. For this reason, the economiser must be designed to be smaller for higher outlet temperatures of the regenerative feed water heating. Increasing feed water temperatures entering the steam generator make the transferable flue

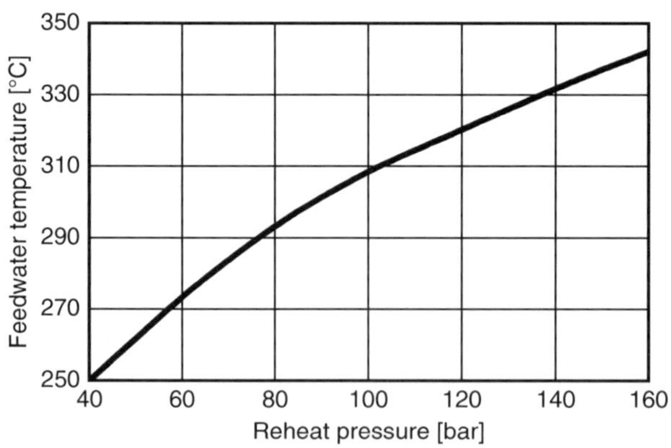

Fig. 4.56 Feed water temperature as a function of the reheat pressure (Rukes et al. 1994)

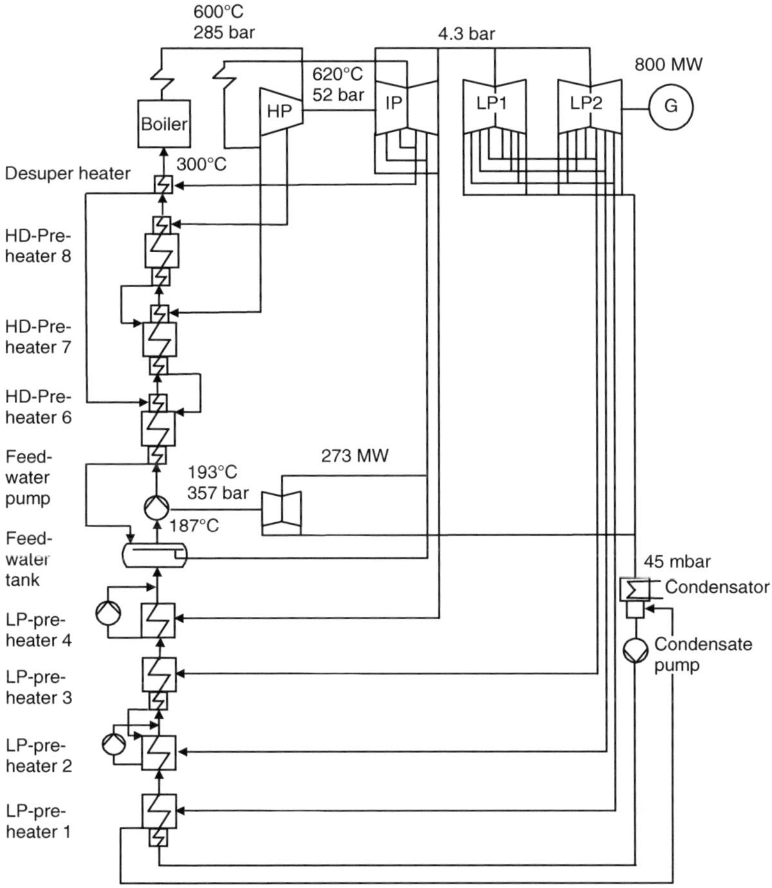

Fig. 4.57 Heat flow diagram of a thermal power plant with advanced steam conditions and nine-stage feed water heating (data from Tremmel and Hartmann 2004)

gas heat in the economiser decrease, which can then be used only to preheat the combustion air.

In designing a power plant, after the feed water heating outlet temperature is defined, further optimisation is only possible within the feed water heating chain. The design should, in this process, provide for the smallest possible temperature difference between the heating medium, i.e. the extracted steam, and the feed water to be heated. By increasing the number of heaters while keeping the same outlet temperature, smaller temperature rises for the individual stages result. This helps to achieve a better adaptation of the temperatures of the heat-absorbing to the heat-dissipating heat transfer medium – water flow and extraction steam flows – and thus to minimise the exergy losses. The improvement in efficiency of each additional stage, as shown in Fig. 4.59, is positive but decreasing, so that a point is reached where installation of yet another stage cannot be justified economically (Eichholtz et al. 1994).

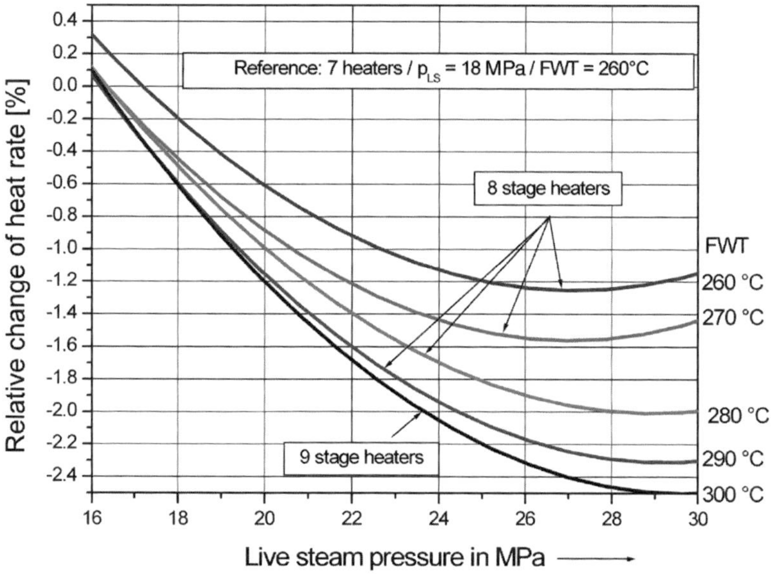

Fig. 4.58 Effect of the live steam pressure and the feed water temperature on the heat rate (Klebes 2007)

Fig. 4.59 Influence of the number of stages on the net efficiency, at constant outlet temperature (Eichholtz et al. 1994)

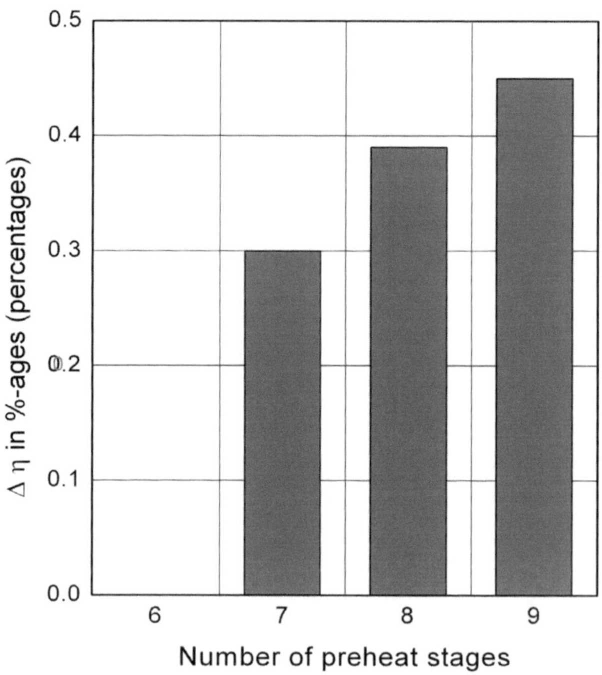

The temperature differences between the heat-dissipating and the heat-absorbing flows in a preheating stage are characterised by the so-called terminal temperature difference (TTD), which is defined as the smallest temperature difference between the two mediums. At the transition to small TTDs, larger heating surfaces and hence heavy and expensive plant components are required. A compact construction is the result when counterflow heat exchangers are used.

Preheaters are usually designed as shell-and-tube heat exchangers. The extracted steam and the feed water are segregated from each other by a heat exchanger surface, which consists of tube bundles. The sensible heat of the steam can be utilised in so-called desuperheaters. The sensible heat of the condensate can be utilised in condensate coolers, which can be mounted either by integration into the preheaters or separately. The desuperheater, with respect to the feed water, is mounted after the preheater(s). This way, the feed water can be heated to a higher temperature than is possible with the condensing preheater. The condensate cooler, with respect to the feed water, is mounted before the preheater.

The most reasonable solution in terms of thermodynamics is to mix, without cooling, the condensate in the preheater with the feed water. This method is not used for HP preheaters because the high feed water pressure requires a complex system of pumps, pipes and fittings. Thermodynamically, it is therefore a compromise to subcool the condensates and to let them flow into the next lowest preheat stage. In configurations with multistage LP preheaters, it is usually economical to pump the condensates of one or several preheat stages back into the condensate flow.

In a direct-contact heater, the heat of the extracted steam is transferred to the feed water by mixing and condensation of steam in water. Given its low terminal temperature difference, the direct-contact heater has thermodynamic advantages. However, because the container is under the pressure of extraction, the entire condensate flow has to be pumped to reach the corresponding pressure level. Because of the necessary pumps, direct-contact heaters are only used in the feed water tank for deaeration.

The common values for the terminal temperature differences of regenerative heaters of modern hard coal power plants are (STEAG 1988) as follows:

- Desuperheater 25 K
- Condensation equipment 2 K
- Condensate cooler 7 K

4.4.1.3 Lower Heat Dissipation Temperatures – Optimisation of the "Cold End"

The portion of the supplied heat which cannot be converted into mechanical work remains as condensation heat in the turbine exhaust steam and is discharged to the environment. In condensation power plants, the exhaust steam temperature of the turbine is about 30°C. At this temperature it is not possible to extract further heat due to the lack of heat sinks.

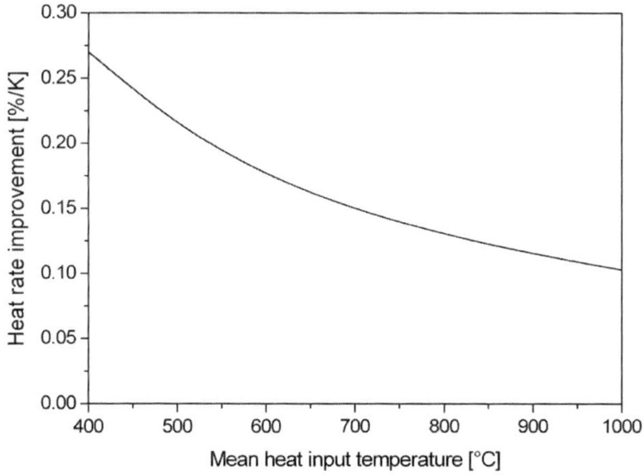

Fig. 4.60 Impact of a heat dissipation temperature reduction of 1 K

Besides the mean heat input temperature into the steam generator, the mean heat dissipation temperature is another factor which determines the thermal efficiency of the cycle. This temperature has to be chosen to be as low as possible in order to maximise the total efficiency. Low exhaust-steam temperatures and pressures in the condenser can be set by low temperatures of the cooling medium. The steam can be further expanded to the low exhaust steam pressure by the final LP blading. More heat is converted into mechanical work and thus the waste heat cut down by this heat fraction.

The heat dissipation temperature has an impact on the efficiency, which increases in strength when the heat input temperature is lower. This correlation is shown in Fig. 4.60 for the Carnot cycle, with a mean heat dissipation temperature of 30°C, corresponding to the condensation temperature. These fundamental correlations also hold true for other thermal power processes. Therefore it is evident that in a pure steam process, in comparison to a combined-cycle (gas and steam turbine) process, a higher efficiency increase can be achieved by improvements at the cold end (Johänntgen 1998). For the reference power plant, with a mean heat input temperature of 376°C, a decrease in the condensation temperature of 1 K diminishes the heat rate by 0.29% in the ideal case.

In a given turbine unit, the steam outlet velocity rises with an increase in the specific volume, i.e. when the condenser pressure decreases. Compared to the isentropic expansion, changes in the condenser pressure cause less change in the heat rate. With the condenser pressure decreasing, the losses increase through the kinetic energy of the exhaust steam, due to the rising outlet velocity. If sonic velocity is reached at very low condenser pressures, a further decrease of the condenser pressure does not improve the efficiency (Adrian et al. 1986). The losses in the exhaust steam are taken into account by the internal efficiency of the turbine.

The optimisation of the cold end must therefore involve not only the design of the cooling circuit but also the choice of the low-pressure turbine. In order to make use of the efficiency advantage of low condenser pressures, it is necessary to enlarge the exhaust steam cross-section of the LP turbine. As well as developing and

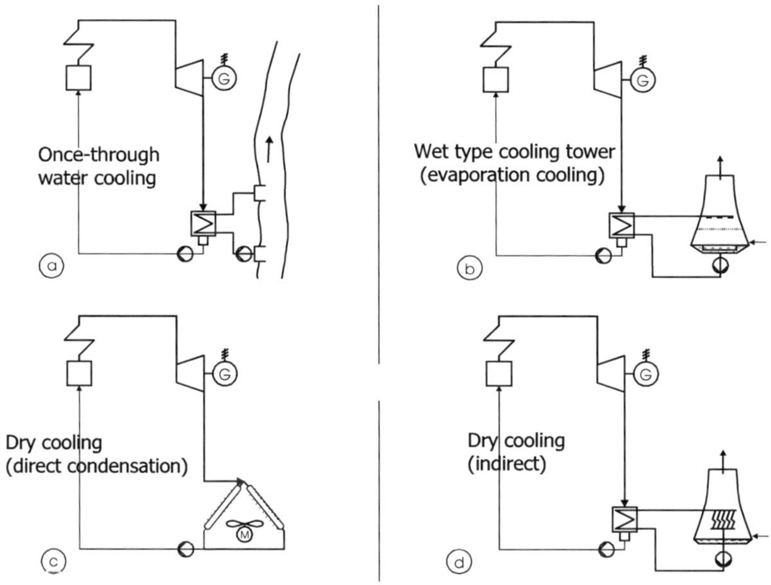

Fig. 4.61 Cooling systems in power plant technology (Baehr 1985)

utilising greater LP last-stage blade lengths – today, last-stage blades are manufactured with lengths up to 1,400 mm, at 3,000 r/min (revolutions per minute) (Neft and Franconville 1993) – a larger outlet cross-section can be obtained with two, three, or four LP turbine components mounted on a shaft. Increasing the number of LP turbines, however, entails step changes in the costs for turbines and turbine houses (STEAG 1988; Weber et al. 2005).

The temperature of heat dissipation is set by the cooling method. Schematics of the cooling systems are shown in Fig. 4.61 (Baehr 1985). Basically, there are three cooling systems to dissipate the waste heat arising in the condenser to the ambient air:

Once-Through Water Cooling

When fresh-water or once-through water cooling with river or seawater is used, the heat in the condenser is directly transferred to the cooling medium. Once-through cooling is simple and effective but can be utilised only at locations where there is fresh water available in sufficient quantities and the inevitable temperature rise ecologically justifiable. In Germany, river water temperatures have an annual average of 12°C; power plants at coastal locations in Denmark are based on a mean seawater temperature of 10°C.

Back-Cooling of Cooling Water Through Evaporation

In Germany, new plant designs mostly incorporate closed-circuit cooling water systems with natural-draught cooling towers. In such systems, the waste heat is initially transferred to the cooling water in the condenser and then backcooled in a cooling tower by heat dissipation to the cooling air. In this process, water is lost through evaporation and has to be replaced. The cooling level theoretically achievable with a wet-type cooling tower is determined by the wet-bulb temperature.[1] This temperature depends on the condition of the air and may lie below the cold inlet air, because of the extraction of evaporation heat (Berliner 1975; Schmidt et al. 1977). Given an annual average temperature of air of 8.5°C (Germany) and a relative air humidity of 75%, the resulting theoretically possible cooling is 6.6°C (STEAG 1988). Though this temperature is below the annual average of rivers, the cold water temperature that is economical, and therefore used, in back-cooling is around 15–20°C.

Dry Cooling

In direct dry cooling, the condenser is directly cooled by ambient air. In indirect dry cooling, an additional water circuit is used, and the warmed cooling water is cooled again in an air/water heat exchanger. For dry cooling systems, it is the dry bulb temperature that sets the temperature difference between the saturated air and the (approach) cooling water, whereas the theoretical limit for wet cooling towers is set by the lower wet-bulb temperature. Depending on conditions at the location, the difference between the dry and wet-bulb temperatures can amount to 15°C (at high temperatures and low air humidities). The poor heat transfer in air requires large heat exchange surfaces and therefore raises the economically achievable cold water temperature. Since, in contrast to wet cooling, dry cooling uses only convection, an air mass flow is necessary which is four times higher than the one in a wet cooling tower. These factors lead to higher exhaust steam temperatures and in consequence higher average heat dissipation temperatures, compared to evaporation cooling. Dry cooling is used only where the additional water required for wet cooling is not available. Indirect dry cooling involves investment costs that are about three times as high as a wet cooling system (Henning 1985).

Hybrid-Type Cooling

In hybrid-type cooling towers, both wet cooling and dry cooling are used. This method combines the advantage of the high cooling efficiency of wet cooling with the advantage of dry cooling, i.e. the absence of water vapours (Sauer 1984). In the variant usually used, the air flow is divided. One part of the flow is used for dry cooling, the other for wet cooling. By mixing the partial air flows, one obtains a wet vapour-free cooling tower exhaust – i.e. the exhaust is not visible. The water to be cooled is first conducted through the dry section and then through the wet section.

[1] The wet bulb temperature is the temperature measured by a moist thermometer or psychrometer. The thermometer is wrapped with a moist fabric. Water evaporates depending on the humidity and temperature of the air. The lower the air humidity and the higher the temperature, the higher the evaporation heat and hence the difference between dry and wet-bulb temperatures. The wet-bulb temperature is used in meteorology to determine the relative air humidity.

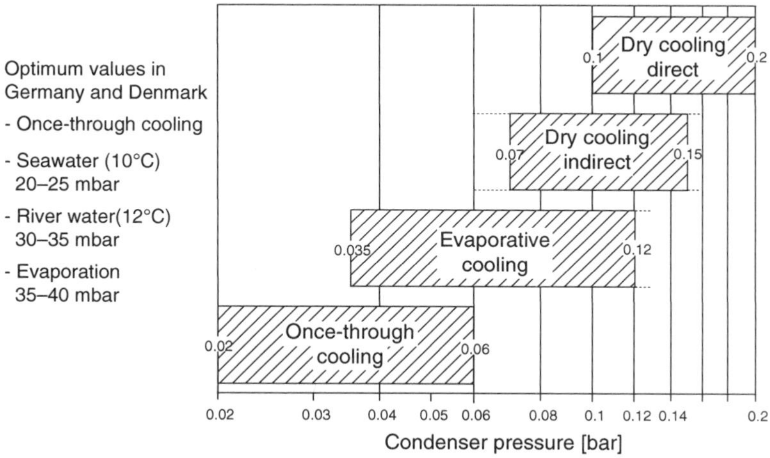

Fig. 4.62 Achievable condenser pressures in different cooling systems (Baehr 1985)

Wet cooling is usually exclusively used during summer operation, with the mixed use occurring in winter. The investment costs of this technique amount to three times as much the costs for a wet cooling tower, and its cooling characteristics resemble that of the wet-type cooling tower (Henning 1985).

Figure 4.62 shows the exhaust steam pressures achievable by the different cooling methods. It becomes evident that the chosen cooling technique has a substantial influence on the condensation temperatures and exhaust steam pressures. The ranges given in Fig. 4.62 are functions of the location-dependent air and water temperatures. Systems using once-through cooling thus offer favourable, systems with dry cooling unfavourable conditions for attaining a high thermal efficiency. Evaporative cooling, in general, involves higher condensation temperatures than once-through cooling, though clearly lower temperatures than dry cooling systems (Baehr 1985). In Denmark, condenser pressures between 20 and 25 mbar are achieved in advanced steam cycles with seawater cooling at an annual average of about 10°C. Reports on power plants with wet cooling towers mention condenser pressures between 35 and 40 mbar (Meier 2004; Lambertz and Gasteiger 2003; Tremmel et al. 2006; Mandel and Schettler 2007; Billotet and Johänntgen 1995; Eichholtz et al. 1994). The reference values for river water cooling in Germany range around 30 mbar.

The impacts of the condenser pressure on the net efficiency is shown in Fig. 4.63 for a power plant with conventional and with advanced steam conditions (Adrian et al. 1986; Kjaer 1993). Evaporative cooling, compared to seawater cooling, has a disadvantage in efficiency of about 1–1.5%, yet an advantage of greater than 1% compared to dry cooling.

The seasonal fluctuations of water and/or air temperatures have a direct effect on the exhaust steam quality in the condenser and hence on the thermal efficiency too. Figure 4.64 shows the yearly trend of cold water temperatures for the cases of seawater cooling and evaporation cooling (Johänntgen 1998).

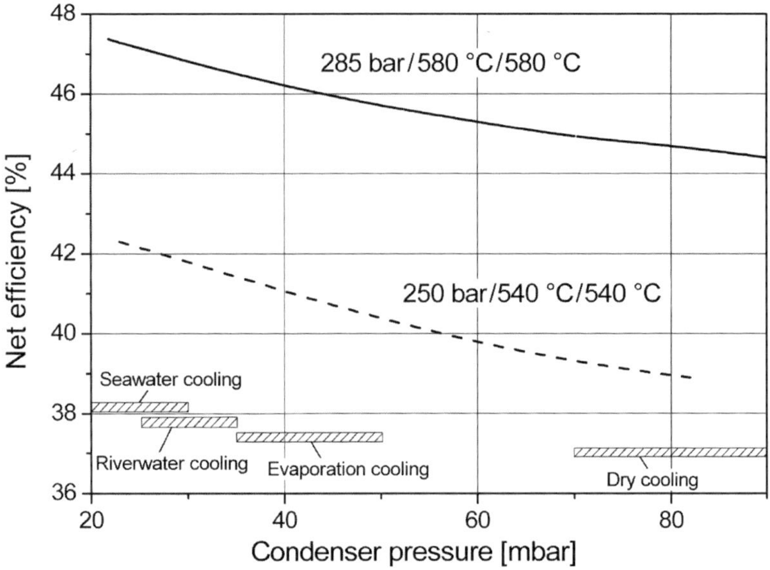

Fig. 4.63 Impact of the condenser pressure on the net efficiency (Adrian et al. 1986; Kjaer 1993)

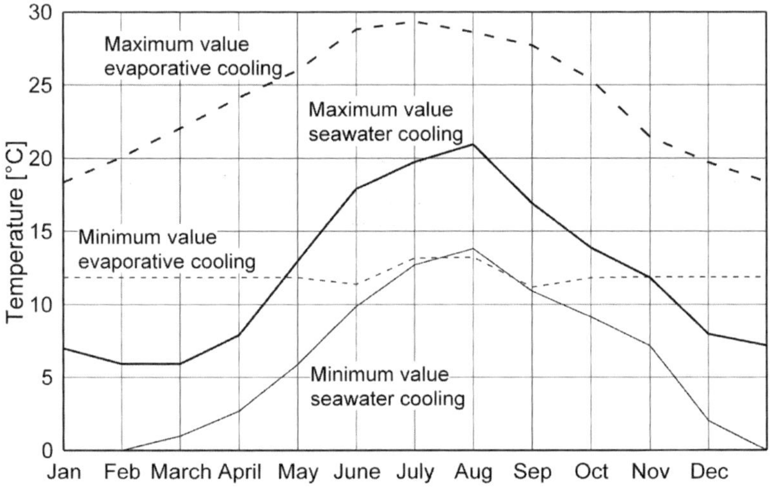

Fig. 4.64 Yearly trend of cold water temperatures (Johänntgen 1998)

Lower exhaust steam pressures in winter have less of an effect on the efficiency, however, than the rise of the condenser pressure in summer, because the outlet loss of the turbine increases with descending pressure. Exhaust steam qualities which are lower than those designed for can also be limited by the allowable exhaust moisture. Wet cooling towers might also confer restrictions on the cold water temperature, for example that they should not fall below 12°C, to prevent icing (Adrian et al. 1986). A temperature rise of 22°C to a level 30°C above the design temperature of 8°C of a

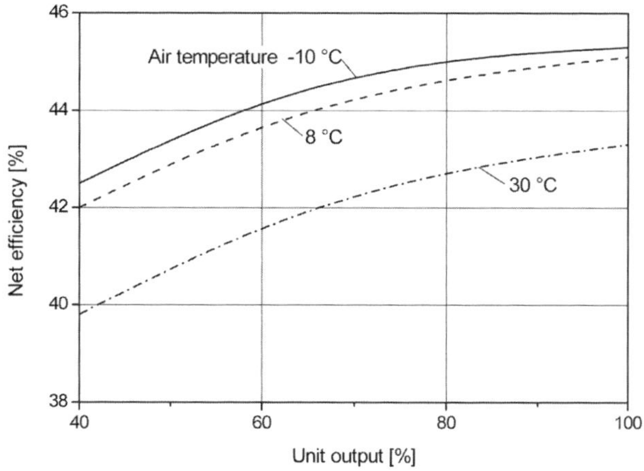

Fig. 4.65 Influence of ambient conditions on efficiency (Eichholtz et al. 1994)

natural-draught cooling tower deteriorates the efficiency of a modern power plant by 1.8%. Ambient temperatures of $-10°C$ yield an improvement of only about 0.2% (see Fig. 4.65) (Eichholtz et al. 1994). The impact depends on the turbine design (Weber et al. 2005). The conditions on site and the legislation concerning water rights and urban planning and building laws set criteria which narrow the choice of thermodynamically reasonable cooling techniques at the cold end of the power plant. In Germany, natural-draught cooling towers for waste heat dissipation and back-cooling of the heat carrier, i.e. cooling water, have become standard. With all cooling systems, the difference between the condensation temperature and the cooling medium temperature for the heat dissipation has to be kept as small as possible.

Back-Cooling by Natural-Draught Cooling Tower

Figure 4.66 presents the schematic diagram and the design data of a closed-circuit system with a natural-draught cooling tower for a 720 MW hard coal power plant (Baehr 1985).

In the condenser, the cooling water gets heated from an inlet temperature $t_{W1} = 20°C$ up to 34.5°C. For back-cooling, the water is transported to the cooling tower where, over about 10–15 m height, it is sprayed through nozzles that are located around the cooling tower cross-section.

The cooling water falls and disperses, via distribution plates, onto the fill packing, which it flows through, then dropping down into the cooling tower basin. In counterflow to the rising cooling air, the water cools to a temperature of $t_{W1} = 20°C$ both by convection and by evaporation, whereupon it is returned to the condenser. In the example shown in Fig. 4.66, 30% of the cooling efficiency is achieved by convective cooling with air and 70% by evaporation.

Ambient air at a temperature of $t_{A1} = 8.5°C$ and with a relative humidity of 76.3% flows into the cooling tower, where its temperature rises to $t_{A2} = 27.1°C$. By warming and buoyancy of the air, a convective flow forms in the cooling tower – this

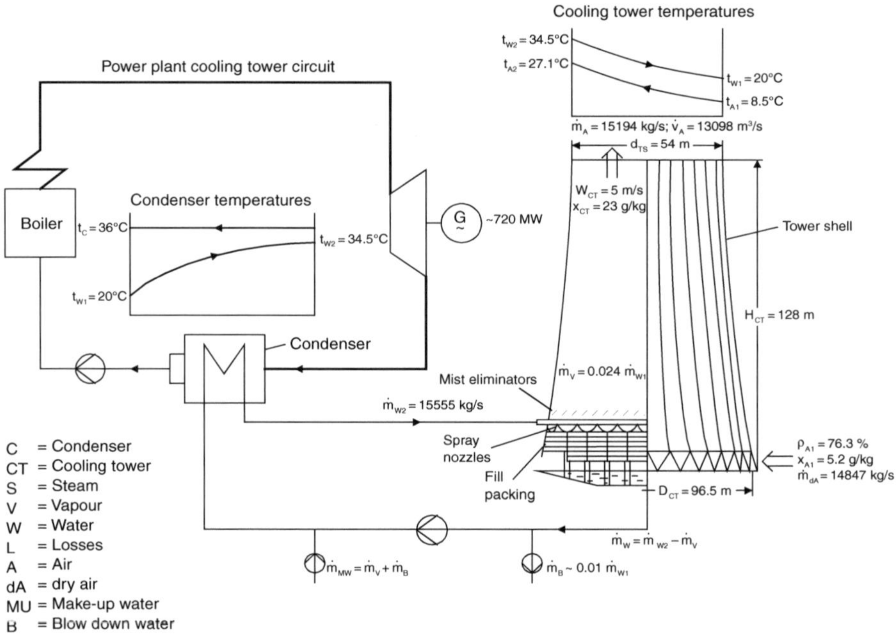

Fig. 4.66 Wet tower cooling circuit with design data for a 720 MW hard coal fuelled power station (Baehr 1985)

defines a natural-draught cooling tower. With increasing humidity, the flowrate slows. Since the cooling power depends on the air mass flow, the air flowrate can be forced much higher by ventilators (a ventilator cooling tower), the driving power demand of which increases the auxiliary power requirement of the plant.

The wet-bulb temperature of the ambient air t_{WB}, which in the example is 6.6°C, is the physical limit for the mean condensation temperature t_{C} and hence for the efficiency improvement at the cold end. For plants with wet cooling towers, the difference between the wet-bulb temperature of the ambient air and the economic temperature, or the mean condensation temperature (here 36°C) results from (see Fig. 4.67) (Odenthal and Spangenmacher 1959)

- The cooling range ($t_{\mathrm{W2}} - t_{\mathrm{W1}}$): This is the temperature rise of the cooling water in the condenser from t_{W1} to t_{W2}, which is determined by the cooling water mass flow with a given heat dissipation. In the cooling tower, the cooling water is cooled back to its temperature t_{W2} before entry to the condenser. In the given example, the difference is $t_{\mathrm{W2}} - t_{\mathrm{W1}} = 14.5$°C.
- The terminal temperature difference (TTD) of the condenser: In the example, the difference is $t_{\mathrm{C}} - t_{\mathrm{W2}} = 1.5$°C.
- The approach $t_{\mathrm{W1}} - t_{\mathrm{WB}}$ of the cooling tower: This is the temperature difference between the temperature of the backcooled water and the theoretically possible cold water temperature, which equals the wet-bulb temperature. In the example, the difference is $t_{\mathrm{W1}} - t_{\mathrm{WB}} = 13.4$°C.

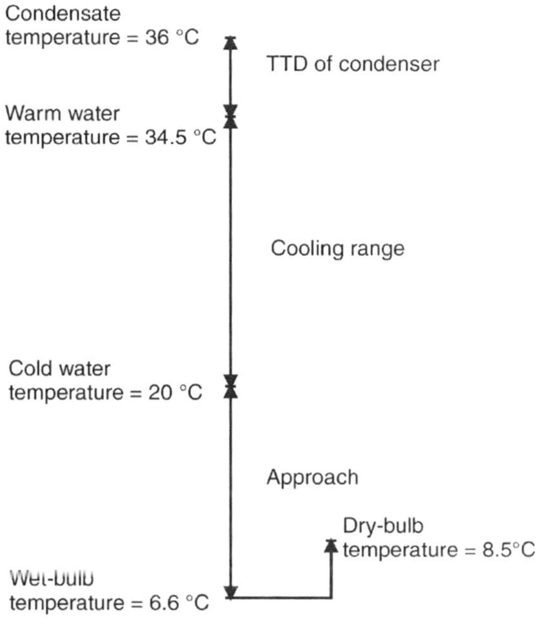

Fig. 4.67 Temperature relations in circuit cooling systems by wet cooling tower

Diminishing the cooling range, the approach or the TTD of the condenser by 1 K results, for each of these parameters, in an equal lowering rate of the condensation temperature (STEAG 1988).

A smaller cooling range is achieved by a greater cooling-water mass flow. The lower water outlet temperature after the condenser then decreases the condensation temperatures correspondingly, with the same TTD maintained.

A greater cooling-water mass flow requires, for the heat and mass transfer, a greater surface for the cooling water to flow down, which is achieved by appropriate inserts, increasing the surface area. In the case of natural-draught cooling towers, enlarging the transfer surfaces as a rule involves the enlargement in height and diameter of the body as well. In the case of ventilator cooling towers, the power demand of the ventilators increases. In the condenser, narrowing the cooling range causes a reduction of the mean logarithmic temperature difference so that, at the same TTD, larger condenser surfaces are needed. The cooling range values common in Germany are between 16 and 10 K, the latter holding true for plants currently in planning.

Small approaches in the cooling tower can be achieved with larger transfer surfaces. In the extreme case of the ideal cooling tower – which only exists theoretically – the water is cooled down to the wet-bulb temperature, and the approach is then $t_{W1} - t_{WB} = 0$. Such an ideal cooling tower has to function by counterflow and has an infinitely large transfer surface (Klenke 1966). Commonly, approaches are between 8 and 12 K.

Smaller terminal temperature differences (TTDs) in the condenser are achieved with larger condenser surfaces. Commonly, condenser TTDs are about 1–2 K.

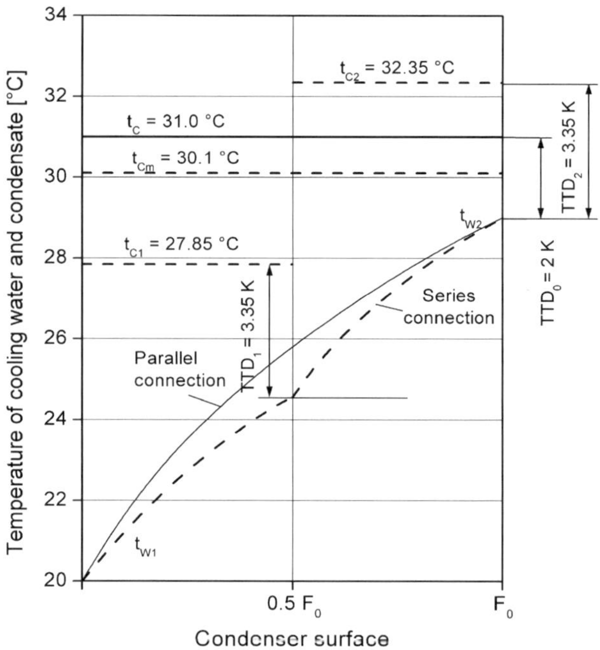

Fig. 4.68 Thermodynamic comparison between parallel- and series-connected partial condensers, both with the same condenser surface (STEAG 1988)

The above-described measures to reduce the condenser temperature necessitate additional investments in the LP turbine, the condenser and the cooling tower.

The reduction of the mean condenser temperature, and hence of the heat rate, while maintaining the same condenser surface area, is possible by implementing water-side series connections of partial condensers. Figure 4.68 demonstrates the advantage of such a configuration in comparison to the often-found parallel connections of partial condensers (STEAG 1988).

Water losses arising through evaporation and blowdown have to be balanced out by additional water. Water loss through evaporation depends on the humidity in the air. Blowdown is necessary in order to prevent minerals contained in the cooling water from accumulating. The entire additional demand for water to account for these losses lies in the order of magnitude of 2–3% of the cooling water mass flow (Baehr 1985).

Pollution of the cooling air and residual contamination of the pre-treated cooling tower make-up water lead to foul deposits in the cooling circuit, which eventually settle at the bottom of the cooling tower basin. This cooling tower slurry is collected over long operating periods and cleaned up during an outage, after drainage of the basin. Until the slurry settles, it is carried along in the cooling cycle. In consequence, deposits form on the inside of the condenser tubes, which deteriorate to a considerable extent the heat transfer. An effective remedial action is constant condenser cleaning by a service system, which carries a number of calibrated cleaning bodies that pass through the tubes, such as sponge rubber balls.

The economic efficiency of the condenser temperature depends on the degree of the utilisation of the power plant and the fuel price on one hand and on the necessary additional investment for an efficiency increase on the other (see also Sect. 4.3.1). The condenser pressure for a mid-range plant, with an operation factor of 4,000 full-load h/annum and a unit size of 500 MW_{el}, in the 1980s ranged between 0.04 and 0.05 bar (STEAG 1988). Recent power station plans provide for values between 45 and 30 mbar (Meier 2004; Lambertz and Gasteiger 2003; Tremmel et al. 2006; Mandel and Schettler 2007; Billotet and Johänntgen 1995; Eichholtz et al. 1994).

4.4.2 Reduction of Losses

4.4.2.1 Internal Turbine Efficiency and Losses

In the expansion process in the turbine, the steam is accelerated, and its kinetic energy converted into mechanical work by impulse transfer onto the rotating blades. The measure for the quality of the conversion into mechanical work is the internal turbine efficiency $\eta_{i,T}$, which indicates the difference between real and loss-free isentropic expansions. For the real thermal cycle efficiency η_{th}, the following applies:

$$\eta_{th} = \eta_{th,0} \cdot \eta_{i,T} \tag{3.31}$$

where $\eta_{th,0}$ represents the thermal cycle efficiency at loss-free expansion.

About two thirds of the total losses occur over the blade stages. The HP first stages (of the turbine), in particular, and the LP last stages (of the turbine) are the areas of the turbine incurring the highest losses. The losses arise through fluid friction in the channels, friction of the rotating blades in the surrounding steam, steam leakages from rotating and fixed parts and through steam moisture in the last stages (Strauß 2006).

The greatest single loss, in the order of magnitude of about one tenth to one third of the total loss, is the outlet loss. It comes about because of the kinetic energy of the exhaust steam. Further losses occur in the inlet valves and in the cross-over pipes (Adrian et al. 1986).

An exhaust steam diffuser partly recovers kinetic energy from the exhaust steam exiting at high velocity from the last blading. The kinetic energy is converted into pressure energy in the diffuser, which is located between the last turbine blades and before the condenser, and partially compensates for the pressure losses arising on the way to the condenser. With a constant condenser pressure, an exhaust steam diffuser brings about a lower pressure after the last turbine stage and in consequence more power is produced in the turbine than would be the case without the diffuser (Schröder 1968).

Besides by an exhaust steam diffuser, it is possible to influence the outlet loss by the exhaust steam velocity, which, at a given steam mass flow and a given condenser

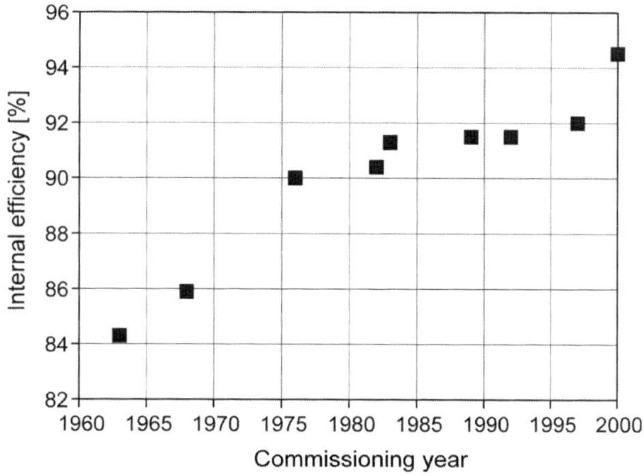

Fig. 4.69 Development of the internal efficiencies of steam turbines (Billotet and Johänntgen 1995)

pressure, can only be varied by means of the cross-sectional area of the turbine outlet. Because of the limited blade length of the last LP stage, the outlet surface can only be enlarged by the number of the LP turbines. The last stages and the exhaust steam cross-section are designed in combination with the heat extraction (see Sect. 4.4.1.3).

Turbine improvements have contributed substantially to the increases in efficiency of modern power plants. Three-dimensional calculations (i.e. computer modelling) of flow processes reveal the potential for reducing the flow losses, and modern manufacturing technologies make it possible to build complex blade geometries (Nowi and Haller 1997; Oeynhausen et al. 1996). The modernisation of the turbine of existing power plants is an effective means to increase the efficiency. Figure 4.69 shows the internal turbine efficiencies for existing and planned power plants (Billotet and Johänntgen 1995).

4.4.2.2 Steam Generator Losses

In the steam generator, or boiler, the chemically bound energy of the fuel is converted into thermal energy of the flue gas and then transferred to the steam – water cycle. The efficiency of the energy conversion is designated as the steam generator or boiler efficiency η_B, and the arising losses are called the boiler losses. Referring to the calorific value of the fuel, the steam generator efficiency of modern hard coal fired furnaces amounts to 94%, while brown coal fuelled furnaces have an efficiency of around 90%.

The losses consist of the following:

- Loss through unburned matter (κ_U)
- Loss through sensible heat of the slag (κ_S)
- Flue gas loss (κ_{FG})
- Loss through radiation and convection of the external surfaces of the boiler (κ_{RC})

Losses through unburned matter are well below 1%. Of these losses, a differentiation is made between unburned gas, unburned matter in the slag and unburned matter in the fly ash. Where a carbon content of less than 5% is required for the use of the fly ash as a by-product, the respective maximum loss of unburned matter in the fly ash will be, for instance, 0.5% for a coal type with an ash content of 10%. Typical losses through unburned matter range around 0.3% (Riedle et al. 1990).

Heat losses due to radiative, conductive and convective transfer to the environment by the steam generator are below 1% and diminish further as the power rating of the steam generator increases. The losses of brown coal fuelled furnaces are significantly higher than hard coal fuelled furnaces because, at the same output, the steam generator has a considerably larger external surface area. Hard coal fuelled furnaces typically have heat losses around 0.3% (Billotet and Johänntgen 1995).

Ash is predominantly removed in the electrostatic precipitator (ESP) as fly ash, though part of it stays in the furnace as slag and is typically removed while in a hot state. The sensible heat of the slag, when unused, results in a portion of the boiler losses. In dry-bottom furnaces, the amount of the so-called hopper ash is about 10% of the total ash mass flow, and the respective loss is below 0.4% of the calorific energy input in hard coal fuelled furnaces. In slag-tap furnaces, the loss by sensible heat is higher, because either all or a large portion of the ash (depending on the degree of retention and the fraction of the re-injected ash) runs off as liquid slag with a high temperature.

The principal loss of the steam generator occurs because the flue gas cannot be cooled down to ambient temperature. After the exhaust steam heat loss, this is the most major loss in a power plant.

Efforts to increase the steam generator efficiency concentrate on reducing the flue gas heat loss. This loss depends both on the flue gas outlet temperature of the steam generator (after the air heater) and on the flue gas mass flow. Figure 4.70 shows the

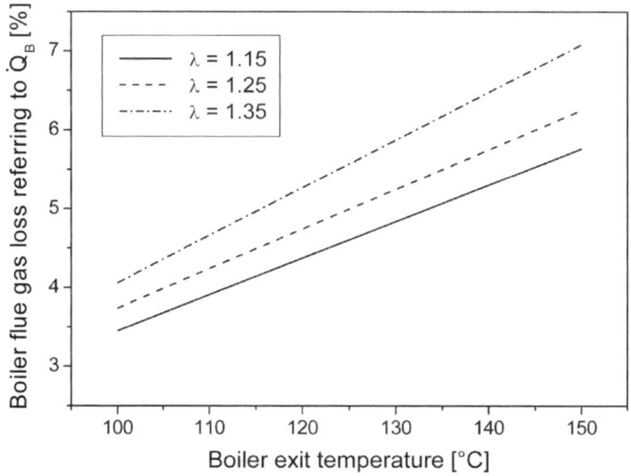

Fig. 4.70 Boiler loss as a function of the boiler exit temperature and air ratio, for hard coal firing (Riedle et al. 1990)

Table 4.2 Boiler losses for the reference power plant and for a new plant

	Reference power plant	New plant
Air ratio	1.3	1.15
Flue gas temperature	130°C	110°C
Flue gas losses	5.3%	3.8%
Boiler radiation	0.25%	0.3%
Loss through unburned matter		
Fly ash	0.2%	< 0.3%
Hopper ash	0.1%	< 0.2%
Sensible heat		
Fly ash	0.02%	0.03%
Hopper ash	0.04%	0.04%
Total boiler loss	**5.9%**	**4.6%**

flue gas loss for hard coal firing as a function of the excess air coefficient and the boiler exit temperature (Riedle et al. 1990). Table 4.2 compares the boiler losses of the reference power plant and a planned power plant.

The flue gas outlet temperature of the boiler has a limit that depends on the condensation of sulphuric acid. Both in the air heater and in the downstream ESP, the temperatures should not fall below this minimal flue gas temperature, which lies above the dew point of sulphuric acid, in order to prevent corrosion and fouling (Müller-Odenwald et al. 1995).

The dew point of sulphuric acid changes with the flue gas contents of sulphur trioxide (SO_3) and water vapour (see Fig. 4.71) (Bauer and Lankes 1997). Part of the sulphur dioxide formed in the combustion is converted to SO_3 in the flue gas path.

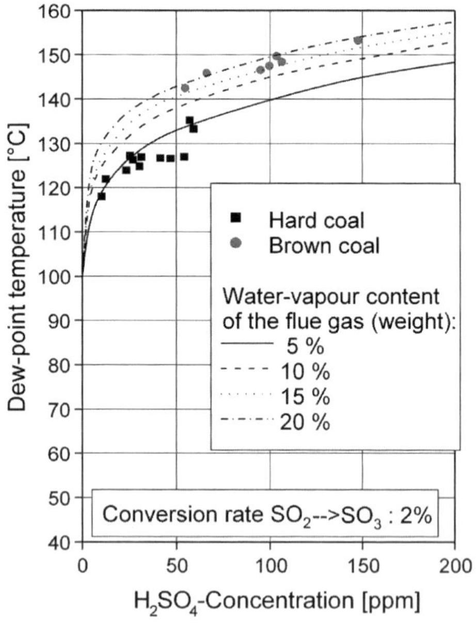

Fig. 4.71 SO_3 dew point of flue gases (Bauer and Lankes 1997)

The SO_3 content correlates to the SO_2 partial pressure in the flue gas and also to the sulphur content of the fuel. The conversion rate and the SO_3 content depend, furthermore, on the excess air and the combustion temperature. A high level of excess air supports the conversion of SO_2 to SO_3, while a high combustion temperature has the opposite effect (Baehr 1985). Typical conversion rates are around 2%. A catalyst for NO_x reduction inserted between the economiser and the air heater may act to form additional SO_3 (Maier et al. 1992). Flue gas temperatures after the air heater, depending on the sulphur content of the hard coal, today commonly range between 120 and 140°C. Because of the resulting higher dew point of sulphuric acid, higher water vapour contents in brown coal firing limit the flue gas temperature between 140 and 170°C, depending on the sulphur content.

For the above-mentioned flue gas temperatures, local temperatures at the cold end of the air heater are below the acid dew point temperature. The lowest plate temperature at the cold end is roughly the arithmetic mean of the flue gas outlet temperature and the air inlet temperature. For the reference power plant, with a 130°C flue gas outlet temperature and a 45°C air inlet temperature, the lowest plate temperature, of slightly less than 90°C, lies well below the acid dew point. In consequence, SO_3 condenses to sulphuric acid in the flue gas, which is taken up by the dust particles. Measurements in plants in service show that this way, 80–90% of the SO_3 can be removed from the flue gas, so the acid dew point for the subsequent equipment decreases. SO_3 condenses into sulphuric acid until it reaches a degree where the acid dew point correlates with the local plate temperature. The temperature at the boiler exit thus determines the SO_3 emissions after the air heater. The decrease of the heating surface temperature and the accompanying H_2SO_4 condensation in the air heater is limited to the point when a condensate film forms on the plates. This causes fouling which cannot be cleaned by soot-blowing. In addition, the corrosive attack becomes more severe with increasing acid condensation.

Minimum achievable flue gas outlet temperatures, achievable through the use of a regenerative air heater, strongly depend on the coal. In some power plants, flue gases are cooled down to 110–115°C, while in other plants, even temperatures of 125°C cause fouling at the cold end. An important part in this process is played by the contents of CaO and MgO in the ash, which have an effect against fouling. A fouling temperature as a function of the ash composition can be defined, below which the local temperature should not fall (see Fig. 4.72).

In consequence, a steam generator designed for low flue gas temperatures restricts the range of coal types.

The flue gas mass flow, as a further parameter influencing the flue gas loss, is augmented by the excess air and also by leaks in the air and flue gas train of the steam generator. This increases the flue gas mass flow to an amount greater than the theoretical air demanded by the stoichiometry of the combustion. For balance-draught furnaces, which are commonly in operation today, design calculations take into account an air leak mass flow between the burners and the economiser of about 1.5% of the total combustion air flow (Adrian et al. 1986). Additional leaks occur in the regenerative air heater, which raise the flue gas mass flow by about 5% of the air flow (see Sect. 4.3.5.6).

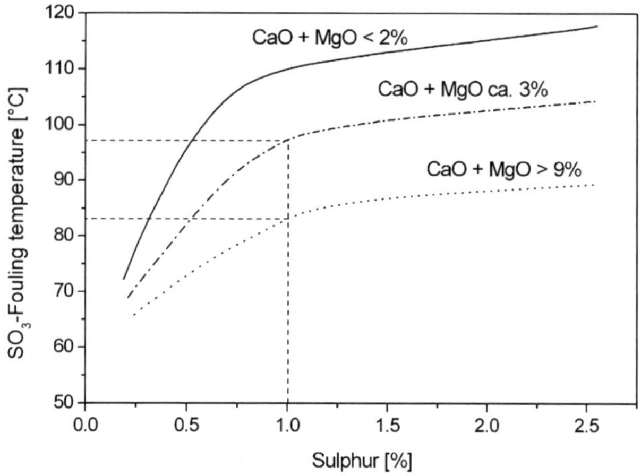

Fig. 4.72 SO_3 fouling temperature as a function of sulphur content and CaO + MgO content (Müller-Odenwald et al. 1995)

As an alternative to the less expensive regenerative air heaters, it may also be possible to use recuperators to preheat the air. The impermeability (i.e. the smaller leakage losses) of these could reduce the flue gas loss and the auxiliary demand of the forced-draught fan.

Furnaces are operated at excess air to achieve the required complete combustion. Lower air ratios do reduce the flue gas loss; however, an excessively low excess air rate causes losses through unburned gases and unburned carbon in the fly ash. So there is an optimum with a minimum boiler loss at a definite air ratio. Lower excess air rates are possible if complete combustion can be guaranteed by other measures, such as increasing the milling fineness (Baehr 1985).

Slag-tap firing systems, with their higher combustion temperatures, allow the setting of lower excess air levels than dry-bottom furnaces. In the past, this level was between 1.2 and 1.35 in dry-bottom furnaces and between 1.1 and 1.2 in slag-tap furnaces. In new and planned power plants, the air ratios for brown coal fuelled furnaces are 1.15, while for hard coal fired dry-bottom furnaces, 1.15 (volatile matter content greater than 25%) or 1.18 (volatile matter content smaller than 25%), and for slag-tap furnaces, the value is 1.1 (Riemenschneider 1995). Besides its effect on flue gas loss, the excess air raises the auxiliary demand of forced-draught (FD) and induced-draught (ID) fans.

Flue gas losses, together with the relationships described above, depend on the firing system. Complete combustion at high temperatures in slag-tap furnaces allows the setting of low air ratios such that the conversion into SO_3 is reduced and hence lower flue gas temperatures can be set. This advantage in the efficiency of slag-tap furnaces, however, is depleted through the loss of the sensible heat of the slag for ash contents of more than 20%.

Compared to a hard coal with a moisture of 10% or so, the flue gas mass flow of a brown coal fuelled furnace, with 50–60% moisture, is larger by more than 20 to over 30%. This fact, in addition to the higher allowable flue gas temperature, explains the high flue gas losses in pulverised brown coal firing, which can be twice as high as those in hard coal firing. The excess air is occasionally made use of to set a mass flow necessary for the heat transfer of the convective heating surfaces. In part-load operation, often a higher excess air is chosen, in order to shift the heat flux distribution in favour of the convective heating surfaces.

4.4.2.3 Loss Through Reheating

National laws impose legal restraints on temperatures at which flue gas may be discharged into the environment. Examples include the USA, Japan and several European countries, including Germany.

Wet scrubbing processes for flue gas desulphurisation cool the flue gases down below the exhaust gas temperature of the steam generator. After flowing through the steam generator, the air preheater and the ESP, the flue gas is conducted to the scrubber. Cleaned of remaining dust and desulphurised, the flue gas – now termed clean gas – leaves the FGD unit via a droplet separator. The flue gases during wet scrubbing cool from the flue gas temperature after the air preheater down to the operating temperature of the FGD, and at this temperature they are saturated with water.

Up to 2004, according to the German Ordinance on Large Combustion Plants (*13.BImSchV* in German), the clean gas had to be dissipated into the open air via a stack at a temperature higher than 72°C. The intention of this high temperature was to raise the plume of the flue gas after it had left the stack to ensure its spread and wide distribution. If the operating temperature of the FGD was lower than the required 72°C, the flue gas had to be reheated (Adrian et al. 1986; 13.BImSchV 2004).

In existing power stations in Germany, in many cases the operating temperature of the FGD is lower than 72°C, and so a regenerative heat exchanger (RHX) is used for reheating. In such heat exchangers, the flue gas heat of the un-desulphurised and cooled flue gases (raw gas) is transferred to the cleaned flue gases (clean gas) (see Fig. 4.73). In this process, the raw gas is cooled from about 110 to 130°C down to 70–75°C, while the clean gas in turn is heated from 45 to 50°C up to about 80–90°C.

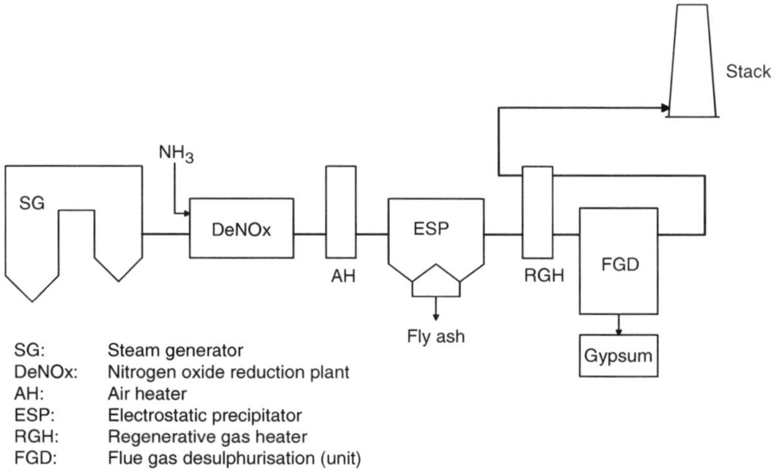

Fig. 4.73 Configuration of the catalyst for high-dust and reheating after FGD

To vaporise the remaining droplets after the FGD unit, it is possible to raise the flue gas temperature by about 5°C by means of an ash-free fuel. An alternative is to return part of the flue gas flow from the back to the front of the rotary gas heater, in order to ensure the prevention of possible build-ups of deposits on the heating surfaces of the rotary gas heater. Adding energy also adds loss: even when reheating does not require additional heat, there is still a higher need for electricity for in-plant use due to the pressure losses of the heat exchangers.

If the flue gas was fed into the cooling tower, the requirement of a flue gas temperature of 72°C was not applicable. The stack in this case was not needed at all or could be reduced to a stack for starting up. This method was applied for the first time in the Völklingen model power plant (Ernst et al. 1986). Today, it is used in all newly built plants with evaporative cooling. The spread of the great volumetric flow of the cooling tower ensures a comparatively better distribution than when the flue gas is discharged via a stack. Feeding the flue gas into the cooling tower offered an advantage in efficiency of 0.1–0.15%, taking into account the lower in-plant electricity demand in comparison to reheating (Billotet and Johänntgen 1995). Where additional energy was needed for reheating, the efficiency increase of feeding the flue gas into the cooling tower was greater than compared to discharge via a stack.

In Germany, the Ordinance on Large Combustion Plants was modified in 2004 to incorporate European law. The requirement of a flue gas chimney inlet temperature of 72°C is no longer included (13.BImSchV 2004). Instead, it has to be proven that the additional contribution of the plant to the local pollutant load will be below certain limits. New power plants currently under construction with once-through cooling estimate a flue gas temperature at the chimney inlet of about 50°C, which is basically the temperature of the FGD. Because of the condensation of saturated flue gas droplets, the chimney has to be designed for wet operation.

The same problems as in the past for reheating after desulphurisation may also occur in NO_x control. Since the $DeNO_x$ process needs a catalyst temperature between 300 and 350°C, the $DeNO_x$ reactor is usually arranged in a high-dust configuration, after the economiser and before the air heater (see Fig. 4.73). In exceptional cases, where a high-dust configuration is not possible, the $DeNO_x$ reactor is arranged in a low-dust configuration, with the unit placed after the dust control and desulphurisation units (see Fig. 4.74). This set-up may be chosen when there are space limitations or when the flue gas composition is such that a short lifetime of the catalysts is expected. In slag-tap firing, with ash recirculation from the ESP to slag-tap furnace, the accumulation of arsenic compounds would, in the case of a high-dust configuration, eventually poison the catalysts associated with the corresponding reactivity loss, so the low-dust configuration is often applied in this firing system.

As previously described, the flue gas temperatures after the flue gas desulphurisation unit are at around 50°C, so it is necessary to reheat the flue gas up to the temperature of about 320°C required for the catalytic flue gas $DeNO_x$ process. The reheating process then usually combines a regenerative preheater and direct reheating by means of natural gas or indirectly by hot steam. A small direct reheating

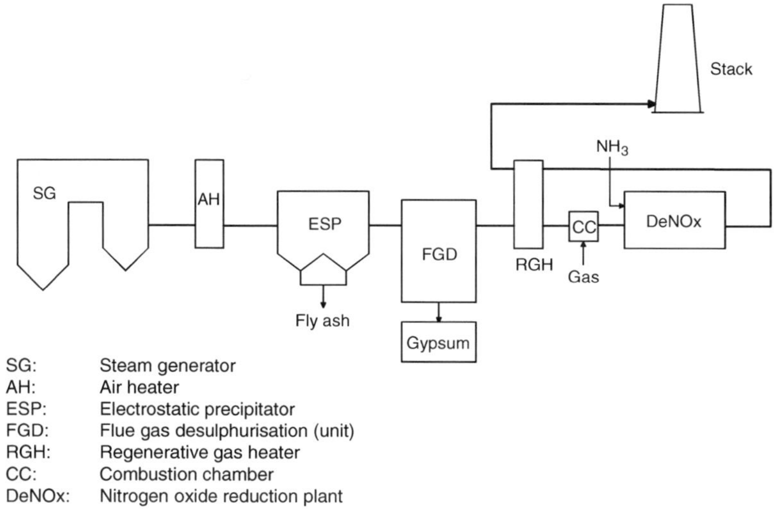

Fig. 4.74 Configuration of the catalyst for low dust

step of 30°C or so by means of primary energy is necessary to compensate for the heat losses from the regenerative heat exchanger (Maier et al. 1992). This need for primary energy lowers the total efficiency by about 1%.

4.4.2.4 Advanced Flue Gas Heat Utilisation

The flue gas temperature after the air heater is chosen in the design such that it does not drop below the acid dew point at any stage before the desulphurisation unit. As it travels from the air heater to the FGD unit, the flue gas temperature drops slightly because of heat losses. The flue gases are cooled down by water spraying from this temperature to the FGD operating temperature, which is 50–80°C in hard coal firing and 90–120°C in brown coal firing (Bruckmann and Hesel 1996).

The designs of advanced power plants make use of this low-temperature heat. The usable temperature gradient is the difference between the flue gas temperature at the boiler exit and the temperature level which the flue gas is cooled down to in the desulphurisation process.

This development utilises an additional heat exchanger to tap low-temperature heat, which is fed to the thermodynamic cycle of the power plant. Due to its low temperature level, the waste heat is transferred either to a low-temperature condensate or to combustion air that has not yet been preheated. By preheating a partial condensate flow in the LP area, an increase of 0.25% in efficiency is obtained for hard coal firing – which is countered by considerable expenditure in equipment and therefore not considered economical (Eichholtz et al. 1994). For brown coal firing, an increase in efficiency of up to 0.7% is reported (Bauer and Lankes 1997).

Figure 4.75 shows a schematic diagram of the configuration for transferring low-temperature flue gas heat to the high-pressure feed water for pulverised hard coal

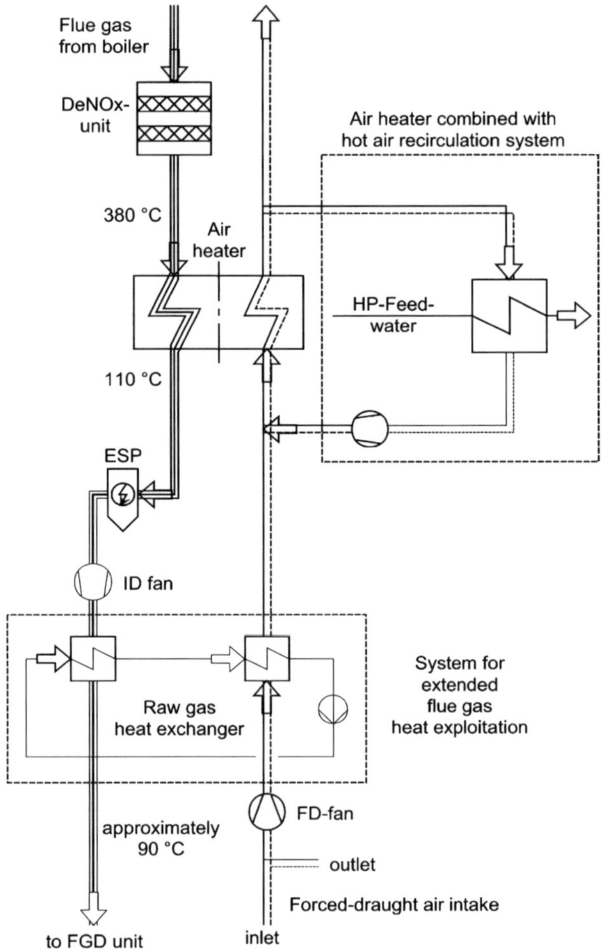

Fig. 4.75 Configuration for extended flue gas heat utilisation (Billotet and Johänntgen 1995)

firing. The flue gas heat is first transferred to the cold combustion air flow in a heat exchanger and then to the HP feed water via a hot-air recirculation system (Billotet and Johänntgen 1995). A similar configuration is used for the power plant Niederaußem K (Tippkötter and Scheffknecht 2004; Lambertz and Gasteiger 2003). Due to the transfer of low-temperature heat of the flue gas to the air, less flue gas heat is required in the regenerative heat exchanger for final air preheating. Therefore, about one third of the flue gas bypasses the regenerative air preheater, and the heat from this portion of the flue gas is transferred by flue gas/water heat exchangers to the low- and high-pressure feed water preheaters.

Advanced flue gas heat utilisation can increase the net efficiency of a hard coal fired power station by 0.6% (Billotet and Johänntgen 1995). In brown coal firing, given the higher flue gas temperature of about 170°C, and the higher flue gas mass flow, increases of up to 1.6% can be achieved (Bauer and Lankes 1997; Bruckmann and Hesel 1996).

In order to prevent the flue gas temperature falling below the dew point when travelling between the air heater outlet and the FGD unit, the low-temperature heat exchanger should be mounted immediately before the FGD unit. The heat exchanger, located where temperatures range close to the sulphuric acid dew point, has to be made of corrosion-resistant material and be equipped with cleaning mechanisms. Such systems have already been developed for flue gas reheating after desulphurisation and so are available today. Shell-and-tube heat exchangers with tubes made of synthetic materials such as perfluoroalkoxy (PFA), polytetrafluoroethylene (PTFE) or polyvinylidene fluoride (PVDF) are suitable to use. The heat exchanger casing itself is made of steel coated with a synthetic material (Suhr 1992).

4.4.2.5 Other Types of Losses

Pipework Losses

The hot HP steam produced in the steam generator is fed to the HP section, and the hot reheater steam is fed to the intermediate-pressure section (IPS) of the turbine – both via pipes with lengths of 100–150 m. Even for the highest capacities, it is common to use two live steam pipes for each run. The energetic heat losses of these pipes, which connect the steam generator to turbine, have an efficiency η_{pipe}, which is around 0.997 (Köhn 1993).

Besides heat losses, pressure losses also cause a temperature decrease and thus lead to exergy losses of the steam between the steam generator and the turbine inlet. The temperature decrease caused by throttling does not have an effect on the pipe loss η_{pipe}, but on the thermal cycle efficiency. The temperature of the live steam decreases between 5 and 2°C in total; the pressure loss ranges between 10 and 5 bar. The latter values hold for new plants and for power plants in the planning stage.

Generator Loss

In the generator, the mechanical energy is converted into electrical energy. The efficiencies of large generators, in the range of 700–900 MVA, with a water-cooled stator and a hydrogen-cooled rotor, are around 98.7%. For this configuration, there are only small possible efficiency increases. An efficiency increase to about 99.4% is expected to be possible with a superconducting rotor winding. In addition to the efficiency increase, a smaller space requirement is expected as well (Riedle et al. 1990).

Losses of the Main Transformer

Main transformers transform the electrical voltage of the power output supplied to the electrical network – from 21 to 220 or 380 kV or higher, in order to keep the network losses low. They are usually designed as three-phase transformers. The transformer efficiency, as the ratio of the supplied energy to the energy absorbed, lies, with large main transformers, at rated power, in the range of 99.6–99.7% (Adrian et al. 1986).

4.4.3 Reduction of the Auxiliary Power Requirements

The auxiliary power requirement of a power plant is the sum total of the electrical and the mechanical power demand for driving ancillary and auxiliary systems. The electrical auxiliary power requirement $P_{\mathrm{aux,el}}$ is comprised of the input power of all machines driven by electric motors, such as fans (FD and ID fans and primary air fan), pumps (the feed water pump if it has an electrical drive, and the condensate, cooling-water and scrubbing slurry pumps), mills, and other power for electrical devices and equipment such as the transducer and the electrostatic precipitator (ESP) for dust removal. Mechanical power is employed if the feed water pump is driven by steam turbine instead of electrically:

$$P_{\mathrm{aux}} = P_{\mathrm{aux,el}} + P_{\mathrm{aux,m}} \tag{4.20}$$

The auxiliary power demand efficiency can be calculated using the auxiliary power demand and the power of the turbine (including the power of a turbine-driven feed water pump):

$$\eta_{\mathrm{aux}} = \frac{P_{\mathrm{aux}}}{P^{*}_{\mathrm{Gen}}} = \frac{P_{\mathrm{aux}}}{P_{\mathrm{Gen}} + P_{\mathrm{aux,m}}} \tag{4.21}$$

About 6–10% of the gross electrical power of a power plant is needed for the auxiliary power of the plant, meaning an auxiliary power demand efficiency of 90–94%.

The electrical auxiliary power demand is made of "useful" power plus losses. The useful or net power is roughly between 90 and 99%, while the losses correspond to between 10 and 1% of the power demand. Process-engineering improvements reduce both the useful power demand and the losses; improvements of the auxiliaries only diminish the losses. In the example of air staging for the reduction of NO_x formation, the useful power demand and losses of auxiliaries are cut down, thus the auxiliary power requirements diminished. The aim of optimisation measures for drives and machines is to reduce the losses.

The auxiliary power demand of different main and auxiliary devices are described in Table 4.3. For the reference power plant and for a new plant, the auxiliary power values reported are given in the table.

The feed water pump has the highest power demand. With the live steam pressures common today for large hard coal power plants, the power of the boiler feed pump is between 2.5 and 3.5% of the rated generator capacity. The power demand depends on the pressure rise and the feed water mass flow. Approximately, by

Table 4.3 Auxiliary power requirement breakdown for the reference and a new power plant

Live steam pressure	Reference power plant 190 bar	New plant 250 bar
FD fan	0.8%	0.5%
ID fan	1.2%	0.8%
Mills and mill fan	0.4%	0.4%
Coal handling, ash removal, dust removal	0.3%	0.3%
Boiler feed pump	2.7% (260 bar)	2.9% (310 bar)
Condensate pump	0.3%	0.2%
Cooling water pump	0.7%	0.6%
FGD unit	2.0%	0.8%
Other equipment	1.0%	0.7%
Total	9.4%	7.2%

neglecting the impact of the efficiency on the feed water mass flow, the result is:

$$\frac{p_{\text{feed}}}{P^{*}_{\text{Gen}}}\,[\%] = \frac{p_{\text{feedpump}}}{100} \tag{4.22}$$

The pressure after the feed water pump is higher than the live steam pressure by an amount equal to the sum of the pressure losses of the HP feed-heating train, the steam generator and the main steam pipe. Table 4.4 compares the pressure losses of the reference power plant to those of an advanced thermal power plant, both power plants being once-through steam generator types. It can be observed that pressure losses are lower for the advanced plant in all areas. An exception is the superheater, the pressure losses of which increase due to the higher live steam temperatures. The pressure losses in the HP area have a large impact on the power demand of the feed water pump, but have no impact on the steam generator efficiency and the thermal efficiency of the cycle, if one assumes the same live steam conditions before the turbine. Pressure losses in the reheater area only have an impact on the thermal efficiency, via the temperature decrease associated with throttling. Reducing the power of the boiler feed pump is a means to contribute to the efficiency increase.

Table 4.4 Pressure losses of the reference power plant and of an advanced thermal power plant

750 MW PP unit	Reference PP (bar)	Advanced thermal PP
Pressure after feed pump	265	320
Pressure before turbine	190	270
Δp HP heater	8	6
Δp economiser	3	2
Δp evaporator	25	10
Δp superheater	20	25
Δp live steam pipe	19	8
Δp Total	75	50

The feed pump input power also depends on the design of the steam generator. In once-through systems, the water-side losses in the steam generator are higher than in natural-circulation systems. The feed pump for the feed water flow control is a variable-speed type. The feed water capacity is disproportionately reduced because of these losses, in particular during part-load operation with sliding-pressure control.

Feed pump configurations for hard coal power plants today are usually either

- a variable-speed turbine-powered 100% duty pump and an electrically driven 50% duty pump or
- two 50% duty pumps with variable-speed electrical drives.

Driving the main duty pump by means of a condensation branch turbine – with its own condenser fed from one of the lower tapping points – is the common technique in large power plant units. The order of magnitude of its capacity is not limited. The turbine can be coupled directly to the pump and has little loss if the operating regime does not deviate too much from the design conditions.

The following are typical options for the electrical drive of a variable-speed feed pump:

- A three-phase current induction motor (cage induction motor with unregulated constant motor speed). The number of revolutions is adjusted by a loss-inducing variable-speed hydraulic coupling.
- A synchronous motor supplied by an electronic power converter. The number of revolutions is regulated through adjustment of the supplied power.
- A wound-rotor induction motor (variable-speed three-phase current induction motor). The slip to synchronous speed is adjusted by a modifiable resistor in the rotor circle. The slip energy can be recovered and fed back into the network.

Turbine drives are adequate for configurations with a single feed pump. They have a lower auxiliary power requirement in the upper load section of base-load power plants compared to electrical feed pump drives.

Power plants in mid-range duty operation are very often equipped with electrical- as well as turbine-driven pumps. They have a comparably low auxiliary power requirement in part-load operation of less than 50% of full load. High-range plant units, above 500 MW or so, must then be equipped with two 50% duty auxiliaries, because the current power maxima are not sufficient for a single auxiliary. A minimum total heat rate is achieved if, in operation with a load above 50–60% of the plant output, a turbine-driven pump is used and, at output rates below 50%, a motor-driven pump.

Other large-size pumps in a thermal power plant are the circulating pumps of the boiler, the condensate pumps and the cooling water pumps. For a 700 MW unit, the capacity of the circulating pump of a once-through steam generator is about 1 MW. The two 50% condensate pumps together need about 2 MW. The input power of the cooling water pumps, depending on the cooling technique, ranges around several MW. The FD and ID fans are next largest in size after the feed pump.

The FD and ID fans are driven almost exclusively by constant-speed three-phase current induction motors. The control features are the adjustable inlet guide vanes or the variable pitch blades of the fans. Inlet vane control is also called variable-pitch control. In countries of the west European continent, large FD and ID fans are preferentially designed and manufactured as variable-pitch axial-flow fans. They feature a relatively wide control range, with high fan efficiencies. Efficiencies of inlet-vane-controlled or pre-rotation-controlled axial-flow fans are lower, as are those of radial-flow fans.

Radial-flow fans have a slightly higher efficiency at the design point, yet a distinguishably lower one in part-load operation. German large-size plants almost exclusively use axial ID and FD fans because of their better part-load efficiencies.

4.4.4 Losses in Part-Load Operation

4.4.4.1 Impact of the Operating Regime of the Steam Generator and Turbine

The output of the turbine is controlled either by enlarging the turbine inlet cross-section (constant-pressure operation) or by modifying the pressure (sliding-pressure operation) (see also Sect. 4.2.4). The different control modes diminish the efficiency in part-load operation to varying extents. Figure 4.76 shows the impacts of the control modes on the heat rate of the turbine, without taking the feed pump work into consideration. Both in constant-pressure operation with throttle control and in sliding-pressure operation, the heat rate increases with decreasing load to a far greater extent than in constant-pressure operation with governing control.

In sliding-pressure control, the lower live steam pressure in part-load operation causes a decrease of the mean heat-input temperature and hence the thermal efficiency.

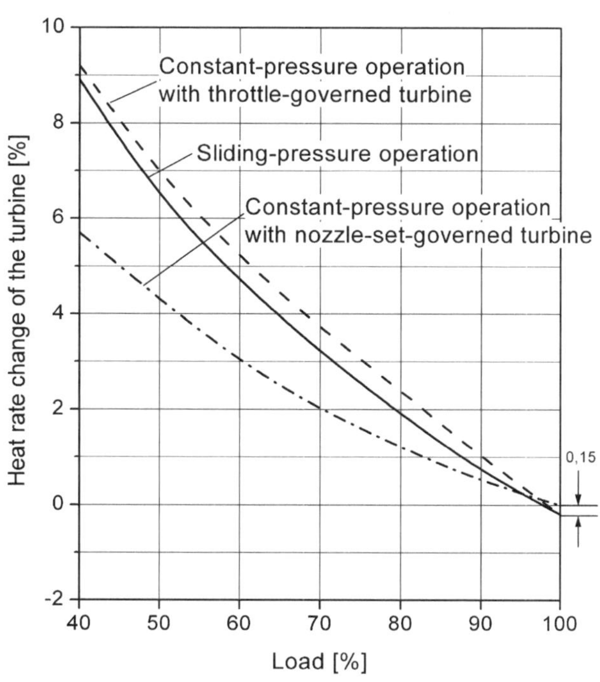

Fig. 4.76 Specific heat rate of the turbine generator as a function of the output, with different control modes (without feed pump capacity) (Baehr 1985)

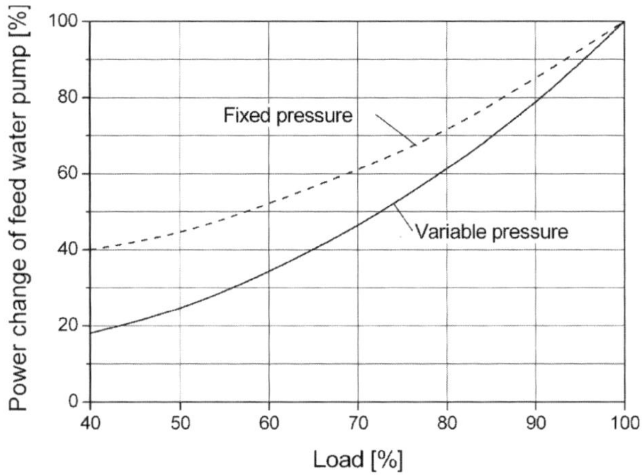

Fig. 4.77 Load dependence of the boiler feed pump power in sliding- and constant-pressure operation (Baehr 1985)

In constant-pressure operation with throttle control, the mean temperature of heat addition changes only a little through the range of the load. The efficiency deteriorates because of the throttling of the steam flow, and the temperature decrease associated with throttling has an additional negative effect.

In constant-pressure operation with governing control, the deterioration of the efficiency is less severe than for throttling control, because only one valve is ever opened partly, while the others are either closed or opened totally. Hence, the throttle loss affects only a part of the flow (Traupel 2001).

The power used by the boiler feed pump decreases both in constant-pressure and in sliding-pressure operation as a consequence of the part-load regime (see Fig. 4.77). It decreases more strongly in sliding-pressure operation because the pump pressure diminishes with the feed water flow for partial loads.

In taking the auxiliary power demand of the feed pump into account additionally to the heat rate of the turbine, one gets a comparison as shown in Fig. 4.78. The efficiency disadvantage of sliding-pressure operation turns into a small advantage compared to constant-pressure operation with governing control. Modified sliding pressure involves a higher heat rate than natural sliding pressure, which is due to the slight throttling of the valves.

At rated load, both constant-pressure operation with throttling control – i.e. the steam flow not being throttled – and sliding-pressure operation have an efficiency advantage, because the turbine control wheel necessary for constant-pressure operation with the nozzle-set governing has a lower HP turbine stage efficiency.

4.4.4.2 Example for the Reference Power Plant

In part-load operation, the efficiency of the reference power plant diminishes because the efficiencies of the plant components change. Besides showing the total efficiency η_{tot}, Fig. 4.79 shows the course of the thermal efficiency η_{th}, the steam generator

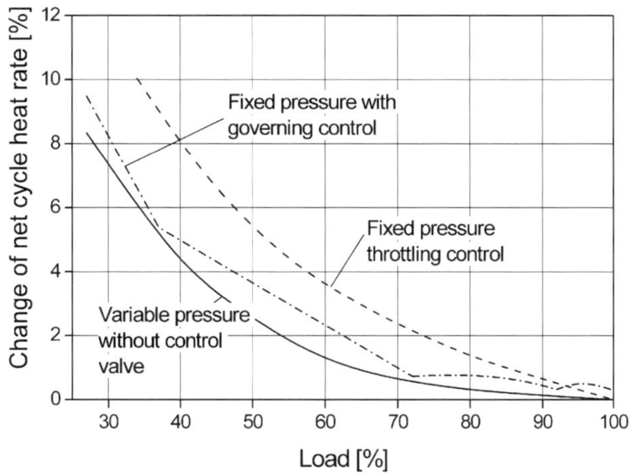

Fig. 4.78 Net heat rate changes with different control modes (Adrian et al. 1986)

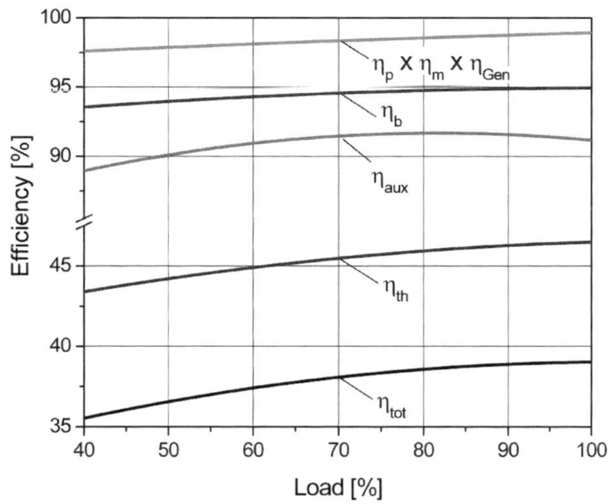

Fig. 4.79 Efficiencies of the reference power plant during part-load operation

efficiency η_{StG} and the auxiliary power demand efficiency η_{aux}, as well as the product of the pipework, mechanical and generator efficiencies $\eta_{pipe}\ \eta_{mech}\ \eta_{gen}$.

The thermal efficiency depends on the live steam conditions before the turbine, on the internal turbine efficiency and on the conditions in the condenser. In the reference power plant, operated at sliding pressure, the mean temperature of the heat input decreases as pressure decreases, even if the temperatures before the turbine are kept constant. In part-load operation, the pressure in the condenser decreases as a consequence of the diminishing heat flux to be dissipated. This advantage of the lower temperature of heat dissipation, however, cannot balance out the disadvantage of the lower temperature of the heat input.

The losses of the steam generator depend only slightly on the load. Though the flue gas temperatures fall with the shift from convective to radiative heat transfer,

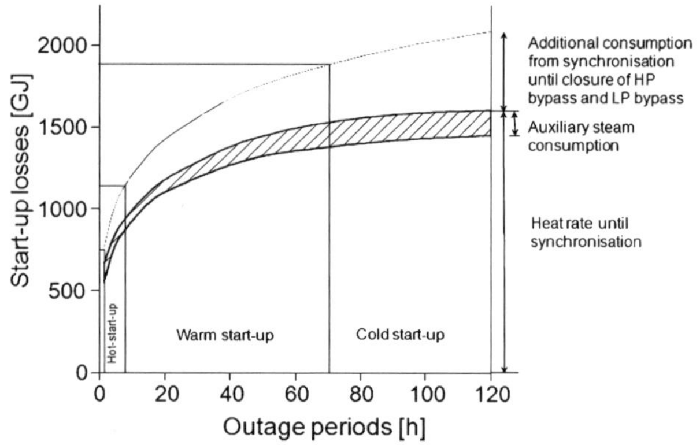

Fig. 4.80 Start-up losses of a 700 MW power plant unit as a function of outage periods (Adrian et al. 1986)

the advantage is nullified by the higher air ratio for the reheater temperature control. Since the power requirement of the auxiliary drives, except for the regulated ones, does not decrease with the load, the auxiliary power requirement with respect to the turbine output increases.

For the reference power plant under consideration in Fig. 4.79, the total efficiency drops from 38.8% at full load to 35.5–40% load.

4.4.5 Losses During Start-Up and Shutdown

Energy supplied to the power plant during start-up heats the steam generator or is lost as waste heat via the stack and condenser. Thermal energy stored in the boiler is lost during shutdown, unless it acts as the standby service until the next (warm) start-up.

The start-up losses continue at their full extent until the power plant is in parallel connection with the electricity network and then diminish accordingly as the live steam and the reheat steam flows are taken up by the turbine, and the steam flow bypasses of the high-pressure, intermediate-pressure and low-pressure turbines are closed (see Sect. 4.2.4.4).

The guideline start-up time for an outage of 48 h of a hard coal fuelled unit is about 4–5 h, for an outage of 8 h, about 2 h, and for an outage of 30 min, about 1 h (STEAG 1988).

The losses occurring during start-up and shutdown depend on

- the capacity of the power plant unit,
- the construction type of the steam generator and
- the period of the preceding outage (Adrian et al. 1986).

The start-up losses decrease with increasing unit capacity. The longer the period of an outage, the higher the losses of the following start-up. Figure 4.80 describes start-up losses of an existing 700 MW power plant unit. The start-up losses of power plants designed today are well below these levels. For a design 400 MW unit, for

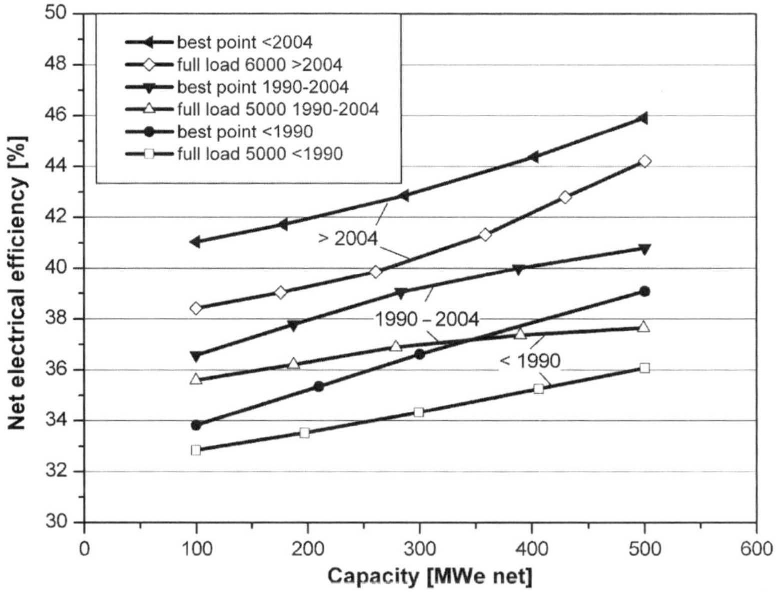

Fig. 4.81 Design and operation efficiencies (data from Theis 2005)

instance, losses of 400 GJ after an 8 h outage are reported. So the specific start-up loss of 2 GJ/MW (see Fig. 4.80) is reduced to about 1 GJ/MW (Kotschenreuther et al. 1993).

4.4.6 Efficiency of Power Plants During Operation

When discussing efficiencies, it has to be pointed out that normally only the best efficiencies at full load are mentioned. In Fig. 4.81, efficiencies which are achieved during operation are compared, for German power plants, to the respective best design efficiencies for different ages and thermal capacities. The efficiency during operation of power plants is about 2–3% (in absolute terms) lower than the design efficiency. With increasing age of the power plant and decreasing thermal capacity, the difference is higher. The lower efficiency during operation is due to the lower part-load efficiency and the losses during start-up and shutdown, as discussed in Sects. 4.4.4 and 4.4.5.

4.4.7 Fuel Drying for Brown Coal

Moist fuels, compared to dry fuels, involve some efficiency disadvantages for a power plant. The high moisture content of brown coals of 50–60% by weight increases the flue gas mass flow and the allowable boiler exit temperature. The steam generator efficiency diminishes as there are higher flue gas losses. The efficiency is about 90%, whereas hard coal fuelled furnaces have rates of 94–95%.

Brown coal needs a pre-drying stage for combustion-engineering reasons. The moisture of German brown coal is almost entirely inherently bound in capillaries, so thermal drying is needed to evaporate the moisture. In a conventional brown coal power plant, moist raw brown coal is milled and dried in a beater-wheel mill. The drying process utilises a 1,000°C hot, recirculated flue gas partial flow to provide the heat. The simultaneous milling accelerates the evaporation heat exchange with the hot flue gas. The dried and milled brown coal powder is injected into the firing together with the milling vapours. The milling – drying, which is state-of-the-art technology for brown coal firing, causes efficiency loss due to the high temperature of the drying medium and, compared to hard coal firing, 1–2% lower thermal cycle efficiencies. This can be explained exergetically by the decline of the temperature level from about 1,000°C to about 150°C and the associated exergy losses. External hot gas drying with the same flue gas and water-vapour temperatures would be unfavourable in exergetic terms as well.

The exergetically unfavourable drying, with its low cycle efficiency, and the influence of the fuel moisture on the flue gas loss have the result that the net efficiency of a brown coal power plant, at comparable steam conditions, may be up to 3% lower than the net efficiency of a hard coal power plant, depending on moisture and sulphur contents. The high temperatures of the partial flue gas flow are not necessary for drying the brown coal and achieving the required moisture content. However, they offer the advantage of a compact construction size and high drying capacities.

The disadvantage of the low cycle efficiency of brown coal (as opposed to hard coal) can be compensated or even overcompensated by fuel drying using low-exergy drying media. Higher cycle efficiencies can be achieved if media with low temperature and exergy are utilised for the drying. Brown coal drying needs only temperatures of somewhat above 100°C, so low-temperature flue gases or extraction steam are possible drying means.

External drying with separate discharges of water vapours and flue gases can diminish the flue gas losses. Separate discharges allow the use of lower temperatures for both the water vapour and the flue gas, due to the low acid dew point.

A further efficiency increase is possible if the evaporation heat of the water vapour can be recovered in the power-generating process. This way, the gross calorific value can be tapped for exploitation. This kind of efficiency increase is only worthwhile for fuels with an adequate difference between the higher and lower heating values or with a high fuel moisture content.

In terms of drying technology, there is a distinction between convection and contact dryers.

Convection dryers transfer the energy of hot gases, necessary for vaporising the moisture, to the substance to be dried in a direct mass and heat transfer. In this process, the gas cools and takes up the evaporated moisture. In contact driers, the energy necessary for vaporisation is captured via heating surfaces (Kallmeyer and Wick 1997; Böcker et al. 1992; Klutz and Holzenkamp 1996).

4.4.7.1 Warm-Gas Drying

Flue gases, extracted after the economiser at 350°C and cleaned of coarse particles, cool down to temperatures above the dew point (140–170°C) during convective drying of moist brown coal in a beater-wheel mill. This temperature, 350°C being low in comparison to the conventional drying temperature of 1,000°C, makes it necessary to use the entire flue gas flow for the drying. Following the removal of the coal particles, the flue gases are conducted into the FGD unit. Instead of a flue gas air heater, a steam air heater is used to preheat the combustion air. This concept of warm-gas drying improves the total efficiency by about 2% compared to hot-gas drying or combined drying and pulverising (Zimmer). The acid dew point should not fall below, in order to prevent corrosion. The drawback there is that the condensation heat of the vaporised water from the flue gas/water vapour mixture cannot be gained isothermally but only to the dew point temperature, and hence is less attractive in terms of thermodynamics. Given that the efficiency potential is thus limited, this method is currently less attractive for integration in the power plant cycle (Kallmeyer and Wick 1997).

4.4.7.2 Drying by Extraction Steam

For the drying of brown coal, the condensation heat of the extraction steam from the turbine can be used. The available techniques are using tubular or fluidised bed dryers (Böcker et al. 1992).

In technical terms, tubular dryers are deemed to be contact dryers. These dryers dominate in the field of the production of dried brown coal. With capacities of up to 50 t/h of raw brown coal, they are used for drying from 60 to 12% moisture. A shell-and-tube heat exchanger is mounted in a slightly inclined rotary cylinder. Extraction steam condenses on the outside of the tubes, transferring the condensation heat for drying to the coal, which flows through the tubes. Air flowing through the tubes with the coal takes up the evaporated fuel moisture. As for warm gas drying, the exploitation of the condensation heat of the evaporated water in the air/water mixture is not attractive from the thermodynamic point of view (Böcker et al. 1992).

Fluidised bed dryers are particularly suited to being incorporated into the power plant cycle. They are a drying technology preferentially used for very moist granules; they stand out due to their very good heat transfer.

A portion of the water vapour from the turbine is recirculated in order to create a pure steam atmosphere in the fluidised bed. This way, the water vapours of the steam-fluidised bed can be condensed almost isothermally, and their evaporation heat can be recovered (Böcker et al. 1992). Fluidised bed dryers can be designed as both convection and contact dryers. In convection drying by fluidised bed, the heat of the extraction steam is transferred to the fluidising medium in an external heat exchanger. Contact dryers transfer the condensation heat of the extraction steam via the in-bed heat-transfer surface placed inside the fluidised bed (see Fig. 4.82). Contact drying offers advantages over convection drying, particularly when high moisture evaporation capacities are required (Klutz and Holzenkamp 1996). It is reported

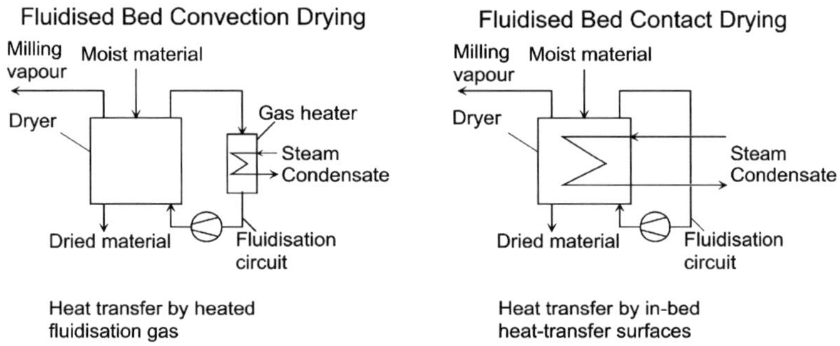

Fig. 4.82 Fluidised bed configurations with convection and contact drying (Klutz and Holzenkamp 1996)

that throughput rates in the order of magnitude of up to 350 t/h of raw brown coal (180 t/h dried brown coal) are possible (Klutz et al. 1996). A steam-fluidised bed with a throughput of 45 t/h of raw coal is in service in Australia (Schmalfeld and Twigger 1996).

The efficiency increase (without water vapour exploitation, which is explained in the following passage) with this kind of technique is in the order of magnitude of 2–3%, depending on the pressure where the extraction steam is taken.

4.4.7.3 Exploitation of the Condensation Heat of the Water Vapours

An additional efficiency increase is possible if the condensation heat of the vapours from the fluidised bed dryer can be utilised. The energetic exploitation, however, requires a drying of the brown coal in a pure steam atmosphere, so that the water vapours can be condensed without an interfering influence from air or flue gas. Using this method, an additional increase of 2–3% can be achieved.

A schematic diagram of the *WTA process* (*Wirbelschichttrocknung mit Abwärmenutzung*: fluidised bed drying with internal waste heat exploitation) is shown in Fig. 4.83 (Klutz et al. 1996). At a temperature of about 100–120°C, the bed is fluidised with water vapour from the drying process. To obtain a residual moisture of the brown coal of 12%, it is necessary to have a fluidised bed temperature around 110°C. After a dust removal stage, the water vapours formed during drying are compressed in a vapour compressor to about 4 bar and then condensed in an in-bed heat-transfer surface to dry the fluidised brown coal bed. In the fluidised bed heat exchanger, the superheat and the condensation heat of the vapours are transferred to provide the evaporation heat of the brown coal moisture. After condensation the heat of the condensate is used to preheat the coal. Another portion of the water vapours is recirculated as a fluidising medium.

For the demonstration of the WTA process, a pilot-scale plant for drying 53 t of raw brown coal per hour was in service from 1993 to 1999. The plant was designed for a grain size of raw brown coal of up to 6 mm. In a further demonstration step, the

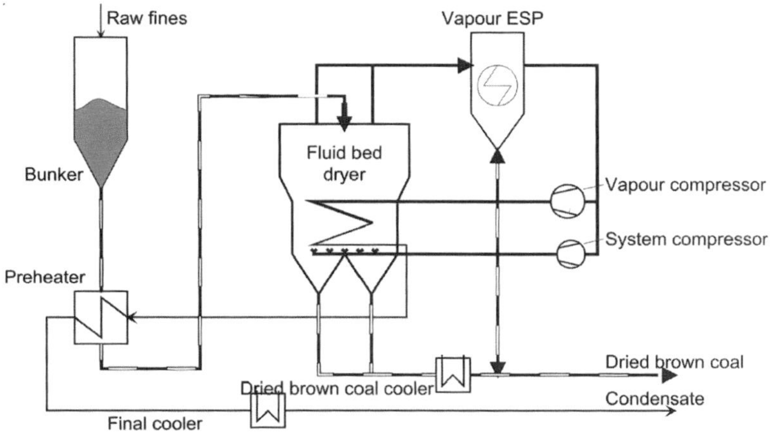

Fig. 4.83 Schematic diagram of WTA-drying – fluid bed drying with internal waste heat exploitation (Klutz et al. 1996)

coarse grain WTA process was scaled up to 170 t/h of raw coal and tested between 2000 and 2002 (Ewers et al. 2003).

For this second stage of development, the focus was put on fine grain drying, because of economic benefits in comparison to coarse grain drying. Fine grain drying substantially improves the heat transfer in the fluidised bed and reduces the amount of vapour required for fluidisation. This results in a much more compact design, smaller heat exchangers and lower flowrates. The specific investment costs are estimated at 70€/kW. Additionally, fine grain drying may remove the need for milling of dried coal, so that dried coal will be able to be fed directly to a pulverised coal fired power station. A WTA fine grain drying plant with a throughput of 30 t/h raw coal per hour was operated between 2000 and 2004 (Ewers et al. 2003; Klutz et al. 2006). To demonstrate commercial-scale application maturity, a WTA prototype with an output of 110 t/h dried brown was operated in 2008/2009. The dried brown coal is fed to the power plant Niederaußem K, delivering up to 30% of the fuel input. It is expected that a 1,100 MW power plant fired exclusively with dried brown coal will require between four and six drying lines, depending on the moisture content of the fuel (Schwendig et al. 2006).

According to reports, the WTA process has a 5.5% efficiency advantage over brown coal fired power plants with hot gas drying processes. Figure 4.84 shows the relative improvement in efficiency, which is a function of the moisture of the raw fuel. The theoretical, maximum improvement is given by the complete drying of the raw coal down to 0% moisture. The actual improvement of the WTA process is lower, due to the residual moisture of the dried brown coal of 12%, and depends on the integration of the process (Schwendig et al. 2006).

In addition to the above-described exploitation of the water vapours for drying, they can also produce power in a vapour turbine or, in the power generation cycle, preheat a feed water partial flow at low temperatures (Elsen et al. 1996).

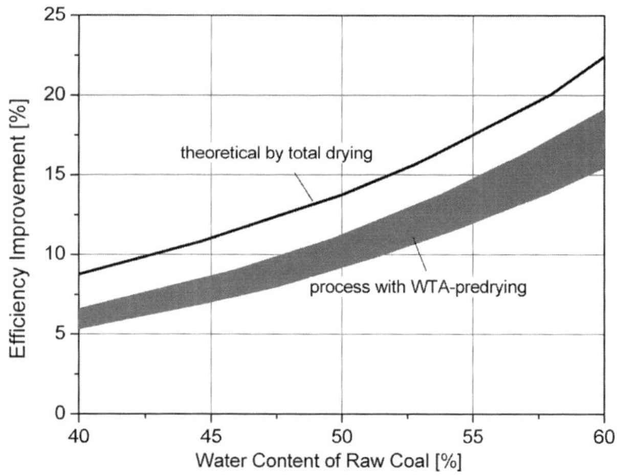

Fig. 4.84 Efficiency improvement by pre-drying (Schwendig et al. 2006)

4.5 Effects on Steam Generator Construction

Looking at their history, it is evident that the transition to live steam conditions above 540°C and 260 bar and reheater temperatures above 540°C was made more than 50 years ago. As early as 1951, in Germany, a plant with a superheater temperature of 600°C was put into service for the first time. Even though at that time there was little experience of using most high-temperature materials, a great number of plants with supercritical steam conditions were built in the 1950s. The peak of the development was reached with the construction of the Eddystone power plant in the USA in 1954, with an output of 325 MW_{el} and live steam conditions of 650°C/650°C/345 bar.

However, the steam generators used in those power plants had several features which at that time were favourable for the application of high steam conditions.

- Compared to modern steam generators with steam-generating capacities up to 2,500 t/h, the steam outputs then ranged around 200 t/h.
- The firing systems, as a rule, were slag-tap furnaces, with furnace walls made of single, vertically arranged tubes with a refractory-lined wall behind them. Today, fully welded, gas-tight membrane walls are constructed with external heat insulation and helically wound wall tubes (Fig. 4.85). As well as the multi-pass type construction, the single-pass type has become generally accepted, often being built in central Europe. The ash is removed from the firing in a dry state.
- Live steam conditions of 600°C/600°C/350 bar, mostly applied in the chemical industry, usually operated at base load for steam and electricity supply. Depending on the fuel prices in a given country, coal-fired furnaces are also built today for mid-range load duty.

The increasing steam outputs and higher steam conditions imply that, for large steam generator components such as HP outlet headers – even when parallel tube

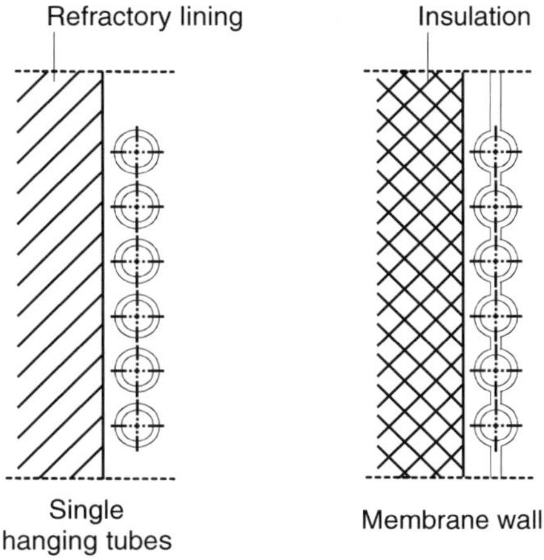

Fig. 4.85 Furnace wall construction of a refractory-lined and fully welded boiler

bundles are used – a large diameter and hence a large wall thickness have to be chosen. This restricts the allowable rate of temperature change, which in turn limits the use of mid-range load-type power plants with their daily start-up and shutdown processes.

For the design and construction of the steam generator walls as membrane walls, instead of the refractory lining with single tubes mounted in front of them, as was common in earlier times, heat-resistant alloy materials are necessary. These materials are such that they do not need any heat treatment after welding at the construction site or after welding repairs at the power plant site. These requirements are met by steel type 13 CrMo 4 4, which was until 2000 the highest-grade alloy steel used for membrane walls. Under high steam conditions, however, the limit of this material is reached. Newly developed membrane wall steels allow higher temperatures for future power plants.

High-temperature corrosion on the gas-swept side and high-temperature oxidation, or scaling, on the steam-swept side of the final superheater heating surfaces augment the requirements of the material to be selected.

The features of advanced steam generators today are

- high capacities,
- fully welded membrane walls,
- a dry-bottom firing arrangement with a large furnace for low NO_x emissions,
- suitability for mid-range load (minimum load at 35–40%) and base load operation,
- a once-through steam generator,
- a single-pass boiler (predominantly in Germany),
- sliding-pressure operation and
- stress-consistent start-up and shutdown processes and fast load changes.

With increasing live steam conditions in large-scale steam generators, the designs of the membrane wall and final superheater heating surfaces and the HP outlet header approach maximum stress limits of the material (Stamatelopoulos and Weissinger 2005; Scheffknecht et al. 2002; Chen and Scheffknecht 2003a; Kather and Scheffknecht 1997). Figure 4.86 illustrates how material developments of steels determine and limit the application of higher steam conditions. This will be discussed in the following sections in regard to the membrane wall, superheater and thick-walled components. The development and qualification of new materials for future power plants with steam conditions up to 350 bar and 750°C is ongoing in various research projects in Europe, Japan and the USA (Blum et al. 2007; Blum and Vanstone 2005; Chen and Scheffknecht 2003b; Chen et al. 2005; Husemann 2003; Kjaer et al. 2002; Viswanathan et al. 2005a, b).

Table 4.5 shows the chemical composition of conventional and new steam generator steels (Scheffknecht et al. 1996; Chen and Scheffknecht 2003b).

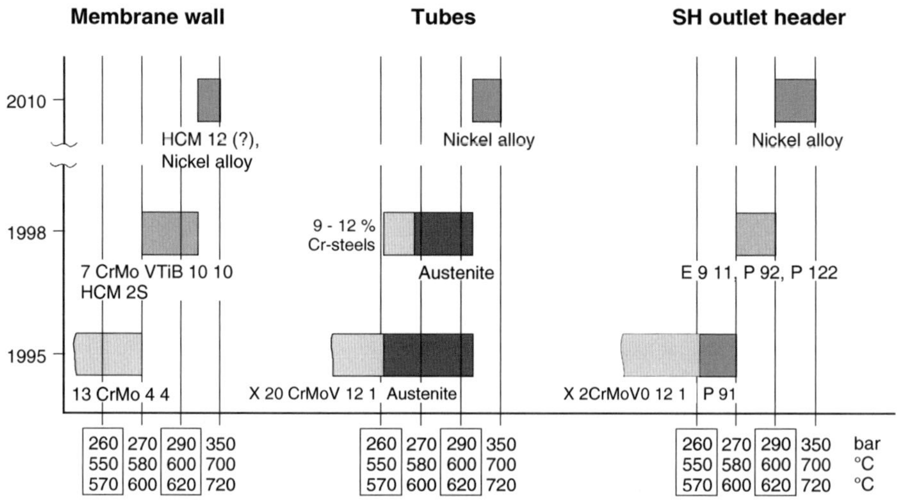

Fig. 4.86 Development of steam conditions and steam generator materials (*Source*: Alstom Power)

4.5.1 Membrane Wall

Higher live steam pressures, live steam and reheater temperatures and higher inlet temperatures of the feed water result in higher compressive and thermal stresses on the evaporative system. These stresses can be reduced by process-engineering measures. The above parameters all cause – to varying extents – an increase of the steam outlet temperature of the evaporator. A higher compressive stress is a result only of the higher live steam pressure. While the effect of higher inlet temperatures of the feed water on the evaporator outlet temperature (the same temperature rise) is some-

Table 4.5 Chemical composition of boiler steels (*Source*: Alstom Power and additions)

Steel	C	Si	Mn	P Max.	S Max.	Cr	Ni	Co	Mo	Al	Cu	V	W	Ti	Nb	Other
15 Mo 3 (1.5415)	0.12–0.20	0.10–0.35	0.40–0.80	0.035	0.035	–	–		0.25–0.35	–	–	–	–	–	–	–
13 CrMo 4 4 (1.7335)	0.10–0.18	0.10–0.35	0.40–0.70	0.035	0.035	0.70–1.10	–		0.45–0.65	–	–	–	–	–	–	–
10 CrMo 9 10 (1.7380)	0.08–0.15	Max. 0.50	0.40–0.70	0.035	0.035	2.00–2.50	–		0.90–1.20	–	–	–	–	–	–	–
15 NiCuMoNb 5 (1.6368)	0.10–0.17	0.25–0.50	0.80–1.20	0.030	0.025	Max. 0.30	1.00–1.30		0.25–0.50	Max. 0.050	0.5–0.8	–	–	–	0.015 0.045	N: Max. 0.020
X 20 CrMoV 12 1 (1.4922)	0.17–0.23	Max. 0.50	Max. 1.00	0.030	0.030	10.0–12.5	0.30–0.80		0.8–1.2	–	–	0.25–0.35	–	–	–	–
X 10 CrMoVNb 9 1 (T/P91) (1.4903)	0.08–0.12	0.20–0.50	0.30–0.60	0.020	0.010	8.0–9.5	Max. 0.40		0.85–1.05	Max. 0.040	–	0.18–0.25	–	–	0.06–0.10	N: 0.030–0.070
7 CrMoVTiB 10 10 (T24) (1.7378)	0.05–0.10	0.15–0.45	0.30–0.70	0.020	0.010	2.20–2.60	–		0.90–1.10	Max. 0.020	–	0.20–0.30	–	0.05–0.10	–	N: Max.0.01 B: Max. 0.0007
HCM 25 (T23)	0.04–0.10	Max. 0.50	0.30–0.60	0.030	0.010	1.90–2.60	–		Max. 0.30	Max. 0.030	–	0.20–0.30	1.45–1.75	–	0.02–0.08	N: Max.0.03 B: Max. 0.0006
HCM 12	Max. 0.14	Max. 0.50	0.30–0.70	0.030	0.030	11.0–13.0	–		0.80–1.20	–	–	0.20–0.30	0.8–1.2	–	Max. 0.20	–
HCM 12A (P122)	Max. 0.15	Max. 0.40	Max. 0.70	0.030	0.020	10.0–12.0	Max. 0.70		0.20–0.60	Max. 0.040	Max. 2.5	0.15–0.30	1.5–2.5	–	0.02–0.10	N: 0.02–0.10
X 10 CrWMoVNb 9 2 NF 616/P92 (1.901)	Max. 0.15	Max. 0.50	Max. 1.00	0.020	0.010	8.0–13.0	–		Max. 1.0	–	–	0.10–0.30	1.5–2.5	–	Max. 0.10	N: 0.02–0.10 B: Max. 0.01
E 911 (1.4905)	0.10–0.18	0.5	0.5	0.010	0.005	10.0–12.0	0.1–1.0		1.0	Max. 0.010	–	0.2	1.0	–	Max. 0.06	N: Max. 0.07
NF12	0.08	0.05	0.5	–	–	11.0	0.5		0.15	–	–	0.2	2.6	–	0.07	N: 0.5, B: 0.004 Co: 2.4–2.8
X3 CrNiMoN 17 13 (1.4910)	Max. 0.04	Max. 0.75	Max. 2.00	0.035	0.015	16.0–18.0	12.0–14.0		2.0–2.8	–	–	–	–	–	–	N: 0.10–0.18
X8 CrNiMoNb 16 16 (1.4981)	0.04–0.1	0.3–0.6	Max. 1.5	0.035	0.015	15.5–17.5	15.5–17.5		1.6–2.0	–		–	–	–	10xC 1.2	–
TP 347H FG	0.04–0.10	Max. 0.75	Max. 2.00	0.040	0.030	17.0–20.0	9.0–13.0		–	–	–	–	–	–	8xC	–
NF 709	0.04–0.10	Max. 0.75	Max. 1.50	0.030	0.010	19.0–22.0	23.0–27.0		1.0–2.0	–	–	–	–	0.02–0.20	0.10–0.40	N: 0.10–0.20 B: Max. 0.007
HR 3C	0.04–0.0	Max. 0.75	Max. 2.00	0.030	0.030	24.0–26.0	17.0–23.0		–	–	–	–	–	–	0.20–0.60	N:0.15–0.35
Super 304 H	0.04–0.10	1.0	Max. 2.00	0.040	0.030	18.0–20.0	8.0–11.0		–					–	0.3–0.6	N, Cu
Alloy 617	0.05–0.1	Max. 0.2	Max. 0.2	0.010	0.010	20.0–23.0	Rest	10.0–13.0	8.0–10.0	0.70–1.40	Max. 0.50			0.2–0.5	–	Fe < 2.00

what obvious, the impact of the higher live steam and reheater conditions can be explained by the higher live steam enthalpy h_{LS}. At the same boiler power output $\dot{Q}_B$

$$\dot{Q}_B = \dot{m}_{LS}\,(h_{LS} - h_{FW}) \tag{4.23}$$

the steam mass flow decreases correspondingly. Given that the heat transferred in the furnace remains constant at a fixed furnace outlet temperature, the outlet temperature from the evaporator rises.

A higher steam pressure, in addition, causes a higher boiling temperature in the evaporator and, with the decrease of the enthalpy of evaporation, higher outlet temperatures of the evaporator, provided there is the same heat input in each case. This is clarified in Fig. 4.87. A simultaneous temperature increase of 10 K at both the HP and the reheater outlet makes the steam temperature in the wall rise by about 7 K, and a pressure increase of 10 bar raises the steam temperature by about 3 K. Because of the greater effect on efficiency, higher live steam temperatures should be preferred to higher live steam pressures if the limit of the membrane walls is reached.

The allowable outlet temperature of the evaporator can be limited by the available materials. Figure 4.88 shows the creep rupture strength for conventional and advanced membrane wall materials. Besides the necessary strength, membrane wall materials have to meet the requirement of being weldable without post-weld heat treatment. In order to avoid hydrogen-induced stress corrosion in the areas of the membrane wall subject to heat loads, post-weld heat treatment is necessary when the hardness of about 350–400 HV 10 is exceeded (Kather and Scheffknecht 1997).

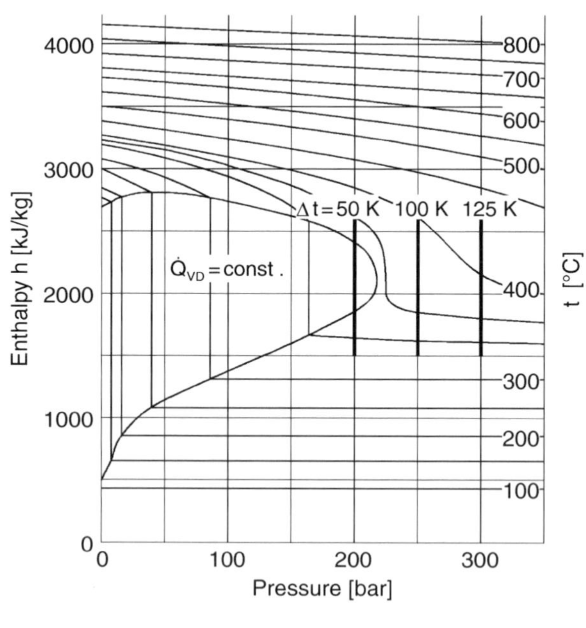

Fig. 4.87 Heat-up in the evaporator as a function of the pressure: $h - p$ diagram (Riemenschneider 1995)

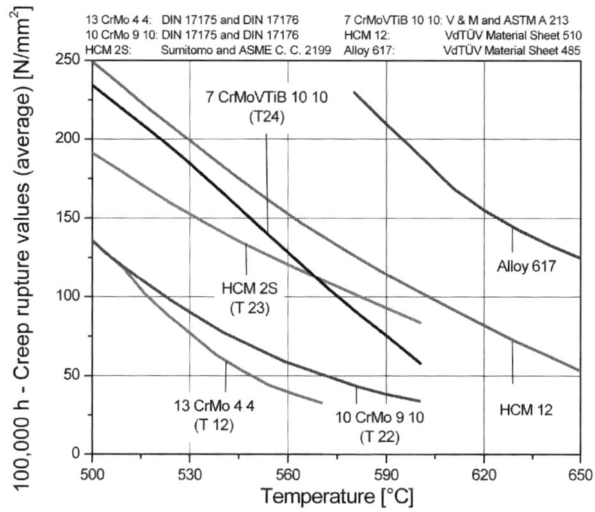

Fig. 4.88 Creep Strength for membrane wall materials (*Source*: Alstom Power)

Post-weld heat treatment is feasible in the workshop, though not during installation, assembly or repair works at the power plant site. For this reason, only materials which do not require post-weld heat treatment are suitable for membrane walls.

This requirement is met by steel-type 13 CrMo 4 4, which was the preferred material until 2000 for steam power plants. With high steam conditions, however, the limit of this material is reached. At the usual temperature additions, an outside tube diameter (OD) of 38 mm and a wall thickness of 6.3 mm, an evaporator outlet temperature of about 435°C is allowable for the membrane wall steel-type 13 CrMo 44 (see Fig. 4.89). By choosing a greater wall thickness, and with mechanical relief from the secondary load of the membrane wall, the allowable temperature can be increased up to a maximum of 460–470°C (Riemenschneider 1995; Scheffknecht et al. 1996). With the material 10 CrMo 9 10 (T22), the steam parameters cannot be increased considerably.

A substantial improvement of the creep strength, with limited hardness levels in as-welded conditions, can be reached by the newly developed, ferritic, 2–2.5% chromium steels. The typical examples of this group of steels are the material HCM 2S (T23) and 7 CrMoVTiB 10 10 (T24). By the use of these steels, the steam temperature limit at the water wall can be increased by approximately 50 K in comparison to the conventional 13 CrMo 4 4 steel. Substantial testing of this steel has been carried out in different research programmes and the steels will be used as the membrane wall material for power plants currently under construction. For a furnace outlet temperature of 1,250°C, steam conditions of 300 bar/640°C can be realised (Chen and Scheffknecht 2003b; Stamatelopoulos and Weissinger 2005).

For very high live steam conditions such as 375 bar and 700°C, with maximum metal temperatures of above 600°C, the 2–2.5% chromium steels are no longer adequate because of their limited creep strength and the lower corrosion and oxidation resistance. In order to meet the strength and corrosion requirements of very high

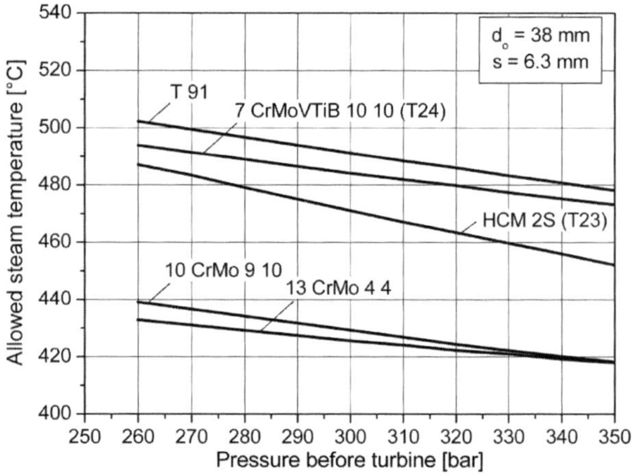

Fig. 4.89 Allowable evaporator outlet temperature for various materials as a function of the pressure before turbine (*Source*: Alstom Power)

live steam conditions, higher-alloy steels are needed. Austenitic steels cannot be used because of their large thermal expansion coefficients, which inhibit the use of a welded joint between austenitic and ferritic water-wall sections. Candidate materials are 9–12% martensitic chromium steels. The most promising candidate was the HCM 12 steel, which has a higher creep strength and corrosion and oxidation resistance, and the hardness in the in-welded condition is the lowest amongst the 9–12% chromium steels (Chen 2003). Unfortunately a major reduction of long-term creep rupture strength at temperatures above 550°C is expected for many 10–12% chromium steels, including HCM 12, due to changes in the microstructure. New routes for the development of iron-based materials for furnace walls have to be found, otherwise the only alternative would be nickel-based superalloys for the hottest part of the furnace wall (Blum et al. 2007). Nickel-based alloys, which do not require post-weld treatment, are potential materials for the water wall. A well-known alloy is the material Alloy 617, which has the advantage of high rupture strength, high corrosion and oxidation resistance and heat expansion coefficients similar to martensitic steels. However, the Ni-based alloys are much more expensive; in comparison to HCM 12, Ni-based alloy tubes are roughly 10 times as expensive (Blum et al. 2007).

Significant efforts are ongoing for the development of heat-resistant steels and their qualification for use, so that higher steam temperatures may be used in membrane wall tubes of evaporators. As an alternative, process-engineering methods which avoid or limit higher evaporator outlet temperatures for high steam conditions have been investigated, such as

- increasing the furnace outlet temperature by using a smaller-sized furnace,
- heat shifting by flue gas recirculation,
- adding additional heating surfaces in the furnace,
- extracting heat from the membrane wall and transferring it to the reheat cycle,
- decreasing the size of the economiser,

- decreasing the feed water temperature,
- minimising the spray attemperator water mass flow into the HP system, in order to raise the cooling water – steam mass flow through the steam generator wall, and
- reducing the pressure losses in the evaporator and the superheater.

A substantial influence on the evaporator outlet temperature is exerted by the furnace outlet temperature. As a rule, this temperature is set to range around 50°C below the ash deformation temperature of the coal. Relatively low furnace outlet temperatures are chosen to allow the firing of a broad range of coal types. This prevents slagging on the first convection heating surfaces, even for coals with a low deformation point. The furnace outlet temperature, at the same time, establishes the heat absorption of the evaporator in the furnace, and hence the evaporative capacity. Reduced dimensions of the furnace, with the consequence of a higher furnace outlet temperature, decrease the evaporator outlet temperature. An increase of the furnace outlet temperature from 1,250 to 1,300°C, for instance, leads to a decrease of the evaporator outlet temperature by about 16°C. So, under advanced steam conditions, efforts are made to set the furnace outlet temperature as high as possible. This, however, may result in a restriction of the usable coal types in the firing. Figure 4.90 shows the impact of the furnace outlet temperature on the wall exit temperature for different steam conditions.

At the same furnace outlet temperature, flue gas recirculation shifts the heat absorption from the furnace to the convective heat exchanger surfaces by the use of a higher flue gas mass flow. The evaporator outlet temperature drops as a result of this process. When designing for flue gas recirculation, the flue gas duct cross-

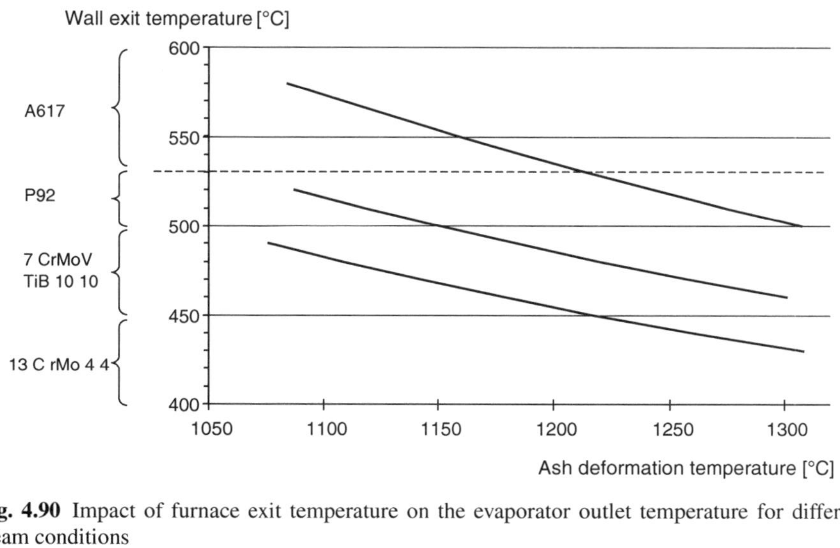

Fig. 4.90 Impact of furnace exit temperature on the evaporator outlet temperature for different steam conditions

section has to be enlarged in order to not exceed the allowable flue gas velocity in the convective heating surface section.

The plant efficiency is diminished by the power consumption of the flue gas recirculation process. Hot flue gas recirculation and cold flue gas recirculation are process options that have to be weighed up. In the cold variant, the flue gas is drawn off after the ESP and recirculated. The disadvantage is the larger air heater surface area. In hot flue gas recirculation, the flue gas is extracted before the air heater and re-injected in the burner area. The higher volumetric flow, due to the higher flue gas temperature, requires a greater power demand. In addition, the dust load of the gas may cause problems. A 10% cold flue gas recirculation lowers the outlet temperature of the steam by 20°C (Heiermann et al. 1993). Design concepts for advanced power plants do not include flue gas recirculation. An increase of the excess air does shift heat to the convective section, but it increases the flue gas losses because of the higher mass flow, and thus should not be incorporated.

Arranging additional heating surfaces in the furnace to remove load from the evaporator is a difficult construction task. Wall heating surfaces mounted in front of the furnace wall result in different ductile behaviour in different parts of the walls. In principle, only heating surfaces exposed to temperatures lower than the mean temperature in the evaporator should be chosen for additional heating surfaces, due to the material restrictions. Only the first, and colder, reheater section could hence be utilised as a wall heating surface. Since, however, those heat exchanger sections are missing in the convective section of the furnace, the other heating surfaces are made larger.

Plate heating surfaces, which are already able to be used at relatively high temperatures of about 1,400°C at the furnace outlet, also reduce the furnace height and are thus suited for removing load from the evaporator. With a wide pitch and the tubes of one plate element mounted tightly next to each other, these heating surfaces are unlikely to experience slag build-up. These types of heating surface, however, can only be taken into consideration for two-pass boilers.

Figure 4.91 shows a concept for transferring heat from the evaporator to the cold reheat steam. Steam is taken from the pipes at the end of the helical winding out of

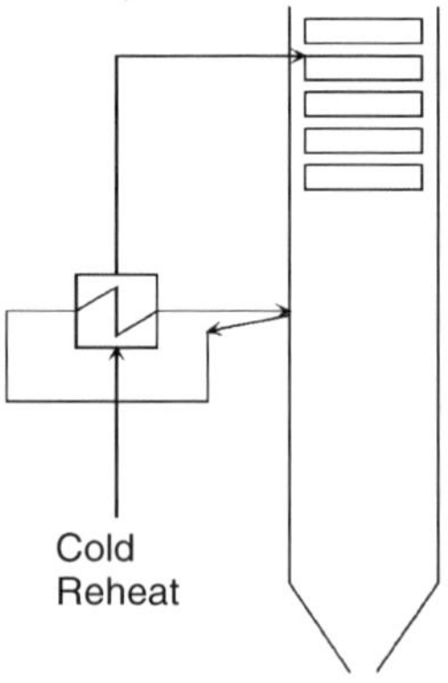

Fig. 4.91 Heat transfer from HP steam to cold reheat steam

the walls, cooled by cold reheat steam and then re-injected into vertically arranged evaporator pipes.

An economiser designed for a smaller temperature rise or, in the extreme case, is not included at all restricts the effect of the higher feed water inlet temperature on the evaporator. Higher flue gas losses can then only be prevented by increasing the dimension of the reheater or the air heating. Greater dimensions of the air heater, due to the higher air preheating temperature at the same furnace outlet temperature, result in an increase of the heat absorption in the evaporator, and thus have to be avoided. For power plants with advanced steam conditions, a larger reheater is taken into consideration for this reason. A drawback, though, may arise in this case because of a deterioration of the convective characteristics of the reheater, so either the injected mass flow would have to be increased or the reheater temperature would drop at part-load operation, if remedial action is not taken by other measures (see Sect. 4.3.5.6).

The reduction of the pressure losses in the evaporator and the superheater results in lower compressive stresses on the evaporator at the same live steam pressure before the turbine. If the maximum load of the membrane material is reached, smaller pressure losses allow higher live steam pressures and therefore a higher efficiency. According to Table 4.4, the predominating pressure loss in advanced steam generators occurs in the superheater. The potential for reduction of superheater pressure losses is only small. Superheater tubes with a larger diameter are more complicated and expensive if the material needed is austenite (Heiermann et al. 1993).

The size of the steam generator also has consequences for the membrane wall design. In small steam generators, a minimum residence time in the furnace is required to ensure primary NO_x control. Burnout entails a relatively low furnace outlet temperature, and hence the problem of an increased steam temperature at the wall outlet. In large steam generators, the dead load can cause the main forces to be in the longitudinal direction of the tube instead of the direction of the tube circumference, because of the inside pressure of the tube. To reduce the total load, the dead load has to be supported by additional weight-bearing components, in order not to be forced to restrict the pressure.

These construction and process-engineering measures only have to be considered if the limits of the available membrane wall materials have been reached. Figure 4.92, for the example of material type 13 CrMo 4 4 and hard coal fired furnaces, shows how the live steam conditions depend on the furnace outlet temperature as the essential design parameter. For coals that allow high furnace outlet temperatures of 1,250–1,300°C, the resulting maximum live steam conditions are up to 300 bar and 600°C. If lower furnace outlet temperatures are required, it is reasonable to decrease the pressure and keep the live steam temperature.

Figure 4.93 shows the limits of the new membrane wall steel 7 CrMoVTiB 10 10 (Lorey and Scheffknecht 2000), which is considered in the design of all new plants. With live steam conditions of the currently planned power stations of up to 300 bar, 600°C and 620°C, the limits of the new membrane wall material are

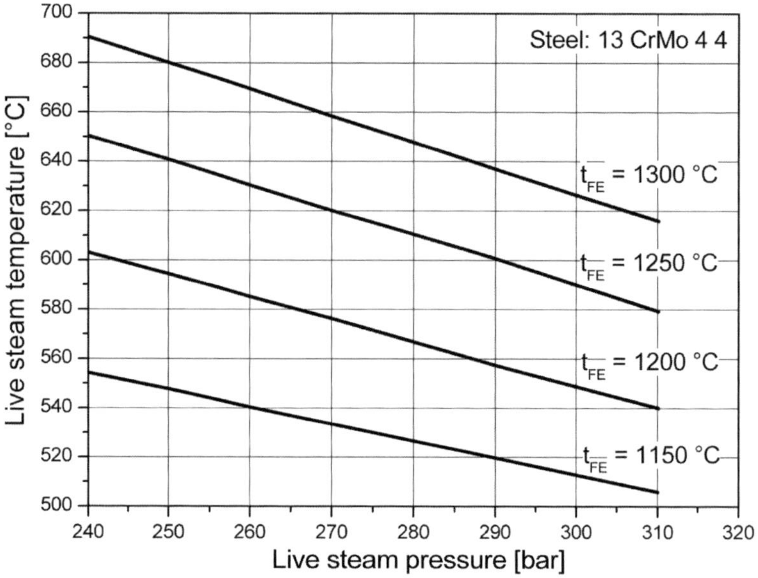

Fig. 4.92 Maximum steam parameters for membrane wall material type 13 CrMo 4 4 (hard coal LCV = 26.1 MJ/kg, feedwater inlet temp. 290°C, reheater temp. = IIP temp. +20 K) (*Source*: Alstom Power)

only reached at low furnace exit temperatures of 1,150°C. With the new material, process-engineering measures are not required.

In Fig. 4.94, the evaporator outlet temperatures for power plants built in Germany, or being planned, are shown in a $h - p$ diagram. In all existing power plants, it was possible to use the well-known material 13 CrMo 4 4 for the membrane walls (see Table 4.7), as the evaporator outlet temperature could be limited to 460°C. On the other hand, it is obvious that power plants which are currently planned require the new 2–2.5% chromium steels, because the evaporator outlet temperature ranges around 480°C. The new 2–2.5% chromium steels can be used for temperatures up to approximately 500°C. For the 700°C power plant with evaporator outlet temperatures of 550°C, new nickel-based alloys are required.

4.5.2 Heating Surfaces of the Final Superheater

The final superheating surfaces are the convective heating surfaces, which are subject to both the highest steam and the highest flue gas temperatures. X 20 CrMoV 12 1, the martensitic steel type commonly used for conventional steam conditions, has proven very worthwhile for steam temperatures up to about 550°C. The corresponding tube wall temperatures are in the range of 600°C. With ever higher temperatures, the creep rupture strength of the material diminishes and cannot be balanced out by thicker tube walls, because this would make the temperature differences

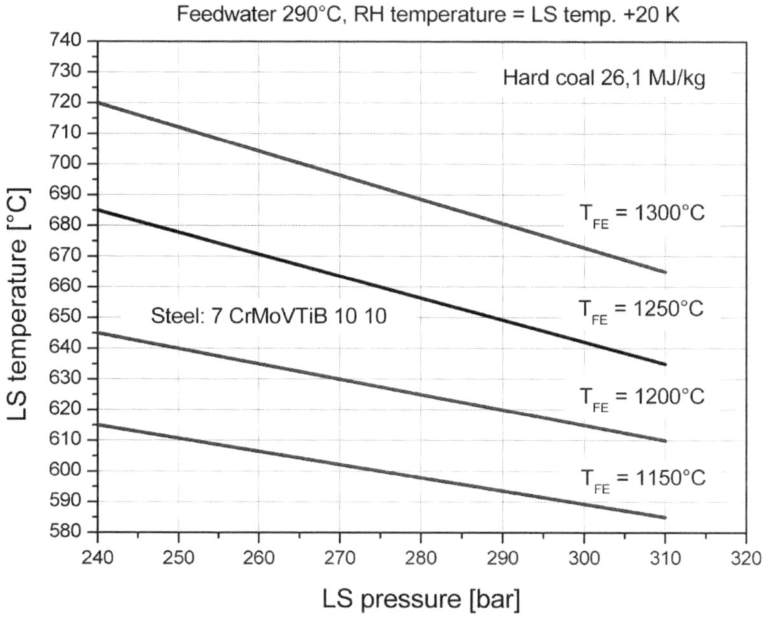

Fig. 4.93 Maximum steam parameters for membrane wall steel 7CrMMoVTiB 10 10 (Lorey and Scheffknecht 2000)

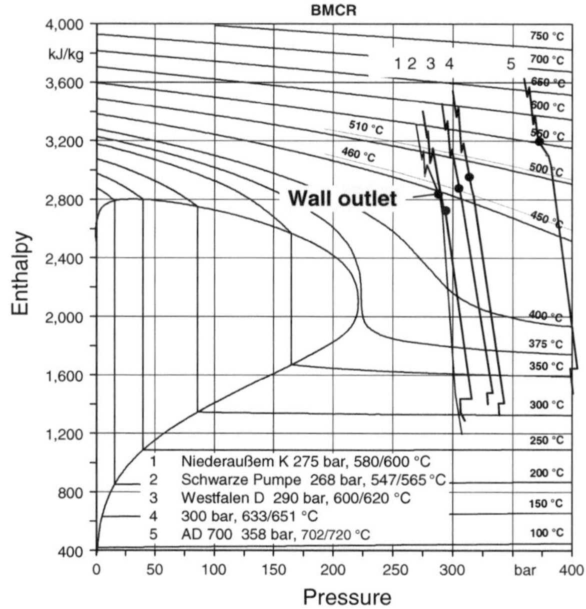

Fig. 4.94 Design of a conventional and of a high-temperature steam generator: $h - p$ diagram (*Source*: Alstom Power)

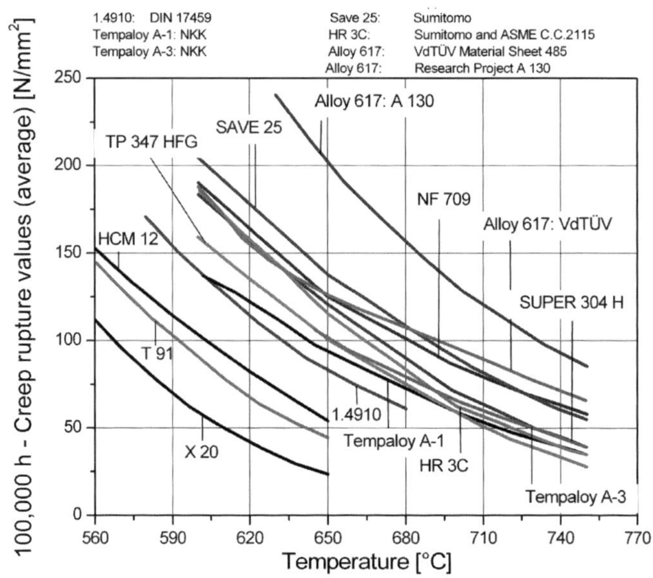

Fig. 4.95 100,000 h mean values of creep rupture for superheater and reheater materials (*Source*: Alstom Power)

between the inside and outside walls too great. Higher steam conditions, for this reason, require a transition to materials with a higher strength.

Figure 4.95 shows the creep rupture strength for 100,000 h of new martensitic materials (T91, HCM 12), of well-established (X3 CrNiMoN 1713, X8 CrNiMoNb 1616) and newly developed austenitic materials (TP 347H FG, HR 3C, Super 304H) and the nickel-based steel Alloy 617. Martensitic material X 20 CrMoV 12 1 is included for comparison.

Figure 4.96 shows, for the different materials, the limits of applicable steam conditions, taking only strength into account. When the steam temperatures are chosen only moderately higher than those possible for X 20 CrMoV 12 1, it is possible to use the martensite-type HCM 12 as an alternative. The martensitic steel T91, however, performs worse under high-temperature corrosion attack due to its low chromium content and should therefore not be used as a heating surface material for superheaters.

For higher steam temperatures, above 560–570°C, higher-alloyed austenitic steels have to be used. The austenitic-type steel X8 CrMoNb 16 16 (not shown in Fig. 4.96), which was used successfully in many high-temperature power plants in the 1950s in Germany and which has a chromium content of 16%, shows the same design strength as X 20 CrMoV 12 1 at temperatures higher by about 70°C. In the group of 18% chromium steels, the austenitic steels TP 347H FG and Super 304H are new developments. Higher-alloyed austenitic steels with chromium contents above 20%, like the newly developed steels HR 3C, NF 709 and SAVE 25,

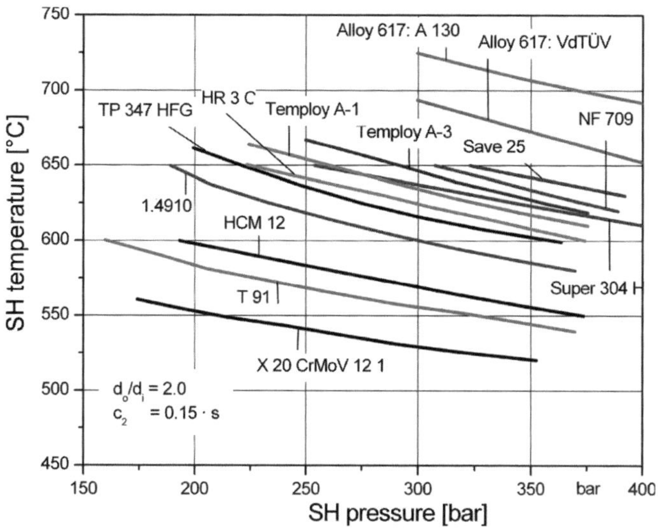

Fig. 4.96 Limits for high-temperature tube materials (*Source*: Alstom Power)

allow steam conditions in the range of approximately 620–630°C at 300 bar. The specific tube costs for the new steels Super 304H, TP 347H FG and HR 3C exceed the conventional steel X 20 CrMoV 12 1 by 28, 41 and 89%, respectively (Lorey and Scheffknecht 2000). Intensive work is continuing to develop a suitable austenitic tube material with a 100,000 h rupture strength of around 100 MPa at 700°C, a material temperature corresponding to 650°C steam temperature. The materials should demonstrate a flue gas corrosion resistance better than a 2 mm metal loss during an exposure of 200,000 h (Blum et al. 2007).

For very high steam conditions, such as 350 bar and 700°C at the boiler outlet, austenitic steels are not adequate, because of insufficient creep strengths. The well-examined nickel-based Alloy 617 is a possible candidate for these conditions. Further nickel-based alloys are being developed for very high temperatures, with the aim of achieving a 100,000 h creep strength value of 100 MPa at 750°C (Chen 2003; Blum et al. 2007).

The design for high steam parameters of the final superheating surfaces, besides considering the strength of the material, also has to consider the resistance of the material to gas-side high-temperature corrosion and steam-side scaling.

In pulverised hard coal fired furnaces, high-temperature corrosion is in particular caused by molten salts such as alkali iron(III) sulphates, $Na_3Fe(SO_4)_3$ and $K_3Fe(SO_4)_3$ (see also Sect. 5.10.4). In the combustion of the pulverised coal, the alkalis sodium and potassium are released in a gaseous state, which then react with SO_3 either in the flue gas or on the tubes and form sodium and potassium sulphates with a low melting point. These sulphates precipitate on the tubes together with other ash components, which then, with iron oxides and SO_3, form the sulphate complexes (Na_3 or K_3)$Fe(SO_4)_3$, the melting points of which are about 590°C.

Figure 4.97 shows the corrosion rate for an austenitic material as a function of temperature (Plumpley and Roczniak 1988; Apblett 1973). The diagram also shows the concentrations of the sulphate complexes and their changes of physical state. At about 580°C, the corrosion rate begins to rise markedly until it reaches the maximum value at about 660°C. In this range the sodium and potassium complexes are present in molten form. In the range above 700°C, these become unstable and evaporate. The material wear rate starts to decrease as a consequence. The corrosion rates of ferritic and martensitic materials are considerably higher at the same temperatures than the rates of the austenitic materials shown in Fig. 4.97.

The location and the level of the maximum corrosion rate depend on the composition of the coal. Prime determining factors of the coal's corrosiveness are the content of volatile alkalis and the SO_3 content of the flue gas. Larger corrosion rates for chlorine-containing coal types are put down to the fact that a higher chlorine content is favourable for the formation of volatile alkalis. CaO and MgO in the coal ash reduce corrosion. CaO or MgO addition to the fuel at a controlled rate can be used to reduce high-temperature corrosion.

Besides the surface temperature of the tube, the flue gas temperature is another important factor affecting the corrosion rate. In the range of the surface temperatures that are associated with high steam conditions and which still lie below the maximum corrosion rate, higher flue gas temperatures lead to higher corrosion rates, because the conditions are favourable for the complexes to melt (see Fig. 4.98).

The corrosion rate depends a lot on the tube material used. From Fig. 4.99 it can be inferred that the resistance of the material against high-temperature corrosion is primarily influenced by the chromium content. The higher the chromium content, the higher the resistance against high-temperature corrosion. With their elevated chromium content, austenites are more corrosion-resistant than ferrites and martensites and should in this respect be given preference for use in the high-temperature area.

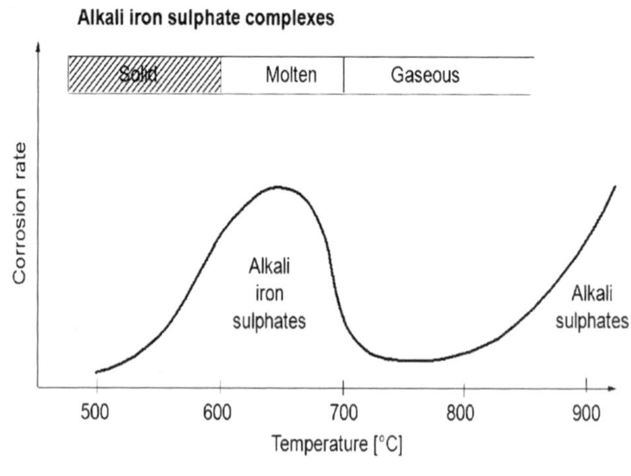

Fig. 4.97 Weight loss of austenitic materials due to high-temperature corrosion, and physical state of corrosive sulphates as a function of temperature

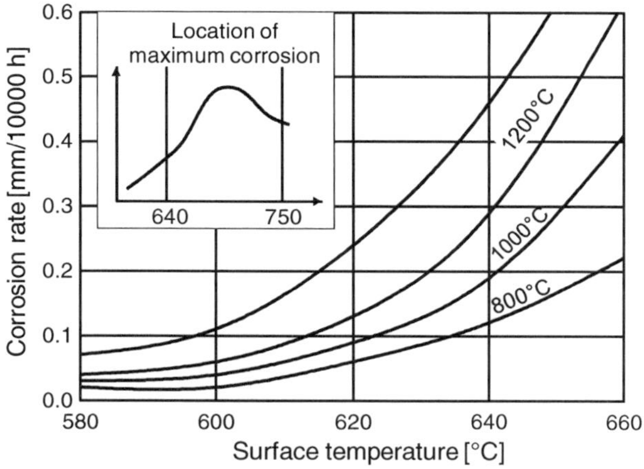

Fig. 4.98 Gas-side corrosion rate as a function of flue gas and wall temperatures (Heiermann et al. 1993)

This advantage of austenites as opposed to ferritic materials, however, applies only in base-load operation of the steam generator. In operating regimes with daily start-ups and shutdowns, in contrast, austenitic steels may show adverse corrosion characteristics compared to ferritic ones, because the austenitic tube material and the oxidic protective layer have differing thermal expansion coefficients. When temperature changes occur, the protective layers flake off, so the tube material becomes subject to a stronger corrosive attack.

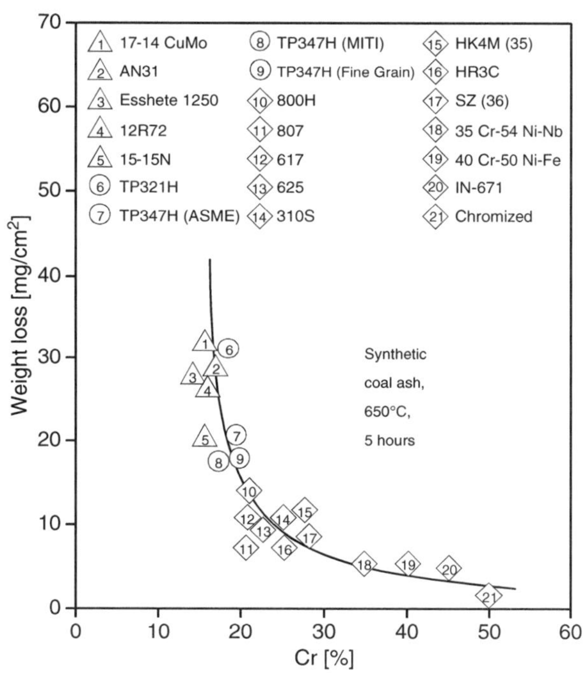

Fig. 4.99 Influence of the chromium content on high-temperature corrosion (Heiermann et al. 1993)

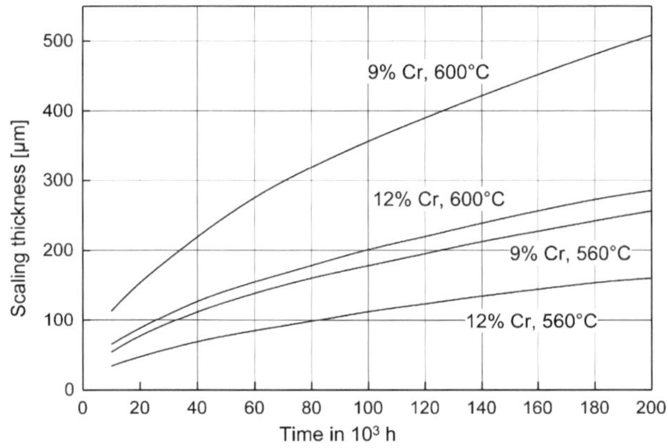

Fig. 4.100 Scaling thicknesses for different chromium contents of a material and different live steam temperatures (Heiermann et al. 1993)

Further measures to control high-temperature corrosion are

- the use of composite tubes or coated tubes,
- the replacement of tubes that are particularly at risk after a defined operating period and
- the application of metal or ceramic half-pipe sections to protect tubes especially at risk, such as the row of tubes of a heating surface bank which are exposed to the flow first. The surface temperature of these protective half-pipes in this case is clearly above that which causes the maximum permissible corrosion rate. The potential to use them, however, remains restricted to relatively few areas in the steam generator, because they considerably impair the heat transfer.

The last reheater stage, with its high steam temperature (usually higher than the reheater temperature due to the lower reheat pressure) and low heat transfer coefficient, is the heating surface most at risk of high-temperature corrosion. It should therefore be arranged downstream of the final superheater on the flue gas side, which makes keeping constant reheater temperatures at part-load operation more difficult.

Another problem arising in high-temperature operations is high-temperature oxidation, or scaling, of the tubes on the steam side. As a consequence of the higher steam temperatures, the scaling rate on the inside of the heated tubes rises, deteriorating the heat transfer and raising the tube wall temperature. The higher temperature diminishes the creep rupture strength of the materials and increases the fireside corrosion rate.

The *scaling constants* of the various materials help to determine scale thickness values as functions of time (Fig. 4.100). These values, in turn, are used to estimate the increases of the tube wall temperatures (Fig. 4.101). Higher chromium contents diminish the effects of high-temperature oxidation.

At the power plant Niederaußem K, with steam conditions of 580°C/600°C (SH/RH), unstable layers of magnetite oxide inside the tubes resulted in tube failures. The austenitic steel X3 CrNiMoN 17 13 (1.4910) was replaced partly by

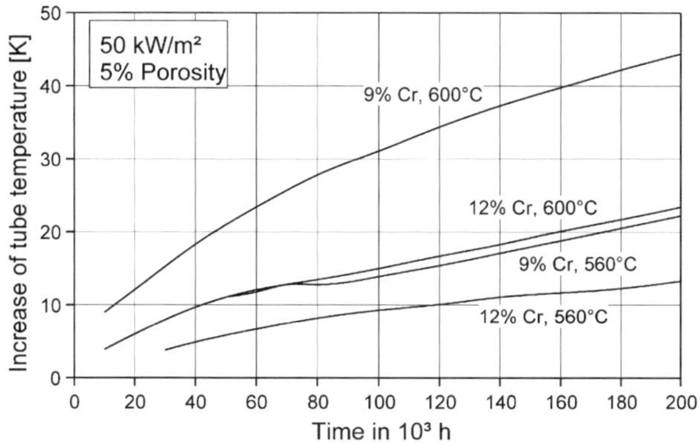

Fig. 4.101 Increase of tube wall temperatures for different chromium contents of the material and different live steam temperatures (Heiermann et al. 1993)

an austenitic steel with a 22% chromium content (DMV310N) (Tippkötter and Scheffknecht 2004; Schlenkert et al. 2006).

4.5.3 High-Pressure Outlet Header

As a thick-walled construction component, the high-pressure outlet header (HP outlet header) limits the rate of load change. An increase of the steam parameters while using the same material would entail a greater wall thickness, and thus would lead to a lower allowable rate of load change. The wall thickness must thus be limited by an appropriate choice of material.

Figure 4.102 shows the creep rupture strength of different materials for thick-walled components. The conventional steel for this application has been for a long time the martensitic steel X 20 CrMoV 12 1. Extensive operating experience is available for the steam temperature range up to 560°C. With an increase of the steam parameters, the limit of X 20 CrMoV 12 1 is reached. The next material adopted was P91, which can also be said to be well examined. In a further development of the 9–12% Cr steels, the creep ruptures were further improved by the addition of tungsten, which led to the development of the steels E911, P92 and P122 (HCM 12A) (Chen and Scheffknecht 2003b). Further improvements of 9–12% Cr steels on top of these developments have proven to be very difficult in the last two decades. Short-term data demonstrated a major improvement, but longer term data showed a dramatic drop in the strength of the steel. As yet, no improvement on the steel P92 has been able to be made (Blum et al. 2007; Blum and Vanstone 2005).

For very high steam conditions of 375 bar and 700°C, nickel-based alloys like Alloy 617 and Alloy 740 are needed. For such high steam parameters, the load change rate of the thick-walled component is very low. In such cases, the power plant should preferably be operated as a base-load unit.

Figure 4.103 shows the necessary wall thickness of the final SH outlet header for different materials at a pressure of 300 bar. If the wall thickness is limited to 100 mm, which represents an outer to inner diameter ratio of 2, the steam parameters are approximately 300 bar/620°C for the steel P92 or P122 and 300 bar/700°C for

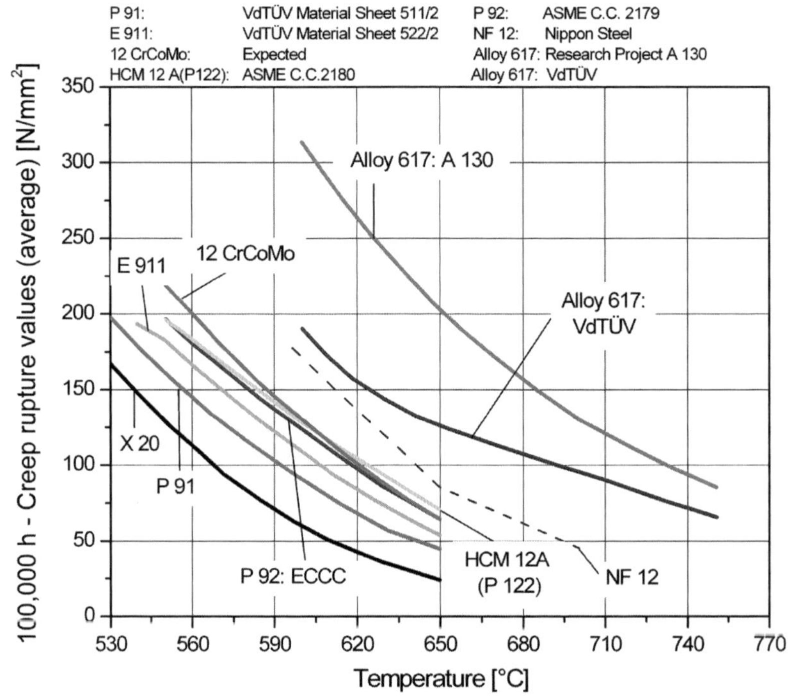

Fig. 4.102 100,000 h creep rupture strength for pipe and header materials (*Source*: Alstom Power)

Alloy 617 (based on current VdTÜV values). An upgrading to the Ni-based Alloy 617 can reduce the wall thickness by about 30% (Chen and Scheffknecht 2003b; Hahn and Bendick 2006). The layout of the plant may, to a limited extent, allow a smaller wall thickness, in addition to the selection of materials. By increasing the number of parallel steam runs, it is possible to reduce the diameter of the header and thus the wall thickness.

High-pressure outlet headers can become limiting components in power plants with advanced steam conditions, because technical regulations result in a greater ratio of the outside to the inside diameter compared to the ratio for live steam pipes. For the live steam pipes, the same materials are considered as for the HP outlet headers. Due to the smaller wall thickness/diameter ratio, the design limit, as opposed to the outlet headers, is not reached.

4.5.3.1 Impacts on the Turbine

In contrast to the steam generator and the tubes, the problems in the construction of steam turbines for extreme steam conditions arise almost exclusively from the temperature (Adrian et al. 1986). The critical components are those swept by hot steam – that is, the cast valve chests and the HP and IP (intermediate-pressure) cylinders, the forged HP and IP rotors and the blading of the first HP and IP stages. The high-temperature power plants in the 1950s and 1960s used austenitic steels for

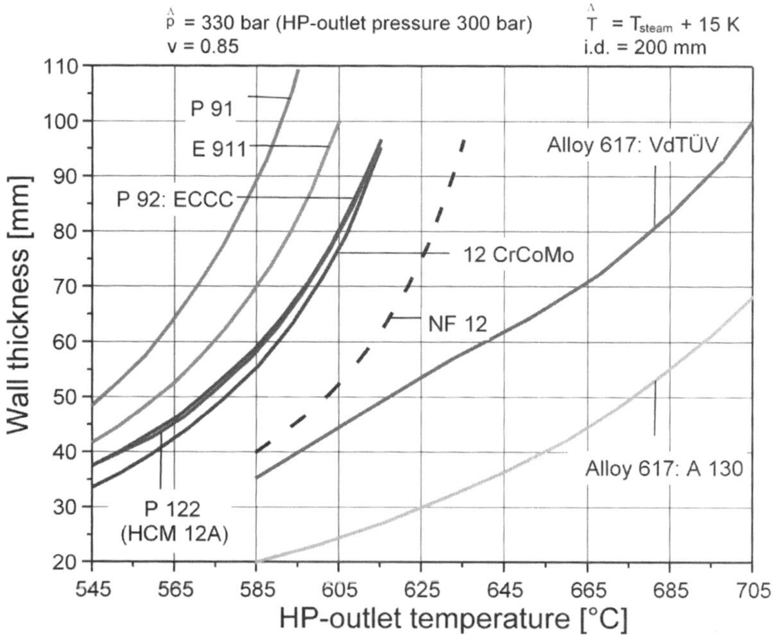

Fig. 4.103 Wall thickness of header materials for different steam conditions (*Source*: Alstom Power)

the critical turbine components such as the valve chest and rotor forgings; however, experience with these steels was unsatisfactory. The high coefficients of thermal expansion and low thermal conductivities, as well as the low yield strengths, mean that thermal cycles cause high thermal gradients and stresses in thick section components. As a consequence, the use of austenitic steels for advanced steam conditions is now avoided. The increase of steam conditions in the turbine section was only possible with the introduction of the use of martensitic 9–11% chromium steels. The 10% chromium steel X14 CrMoVNbN10 2 as a standard for rotor materials today allows steam conditions between 585 and 610°C. Within different phases of the European COST programme, work is ongoing to develop martensitic steels for temperatures up to 630 and 650°C, by the increase of the chromium content to 11% and the addition of Co and B to increase the oxidation resistance and the creep rupture strength. The temperature at which the 100,000 h rupture creep strength is around 100 MPa is a reasonable indicator of the maximum application temperature as a steam turbine material. To support the development of steam turbines for temperatures greater than 700°C, development of nickel-based alloys for use in critical turbine components is currently in progress in European programmes such as AD 700. The timeframe to develop a new, completely reliable, mature turbine steel is about 12 years (Blum and Vanstone 2005; Blum et al. 2007; Tremmel et al. 2006; Oakey et al. 2003; Wichtmann et al. 2005).

4.5.4 Furnaces Fuelled by Dried Brown Coal

The combustion of pre-dried brown coal requires a new design for the steam generator. The combustion temperatures of dried brown coal are considerably higher than the temperatures of raw brown coals with high moisture contents. By drying a Rhenish coal from a raw-coal moisture content of 54% to 12% residual moisture, the calorific value increases from 9–10 MJ/kg to 19–20 MJ/kg. The doubling of the calorific value makes the adiabatic combustion temperature rise to 2,020°C; with raw brown coal, it lies at 1,560°C. The removal of water vapour increases the flue gas temperatures above the burners to an average of about 1,550°C. In furnaces fuelled by raw brown coal the temperatures are lower by about 300°C.

The drying process has no influence on the ash properties, so for dried brown coal fired furnaces, similar furnace outlet temperatures have to be achieved. These range around 1,050°C for Rhenish brown coals. Non-symmetries, which are inevitable in raw brown coal, can be reduced by a more symmetric feeding of the fuel. This results in a more uniform flue gas temperature distribution at the furnace exit, allowing a slight increase of the furnace exit temperature of approximately 20 K (Ewers et al. 2003). Furnaces for dried brown coal have slimmer constructions, because the flue gas mass flow gets reduced by the water vapour fraction. The furnace height – set by the need for flue gas cooling – is higher with lower fuel moisture contents (see Fig. 4.104). The convective heating surfaces are smaller for higher combustion temperatures and smaller flue gas mass flows.

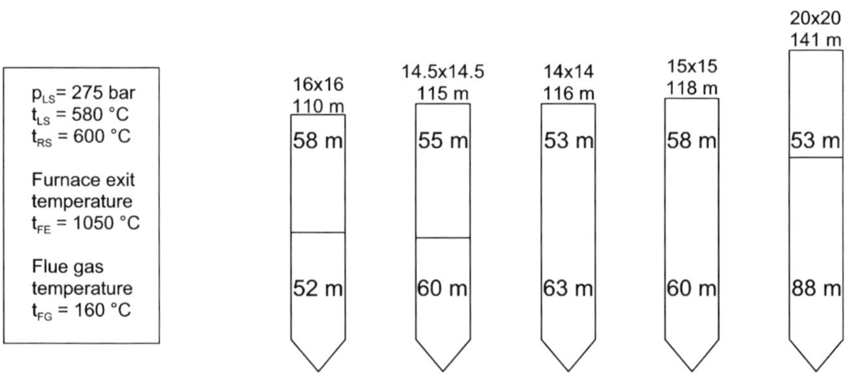

Thermal output [MW]	1000	1000	1000	1000	2000
Coal	RBC	DBC 30	DBC 12	DBC 12	DBC 12
Recirculation [%]	-	-	-	15	-
Wall temperature	435	488	510	465	510
Gain in efficiency	-	2.8% Points	4% Points	3.8% Points	4% Points

Fig. 4.104 Influence of the brown coal drying degree on steam generator dimensions (Riemenschneider 1995)

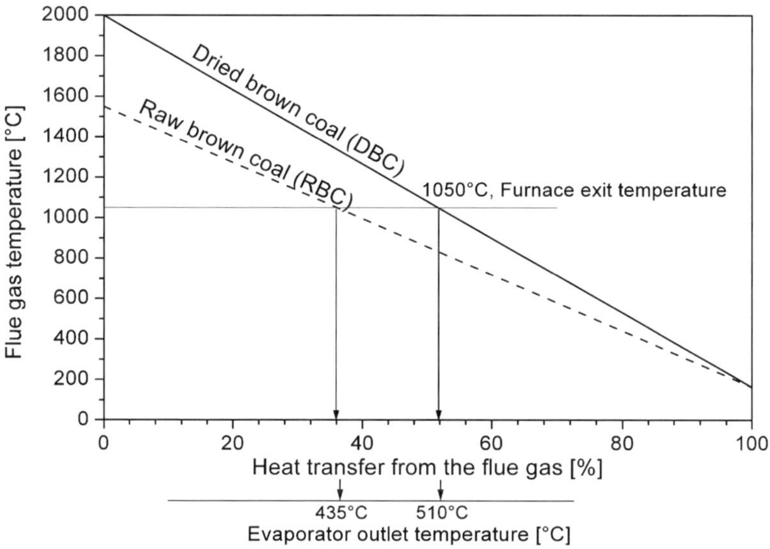

Fig. 4.105 Heat absorption in the membrane wall in raw brown coal and dried brown coal firing systems (1,000 MW_{el}, 275 bar, 580°C, 600°C (Pollack and Heitmüller 1996)

The heat flux in the combustion chamber increases substantially. The consequence of this is very high steam temperatures of more than 500°C developing in the enclosing walls of the furnace (see also Fig. 4.105) (Pollack and Heitmüller 1996). This wall temperature was not acceptable for the membrane wall material 13 CrMo 4 4 (which was the available steel when the study was done), but also too high for the limits of the new membrane wall steels like HCM 2S (T23) or 7 CrMoVTiB 10 10 (T24). Additionally, an increase of live steam conditions further raises the evaporator outlet temperature. Compared to raw brown coal, the smaller flue gas mass flow of dried brown coal is a disadvantage. Compared with hard coal fired furnaces, the low furnace outlet temperature which is required for the fuel brown coal is of disadvantage.

The process-engineering measures discussed in Sect. 4.5.1 provide possible solutions. By recirculating 20% of the total flue gas flow it is possible to keep the wall at 460–470°C, which is comparable to hard coal firing. Flue gas recirculation diminishes the efficiency by 0.2% (Riemenschneider 1995). This temperature range could be met by the steel 13 CrMo 4 4, provided measures such as the mechanical relief of loads are implemented; however, the new steel 7 CrMoVTiB 10 10 (T24) will provide more flexibility. Flue gas recirculation is also required to limit the size of the furnace.

A steam generator fuelled by dried brown coal is in principle realisable technology. However, detailed investigations concerning emissions, corrosion, fouling and slagging have to show to what extent a less conservative and thus more economical design would be possible (Riemenschneider 1995). In 2008, large-scale tests started in the 1,100 MW unit Niederaußem. Up to 110 t/h of dry brown coal from a WTA dryer prototype are fired in the boiler, corresponding to 30% of the furnace capacity. From about 2015 on, the dry brown coal fired power plant will presumably be the new standard for brown coal fired power stations (Schwendig et al. 2006).

4.6 Developments – State of the Art and Future

In the field of steam production, the recent years have seen the beginning of a rapid development towards enhancing the efficiency. The characteristic of this development is the transition to higher live steam conditions with live steam temperatures of more than 540°C and live steam pressures of more than 250 bar. The higher live steam temperatures make the thermal cycle efficiency increase, because of the higher mean temperature of the heat input. Potential for optimisation is also available at the cold end of the steam cycle. Depending on the boundary conditions – once-through water cooling or evaporative cooling – the condensation pressure is reduced when optimising (in regard to cost and effectiveness) the cooling circuit.

Besides increasing the thermal efficiency, cutting down the auxiliary power demand by the use of low-loss feed pump drives and by reducing the various losses also contributes substantially to the enhancement of the individual efficiencies. Measures aimed at reducing the boiler losses focus on diminishing the flue gas temperatures and at reducing the flue gas mass flow. Today, it is possible to achieve steam generator efficiencies for hard coal power plants of up to 95%; for brown coal, the values reach around 90%. Further developments in turbine engineering also help to advance the power plant technology by increasing the efficiency.

When discussing efficiencies, it has to be pointed out that the average (best value) efficiencies of existing power plants are considerably lower in comparison to the state of the art, depending on the age of the existing power plants. Figure 4.106 compares average design efficiencies of existing power plants in different parts of the world. Furthermore, these efficiencies are normally the maximum design values at full load. The measured efficiencies for power plants in operation are lower, as can be seen in Fig. 4.81.

4.6.1 Hard Coal

In the following sections, the following stages of development of hard coal fuelled power plants shall be discussed:

A Power plants with conventional steam conditions (160–200 bar/540°C/540°C)
B Power plants with raised steam conditions utilising known materials (250 bar/540–560°C/560°C)

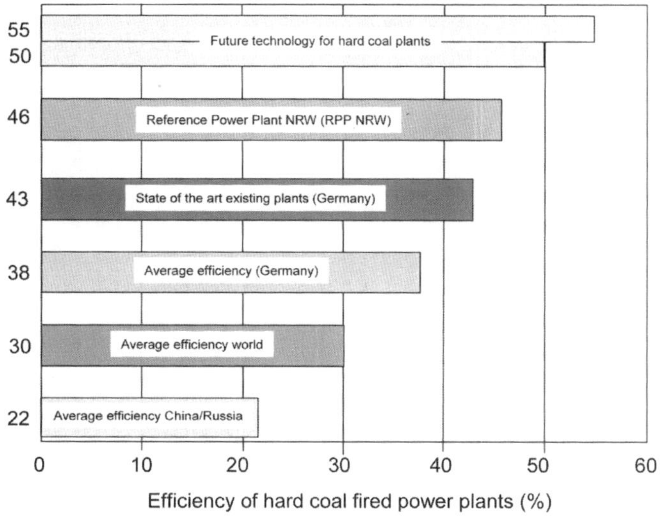

Fig. 4.106 Average efficiency of hard coal fired power stations in different regions (Meier 2004)

C Power plants with high steam conditions employing austenitic materials for the final-stage superheater and available new steel types for thick-walled parts (270–290 bar/580°C/600°C)

D Power plants with the highest steam conditions employing austenitic materials for superheaters, and new ferritic steel types for thick-walled parts and the evaporator (300–330 bar/600–630°C/600–630°C)

E Power plants with steam conditions up to 350 bar/700–750°C, with future materials

Figure 4.107 gives a breakdown of the efficiencies of the individual development stages for hard coal power plants. In the long term, an increase of the efficiency to more than 50% can be expected, which will require the development of suitable materials. Table 4.6 describes the materials used and to be used for boilers.

Table 4.6 Materials required for steam generator advancements

	Evaporator	Superheater	Thick-walled components	Steam parameters [bar/°C/°C/°C]		*Commissioning year*
A	13 CrMo 4 4	X20	X20	180/540/540	39	1985
B	13 CrMo 4 4	X20*	X20*	250/540/560	43	1993
C	13 CrMo 4 4*	Austenite	P91*	270/580/600	45	2000
D	HCM 2S* 7 CrMo...	Austenite*	Ferrite (P92, NF 616, NF 12) or austenite*	300/600/620	47	2010
E	HCM 12	Nickel alloy	Nickel alloy	350/700/700	50	2020

* Material stress limit reached.

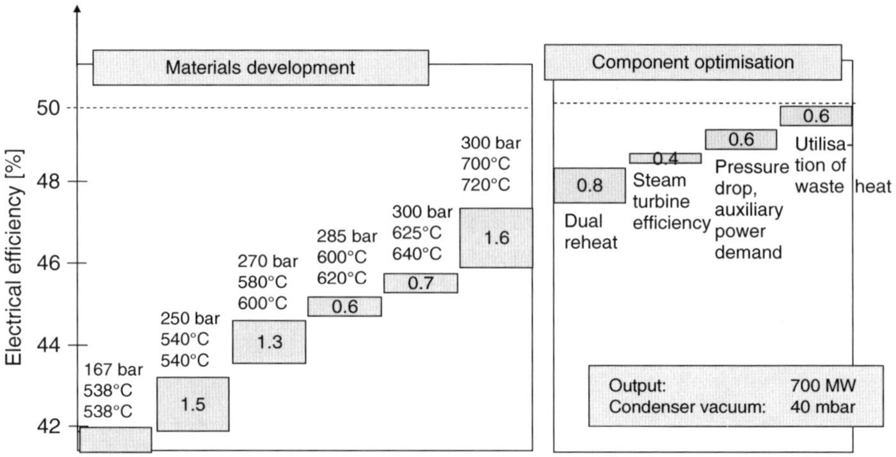

Fig. 4.107 Efficiency development in hard coal fired power stations (Rukes 2002)

A Power Plants with Conventional Steam Conditions

The reference power plant described in Sects 4.1 and 4.3 – a heat-flow diagram is presented in Fig. 4.28 – is a thermal power plant typical of Germany with conventional steam conditions that, until the end of the 1980s, were considered as the economic optimum. Typical efficiencies were somewhat below 40%. This corresponds to the average efficiency of existing power plants in Germany. Power plants with higher steam conditions have been designed and built since, here and there at first in Germany and Denmark, and then in other countries. Operating data and conclusions made from experience are now available.

B Power Plants with Raised Steam Conditions Utilising Known Materials (250 bar/540–560°C/560°C)

Still employing the known creep-resistant steel types – 13 CrMo 4 4 for the evaporator, X 20 CrMoV 12 1 for the superheater and thick-walled components – the steam conditions are raised through process-engineering modifications and by exhausting design reserves. A requirement for this is the homogeneous distribution of steam temperatures over the heating surfaces and of flue gas temperatures over the flue gas cross-sections. Temperature asymmetries of the flue gas are avoided by adequate control of the firing rate; of the steam, by adequate configuration of the heating surfaces or by a limited heat-up of the stages. The material limits of thick-walled components and final-stage superheating surfaces are then reached.

Representing this power plant development, Table 4.7 lists several plants: for hard coal the German Staudinger 5 and Rostock power stations and the Danish Esbjerg 3 power station, and for brown coal, the "Schwarze Pumpe" power station in Brandenburg, Germany.

Table 4.7 Data concerning various advanced steam power plants (Billotet and Johänntgen 1995; Breuer et al. 1995; Eichholtz et al. 1994; Lambertz and Gasteiger 2003; Meier 2004; VGB 2004; Spliethoff and Abröll 1985; Tippkötter and Scheffknecht 2004; Köhn 1993; Kjaer 1993; Vattenfall 2007)

Power Plant	Bexbach I/D	Staudinger/ D	Esbjerg 3/DK	Hessler/ D	Bexbach II/D	Aalborg/ DK	Referenzk-raftwerk NRW/D	Moorburg/ D	Neurath/ D	Schwarze Pumpe/D	Lippendorf D	Niederaußem K/D
Commissioning	1983	1992	1992	Planning 1994	Planning 1995	1998	Planning 2004	2010		1997	1999	2002
Coal					Hard coal						Brown coal	
LHV [MJ/kg]		27					25	25.1	8.5		10.5	9.2
P_{gross} [MW]	742	509	420	700	750	385	600	840	600	740	933	1,012
P_{net} [MW]	705	553		732			555.5	779	562		870	965
p_{LS} [bar]	190	250	251	275	250	290	285	276	176	260	268	275
t_{LS} [°C]	530	540	560	580	575	582	600	600	530	545	554	580
p_{RH} [bar]	37		56	50		74	60	51			52	60
t_{RH} [°C]	530	560	560	600	595	580	620	610	530	560	583	600
p_{RH2} [bar] t_{RH2} [°C]	–	–	–	–	–	19/580	–	–	–	–	–	–
$t_{feed\ water}$ [°C]	250	270	275	300	290	290	303	293		270	270	295
n_{stages}	7	7		9	9	9		9		7	8	10
$t_{furnace\ exit}$ [°C]	1,110		1,260	1,250	1,200	1,300				965	1,000	1,050
Flue gas exit temp. [°C]	130	125	104	125	110	100	115	115	142	170	170	160
Air ratio	1.3	1.2			1.15	1.17	1.15	1.17			1.15	1.15
Heat shift [°C]	–	–	–	–	110/90	–		–	–	–	170/130	160/100

Table 4.7 (continued)

Power Plant	Bexbach I/D	Staudinger/ D	Esbjerg 3/DK	Hessler/ D	Bexbach II/D	Aalborg/ DK	Referenzk-raftwerk NRW/D	Moorburg/ D	Neurath/ D	Schwarze Pumpe/D	Lippendorf D	Niederaußem K/D
Cold water [°C]	16.6	18	10	17.7	15	10		12	21.5	17.5	16.4	14.7
Vacuum [mbar]	55	38/52	23	37	30/43	23	45	27.4	66	42	38	28.5/35.5
Cooling range [°C]	15.9				12			5.9			10.2	10.3
Heat extraction [MW]	–	300	–	–	160	–		225	–	120		
Flue gas extraction	Chimney	Cooling tower	Chimney	Cooling tower	Cooling tower	Chimney	Cooling tower	Chimney	Cooling tower	Cooling tower	Cooling tower	Cooling tower
Efficiency												
Auxiliary power [%]	9.4	8.1	7.8		7.1	7.2		7.2			6.8	7.3
η_B [%]	94.0	94.5	95.8	94.4	95.4	95.4	95.0			89.8	90.6	94.4
η_{tot} [%]	38.7	43	45.3	45	46.3	47	45.9	46.5	35.5	40.4	42.3	43.2
Materials												
Evaporator	*	*	*	*	*	*	**	**	*	*	*	*
Final superheater	X20	X20	X20	Aust.	Aust.	Aust.	Aust TP 347H FG	Aust Super 304H	X20	X20	Aust.	Aust.

* 13 CrMo 4 4
** 7 CrMoVTiB 10 10

The plant currently (2009) featuring the highest efficiency amongst the hard coal fuelled thermal power plants of the public power supply in Germany is the 553 MW Rostock power plant unit with 43.2%, which went into operation in 1994. The efficiency is achieved by high live steam conditions (262 bar/545°C/560°C), by feed water preheating up to 270°C and by seawater cooling. The 553 MW power plant Staudinger 5, which went into operation in 1992, is similar in design and steam conditions and achieves an efficiency of 42.7%, with an optimised evaporative cooling system. The losses of the boiler and the turbine were diminished and the auxiliary power demand reduced by utilising low-loss drives for the main feed pump and condensate pumps, as well as more effective fans. Still employing conventional materials, the design limitations were set by the final-stage superheater material, X20; higher temperatures would have required the use of austenitic materials (Rukes et al. 1994; E.ON 2006).

The Danish plant of Vestkraft has roughly comparable thermodynamic conditions and reaches a noticeably deeper vacuum due to the once-through cooling with cold seawater. Cooling with seawater helps to achieve a vacuum of 22 mbar at an annual average of 10°C, whereas inland in Germany, it is only possible to achieve cold water temperatures of 15–18°C by evaporative cooling, which correspond to a vacuum of 36–42 mbar. By seawater cooling and the rather lower steam generator flue gas temperature of 104°C, which, however, requires low-sulphur coal, it is possible to achieve an efficiency which is better by 2%.

C Power Plants with High Steam Conditions Employing Austenitic Materials for the Final-Stage Superheater and Available New Steel Types for Thick-Walled Parts (270–290 bar/580°C/600°C)

For power plants designed before 2000, net efficiencies of 45–46% could be achieved (Billotet and Johänntgen 1995; Kotschenreuther et al. 1993; Eichholtz et al. 1994; Kjaer 1993). The respective live steam conditions of 270–290 bar and 580°C and the reheat temperature of 600°C allowed the use of the approved 13 CrMo 4 4 material for the evaporator wall, whereas the superheating surfaces required austenitic materials. Only the development of the ferritic steel P91 for thick-walled components made it possible to employ advanced steam conditions. Helically wound tubing for the evaporators in the furnace section was used.

In the planning of the 700 MW Hessler power plant, an efficiency was calculated that ranged around 45% (Eichholtz et al. 1994).

The Danish Aalborg power plant achieves an efficiency of 47%, applying double reheating, cooling with seawater and low boiler-outlet temperatures. This plant still holds the world record for efficiency of a coal-fired steam power plant today (2009).

D Power Plants with the Highest Steam Conditions Employing Austenitic Materials for Superheaters, and New Ferritic Steel Types for Thick-Walled Parts and the Evaporator (300–330 bar/600–630°C/600–630°C)

Power plants employing austenitic materials for superheaters, and new martensitic steel types for thick-walled parts and the evaporator, are predicted to raise efficiencies up to 48%. This is made possible by raising the steam conditions up to 300–330 bar and 630°C/630°C and advancing the process and the single components (Kjaer 1994). The transition to higher steam conditions requires higher-duty steel types both for the evaporator wall and for thick-walled components. For the evaporator, new steels like HCM 2S and 7 CrMoVTiB 10 10 are now available. For thick-walled components, steel P91 becomes unsuitable, and the newly developed steel P92 has to be used. The live steam conditions mentioned mean that the stress maximum occurs on the evaporator walls, the final-stage superheater and the thick-walled components. Turbine shafts and cylinders, as well, need to be built with austenitic materials. From power plants used in the chemical industry, data based on experience with turbines at live steam conditions of more than 300 bar and temperatures up to 625°C is available.

A number of power plants are today (2009) under construction in Germany (Datteln, Karlsruhe, Moorburg and Westfalen). The current maximum achievable steam conditions with presently available steels are considered to be 300 bar and 600°C/620°C. The design of these power plants will be similar to the design of the reference power plant North Rhine-Westphalia (RPP NRW), which is given in Table 4.7. The RPP NRW achieves an efficiency of 45.9% with live steam conditions of 285 bar and 600°C and reheat steam temperatures of 620°C. The design study showed that by increasing the feed water preheating up to 320°C, decreasing the condenser pressure to 35 mbar and increasing the flue gas heat utilisation, it is possible to raise the efficiency to 47.3%. The power plant in Moorburg has a design efficiency of 46.5% (Meier 2004; VGB 2004; Michel 2006; Klebes 2007; Schmitz 2007; Willeke 2007; Mandel and Schettler 2007; Then et al. 2007; Vattenfall 2007).

E Power Plants with Steam Conditions up to 350 bar/700–750°C, with Future Materials

Initiatives have commenced to further increase the live steam temperatures up to 700–750°C and the live steam pressure up to 350 bar (Blum et al. 2007; Bauer et al. 2003). For a power plant with a single reheat cycle cooled by a wet cooling tower, net efficiencies are in the range of 50–51%, and 53–54% if it is based on a double reheat cycle cooled by seawater (Hoestgaard-Jensen et al. 2003). The 700°C technology is expected to be mature after 2010 and long-term targets are net efficiencies above 55%, based on maximum steam temperatures in the range of 800°C (Kjaer et al. 2002). In Germany, E.ON has announced it will build a 550 MW power plant with 365 bar/705°C/720°C and an efficiency of about 50%, which is expected to be operational in 2014 (Bauer et al. 2008).

In order to run such a high-temperature process, new materials and construction procedures have to be developed and approved for the steam generator, turbine and piping. The testing of the critical components for the 700°C technology was the reason for launching the European Research Project AD 700 in 1998. Major targets were the development of austenitic steels and nickel-based superalloys for the hottest parts of boilers, steam lines and turbines, and the development of boiler and

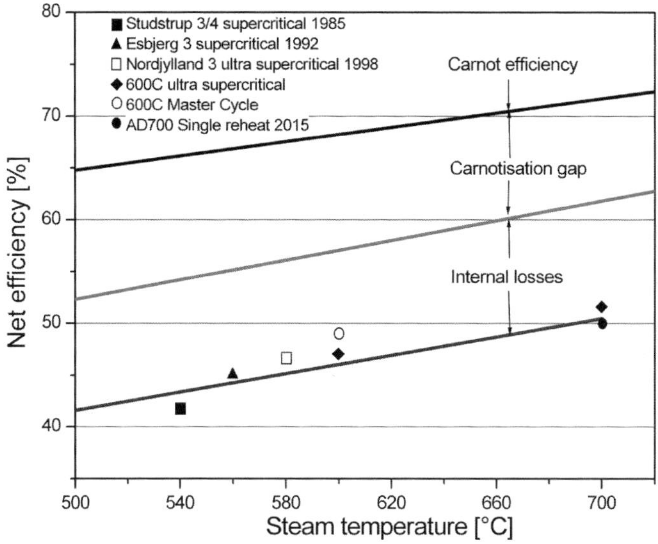

Fig. 4.108 Net efficiency of seawater-cooled supercritical power plants (Kjaer and Drinhaus 2008)

turbine designs. The materials under consideration for the evaporator are the HCM 12 steel or various nickel-based alloys. For the convective heating surfaces and for thick-walled components, nickel-based alloys are the only choice (Blum et al. 2007; Bauer et al. 2003; Chen et al. 2005). When applying the new materials, problems to do with wall thickness, high-temperature corrosion and steam oxidation have to be taken into account. Besides the mechanical properties and the workability, the costs have to be borne in mind when choosing the material, considering the high quantities needed of it in a thermal power plant.

Figure 4.108 shows the net efficiency of a number of seawater-cooled supercritical power plants versus the maximum steam temperature and the ideal Carnot efficiency. It is obvious that higher process temperatures drive the Carnot and the net efficiency upwards. Over the past 20 years, materials development has resulted in an increase of live steam and reheat temperatures by 60 K to about 600°C, corresponding to a heat rate improvement of 3%. However, further steel-based improvements are not expected for the next 10–15 years as no new candidates replacing P92 are in sight, and presently only nickel-based alloys seem to allow steam conditions of 700°C.

Figure 4.108 divides the gap between the Carnot and the net efficiency into two gaps of almost equal size, one reflecting the lack of thermodynamic completeness of the super critical water/steam cycles (the Carnotisation gap) and the other internal losses (the internal losses gap). The dividing curve has been calculated by setting all equipment efficiencies at 100% and parasitic losses at zero and calculating the net efficiency of the supercritical power plants. Both gaps are about 10% for a steam temperature of 600°C. Whereas higher steam temperatures raise the Carnot efficiency, optimisation of the thermodynamic cycle with a higher live steam pressure, a higher feed water preheating temperature and double reheating, etc. can reduce the Carnotisation gap. It is expected that the optimisation of the Carnotisation gap would provide limited potential for efficiency increase. Efficiency improvements of components and reductions of the auxiliary power demand are already quite advanced and only minor improvements are likely to be made (Kjaer and Drinhaus 2008).

4.6.2 Brown Coal

A considerable increase of efficiency could be achieved as well for brown coal. After the 600 MW units built after 1972 in the Rhenish mining area, with efficiencies around 35.5% (Heitmüller et al. 1996), it was possible to achieve an efficiency of 40% for the Schwarze Pumpe power plant (2 × 800 MW), at live steam conditions comparable with the power plant Staudinger 5 (Lauterbach et al. 1993). The Lippendorf power plant (2 × 930 MW), in service since 1997, reaches an efficiency of 42.3%, with again higher steam conditions (Breuer et al. 1995). The operating conditions of the Niederaußem power plant, put into operation in 2002 in the Rhenish mining area, can be compared to those of the plant design of the Hessler plant and achieves an efficiency of 43.2% (Heitmüller et al. 1996; Böcker and Hlubek 1995; Lambertz and Gasteiger 2003). Brown coal fired power stations with similar steam conditions are planned (Kehr et al. 2005). Comparing the power plants in the Rhenish mining area with those in East Germany, it has to be considered that the East-German brown coal types, having a higher fouling tendency, limit the furnace-outlet temperature to values between 950 and 1,000°C, whereas the Rhenish brown coal allows furnace-outlet temperatures of around 1,050°C, and thus favours the application of higher steam conditions (see Table 4.7).

A pre-drying stage for the moist brown coal is a point of discussion as well. Such a stage would increase the efficiency by 2%, so that brown- and hard coal power plants would in the end feature more or less the same overall efficiency. If the drying method additionally exploited the condensation heat of the water vapours, the efficiency could be increased by about 4–6% to eventually reach 47–49% (280 bar/580°C, 600°C) (Ewers et al. 2003).

References

13.BImSchV. (2004). 13. Verordnung zur Durchführung des Bundesimmissionsschutzgesetzes (Verordnung über Großfeuerungsanlagen, 13. BImSchV) vom 23.7.2004, BGBl. I Nr. 37, p. 1717, from http://www.gesetze-im-internet.de/bimschv_13_2004/index.html.

Adrian, F., Quittek, C. and Wittchow, E. (1986). Fossil beheizte Dampfkraftwerke. Handbuchreihe Energie, Band 6, Herausgeber T. Bohn. Technischer Verlag Resch, Verlag TÜV Rheinland

Apblett, W. R. (1973). *Corrosion in coal combustion systems*. Corrosion and erosion of coal conversion systems materials.

Baehr, H. D. and Kabelac, S. (2006). Thermodynamik: Grundlagen und technische Anwendungen. Berlin, Heidelberg, Springer.

Baehr, R. (1985). Konzeption und Aufbau von Dampfkraftwerken, Handbuchreihe Energie, Band 5, Herausgeber T. Bohn. Technischer Verlag Resch, Verlag TÜV Rheinland.

Bauer, F., Stamatelopoulos, G., Vortmeyer, N. and Bugge, J. (2003). Driving coal-fired power plants to over 50% efficiency. *VGB PowerTech* **83**(12): 97–100.

Bauer, F., Tschaffon, H. and Hourfar, D. (2008). Role of 700°C technology for the carbon-low power supply. *VGB PowerTech* **88**(4): 30–34.

Bauer, G. and Lankes, F. (1997). Erhöhung des Blockwirkungsgrades durch verbesserte Abgaswärmenutzung. *VGB Kraftwerkstechnik* **77**(5): 376–381.

Berliner, P. (1975). Kühltürme. Grundlagen der Berechnung und Konstruktion. Berlin, Heidelberg, New York, Springer.

Billotet, T. and Johänntgen, U. (1995). Kraftwerk Bexbach II – Fortschrittliche Stromerzeugung aus Steinkohle. *VGB Kraftwerkstechnik* **75**(1): 21–25.

Blum, R. and Vanstone, R. W. (2005). Materials Development for Boilers and Steam Turbines Operating at 700°C. Proceedings of the 1st International Conference on Super-High Strength Steels, 2–4 November 2005, Rome, Italy, Associazione Italiana di Metallurgia.

Blum, R., Kjaer, S. and Bugge, J. (2007). Development of a PF Fired High Efficiency Power Plant (AD700). Proceedings of the Riso International Energy Conference "Energy Solutions for Sustainable Development", Denmark, 2007 Riso-R-1608 (EN).

Böcker, D. and Hlubek, W. (1995). Umwelt- und Klimaschutzprogramm für die rheinische Braunkohle. *ET* **45**(3): 119–127.

Böcker, D., Klöcker, K. J. and Klutz, H.-J. (1992). Verfahren zur Trocknung und Mahlung von Braunkohle. *BWK* **44**(7/8): 315–322.

Breuer, H., Kehr, M. and Gerstmeyer, R. (1995). Darstellung und Entwicklung der neuen Braunkohlekraftwerke der VEAG. Beispiel des Gemeinschaftskraftwerkes Lippendorf. *VGB Kraftwerkstechnik* **75**(2): 93–100.

Bruckmann, W. and Hesel, U. (1996). Exergiesteigerung durch Schaltungsänderung der Wärmeauskopplung bei Kohlekraftwerken. *VDI-Tagung. "Entwicklungslinien der Energie- und Kraftwerkstechnik"*. Siegen.

Chen, Q. and Scheffknecht, G. (2003a). Neue Dampferzeuger- und Rohrleitungswerkstoffe und ihre Anwendungsgrenzen. VGB Konferenz "Kraftwerke im Wettbewerb – Technik, Betrieb und Umwelt 2003". Köln.

Chen, Q. and Scheffknecht, G. (2003b). New boiler and piping materials. *VGB PowerTech* **83**(11): 91–98.

Chen, Q., Stamatelopoulos, G., Helmrich, A., Kjaer, S. and Stolzenberger, C. (2005). Investigations on oxidation and corrosion characteristics of the advanced boiler materials at steam temperatures up to 720°C, Proceedings of the 1st International Conference on Super-High Strength Steels, 2–4 November 2005, Rome, Italy, Associazione Italiana di Metallurgia.

Doležal, R. (1990). Dampferzeugung: Verbrennung, Feuerung, Dampferzeuger. Berlin, Heidelberg, New York, Springer.

E.ON (2006). www.kraftwerk-rostock.de.

Effenberger, H. (2000). Dampferzeugung. Berlin, Heidelberg, Springer.

Eichholtz, A., Hourfar, D. and Kübler, D. (1994). Der 700-MW-Steinkohleblock der VEBA Kraftwerke Ruhr AG in Gelsenkirchen-Heßler. *VGB Kraftwerkstechnik* **74**(1): 25–29.

Eitz, A. W. (1996). Kraftwerkstechnik in Deutschland – gestern, heute und morgen. Fachvortrag anläßlich der 75-jährigen Jubiläumsfeier der Firma EVT.

Elsen, R., Blumenthal, U., Wick, W. and Ewers, J. (1996). *Wirkungsgradverbesserung neuer Braunkohlekraftwerke durch vorgeschaltete Trocknung*. VDI-Berichte 1280, pp. 149–164.

Ernst, G., Leidinger, B., Natusch, K., Petzel, H. and Scholl, G. (1986). Kühlturm und Rauchgasentschwefelungsanlage des Modellkraftwerkes Völklingen. VDI-Fortschrittberichte, Reihe 15, Nr. 45.

Ewers, J., Klutz, H.-J., Renzenbrink, W. and Scheffknecht, G. (2003). The development of pre-drying and BoA-Plus technology. *VGB PowerTech* **83**(11): 60–65.

Franke, J., Cossmann, R. and Huschauer, H. (1995). Benson-Dampferzeuger mit senkrecht berohrter Brennkammer: Praxisnaher Großversuch belegt Auslegungssicherheit. *VGB Kraftwerkstechnik* **75**(4): 353–359.

Franke, J., Köhler, W. and Wittchow, E. (1993). Verdampferkonzepte für Benson-Dampferzeuger: Heutiger Stand und neue Entwicklungen. *VGB Kraftwerkstechnik* **73**(4): 352–361.

Hahn, B. and Bendick, W. (2006). Stand der Werkstoffentwicklungen für neue Kraftwerke, Möglichkeiten für deren Nutzung für die Ertüchtigung von Altanlagen. *VGB PowerTech* **86**(6): 46–55.

Heiermann, G., Husemann, R.-U., Hougard, P., Kather, A. and Knizia, M. (1993). Dampferzeuger für fortgeschrittene Dampfzustände. *VGB Kraftwerkstechnik* **73**(8): 678–689.

Heitmüller, R. J., Kallmeyer, D., Kückelhaus, K. and Wick, W. (1996). Neue Braunkohleblöcke mit hohem Wirkungsgrad bei RWE Energie AG. Düsseldorf VDI-Berichte 1280, pp. 77–90.

Henning, H. (1985). Stand und Entwicklung im Kühlturmbau. *Technische Mitteilungen* **78**(10): 511–524.

Hoestgaard-Jensen, P., Bauer, F. and Kjaer, S. (2003). Emax: VGB initiative for clean coal technologies. *VGB PowerTech* **83**(5): 92–95.

Husemann, R.-U. (2003). Development status of boiler and piping materials for increased steam conditions. *VGB PowerTech* **83**(9): 124–128.

Johänntgen, U. (1998). Kühlwasserseitige Bypass-Schaltung an Kondensationsanlagen. *VGB Kraftwerkstechnik* **78**(3): 57–64.

Kallmeyer, D. and Wick, W. (1997). Entwicklungsschritte zum trockenbraunkohlebefeuerten Kraftwerk. *VGB Kraftwerkstechnik* **77**(4): 300–303.

Kather, A. and Scheffknecht, G. (1997). Neue Werkstoffe und Dampferzeugerbau. *BWK* **49**(11/12): 62–67.

Kefer, E., Köhler, W. and Wittchow, E. (1990). Wärmeübergang und Druckverlust in Dampferzeugerrohren: Forschung und Anwendung. *VGB Kraftwerkstechnik* **70**(10): 827–832.

Kehr, M., Elsen, R. and Ewers, J. (2005). Aktuelle Neubauprodukte und Weiterentwicklungen in der Kohlekraftwerkstechnik. *VGB PowerTech* **80**(4): 30–35.

Kjaer, S. (1990). Kohlestaubgeförderte Kraftwerksblöcke mit fortgeschrittenem Wasser/Dampfprozess. *VGB Kraftwerkstechnik* **70**(3): 201–208.

Kjaer, S. (1993). Die zukünftigen 400 MW-ELSAM-Blöcke in Aalborg und Skaeraek. *VGB Kraftwerkstechnik* **73**(11): 933–940.

Kjaer, S. (1994). Advanced steam cycles. *IFRF-Conference "Advanced Power Generation", TOTEM 10*. Veldhoven, NL.

Kjaer, S. and Drinhaus, F. (2008). Presentation of a modified double reheat cycle. 40. Kraftwerkstechnisches Kolloquium, 14–15 Oktober 2008, Vortrag V 5.5, Dresden.

Kjaer, S., Klauke, F., Vanstone, R. W., Zeijseink, A., Weissinger, G., Kristensen, P., Meier, J., Blum, R. and Wieghardt, K. (2002). The advanced supercritical 700°C pulverised coal-fired power plant. *VGB PowerTech* **82**(7): 46–49.

Klebes, J. (2007). High-efficiency coal-fired power plants based on proven technology. *VGB PowerTech* **87**(3): 80–84.

Klenke, W. (1966). Die Kühlturmkennlinie als Mittel für die Beurteilung von Kühltürmen. *BWK* **18**(3): 97–105.

Klutz, H. J. and Holzenkamp, M. (1996). Verfahren und Anlagen zur Braunkohletrocknung. Düsseldorf, VDI-Berichte Nr. 1280, pp. 91–105.

Klutz, H.-J., Klöcker, K. J. and Lambertz, J. (1996). Das WTA-Verfahren als Vortrocknungsstufe für moderne Kraftwerkskonzepte auf Basis Braunkohle. *VGB Kraftwerkstechnik* **76**(3): 224–229.

Klutz, H.-J., Moder, C. and Block, D. (2006). WTA-Feintrocknung, Baustein für die Braunkohlekraftwerke der Zukunft. *VGB PowerTech* **86**(11): 57–61.

Knizia, K. (1966). Die Thermodynamik des Dampfkraftprozesses. Berlin, Springer.

Köhn, G. (1993). Staudinger 5 – Auslegung und Erfahrungen aus Inbetriebnahme und erster Betriebszeit. *VGB Kraftwerkstechnik* **73**(11): 927–932.

Kotschenreuther, H. and Klebes, J. (1996). Die Marktentwicklung bestimmt die zukünftigen Kraftwerkskonzeptionen in Europa. Düsseldorf, VDI-Berichte Nr. 1280, pp. 39–52.

Kotschenreuther, H., Häuser, U. and Weirich, P. H. (1993). Zukünftige wirtschaftliche Kohleverstromung. *VGB Kraftwerkstechnik* **1**: 22–28.

Kral, R., Schröder, S. and Zipfel, T. (1993). Versuche mit einem senkrecht berohrten Benson-Verdampfer in einem 160-t/h Dampferzeuger. *VGB Kraftwerkstechnik* **73**(9): 793–797.

Lambertz, J. and Gasteiger, G. (2003). BoA Niederaußem: Konzeption und Betriebserfahrungen. *VGB PowerTech* **83**(5): 82–87.

Lauterbach, K., Mandel, H., Haase, H. and Kotschenreuther, H. (1993). Konzepte für 800-MW-Turbinen der VEAG im Kraftwerk Boxberg und im Gemeinschaftskraftwerk der VEAG/ESPAG am Standort Schwarze Pumpe. *VGB Kraftwerkstechnik* **73**(6): 500–508.

Ledinegg, M. (1966). Dampferzeugung, Dampfkessel, Feuerungen. Wien, New York, Springer.

Lehmann, L., Klein, M. and Wittchow, E. (1996). Der Block 4 im Kraftwerk Heyden: Referenzanlage für moderne 1000 MW-Steinkohlekraftwerke. *VGB Kraftwerkstechnik* **76**(2): 85–94.

Lorey, H. and Scheffknecht, G. (2000). Zwangdurchlauf-Dampferzeuger mit hohen Dampfparametern, Entwicklungsstand und Entwicklungstendenzen. *BWK* **52**(1/2): 42–48.

Maier, H., Bilger, H. and Mayer, G. (1992). Betriebserfahrungen mit der SCR-Stickoxidminderungstechnik. *VGB Kraftwerkstechnik* **72**(9): 796–800.

Mandel, H. and Schettler, H. (2007). Die Neubaublöcke von Vattenfall Europe mit ihren technischen Herausforderungen. *VGB-Konferenz, "Kraftwerke im Wettbewerb"*. Hamburg.

Meier, H. J. (2004). Reference Power Plant North Rhine-Westfalia (RPP NRW). *VGB PowerTech* **84**(5): 76–89.

Michel, M. (2006). Kraftwerksneubau der E.ON Energie AG in Deutschland. *VGB PowerTech* **86**(4): 46–51.

Müller-Odenwald, H., Demuth, J. and Farwick, H. (1995). Der Luftvorwärmer – eine Komponente zur Emissionsminderung ($CO_2 + SO_3$). *VGB Kraftwerkstechnik* **75**(11): 972–979.

Neft, H. G. and Franconville, G. (1993). Neue Dampfturbinenkonzepte für höhere Eintrittsparameter und längere Endschaufeln. *VGB Kraftwerkstechnik* **73**(5): 409–415.

Nowi, A. and Haller, B. R. (1997). Wirkungsgradentwicklung bei Dampfturbinen. *VGB Kraftwerkstechnik* **77**(7): 551–555.

Oakey, J. E., Pinder, L. W., Vanstone, R., Henderson, M. and Osgerby, S. (2003). Review of status of advanced materials for power generation. Report Nr. COAL R224, DTI/Pub URN 02/159, 2003.

Odenthal, A. and Spangenmacher, K. (1959). Der Kühlturm im Dampfkraftprozeß. *BWK* **11**(12): 556–652.

Oeynhausen, H., Drosdziok, A. and Deckers, M. (1996). Dampfturbinen für die neue Kraftwerksgeneration. *VGB Kraftwerkstechnik* **76**(12): 974–979.

Plumpley, A. L. and Roczniak, W. R. (1988). Coal ash corrosion field testing of advanced boiler tube materials. Proceedings of the 2nd EPRI International Conference on "Improved Coal Fired Power Plants" Palo Alto, CA.

Pollack, M. and Heitmüller, R. J. (1996). Trockenbraunkohle-gefeuerte Dampferzeuger. VDI-Berichte 1280, pp. 133–148.

Reuter, F. D. and Hönig, J. (1988). Beeinflussung der Kesselauslegung durch Einbau einer SCR-Anlage. *VGB Kraftwerkstechnik* **68**(2): 118–124.

Riedle, K., Rukes, B. and Wittchow, E. (1990). Die Erhöhung des Kraftwerkswirkungsgrades in der Vergangenheit und in der Zukunft. *VGB Konferenz "Kraftwerkstechnik 2000"*. Essen, VGB TB 120.

Riemenschneider, G. (1995). Entwicklungspotentiale bei stein- und braunkohlegefeuerten Dampferzeugern. *BWK* **47**(7/8): 296–302.

Rukes, B. (2002). Perspectives of fossil power technology. *VGB PowerTech* **82**(10): 71–76.

Rukes, B., Vollmer, W. and Wittchow, E. (1994). Abgestimmte Auslegung von Dampfturbine und Dampferzeuger. *VGB Kraftwerkstechnik* **74**(5): 405–411.

Sauer, E. (1984). Abwärmetechnik, Kühlsysteme, Umweltprobleme, Abwärmenutzung. Köln, Verlag TÜV Rheinland.

Scheffknecht, G., Chen, Q. and Kather, A. (1996). Dampferzeuger mit fortschrittlichen Dampfparametern. VDI-Berichte Nr. 1280, pp. 335–353.

Scheffknecht, G., Stamatelopoulos, G. and Lorey, H. (2002). Moderne Kohlekraftwerke. *BWK* **54**(6): 46–50.

Schelenkert, J. U., Rüsenberg, D., Bruch, U. and Niepel, A. (2006). *Erfahrungen beim Betrieb des Dampferzeugers Block K im Kraftwerk Neideraußem der RWE Power*. VGB-Fachtagung Dampferzeuger, Industrie- und Heizkraftwerke, 4–6 September 2006, Würzburg.

Schmalfeld, J. and Twigger, C. (1996). Erfahrungen mit dem Betrieb der Dampfwirbelschicht-Trocknungsanlage Loy Yang, Australien. Düsseldorf, VDI-Berichte Nr. 1280, pp. 107–117.

Schmidt, E., Stephan, K. and Mayinger, F. (1977). Technische Thermodynamik. Band 2, Mehrstoffsysteme und chemische Reaktionen. Berlin, Heidelberg, New York, Springer.

Schmitz, E. (2007). Energiepolitische Perspektiven für Europa und Deutschland sowie die Rolle der Steinkohle im Energiemix. VGB-Konferenz *"Kraftwerke im Wettbewerb"*. Hamburg.

Schröder, K. (1968). Große Dampfkraftwerke, Band 3a: Die Kraftwerksausrüstung. Berlin, Heidelberg, New York, Springer.

Schwendig, F., Klutz, H. J. and Ewers, J. (2006). Das Trockenbraunkohle befeuerte Kraftwerk. *VGB PowerTech* **86**(12): 51–57.

Smith, J. W. (1996). Babcock & Wilcox Company Supercritical (Once Through) Boiler Technology. Barberton, OH, Firmenschrift Babcock & Wilcox BR-1614.

Sobbe, W. (2004). Shell boilers with a steam output of up to 54 t/h. *VGB PowerTech* **84**(1–2): 77–79.

Spliethoff, H. and Abröll, G. (1985). Das 750 MW-Steinkohlekraftwerk Bexbach. *VGB Kraftwerkstechnik* **65**: 346–362.

Stamatelopoulos, G.-N. and Weissinger, G. (2005). *Die nächste Generation von Steinkohlekraftwerken*. VGB-Konferenz "Kraftwerke im Wettbewerb 2005, Ordnungspolitik, Markt und Umweltschutz", Potsdam.

STEAG (1988). Strom aus Steinkohle: Stand der Kraftwerkstechnik. Berlin, Heidelberg, New York, London, Paris, Tokyo, Springer.

Strauß, K. (2006). Kraftwerkstechnik: Zur Nutzung fossiler, nuklearer und regenerativer Energiequellen. Berlin [u.a.], Springer.

Stultz, S. C. and Kitto, J. B. (1992). Steam, Its Generation and Use. Barberton, OH, The Babcock & Wilcox Company.

Suhr, L. (1992). Wärmeauskopplung aus Kraftwerksabgasen mit korrosionsfesten Wärmetauschern zur Kondensat- und Luftvorwärmung. *VGB Kraftwerkstechnik* **72**(8): 687–690.

Theis, K. A. (2005). Kurzbericht über die Tätigkeit der VGB PowerTech in 2004/2005. *VGB PowerTech* **85**(9): 37–69.

Then, O., Wüllenweber, H.-J. and Kleinhörster, B. (2007). Neue Kohlekraftwerke bei RWE Power – Technologien und Herausforderungen. VBG-Konferenz "Kraftwerke im Wettbewerb", Hamburg, 25–26. April 2007.

Tippkötter, T. and Scheffknecht, G. (2004). Betriebserfahrungen mit dem neuen BOA-Block und Ausblick. *VGB PowerTech* **84**(4): 48–55.

Traupel, W. (1982). Thermische Turbomaschinen, Band 2, Geänderte Betriebsbedingungen, Regelung, Mechanische Probleme, Temperaturprobleme, 3. Auflage. Berlin, Heidelberg, New York, Springer.

Traupel, W. (2001). Thermische Turbomaschinen, Bd. II, Strömungsmaschinen. Berlin, Springer.

Tremmel, A. and Hartmann, D. (2004). Efficient steam turbine technology for fossil fuel power plants in economy and ecologically driven markets. *VGB PowerTech* **84**(11): 38–43.

Tremmel, A., Mander, H., Klauke, U. and Brandt, C. (2006). Modernste Turbinentechnologie mit höchsten Dampftemperaturen für das Kraftwerk Boxberg R. *VGB PowerTech* **86**(12): 71–75.

UCTE (2004). UCTE operation handbook, P1 load-frequency control and performance, from http:/www.ucte.org.

Vattenfall (2007). Antrag auf Genehmigung nach dem Bundes-Immissionsschutzgesetz für die Errichtung und den Betrieb des Kraftwerkes Moorburg Stadt Hamburg.

VDN (2007). TransmissionCode 2007, Netz und Übertragungsregeln der deutschen Netzbetreiber, from http://www.vdn-berlin.de.

Verbundgesellschaft, D. (1991). Das versorgungsgerechte Verhalten der thermischen Kraftwerke. Heidelberg, DVG.

Verbundgesellschaft, D. (1996). Anforderungen an die Primärregelung im UCPTE-Verbundbetrieb. Heidelberg, DVG.

VGB (1995). Kraftwerke 1970 bis 1995 – Wege und Meilensteine. Ein Überblick anläßlich des 75jährigen Bestehens der VGB. Essen.

VGB (2004). Konzeptstudie Referenzkraftwerk Nordrhein-Westfalen.

Viswanathan, R., Henry, J. F., Tanzosh, J., Stanko, G., Shingledecker, J., Vitalis, B. and Purgert, R. (2005a). U.S. program on materials technology for ultra-supercritical coal power plants. Journal of Materials Engineering and Performance. **14**(3): 281–291.

Viswanathan, R., Purgert, R. and Rao, U. (2005b). Materials technology for advanced coal power plants. Proceedings of the 1st International Conference on Super-High Strength Steels. A. I. d. Metallurgia. Rome, Italy.

Wauschkuhn, A. (2001). Wirtschaftlichkeitssteigerung thermischer Kraftwerke unter Berücksichtigung der Prozessdynamik. University of Stuttgart. Doktorarbeit, Shaker, Aachen.

Weber, F., Alzinger, R. and Maschke, U. (2005). Optimierungspotential für Kraftwerke zur Stromerzeugung – Maßnahmen am kalten Ende. *VGB PowerTech* **85**(9): 85–87.

Wichtmann, A., Deckers, M. and Ulm, W. (2005). Ultra-supercritical steam turbine turbosets – best efficiency solution for conventional steam power plants. *VGB PowerTech* **85**(11): 44–49.

Willeke, A. (2007). Das Steinkohlekraftwerk Datteln 4 – Start der E.O.N.-Konvoi-Projekte in der 1.100-MW-Klasse. VGB-Konferenz "Kraftwerke im Wettbewerb", Hamburg, pp. 25–26 April 2007.

Wittchow, E. (1982). Trommelkessel oder Durchlaufkessel: Einfluß des Verdampfersystems auf die Auslegung und das Betriebsverhalten der Anlage. *VGB Kraftwerkstechnik* **62**(5): 346–356.

Wittchow, E. (1995). Weiterentwicklung des Benson Dampferzeugers. Düsseldorf, VDI-Berichte Nr. 1182, pp. 121–136.

Zehtner, W. (2009). Betriebsoptimierung von Steinkohlekraftwerken durch Simulation. München, TUM. Disserationsschrift.

Zimmer, C. and Leithner, R. (1995). Wirkungsgradverbesserungen eines konventionellen Kraftwerkes durch Braunkohleaustrocknung mit Abgas. *BWK* (47), Nr. 3, pp. 78–81.

Chapter 5
Combustion Systems for Solid Fossil Fuels

Coal firing systems are comprised of the sub-systems of fuel supply and preparation, fuel and combustion air transport and distribution, the furnace for releasing the heat from the fuel and flue gas cleaning.

The systems used for combusting solid fossil fuels are as follows:

- Grate firing
- Fluidised bed firing
- Pulverised fuel firing (Stultz and Kitto 1992; Strauß 2006; STEAG 1988; Doležal 1990; Günther 1974; Gumz 1962; Görner 1991)

Table 5.1 compares the advantages and disadvantages of different combustion systems. Figure 5.1 gives the characteristic gas and solid fuel flow velocities, pressure losses and heat transfer coefficients of each of the combustion systems.

In a grate firing system, the solid fuel lies in a bulk bed on a moving grate. The fuel burns with the combustion air which is blown through the grate bars and through the bulk. At low flow velocities, single coarse coal particles with sizes up to 30 mm (approximately the size of a nut) remain in the coal layer on the grate. Notable quantities of solids are not entrained. Because of the limited capacity of this furnace type, coal-fired grates are only used for industrial and thermal power plants of small capacity. Grate firing is the preferred system for ballast-containing fuels such as waste, or for solid industrial wastes, or biomass, because no or minor fuel preparation is required.

In fluidised bed firing, the solid fuel is fluidised and burns while in a gas – solid suspension. The fluidising medium also provides the oxygen for the oxidation of the fuel. With the lower flow velocities of the bubbling fluidised bed (BFB), only the fine-grained ash from the fluidised bed is entrained in the gas after burnout and abrasion of the coal. Coarse-grained ash accumulates in the fluidised bed, from where it is removed. With the higher flow velocities of combustion air and combustion gases of the circulating fluidised bed (CFB), the entire solid flow in the furnace is entrained and circulated. The circulating fluidised bed occupies the entire furnace volume. In both systems, the solids stay in the furnace appreciably longer than the gas flow.

H. Spliethoff, *Power Generation from Solid Fuels,* Power Systems,
DOI 10.1007/978-3-642-02856-4_5,

Table 5.1 Comparison of grate, fluidised bed and pulverised fuel firing systems

Grate firing systems	Bubbling fluidised bed (BFB) and circulating fluidised bed (CFB) firing systems	Pulverised fuel firing systems
Advantages	Advantages	Advantages
– Relatively minor fuel preparation requirement – Clear design – High process availability – simple operation – Low auxiliary power demand	– Relatively minor fuel preparation requirement – Flue gas cleaning consists only of particulate collection	– High process availability – Large capacities – High power density – Good burnout – Utilisable ash
	Disadvantages of BFB and CFB	Disadvantages
– Low NO_x emissions (e.g. bituminous coals $< 400 mg/m^3$) – Partial desulphurisation by limestone addition	– High limestone demand for sulphur capture – Ash not utilisable without further preparation	– Relatively major fuel preparation requirement – Flue gas cleaning needed for particulates, SO_2 and NO_x
Disadvantages	Advantages of CFB against BFB	
– High combustion losses of 2–4% unburnt carbon – High flue gas temperatures due to limited air preheating – Unsuitable for fine-grained fuels	– Better burnout – Lower limestone demand for sulphur capture – Lower emission values – No in-bed heating surfaces at risk of erosion – Better power control	

In pulverised fuel firing systems, the coal particles are carried along with the air and combustion gas flow. Because particles are entrained in the gas flow, this firing type is also known as entrained-flow combustion. Pulverised fuel and combustion air are injected into the firing via the burner and mixed in the furnace. With a fine raw coal milling degree and high combustion gas flow velocities, particle and gas residence times are almost equal. The combustion of the pulverised coal/air mixture being a rapid process distributed over the entire furnace makes it possible to achieve higher capacities than grate or fluidised bed firing systems.

The choice of the firing system depends on the properties of the fuel and on the steam generating capacity (Strauß 2006). Combustion systems for solid fuels are offered on the market with the capacities shown in Table 5.2:

Table 5.2 Output ranges of firing systems

Firing system	Output range [MW_{th}]
Pulverised fuel firing	40 up to 2,500
Bubbling fluidised bed firing	up to 80
Circulating fluidised bed firing	40 up to 750
Grate firing	2.5 up to 175

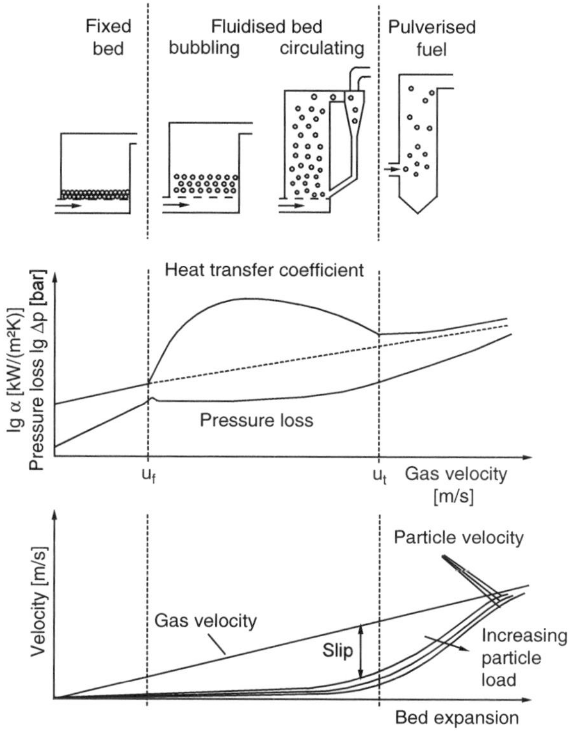

Fig. 5.1 Distinctive features of firing systems (Görner 1991)

5.1 Combustion Fundamentals

The purpose of the combustion process is to release by oxidation the energy which is chemically bound in the fuel and to convert it into sensible heat.

The heterogeneous combustion process of solid fuels is more complex than the homogeneous combustion of gaseous fuels. Solid fuels such as coal are composed of different fractions of organic matter and minerals. As the fuel heats up in the furnace, the pyrolysis of the organic matter starts. In this process, volatile intermediate products such as hydrocarbons, carbon oxides, hydrogen, sulphur and nitrogen compounds and residual char (as a solid intermediate product) are generated. Ignition begins the combustion process. Prerequisite for ignition, besides a sufficiently high temperature, is the forming of a burnable mixture. Under these conditions, the volatile matter and the residual char combust together with the oxygen of the combustion air. Figure 5.2 schematically presents the combustion process of coal in pulverised fuel firing.

The combustion of solid fuels evolves in the partial processes of (Doležal 1990; van Heek and Mühlen 1985)

- drying,
- pyrolysis,
- ignition,

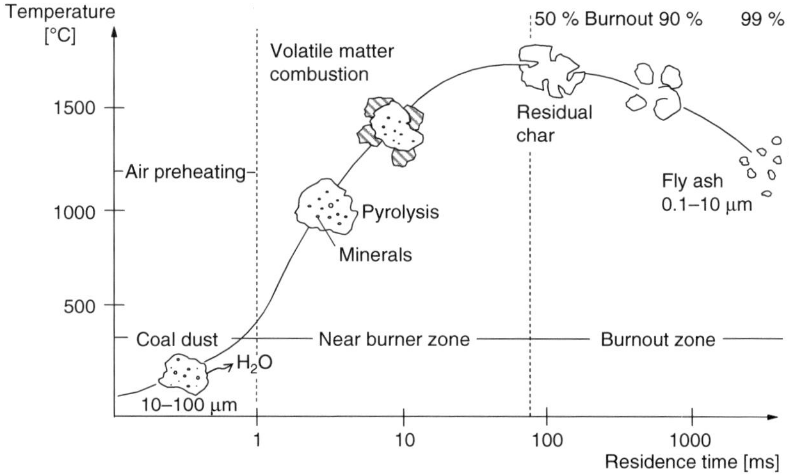

Fig. 5.2 Schematic drawing of the combustion process in pulverised fuel firing

- combustion of volatile matter and
- combustion of the residual char.

The first two partial processes are a thermal decomposition as a consequence of the heating up of the fuel. The quantity of heat necessary to heat the fuel up to ignition temperature is transferred mostly by convection. In pulverised fuel firing, for example, hot flue gas is admixed in the near-burner zone, while in a fluidised bed, the heat is transferred by particles of solid matter. In grate firing systems, heating up is carried out by means of refractory-lined hot walls transferring the heat to the fuel by radiation.

In the last two partial processes – combustion of volatile matter and combustion of residual char – the organic matter is converted chemically. Conversion is divided into homogeneous and heterogeneous reactions. The partial processes do not necessarily run one after the other but, depending on the firing type, may overlap. Table 5.3 provides an estimate of the necessary time for each of the partial processes. It is evident from the table that the total combustion time of all firing systems is determined by the combustion of the residual char.

In the following, the partial processes of solid fuel combustion are discussed in more detail.

5.1.1 Drying

Water can adhere both to the particle surface and to the pores inside the coal particle. As the fuel heats up in the furnace, water begins to vaporise (at temperatures above 100°C). At temperatures up to 300°C, the vaporised pore water becomes desorbed or released. Besides water vapour, other gases such as methane, carbon dioxide and

Table 5.3 Partial processes of coal combustion in firing systems

Firing system	Particle diameter [mm]	Heating rate [K/s]	Drying and pyrolysis period [s]	Time of volatile matter combustion [s]	Time of residual char combustion [s]
Fixed bed firing	100	10^0–10^2	ca. 100	Determined by release and mixing with combustion air	>1,000
Fluidised bed firing	5–10	10^3–10^4	10–50		100–500
Pulverised fuel firing	0.05–0.1	10^4–10^6	<0.1		1–2

nitrogen, which have formed during the coalification process, outgas as well (van Heek 1988).

Depending on the combustion system, the firing is capable of drying fuels with different moisture contents. Whereas grate or fluidised bed firing systems can be fed with moisture-containing fuels without further treatment, for pulverised fuel firing the fuel is predried in mills in order to ensure a fast combustion process within the available residence time.

5.1.2 Pyrolysis

The decomposition of the organic coal substance and the formation of gaseous products during the heating of the coal are termed devolatilisation or pyrolysis (van Heek and Mühlen 1985; Zelkowski 2004; Rüdiger 1997; Klose 1992).

Devolatilisation of volatile matter by cracking of compounds of organic coal structures starts at temperatures above 300°C. In a temperature range up to about 600°C, tars (liquids at lower temperatures) and gaseous products are formed. The gases consist of carbon dioxide (CO_2), methane (CH_4) and other, lighter hydrocarbons such as C_2H_6, C_2H_4 and C_2H_2. Tars are complex hydrocarbon compounds, in their organic structure similar to the base fuel, which evaporate from the coal substance at temperatures between about 500 and 600°C (Solomon and Colket 1979).

The particle form remains almost unchanged up to temperatures of about 400°C. Above this temperature, the coal particle begins to soften. The tars and gases formed inside the coal can swell the particle at temperatures reaching slightly above 550°C. The particle solidifies into the so-called semi-char which has a cavity structure with a distinct pore system and an enlarged surface area (van Heek and Mühlen 1985).

Further heating, above about 600°C, converts the semi-char into char, releasing mainly carbon monoxide and hydrogen in the process (Anthony and Howard 1976). With rising temperatures, light gas components such as hydrogen and carbon monoxide, as well as soot, form from the tar compounds.

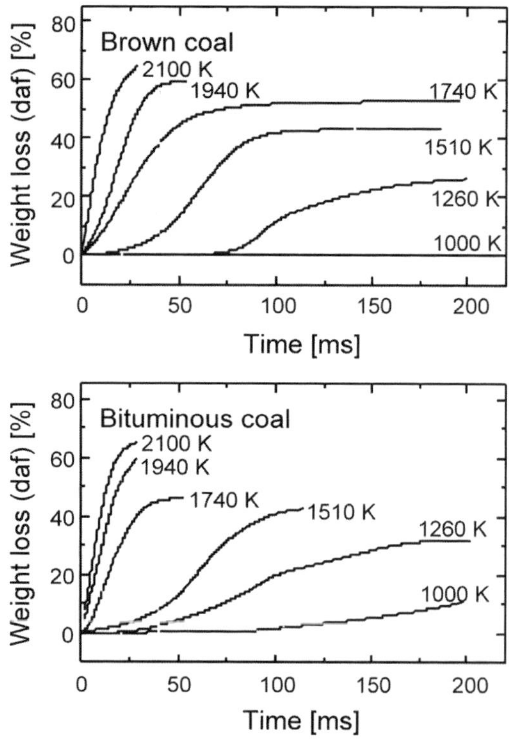

Fig. 5.3 Impact of temperature and residence time on weight loss during pyrolysis (Kobayashi et al. 1977)

The fraction and the composition of the volatile components and the history of their release depend on the coal type, the grain size, the heating rate and the final temperature of the heating. As the heating rate and the coalification degree increase, the devolatilisation maxima of the components shift towards higher temperatures.

The yield of volatile matter increases with rising end temperatures. Figure 5.3 shows the weight losses of a hard and a brown coal determined during pyrolysis at short residence times and high heating rates (Kobayashi et al. 1977).

The volatile matter content determined at high temperatures and heating rates of entrained-flow reactors may amount to 1.1–1.8 times the content detected in proximate analysis (Sayre et al. 1991). For coals with a strong tar release, in particular, the yields of volatile matter are significantly higher, because the conditions of the entrained-flow reactor impede the decomposition of the tar into char and gas.

Figure 5.4 shows the composition of the volatile matter as a function of the temperature during the pyrolysis of a hard and a brown coal (Smoot and Smith 1985). In the pyrolysis of the hard coal, the tar products predominate, whereas CO and water comprise the larger fraction of the volatile matter for the brown coal. At higher temperatures, stable compounds form increasingly, while the tar fraction decreases.

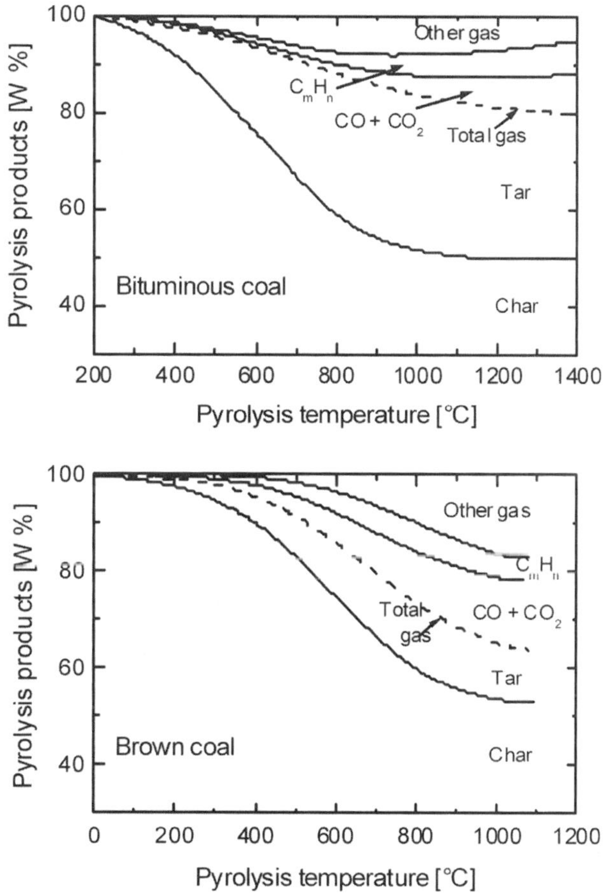

Fig. 5.4 Distribution of products of pyrolysis of a brown and of a hard coal (Smoot and Smith 1985)

5.1.3 Ignition

Ignition begins the process of combustion. The ignition temperature is defined as the temperature above which combustion evolves independently. At temperatures below the ignition temperature, the heat released during fuel oxidation is dissipated to the environment, so the temperature does not rise notably. Only at or above the ignition temperature does the reaction velocity reach a rate where the amount of heat released exceeds the amount dissipated to the surroundings. Thus the reaction is accelerated, so a stable combustion can be maintained (Doležal 1990).

In the combustion of solid fuels, both the volatile components and the residual char have to be ignited. The volatile components ignite as soon as they form a combustible mixture with the combustion air and the ignition temperature of the mixture is either reached or exceeded. The residual char particle, in order to ignite, has to reach or surpass its ignition temperature and receive sufficient oxygen at its surface (Zelkowski 2004). The ignition temperatures of the combustible mixture of

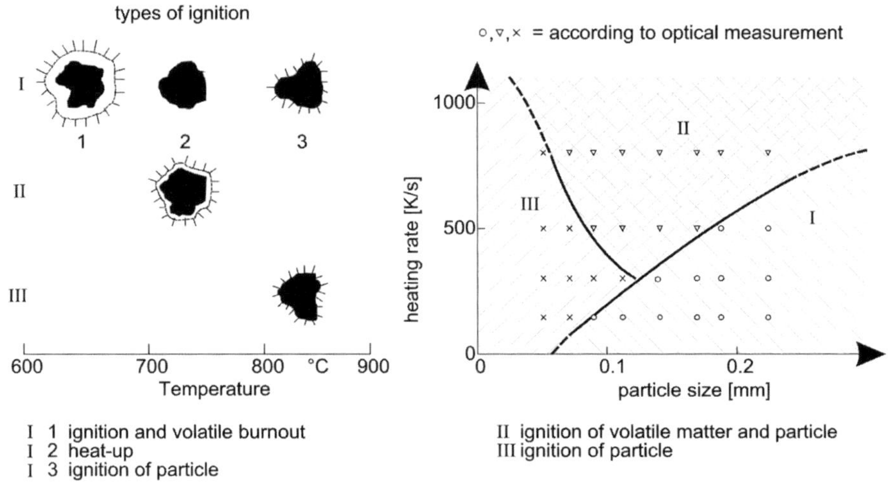

Fig. 5.5 Ignition mechanism as a function of the heating rate and the particle size for a high-volatile bituminous coal (hvb) (Stahlherm et al. 1974)

volatile matter and combustion air range between 500 and 700°C, while the ignition temperatures of the residual char particle lie above 800°C.

In coal combustion, the history and sequence of ignition processes above all depend on the heating rate and the particle size. The impact of these two parameters on the ignition mechanisms in the combustion of a high-volatile bituminous coal, determined at a laboratory-scale plant, is demonstrated in Fig. 5.5 (van Heek and Mühlen 1985; Stahlherm et al. 1974; Stahlherm 1973).

During slow heating and with coarse particles, the volatile components are first released, then ignite in the near-particle zone and then burn out. Devolatilisation and volatile matter combustion result in a gas atmosphere that envelops large particles, thus impeding the diffusion of oxygen to the particle, which can ignite only after the volatile matter has burned up (ignition mechanism I).

Coarse particles and high heating rates favour the simultaneous ignition of volatile matter and residual char (ignition mechanism II). Pyrolysis reactions shift towards higher temperatures, with the ignition temperature of the particle changing to a lesser extent. This way, the ignition of the particle is possible even before all the gases are burned completely.

With very small particles, ignition happens directly at the particle surface. Given the great surface-to-volume ratio, these particles are rapidly heated up, so the ignition temperature of the particle is reached even before an ignitable mixture has formed around the particle (ignition mechanism III) (Stahlherm et al. 1974).

Besides the high-volatile bituminous coal analysed in Fig. 5.5, a low-volatile anthracite coal was investigated as well. At the same conditions, ignition took place at the particle surface (Stahlherm et al. 1974).

For coarse-grained coal in grate firing, the volatile matter ignites first, whereas medium-sized coal particles and higher heating rates in fluidised bed firing promote the simultaneous ignition of volatile matter and particle. High heating rates and small particle sizes in pulverised fuel firing make low-volatile bituminous (lvb) coals ignite at the particle, whereas high-volatile bituminous (hvb) coals show a simultaneous ignition of both volatile matter and particle.

The ignition temperature, in solid fuel combustion, depends not only on the fuel characteristics, such as the volatile matter, moisture and ash contents, and on the physical structure, such as the particle size and the inner surface of the coal, but also on the combustion conditions of the firing system (heating rate, dust and gas concentrations, etc.). Depending on the fraction of volatile components, the ignition temperature is high for lean fuels and char and low for higher volatile fuels. The temperature decreases with increasing fineness of the fuel (STEAG 1988; Doležal 1990). Figure 5.6 gives reference values as a function of the volatile matter content and oxygen concentration for the design of pulverised coal firing systems (Zelkowski 2004).

The ignition velocity – which is understood as the velocity of flame propagation in the mixture – has a clear dependence on the volatile components, the ash content and the primary air mixture in the case of a hard coal flame, as in Fig. 5.7. The ignition velocity always reaches a maximum depending on the primary air fraction. At low air ratios, the oxygen in the primary air is not sufficient to combust the volatile matter in the near-burner zone. With a stronger primary air flow, the primary air which is not needed for the combustion of the volatile matter serves to decrease the flame temperature. In both cases, the ignition velocity decreases. A higher ash content also has a delaying effect on ignition. The ignition velocity is a crucial parameter for the burner design for two reasons. On the one hand, the burner throat velocity has to be notably higher than the ignition velocity in order to surely prevent the flame from flashing back. On the other hand, to have a stable flame front, it has to be ensured that zones form where the flow velocity is equal to the ignition velocity (Doležal 1990).

In pulverised fuel firing, the coal as well as the carrier gas flow (consisting of primary air and vapours) has to be preheated – starting from classifier temperature (i.e. the temperature in the mill) – to values equal to or higher than the ignition

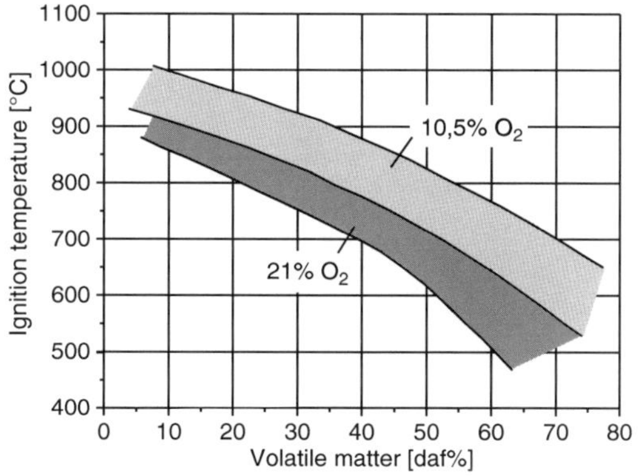

Fig. 5.6 Ignition temperature as a function of the volatile matter (Zelkowski 2004)

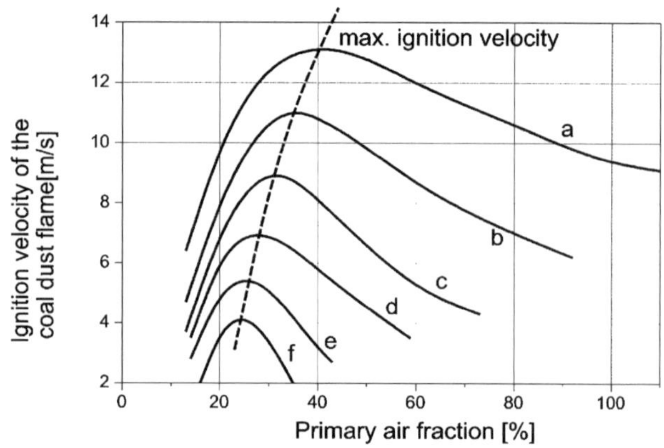

Fig. 5.7 Ignition rate as a function of the primary air fraction (Doležal 1990)

	a	b	c	d	e	f
Volatiles %	30	30	30	20	30	15
Ash %	5	15	30	5	40	5

temperature. For this reason, only the amount of primary air that is necessary for the combustion of the volatile matter should be fed.

5.1.4 Combustion of Volatile Matter

The homogeneous combustion of the volatile components is characterised by a very high reaction velocity, so that the burning time is essentially determined by their release and mixing with air.

The highest concentrations of volatile components develop on the particle surface, the concentration diminishing with increasing distance from the particle. The volatile matter combustion stabilises into a diffusion flame in areas where there is a stoichiometric concentration of volatile matter and oxygen. The diameter of a flame enveloping a particle is about three to five times the diameter of a particle (Zelkowski 2004). In pulverised coal combustion, the volatile matter combustion processes of the individual particles combine so they can be considered a coherent gas flame.

5.1.5 Combustion of the Residual Char

The volatile matter having been released from the particle, it remains a porous structure consisting almost only of carbon and ash. The carbon, at a sufficiently high particle surface temperature, is oxidised by oxygen, carbon monoxide, carbon dioxide and water vapour.

At the same temperature, the reaction velocity of the heterogeneous combustion of solid residual char with oxygen is orders of magnitude lower than the homoge-

Fig. 5.8 Combustion process of a char particle

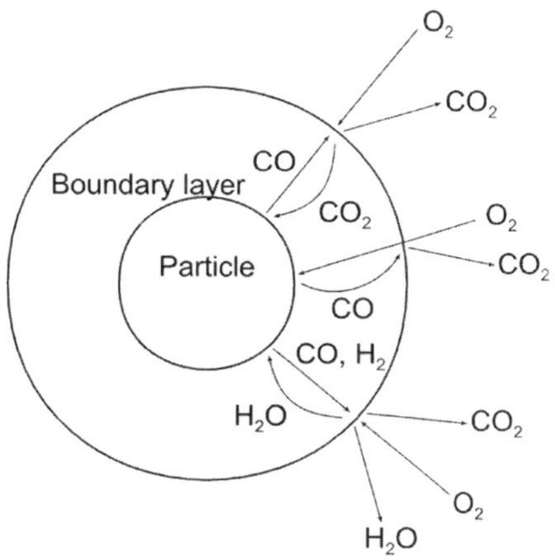

neous volatile matter combustion. Residual char combustion therefore determines the total combustion time and is decisive for the design of firing systems.

Figure 5.8 schematically shows the course of residual char combustion of a single particle. At the surface or inside the particle, the heterogeneous oxidation of the carbon takes place with oxygen, carbon dioxide and water vapour as oxidants:

$$C + {}^1\!/\!{}_2O_2 \leftrightarrow 2CO \tag{5.1}$$

$$C + CO_2 \leftrightarrow 2CO \text{ (Bouduard reaction)} \tag{5.2}$$

$$C + H_2O \leftrightarrow CO + H_2 \text{ (heterogeneous water–gas reaction)} \tag{5.3}$$

Today it is considered proven that directly at the particle surface, initially only a conversion to carbon monoxide takes place, either by combustion (5.1) or by gasification (5.2) and (5.3) (Zelkowski 2004). Around the coal particle, a gaseous atmosphere consisting of the combustion products CO and H_2 and the oxidants O_2, CO_2 and H_2O forms. The oxidants have to diffuse to the particle surface through this laminar boundary layer and, vice versa, the combustion products from the particle to the environment.

The following homogeneous oxidation

$$CO + {}^1\!/\!{}_2O_2 \leftrightarrow CO_2 \tag{5.4}$$

$$H_2 + {}^1\!/\!{}_2O_2 \leftrightarrow H_2O \tag{5.5}$$

takes place in the surrounding boundary layer.

In heterogeneous reactions, the conversion velocity $\mathrm{d}m_C/\mathrm{d}t$ of the carbon mass m_C of a coal particle is proportional to the reacting surface A, to the reaction velocity k_{tot} and to the oxygen partial pressure pO_2 in the environment of the particle:

$$\frac{\mathrm{d}m_C}{\mathrm{d}t} = Ak_{tot}pO_2 \tag{5.6}$$

Given that besides the chemical kinetics, the mass transport processes also exert an influence on the burning process, the conversion velocity of the residual char combustion is limited by the slowest one of the participating processes. Which of the partial processes determines the conversion velocity in the end depends essentially on the reaction temperature.

As a function of the temperature, a distinction is made between three areas. In each, either

- the chemical reaction,
- the pore diffusion or
- the boundary layer diffusion

determines the velocity. The three areas are shown in an Arrhenius diagram in Fig. 5.9. In this diagram, the natural logarithm of the reaction velocity is plotted over the reciprocal of the absolute temperature.

In the chemical reaction (area I), the oxygen can at first, at low temperatures, sufficiently quickly reaches the inside of the char residue via the finely branched pore system without undergoing notable conversions. Thus the concentration of oxygen is equal to the concentration in the free gas atmosphere, as shown in Fig. 5.10. Only the chemical reaction of the oxygen with the carbon surface of the pores influences the combustion velocity.

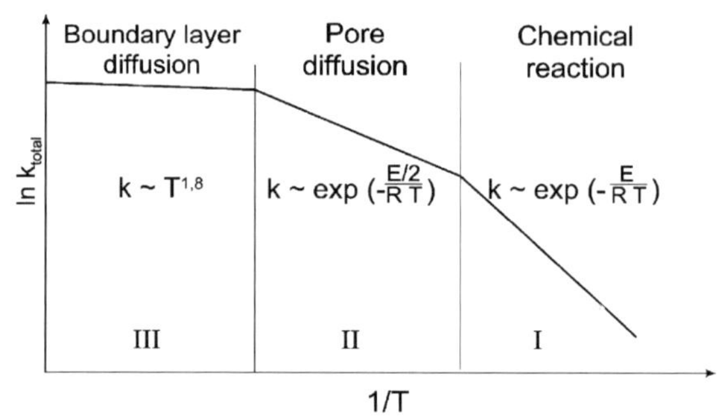

Fig. 5.9 Arrhenius diagram of char combustion

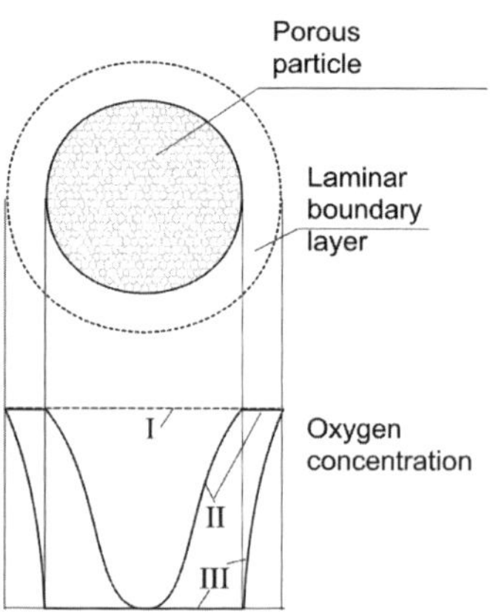

Fig. 5.10 Oxygen concentration profile around a char particle

In pore diffusion (area II), the velocity of the chemical reaction increases with rising temperatures. In the inside of the char residue, the oxygen molecules get depleted so that a concentration drop from the fringe to the centre of the particle develops. The burning velocity in this area depends on how fast oxygen can be supplied by pore diffusion.

In boundary film diffusion (area III), at still higher temperatures, oxygen is no longer able to penetrate into the pores. The gradient of the oxygen partial pressure shows that the combustion process takes place only on the outer surface of the particle. The particle is enveloped by a laminar boundary layer and the conversion velocity is determined by the diffusion of the oxygen through this layer.

The total velocity is the result of the single reaction velocity constants:

$$k_{\mathrm{tot}} = \frac{1}{\dfrac{1}{k_{\mathrm{diff},b}} + \dfrac{1}{k_{\mathrm{diff},p}} + \dfrac{1}{k_{\mathrm{chem}}}} \tag{5.7}$$

The temperature zones shift depending on the particle size and the coal type. Whereas pore and boundary layer diffusion determine the reaction velocity at temperatures above a level of 1,450°C or so for coal particles of 20 μm, this holds true even at 1,150°C in the case of larger particles of 200 μm.

During the combustion process, the relative ash fraction in the coal particle increases. An ash layer enveloping the remaining combustible matter develops, so the oxygen has to penetrate this ash cover. Given that as the burning process proceeds, the ash cover grows thicker, the combustion velocity gradually decreases. The more retarded the combustion is, the more ash and the less pores the fuel

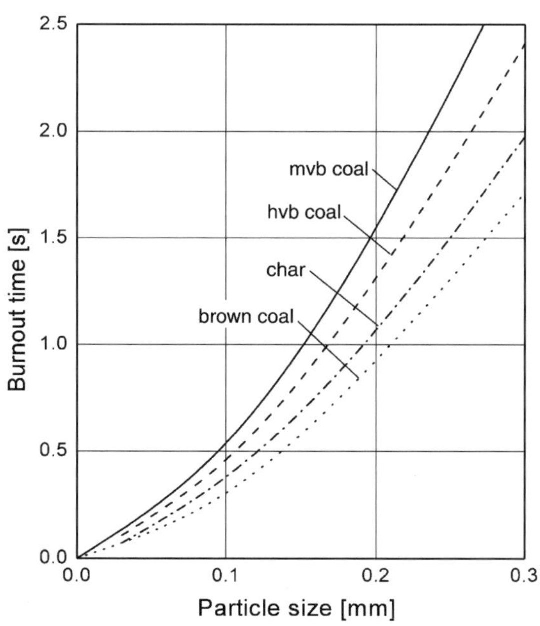

Fig. 5.11 Burn times for pulverised coal as a function of particle size ($t = 1{,}300°\mathrm{C}$, $\lambda = 1.2$) (hvb: high-volatile, mvb: medium-volatile) (Gumz 1962)

contains (Zelkowski 2004). The pyrolysis process preceding the char combustion has a positive effect on the burnout. Depending on the volatile matter content, a more-or-less marked cavity structure is formed in the char during pyrolysis. This structure considerably enlarges the surface available for the chemical reaction in the raw coal particle (Rüdiger 1997; Spliethoff 1995). Coals with a higher volatile matter content burn faster because the respective residual char gets a much larger surface area through pyrolysis than the residual char of a low-volatile bituminous (lvb) coal. Figure 5.11 shows the combustion time of different coals at a temperature of 1,300°C (Gumz 1962).

5.2 Pollutant Formation Fundamentals

5.2.1 Nitrogen Oxides

Different mechanisms during the combustion of fossil fuels cause the formation of NO and NO_2, which, combined, are termed NO_x (nitrogen oxides). Nitrogen oxide emissions from power plants are composed of about 95% NO and 5% NO_2 but are calculated simply as NO_2. This is because nitrogen monoxide (NO) formed inside the flame is converted into NO_2 in the flue gas path after the furnace as temperatures fall below 600°C, as well as in the atmosphere (Jacobs and Hein 1988).

Because emission regulations prescribe measurement of the sum of NO and NO_2, the term NO_x emissions will always be used when discussing emissions in this text. In the context of combustion engineering measurements, the nitrogen oxides at the furnace exit will also be termed NO_x emissions, regardless of whether they are further reduced by secondary measures. However, if nitrogen oxide concentrations at a specific location within the combustion process are considered, the designation will be NO concentrations or NO_x concentrations, if NO and NO_2 are measured.

In the combustion of fossil fuels without organically bound nitrogen, emissions of nitrogen oxides, formed at high combustion temperatures from nitrogen of the combustion air, can in most cases be limited to allowable values by combustion engineering measures. If nitrogenous fuels and low combustion temperatures are used, nitrogen emissions are mainly formed out of the fuel nitrogen, if present. During combustion, the fuel nitrogen is converted partly or totally into nitrogen oxide.

In pulverised coal combustion, nitrogen oxides can be formed by three different mechanisms (de Soete 1981; Leuckel 1985; Warnatz 1985; Wolfrum 1985):

- Thermal NO formation
- Prompt NO formation and
- NO formation out of the fuel nitrogen

Figure 5.12, in a simplified way, describes the pathways of reaction and Fig. 5.13, for the different formation mechanisms, shows the NO_x emissions at the furnace exit as a function of the furnace temperature (Pohl and Sarofim 1976; Zelkowski 2004).

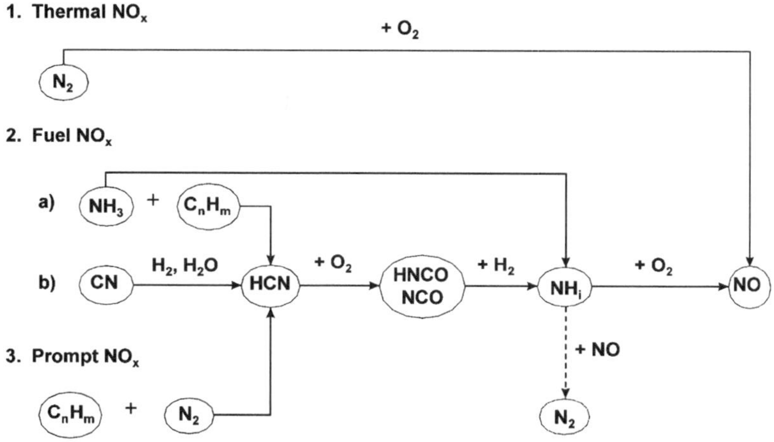

Fig. 5.12 NO_x formation mechanisms

5.2.1.1 Thermal NO Formation

Thermal NO forms from molecular nitrogen in combustion air, following the Zeldovich mechanism (Zeldovich 1946). At high temperatures, oxygen molecules break apart. The resulting oxygen atoms react with the molecular nitrogen to form nitrogen monoxide and atomic nitrogen:

$$O + N_2 \leftrightarrow NO + N \tag{5.8}$$

The conversion process starts at temperatures above 1,300°C and the conversion rate increases exponentially with the temperature. The conversion is proportional to

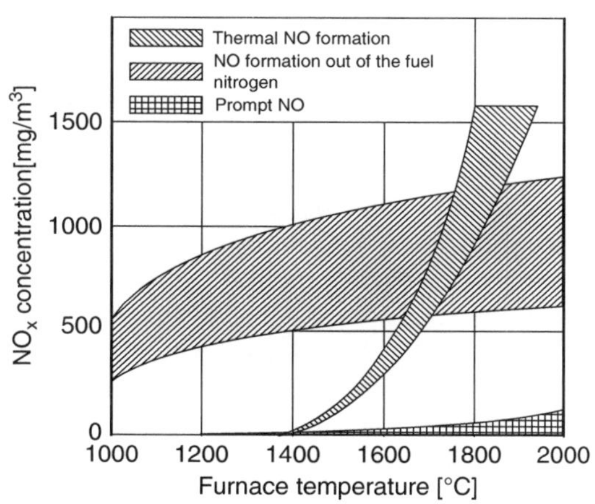

Fig. 5.13 NO_x emissions in coal combustion (Zelkowski 2004)

the concentration of atomic oxygen. The formed nitrogen atom in turn reacts with an oxygen molecule:

$$N + O_2 \leftrightarrow NO + O \quad (5.9)$$

Under oxygen-deficient conditions, NO formation primarily evolves via the following reaction:

$$N + OH \leftrightarrow NO + H \quad (5.10)$$

For pulverised coal-fired furnaces with dry ash removal, the fraction of thermal NO in NO_x emissions is reported as 20% or so (Blair et al. 1978); furnaces with molten ash removal may have a higher percentage (Bertram 1986).

5.2.1.2 Prompt NO Formation

Prompt NO, a notion introduced by Fenimore (1970), describes a mechanism where, in an early phase in the flame front, molecular nitrogen is converted into NO via intermediate products with hydrocarbon radicals participating. The starting reaction evolves as follows:

$$CH_i + N_2 \leftrightarrow HCN + N \quad (5.11)$$

The intermediate products formed in the process can then be oxygenated to form NO via different reactions. In industrial combustion systems, prompt NO plays a minor part. In pulverised coal combustion, the estimated amount of prompt NO is less than 10 ppm.

5.2.1.3 NO Formation from Fuel Nitrogen

Coal has a 0.5–2% fuel nitrogen content, part of which can be converted to NO in the combustion process. In the case of a complete conversion of the fuel nitrogen, a high-volatile hard coal with a nitrogen content (daf) of 1.5% would produce NO_x emissions of 4,500 mg/m^3 at 6% O_2. The conversion rates of fuel nitrogen to NO in industrial furnaces are between 15 and 30%. The quantity of NO formed this way depends on the nitrogen content of the coal, the air ratio, the temperature and other parameters characterising the course of combustion. NO from fuel nitrogen, in comparison with thermal NO, is formed even at temperatures lower than 1,300°C and the reactions run at a higher velocity.

The current state of knowledge is that in pulverised coal combustion with fast devolatilisation of the coal particles, part of the fuel nitrogen is released together with the volatile matter and the remaining part stays in the residual char (see Fig. 5.14). The nitrogen oxides from the volatile fuel nitrogen and from the residual

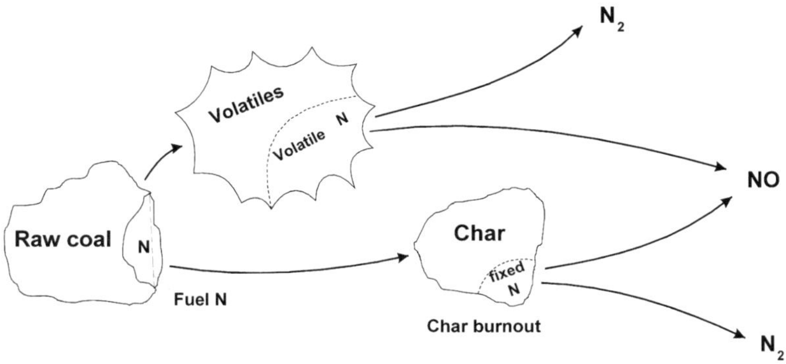

Fig. 5.14 Distribution of the fuel nitrogen during pyrolysis

char nitrogen are formed by different pathways of reaction. Nitrogen oxide formation from fuel nitrogen in pulverised coal combustion depends on

- the devolatilisation of the fuel nitrogen,
- the formation of NO from the residual char nitrogen and
- the formation of NO from the nitrogen of the volatile matter (Glarborg et al. 2003).

Devolatilisation of the Nitrogenous Components

The nitrogen in the coal is partly released through devolatilisation, together with the volatile components, in the form of nitrogen compounds of the amine class (N—H, e.g. NH_3) or the cyanogens class (C=N, e.g. HCN). The fractions of the fuel nitrogen getting released with the volatile matter and the quantity remaining in the residual char are values that essentially depend on the pyrolysis temperature and the coal type.

At low pyrolysis temperatures, the nitrogen mainly remains in the residual char. At high temperatures of 1,300–1,500°C, typically occurring in flames, 70–90% of the fuel nitrogen may be released, according to studies by different authors (Blair et al. 1978; Wendt 1980). Notable quantities of nitrogenous components devolatilise only after a mass loss of the coal of 15%; afterwards the release of fuel nitrogen, in flow reactors, develops proportionally to the total weight loss of the coal (Pohl and Sarofim 1976).

With decreasing coalification, the fraction of volatile fuel nitrogen released as NO_x decreases at a constant pyrolysis temperature. The coalification degree also has an influence on the distribution of the gaseous nitrogen compounds. Results of investigations into air staging revealed that HCN is the dominating nitrogen component in the primary zone for hard coals with a low volatile matter content, while for high-volatile hard coals and for brown coals, a larger fraction of NH_3 was found (Chen et al. 1982b; Wendt and Dannecker 1985; Di Nola et al. 2009; Di Nola 2007).

NO Formation from Residual Char Nitrogen

The conversion rates of residual char nitrogen to NO are low – the percentage is at 10–25% (Pohl and Sarofim 1976; Song et al. 1982). This fact is put down to the indirect reduction of NO on the coal particle surface. In contrast to the formation of nitrogen oxide from volatile nitrogen, heterogeneous nitrogen oxide formation can be influenced only to a limited extent (Pohl et al. 1982; Schulz 1985). Influence on the conversion rates is exerted by the flame temperature, the air ratio and the characteristics of the char. With higher temperatures, the formation of NO from residual char nitrogen decreases (Pohl and Sarofim 1976; Song et al. 1982). Conversion rates of residual char nitrogen to NO of less than 10% were measured in combustion in reducing conditions (Pohl and Sarofim 1976).

NO Formation from Volatile Fuel Nitrogen

In pulverised coal combustion, the conversion of volatile fuel nitrogen to NO may reach considerably higher rates than the conversion of residual char nitrogen. The rate strongly depends on the combustion conditions and can be reduced effectively by primary measures such as air staging. Essential parameters pertaining to the conversion into NO are the air ratio, the concentration of nitrogen in the gas phase and the temperature (Fenimore 1976, 1978). The fuel nitrogen released by devolatilisation can be oxidised to NO or decomposed to molecular nitrogen by reduction mechanisms. Combustion engineering measures can particularly help to reduce NO formation from volatile fuel nitrogen, to the extent that, according to the opinion of several authors, the NO formation from residual char establishes a limiting value to the total NO_x emissions which cannot be further reduced by air staging measures (Mechenbier 1989; Wendt 1980; Spliethoff and Hein 1997).

In industrial firing systems, the conversion of total fuel nitrogen to NO is about 30%; by means of primary measures like air staging it is possible to achieve conversions as low as 5%.

5.2.1.4 NO-Reducing Mechanisms

During the process of the combustion, it is possible to reduce nitrogen oxides that form. A difference is made between

- heterogeneous reduction and
- homogeneous reduction.

Heterogeneous reduction is the reduction of residual char which has not yet undergone reaction. The very low level of NO formed from residual char has to be put down to the reduction of NO on the surface of the coal particle. Heterogeneous reduction plays an important part when there are high loads of pulverised coal with a large fraction of unburned matter, as in fluidised bed or grate firing systems (Schulz 1985).

In pulverised coal combustion, heterogeneous reduction is of minor importance (Glass and Wendt 1982). On the one hand, the particle load outside the flame zone is low and, on the other hand, heterogeneous reduction needs a high degree of activation energy. The ratio of homogeneous to heterogeneous reduction rates is more or less 100 to 1 in pulverised coal combustion (Schulz 1985).

Homogeneous reduction plays the essential part in the context of combustion engineering measures for NO_x reduction. However, reduction mechanisms should not be considered separately but in correlation to the possible ways of formation. The homogeneous formation and reduction mechanisms are combined in Fig. 5.15. This simplified reaction diagram is also denoted as the fuel N mechanism.

Figure 5.15 shows the NO formation and reduction pathways of homogeneous nitrogen components for all combustion zones and conditions. The effective reaction processes that occur will depend on the combustion conditions, possibly differing from zone to zone in the combustion. Efficient NO reduction by combustion engineering measures can be achieved by setting in each of the zones those combustion conditions which promote the decomposition and prevent the formation of NO.

Homogeneous NO formation and reduction can be divided into the following major reactions:

- Conversion of HCN to NH_i
- Conversion of NHi to N_2 or NO
- NO decomposition by CH_i

Conversion of HCN to NH_i

HCN is converted to NH_i both under fuel-lean and under fuel-rich conditions (Haynes 1977; Just and Kelm 1986). The reaction velocity of the conversion of cyanide species into NH_i increases with rising temperatures and higher excess-air ratios (Eberius et al. 1981). The conversion of cyanide radicals to NH_i is slow and therefore determines the velocity (Fenimore 1976; Just and Kelm 1986). High hydrocarbon concentrations impede the HCN decomposition, which only takes place after the hydrocarbon radicals have been consumed (Fenimore 1978).

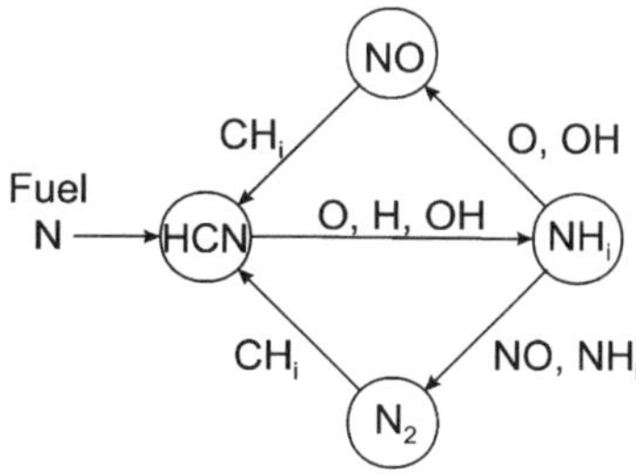

Fig. 5.15 Homogeneous formation and reduction mechanisms

Conversion of NH_i to N_2 or NO

The NH_i compounds originating from the decomposition of HCN either react with NO to form N_2

$$NH_i + NO \rightarrow N_2 + \text{products} \tag{5.12}$$

or are oxygenated to NO under excess-air conditions that arise at the latest when burnout air is added following an air-deficient zone:

$$NH_i + O_2 \rightarrow NO + \text{products} \tag{5.13}$$

Besides the decomposition of the NH_i species via NO, self-decomposition of the NH_i compounds is possible as well. Thus the conversion of the ammonia species into NO or N_2 primarily depends on the fuel – air ratios. In air-deficient zones, the ammonia radicals that are present are mostly decomposed, leaving N_2; in excess-air zones, at the common firing system temperatures of more than 1,000°C, they are oxidised to form NO.

Within a small range of temperatures, between 900 and 1,000°C, and while also in excess-oxygen conditions, nitrogen oxides are decomposed via ammonia radicals (Wolfrum 1985). These conditions exist in such cases as that of ammonia addition in a 900–1,000°C hot flue gas flow with excess air or when there is burnout air addition at the end of a reduction zone containing ammonia radicals in air- or fuel-staged operation. The location of the temperature window depends on the flue gas concentrations of O_2, CO, H_2 and H_2O. The reaction times are some hundredths of seconds (Hemberger et al. 1987).

NO Reduction by CH_i

Besides the decomposition of NO via NH_i species, it is also possible for NO to be decomposed via hydrocarbon radicals to form HCN (Wendt 1980; Chen et al. 1982a; McCarthy et al. 1987; Myerson 1974):

$$NO + CH_i \rightarrow HCN + \text{products} \tag{5.14}$$

The decomposition reactions via hydrocarbon radicals are 10–100 times faster than the conversion from HCN into NH_i (Just and Kelm 1986). The decomposition by hydrocarbon radicals is also termed the NO recycle mechanism, because already-formed NO re-enters the fuel N mechanism.

When taking technical measures to reduce NO_x emissions, NO reduction mechanisms through ammonia or hydrocarbon radicals are those that diminish NO_x emissions most significantly. While in air-staged combustion, NO is reduced mainly through NO decomposition by NH_i compounds, fuel staging additionally makes use of NO decomposition through hydrocarbon radicals.

For an effective reduction by means of fuel staging, the objective to be attained is the complete decomposition of the nitrogen oxides through hydrocarbon radicals. As the decomposition reactions via CHi radicals run very quickly, the decomposition rate of nitrogen oxides is determined by how fast and complete the admixture of the hydrocarbon-containing reduction fuel is. The reaction conditions should be favourable for the slow conversion of HCN to NH_i, with high temperatures and low hydrocarbon concentrations, in order to completely decompose HCN to N_2.

5.2.2 Sulphur Oxides

Coal is a fuel which contains sulphur, the major fraction of which is converted into sulphur dioxide during combustion. The sulphur content of coal may be up to 8%, but usually the fraction is below 2%. Accordingly, as an example, if there is a fuel sulphur to SO_2 conversion rate of 90%, with a hard coal having a sulphur content of 1%, the resulting SO_2 emission level is 1.6–1.7 g/m^3.

The sulphur can exist in different forms in the coal, for instance, as follows:

- organic sulphur which is bound in the organic coal structure;
- sulphides, which originate from the mineral impurities such as pyrite (iron sulphide (Fe_2S)) or marcasite;
- sulphates, which are found in particular in younger hard coals and brown coals ($CaSO_4$, Na_2SO_4);
- elemental sulphur (Gumz 1962; Morrison 1986).

Pyrite and organic sulphur dominate in coals. Sulphate sulphur, like gypsum or iron sulphate, usually has a fraction of the total sulphur less than 0.1%; the fraction of elemental sulphur is smaller than 0.2% (Morrison 1986).

The relative distribution of pyrite and organic sulphur depends on the coalification degree. While most of the sulphur is bound organically in younger fuels, like brown coal, the fraction of organic sulphur in the total sulphur content of hard coals ranges between 40 and 80% (Morrison 1986). The organic sulphur is less stable than the inorganic type. It is released as H_2S as early as in the devolatilisation phase, together with the volatile components (Zelkowski 2004). Both the pyrite and the organic sulphur participate in the combustion and are oxygenated to sulphur dioxide, SO_2. Another oxidation, forming sulphur trioxide (SO_3), does occur, but the fraction is small due to the short residence time in industrial firing systems (Hein and Schiffers 1979).

If the coal ash contains alkalis or alkaline earths, sulphur dioxides can be captured in the ash. However, this type of capture needs low temperatures, such as arise in brown coal combustion due to the high-moisture load (STEAG 1988). In pulverised hard coal combustion, the conversion of the fuel sulphur into SO_2 reaches a relatively high rate of between 85 and 90% – and is more or less independent from the combustion conditions (Morrison 1986).

5.2.3 Ash formation

Solid fuels contain inorganic mineral matter and inorganic elements, which can be bound organically in the coal or present in the form of simple salts. At high temperatures in the combustion process, these constituents undergo chemical and physical transformations to form ash.

Mineral matter in coal commonly includes alumino-silicate clays, silicates, carbonates and disulphides as major components. According to its association with the coal particle, it can be classified into two groups, namely included minerals and excluded minerals. Included minerals refer to those locked inside the coal matrix and generally have smaller sizes. Excluded minerals are those liberated from the coal completely during crushing, grinding and milling processes and are relatively large.

As part of the coal preparation process, a portion of the excluded minerals can be separated from the mined coal. Smaller or larger fractions, however, remain dispersed in the coal. If as-mined coal is used directly in power plant furnaces, as in the case of brown coal, the mineral components remain in the coal completely. In the case of hard coal, the preparation process separates the coal into high-grade coal, with some 10% of mineral components, low-grade or high-ash coal, with about 30–40% of mineral components, and overburden, with a small percentage of residual coal. Hard coal power stations commonly use high-grade coal.

Organically bound inorganic elements such as Na, K, Ca and Mg, which are distributed within the coal macerals, are commonly found in lower rank coals. In the lowest rank coals, these elements can comprise up to 60% of the total inorganic content. However, they only represent a very small proportion in high-rank coals (Wu 2005). In high-rank coals, sodium and potassium are either in the form of water-soluble chlorides or alumino-silicates (Heinzel 2004).

Figure 5.16 shows a diagram of the mechanisms of ash formation (Beer 1988). In the combustion of pulverised coal, the first partial process is fragmentation, where several particles originate from one single coal particle. Through the burnout of the combustible matter surrounding the mineral components, finely distributed ash components reach the particle surface. With the carbon burnout increasing, the molten ash components sticking to the coal structure merge into ever-larger particles on the shrinking coal particle. In pulverised coal combustion, ash particles with a size of 1–20 μm develop this way.

Part of the ash may vaporise at high temperatures. The extent of vaporisation is affected by the char particle temperature. For example, about 1% of the ash of a hard coal vaporises at temperatures of 1,400–1,600°C in the pulverised coal flame. The vaporised ash particles condense in the process of cooling and form very fine dust particles in the range of 0.02–0.2 μm (also known as aerosols) by nucleation, which

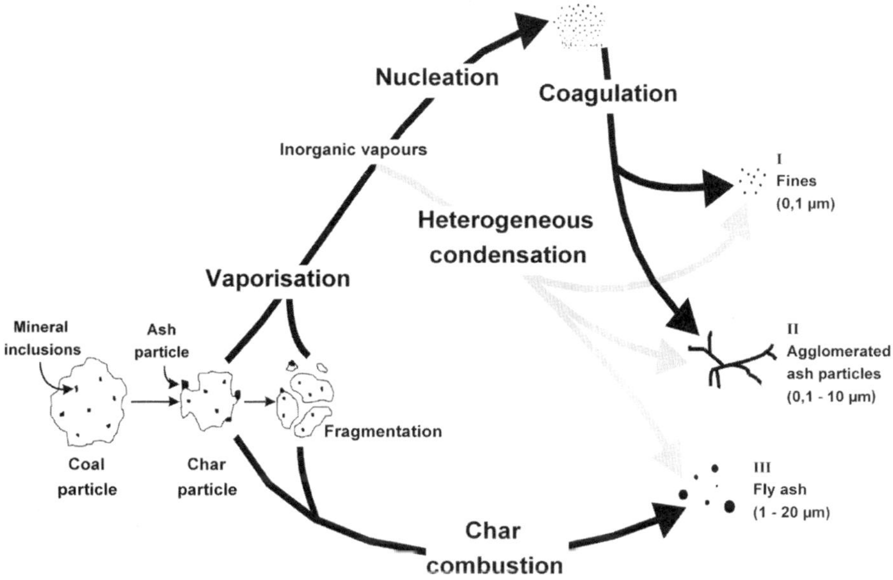

Fig. 5.16 Formation of fly ash in pulverised coal combustion (Beer 1988)

in turn can coagulate. A possible additional process is condensation on available ash particles and on the furnace walls (Beer 1988; Sarofim et al. 1977; Amdur 1986).

Because of the different mechanisms of flue dust formation described above, various authors observe a bimodal distribution of the dust of the cleaned gas with maxima between 0.1 and 0.5 μm and between 1 and 5 μm (Kauppinen and Pakkanen 1990).

Fine dusts may cover more than 99% of the total surface of the fly ash. With their ability to take up gaseous and vaporous pollutants, they have an especially harmful effect on health. The distribution of trace elements, such as heavy metals, over the different particle fractions is a particularly interesting factor in view of the limited removal effect of dust collectors. A general phenomenon to be found with small particles is the accumulation of metal components in the dust (Laskus and Lahmann 1977; Albers et al. 1987).

The ash content of the coal, the combustion system and the combustion conditions all exert an influence on both the quantity of discharged dust and the particle distribution of the fly ash. Table 5.4 shows typical contents of fly ash and Fig. 5.17 plots the particle size distribution relating to different combustion systems (Soud 1995).

In the commonly used pulverised fuel firing system with dry ash removal, 70–90% of the ash is released from the firing as fly ash, while some 10–30% is removed as coarse-grained or even coarse-graded hopper ash, mostly originating from ash deposits. Finely milling the coal will likewise produce a relatively fine fly ash, with a mean diameter of about 30 μm. In slag-tap firing, the fly ash fraction is low because of the primary removal of molten ash. In large slag-tap furnaces, the

Table 5.4 Dust content of firing systems

Firing system	Dust content after firing [g/m^3]
Pulverised fuel firing	5–30
Grate firing with spreader stoker	2–5
Grate firing	1–3
Cyclone firing	0.5–1.5

fly ash amounts to about 50%, while it ranges around 10–30% in cyclone slag-tap furnaces. Given the rotating pattern of the gas flow, only the coarse particles gather on the cyclone wall, while the small ones are carried out of the cyclone with the gas. The fly ash of a cyclone firing system, considering its particle size distribution, therefore features a considerably finer dust than the ash of a dry-bottom firing system. In grate firing systems, the fly ash fraction is only about 40% due to the coarse fuel, the rest is extracted as bottom ash. The fly ash is significantly coarser than the average ash in pulverised fuel firing. Grate firing systems with a spreader stoker feature a higher flue dust fraction.

In circulating fluidised bed firing, the total ash flow is carried out from the furnace, so needs a dust collecting unit.

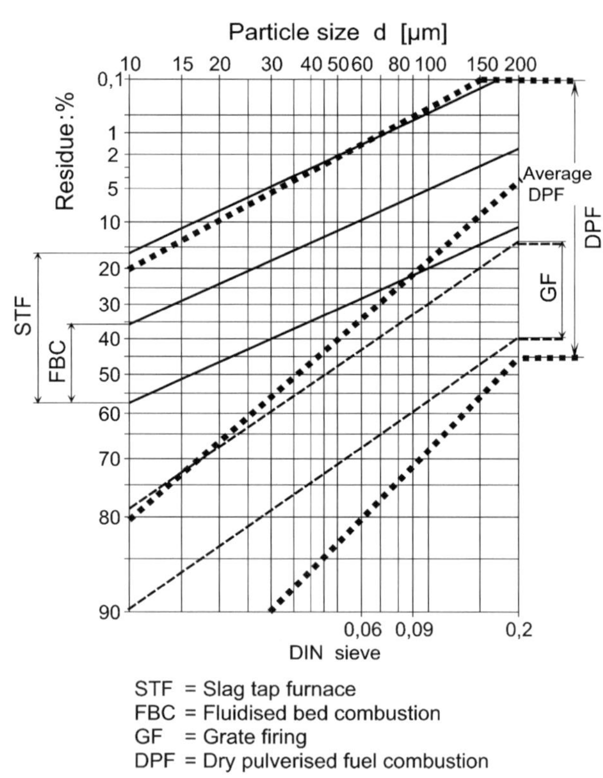

Fig. 5.17 Particle size distribution of fly ashes relating to different combustion systems (*Source*: Alstom Power)

The data on the amount of dust and the properties of the ash are of great importance for the design of the secondary ash removal system (Stultz and Kitto 1992; Klingspor and Vernon 1988; Soud 1995).

5.2.4 Products of Incomplete Combustion

The purpose of the combustion process is the complete conversion of the fuel to transform the bound fuel energy into the sensible heat of the flue gas. Incomplete conversion causes loss and produces emissions of

- carbon monoxide,
- hydrocarbons and
- soot (Baumbach 1990).

In general, the emissions from incomplete combustion in large-scale firing systems stay below the prescribed limiting values. Higher emission levels arise in small plants, in particular, where the combustion process is transient. The combustion techniques under consideration in this text – pulverised fuel, fluidised bed and grate firing – during stationary operation feature high fuel conversion rates and complete combustion.

The completeness of the combustion is influenced by the combustion control, the temperature and the residence time. The design of a combustion plant has to be such that the fuel, depending on the temperature, remains in the furnace sufficiently long: the higher the temperature, the faster the oxidation reactions of the fuel.

CO in common firing systems always forms as an intermediate product of the combustion, which in the course of the combustion process is almost completely converted to CO_2. Typical CO emissions in pulverised fuel firing are below $50\,mg/Nm^3$. CO is also used as a reference value for emissions of hydrocarbons. Soot rarely develops in the combustion of solid fuels in firing systems operated at excess air. It is virtually undetected as a solid matter combustion residue in the ash.

The emissions from incomplete combustion also have to be considered in the context of other kinds of emissions. For instance, with lower air ratios of the combustion process, NO_x emissions decrease and CO emission increases.

When measures for nitrogen oxide reduction are taken, it can be observed that the burnout partly deteriorates and CO emission rises. This rise can be counteracted by a longer residence time in the burnout zone or by a finer milling. Newly developed concepts of nitrogen oxide abatement, which will be considered in Sect. 5.7, show that a reduction of NO_x emissions is not necessarily associated with a deterioration of the burnout. By setting high temperatures, for instance, both the combustion course and nitrogen oxide reduction can be accelerated.

5.3 Pulverised Fuel Firing

The basic idea of a firing system using pulverised fuel is to use the whole volume of the furnace for the combustion of solid fuels. The fuel is milled to the size of a fine grain, mixed with air and burned in the flue gas flow. Because the pulverised fuel is carried through the furnace within the residence time of the combustion gas flow, the burning time is limited to a short period. Fuel drying and milling give the fuel the prerequisites for rapid ignition and fast burnout of the fuel. Compared to grate or fluidised bed firing systems, pulverised fuel firing gives larger power densities. The relatively small mass of fuel inside the furnace provides good controllability to these firing systems, while the disadvantage lies in the need for a high degree of fuel preparation. In the power station sector, pulverised fuel combustion is the predominate system. In Europe, it is used in power plants of up to 2,300 MW_{th} (Strauß 2006; STEAG 1988; Doležal 1990), while globally the largest capacities built are around 4,000 MW_{th} (Stultz and Kitto 1992).

5.3.1 *Pulverised Fuel Firing Systems*

Almost all coal types, from anthracite to lignite coal, can be combusted using pulverised fuel firing. The firing system, however, must be designed to take into account the fuel characteristics, such as the calorific value and the volatile matter, ash and moisture contents. This includes not only the burner system and the furnace in question but also the coal preparation and by-product utilisation and disposal processes.

Pulverised fuel firing systems are differentiated according to

- the state of the ash and the kind of ash removal from the furnace (dry, molten) and
- the fuel dust system that blows the pulverised fuel directly or indirectly into the furnace.

Pulverised fuel firing systems, in large capacity steam generation plants, are usually designed as dry-bottom furnace types with dry ash removal, and less frequently as slag-tap furnaces. Slag-tap firing systems melt the fuel ash and remove it in a molten state as slag. Dry-bottom furnaces are suitable for almost all kinds of fuel. Slag-tap furnaces are beneficial for high-ash or low-volatile coals. When deciding upon using slag-tap firing, another important criterion may be the utilisation of the ash, as the granulated material produced may be more useful. Because of the capital and maintenance cost drawbacks, slag-tap furnaces are rarely built today.

Depending on the ash and moisture contents of the raw coal, the pulverised coal is blown in either directly or indirectly. Figure 5.18 shows the injection systems used in hard-coal and brown coal-fired furnaces. The applicability of hard-coal firing systems as a function of the volatile matter and ash contents of the raw coal is given in Fig. 5.19 and for pulverised brown-coal firing systems as a function of the moisture and ash contents in Fig. 5.20.

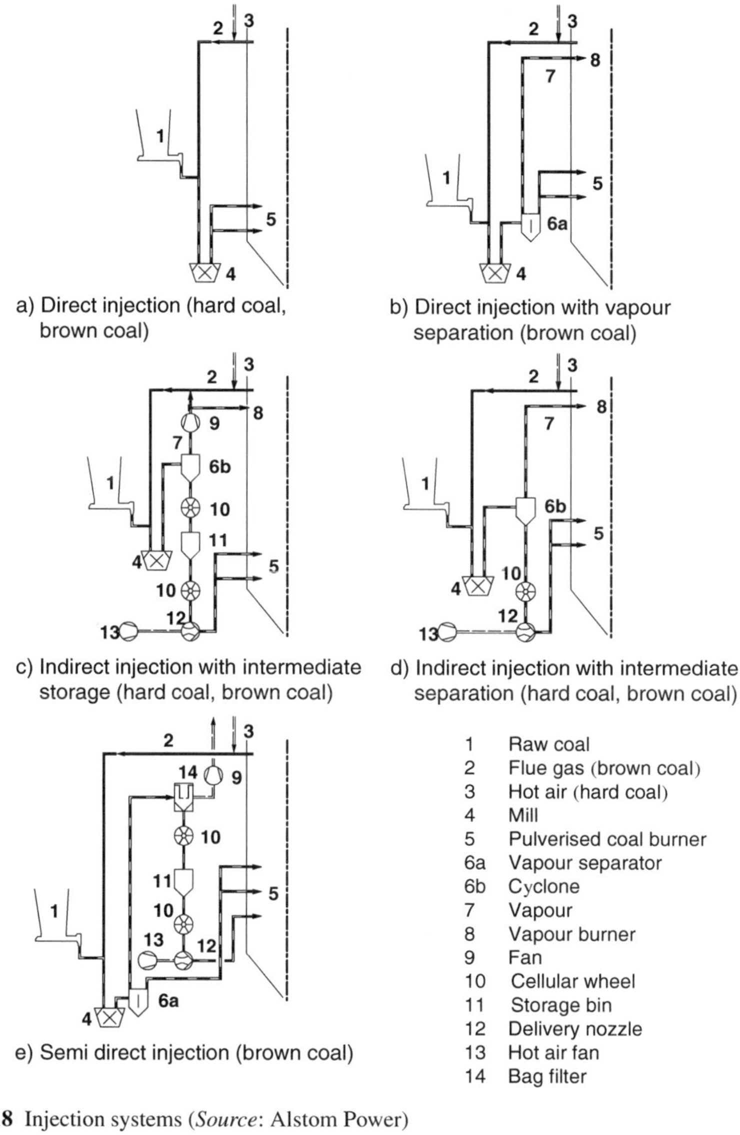

Fig. 5.18 Injection systems (*Source*: Alstom Power)

The technique of direct pulverised fuel injection (Fig. 5.18a) is such that the dried and pulverised coal is blown into the firing by primary (or transport) air and the milling vapours. In design, the general preference is for hard-coal and brown-coal firing systems to have direct pulverised fuel injection, since this system is relatively simple and cost-efficient. It needs a smaller number of components, less expenditure for monitoring and measurement and less auxiliary power.

An excessively high-fuel moisture content is likely to impair a stable ignition. For this reason, the system of direct injection for high-moisture brown coals is only

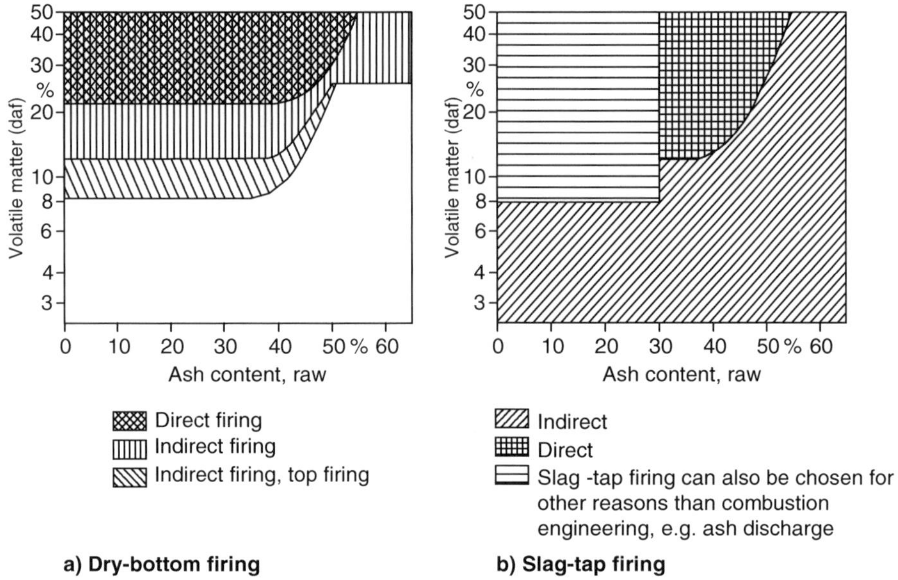

Fig. 5.19 Applications of pulverised hard-coal firing systems as a function of volatile matter and ash contents (*Source*: Alstom Power)

suitable for calorific values somewhat higher than 5.5 MJ/kg, or, depending on the ash content, for moisture contents below 50–65% (Lehmann 1990). In order to raise the ignition stability of fuels of higher moisture contents, the dust concentration is increased by segregating the milling vapours from the fuel flow that is channelled

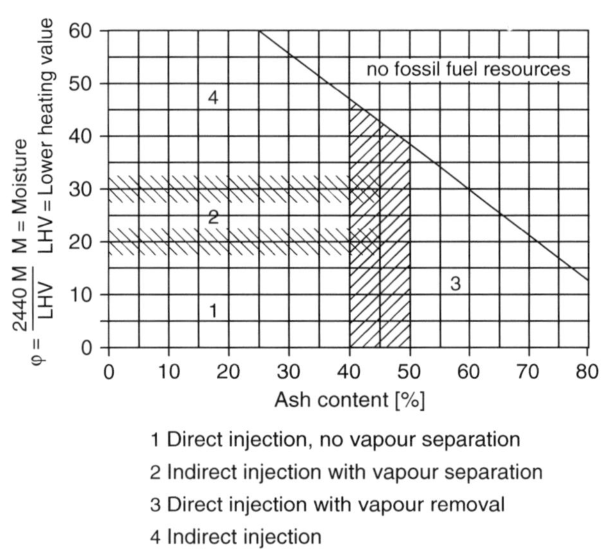

Fig. 5.20 Applications of pulverised brown coal firing systems as a function of moisture and ash contents of the fuel as mined (*Source*: Alstom Power)

to the main burners. The vapours, to burn their residual dust content, are fed into the furnace above the main combustion zone. Figure 5.18b shows this kind of direct injection with vapour separation.

If, in the case of higher moisture contents, injecting the vapours would be disadvantageous, the vapours are cleaned of dust separately by ESP or bag filter (Fig. 5.18e). The resulting super-fine dust, accounting for as much as 30% of the total fuel heat, is blown in by air, which raises the flame stability of the main burners. This system of semi-direct firing is used for fuels with calorific values below 5.5 MJ/kg and moisture contents of more than 70% (Lehmann 1990).

In the combustion of high-ash hard or brown coals, intermediate storage – the so-called bin-and-feeder system (Fig. 5.18c) – or intermediate separation (Fig. 5.18d) is used to achieve a high dust saturation and a high primary mixture temperature. In both systems, the pulverised coal, after milling, is separated from the milling vapours and, after the intermediate steps, transported pneumatically to the furnace. The difference is that in intermediate separation, the separated pulverised coal is directly transported to the burners by a particular transport gas flow, whereas the bin-and-feeder system stores the separated coal powder, charges the carrier air with it via a pulverised coal feeder and feeds it to the firing independently to the raw coal feeding system. When pulverised coal firing was first used, the bin-and-feeder system was used for all coal types in order to be able to continue the operation of the combustion plant if the mills were not operating. A bin-and-feeder system, however, has a higher capital cost compared to an intermediate removal system (STEAG 1988). Bin-and-feeder and intermediate removal systems are used both with low-volatile, high-ash hard coals and with high-volatile, high-ash lignite coals.

5.3.2 Fuel Preparation

Pulverised fuel combustion requires a preparation step in order to completely burn the fuel in the furnace within a short residence time (typically between 2 and 5 s). Combustible coal powder is dry and fine-grained and is the product of the preparation of more-or-less moist, coarse-grained run-of-mine coal. The preparation of the fuel for combustion consists of the steps of crushing, milling, drying and classification. The necessary milling fineness depends on the types of fuel and the firing system. Figure 5.21 shows the requirements for the milling of the fuel as a function of the volatile matter content for dry-bottom and slag-tap firing systems.

5.3.2.1 Drying

The process steps of milling and drying are combined in a simultaneous drying – grinding process to reduce the amount of time required to complete the drying.

During drying, the moisture in the coal is vaporised and removed by the carrier air or carrier gas flow. The drying heat is supplied together with the carrier gas flow. The control variable is the classifier temperature and the manipulated variables are the hot and cold gas flows. In direct injection systems, the classifier temperature sets

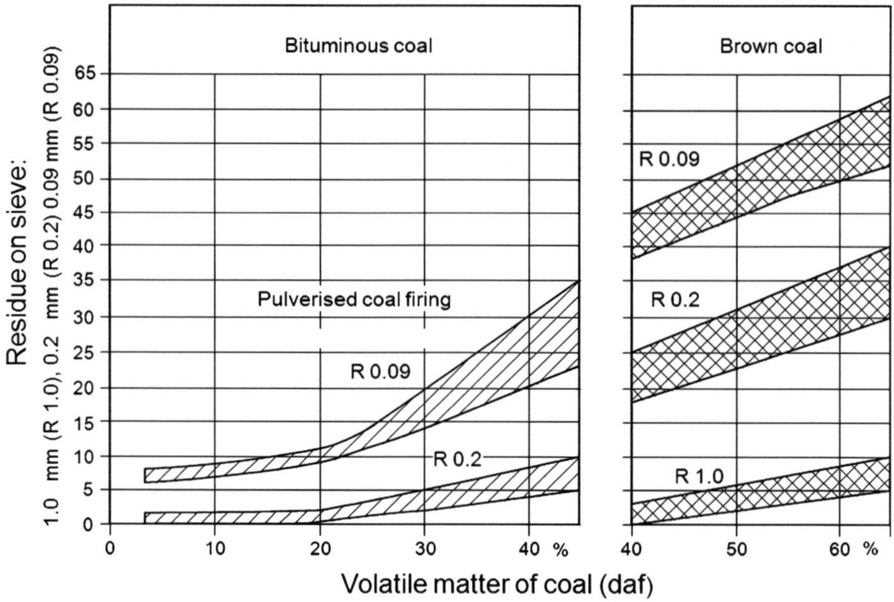

Fig. 5.21 Requirements for milling (*Source*: Alstom Power)

the temperature of the primary air mixture. The drying medium is either hot air or flue gas.

Hard coal is generally dried by hot air to a residual moisture content of 1%. Brown coal types with moisture contents up to 60% and above require great quantities of heat for drying. The heat needed for brown coal is around 16% of the fuel heat, while hard coal needs only around 3%. For this reason, the only drying medium considered for brown coal is hot flue gas, which is extracted from the furnace at high temperatures of around 1,000°C. The residual moisture of brown coal after drying ranges between 12 and 18% (Doležal 1990).

5.3.2.2 Milling

The grinding of coal is performed via one of the following different methods (Zelkowski, 2004):

Gravity Mills (Ball mills)

In an armoured drum rotating slowly around its horizontal axis, the coal is ground by crunching and grating in a milling bed of steel balls. These mills are used for the production of an especially fine powder and for grinding hard types of coal (Fig. 5.22).

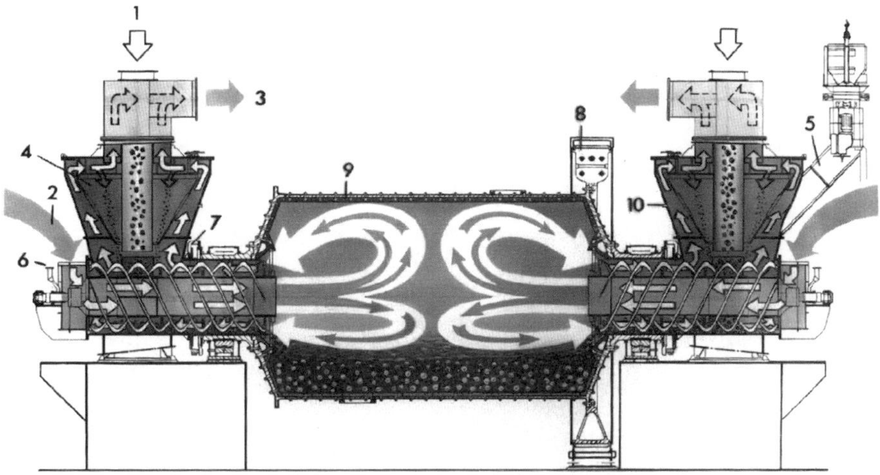

Fig. 5.22 Schematic drawing of a ball mill (*Source*: Alstom Power)

Applied-Force Mills (Bowl Mills, Roller Mills)

In applied-force mills, coal is comminuted and ground by pressure. On a motor-driven grinding table with a vertical axis of rotation, the coal lies in a bed which is passed over by two or three grinding rollers at the circumference. The grinding rollers, hinged and rotatable on swing hammers, are unpowered but are pressed against the bed by spring elasticity or a hydraulic system. The run-of-mine coal is fed to the table centrally from above. Figure 5.23 shows the schematic diagram of a roller mill. Applied-force mills are often used for hard coals.

Beater Mills

In beater mills, the incoming coal is caught by rapidly circulating beaters which are fixed at the perimeter of a rotor and comminuted by the impact of the beater against the armoured mill housing. A differentiation is made between beater mills and beater-wheel mills. Beater mills have a rotor equipped with rigid beater arms on which movable beater tips are mounted. Beater-wheel mills are equipped with the so-called impact plates mounted on a circulating wheel. Beater-wheel mills, like beater mills, have a ventilating effect – they transport the pulverised coal and carrier gas to the burners. Beater-wheel mills are used for the grinding of brown coal. They are usually prefaced by a beater mill for primary crushing (Fig. 5.24). Beater mills as well as gravity and applied-force mills are used in hard coal firing plants.

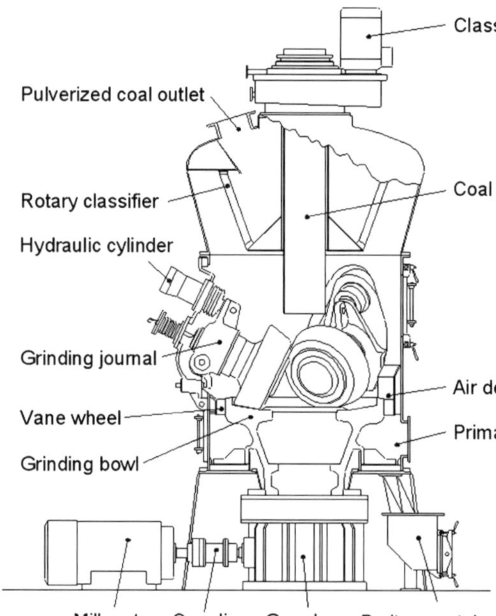

Fig. 5.23 Schematic drawing of a bowl mill (*Source*: Alstom Power)

5.3.2.3 Classifiers

Classifiers separate combustible fine dust from coarse dust, the latter being returned to the mill. Static classifiers installed at the mill outlet or after the mill have little selectivity. If there are high dust fineness requirements, centrifugal classifiers are used.

5.3.3 Burners

After the preparation processes of fuel drying and milling and the injection of the dust – air mixture into the furnace, the combustion process starts, with ignition and the mixing of the dust – air with the remaining combustion air. The milling system, furnace and burners have to be designed to ensure the reliable ignition and complete combustion of the fuel.

The pulverised coal, transported from the mills by the carrier gas, known as primary air, is blown into the furnace via the burners together with the remaining combustion air, known as secondary air. The purpose of the burner is to produce in the furnace an adequate flow pattern for mixing, ignition and combustion. The method of injection has a substantial impact on the near-burner area, in particular, and thus on ignition and pollutant formation. Aside from this, the burner design and installation also influence the downstream end of the process. While the design of pulverised fuel combustion in the past used to follow only the objectives

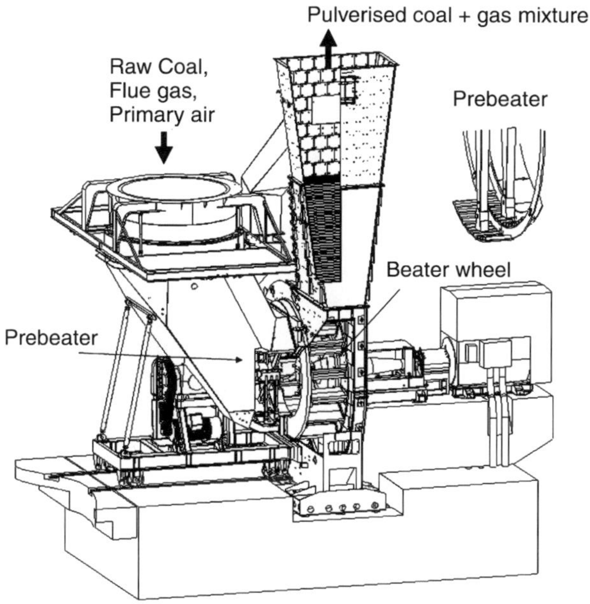

Fig. 5.24 Schematic drawing of a beater-wheel mill with a primary beater stage (throughput raw lignite ca. 170 t/h, ventilation $535,000\,\mathrm{m}^3/\mathrm{h}$, diameter of Wheel 4,300 mm) (*Source*: Alstom Power)

of stable combustion and complete burnout of the fuel, today additional requirements such as low emission levels or usability of the combustion residues have to be met.

For a stable ignition, the coal powder – air mixture has to be heated up to the ignition temperature within a very short time. The supply of heat by radiation is only a relevant mechanism at very high temperatures, for instance, in slag-tap firing. A substantial fraction of the ignition energy has to be provided by recirculation of hot flue gases. The basic aim is to first raise to ignition temperature the primary mixture alone and then to admix the secondary air only after ignition. The ignition can be positively influenced by finer milling, a high air preheating temperature and a high-dust saturation in the primary air mixture.

Figure 5.25 shows the flow fields of a jet burner and a vortex burner. A jet burner usually injects secondary air via two inlets and primary air through a jet in between these, the primary jet loaded with pulverised coal. With hard coal, the primary jet is injected at about 18–22 m/s, and with brown coal, at about 10–14 m/s. The secondary air is injected at a considerably higher velocity, i.e. 40–80 m/s, thus it defines the flow field. The secondary jet sucks hot flue gas from the furnace and mixes it with the primary air jet. The distance between the pulverised coal jet and the secondary air jets has to be such that the coal particles ignite before mixing with the secondary air. The heating to ignition temperature is affected by the recirculated

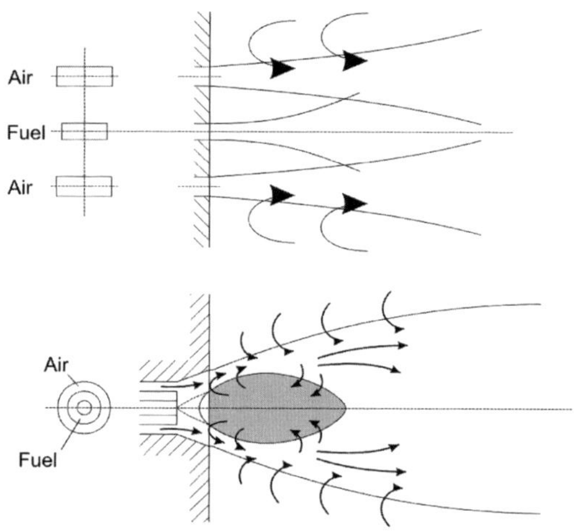

Fig. 5.25 Flow fields of a jet burner (*above*) and a swirl burner (*below*)

hot flue gas. Ignition for a jet burner occurs at a distance of 0.8–1.5 m from the burner (Strauß 2006; Lehmann 1996; Adrian et al. 1986).

Swirl burners inject the primary and secondary air jets into the furnace via concentrically arranged annular nozzles. The secondary air is fed by the outer annular nozzles at high velocities (30–50 m/s); the primary air is injected through the centre nozzle at a velocity of about 18–25 m/s. An adjustable cascade which passes air axially is a possible additional device to give the secondary air an added swirl. The swirling of the secondary air, the cone-shaped widening of the burner tip and the interior barrier baffle plate create an inner recirculation zone which returns hot flue gases from the area of complete combustion to the flame core. Ignition takes place in the shear zone of the opposed flows of primary and secondary air immediately next to the burner. Because of the intensive mixing of primary and secondary air, these burners historically featured a secure and excellent flame stability, but also a high level of nitrogen oxide emissions. By dividing the secondary air flows and delaying the mixing, it was possible to significantly reduce the NO_x emissions.

The impact of the burner design on NO_x emissions will be explained in Sect. 5.7.1.2 in the context of emissions of nitrogen oxides.

5.3.4 Dry-Bottom Firing

Dry-bottom firing, suitable for both hard and brown coal, is a widely used power plant technology. In this firing system, the ash is designed to leave the furnace in a solid state. The broad temperature range of the flame core lies between 1,000 and 1,600°C, depending on the coal and the burner type.

The flame temperature considered sufficient is that which ensures a stable ignition and a sufficiently fast and complete process of combustion. In the flame centre during dry-bottom firing, it is possible for ash particles to melt. Therefore, it has to be ensured that the ash particles in this state do not coagulate, agglomerate and cause slagging on the furnace walls. Otherwise, in an extreme case, the dry-bottom firing could turn into slag-tap firing. The ash is removed from the furnace in a dry state, either as slag or as fly ash. The coarse particles of the slag, which develop through the sticking and sintering of ash in the furnace, fall into the furnace hopper, while the fly ash is carried out with the flue gas and is removed in the electrostatic precipitator. The fraction of slag amounts to 10–15% of the total ash.

Dry-bottom firing is suitable for a broad range of coal types:

- Those with ash in fuel up to 50% (dry)
- High- and medium-volatile coals with volatile matter contents higher than 20% (daf) (Doležal 1990)

The burner configurations used in dry-bottom firing systems are tangential firing, frontal firing and opposed firing (see Fig. 5.26). The residence time for hard coal, in the entire space of the furnace, lies at about 4–5 s; in the space from above the upmost burners to the furnace top, it ranges around 2 s (Adrian et al. 1986). The lower the volatile matter content and the less reactive the fuel, the higher the furnace temperature and/or the longer the residence time have to be in order to achieve complete burnout.

In tangential firing systems for hard coals, the jet burners are mounted in the furnace corners and oriented towards a fire circle to achieve the longest possible flame. Figure 5.27 shows the burner arrangement of a 900 MW_{el} power plant. The furnace is equipped with six burner levels, the four corner burners of each level being supplied with pulverised coal by one mill. In the case of the 900 MW_{el} plant, two burners are arranged in one compartment, as shown in Fig. 5.27. Every burner unit consists of two pulverised coal nozzles and one lower, central and upper air

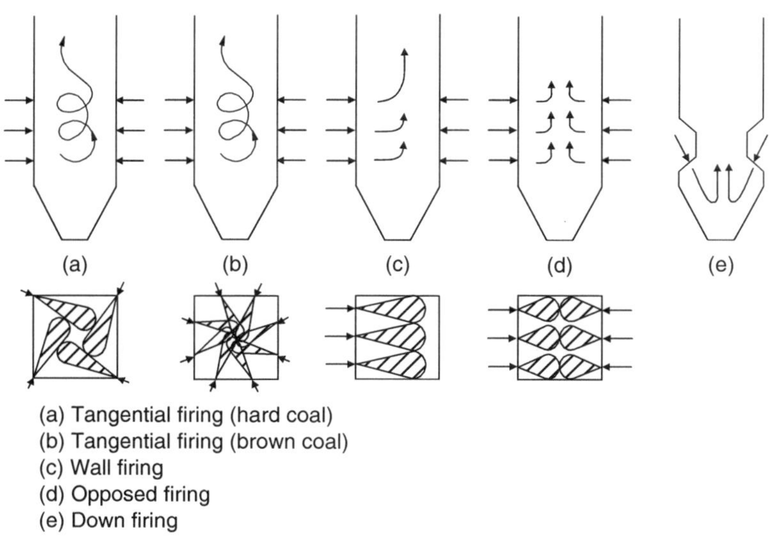

Fig. 5.26 Burner configurations of dry-bottom firing systems (Soud and Fukasawa 1996)

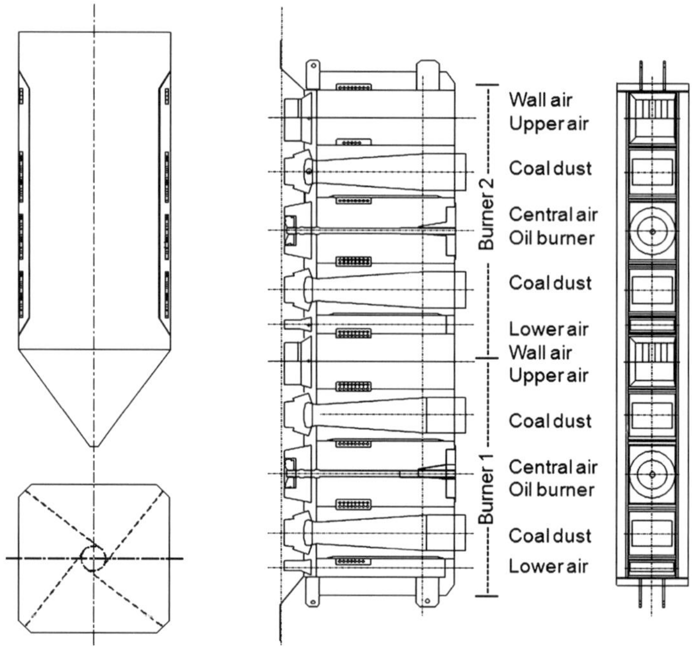

Fig. 5.27 Jet burners for a tangential hard coal firing (*Source*: Alstom Power)

nozzle each. The central air nozzle has one integrated oil burner. The separation of fuel and air nozzles and the parallel channelling of the fuel and air flows result, in comparison to vortex burners, in a delayed mixing of fuel and combustion air and in consequence to less formation of nitrogen oxides.

Brown coal-fired furnaces are predominantly designed as tangential firing systems with jet burners. The burners in this case are positioned in vertical stacks in the side walls. In contrast to hard coal firing, each stack of burners is supplied by a separate mill. One reason for this is the combined drying – crushing, using hot flue gases, which requires a mechanical draught mill for the transport of the primary air mixture. With the limited pressure increase of the beater-wheel mills, the supply of one burner level by a single mill is not possible, because the pressure loss of the pipelines, which are of differing lengths, would have to be balanced out by flow restrictors which would cause too high a pressure loss. The connection to a vertical burner row, however, results in asymmetries in the firing.

The high-moisture content and the flue gas recirculation via the drying – crushing unit are two factors why the dimensions of the furnace are larger for brown coal than that of a hard-coal firing system. For the drying – crushing process, the flue gases are extracted at 1,000°C and fed to the mill. The ground fuel, in the most frequently applied direct injection method, is fed to the furnace via the burners along with the milling vapours.

The milling degree of brown coal is relatively coarse. If a mill without a classifier is used, the residues on the 1 mm sieve may amount to a fraction of 12%. In consequence, the fraction of unburned matter in the residues increases proportionally to the milling degree. Therefore, the design in various plants includes a secondary combustion grate below the furnace hopper in order to diminish the losses through residual material (Lehmann 1996).

Hard coal-fired furnaces with a wall or opposed firing configuration work, as a rule, with swirl burners. Swirl burners, as opposed to jet burners, can also be operated as individual burners, so they allow a greater freedom with respect to the configuration of the furnace. The modifications to swirl burners with regard to NO_x reduction are described in Sect. 5.7.1.2.

The down firing type shown in Fig. 5.26e is often used for the combustion of low-volatile coal types. The injection against the main direction of the flow creates a longer residence time of the fuel. To achieve higher temperatures, it is possible to apply refractory lining in part of the furnace (Stultz and Kitto 1992). Further measures, such as intermediate removal, as described in Sect. 5.3.1, are taken when the fuel is a low-volatile coal.

5.3.5 Slag-Tap Firing

Achieving the highest possible degree of ash retention in the firing was the objective of the development, and the reason for the spread, of slag-tap firing technology in the 1960s. The temperatures in the furnace have to range between 100 and 200°C above the ash fluid temperature to be able to remove the ash in a molten state. In large-volume slag-tap boilers, the combustion of the coal dust takes place in the flow, as in dry-bottom firing, whereas in cyclone slag-tap boilers, the coal particles burn on a slag layer on the wall of a cyclone (Doležal 1990, 1961).

In contrast to firing types with dry ash removal, in slag-tap firing the heat release and the heat transfer by radiation are separate. The slag-tap chamber has the function of burning the fuel and retaining the ash at a sufficiently high temperature. Any kind of heat dissipation is unwanted in the chamber. Only after the chamber are the flue gases cooled down by radiating and convecting heat (Adrian et al. 1986).

5.3.5.1 Large-Volume Slag-Tap Boilers

In large-volume slag-tap boilers the pulverised fuel burns in the slag-tap furnace, which is designed to achieve melting and complete burnout at high temperatures. The residence time in the slag-tap furnace is typically 1 s or shorter. The volumetric heat release rate of the slag-tap furnace ranges between 0.5 and roughly 1 MW/m^3 (Doležal 1990).

Among the construction types which have been developed for slag-tap boilers, the U-furnace has become the most widely accepted. The cross section of such a large-volume slag-tap boiler is shown in Fig. 5.28. The high temperatures of 1,400–1,600°C in the slag-tap furnace make the ash melt, which partly precipitates

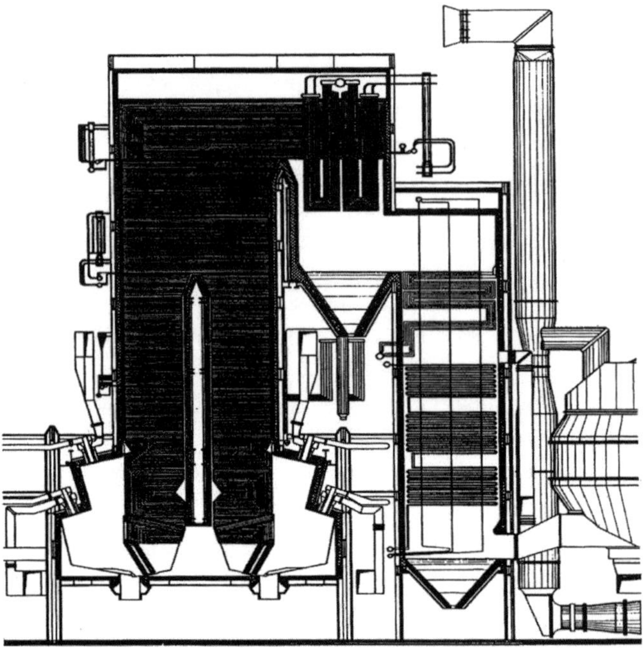

Fig. 5.28 Divided slag-tap furnace

on the walls. The cooling effect of the evaporator tubes solidifies the molten slag closest to the wall side along the walls and bottom of the furnace. The molten slag runs down this insulating layer, is collected at the bottom and routed to a water bath where it granulates.

The combustion process is accelerated by the use of finely pulverised coal, swirl burners and strong air preheating – together they can achieve the high temperatures desired in the slag-tap furnace. By applying ceramic refractory material to the evaporator tubes, held by studs welded onto the tubes, the heat extraction from the slag-tap furnace is diminished, including when a slag layer is absent (for example, during start-up). The refractory material at the same time forms the base for the insulating slag layer. The purpose of the refractory material and the insulating slag layer is to protect the furnace tube walls against corrosion. Figure 5.29 schematically shows the refractory lining. The slag screen forms the boundary of the slag-tap furnace. The function of this screen is to abate heat emission from the slag-tap furnace and to separate molten and softened ash particles from the flue gas.

In slag-tap firing, the molten state removal of ash must be ensured even in part-load operation. At minimum load, there may be the risk of the slag solidifying on the bottom or in the slag discharge mouth. The ash would then accumulate on the furnace bottom and melt out only with a load increase. A slag-tap firing system, in thermal terms, should therefore be designed such that, even at minimum load, sufficient temperatures are achieved to guarantee a satisfactory ash flow. This has

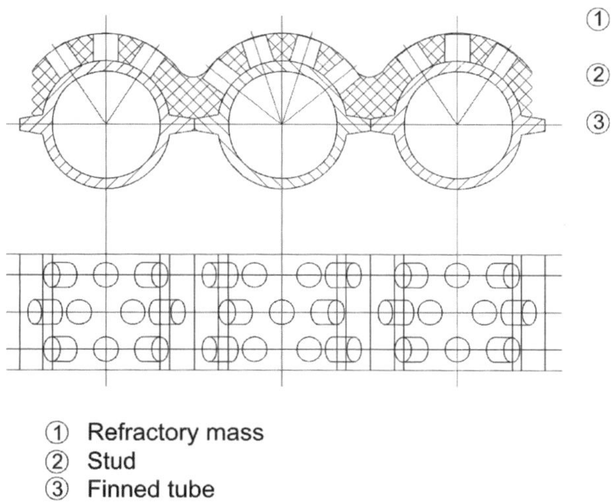

Fig. 5.29 Studding and refractory lining of the slag-tap furnace walls (Doležal 1990)

the consequence, however, that the temperatures in the upper output ranges rise considerably, thus provoking the volatilisation of the ash. Slag-tap firing systems with intensive combustion, using a suitable coal type, allow operation with molten state removal down to around 30% of the maximum power output, so in this respect they are indeed equal to dry-bottom firing systems (Doležal 1990). In the case of the slag-tap furnace shown in Fig. 5.28, the molten state removal at part-load operation is made easier by the division of the furnace chamber into two, with a two-level arrangement of the burners. At minimum load, the bottom burners are operated to intensively heat the bottom section of the furnace. At part loads below 50%, the furnace is operated with one chamber only.

To a certain extent, the insulating slag layer of the furnace has a self-regulating effect. If the temperature in the slag-tap furnace drops at part load, the slag layer grows, so the heat extraction from the furnace diminishes, counteracting the temperature drop.

Primary ash retention in slag-tap firing – understood as the ratio of the ash removed in molten form to the fuel ash input – amounts to 40–60% depending on the coal and the firing type. The recirculation of the filter ash into the slag-tap furnace is a possible means to transform the total fuel ash into slag granulate. The ash which becomes granulated in the water bath can be made use of, for instance, in the construction industry as filling material, or as a gravel substitute in road construction. Its specific volume is only one-third that of fly ash.

As mentioned previously, a possible consequence of the high combustion temperatures in a slag-tap furnace is the partial volatilisation of certain ash components. The gaseous ash components precipitate on the convective heating surfaces in the form of very persistent fouling deposits. In contrast to dry-bottom firing systems where, if anything, only the alkalis and the sulphides volatilise, it is possible that the very high temperatures of more than 1,800°C in slag-tap firing systems cause the volatilisation of silicon as well (Doležal 1961).

One advantage of a slag-tap boiler in comparison to a dry-bottom boiler is a higher steam generator efficiency. Due to the high temperatures, it is possible to operate this firing at a lower air ratio (1.05–1.15) than dry-bottom firing. In addition, the low SO_3 content and the associated low acid dewpoint in the flue gas allow lower boiler exit temperatures in slag-tap firing systems. A drawback is that heat is lost through the removal of the hot molten slag, the magnitude of which depends on the ash content of the coal. For low volatile coal types, the burnout is significantly better than in dry-bottom firing. The resulting losses are one parameter for consideration in comparison with a dry-bottom firing system (Fig. 5.30).

Today, one application of slag-tap firing considered particularly apt is for the combustion of low-volatile coals, because the slag-tap furnace in this case ensures complete combustion in a more efficient manner than the dry-bottom furnace. In the past, however, the combustion of medium- and high-volatile hard coals in slag-tap furnaces was also successful. For coals with a very high ash content, a lower efficiency compared to dry-bottom firing arises through the heat loss of the slag. With respect to ash fluid temperatures, a broad range of fuels can be burned in slag-tap firing. Coal types have been used, for instance, with a fluid temperature higher than 1,600°C. However, the necessary temperatures for acceptable operation, in this case above 1,800°C, may provoke ash volatilisation and heat exchanger fouling.

The essential disadvantages of slag-tap firing in comparison to dry-bottom firing systems are the higher capital costs and the maintenance necessary for the restoration of the slag-tap furnace refractory lining and the wall and slag screen heating surfaces. Slag-tap firing systems involve about 10% higher capital costs than dry-bottom systems. Another disadvantage is the higher nitrogen oxide emissions of the slag-tap furnace, but it is possible for slag-tap firing to achieve a reduction of NO_x emissions to between 800 and 1,000 mg/m^3 and should be possible to further reduce these emissions by means of the methods described in Sect. 5.7. Slag-tap firing systems are justified when the given coal type requires appropriately high

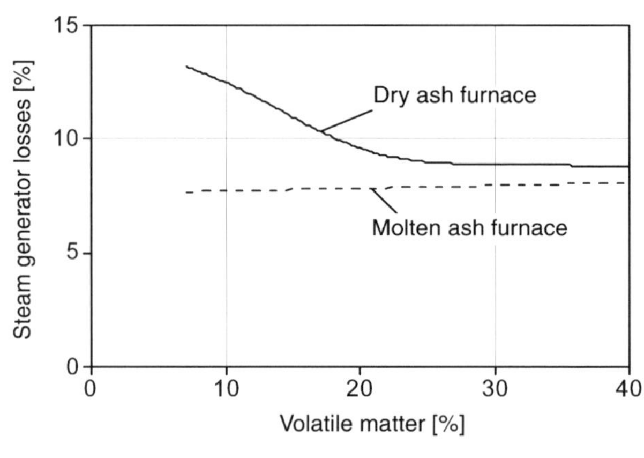

Fig. 5.30 Steam generator losses of slag-tap and dry-bottom firing systems (Doležal 1990)

temperatures for the complete combustion or where the ash can only be utilised or disposed of in granulate form (Kather 1995).

The first slag-tap firing system was put into service in 1934 in Czechoslovakia (Doležal 1954). In Germany, the development of this technology continued after WW II. Large-scale slag-tap boilers were scarcely used in other countries. The advantages of slag-tap firing technology which resulted from the development in Germany in turn resulted in 75% of the use of the technology being in Germany by 1970. Due to the disadvantages mentioned above, however, slag-tap firing systems were rarely built in the time that followed.

5.3.5.2 Cyclone Furnaces

Cyclone furnaces differ from the previously described large-volume slag-tap furnaces by a higher volumetric heat release rate, which results from the use of much coarser coal particles and a higher primary ash retention. The combustion process takes place in a (usually horizontally installed) cyclone which typically measures 1.8–3 m in diameter. The volumetric heat release lies between 4 and 8 MW/m^3. The cyclone, as with the large-scale slag-tap furnace, is lined with a ceramic material to reduce heat absorption. Due to the small cyclone volume, only about 10% of the total heat is transferred to the cyclone heating surface of the steam – water cycle (Stultz and Kitto 1992).

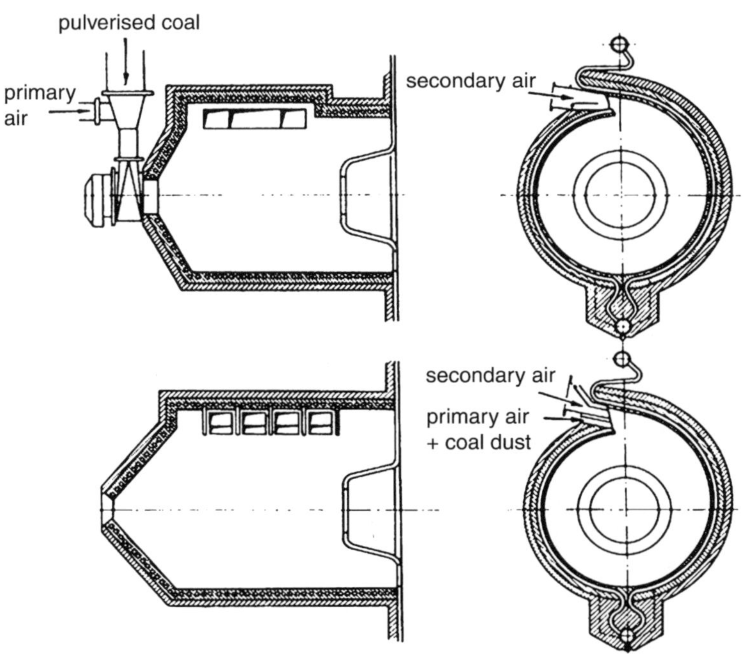

Fig. 5.31 Cyclone construction types (Doležal 1961)

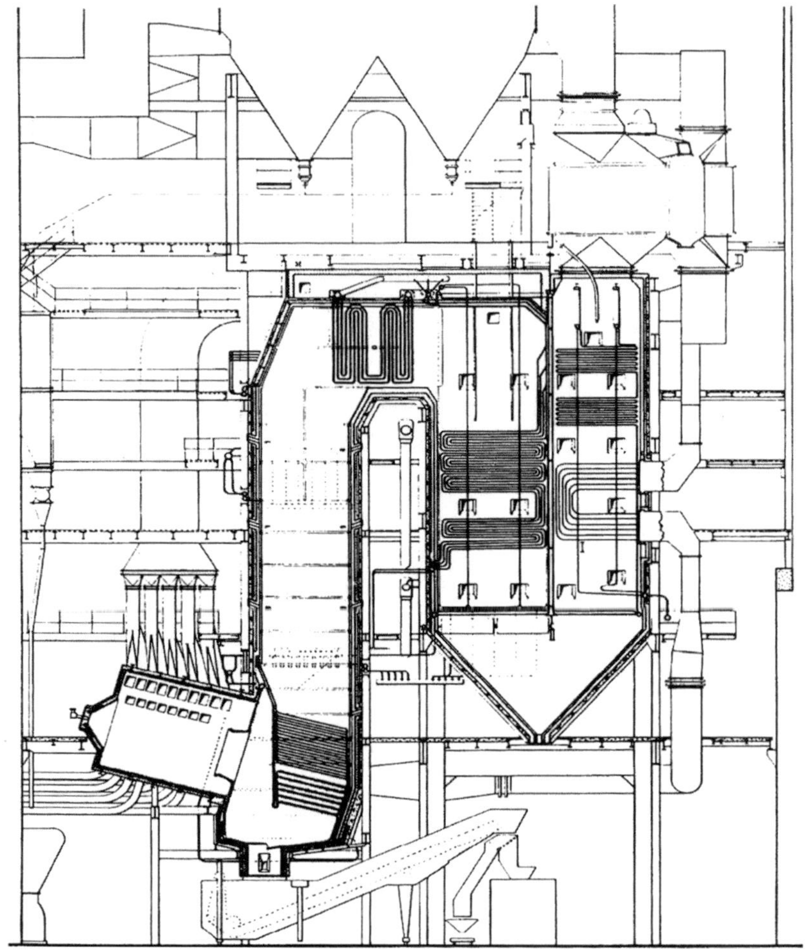

Fig. 5.32 Steam generator with cyclone furnace (Doležal 1961)

In the cyclones shown in Fig. 5.31, the tangential injection of the secondary air creates a rotary flow which hurls the coal, fed either axially (American construction) or tangentially (German construction), against the wall. The coarse coal then burns (i.e. to completion) while attached to the molten ash-covered cyclone wall. The residence time of the particles exceeds the time of the gas many times over. The rotary flow has, as a result, a cyclone removal efficiency of up to 90%. The hot flue gases, with temperatures from 1,600 to 1,800°C, exit the cyclone at a high velocity and are cooled in a radiation duct (see Fig. 5.32). An effective size reduction of this radiation duct can be achieved by inserting plate elements and separating walls, because the duct has no combustion engineering function but only serves to exchange heat (Doležal 1954). The slag at the cyclone base flows towards the furnace bottom and from there through a drain into a water bath. In a power station, several cyclones are usually installed next to and above each other to achieve greater outputs. In one particular 1,100 MW_{el} plant, for instance, there are 23 cyclones (Stultz and Kitto 1992).

Sufficient coal preparation is achieved by a crusher which mills the coal to a size smaller than 4 mm, with a mean particle diameter of 0.5 mm. The lesser milling work requirement of cyclone firing systems, however, is counterbalanced by the higher power demand of the air fans. A fan pressure of 0.5–1 bar is needed due to the high pressure loss in the cyclone. The cyclone furnace requires a lower auxiliary power, in comparison to dry-bottom furnaces, only when the feedstock is a high-ash coal type (Stultz and Kitto 1992).

5.4 Fluidised Bed Firing Systems

Fluidised bed firing (BFB) technology was industrially applied for the first time in the 1920s by Winkler for the gasification of coal. The development of bubbling fluidised bed combustion began in the 1960s, resulting in the first commercial applications at the beginning of the 1970s, with capacities of up to 20 MW_{th}. The capacity of bubbling systems have increased since then – today, the biggest are the Shawnee (USA) plant, with an electrical output of 160 MW_{el}, which went into service in 1988, and the 350 MW_{el} Takehara plant in Japan, which started up in 1995. However, these BFB demonstrations do not seem to have led to any follow-up installations. This is probably because circulating fluidised bed combustion (CFBC) has come to dominate the larger scale applications.

At the end of the 1970s, circulating fluidised bed techniques were developed as an alternative to bubbling systems and have superseded them more and more since. With technological advances, the achievable unit size has increased steadily over the past decades. Today, CFBC units in operation range in size from a few MW_{th} to 300 MW_{el}. In 1995, a circulating fluidised bed furnace with an output of 259 MW_{el} was put into service in Gardanne, Provence-Côte d'Azur (France) and in 2001 the JEA plant of 2×300 MW_{el} went into service in Jacksonville/FL/USA. A 460 MW_{el} boiler, which was built in Lagisza (Poland), is currently the world's largest CFBC unit (Goidich et al. 2006). A capacity increase of up to around 800 MW_{el} is considered feasible within the medium-term future (Hotta and Venäläinen 2006). Figure 5.33 shows the development of the thermal capacity of bubbling and circulating fluidised bed furnaces installed worldwide (Wu 2006; Koornneef and Junginger 2007).

The way a fluidised bed works is shown in Fig. 5.1. A fluidised bed consists of a packed bed of fuel particles above a grid through which air can be passed upwards to the bed. According to the velocity of the air, the bed is said to have one of three distinct stages of fluidisation – fixed bed, bubbling fluidised bed or circulating fluidised bed.

At low gas velocities, the gas simply flows through the packed bed without disturbing the particles significantly. The bed acts as a porous media and is called a fixed bed. As the gas velocity increases, the gas particle drag forces compensate for

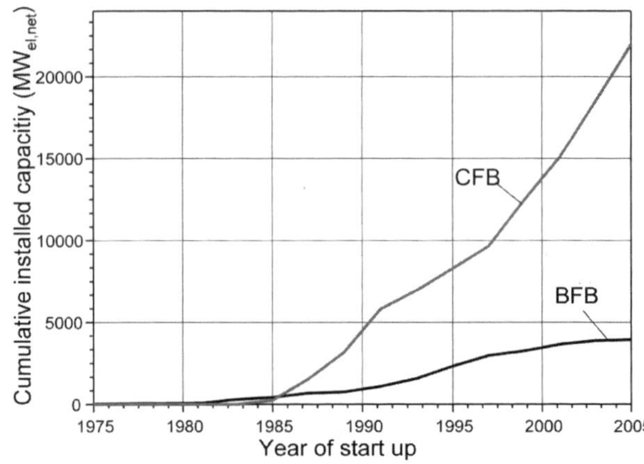

Fig. 5.33 Installed capacities of bubbling and circulating fluidised bed furnaces; data from Koornneef and Junginger (2007)

the bed weight and the point of minimum fluidisation is reached. The inter-particle distances increase, the bed expands and the particles appear to be suspended in the gas. When the gas velocity exceeds the minimum fluidisation velocity, the excess gas passes through the bed as bubbles and the remainder leaks through the bed material. The bed is then called a bubbling fluidised bed and the furnace above the bed is defined as the freeboard. Bubbling fluidised beds are normally operated at gas velocities several times higher than the minimum fluidisation velocity. The passage of the bubbles result in an intensive mixing of the bed particles, although the particles remain in close contact and are not carried upwards to a significant degree. The height of the fluidised bed, and hence the distance between each of the single particles – the void fraction – is proportional to the throughput of air. The necessary air pressure is higher than in grate firing systems and remains constant in fluidised bed firing even at increasing velocities of the air. Only when the velocities exceed the particle free-fall velocity are the bed particles entrained in the gas stream and lifted out of the bed. The gas velocity at this point is known as the particle terminal velocity. The fluidisation and the terminal velocities limit the air velocity range of a furnace using a bubbling fluidised bed. The range for circulating fluidised beds lies above the terminal velocity, in the transition range to pneumatic transport (Wu 2003; Doležal 1990).

The temperatures of fluidised beds are limited to between 800 and 900°C, on the one hand, in order to ensure the prevention of the sticking of ash particles and, on the other hand, to achieve an optimum SO_2 capture.

5.4.1 Bubbling Fluidised Bed Furnaces

The bubbling fluidised bed features a defined bed surface and a high solid matter density. The gas velocities range between 1 and 2 m/s. The fluidised bed, consisting of 96% inert material and limestone, has a height of 1–1.5 m. Fine particles are carried out of the bed and removed in a downstream dust collector. Recirculating the filter ash into the furnace limits the unburned fuel loss through unburned combustibles (Strauß 2006). Without ash recirculation, the resulting burnout degrees with hard coal are 80–90% (STEAG 1988).

Another feature of the bubbling fluidised bed is the in-bed heat transfer surface installed inside the fluidised bed. Despite the small temperature gradient, heat flow rates arise which are usually found only in pulverised fuel firing furnaces ($q_w = 300\,kW/m^2$). About 50% of the total heat is transferred by the in-bed heat exchanger, the remaining heat being transferred via the downstream heat transfer surfaces. Erosion and corrosion of the in-bed heat transfer surfaces are a problem of the bubbling fluidised bed, and one explanation for it being superseded by circulating systems. To limit the corrosion hazard caused by the reducing atmosphere, air staging is not applied.

Lower velocities in part load operation reduce the heat transfer only a little. In order not to lose too much heat during part load operation, it is necessary to diminish the heat transfer of the in-bed surfaces. Adjustment to the steam generation process can be performed either by reducing the bed height or by taking furnace modules of a fluidised bed out of service. The bed height reduction and the functional change of the in-bed heat transfer surfaces into gas/steam heating exchangers have the effect of cooling down the freeboard temperatures and raising the emissions.

The design capacity of a bubbling fluidised bed furnace can only be increased by the bed surface; the bed height, being limited by the pressure loss, cannot be a variable design parameter (Bunthoff and Meier 1987). The cross-sectional area heat release rate of bubbling fluidised beds, from 1.2 to $1.6\,MW/m^2$, lies in the same order of magnitude as a grate firing system. Ever-larger bed surfaces, however, make the fuel and air distribution more difficult, as eventually air bubbles may come through. The upper capacity limit of a module is $80\,MW_{th}$ or so (Strauß 2006). Figure 5.34 schematically shows the construction of a bubbling fluidised bed furnace.

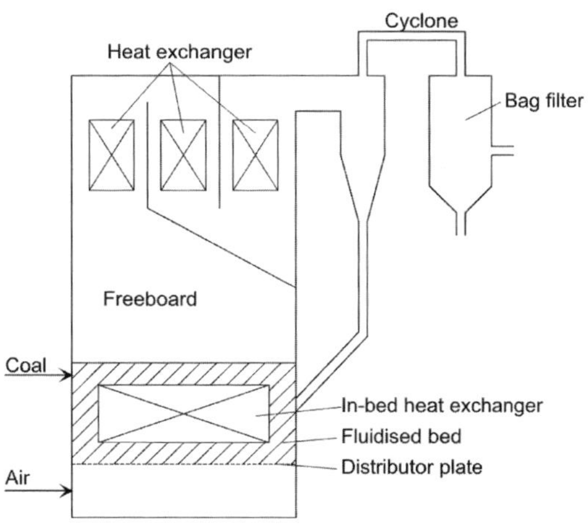

Fig. 5.34 Schematic of a bubbling fluidised bed firing system

Bubbling fluidised bed technology today is used either in cases of low capacities or where fuels are used with low calorific value and hence low adiabatic combustion temperatures. In these cases of application, the in-bed heat transfer surface may be unnecessary. Bubbling fluidised beds are also a preferred solution for biomass. Depending on the lower heating value (LHV) of the biomass, the bed is operated fuel-rich, with stoichiometries of around 0.3, in order to keep the bed temperature around 800°C or lower. With the addition of secondary air in one or several stages, the temperatures should not exceed 950–1,100°C, depending on the ash composition (see also Sect. 6.2.3).

5.4.2 Circulating Fluidised Bed Furnaces

In a circulating fluidised bed, a gas velocity of up to 8 m/s is chosen, so high that most particles are carried up by the gas flow. No definite fluidised bed with a high density of particles is established with these velocities. The solid load continuously decreases with the height.

The fluidising velocity is one of the key parameters for fluidised bed design. Its choice affects most of the other design parameters. Higher fluidising velocities allow the use of a smaller bed cross-sectional area. Circulating fluidised bed furnaces are therefore built more compactly with cross-sectional area heat release rates of 5–7 MW/m^2 (JBDT 1992). The furnace height has to be increased, however, in order to provide a sufficient particle residence time for combustion and sulphur retention and to accommodate heat transfer surfaces. Higher fluidising velocities also increase erosion and fan power requirements (Wu 2006).

In a CFBC, much finer bed particles (with a diameter of 150 μm) are used. This, in combination with a high fluidising velocity, ensures that the particles are entrained in the gas flow and circulated in the system. The actual particle size depends on the fluidising velocity. Feed coal particle sizes typically range from 3 to 6 mm. However this may vary according to the characteristics of the fuel fired. For less reactive high-ash and low-volatile fuels, a smaller particle size is usually selected (Wu 2003).

The residence time of the particles in the fluidised bed depends on the velocity and particle size. Locally inside the fluidised bed, particles may move against the gas flow. This so-called internal recirculation provides for a substantially more intensive and thorough mixing compared to a bubbling fluidised bed. While the solid matter load in the upper part of bubbling fluidised beds amounts to around 50 g/Nm^3, this load in circulating fluidised beds may reach up to 10 kg/Nm^3 (Strauß 2006). The circulating solid matter is removed by a cyclone downstream of the fluidised bed and fed back into the lower part of the fluidised bed by a recirculating duct. The essential advantage of the circulating fluidised bed lies in the uniform temperature of the total cycle, a consequence of the large mass of circulating solid matter with a high heat capacity.

A decisive factor in the combustion and in the desulphurisation process is the contact time between solid matter and gas. While the contact, in the case of bubbling

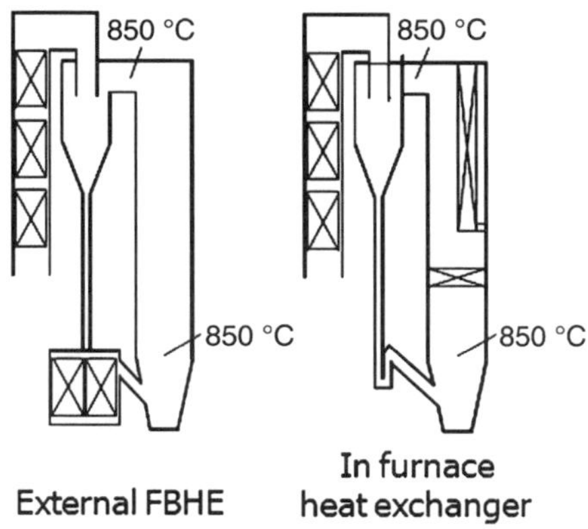

Fig. 5.35 Circulating fluidised bed systems

fluidised beds, is limited to 0.5 s, due to the height of the fluidised bed, the contact is prolonged to 4–5 s and more in circulating fluidised beds because of the use of the entire height of the furnace (Takeshita 1994). The longer contact time and the intensive mixing reduce the limestone demand for desulphurisation.

In-bed heat transfer surfaces are not used for heat extraction from the combustion chamber of circulating fluidised beds. The heat is transferred via the furnace walls, built-in plate heating surfaces and heat transfer surfaces that cool the recirculated ash. In all systems, the fluidised bed furnace is built of water/vapour-cooled membrane walls, with the bottom part refractory-lined. Heat is transferred to the furnace walls or plate heating surfaces mainly by particle convection. Depending on the solid matter load, the resulting heat transfer coefficients range between 230 and 280 W/m^2 K. An additional fraction of the released fuel heat is transferred to the steam – water cycle through convective heat transfer surfaces from the hot flue gases having left the cyclone. The remaining usable fraction is used to preheat the combustion air.

As the heat transfer via the furnace walls is not sufficient to extract the heat, additional heat exchangers are required either within the furnace or after the furnace in the ash recirculation stream ("external" heat exchangers). The arrangement of these heat exchangers was one of the most obvious differences in the design of different boiler manufacturers in the past (see Fig. 5.35).

5.4.2.1 Systems with External Fluidised Bed Heat Exchangers

The characteristic of this arrangement, which was originally developed by Lurgi (now Lurgi Lentjes), is the fluidised bed heat exchanger for cooling the externally recirculated ash flow. Part of the solids collected by the cyclone, at temperatures of 845–900°C, is diverted into the circulating fluidised bed via an ash discharge valve. A series of heat exchanger bundles, which can perform superheater, reheater and/or evaporator duties, can be located in the bed. The solids are fluidised with the air and cooled down to temperatures around 600°C, then returned to the lower furnace. The recirculation of cold ash allows low-ash recirculation rates when setting the fluidised bed combustion temperature of 850°C. The fluidised bed velocity ranges between 6 and 8 m/s.

A fluidised bed heat exchanger (FBHE) has a relatively high rate of heat transfer from the hot solids to the tube bundles. With a low fluidising velocity, typically less than 0.3 m/s, and fine particle sizes (about 200 μm), the potential for erosion of the tubes is eliminated. As the heat exchanger is fluidised with air and not exposed to corrosive elements in the flue gas stream, the potential for corrosion is also minimised. Additionally, with the ash flow control valve, one can control the heat transfer to the immersed bundles. This in turn controls the furnace temperature (Takeshita 1994; Wu 2006).

5.4.2.2 Systems with Plate Heat Exchangers

This circulating fluidised bed firing technique, originally developed by the Ahlström Company, Finland (now Foster Wheeler), with gas velocities from 6 to 8 m/s, is in the category of classical circulating fluidised bed types. The circulated ash is removed outside the furnace in a refractory-lined cyclone and fed again, uncooled, to the fluidised bed. The temperature of the solid matter – flue gas mixture at the exit of the furnace ranges between 800 and 900°C, thus corresponding to the combustion temperature. The heat is transferred inside the furnace via the furnace wall and via additional platen heating surfaces in the upper section of the furnace and outside the furnace in downstream convective heat transfer surfaces. In order to maintain the fluidised bed temperatures, high-ash recirculation rates are required. As the boiler size increases, the furnace surface to volume ratio decreases and it then may not be possible to perform all the required heat exchange in the furnace and back passage. Hence an external fluidised bed heat exchanger has to be used for boilers in the 300 MW_{el} range or higher.

5.4.2.3 Solid Separation Systems

The most commonly applied separation systems are cyclones consisting of a steel shell lined with heat- and erosion-resistant refractory material. However, these thick multi-layer linings often require high maintenance efforts. This has led to the development of water- or steam-cooled cyclones, which are lined with a thin layer of refractory material held in place by a dense pattern of metal studs.

In order to achieve a more compact design of the CFBC, Foster Wheeler has developed a compact separator integrated with the furnace. The design is still based on centrifugal separation but has flat walls, thus simplifying fabrication and construction. Gas with entrained solids enters the separator through a tall and narrow opening and exits from the top. A swirling imparted to the gas flow causes solid separation.

Another possibility for achieving a more compact design is to recirculate the ash inside the furnace (Maryamchik and Wietzke 2005). Impingement separators in the form of a U-beam mounted in the upper furnace section return the ash within the

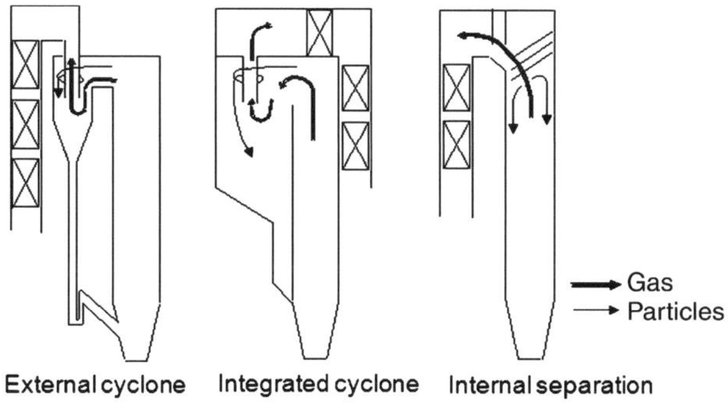

Fig. 5.36 Particle separation configurations

furnace, allowing the external recirculation to be omitted for extreme cases (see Fig. 5.36).

A decisive factor in ensuring the complete combustion of the coal in fluidised bed firing is ensuring that the furnace residence time of the coal particles is longer than the burnout time. Depending on the fluidising velocity, larger particles stay in the fluidised bed, while smaller ones are carried out. The correlations are shown in Fig. 5.37 (Michel 1992). For brown coal, for all particle sizes, the time required for burnout is less than the residence time in the freeboard, so that, if it was for the sake of the burnout, the process could be run without recirculation. For hard coal, the residence time in the freeboard is only sufficient for the burnout of very small particles. Particles between 0.04 and 0.8 mm have to be recirculated to achieve a complete burnout. However, given that the cyclone can only partly separate the

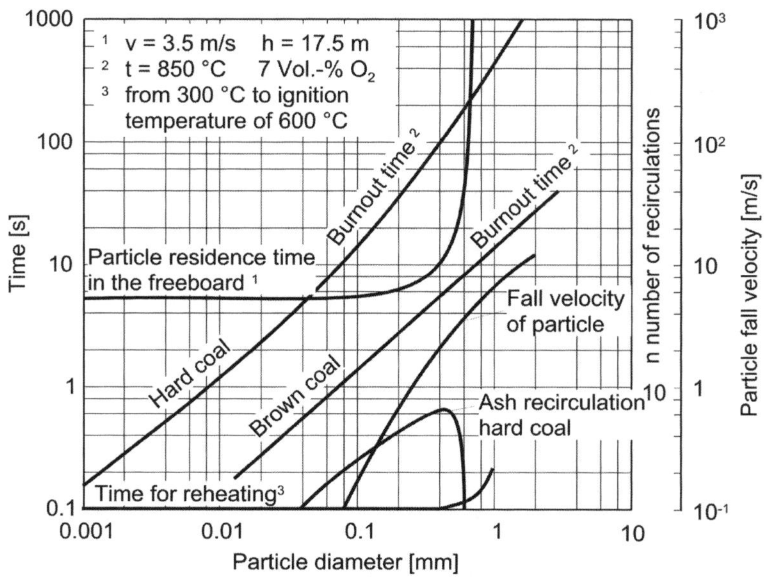

Fig. 5.37 Particle burnout behaviour (Michel 1992)

particles to be recirculated, the result for hard coal is a loss due to unburned combustibles in the fly ash.

The fly ashes of hard coal-fired circulating fluidised bed furnaces in general have high carbon contents, typically of 30%. Measures to increase the removal efficiency of the recirculating cyclones, in particular for small particles, can reduce loss through unburned combustibles.

Start-up times of circulating fluidised bed systems are determined by the refractory lining of the recirculating cyclone and the ash flow ductwork. Because the heating-up may only proceed at a maximum temperature change of 50–80°C per hour, long times for the start-up from a cold state (cold start; outage > 72 h) with auxiliary fuels are needed which, accordingly, cause start-up losses. Hot starts (outage < 8 h) or load changes are assisted by the great heat storage of the fluid bed and the refractory lining, and the start-up times from a hot state or from rates of load change are comparable to pulverised fuel firing systems (VGB 1997). For warm starts (outage 8–72 h) the start-up time ranges in between and depends on the outage time.

The load is controlled by modifying the fuel feed rate and the air flow to the steam generator. By means of supplementary ash from the bed ash storage, the circulating ash flow can be changed and the heat extraction adjusted according to the firing rate in order to set the desired fluidised bed temperature. This procedure manipulates the impact of the solid matter load on the heat transfer coefficient. In circulating fluidised bed systems with external fluid-bed heat exchangers and a high circulating ash flow, the fluidised bed temperature is controlled by modifying the ash recirculation temperature (Stultz and Kitto 1992).

The circulating has almost superseded the bubbling fluidised bed because of its numerous advantages in the mid-load and upper load range of fluidised bed furnaces. The main advantage of the fluidised bed firing system, as opposed to other combustion technologies, lies in the fact that it is able to meet emission control standards without additional desulphurisation and $DeNO_x$ plants.

Due to the high-ash load in circulating fluidised bed furnaces, the heat transfer surfaces are subject to increased erosion. However, satisfactory solutions for this problem have been achieved by taking measures such as installing the tubes parallel to the direction of flow and coating areas threatened by erosion with ceramic material. Another problematic issue is the utilisation of the combustion residues. The residual material, consisting of inert material, fuel ash, additive(s) and products of the desulphurisation process, is not suitable for immediate utilisation. Table 5.5 draws a comparison between circulating fluidised beds and pulverised fuel firing systems (VGB 1997).

5.4.2.4 Future Developments

Considerable efforts are continuing to further improve the performance of CFBC boilers and to scale up the technology to 600–800 MW_{el}. As the unit size increases, the furnace depth remains constant while the furnace width increases and cyclones are added as required. A limit is reached when the unit size approaches 300 MW_{el} or when four cyclones are required. For larger electrical capacities, a deeper furnace

Table 5.5 Comparison between circulating fluidised bed firing (CFBF) and pulverised fuel firing systems (PFF)

	CFBF	PFF
Capacity range	600 MW_{el}	1,000 MW_{el}
Fuel range	Wide	Limited
Fuel flexibility	Great	Moderate
Space required	Little	Moderate
Efficiency	High	High (higher than CFBF)
Availability	>90%	>90%
Investment costs	90% (of PFF)	100% (of which ca. 10% $DeNO_x$, 15% FGD unit)
Operating costs	High limestone consumption (Ca/S = 1.5–2.7) ash disposal landfill, mining (0–50 € per ton)	Low limestone consumption (Ca/S = 1.05) ash utilisation as building material, revenue up to 10 € per ton utilisation of gypsum, revenue up to 5 € per ton

is used and cyclones are arranged in parallel at both sides of the furnace. Depending on the manufacturer, the furnace either still has one fluidising grid or is split to form a dual grid (Wu 2006; Stamatelopoulos and Weissinger 2005a, b; Goidich et al. 2006).

5.5 Stoker/Grate Firing Systems

The stoker-fired furnace is considered the oldest combustion system for solid fuels. At the beginning of the 20th century, mechanical stoker firing was the only available combustion technology for coal. Having limited capacities, it was increasingly superseded in the 1920s and 1930s by pulverised fuel firing (Stultz and Kitto 1992). Stoker-fired furnaces are used in the capacity range from 0.3 up to 150 MW_{th} in industrial and thermal power plants (JBDT 1985). In this range, they compete with fluidised bed furnaces, which is why these furnace types have become a seldom-applied technology for fossil fuels. In waste incineration, in contrast, stoker firing is almost the only technology and it is suited for biomass as well.

The type of stoker firing is chosen depending on the properties of the fuel feedstock. For coals, the system commonly used is the travelling grate stoker; for high-ash fuels like biomass and waste, firing systems with a mixing function, such as pusher-type grate firing systems, are used.

5.5.1 Travelling Grate Stoker Firing

Characteristic of the travelling grate stoker firing is the travelling of a glowing fuel bed on the upper track of a circulating grate belt. The coal glides from the coal bunker onto the grate and forms a bulk layer which is then heated by radiation from the furnace and thereby dried, then devolatilised and ignited, radiating heat back

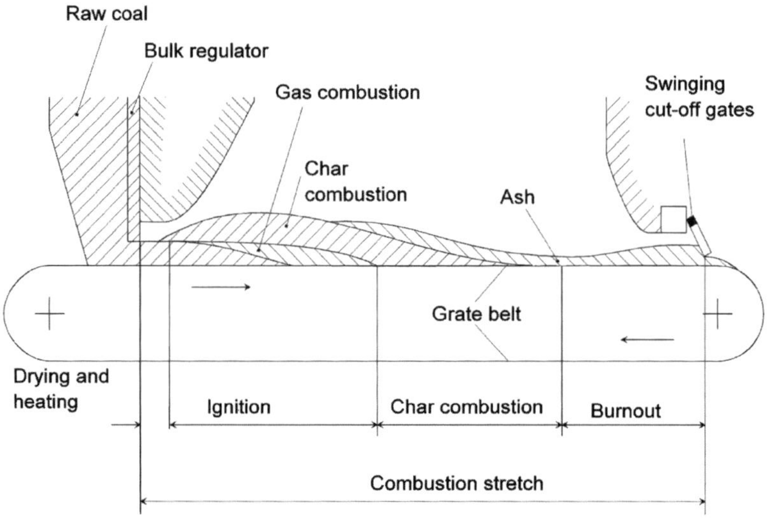

Fig. 5.38 Combustion procedure for a travelling grate (Adrian et al. 1986)

into the furnace. After combustion of the solid coke residue, the ash dammed up by swinging cut-off gates at the end of the grate falls into the ash hopper (see Fig. 5.38). The coal bulk layer resting and burning on the travelling grate is not poked (Strauß 2006; Doležal 1990; Lehmann 1996; Adrian et al. 1986).

The travelling grate is composed of a great number of parallel grate bars mounted in rows one behind the other and linked by joints. The major part of the combustion air, the primary air, is blown into the layer from below, cooling the grate at the same time. Combustion air preheating is therefore limited to about 150°C.

Stoker-fired furnaces are suitable for the combustion of lump fuels – fines in the fuel are not desirable because they may fall through the grate clearances, thus increasing the unburned fuel loss. Strongly caking coal is little suited to combustion on travelling grates – it forms big, very slowly burning coke cakes which hamper the homogeneous distribution of combustion air throughout the fuel bed. Lightly caking coal, in contrast, is desirable because the forming of small lumps that cling together prevents the discharge of fines without obstructing the air flow through the bed (Adrian et al. 1986).

The fuel bed height is adjusted according to the combustion characteristics of the feedstock. It depends on the volatile matter content and the grain characteristics of the fuel. Figure 5.39 provides reference values for the bed height to be chosen (Adrian et al. 1986). For finer fuel grains and a higher volatile matter content, the bulk height is reduced in order to ensure the passage of air.

A good burnout is achieved by sufficient residence times for the solid fuel on the grate and for the gases in the furnace. For optimum conditions it is necessary that the bed height, the grate forward movement and the course of burnout are co-ordinated. The velocities of the grate forward movement vary from 1.5 to 15 m/h. Too slow a grate movement along with too high a bed results in excessively high

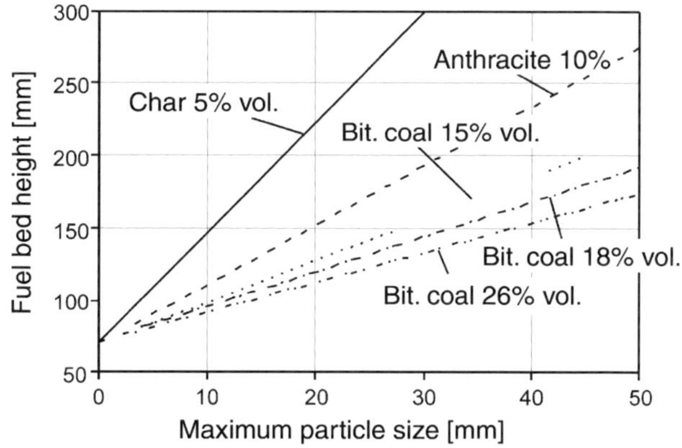

Fig. 5.39 Bed height of hard coal on travelling grates (Adrian et al. 1986)

concentrations of unburned gases above the devolatilisation zone. The lower limit of the grate velocity is a consequence of the danger of a flashback into the feed hopper. Too fast a movement and too low a bed height, in turn, may break off the ignition.

The load of a travelling grate stoker is controlled mainly by adjusting the grate forward movement, as the bed height as a manipulated variable is too slow in responding (Strauß 2006).

The ratio of the fuel heat release to grate surface area is termed the grate load. The maximum grate load is limited by the combustion and slagging behaviour of the coal. When there is too high a grate load, the consequence may be slagging on the grate and hence operational malfunctions. Typical grate loads lie between 1 and $2\,\mathrm{MW/m^2}$.

The maximum output of a travelling grate stoker-fired furnace is determined by the grate surface area. These areas have a size of up to $100\,\mathrm{m^2}$, yielding heat outputs up to $125\,\mathrm{MW_{th}}$ (JBDT 1985).

For fuels with fines, feeding is better executed by spreader stokers which throw the fuel onto the grate mechanically (see Fig. 5.40), rather than by hoppers. This way, finer fuel particles already start to burn while in the air. The grate load may thus be greater by 50% compared to hopper feeding (Strauß 2006). The fuel in this case ignites via the base fire on the grate. As part of the fuel is carried out of the furnace, it is reasonable to recirculate the fly dust in order to limit the loss through unburnt carbon. This firing system represents a transition to a pulverised fuel firing system. Achievable thermal outputs reach a maximum of $175\,\mathrm{MW_{th}}$.

5.5.2 *Self-Raking Type Moving-Grate Stokers*

For high-ash, low-calorific fuels, travelling grates are not an adequate technology, because the movement of the grate only serves for fuel feeding and for the transport of fuel and ash. Difficulties arise igniting such fuels and during burnout. Such types

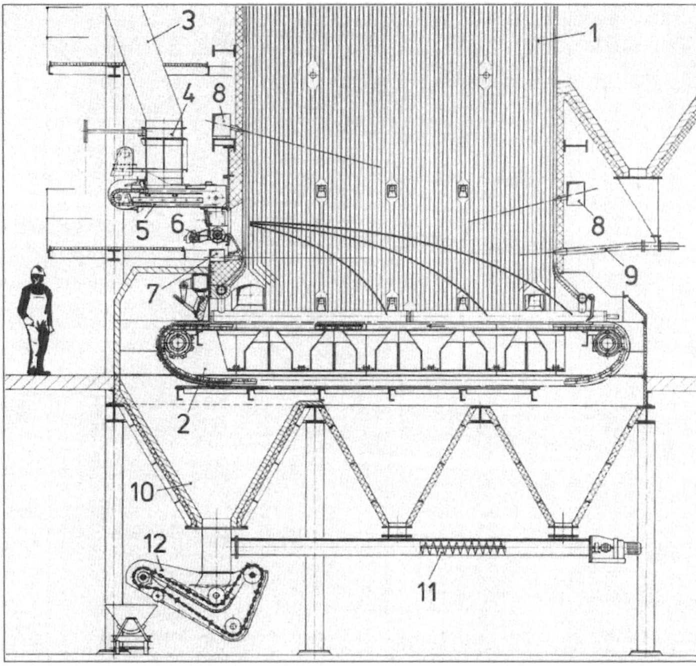

1	Steam generator	2	Louvre type travelling grate
3	Chute	4	Shut off gate
5	Feeder	6	Spreader
7	Carrying air	8	Secondary air
9	Coke fines feed back	10	Ash hopper
11	Screw conveyor	12	Submerged scrapper conveyor

Fig. 5.40 Travelling grate stoker firing with a spreader stoker (*Source*: Alstom Power)

of fuel have to be constantly broken up by poking/raking in order to ensure the access of air and to subject all of the fuel to the radiant heat.

A difference is made between pusher-type grates where the rams move the fuel in the direction of the proceeding combustion and reciprocating grates where the rams move the fuel against the direction of combustion. The applications of different grate variants are more thoroughly discussed in Sect. 6.4.2.

Figure 5.41 shows a schematic drawing of a pusher-type grate firing system. Every second row of grate bars is fixed, while the bar rows mounted in-between make a pushing movement, forced by a hydraulic cylinder. The coal is transported by this pushing, burning in a layer roughly 500 mm thick. The reciprocal row movements of the grate poke the coal and thus break the coal cakes. The grate load here is about $1\,\mathrm{MW/m^2}$ of grate surface area. The necessary power of the grate drive is higher than the power used by a travelling grate. This construction type is suitable for moist brown coals, caking hard coals, wood and waste. With respect to the grains, this construction type is less delicate than the travelling grate.

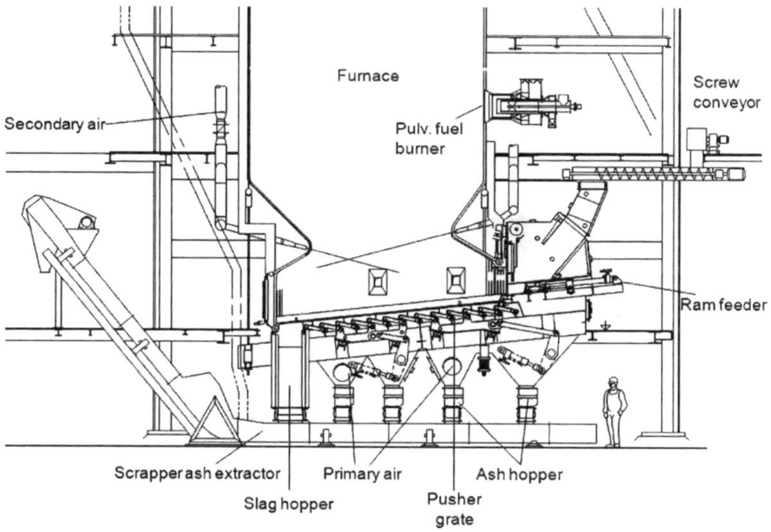

Fig. 5.41 Pusher-type grate firing for biomass/sludge (*Source*: Alstom Power)

5.5.3 Vibrating-Grate Stokers

In this furnace construction, water- or steam-filled tubes connected in a web formation form a flat-inclined grate surface. The combustion air is fed through air slots in the webs. By a short rocking movement of the whole-grate track, triggered in defined intervals, the coal on the inclined grate is transported, poked and evened out, and burned-out slag is transported into the slag hopper at the grate end.

The size of load changes are determined by the vibrating or rest frequency (Adrian et al. 1986). Suitable fuels for a vibrating-grate stoker have a calorific value above 20 MJ/kg, ash contents up to 20% and more than 16% volatile matter. When suitable coal types are used, grate loads up to $1.5\,MW/m^2$ in continuous operation are possible. The water-cooled grate surface also allows operation with low-ash coal at slight excess air. Wood can be burned combined with coal.

5.6 Legislation and Emission Limits

Air pollutant emission control for solid fuel-based power generation has been introduced and adopted in many countries. Emphasis historically began with the drive to reduce emissions of particulate matter (PM), followed by the acid rain precursors sulphur dioxide (SO_2) and nitrogen oxides (NO_x), and more recently mercury (Hg). Different legislative controls such as emission limits/standards, BAT (Best Available Technology), cap and trade, integrated pollution prevention and control (IPPC), fines, taxes and levies are adopted in different countries. Emission limits/standards are simple fixed limit values for a source or source type and are applied in most countries, for example, across Europe. Including BAT within international or national legislation requires the application of the newest technology. Integrated pollution prevention and control (IPPC) moves away from fixed standards for different emissions to a broader, integrated and preventive approach and includes various emissions, but also wider issues such as energy efficiency and the minimisation of waste. Cap and trade systems allow trading of total emissions of a single pollutant (Nalbandian 2004; Sloss 2003).

The main EU policies which are relevant for power stations are the Integrated Pollution Prevention and Control Directive (IPCC), the National Emissions Ceiling Directive (NECD) and the Large Combustion Plants Directive (LCPD).

The EU IPCC directive of 1996 required the introduction of an integrated environmental licensing system which applies to combustion installations greater than 50 MW_{th} and had to be implemented by the member states by 1999. The IPCC directive includes a large number of air pollutants such as SO_2, NO_x, CO, VOC, metals, particulate matter, chlorine, fluorine, dioxins and furans and specifies that best available techniques should be installed.

The National Emissions Ceiling Directive (2001/81/EC) set limits for each member state for the SO_2, NO_x, VOC and ammonia for the year 2010 (Sloss 2003).

The LCPD establishes emission limits for sulphur dioxide (SO_2), nitrogen oxides (NO_x) as well as emissions of fine and coarse particulate matter (PM), for all existing and new plants with a thermal capacity more than 50 MW. New combustion plants must meet emission limit values (ELVs) as given in Table 5.6 (Nalbandian 2007). The LCPD does not include waste fuels which are covered by the Waste Incineration Directive, but does include biomass fuels such as agricultural and forestry residues, waste from the paper and pulp industry and wood wastes, except those containing halogenated organic compounds or heavy metals as a result of treatment.

The EU directives were implemented by national legislation in member countries, with the compliance mechanism left largely optional. Therefore specific legislation varies from country to country.

In Germany emissions from large combustion plants with thermal outputs in excess of 50 MW_{th} are regulated by the Ordinance on Large Combustion Plants (13.BImSchV 2004). The emission standards are given in Table 5.7. For plants with thermal capacities between 1 and 50 MW_{th} emission limits are given in the Technische Anleitung Luft (TALuft 2002).

Waste incineration plants must comply with the stricter limits of the Ordinance on Incinerators for Waste and Similar Combustible Materials (17.BimSchV 2003), which is in line with the European Waste Incineration Directive. When waste fuels are co-fired in coal-fired power plants up to a fraction of 25% (based on heat input), the emission limits are calculated based on the emission limits of the two ordinances. This calculation, which is known as the mixing rule, takes account of the calorific values, flue gas volumes and other data. In some cases, the limits for waste fuels include pollutants not included in the Ordinance on Large Combustion Plants.

Table 5.6 Emission limits of the EU Large Combustion Plant Directive (Nalbandian 2007)

Revised 2001 LCPD emission limit values (ELVs) for new plants

Pollutant	Fuel type	ELVs (mg/m^3) 50–100 MW_{th}	100–300 MW_{th}	> 300 MW_{th}
SO_2	Biomass	200	200	200
	General case	850	200	200
	Liquid	850	400–200 (linear decrease)	200
	Gaseous	35 – in general 5 – liquefied gas 400 – low calorific gases from coke oven 300 – low calorific gases from blast furnace		
NO_x	Solid – biomass	400	300	200
	Solid – general case	400	200	200
	Liquid	400	200	200
	Gaseous			
	– Natural gas	150	150	100
	– Other gas	200	200	200
	Gas turbines			
	– Natural gas	50	50	50
	– Other gases	120	120	120
	– Liquid fuels	120	120	120
Particulate matter	Solid	50	30	30
	Liquid	50	30	30
	Gaseous	5 – as a rule 10 – blast furnace gas 30 – steel industry gases that can be used elsewhere		

5.7 Methods for NO_x Reduction

Measures to reduce nitrogen oxide can be divided into two categories: combustion engineering methods and methods applied downstream of the furnace. For this categorisation, the furnace is defined as the space where the fuel is burned. The different methods of NO_x reduction can be classified as follows (Spliethoff 2000; Spliethoff and Hein 2002; Wu 2002):

Combustion Engineering Measures

- Measures applied at the burner
  - flue gas recirculation
  - air staging
  - fuel staging

Table 5.7 Emission standards for solid fuels in Germany (17.BimSchV 2003; 13.BImSchV 2004)

	13 BImSchV (2004)		17. BImSchV
	Plant size (MW_{th})	Emission limit (mg/Nm^3)[a]	Emission limit (mg/Nm^3)[b]
Particulate matter	>50	20	10/30
CO	50–100	150	50/100
	>100	200	
SO_2	50–100	850	
	50–100 fluid. Bed	350 (75% retention)	50/200
	>100	200 (85% retention)	
	Biomass >50	200	
NO_x	50–100	400	
	>100	200	
	Biomass 50–100	350	
	Biomass 100–300	300	
	Biomass >300	200	200/400
	Wood 50–300	250	
	Wood >300	200	
	FB, 50–100, no wood	300	
	FB, >100, no wood	200	
HCl			1/4
HF			10/60
Organic compounds as total C			10/20
Hg		0.03	0.03/0.05
Cd, Th		0.05	0.05
Sb, As, Pb, Cr, Co, Cu, Mn, Ni, V, Sn		0.5	0.5
As, benzopyrene, Cd, Co, Cr, Co, Cd		0.5	0.05
Dioxins, furanes		0.1 ng/Nm^3	0.1 ng/Nm^3

[a] 6% O_2, daily average for gaseous emissions
[b] 11% O_2, daily average/half hour average for gaseous emissions

- Measures applied in the furnace
  - air staging
  - fuel staging
  - addition of a reducing agent

Downstream Flue Gas Cleaning Measures

- selective non-catalytic reduction (SNCR)
- selective catalytic reduction (SCR)

The SNCR method – although applied in the temperature range between 900 and 1,100°C – is classified under downstream flue gas cleaning methods because, generally, combustion is complete at the point where the reducing agents are added.

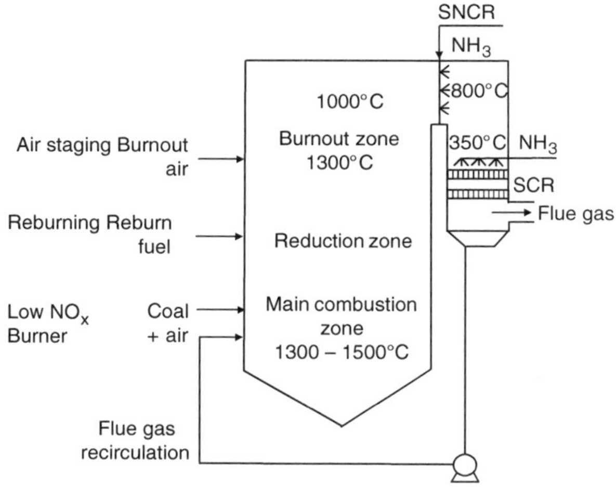

Fig. 5.42 Methods of NO_x reduction

The terms "primary" and "secondary" measures are deliberately avoided in the following because their meaning is not clear. Primary measures could be understood as the sum total of all methods and measures that limit the formation of nitrogen oxides to low values. Secondary measures, in consequence, would be methods that in turn reduce already-formed NO. In the sense of this definition, combustion engineering methods such as fuel staging would also be secondary measures because already-formed NO is reduced. For these reasons, the following material is based on the division into combustion engineering and downstream flue gas cleaning methods, since the corresponding classification is clearly determined by the location of the method applied.

Figure 5.42 schematically shows the current methods of NO_x reduction used in pulverised coal-fired furnaces.

5.7.1 Combustion Engineering Measures

Air staging and fuel staging have both proven to be successful combustion engineering techniques to reduce nitrogen oxide in pulverised coal firing. The effect of these measures relies on the setting of zones of air deficiency where the formation of nitrogen oxides is kept low or already-formed NO can be reduced. Figure 5.43 shows schematically the principle of fuel-staged and air-staged combustion (Spliethoff and Hein 2002).

Air staging divides the combustion course into two zones: a primary zone that restricts the formation of NO by air deficiency and a burnout zone operated at excess air. In fuel staging, reducing conditions are set by adding a reductive or reburn fuel after a first zone which is set nearly stoichiometric or with excess air. The air ratio in fuel staging of the primary or main combustion zone is determined by the requirements of the primary fuel and the furnace.

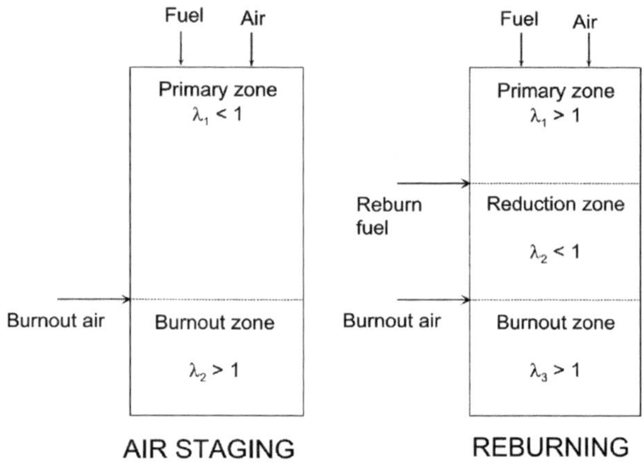

Fig. 5.43 The techniques of air and fuel staging

The techniques of air and fuel staging can be applied both in the furnace and at the individual burner. In the case of air or fuel staging in the furnace, there are clear, locally delimited zones with different stoichiometric conditions. When air or fuel staging is applied at the burner, the zone formation is determined by mixing processes. Both methods are mainly based on the homogeneous reduction of gaseous nitrogen oxides, or their precursor components, under fuel-rich conditions. In air staging the reduction of NO is dependent on the degree of coalification of the fuel: the more that volatile nitrogen is released in the primary zone, the more likely it is that the gaseous nitrogen compounds can be reduced to molecular nitrogen. In fuel staging, the main fuel being combusted and converted into the gaseous phase in the first stage, the NO_x emissions do not depend on the coalification of the primary fuel but on the volatile matter content of the reburn fuel.

In Fig. 5.44, the possible NO formation and reduction mechanisms are related to the three combustion zones – the main load burning or primary zone, the reduction zone and the burnout zone – of fuel staging in a slag-tap furnace, with pulverised coal as the main fuel and coke-oven gas as the reburn fuel. With pulverised coal as the primary fuel at combustion temperatures below 1,400°C, nitrogen oxide in the first stage of fuel staging mainly forms out of the nitrogen content in the coal. There are two distinct pathways of NO formation. As the devolatilisation of the pulverised coal commences, part of the fuel nitrogen gets released with the volatile matter as a consequence of the rapid heat-up of the furnace, while the remaining nitrogen stays with the residual char. Depending on the temperatures and the air/fuel ratios, the volatile nitrogen and the nitrogen remaining in the residual char can be converted to varying degrees of completeness to NO in the main burning zone. In a furnace with molten ash removal, thermal NO formation can add to nitrogen emissions, but combustion engineering measures such as air staging significantly reduce this formation pathway as well (Spliethoff 1992).

In the reduction zone, nitrogen oxides that formed in the main burning zone are reduced by homogeneous reactions. Hydrocarbon radicals formed from hydrocarbon fractions of the reburn fuel affect a fast reduction of the nitrogen oxides into

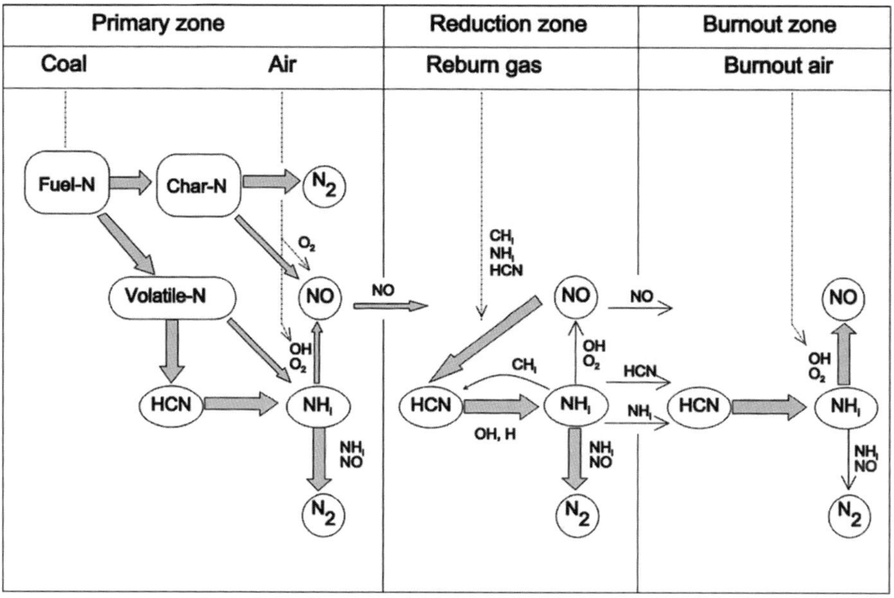

Fig. 5.44 Reactions of nitrogen formation and reduction in fuel staging with pulverised fuel as the primary fuel and gas as the reburn fuel (Spliethoff 1992)

hydrogen cyanide (HCN). Hydrogen cyanide can then be converted into ammonia radicals (NH_i) in a second, slower and thus rate-limiting step (Just and Kelm 1986), and NH_i, depending on the reduction zone atmosphere, can be oxidised to NO or completely decomposed to N_2. Through burnout air addition, the nitrogen components HCN and NH_i that were not decomposed in the reduction zone are oxidised to NO in the burnout zone, i.e. HCN to a major degree, NH_i to a minor degree (Kolb 1990). If burnout air is added in a flue gas temperature range of 900–1,000°C, a further decomposition of NO via NH_i radicals is possible (Hemberger et al. 1987).

5.7.1.1 Investigations at Experimental Plants

Experimental plants provide the potential to vary combustion parameters more widely than it would be possible to within the operational boundary conditions of an industrial plant. By appropriately designing the experimental plant, individual parameters can also be investigated separately from one another (Spliethoff and Hein 2002).

The following sections present investigations at experimental plants in order to demonstrate the potential for NO reduction at the industrial scale. The results presented were obtained mostly from experiments at a semi-industrial 0.5 MW furnace with pulverising equipment and at an electrically heated tube reactor (Spliethoff et al. 1995b). The plan of the electrically heated combustion reactor is shown in Fig. 5.45. The reactor's electrical heating allows the temperatures along the

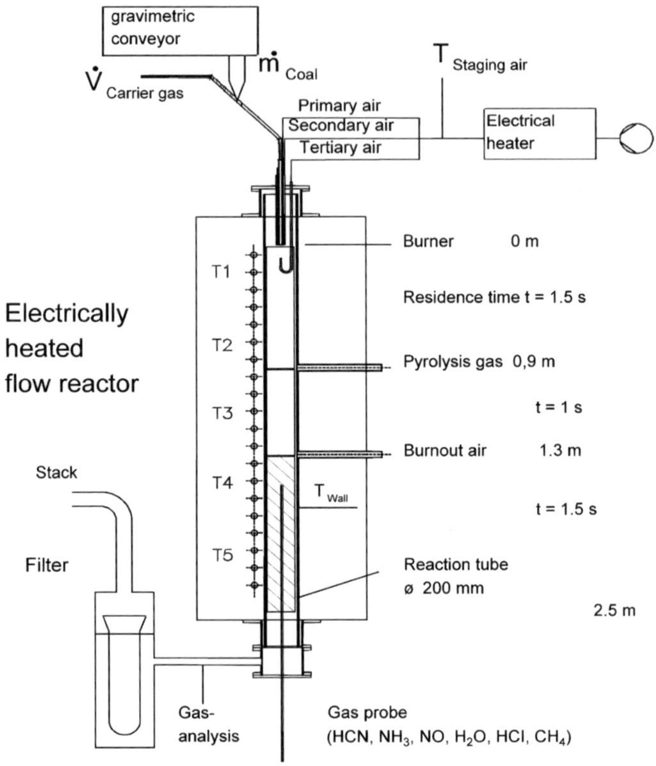

Fig. 5.45 Electrically heated tube reactor (20 kW_{Fuel})

combustion course to be set to facilitate separate investigations of the impacts of individual parameters on the combustion course and pollutant formation. For the methods of air-staged and fuel-staged combustion of hard and brown coals, systematic parameter studies were carried out as functions of temperature, stoichiometry and residence times (Greul et al. 1996b; Spliethoff et al. 1996). At the semi-industrial 0.5 MW plant shown in Fig. 5.46, the temperature – residence time course of the combustion was set within boundary conditions similar to an industrial plant (Maier 1998). Investigations concentrated on mixing phenomena.

Air Staging

In a separate investigation using two different coal types, the NO_x emissions and the NO, HCN and NH_3 concentrations, as well as the nitrogen remaining in the residual char, were measured in the primary zone (Fig. 5.47) (Chen et al. 1982b). At a specific air ratio, termed the optimum air ratio, a minimum NO_x emission level appears. At a higher air ratio, NO is only insufficiently decomposed in the primary zone. Below the optimum air ratio, NO in the primary zone is almost completely

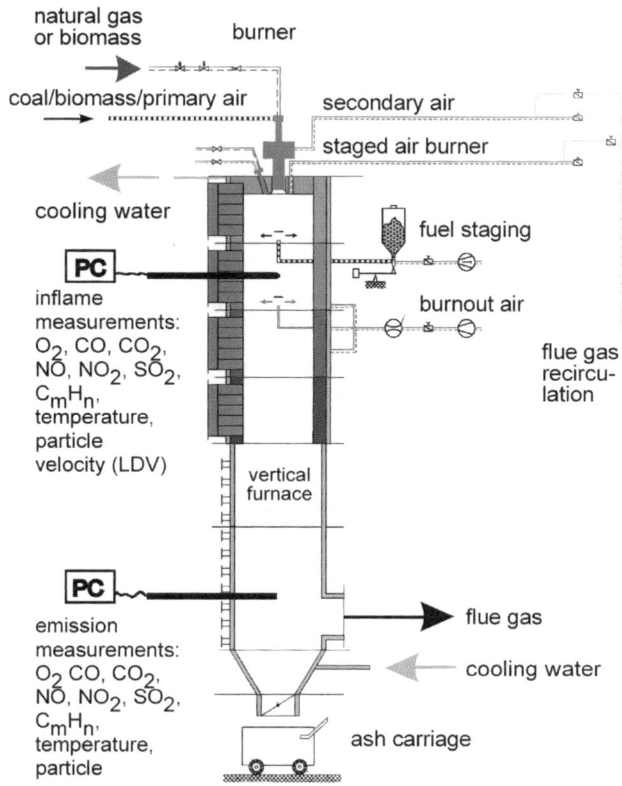

Fig. 5.46 Dry-bottom pulverised-fuel-fired furnace (0.5 MW)

decomposed, but part of the nitrogen stays in the residual char or forms NH_3 and HCN as intermediate components which, when burnout air is added, can again be oxidised to NO. The nitrogen in the residual char cannot be reduced by way of homogeneous gas phase reactions.

In the following sections, the influence of the air ratio, residence time and temperature on NO_x emissions in the primary zone shall be discussed using the example of the investigations at the electrically heated combustion reactor described previously (Spliethoff 2000; Spliethoff and Hein 2002).

The lower the air ratio of the primary zone, the lower will be the NO_x emission level. This correlation persists until the minimum NO_x emission rate develops at a definite air ratio (the optimum air ratio) (Fig. 5.48). Higher and lower air ratios than this cause higher NO_x emissions. At a residence time in the primary zone of 3–4 s and an air ratio of about 0.8, the NO_x emissions drop to the lowest level. At shorter primary zone residence times, the optimum was not attained within the investigated range of air ratios.

Longer residence times in the primary zone significantly reduce the NO_x emissions (Fig. 5.48). However, a prolongation beyond 4 s results in only slightly lower NO_x emissions, so the residence time between 3 and 4 s is considered sufficient.

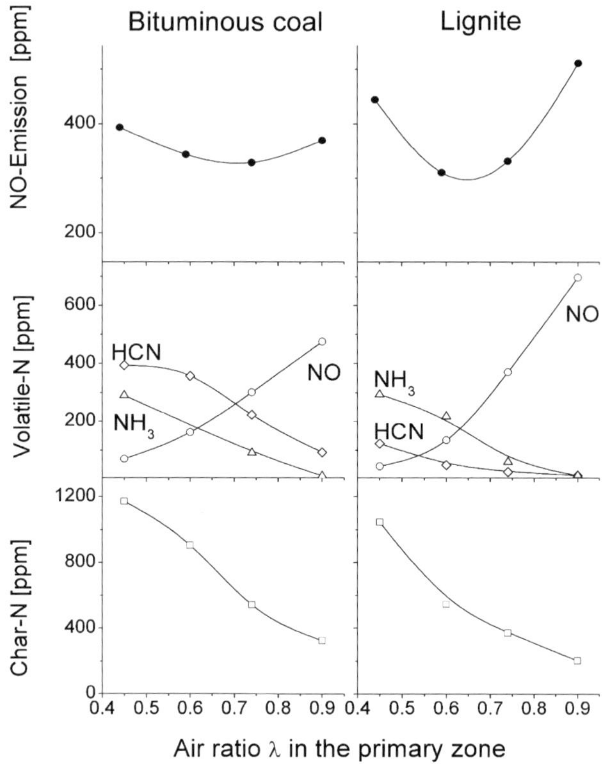

Fig. 5.47 NO_x emissions and nitrogen components in the primary zone (Chen et al. 1982b)

In scaling up the residence time influence on NO_x emissions from this experimental plant to an industrial plant scale, one has to take into account that reducing conditions develop only after the complete reaction of the fuel. The experimental set-up chosen here – the injection of the fuel in the direction of the main flow – results in a shorter effective residence time in the primary zone. In industrial plants,

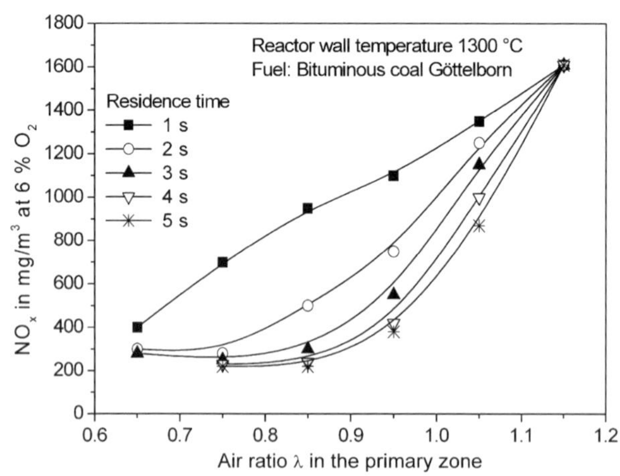

Fig. 5.48 Effect of residence time on a high volatile hard coal

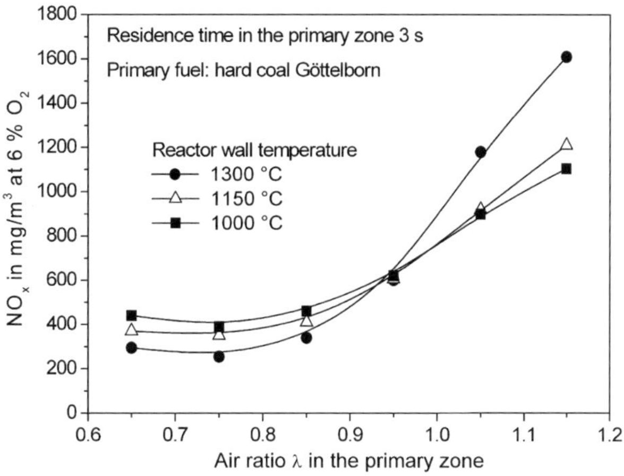

Fig. 5.49 Temperature influence on NO_x formation from a high volatile hard coal

the fuel fed via the burners is mostly injected horizontally into the furnace, and so the flue gases are only gradually led upwards by the main flow. Therefore, the same NO reduction effects are to be expected even at shorter residence times.

The extent of the influence of temperature on NO reduction depends on the flue gas atmosphere. In excess-air conditions, higher temperatures produce higher NO_x emissions, whereas under a deficiency of air, NO reduction is accelerated. The positive temperature effect is shown in Fig. 5.49 for a high-volatile hard coal. This positive impact of the temperature is also valid for brown coal.

The effect of the temperature in air staging can be explained using the example of NO concentrations measured along the combustion course (see Fig. 5.50). In all configurations of the air ratio and temperature, the measurements showed that NO concentrations increased considerably in the immediate burner vicinity. They decreased again in fuel-rich conditions. Higher temperatures lead first to higher NO concentrations in the near-burner area. In air-deficient conditions, though, NO decomposition speeds up so that, at the low air ratio of 0.75, lower NO_x emissions result at a high rather than a low temperature (Spliethoff and Hein 1997).

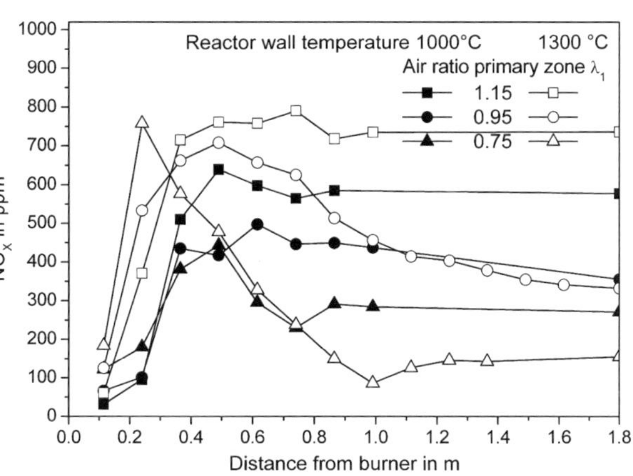

Fig. 5.50 Concentrations along the combustion course at different temperatures and air ratios

Given that the effect of air staging is based on homogeneous decomposition reactions, high-volatile fuels are better suited to NO_x emission reductions via this method than low-volatile ones. Figure 5.51 plots the NO_x emissions achievable at defined residence times and temperatures for fuels with different volatile matter contents. While it suggests that it is possible to stay below the German emission standards when using brown coal at different temperatures and air ratios and with sufficient residence times, these standards cannot be complied with using low-volatile fuels. For hard coals with a volatile matter content of 35–40% daf, it is possible to achieve a value of about 250 mg/Nm^3.

Fuel Staging

The optimum residence time in fuel staging ranges from around 1–1.5 s; a further rise of the residence time results in only slightly lower NO_x emissions. Analogously to air staging, higher temperatures bring about a decrease in NO_x emissions in fuel staging.

Figure 5.52 shows a comparison of the effect of different gaseous reductive agents at a residence time in the reduction zone of 1.5 s (Greul 1997). Among these, methane and the synthetic pyrolysis gas mixtures 1 (8% H_2, 25% CO, 61% hydrocarbons) and 2 (28% H_2, 17% CO, 59% hydrocarbons) show more or less the same NO_x emission levels. With a pyrolysis gas produced from coal, NO_x emissions around 200 mg NO_x/Nm^3 are achieved even in less fuel-rich conditions in the reduction zone. This pyrolysis gas was produced in the entrained flow at a temperature of about 1,000°C and, besides the gaseous components (51% H_2, 18% CO, 27% hydrocarbons), also contains tar vapours such as benzene, toluene, naphthalene and nitrogenous components. The positive effect is put down to the nitrogen components in the tar released after a delay in the reduction zone (Greul et al. 1996b). Without these tar components, a clearly weaker NO_x reduction is the result. Pyrolysis gases produced from other fuels such as straw and sewage sludge also show this characteristic (Rüdiger 1997).

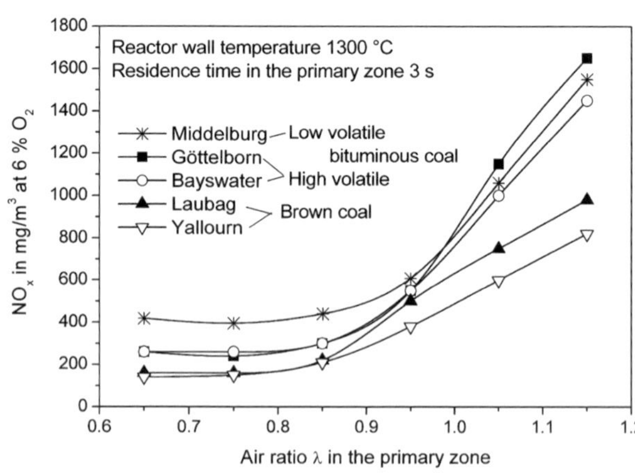

Fig. 5.51 Influence of the coal type in air staging

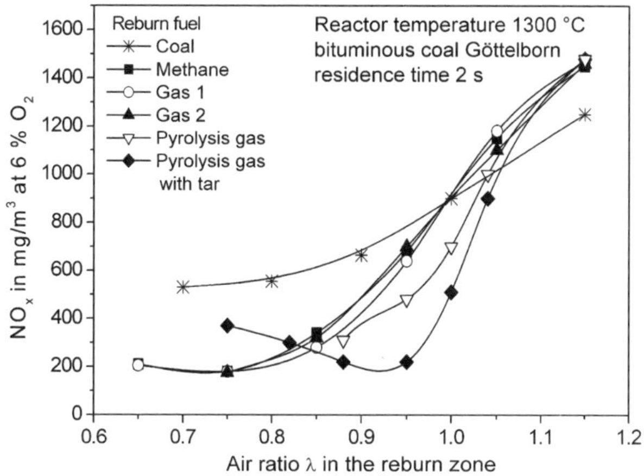

Fig. 5.52 NO_x emissions with different gaseous reduction fuels (Greul 1997)

In the 0.5 MW experimental plant, not only gaseous but also liquid and solid reducing fuels were trialled. The focus was on the effect of the admixture of the fuel and the preparation of the solid fuels on NO_x emissions and burnout. The comparison of the NO_x emissions for a residence time in the reduction zone of 1.5 s using the various admixtures and preparations is shown in Fig. 5.53 (Spliethoff et al. 1995b). Using pulverised coal as the reduction fuel, higher emission levels compared to gaseous or liquid fuels clearly result. Pulverised coal, compared to gaseous or liquid reburn fuels, may also have the disadvantage of deteriorating the burnout quality. Investigations revealed a decline of the total burnout in fuel staging from 99.5 to

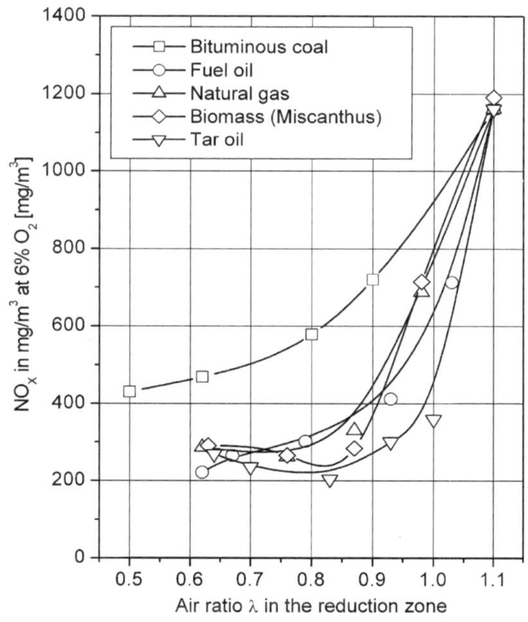

Fig. 5.53 NO_x emissions of gaseous, liquid, and solid reburn fuels (0.5 MW furnace)

around 97%, despite the use of a high-volatile coal and fine milling with 2% residues on the 90 μm screen. By applying gaseous or liquid reburn fuels, a burnout rate with more than 99%, which is nearly complete for experimental plants, was achieved at all parameter settings. The higher reductive effect of tar oil compared with fuel oil or natural gas is put down to the nitrogen compounds in tar oil which, with a delayed release, favour the decomposition of NO (Greul 1997). Comparing to the results from tests using the electrically heated tube reactor, the lower temperatures of the semi-industrial furnace cause roughly a 100 mg higher NO_x emission level. The emissions achieved with the biogenetic reburn fuels, i.e. straw and *Miscanthus*, lie in the order of magnitude that could be reached with gaseous (natural gas) or liquid fuels (fuel oil). They are lower, though, than in the case of pulverised coal as a reburn fuel.

Comparison of Air Staging to Fuel Staging

For both techniques, the same total residence time, given by the size of the firing system, must be used in order to be able to compare the effects. In the case of the NO_x emissions shown in Fig. 5.54, the residence time in the primary zone was 3 s with air staging; with fuel staging, the residence time in the primary zone was set at 2 and 1 s in the reduction zone. The residence time in the burnout zone was 1.5 s with both staging techniques (Spliethoff et al. 1995b).

For brown coal, it is possible with air staging to achieve lower emissions of well below 200 mg/Nm^3, a level not attainable with fuel staging even with brown coal as the primary fuel and natural gas as the reburn fuel. NO_x emissions in fuel staging are lower, however, with medium- and low-volatile coals. Hence it is the case in this scenario that the lower the content of volatile matter, the more favourable the result of the fuel staging in terms of NO_x emissions. Gaseous, liquid or high-volatile solid fuels such as biomass used as reductive agents in fuel staging feature the highest NO_x reduction degrees and have only a small effect on the combustion and the fly ash.

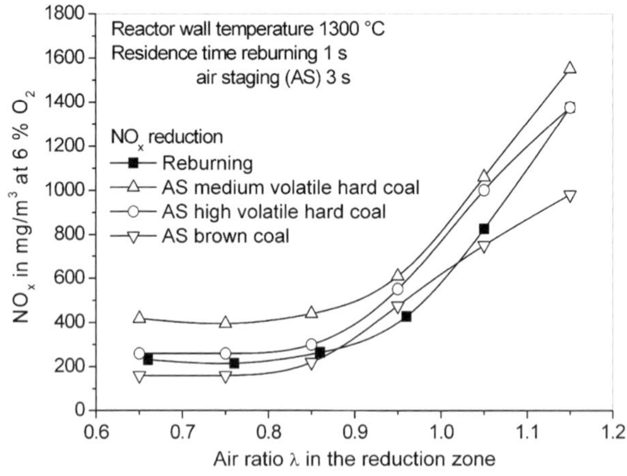

Fig. 5.54 Comparison of NO_x emissions in air staging and fuel staging

Optimisation of NO_x Abatement by NH_3 Addition

For a further reduction of NO_x emissions in air or fuel staging, another option is to utilise nitrogenous reductive agents, mostly ammonia-based. Reducing substances are either added to the air-deficient reduction zone in air or fuel staging or to the excess-air burnout zone (Fig. 5.55). The latter technique corresponds to the SNCR method where, in a narrow temperature range around 900°C, ammonia as a reductive agent selectively decomposes NO_x concentrations. Through the addition of ammonia to the reduction zone, the nitrogen oxides are more quickly decomposed into molecular nitrogen. The test results of fuel staging with ammonia addition show that the effect strongly depends on the method of fuel injection. If NH_3 is injected together with the reburn fuel, ammonia can be oxidised to NO. Ammonia added only after a reducing zone has formed, however, has an NO-reducing effect (Greul et al. 1996a; Spliethoff and Haferkamp 1991).

Figure 5.56 shows the test results of air staging with NH_3 injection into the air-deficient primary zone at an air ratio of 0.95. By ammonia addition, and using air staging, an NO_x emission level below 200 mg/Nm^3 can be achieved at a

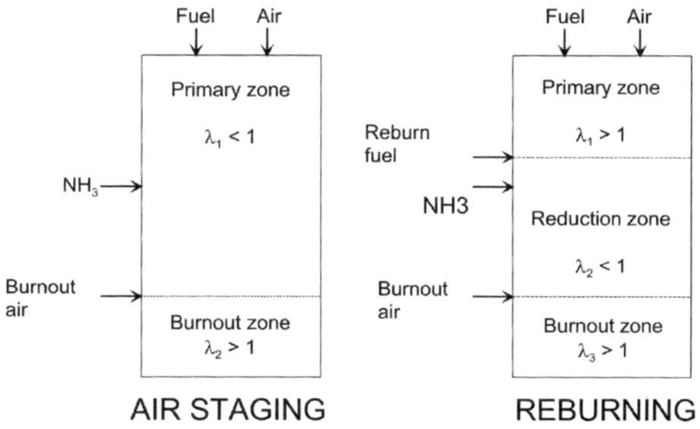

Fig. 5.55 Addition of NH_3 in air and fuel staging

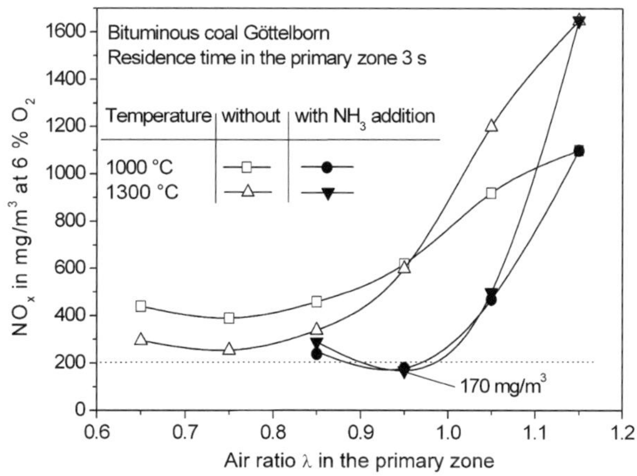

Fig. 5.56 Effect of NH_3 addition on NO_x emissions with air staging

stoichiometric NH_3/NO_x molecular ratio. The reduction here can be put down to the accelerated decomposition of NO under reducing conditions. Due to the high temperatures of 1,300°C in the burnout zone, a selective catalytic reduction is not possible.

5.7.1.2 NO_x Abatement in Pulverised Coal Combustion – State of the Art

Air Staging in Hard Coal-Fuelled Wall Firing Systems

The development of NO_x abatement by air staging at the burner and inside the furnace shall be described by the burner development of the Babcock Borsig Power Company (today Hitachi Power Europe) (Tigges et al. 1996). Similar designs of low NO_x swirl burners have been developed by other manufacturers, which are described in Wu (2002). The vortex burners used at the beginning of the 1970s were designed to achieve a complete and stable combustion by intensive mixing of combustion air and pulverised coal. The burner in Fig. 5.57a shows a coaxial core-air injection, a swirled coaxial secondary air injection and a concentric pulverised coal injection. The intensive mixing is achieved by swirling the secondary air and by dust separators. Given that all of the combustion air is already involved in the combustion process in the near-burner area, high NO_x emissions are an inevitable side-effect.

The first low-NO_x burners (first generation) injected the secondary air after a delay in order to decrease the NO_x emissions (Fig. 5.57b). The delayed admixing on the one hand diminished NO formation due to the lower oxygen available in the primary zone. On the other hand, however, it slowed down the combustion process so that the flame extended, the flame temperatures fell and, in many cases, the unburnt combustible material content increased and the flame became unstable. The burner configuration shown in Fig. 5.57b features a division of the secondary air into two partial flows. The inner swirled secondary air, as well as the unswirled tertiary air, slowly mixes after a delay into the primary combustion zone. This air staging at the burner was often combined with air staging in the furnace. Compared to swirl burners, it was possible with these first low-NO_x burners and air staging in the furnace to diminish the NO_x emissions by about 50%.

Depending on the coal, air ratio and plant type, NO_x emissions could be kept between 450 and 600 mg/Nm^3. The deterioration of the burnout had to be made up for by a finer milling degree, which was achieved by using dynamic classifiers within the mills. The corrosion of the furnace walls in these cases depended on the air ratio of the furnace air staging. It could be limited, for example, by a lateral air curtain.

The fluid dynamics of later, second-generation low-NO_x burners (those used today) are optimised by intensively mixing the pulverised coal with part of the combustion air and completely mixing in, after a delay, the combustion air necessary for total burnout. The high mixedness of the coal powder is achieved by the combination of swirl vanes and by toothed rings at the burner outlet so that single streaks were formed which mixed intensively with the air. This thorough mixing ensures a stable

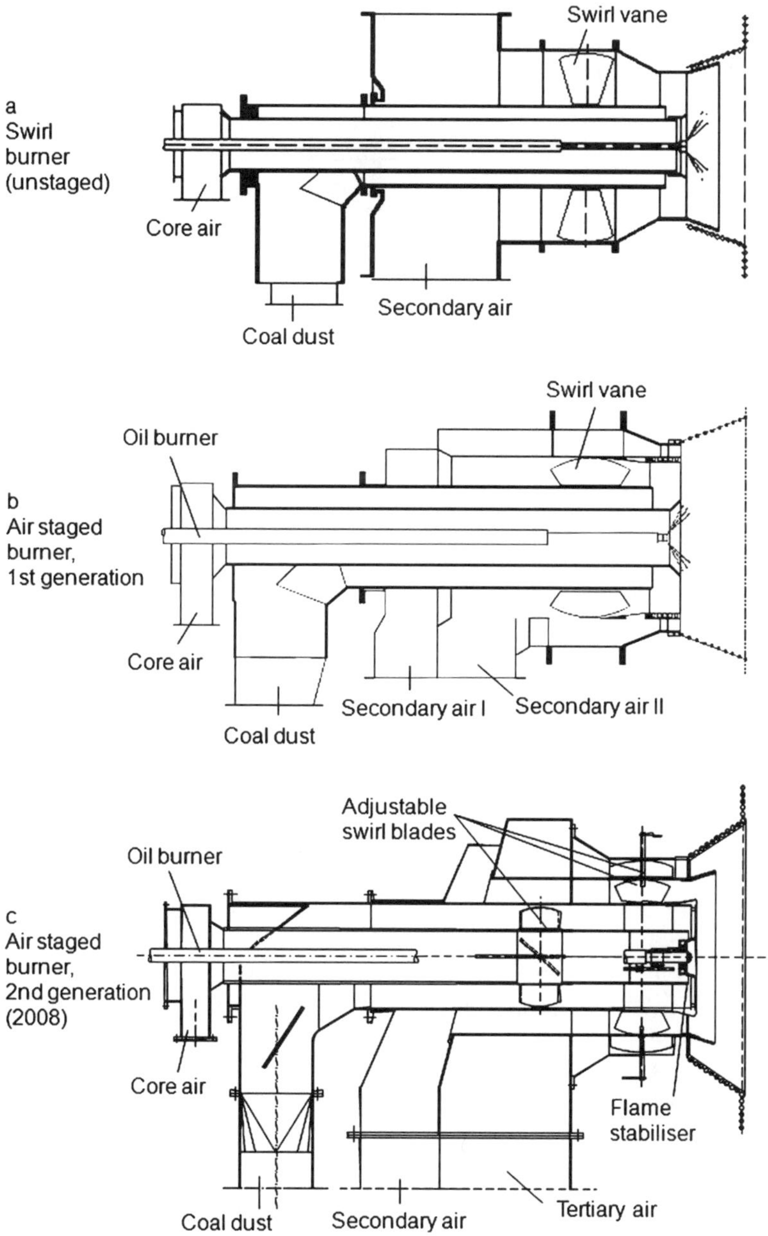

Fig. 5.57 Technological development of the swirl burner (*Source*: Hitachi Power Europe; Tigges et al. 1996; Leisse and Lasthaus 2008)

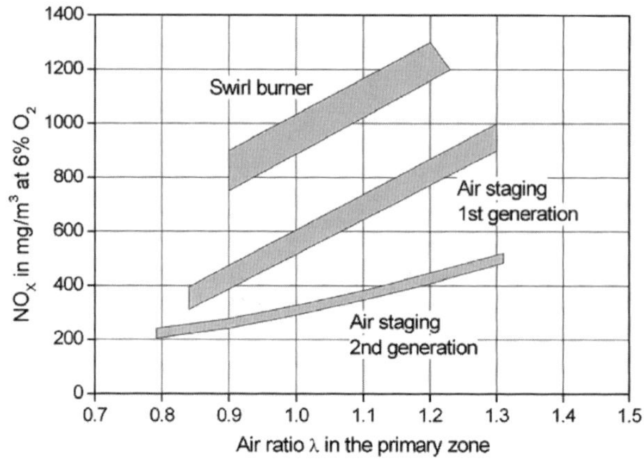

Fig. 5.58 Decrease of NO_x emissions with swirl burners (Tigges et al. 1996; Leisse et al. 1993)

flame with high temperatures at the burner and a rapidly formed reducing zone. The swirling of secondary and tertiary air and the flashings at the burner outlet clearly separates the oxygen-rich outer zone from the inner, reducing zone; see Fig. 5.57c (Leisse and Lasthaus 2008). By means of these burners, NO_x emissions of around 300 mg/Nm^3 are achieved in combination with near-stoichiometric furnace air staging and using high-volatile hard coals (see Fig. 5.58). The higher the content of volatile matter of the coal, the lower the level of the NO_x emissions (Tigges et al. 1996, 1997; Leisse et al. 1993). Due to the burner working principles of rapid mixing and high temperatures in the flame core, burnout is not affected or can even improve. The applicability of furnace air staging can be limited by corrosion of the membrane walls under reducing conditions.

Long years of experience with advanced low-NO_x swirl burners underlines that NO_x emissions of around 300 mg/Nm^3 or lower can be achieved. However, these values require high-volatile bituminous coals and air staging in the furnace. In the 550 MW_{el} power plant Staudinger, 300 mg/m^3 NO_x is achieved with a high-volatile Colombian coal (40% VM daf); in the case of a medium-volatile South African coal (28% VM daf) about 400 mg/m^3 is reported. At the 300 MW_{el} power plant Altbach, NO_x emissions are between 230 and 300 mg/m^3 for a high-volatile coal and 330–430 mg/m^3 for a medium-volatile coal (VGB 2007). In retrofit applications the potential to lower the burner stoichiometry can be limited (Jochum and Reidick 2005).

Air Staging in Hard Coal-Fuelled Tangential Firing Systems

Considerable efforts to further reduce NO_x emissions have also been undertaken for tangential firing systems (Epple et al. 1995). The jet burners and the flue gas recirculation in the furnace that are used in tangential firing technology already produce low NO_x emission levels. While vortex burners, which technically advanced into low-NO_x burners, make it possible to apply air staging both at the burner and in the furnace, the technology of jet burners used in tangential firing uses air staging

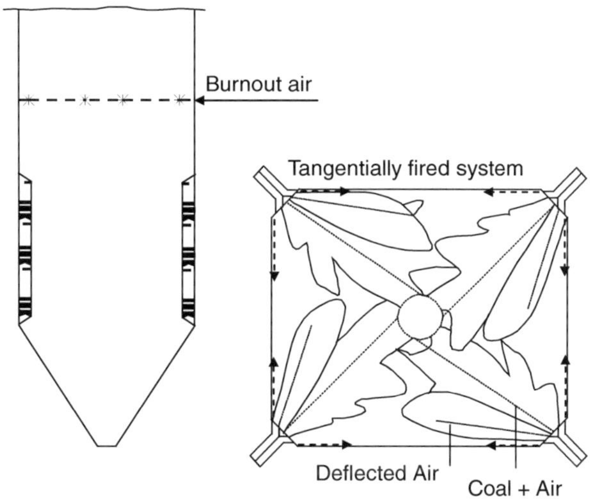

Fig. 5.59 Schematic presentation of air staging (Effenberger 2000)

only in the furnace. Because elongated flames are one of the combustion advantages of using jet burners, these burners are not suited to burner air staging. For the purpose of prevention of the corrosion of the furnace walls, the air is not only staged over the furnace height (axially) but also on the horizontal plane, at an angle to the burner axis (radially). It is achieved by deviating a portion of the combustion air flow towards the walls. The principle of overlapping axial and radial air staging is shown in Fig. 5.59.

An example is as follows: by retrofitting a 770 MW_{el} tangential firing system according to the axial and radial air staging principle, NO_x emissions were diminished from about 850 mg/Nm^3 to values between 250 and 300 mg/Nm^3 throughout the entire load range. The unburnt combustible material increased slightly, but with a 3% loss at ignition within the whole load range, stayed significantly below the value required for utilisation in the concrete industry. Measured O_2 and CO concentrations near the wall suggest that corrosion of the furnace walls, for the present case of a high-volatile hard coal with a high chlorine content up to 0.4%, can be avoided by the applied measures (Brüggemann et al. 1997).

The principle of axial and radial air staging has been applied at the 2x 900 MW_{el} hard coal-fired power station WaiGaoQiao/China. Figure 5.60 shows the measured NO_x emissions as a function of the burner stoichiometry. The Chinese hard coal has a volatile content of 37% (daf) (VGB 2007; Brüggemann 2008).

Air Staging with Brown Coal

After a development and testing phase of several years, combustion engineering measures were retrofitted to pulverised brown-coal furnaces in Germany as the only technology used to comply with the NO_x emission standards (Hein and Kallmeyer 1989; Kallmeyer and König 1987). In new 800–900 MW_{el} brown-coal furnaces, combustion engineering measures are also the only ones taken. Figure 5.61 shows

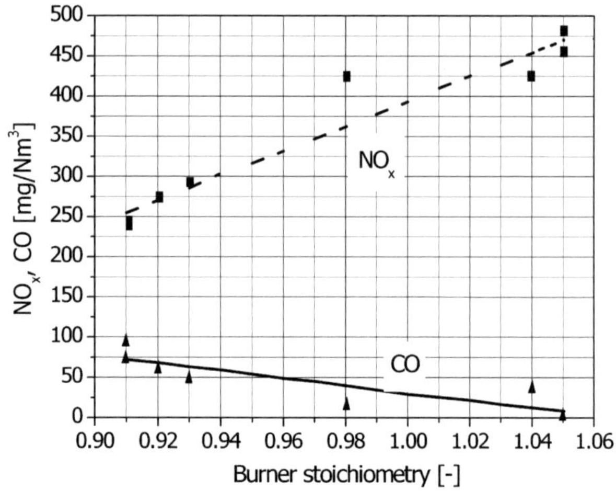

Fig. 5.60 Effect of burner stoichiometry on NO_x emissions when air staging with tangential firing (VGB 2007; Brüggemann 2008)

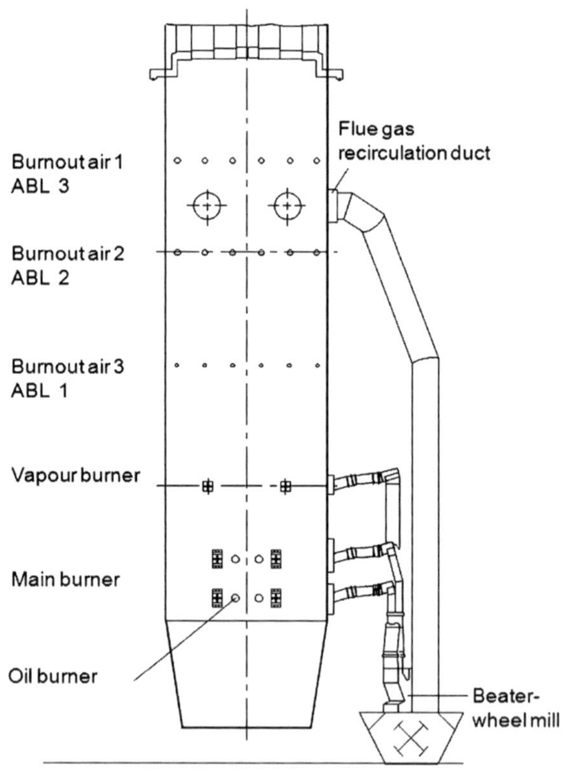

Fig. 5.61 Brown-coal fuelled steam generator with low-NO_x firing (*Source*: Alstom Power)

the furnace of an 800 MW brown coal-fired steam generator equipped with this technology. In this plant, the following measures for NO abatement are applied:

- Reduction of the total excess-air ratio
- Air staging with multi-staged burnout air addition
- Main burning zone volume reduction by fuel compression
- Flue gas recirculation
- Fuel splitting (fuel-rich/fuel-lean burners, vapour burner)

These NO-reducing measures make use of the high reactivity and the high-volatile matter content of brown coal. In contrast to hard-coal furnaces, the fouling behaviour in brown-coal furnaces requires much lower furnace outlet temperatures that, according to the fuel, should range between 950 and 1,100°C. This results in large combustion chambers with long residence times, a favourable feature for NO reduction. The residence time in the reducing primary zone is of key importance for NO abatement. To increase the residence time, the fuel is compressed by charging only the lower burner levels. The air ratio of the primary zone, between 0.9 and 1.0, effects the decomposition of the nitrogen oxides into nitrogenous intermediate products or molecular nitrogen.

The burnout air has to be added in stages and the total excess-air rate reduced in order to prevent reformation of nitrogen oxides. Experience has shown that compliance with the NO_x limit requires a residence time of about 4 s above the main burners (Kather 1995). If in the retrofit case these residence times are not possible it is necessary to apply additional techniques such as flue gas recirculation or milling vapour burners. Flue gas recirculation improves the mixing in the air-deficient zone and thus accelerates the NO decomposition reactions. The techniques of vapour separation and subsequent injection of the vapours above the main combustion zone correspond to the principle of fuel staging because the vapours contain about 30% of the fed pulverised brown coal.

Operational experience of air staging has shown that corrosion is not a problem for power plants fired with Rhenish brown coal. However, eastern German brown coals with a raw coal sulphur content of more than 1.5% have caused severe water – wall corrosion due to high flue gas concentrations of H_2S. In order to avoid the problem, measures were taken to guarantee concentric firing over the cross section of the tangentially fired boiler and to avoid flame impingements normal for brown coal firing, which come about from the non-symmetric firing arrangement, one mill being out of service (VGB 2007; Kahlert et al. 2008).

These NO-reducing measures exploit the well-known burner concepts of brown coal furnaces. The burners have a simple construction, designed as jet burners. One typical example is presented in Fig. 5.62a. The combustion air enters the furnace via the inlets separately from the coal and mixes after a certain delay such that these burners by themselves stage the air addition. The further developed design shown in Fig. 5.62b aims at preventing streaks of coal dust while mixing intensively with part of the combustion air so that a rapid ignition at the burner is achieved.

Currently, research is in progress into a brown-coal burner design following the construction of hard-coal burners (Fig. 5.62c). The air swirlers in this design,

Fig. 5.62 Development of brown-coal burners (*Source*: Hitachi Power Europe; Tigges et al. 1996)

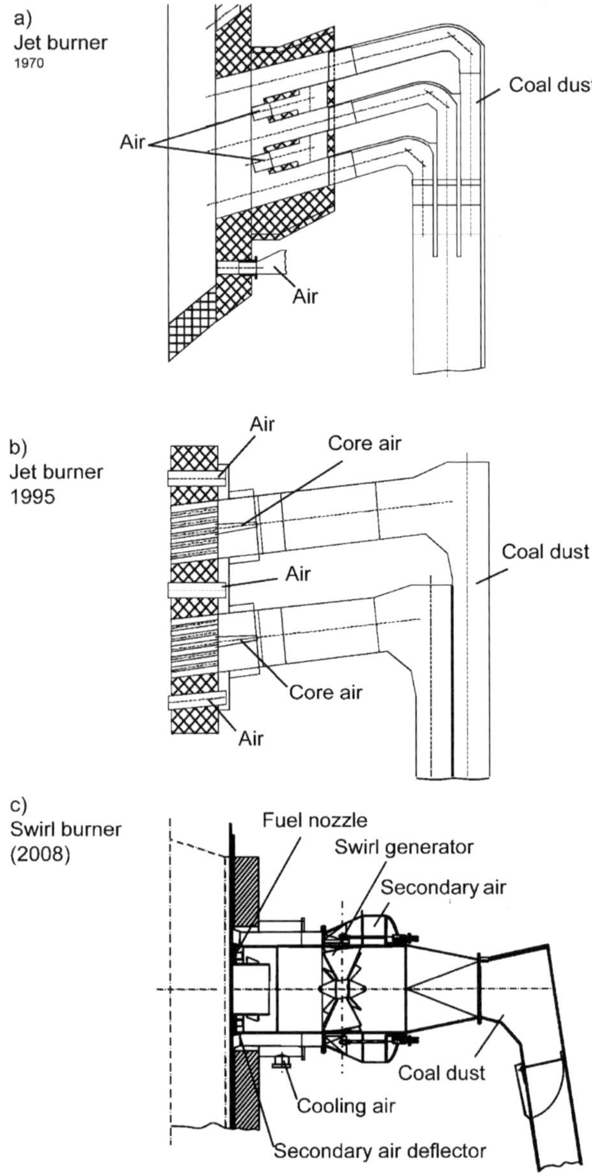

mounted in the pulverised coal duct, cause the coal dust to accumulate on the interior walls, while the stabilisation ring divides the coal flow of the burner into single streaks. The swirl of the air and flashing between primary and secondary air promote a reducing atmosphere in the flame core. Results from operation document a good NO reduction and a stable combustion behaviour (Brüggemann et al. 2003).

Air Staging in Slag-Tap Boilers

Besides air-staged burners, flue gas recirculation is another technique for NO_x reduction in slag-tap boilers, by means of which it is possible to achieve NO_x emissions between 800 and 1,200 mg NO_x/m^3 in furnaces with molten ash removal (Bertram 1986; Biber 1986; Strauß and Thelen 1989). Figure 5.63 shows the NO_x reductions achieved at a 160 MW_{el} slag-tap furnace by the use of air staging at the burner and flue gas recirculation via the mills (Spliethoff 1992).

The application of furnace air staging for NO_x reduction in slag-tap furnaces, however, has its limits, because both the reducing flue gas atmosphere and the liquid slag produced under reducing conditions can attack the refractory lining. Wall air, or an air curtain, to protect the slag-tap furnace walls can be used only to a limited extent because the cool air may hinder the slag flow. In any case, a reducing flue gas atmosphere has to be avoided in the area of the bottom of the slag-tap boiler or of the slag removal. Consequently, in a slag-tap furnace only air staging with short residence times (below 1 s) in the primary zone can be used.

Fuel Staging in Dry-Bottom Furnaces

While air staging is a widely applied technique, fuel staging or reburning is still rarely made use of at an industrial scale for dry-bottom furnaces. Depending on the coalification degree, fuel staging, in comparison to air staging, is advantageous for emissions to some extent for high-volatile hard coals and to a great extent for low-volatile coal types, though it requires more technical expenditure. For large-dimensioned dry-bottom furnaces, the admixture of the reburn fuel is difficult because of its relatively small mass flow. Its homogeneous distribution in the flue gas flow cannot be adjusted adequately. In furnaces with small capacities, fuel staging is a practicable technique because, for them, a complete mixing and homogeneous distribution of the reburn fuel in the flue gas flow is possible. In fuel staging, as opposed to air staging, the corrosion hazard in the furnace is limited to a smaller

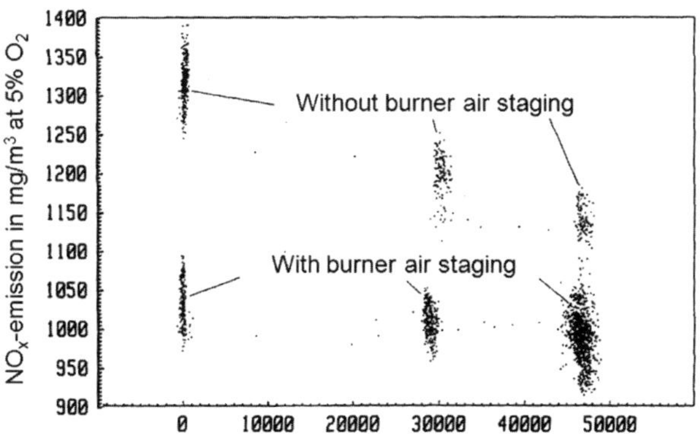

Fig. 5.63 Effect of burner air staging and flue gas recirculation on NO_x emissions (Spliethoff 1992)

wall area due to the reducing flue gas atmosphere, so the need for an air curtain or wall air is limited.

The in-furnace fuel staging technique was applied in an oil-fuelled furnace for the first time in 1982 by Mitsubishi, whose term for it was the MACT method. By adding oil as the reduction fuel above the main combustion zone, it was possible to reduce NO_x emissions by 50% (Murakami 1985; Takahashi et al. 1982). The application of the MACT method in pulverised coal-fuelled furnaces, with pulverised coal as the reburn fuel, turned out to be costly. The further developed version, called the "Advanced MACT Method", did without the addition of reduction fuel. Instead, air-deficient conditions were set in the main combustion zone (Araoka et al. 1987). In the case of pulverised coal-fuelled dry-bottom furnaces, the differences between air and fuel staging become indistinct. In dry-bottom furnaces with several burner planes, the difference between air and fuel staging can hardly be determined, as the firing has alternating zones of excess air and a deficiency of air. By applying a firing regime which consisted of fuel staging at the burner and air staging in the furnace in a hard-coal-fuelled furnace in Germany, it was possible to achieve NO_x emission levels around 400 mg/m^3 (Benesh et al. 2001).

Investigations into classical fuel staging (or reburning) in the furnace, with a reduction zone well separated from the main combustion zone, were carried out in demonstration projects funded by the European Union. The plants involved in the project were a 600 MW_{el} power plant unit in Scotland with natural gas as the reburn fuel and a 320 MW_{el} power plant unit in Italy with pulverised coal as the reburn fuel (Macphail et al. 1997; Bertacchi et al. 1997).

The NO_x emissions could be reduced to ca. 350 mg/Nm^3 by fuel staging with natural gas, with a slight rise of the unburned combustible material (Ghribelli et al. 1999). Air staging tests at this plant, using the same air ratios in the reduction zone, produced both higher NO_x emissions and higher C-contents in the fly ash. The results of the demonstration tests with pulverised coal as reburn fuel were NO_x emission levels between 300 and 370 mg/Nm^3 at a relatively short residence time in the reduction zone of 0.5 s, associated with a rise of the unburned combustible C-content in the fly ash from 6 to 8%. Air staging in the furnace brought about roughly the same level of NO_x emissions at this plant, but significantly higher rates of unburned combustible matter resulted (Bertacchi et al. 1997; Wu 2002).

In the USA, fuel staging was tested using natural gas in several power plants. At a 158 MW_{el} wall-fired furnace and at a 71 MW_{el} tangentially fired furnace, for instance, the NO_x emissions determined ranged around 330 mg/Nm^3 at 6% O_2, corresponding to a reduction of 60–70% (Folsom et al. 1995).

Fuel Staging in Slag-Tap Furnaces

In contrast to dry-bottom furnaces, where the original principle of fuel staging, i.e. with a clear separation of the main combustion zone from the reburn fuel addition, is followed only in some industrial plants, the situation is different for slag-tap furnaces.

In the USA a total capacity of more than 26,000 MW_{el} of cyclone furnaces are operated. Reburning has been judged as an adequate NO_x reduction technique for most of these cyclone furnaces. When reburning is used, the cyclone burners operate within their normal non-corrosive, oxidising conditions, thereby minimising any adverse effects on combustor and boiler performance. Fuel staging has been demonstrated or is operated at several, mostly, smaller plants with natural gas and pulverised coal (DoE 1999). The NO_x emission concentrations achievable with natural gas as the reduction fuel range between 350 and 450 mg/Nm^3 at 6% O_2, with reduction rates of about 55–70% from the high baseline emissions of cyclone furnaces. The necessary fraction of natural gas is between 15 and 20% of the fuel heat input (Folsom et al. 1995; Booth et al. 1991; Farzan et al. 1995). Using pulverised coal, NO_x emission concentrations between 420 and 480 mg/Nm^3 could be reached (Newell et al. 1995).

The economic viability of reburn technology for cyclone NO_x control has been challenged by the application of SCR and air staging technologies. The majority of cyclones are large, base-loaded units, and utilities have chosen to apply overfire air and SCR technologies to meet the more stringent NO_x emission levels (Farzan et al. 2004).

In Germany, fuel staging was investigated with coke-oven gas and tar oil as reduction fuels at a 160 MW_{el} slag-tap furnace. Figure 5.64 shows a sectional view of the firing plant. Under test conditions using coke-oven gas, NO_x emissions could be reduced to 350 mg/Nm^3 at an air ratio of $\lambda = 0.9$ in the reduction zone (referenced to 5% O_2) and after optimisation towards the minimum values of below 300 mg/Nm^3 (see Fig. 5.65) (Spliethoff 1992, 1991). Whereas with coke-oven gas, the same NO_x emissions were detected in the industrial plant and in comparative investigations at an experimental plant, the fuel staging test with tar oil at the 160 MW_{el} slag-tap furnace revealed higher NO_x emissions, of 500 mg NO_x/m^3, than tests at a 0.5 MW test furnace where 200 mg/Nm^3 were detected. The lower reducing effect of tar oil has to be put down to the insufficient admixture of the liquid fuel under test conditions at the industrial plant (Spliethoff et al. 1995a; Spliethoff 1995).

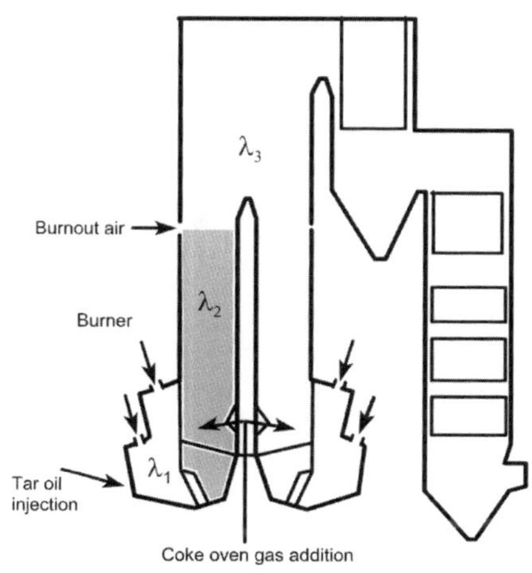

Fig. 5.64 Slag tap furnace Fenne 3

Fig. 5.65 NO_x emissions with different reburn fuels

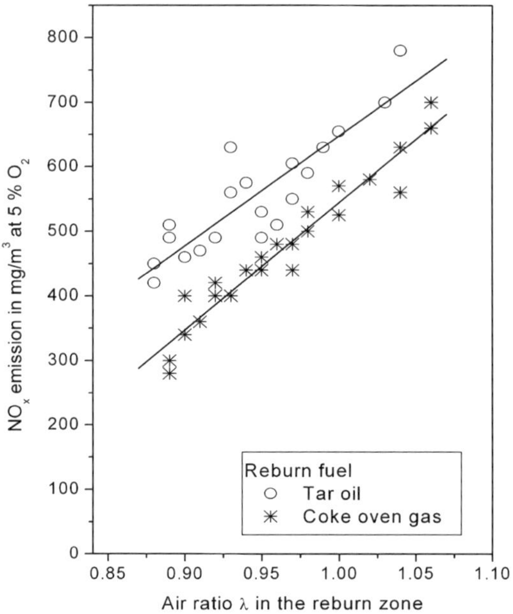

5.7.1.3 NO_x and N_2O Reduction in Fluidised Bed Combustion – State of the Art

Under the conditions of the low combustion temperatures of around 850°C in fluidised bed furnaces, thermal NO does not form. Therefore the nitrogen in the fuel is the main source for nitrogen oxide emissions. The conversion into NO lies below 10%. Contrasting with pulverised fuel combustion, this low conversion percentage has to be put down to the heterogeneous reduction of nitrogen oxides on the pulverised coal.

The most effective measure to limit NO_x emissions in fluidised bed combustion is to decrease excess air. By the utilisation of air staging it is possible with almost all fuel types to meet the limit of 200 mg NO_x/m^3 at 6% O_2. In this process, the combustion air is injected through the distributor plate as primary air and on several levels above the fluidised bed as secondary air. However, the method of air staging is restricted to circulating and stationary fluidised bed furnaces without in-bed heat transfer surfaces, the tubes of which would be corroded in the reducing flue gas atmosphere (Stultz and Kitto 1992).

In stationary fluidised bed combustion, the NO_x emissions change with the temperature; in circulating fluidised bed combustion, the temperature influence is slight.

For several years, N_2O emissions from fluidised bed furnaces have been the subject of critical observation as well. N_2O belongs to the group of greenhouse gases and is a contributor to the decline of the stratospheric ozone layer (Jacobs and Hein 1988). While the N_2O emissions in the combustion of fossil fuels in other firing systems have a level below 10 vpm, the values measured in fluidised bed furnaces reach a maximum of up to 250 ppm (Bonn and Baumann 1991; Bonn et al. 1993).

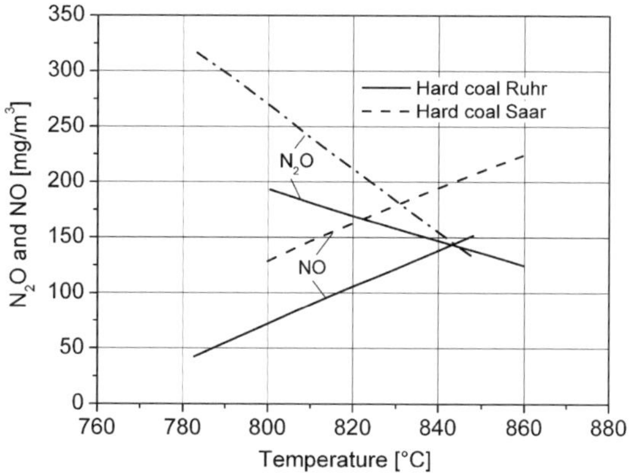

Fig. 5.66 NO and N_2O emissions as a function of the temperature in a fluidised bed furnace (König 1996)

N_2O emissions are dependent on operational conditions and the fuel. With older coal types, higher N_2O emissions were measured than with younger coal types. This can be put down to the binding form of the nitrogen in the coal structure. In the case of hard coal, the fuel nitrogen is mainly released as HCN, which is considered a precursor to N_2O formation. In the case of younger, brown coal, mainly NH_3 from fuel nitrogen is emitted (Bonn et al. 1993; Takeshita et al. 1993).

Comprehensive tests of parameters such as the temperature, coal type, the stoichiometries and the bed material have been carried out at a circulating fluidised bed test facility (König 1996). Given the fuel, the temperatures of the fluidised bed and of the freeboard have a great influence on the formation of N_2O. Figure 5.66 shows the NO and N_2O emissions as a function of the temperature. Due to the lower freeboard temperatures at part load, high N_2O emissions arise at low outputs. If the temperature in the freeboard is sufficiently high, the N_2O emissions remain low. NO_x and N_2O emissions have contrary behaviours as functions of temperature and total air ratio, so the sum of the two emission levels stays more or less constant. In contrast, air staging with an air-deficient operating regime in the fluidised bed involves both lower NO_x and lower N_2O emissions.

5.7.1.4 NO_x Reduction in Grate Firing Systems – State of the Art

The combustion temperatures do not significantly exceed a temperature of 1,300°C because of the relatively high excess air level of 1.4–1.6 (JBDT 1985), so thermal NO formation in grate firing systems can be ignored. Air staging is a state-of-the-art technique. In air staging, only part of the combustion air is injected as primary air to each of the combustion zones. The devolatilised components are burned above the grate at first at air deficiency and, after secondary air addition, to completion. A characteristic of the combustion of the residual coke remaining on the grate after devolatilisation is that little nitrogen oxide arises. This is because the nitrogen oxides formed in the lower coal layers during residual coke combustion, as they flow through the layers lying above, are reduced via heterogeneous decomposition reactions on still-unreacted carbon to molecular nitrogen. The total combustion air, depending on the volatile matter content of the coal, is divided into roughly 70–80% primary air and about 20–30% secondary air.

In the combustion of hard coals, by applying combustion engineering measures such as air staging and flue gas recirculation, it is possible to achieve NO_x emissions below 450 mg/m^3 (Schroth 1985).

5.7.2 NO_x Reduction Methods, SNCR and SCR (Secondary Measures)

Downstream $DeNO_x$ processes following the firing can be classified either as oxidation or as reduction processes. In oxidation processes (which shall not be discussed in detail in this context) NO is oxidised to form NO_2 and then separated by a scrubbing liquid (STEAG 1988).

The downstream $DeNO_x$ processes most frequently applied are reduction techniques, i.e. the methods of

- selective non-catalytic reduction (SNCR) and
- selective catalytic reduction (SCR) (Baumbach 1990; Wu 2002)

In both processes, nitrogen oxides are reduced to molecular nitrogen and water vapour by ammonia. Both methods work selectively, i.e. only a reaction with the nitrogen oxides occurs. In contrast, hydrocarbons as reducing agents in fuel staging do not react selectively because they also consume oxygen.

5.7.2.1 Selective Non-catalytic Reduction (SNCR)

The method of selective non-catalytic reduction uses either ammonia or urea as the reducing agent. The flue gas temperature necessary for optimal reduction lies between 900 and 1,050°C, depending on the flue gas composition and the reducing agent employed. Figure 5.67 shows the influence of the O_2 content on the temperature range under laboratory conditions. By adding further substances such as hydrogen, the range can be expanded to temperatures below 900°C (Wolfrum 1985).

In the optimal temperature range, the reaction triggered by NH_3 addition is

$$4\,NO + 4\,NH_3 + O_2 \rightarrow 4\,N_2 + 6\,H_2O \tag{5.15}$$

or, by urea addition,

$$2\,NO + (NH_2)_2CO + \tfrac{1}{2}\,O_2 \rightarrow 2\,N_2 + 2\,H_2O + CO_2 \tag{5.16}$$

If the temperature is too low, the ammonia slip (the unreacted ammonia in the flue gas) increases; if the temperature is too high, NH_3 burns, forming additional nitrogen oxides. Excess ammonia can react with SO_3, forming ammonium salts which can lead to fouling in the following heat transfer surfaces and the air preheater. Using urea may carry the risk of developing emissions of N_2O.

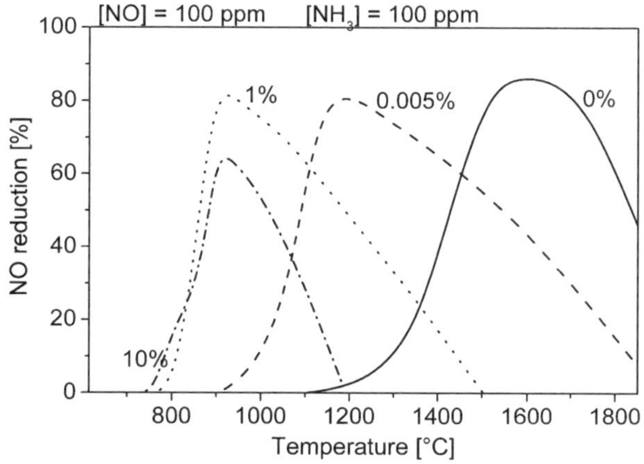

Fig. 5.67 NO reduction as a function of temperature and oxygen content (Wolfrum 1985)

Prerequisite for an effective reduction, besides maintaining the temperature window, is a good mixing of the employed reducing agent with the flue gas. The mixing momentum may be given either through the burnout air or by a recirculated flue gas flow. A greater number of single nozzles improve the thoroughness of the mixing and the distribution. In this process, the most homogeneous possible temperature and concentration distribution in the flue gas flow should be established before injection. Due to the inevitably inhomogeneous distributions of temperatures and flue gas concentrations across their flue gas profiles, it is difficult to attain in industrial plants the reduction degrees achievable in laboratory plants.

An NH_3 dosing rate set higher than an amount corresponding to the NH_3/NO stoichiometry at the injection location causes NH_3 slip, which increases with a higher NH_3/NO stoichiometry. NO_x reduction and NH_3 slip have to be weighed up.

Even though it is possible to achieve removal efficiencies of 70% under favourable basic conditions, values between 30 and 50% are rather typical at stoichiometries up to a maximum of 2 – so as to limit both the consumption of ammonia and the NH_3 slip (Stultz and Kitto 1992; Zellinger and Tauschitz 1989; Staudt et al. 1995; Himes et al. 1995; Hofmann et al. 1989; Gebel et al. 1989).

The SNCR technique is therefore only suitable for combustion plants where relatively low $DeNO_x$ degrees are required. It is mainly applied in furnaces with relatively low thermal outputs and corresponding small furnace cross sections, because an even flue gas flow and comparatively good admixing can be worked with more easily.

5.7.2.2 Selective Catalytic Reduction (SCR)

Selective catalytic reduction processes, similar to SNCR processes, utilise ammonia for reducing NO_x; however, the reduction process runs at significantly lower temperatures, using catalysts which reduce the activation energy. In the flue gas flow, the nitrogen oxides NO and NO_2 and the reducing agent, ammonia, are selectively reduced to nitrogen and water via the following reactions:

$$4\,NO + 4\,NH_3 + O_2 \rightarrow 4\,N_2 + 6\,H_2O \tag{5.17}$$

or

$$2\,NO_2 + 4\,NH_3 + O_2 \rightarrow 3\,N_2 + 6\,H_2O \tag{5.18}$$

The catalyst mediums are made of porous, ceramic basic materials, which provide a large surface area for adsorption. The material usually used is a SO_2-resisting titanium oxide or silicon oxide. Active metal compounds such as tungstic oxide with small additions of vanadium pentoxide or, alternatively, molybdenum, copper or iron oxide (MoO, CuO, FeO) are either mixed with or applied on the basic mass (STEAG 1988). The catalysts are made in the form of pellets, honeycomb catalysts or coated plates (VGB 1996).

Depending on the catalyst used, the reduction can for the most part develop even at temperatures around 100°C (Hannes et al. 1987). For coal-fired steam generators, though, the temperatures for catalytic NO_x reduction have to lie between 320 and 400°C. The upper temperature limit is imposed by the risk of surface sintering, which blocks the access to the pores and the interior surface. The lower temperature limit is determined by the temperature at which, with SO_3 and NH_3, ammonium sulphate and ammonium hydrogen sulphate form, which may provoke fouling and corrosion in the catalyst and the following heat exchanger surfaces (the air preheater). The tendency of the catalyst to oxidise some of the SO_2 into SO_3 increases the formation of sulphates and in addition raises the acid dew point (Frank et al. 2006).

One of the essential prerequisites for effective reduction, besides a homogeneously distributed NO concentration in the flue gas, is an even distribution of flue gas flow, fly ash and the reducing agent over the reactor cross section. For this reason, flow straighteners are arranged in front of the catalysts in the flue gas canal. Ammonia is added through a great number of nozzles with adjustable single flows to achieve a uniform distribution. Unequal distributions are problematic because they can provoke sub- and over-stoichiometric zones which increase the NO_x emissions or NH_3 slip. Dust deposits may clog some of the catalyst sections so that the flow velocity increases in other sections. The consequences of this are higher NO_x emissions and an increased ammonia slip. The higher flow velocity also involves the risk of erosion.

The SCR process uses almost the entire amount of ammonia for the reduction of NO, so high removal degrees are achieved even at a stoichiometric ratio below 1. The NH_3 consumption in the SCR process is therefore significantly lower than in the SNCR process, but yields higher removal degrees.

The design of the catalysts of an SCR process is usually based on an expected removal efficiency after about 2–3 years. In particular, it takes into account the diminishing activity of the catalyst. In various high-dust installations with the catalyst arranged in the flue gas flow before the dust removal unit, losses of activity of the catalyst of about 10–20% within a period of 20,000 h have been determined (Maier 1992; Farwick and Rummenhohl 1993). The aging of the catalysts is put down to various flue gas components, such as heavy metals, alkalis and SO_2 – in

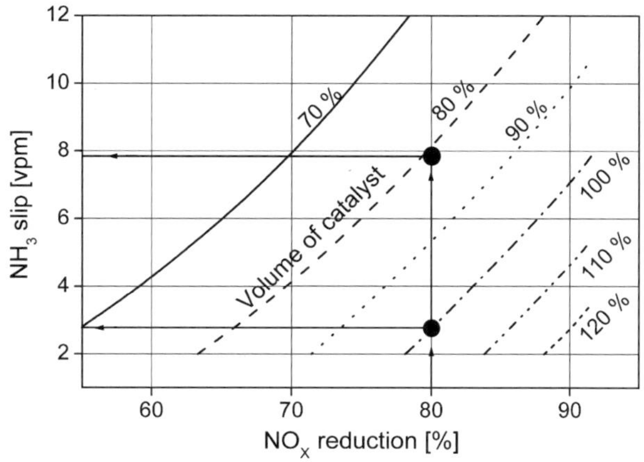

Fig. 5.68 Correlation between NH_3 slip, catalyst volume and NO_x reduction degree (Becker 1986)

technical terms, a poisoning of the catalyst. If the activity falls short of the designed level, it is necessary to replace some catalyst volume to limit the NO_x emissions and the ammonia slip.

In the design process, the necessary catalyst volume is calculated from the required removal efficiency and the level of slip to be maintained (Fig. 5.68). Where higher input concentrations require greater removal efficiencies, it is necessary to provide for a greater catalyst volume, while keeping the ammonia slip at the same level (Becker 1986, 1987).

Due to the inevitable occurrence of inhomogeneous distributions of the reducing agent or flue gas concentrations, the resulting removal efficiency of the SCR process reaches a maximum around 90%. These inhomogeneous distributions result in the forming of zones of ammonia excess or deficiency. In excess zones, ammonia reduces the nitrogen oxides totally. If the NO_x concentration reaches zero in some places, the inevitable consequence is a remainder of NH_3. This surplus of ammonia (NH_3 slip) can only be diminished to a limited extent, even by additional catalyst volume.

The inhomogeneous distributions are, with lower emission control standards, all the more problematic because they make the excess zones, where NO_x forms, more probable. Low NO_x emission control standards, using SCR technology, thus lead to barely controllable problems. Lower NO_x emissions are limited not by the catalyst activity but because of the mixing problems (Frank et al. 2006).

Investigations show that about 70% of the excess ammonia is taken up by the fly ash, possibly resulting in odour nuisances in cases of high ammonia slip. In general, the slip is limited to values of less than 2 ppm so as not to risk the usability of the ash. The catalysts are usually placed after the economiser and before the air heater in what is called a high-dust configuration, i.e. in the dust-laden flue gas flow (see Sect. 4.4.2.3). The thermodynamic design has to ensure that the flue gas temperatures in the catalyst area do not rise higher than 400°C and do not fall below 320°C. The flue gas temperature after the economiser, by installing a fireside or waterside economiser bypass, has to be limited to temperatures above 320°C to prevent catalyst fouling (Reuter and Hönig 1988).

Besides this high-dust configuration, in Germany a low-dust configuration with the placement of the catalyst after the dust collection and flue gas desulphurisation units is used in some plants as well. This arrangement is chosen in cases where, for reasons of space, a high-dust configuration is not possible, or where the flue gas compositions give reason to expect a short lifetime of the catalysts. If a high-dust configuration was applied, for instance, in slag-tap furnaces with fly ash recirculation, an enrichment with arsenic compounds would have the consequence of poisoning the catalyst, entailing a corresponding loss of activity. For this reason, the low-dust configuration is often applied in these furnace types.

The charging of the catalyst with desulphurised and fly-ash-free flue gas prolongs the lifetime of the catalyst in comparison to the high-dust configuration. Lower dust concentrations allow the use of catalysts with smaller pore diameters and larger active surfaces, so more compact designs are possible. The lower SO_2 concentration of the flue gas makes it possible to use a catalyst with higher activity.

The flue gas temperatures after the FGD unit are only around 50°C at first, so it is necessary to reheat the flue gases to a temperature of about 320°C for catalytic NO_x removal. The reheating process mostly consists of a combination of regenerative heating and heating by natural gas or steam. This direct reheating is needed to compensate for the heat losses (about 30°C) of the gas heater and the $DeNO_x$ unit. These heat losses ultimately diminish the efficiency of the boiler and the total efficiency (Maier et al. 1992).

5.7.3 Dissemination and Costs

Combustion engineering measures to reduce NO_x formation are state of the art in modern pulverised coal power stations. Low NO_x burners are currently in operation in more than 800 pulverised coal-fired units of a total capacity of above 295 GW_{el} as a stand-alone measure or combined with over-fire air (OFA) (Nalbandian 2004).

SCR technology has been used commercially in Japan since 1980 and in Germany since 1986 on power stations burning mainly low-sulphur coal and in some cases medium-sulphur coal. The technology has been (since the mid-1990s) and continues to be retrofitted in many existing coal-fired power plants in the USA. SCRs are currently in operation or under construction in more than 315 pulverised coal-fired units of a total capacity of above 130 GW_{el} (Nalbandian 2004). In Germany, power stations with a capacity of 35 GW_{el} are equipped with SCR (Frank et al. 2006).

The first commercial SNCR application in a coal-fired power plant was in 1980. SNCR systems are currently in operation or under construction in around 50 units of a total capacity of approximately 10 GW_{el}. As it is difficult to achieve good mixing in large boilers, almost all the commercial applications to date have been limited to typically less than 200 MW_{el} (Nalbandian 2004).

Table 5.8 Capital and production costs of NO_x reduction measures (data from Wu 2002; Soud and Fukasawa 1996)

Primary measures	NO_x reduction [%]	Capital costs [€[a]/kWe]	Production costs [€cent/kWh$_e$]
Low NO_x burner	30–50	10–15	0.035–0.045
Furnace air staging	50	5–20	0.03–0.05
Low NO_x burner and furnace air staging	60	15–30	0.03–0.06
Coal reburning	50	15–45[c]	0.05–0.15
Gas reburning	50	10–15[b]	0.11–0.2[d]
Flue gas cleaning			
SNCR	50	5–30	0.05–0.13
SCR	70–90	50–80	0.15–0.25

[a]1€ = $1 (2002), [b]no pipeline included, [c]upper value includes pulveriser; [d]depending on difference between gas and coal price.

The capital costs of a high-dust SCR unit after a pulverised hard coal-fired boiler range between 50 and 80 €/kW of electrical power. In relation to the total capital costs of a large pulverised coal-fired power plant, the result is a cost fraction of about 5–8%. The additional power production costs range around 0.2–0.3 € cents/kWh at total prime costs of about 4–6 cents/kWh. The costs for a low-dust configuration are about twice as high as a high-dust configuration.

The costs for SNCR, from 5 to 30 €/kW, are significantly lower than the expenditures for SCR. The large spread is a result of the expenditure necessary for the adjustment of local temperature and concentration conditions at differing capacities and the associated control. High $DeNO_x$ degrees, in particular, result in greater expenditures (Himes et al. 1995).

Table 5.8 draws a comparison between reduction potential and necessary capital costs of different NO_x reduction measures.

5.8 SO_2-Reduction Methods

Desulphurisation methods may be categorised into

- methods to reduce the sulphur content of the fuel and
- methods to desulphurise the flue gas.

The flue gas desulphurisation methods distinguish between

- additive injection in the furnace or in the flue gas ducts (dry) and
- downstream desulphurisation processes (semi-dry, wet).

5.8.1 Methods to Reduce the Sulphur Content of the Fuel

The reduction methods involving the ash and sulphur contents in the coal usually employ physical separation processes which make use either of the density difference between the combustible and the mineral substances or of differing surface qualities. The methods used in coal preparation are wet processes. In Germany, about 95% of mined coal is treated in preparation processes. The costs of these physical preparation processes range between 1.5 and 2.5 €/tonne of fuel. During this preparation, however, only part of the pyritic sulphur is removed (Vernon and Jones 1993). The effectiveness of the separation process depends on the coal properties but is never sufficient to comply with emission control standards. It is, for instance, possible with some coal types to diminish the sulphur content from 1.3 to 1% (Chugtai and Michelfelder 1983).

A more complete removal of sulphur, including the organic fuel sulphur, is possible by chemical and biological processes, but they are relatively expensive and do not yet meet the state of the art (Vernon and Jones 1993).

5.8.2 Methods of Fuel Gas Desulphurisation

Sulphur oxides in flue gases can be captured by different alkali and alkaline earth compounds and, principally, by all kinds of metal oxides as well, hence there being numerous variants of SO_2 reduction in flue gases (STEAG 1988).

Desulphurisation methods today mainly employ processes based on limestone ($CaCO_3$) or lime (quicklime (CaO) or hydrated lime ($Ca(OH)_2$)). The processes work in dry, semi-dry or wet conditions. In dry conditions, the desulphurisation medium and the products of desulphurisation are solids, in wet conditions both are liquid. In a semi-dry process, the medium is added in a liquid state and the products are solids. In contrast to alkali compounds, which are more active in SO_2 capture, lime has the advantages of being more cost-effective, available in great quantities in nature, and of forming a by-product (gypsum) which is reusable.

The application of alkali compounds or metal compounds is a potentially viable option for small plants where the capital costs would be relatively small compared to the operating expenses.

Figure 5.69 shows the different possible applications for additives in desulphurisation processes (Vernon 1990). For dry injection, possible devices are the burners or special injection nozzles above the furnace, upstream of the economiser or the ESP. In downstream FGD units, the capturing agent is added in a suspension with water.

5.8.2.1 Additive Injection in the Furnace

With respect to process equipment, the injection into the flue gas flow of powdery calcium-based additives is a relatively simple variant of an SO_2 reduction process. Only installations for the preparation, storage, dosage and transport of the additive

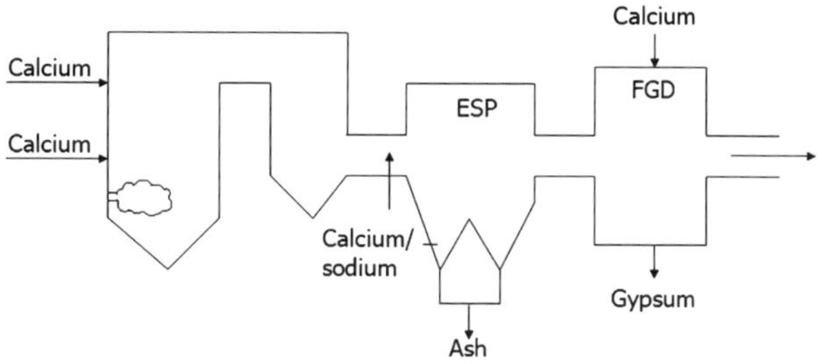

Fig. 5.69 Locations of additive injections for flue-gas desulphurisation

are needed. The dry additive is injected into the furnace or into the flue gas ducts, then the products of desulphurisation and the unreacted additive have to be removed in the dust collector together with the fly ash. The injected additive increases the load on the dust collector plant. Another drawback may be the limited usability of the resulting mixture of fly ash, calcium sulphate and limestone. In the case of dry additive injection into the flue gas flow, Ca/S ratios which are higher than the stoichiometric additive ratio are necessary.

Calcium-based additives are the source from which calcium oxide, CaO (quicklime), forms in a first step (calcination) under heat. If $Ca(OH)_2$ (calcium hydrate or hydrated lime) or $CaCO_3$ (limestone) are used as additives, the dehydration or decarbonisation process evolves according to the following reactions:

$$Ca(OH)_2 \rightarrow CaO + H_2O \quad (5.19)$$

$$CaCO_3 \rightarrow CaO + CO_2 \quad (5.20)$$

The reactive CaO then reacts with the sulphur oxides, forming sulphates. Calcium oxide, though, also bonds with other flue gas components:

$$CaO + SO_2 + 1/2O_2 \rightarrow CaSO_4 \quad (5.21)$$

$$CaO + SO_3 \rightarrow CaSO_4 \quad (5.22)$$

$$CaO + 2\,HCl \rightarrow CaCl_2 + H_2O \quad (5.23)$$

$$CaO + 2\,HF \rightarrow CaF_2 + H_2O \quad (5.24)$$

The reactions of lime are temperature dependent. The optimum temperature range of a dry reduction process via lime is at gas temperatures of around 850°C. At lower temperatures, the calcination reaction does not run satisfactorily. At too high a temperature (above 1,100°C), the lime sinters and so becomes inactive. In addition, calcium sulphate that has formed may decompose. Figure 5.70 depicts the dependence of the desulphurisation process on the flue gas temperature for a range of powdery additives (Wickert 1963).

Further parameters having an influence on the desulphurisation degree are the particle size and the type of additive. Since the reaction occurs on the surface, a finer grinding of limestone improves the desulphurisation extent. Sulphate already formed on the limestone surface may block further reaction. Hydrated lime

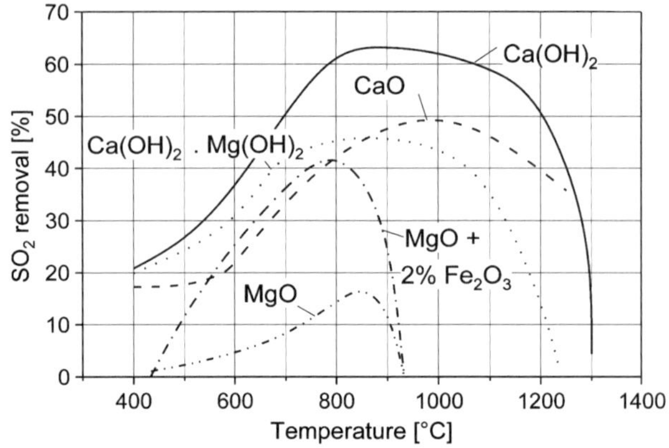

Fig. 5.70 Effect of temperature on the desulphurisation process for a range of additives (Wickert 1963)

($Ca(OH)_2$), which triggers the calcination reaction even at lower temperatures, is more reactive than calcium carbonate (limestone ($CaCO_3$)) (Vernon 1990). Because the equilibrium of the capture reaction cannot be achieved in industrial combustion systems, it is necessary to take the residence time into account as an additional parameter (Hein and Schiffers 1979).

High temperatures and a low loading of ground lime in the flue gas put a limit to SO_2 reduction in the furnace of boilers fired by pulverised hard coal. From the process engineering point of view, the injection of the additive in combination with the coal or via the different combustion air flows is possible. In tests injecting $Ca(OH)_2$ into the external recirculation zone at a Ca/S ratio of 2, it was possible to achieve desulphurisation degrees between 50 and 60%. Higher rates of sulphur removal were only possible with very high amounts of additive. The injection via nozzles above the furnace yielded better results, but this technique, due to the mixing problem, appears suitable only for small steam generators (Chugtai and Michelfelder 1983).

The significantly lower combustion temperatures of pulverised brown coal firing – the maximum temperatures range between 1,100 and 1,150°C – provide much more favourable conditions for in situ desulphurisation. Additionally, brown coal ash already contains a certain fraction of alkaline earths, so desulphurisation partially occurs even without an additive. Figure 5.71 shows the measured desulphurisation rates in brown coal furnaces as a function of the Ca/S ratio (Hein and Schiffers 1979). The cause for the higher capture of $Ca(OH)_2$ or $CaCO_3$ is its spontaneous decomposition upon being fed to the furnace, which yields a large reactive surface of the CaO (Fig. 5.72).

In contrast to pulverised fuel firing, desulphurisation in fluidised bed furnaces can be carried out in the optimum temperature range of desulphurisation. A high concentration of ground lime, a good mixing of the additive in the fluidised bed and, in particular, a long contact time between flue gas and additive have a positive effect on the desulphurisation process. In bubbling fluidised bed furnaces, removal efficiency rates of more than 80% have been achieved at Ca/S molecular ratios of 2–4. In circulating fluidised bed furnaces, higher values of about 90% were achieved at Ca/S ratios of 1.5–2. Apart from limestone, dolomite ($CaCO_3 \times MgCO_3$) is also

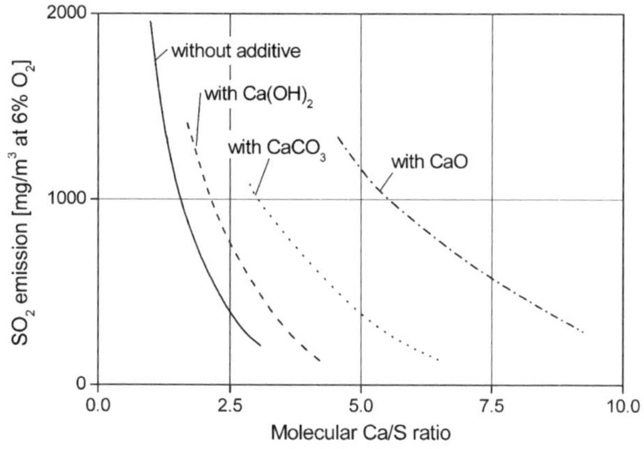

Fig. 5.71 SO_2 emissions as a function of the Ca/S ratio in pulverised brown coal combustion (Hein and Schiffers 1979)

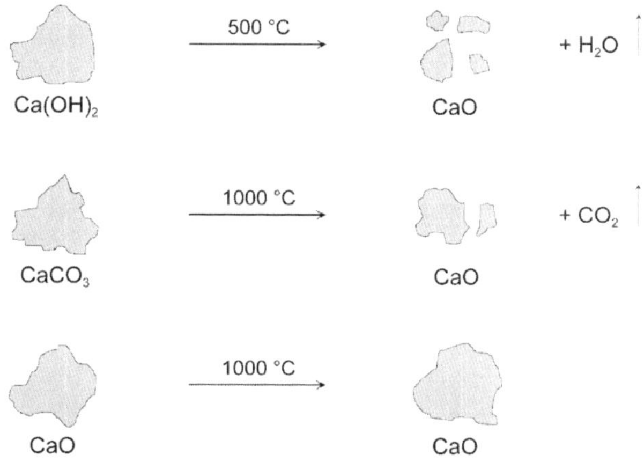

Fig. 5.72 Decomposition of additives with heat

used as an additive in fluidised bed firing. Figure 5.73 shows the desulphurisation rates achieved at temperatures below 880°C in a 110 MW_{el} CFBC as a function of the Ca/S ratio. Removal efficiency rates of 70–90% require Ca/S ratios of 1.6–3.1. For a removal rate of 95%, a Ca/S ratio as high as 4 is necessary (Takeshita 1994).

In grate firing systems, the injection of limestone and slaked lime (calcium hydroxide $Ca(OH)_2$) above the flame zone showed desulphurisation rates of 60% at a Ca/S ratio of 3. In contrast, the addition of the additive to the fuel or secondary air at the same Ca/S ratio yielded a worse result: a desulphurisation rate of 20% for injection with the fuel and of 40% for injection with the secondary air (Hössle 1985).

Additives for desulphurisation can also be added in lower temperature regions. At temperatures below 500°C, however, the degree of desulphurisation by hydrated lime, $Ca(OH)_2$ – which has the highest reactivity among the calcium additives – drops considerably, so for removal rates to be sufficient, high stoichiometric ratios of the additive are required (STEAG 1988). As an example, a stoichiometric ratio

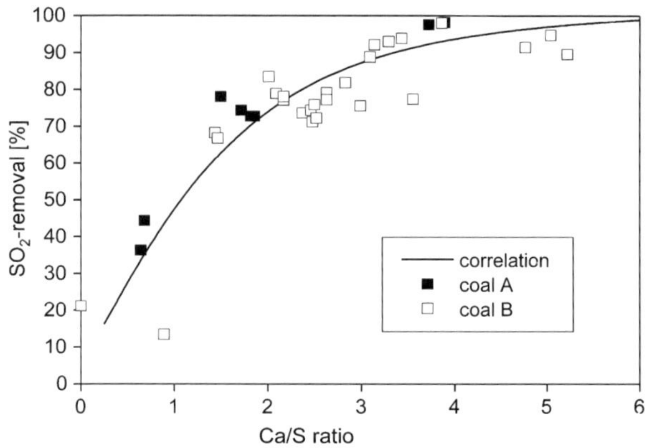

Fig. 5.73 Desulphurisation rate as a function of the Ca/S ratio for a circulating fluidised bed (Takeshita 1994)

of 8 is necessary in order to achieve removal rates of 50% at 150°C. So to efficiently use $Ca(OH)_2$ in the temperature range after the air heater or before the ESP (130–180°C), it is necessary to humidify and/or cool down the flue gas to near dew point. This way, removal rates of 60–80% can be achieved. With this humidification and cooling down of the flue gas, this method becomes a wet process. Besides calcium, which has found widespread use as an additive for flue gas desulphurisation, sodium-based additives are also suitable (Vernon 1990; Nolan 1994).

5.8.2.2 Downstream Desulphurisation (Semi-dry, Wet)

Semi-dry SO_2 removal processes, as well as wet processes, are techniques which achieve a better utilisation of the reacting agents than dry processes. This, above all, is owing to the kinetics, which proceed faster in the liquid. While in wet processes, sulphur dioxide is removed by absorption in the aqueous phase only, semi-dry processes are followed by an additional dry SO_2 capture which works in the same way as in dry processes (STEAG 1988).

Wet Flue Gas Desulphurisation

Wet desulphurisation processes applied at an industrial scale predominantly use lime or limestone. The different processes are usually distinguished by the different scrubber types, the absorbents and the final products. These processes feature high removal rates of between 95 and 99% at an almost stoichiometric quantity of absorbents (Takeshita 1994).

The process, in very simplified terms, has the following reactions following the application of lime (Rosenberg and Oxley 1993):

$$Ca(OH)_2 + SO_2 + H_2O \rightarrow CaSO_3 \cdot 1/2\,H_2O + 3/2\,H_2O \qquad (5.25)$$

and with limestone:

$$CaCO_3 + SO_2 + 1/2\,H_2O \rightarrow CaSO_3 \cdot 1/2\,H_2O + CO_2 \qquad (5.26)$$

and, subsequently, using both additives:

$$CaSO_3 \cdot 1/2\,H_2O + 1/2\,O_2 + 3/2\,H_2O \rightarrow CaSO_4 \cdot 2\,H_2O \qquad (5.27)$$

$CaSO_3$, as a product formed through desulphurisation, is usually oxidised to form $CaSO_4$ in order to create a usable product. This can be done either separately in different reactors, as described by the equations above, or, most commonly, in one reactor, the absorber.

Most desulphurisation plants in Germany and the USA use lime (CaO, $Ca(OH)_2$) or limestone ($CaCO_3$). Today, limestone, being a more cost-effective material, is predominant (STEAG 1988).

Figure 5.74 shows a process schematic using limestone as the additive and which produces gypsum as a final product. Figure 5.75 describes the reactions in a limestone-based process. After the sulphur-containing flue gases have passed the dust removal stage, they are cooled and conducted through the absorption tower. The absorption tower consists of a reaction space, a suspension sump or a washing liquid tank and a demister. The limestone suspension for SO_2 removal is fed to the suspension sump and mixed with the washing liquid. Pumps then transfer the washing liquid through a pipe to the spraying levels which are arranged in the upper part of the absorber vessel. In the counterflow of falling droplets of washing liquid and rising flue gas flow, sulphur dioxide reacts with limestone and the intermediate products are formed. SO_2 is absorbed in the washing liquid and converted into calcium dihydrate by way of complex reactions. Besides SO_2, SO_3, Cl and F are also removed. In the sump of the absorption tower, the limestone dissolves, forming gypsum crystallites. The oxygen necessary for oxygenation comes partly from the flue gases; additionally, air is injected into the suspension. Washing liquid is extracted

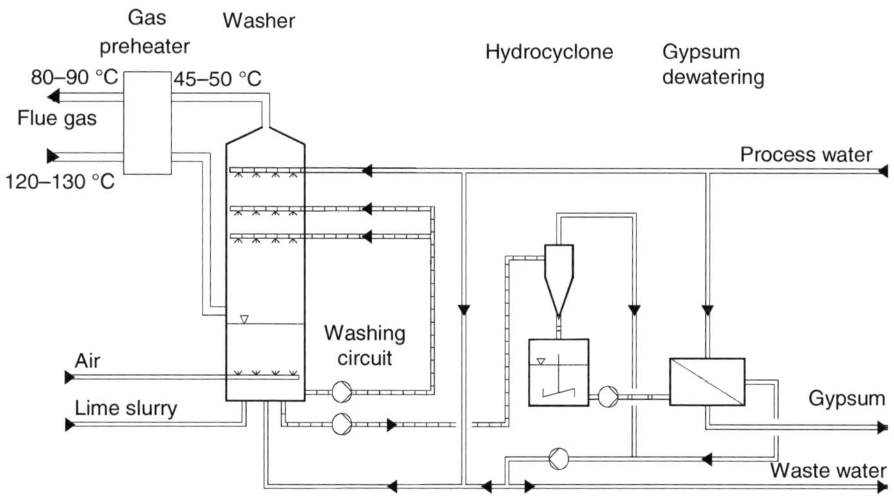

Fig. 5.74 A wet flue gas desulphurisation plant with gypsum production

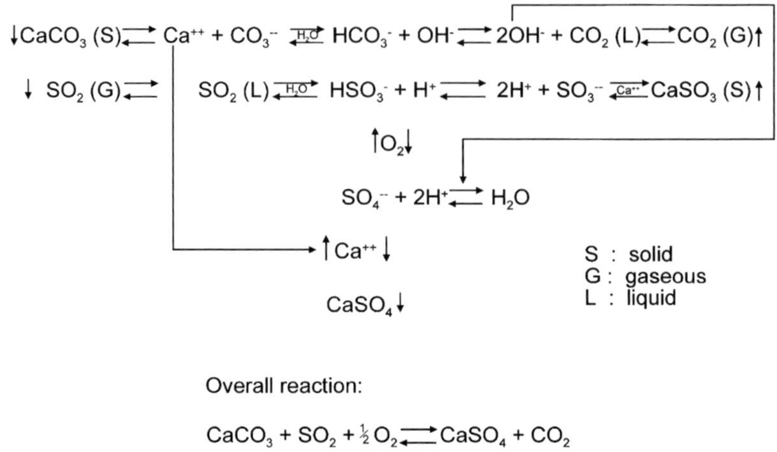

Fig. 5.75 Reaction mechanisms of flue gas desulphurisation by limestone

from the absorber sump, concentrated in a thickener and subsequent hydrocyclones and then dewatered by a vacuum band filter or centrifuges.

Semi-dry Spray Dryer Process

The spray dryer process, as a semi-dry desulphurisation method, usually uses a lime slurry, which is made of quicklime (CaO) and water (Kolar 1995). The lime suspension is sprayed and distributed in the reactor which the flue gases to be cleaned are conducted through. This way, the gas components – SO_2, mainly, and HCl and HF – are provided with, in one passage, a large surface for chemisorption. The capture process of sulphur dioxide runs according to the following elemental formula:

$$Ca(OH)_2 + SO_2 \rightarrow CaSO_3 + H_2O \tag{5.28}$$

The reactor residence time of the gas necessary for absorption and drying may be more than 10 s in this process, which requires correspondingly large vessel dimensions. The temperatures of the flue gases entering the absorption tower range between 130 and 300°C and are cooled by the evaporation of the lime slurry. Care must be taken so that the temperature does not fall below the dew point. The dry desulphurisation products are removed in an ESP or a bag filter. Despite the long residence times of the lime slurry and the flue gas in the absorption tower, the reacting agents are not converted completely. Unreacted $Ca(OH)_2$ mixes with other residual and desulphurisation products. Compared to wet processes, an excess of the added Ca is necessary. At a removal efficiency of 80%, the Ca/S ratio is about 2 (Adrian et al. 1986). In contrast to wet desulphurisation techniques, spray dryer processes do not produce waste water. A drawback is the lesser usability of the product (Vernon and Jones 1993).

The application of spray dry scrubbers is generally limited to flue gases from plants of up to 200 MW_{el} capacity. Larger plants require the use of several modules to deal with the total flue gas flow. This is why in general the technology is used in small- to medium-sized coal-fired power plants. Spray dry scrubbers in commercial use have achieved removal efficiencies in excess of 90% (Nalbandian 2004).

5.8.3 Dissemination and Costs

The application of FGD technology for SO_2 control in pulverised coal power plants began in the early 1970s in Japan and the USA. Western Europe followed in the 1980s. FGD has now become more widespread and is also installed in central and eastern Europe and in Asia. Today, there are over 780 FGD systems installed worldwide, with a total capacity of over 260 GW_{el}. Wet desulphurisation processes are dominant, with a market share of 88%. The semi-dry processes have a share of about 8%. The rest comprises dry processes and other methods (Nalbandian 2004).

The dry additive processes, though requiring only low capital costs, achieve only moderate desulphurisation efficiency rates of up to 60% and are therefore scarcely applied. Wet scrubbers, compared to spray dryers, have about 30% higher capital costs but feature the advantages of a low additive demand and a usable product.

In the USA, the capital cost of wet FGD systems in the early 1990s reached approximately $200/kW. By the late 1990s, capital costs were reduced to $125/kW, which corresponds to the current capital costs quoted in the USA. The cost reductions can be explained by design improvements in using high-velocity absorbers with increased sorbent utilisation. High tower velocities result in a smaller tower cross-sectional area, and hence provide material cost savings. The wet absorber typically represents 40–50% of the cost of an installed FGD system.

The capital costs for a wet desulphurisation plant range between 80 and 120 €/kW_{el} (2005). The additional costs from flue gas desulphurisation are about 0.4–0.6 € cents/kWh at power generation costs of 3–6 cents (Nalbandian 2006).

5.9 Particulate Control Methods

Combustion of solids that contain mineral matter produces ash in the form of particulates in the flue gas. The formation of ash has been discussed in Sect. 5.2.3. for the different firing systems. Particulate concentrations are usually regulated by law for environmental and health reasons; to meet the required emission limits, the following devices are used for their removal from flue gases:

- Mechanical separators
- Electrostatic precipitators
- Fabric filters

The required collection efficiency is determined by the particulate emissions of the firing system and the required emission limit.

5.9.1 Mechanical Separators (Inertia Separators)

Inertia separators are thus called because they remove particulates by making use of the inertia of the particles. Gas and particulates are separated by deflecting the dust-laden gas flow.

Cyclone separators use the centrifugal force of the dust particles. A cyclone consists of a cylindrical vessel with a tangential gas inlet and a gas outlet which is mounted in a central position as part of the vessel cover (Fig. 5.76). A downward vortex forms in the cyclone and particles are collected near the wall and fall to the bottom.

Mechanical separators are in widespread use in industrial plants and are most frequently employed where emission limits are less strict. Their features are a simple construction, a small size, robustness and low operating and construction costs.

The removal rate of simple cyclones is limited to about 90%. As the removal efficiency is dependent on the centrifugal force, the removal rate can be increased either by raising the speed or by reducing the cyclone diameter. Speed increases are ruled out because of the resulting increase in pressure loss. In multicyclone installations, a reduction in diameter is achieved by dividing the gas flow into several small cyclones connected in parallel. Most of the time, though, an increase of the removal efficiency, for instance, by two-stage cyclones or by multicyclones also

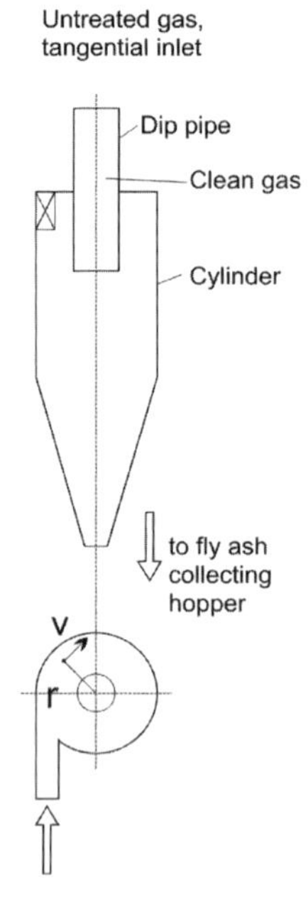

Fig. 5.76 Schematic of a cyclone separator

involves an increase in the loss of pressure. In general, with decreasing particle size, the removal efficiency decreases (Vernon 1990).

5.9.2 Electrostatic Precipitators

In coal-fuelled power plants, particulates are most frequently removed by electrostatic precipitators (ESPs), which reliably meet the strict emission limits of today. ESPs feature high removal rates and low pressure losses and are suitable for large flue gas flows. Their removal rate may be more than 99.8% (Zhu 2003; Soud 1995; Vernon 1990).

In electrostatic filtration the dust particles are electrically charged by and then removed in an electric field (Löffler 1988). The functioning of an ESP is shown in Fig. 5.77. The charging of the dust particles is performed mostly by corona discharge. The electric field is created by applying a high DC voltage of 30–80 kV between the discharge electrodes and the earthed collecting electrodes. The gas is ionised in the vicinity of the discharge electrodes and split into positive gas ions, which remain in the vicinity of the discharge electrode wire, and electrons. The electrons migrate to the collecting electrodes, and on their way are collected by particulates. These dust particles, as charge carriers in this electric field, then migrate towards the collecting electrodes, where they accumulate. Through mechanical rappers the dust is removed, collecting in the bottom hopper, from where it gets discharged.

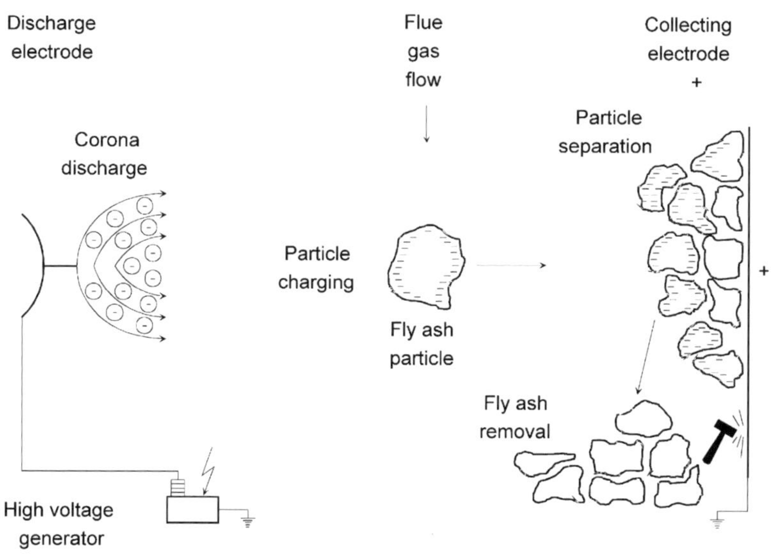

Fig. 5.77 Principles of electrostatic precipitation (Soud 1995)

In the dust collector, the gas to be cleaned is conducted through channels, the edges of which are collecting electrode plates. The width of the channels is usually between 200 and 600 mm (Löffler 1988); in large hard coal-fuelled boilers the width is 300 mm. Discharge electrode wires are suspended parallel to the direction of flow in the centre of each channel, at distances of 300 mm apart. The number of collection channels amount to the width of the ESP; the height of the electrode plates determine the effective ESP height and the length of the channels the effective length.

Because the removal efficiency of the ESP increases with higher voltage, the voltage is set high. This is done by a voltage regulator. The voltage regulator, cycling through each field, checks the sparkover limit for each and resets the voltage somewhat below the sparkover voltage, at which the electrical field breaks down.

The electrical resistance of the ash particles has a great impact on the removal efficiency of the electrostatic precipitator. It has been shown that the specific dust resistance range favourable for the removal process extends from 10^4 to 10^{11} Ω (Löffler 1988). Particles with low resistance and high conductivity, such as carbon, give off their charge before impinging and contacting the collecting electrode. They do not adhere to the collecting electrode but instead migrate with the gas flow somewhat further downstream until they get recharged and travel once more towards the collecting electrode. With too high a resistance, the dust charge carriers are removed only very slowly and an electric field forms over the dust layer covering the collecting electrode. This counteracts the field between the discharge electrode and the collecting electrode, reducing its removal efficiency. The electric field developing over the dust layer may grow so large that gas discharges may occur in the pores of the dust layer and cause an electrical breakdown, which results in a re-entrainment of the particles in the flue gas (Baumbach 1990).

The resistance of the dust depends on the dust composition, the gas composition and the temperature. The electrical resistance of a particle is a product of the resistance of the surrounding gas atmosphere and the surface resistance of the particle. The gas composition has an influence on the resistance of the gas atmosphere around the particle, and the dust composition and potentially condensed components on it have an impact on the resistance on the particle surface (Klingspor and Vernon 1988). Higher water vapour contents and concentrations of SO_3 result in a lower resistance and hence in better removal efficiencies. Coals with lower sulphur contents thus have inherent difficulties in the removal process (see Fig. 5.78). This is put down to the lower sulphuric acid contents on the surface of the dust particle (Klingspor and Vernon 1988). To increase the removal efficiency of problematic dusts, sulphur trioxide, SO_3, is added to the flue gas in some facilities.

The electrical resistance of the fly ash depends on the composition. Alkalis in the ash bring about a better conductivity; alkaline earths, a worse one. Condensed sulphur components and humidity also increase the conductivity, as described above. Also influential are the contents of silicon, aluminium and iron. The sizing of the ESP is strongly influenced by the resistance of the dust. A high resistance, as found with low-sulphur coals, requires bigger and hence more expensive ESPs (Zhu 2003; Wu 2000; Soud 1995; Stultz and Kitto 1992; Klingspor and Vernon 1988).

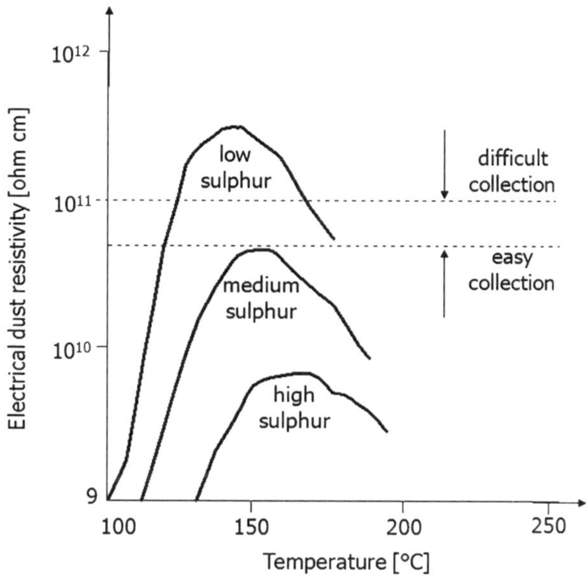

Fig. 5.78 Electrical dust resistance for different coals (Wu 2000)

5.9.3 Fabric Filters

Fabric filters are permeable filtering media on which dust is collected, used during a filtering period and removed at intervals for cleaning. One construction uses filter bags mounted on a supporting structure such as helical springs which separate untreated gas from cleaned gas regions. A multitude of filter bags connected in parallel form one module. Another construction uses pockets instead of bags, where the inside of the pocket corresponds to the inside of the bag. Hence, filters are designated as bag or pocket filters; the term fabric filter is derived from the use of the filter medium.

The parameters which decide the size of construction and the capital costs of a bag filter are the ratio of the filter surface area to the gas flow (air-to-cloth ratio) and the approach velocity towards the filter surface. Typical values for reverse flow cleaning range between 0.008 and 0.011 m/s and for pulse jet cleaning, between 0.015 and 0.2 m/s. Typical dimensions of filter bags with reverse flow cleaning are roughly 300 mm in diameter and 10 m in length; those with pulse jet cleaning have a diameter of 150 mm and a length between 3 and 6 m (Soud 1995; Stultz and Kitto 1992).

The filtering media are woven or non-woven fabrics made from natural or synthetic fibres. Owing to their better resistance to wearing and chemical influences, fabrics of synthetic fibres are preferred today to natural fibres (wool, cotton). The choice of the filtering medium is determined by the properties of the particulates and the operating temperature. Natural fibres can be used up to 110°C or so. The thermal stability of different synthetic fibres ranges between 100 and 280°C; with temperatures up to 260°C it is also possible to use glass fibre fabrics. For temperatures above 300°C, the filtering media have to be made of metal fibres or mineral fibres (Zhu 2003; Löffler 1988).

Separation by filtration with a bag filter is a mature technology which presents an alternative to electrostatic filtration for power plant combustion systems. In the case of a high-dust resistance, which would require a big ESP, bag filters are an economical alternative. In industrial combustion plants and in small plants, bag or pocket, filters are usually used for dust collection. These filters make it possible to achieve collection efficiencies of more than 99.9%, a rate not reached by mechanical dust collectors. The collection efficiencies are relatively independent of the dust load and the properties of the fly ash (Stultz and Kitto 1992; Löffler 1988).

Particle removal, at the beginning of the filtering period, depends on the pore size of the fabric (Löffler 1988). Subsequently, the particles form a dust layer covering the filter surface. The filtering effect of the filtering medium is supported by deep-layer filtration of the dust layer. While the sifting filtration on the surface collects particles to a size of 1 μm, deep-layer filtration is able to remove particles even smaller than 0.5 μm (Baumbach 1990).

Figure 5.79 shows the mounting of the filter bags in a filter casing for different cleaning methods (Soud 1995). A filter unit usually consists of several filter modules. The dust layer formed on the filtering medium has to be cleaned off periodically to limit the pressure loss. The possible cleaning methods are the following:

- Cleaning by a reverse flow of purified gas at low pressure
- Mechanical cleaning through rapping
- Pulse jet cleaning

In pulse jet cleaning, the dust deposits sticking on the outside of the bag filter are cleared away by pulses of compressed air from the inside of the bag. This cleaning-off, a chronologically staggered process in the modules, can be run during plant operation.

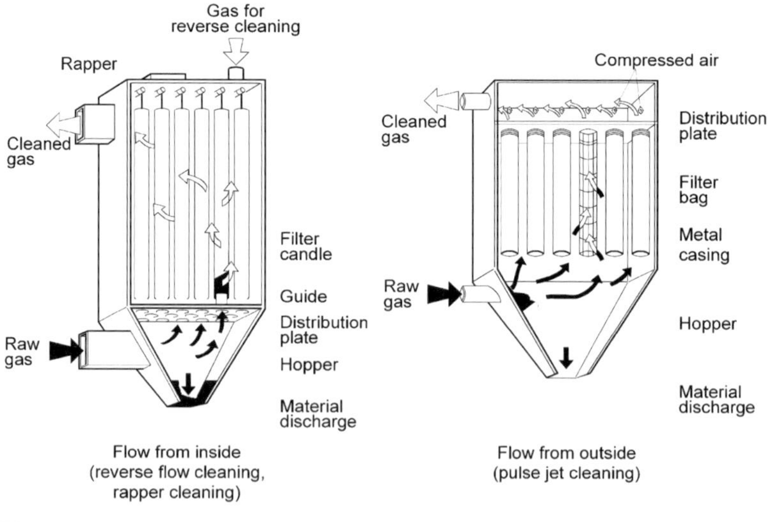

Fig. 5.79 Schematic drawing of a bag filter (Soud 1995)

In cleaning by a reverse flow of purified gas, the dust layer is taken off by reversing the gas flow and transported to the filter bottom. In rapper cleaning, a purging gas flow from the cleaned to the untreated gas also assists the transport of the filter cake which has been cleaned off by rapping. In these two methods, often applied in the USA, cleaning-off during the filtering process is not possible, so that the in- and outflow have to be diverted by means of valves. The filter therefore consists of several filter casings or modules, one casing is taken out of service for cleaning at a time (Soud 1995).

The cleaning of filters is performed in power plants in a temperature range typically between 120 and 180°C. The limit of the lower temperatures in this range is set by the acid dew point, under which dust particles would not stick together. The limit of the upper temperatures is set by the allowable temperatures of the filter medium.

5.9.4 Applications and Costs

The application of the different removal methods depends on the required collection efficiency, the capital costs and the operational costs (pressure loss, maintenance). Table 5.9 presents the collection efficiency as a function of the particle size, comparing the different methods. Mechanical collectors are not efficient at removing small particles, so they are not suitable where strict emission control standards apply.

The pressure losses of electrostatic precipitators (ESPs) are the lowest, whereas the losses of filtration separators and cyclone dust collectors are considerably higher. The different pressure losses influence the power demand of the flue gas fans. The operating costs of a bag filter are higher due to the higher pressure loss and the limited lifetime of the filter bags.

The notable attractions of mechanical dust collectors are both the low capital costs and the low operating costs. They are used where the emission control standards can be complied with without any additional particulate treatment, for example, in small furnaces, which are often only regulated for coarse fly ash particles. The emission control standards laid down for industrial and power plant furnaces can only be complied with by ESPs and fabric filters. The respective costs depend mainly on the size of the plant, the required emission limits and the properties of the dust. Generally, ESPs are more economical for large power plant furnaces and bag filters for industrial plants. The costs are equal at 250 MW_{th}, with lower emission control standards and a high-dust resistance favouring the bag filter (Soud 1995; Takeshita 1995; Vernon and Jones 1993). In the power range of large pulverised coal-fired power plants, electrostatic precipitators are employed almost exclusively (Stultz and Kitto 1992).

Table 5.9 Collection efficiency as a function of particle size (Soud 1995)

	Dust collection			
Separation system	$< 1\,\mu m$	$1\text{–}3\,\mu m$	$3\text{–}10\,\mu m$	$> 10\,\mu m$
ESP	96.5	98.25	99.1	99.5
Bag filter	99	99.75	> 99.95	> 99.95
Multicyclone	11	54	85	95

For large pulverised coal-fired power plants, reports cite capital costs for dust collection of about 30–50 €/kW_{el} (1990), i.e. a fraction of 3–4% of the total capital costs. The additional cost of power generation amounts to about 0.25 € cents/kWh at a total cost of 3–6€ cents/kWh (Takeshita 1995).

5.10 Effect of Slag, Ash and Flue Gas on Furnace Walls and Convective Heat Transfer Surfaces (Operational Problems)

Ash components in solid fuels affect coal-fired furnaces in numerous ways. Because it is incombustible matter content, ash reduces the calorific value. Ash particles in the hot flue gas flow participate in the transfer of radiant heat. On the one hand, ash is employed as a heat carrier; on the other hand, heat is lost through the discharge of ash and slag. In the following sections, the interactions of ash, slag and flue gas in the context of their impact on heating surfaces and the respective heat transfer will be discussed. The ash fraction and the ash properties of solid fuels are decisive parameters for the design of the furnace and are a requirement for estimating ash deposits, erosion and corrosion. The predicted impacts of the ash on the operation of solid fuel-fired furnaces have to be taken into account during the design of the plant. Within this chapter the focus is on ash-related problems in coal-fired furnaces, but the principles are also valid for biomass and waste combustion, which are discussed in Sect. 6.

Deposits may form both in the furnace and on the convective heat transfer surfaces. Slagging refers to molten deposits, while fouling refers to dry solid (i.e. not molten) deposits. Another differentiation which shall be used in the following refers to the location of the deposits. In the area of the furnace and the platen heating surfaces where heat is mainly transferred by radiation, the term used is slagging, and in the area of the convective heat transfer surfaces, the term is fouling. Figure 5.80 depicts the areas of possible slagging and fouling deposits for a single-pass and a two-pass boiler (Couch 1994).

Slagging and fouling are often associated with erosion and corrosion. For example, deposits in many instances cause corrosion damage to or narrowing of the flue gas ducts. Locally narrowed areas raise the flue gas velocity in those areas and thus increase the eroding action. Erosion may impede the formation of protective metallic oxide layers, so that tube walls are attacked by corrosion (Skorupska 1993). Erosive action, though, may also wear away deposits or impede their growth.

Operational disturbances, damage of plant components and resulting failures can be a direct consequence of slagging and fouling:

Impediment of the Heat Transfer

Deposits insulate the heating surfaces, deteriorating the heat transfer from the flue gas to the steam – water cycle. Slagging in the area of the furnace reduces the heat

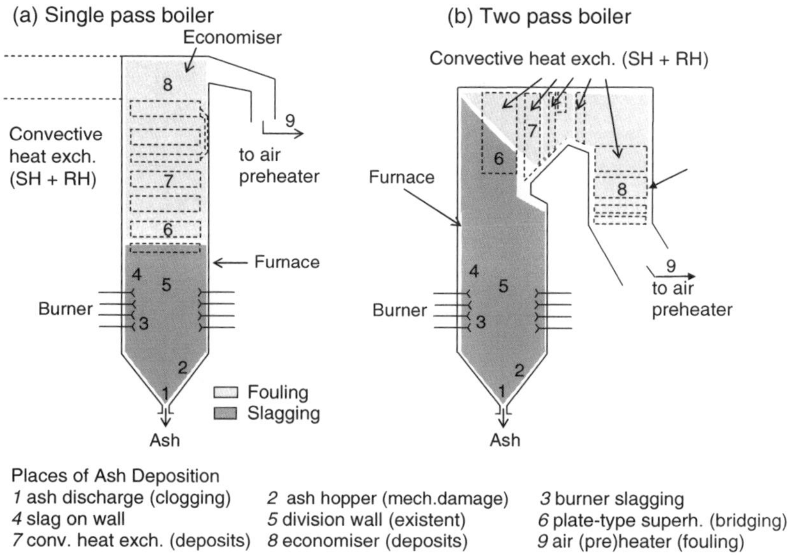

Fig. 5.80 Fouling and slagging in single-pass and in two-pass boilers (Couch 1994)

absorption of the evaporator and shifts heat to the convective heat transfer surfaces. This displacement of heat has to be counteracted by the design of the firing system; as a last remedy the thermal output has to be reduced. A higher heat absorption of the convective heat transfer surfaces has to be balanced out by spray attemperation to limit the maximum steam temperatures. Reheater attemperation in this context has the consequence of lower efficiencies. If, in the case of severe slagging in the furnace, the furnace outlet temperature is exceeded, the power output has to be reduced in order to avoid molten deposits on the convective heat transfer surfaces. Fouling in the convective section may increase the flue gas heat loss.

Narrowing of the Flue Gas Path

In an extreme case, deposits and bridging of solid material make the flue gas path in the convective heat transfer section so narrow that the boiler eventually has to be shut down to remove the deposits manually. The sections of the system affected are the platen heating surfaces at the furnace outlet and the first heat exchanger tube bundles.

Blocking of the Ash Discharge

Deposits that form in the furnace may detach and clog the discharge of the bottom ash via the hopper of the furnace. If necessary the boiler has to be shut down.

Damage to the Steam Generator

The pressure part of the steam generator system may be damaged by large falling slag lumps.

Material Wear through Erosion or Corrosion

Metal heat exchanger materials may be worn through corrosion and erosion. In the case of excessive material wear, it becomes necessary to replace heating surfaces.

5.10.1 Slagging

5.10.1.1 The Process of Slagging

Deposit formation on heat exchangers is determined by three main steps:

- Release and conversion of ash-forming elements
- Transport to the tubes and deposition
- Deposit reactions (Heinzel 2004)

The process of ash formation is described in Sect. 5.2.3. Excluded and included mineral matter, organically bound elements and solid or dissolved salts all undergo transformations to form ash. Ash formation includes the processes of coalescence, fragmentation, fusion, vaporisation and condensation, which can occur sequentially or simultaneously. The results are particles of different sizes and chemistry.

Excluded and included minerals form the major part of the ash; their particle distribution is determined by the coalescence and fragmentation processes. At the high temperatures of pulverised coal combustion, ash particles melt either completely or partially. The physical state of the ash particles at a given place in the furnace is then determined by the cooling process. This process is decisive in the formation of deposits in the furnace.

During combustion, reactive alkalis, either in the form of simple salts or organically bound, are released to form vapour. When cooling down, the alkalis nucleate to form very fine particles or condense on ash particles.

Depending on the size, the particles are transported via different mechanisms to the heat exchange tubes. Larger particles collide with the tubes by impaction, while small particles are transported by turbulent diffusion or thermophoresis (see Sect. 7.2.2).

On clean tubes, deposits form only very slowly at first. If molten ash particles get onto a clean tube, they are cooled very quickly down to temperatures below the deformation point. They flake off and are caught again by the flue gas flow. Gradually, though, a basic layer of very fine particles, which tend to be transported selectively towards walls and tubes, forms. This basic layer may react chemically, either with ash particles with low melting points that come afterwards or with the condensate of vaporised ash components, to form a sticking layer (Zelkowski 2004). The insulating effect of the basic layer makes its surface temperature rise, so that particles impinging on the tube cool to lesser and lesser extents, and eventually hardly at all. With the layer growing in thickness, the exterior temperature rises, and the layer reaches a plastic state on its outside. In consequence, all particles that come into contact with the wall stick to it and the layer thickness continues to grow. If the temperatures on the outside reach the ash fluid temperature, a state of equilibrium develops, where the ash starts to flow down the layer, draining off. The slag layer and its effect on the heat transfer reach a stable state. The dripping slag, however, may cause problems in furnace areas below.

If the ash fluid temperature at the outside slag layer is not reached, for instance, in the cases of high-ash melting temperatures or low furnace temperatures, the deposits continue to grow. Due to the weaker solidification of the first (inner) layers, it is also possible for deposits to fall from the tubes if the total mass becomes large enough. The deposit formation in these temperature ranges is uncontrolled. It can be possible, however, to remove the deposits by using blowers, thus avoiding problems due to solidification and total mass. Changing the load of a plant may intensify the problems if deposits build up at part load and solidify at full load.

Slagging is not only influenced by coal and ash properties but also depends to a great extent on the combustion conditions and the furnace design.

The process of slagging is based on the following conditions (Skorupska 1993; Juniper 1995):

- Ash particles can only form deposits if they can reach the wall. Their trajectories have to be seen in the context of furnace design and the aerodynamics of the burners
- The linkage forces between the walls and particle(s) have to be great enough for the particles to adhere to and not bounce off the wall. For large particles with a high velocity, the necessary linkage forces are greater than for small particles. The linkage forces are a function not only of the softening and melting process but also of the partial vaporisation of mineral components
- If particles have deposited on the wall, the wall – particle linkage forces have to be strong enough so that the particles do not fall off because of their dead weight. The linkage forces between the single particles and between the particles and the wall increase with time due to the diffusion of gaseous compounds into the deposits and through chemical reactions (Raask 1973)

Slagging in the furnace poses a problem only in dry-bottom furnaces. In slag-tap furnaces, the behaviour described above is intentionally made use of for primary ash removal. Suitable feedstocks for slag-tap furnaces are those coal types with low melting points, which would cause slagging in dry-bottom furnaces. Inversely, the fuels suited to dry-bottom furnaces are coals with high melting points, because the low furnace temperatures keep the ash in a solid state, which in turn is not desirable in slag-tap furnaces. Slagging should be limited in slag-tap firing systems, though, to the furnace. In the downstream radiation region, slagging is as undesirable as in dry-bottom furnaces.

In stoker-fired furnaces, slagging may occur on the grate, in the furnace and on the first superheater. While a slight sintering of the burning coal layer is desired to impede the release of fines, slag deposits or cakings hamper the penetration of air through the coal layer, hence have to be avoided. Due to the lower ash load, there is less danger of slag deposits on the furnace walls and superheater surfaces than in pulverised fuel-fired furnaces. Slag problems in stoker firing systems are to be expected, particularly if alkali-rich biomass is used as a fuel.

Fluidised bed furnaces, with their low combustion temperatures, are normally not affected by slagging. However, bed sintering can occur, particularly with alkali-rich biomass as a fuel feedstock. This is discussed in Sect. 6.2.5.

5.10.1.2 Evaluation of the Slagging Behaviour

Ash Fusion Behaviour

Laboratory analysis of ash fusion behaviour yields data for the design of a furnace (namely the furnace outlet temperature) and the first guideline values for the evaluation of the slagging behaviour of the ash. Ash produced in the laboratory at temperatures of about 800°C is used to make a cylindrical, cube-shaped or pyramidal sample body. The changing shapes of the sample body during a slow heating-up are recorded photographically. In accordance with convention, a distinction is made between initial ash deformation, spherical or softening, hemispherical and fluid temperatures. The temperature at which the first changes of shape are discernible is designated as initial ash deformation temperature. The temperature at which the sample body changes from a solid into a plastic state, losing its shape, is termed hemispherical (see also Fig. 2.3). Types of coal ash with high fusion temperatures have less of a tendency towards slagging, since they cool below the ash deformation temperature inside the furnace and so do not stick to the walls. Low fusion temperatures make it likely that slagging occurs. The ash fusion behaviour, however, can only give a limited amount of information about the slagging behaviour (Albrecht and Pollmann 1980), because the ash sample used in the laboratory can only partially represent the composition and structure of deposits in the boiler. However, this method is not suitable for detecting sintering, because the sample body keeps its shape. With sintering, melting only occurs locally on the particle surface.

Investigations into Ash Viscosity

Physical parameters of interest can be determined by viscosity measurements of the molten ash (slag), though these tests are confined to homogeneous ashes. This method was originally applied to evaluate coal types for their suitability for slag-tap furnaces (Stultz and Kitto 1992; Albrecht and Pollmann 1980). Usually, the viscosity is determined by measuring the torque of a platinum plate rotating in the slag. In a defined flue gas atmosphere, the slag temperature is decreased step by step and thus the viscosity determined depending on the temperature. As the slag cools down, a linear decrease of the logarithm of the viscosity correlated to the temperature can be observed initially. Below a certain temperature, it can be seen that, due to the partial crystallisation of high melting temperature components, the viscosity increases more strongly. The viscosity here usually ranges around 250 poise. The temperature at which a viscosity of 250 poise develops is designated as the temperature of critical viscosity, T_{cr} or T_{250}. The end of the plastic phase is reached with the solidification temperature, at a viscosity of about 10,000 poise.

The limiting temperature for a coal for use in slag-tap furnaces is 1,425°C for a critical viscosity of 250 poise. This temperature should not be exceeded, since coals with higher temperatures at T_{250} either require excessively high temperatures in the slag-tap furnace or impair the slag flow. Conversely, it is possible to apply a limit to dry-bottom furnaces and classify coals as slagging if the ash viscosity of 250 poise is reached at lower temperatures (Albrecht and Pollmann 1980). Coals with low temperatures at T_{250} generally have low deformation temperatures as well.

Figure 5.81 shows the viscosity behaviour for two different coal types and various flue gas atmospheres (Stultz and Kitto 1992). Both the temperature at which the plastic behaviour sets in and the temperature range of the plastic behaviour give information about the slagging behaviour. A coal type with a slight slagging tendency features high temperatures T_{cr} and a narrow temperature range of plastic behaviour. Coals with a tendency towards slagging (slagging coals) have low temperatures T_{Cr} and large temperature ranges of plastic behaviour. The impact of the reducing atmosphere should also be taken into account in the case of the coal shown in Fig. 5.81, which has a high Fe content that produces a widening of the temperature range of plastic behaviour. This may be important when considering NO_x-reducing measures in the furnace.

Since methods for the determination of the slag viscosity are costly, the viscosity is usually estimated by calculation, using the chemically analysed ash composition. One approach to establish the viscosity from the ash analysis is the determination

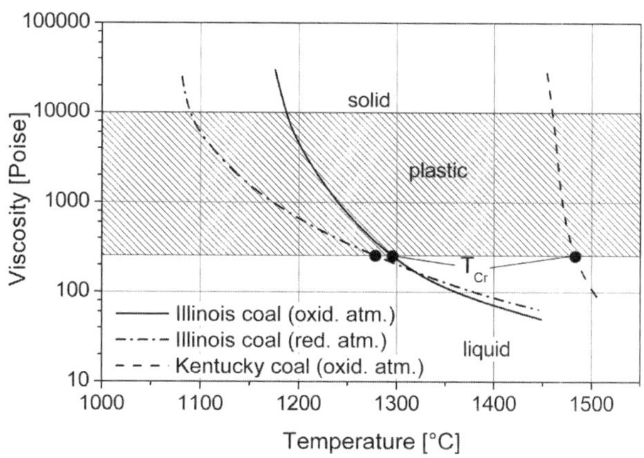

Fig. 5.81 Viscosities of different coal types as a function of the temperature (Stultz and Kitto 1992)

of the basic and the acid ash components in proportion to the total ash or the ratio of basic to acidic components.

Impact of the Ash Composition

In an ash analysis of a coal, its ash components are investigated to determine their elemental composition, usually indicated as oxides of the elements. The ash is produced in the laboratory by slowly heating a coal sample to 815°C with an air supply. Coal ash contains silicon, aluminium, iron, calcium, small amounts of magnesium, titanium, sodium, potassium, phosphorus and sulphur. Ashes of hard coals typically have higher contents of silicon, aluminium and iron; ashes of younger coals have higher contents of calcium, magnesium and in some cases of sodium. Although the ash components are given as oxides, they are actually found in various compounds, i.e. as silicates, oxides and sulphates. Even if there is no clear correlation between the ash composition and the slagging intensity of the different compounds of the ash elements and their conversion processes, it is well known that certain chemical elements in the ash intensify the slagging process. Such elements include sodium, potassium, calcium and iron.

Under the effects of the temperature and flue gas, the ash-forming components can be converted into other compounds which have a lower melting temperature, or, alternatively, different ash compounds can form a eutectic mixture which has a lower melting temperature than the single compounds. Table 5.10 presents a compilation of melting points of compounds and Table 5.11 of melting points of mixtures.

Iron may be found in different compounds such as pyrite (FeS_2), siderite (Fe_2CO_3), haematite (Fe_2O_3), magnetite (Fe_3O_4) and ankerite [$(Ca,Fe,Mg)CO_3$]. Under oxidising conditions, pyrite is converted into Fe_2O_3 and SO_2. Under reducing conditions though, pyrrhotite (FeS), weakly oxidised compounds (FeO) and metallic iron are

Table 5.10 Melting points of compounds in furnaces (Hein 1984)

Compound	Melting point (°C)	Compound	Melting point (°C)
Sulphates		*Oxides*	
$CaSO_4$	1,447	MgO	2,800
Na_2SO_4	884	CaO	2,570
K_2SO_4	1,076	Al_2O_3	2,020
$Na_2K_3Fe_2(SO_4)_6$	552	SiO_2	1,723
		Fe_2O_3	1,566
		FeO	1,369
Sulphides		*Silicates*	
Na_2S	1,175	$Na_2O \cdot 2SiO_2$	874
K_2S	840	$Na_{2O} \cdot SiO_2$	1,089
FeS	1,195	$CaO \cdot SiO_2$	1,544
FeS_2	1,171	$CaO \cdot Al_2O_3 \cdot 2SiO_2$	1,553
		$K_2O \cdot Al_2O_3 \cdot 6SiO_2$	1,150
		$Na_2O \cdot Al_2O_3 \cdot 6SiO_2$	1,118

Table 5.11 Eutectic mixtures with low melting points (Zelkowski 2004; Hein 1984)

System	Melting point (°C)	System	Melting point (°C)
Pure oxides		*Silicates*	
$Al_2O3–SiO_2$	1,590	$SiO_2–Al_2O_3–CaO$	1,165–1,260
$CaO–Fe_2O_3$	1,205	$SiO_2–Al_2O_3–Fe_2O_3$	1,073
CaO-FeO	1,133	$SiO_2–Al_2O_3–K_2O$	750
$CaO–SiO_2$	1,436	$SiO_2–CaO–Na_2O$	725
$Na_2O–SiO_2$	827	$SiO_2–CaO–K_2O$	710
		$CaO–FeO–SiO_2$	1,093
		$CaO–FeO–SiO_2–MgO$	< 1,047
Sulphurous compounds			
$Na_2SO_4–NaCl$	625		
$Na_2SO_4–CaSO_4$	918		
$Na_2SO_4–CaSO_4–K_2SO_4$	845–933		
$Na_2S–FeS$	640		

formed. The compounds formed in reducing flue gas atmospheres lower the ash melting point, whereas haematite, Fe_2O_3, which forms at excess oxygen, raises the ash melting temperature (Stultz and Kitto 1992). Mixtures of different compounds such as FeS and FeO further decrease the melting point.

A similar effect to iron is also found with calcium. The melting temperature depends on the state of oxidation – calcium oxide has very high, calcium sulphide very low melting temperatures. Mixtures of $CaSO_4$ and CaS, as a eutectic, have a melting temperature of 850°C. Slagging caused by calcium predominately occurs in the combustion of brown coal.

Severe slagging, as an example, occurs in the combustion of coals which have rather large fractions of sodium sulphate (Na_2SO_4) and/or sodium chloride (NaCl). Difficulties in the combustion arise because of the low melting temperatures of the eutectic mixture of $NaSO_4$–NaCl, with a minimum of 625°C. Ash components molten on the coal surface may also hamper the supply of oxygen.

Without taking into consideration the various effects of certain ash compounds, the calculation of the melting temperature is often based on a determination of the ratio of basic to acidic ash components. The basic components include iron, alkaline earths and alkalis and among the acidic ones are silicon, aluminium and titanium. Both acidic and basic ashes have high melting temperatures. If basic and acidic ashes get mixed, low melting temperatures are the consequence (Fig. 5.82). The minimum melting temperatures are found in the range of 40–45% basic components (and hence 55–60% acidic), which corresponds to a base/acid ratio of 0.7–0.8. Coals with a base/acid ratio between 0.5 and 1.2 are categorised as slagging coals.

Slagging Indices

In accordance with the correlations described above, the slagging behaviour is predicted using a range of evolving indices, a selection of which are described in Table 5.12. Slagging indices are mostly based on the ash viscosity behaviour, which is either determined by experiment or calculated from the ash composition.

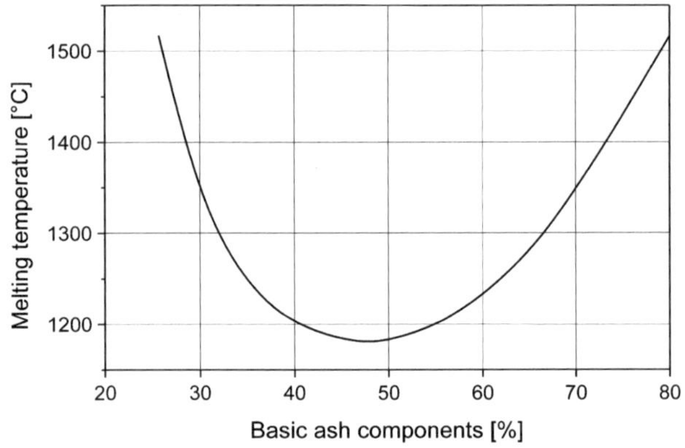

Fig. 5.82 Melting temperature of ash as a function of basic ash components (Stultz and Kitto 1992)

Predictions of the slagging tendency of a coal type by means of these indices, being dependent on the ash composition or the ash fusion behaviour, are only approximations. There are several explanations for this fact:

- The indices have evolved for particular coal types in individual plants, hence are only partially applicable to other coal types and general boundary conditions
- The composition of the sample incinerated in the laboratory does not correspond to the ash composition in a firing system, because in the laboratory, the incineration is done at low temperatures and heating rates, hence does not take into account the conversion of ash at high temperatures and heating rates. The vaporisation and condensation of ash components may have an influence on slagging

Table 5.12 Slagging and fouling indices (Stultz and Kitto 1992; Zelkowski 2004; Juniper 1995; Bals 1997)

Index	Formula	Problem area
Viscosity temperature T_{250}	T_{250} = temperature at $\eta = 250$ poise	$T_{250} < 1{,}400°C$
Calculated viscosity at 1,426°C		$CV_{1,426} < 350$
Multi-viscosity index	$MV = \dfrac{T_{250} - T_{10,000}}{97.5 \cdot F_s}$ $F_s = 10^{(0.00186 \cdot T_{2,000} - 1.933)}$ T_x temperature at $\eta = x$ poise	$MV > 0.5$
SiO_2 ratio	$S_R = \dfrac{SiO_2}{SiO_2 + Fe_2O_3 + CaO + MgO} \cdot 100$	$S_R < 72$
Base/acid ratio	$\dfrac{B}{A} = \dfrac{Fe_2O_3 + CaO + MgO + Na_2O + K_2O}{SiO_2 + Al_2O_3 + TiO_2}$	$B/A > 0.5$ for dry-bottom furnace
Slagging index	$R_S = \dfrac{B}{A} \cdot S_{wf}$ (S sulphur)	$R_S > 0.6$
Fouling index	$R_F = \dfrac{B}{A} \cdot Na_2O$	$R_F > 0.2$

- Even with the same total ash composition, there may be different compounds present in the ash and inhomogeneous distributions of the ash components
- The indices determine the tendency of ash towards deposit formation but cannot predict the method of removal nor destruction of the deposits (Kautz and Zelkowski 1985)

Although the indices cannot offer a safe prediction of the slagging behaviour, they are usually the only available information and, because they are associated with experiences at industrial plants for similar coal types and with investigations at semi-industrial experimental plants, provide a valuable guideline for the design of combustion systems.

Further Development

The prediction of deposits is currently following two distinct principal lines of development. For one line, the aim is to predict the slagging and fouling tendencies by means of improved or advanced experimental methods, while the other line pursues the analysis of deposits of real combustion processes.

- Fusability and sintering tests
  Given the slight predictive power of conventional ash analyses and characteristics, there are intensive efforts ongoing to develop new or improved sintering and fusion analyses for prediction. Usually the samples used in these investigations are not laboratory-produced ashes but real ash samples from experimental furnaces or industrial plants, so that the complex reactions running in combustion processes are taken into account. The assigning of temperatures to significant changes of shape, as practiced in the conventional ash fusion analysis, full of uncertainties, is dropped in favour of a continuous determination of the shrinkage or shape change. The height of the sample body is measured as a function of the temperature. Figure 5.83 as an example shows such a result for various slagging and fouling studies of different brown coals at a 325 MW_{th} pulverised brown-coal combustion plant. The measured changes of height of deposits taken from the furnace, using different coal types, are compared with the measured flue gas temperatures (Heinzel et al. 1997; Heinzel 2004).

  Different methods utilise different sample shapes or in addition subject the sample to pressure. Alternatively or additionally, other properties may be used to determine the clinkering as well. For instance, the electrical or thermal conductivity is a measure for the sintering behaviour of the sample.

  Another method, known as simultaneous thermal analysis (STA), is based on the simultaneous measurement of the weight and temperature during the slow continuous heating of a sample compared to an inert sample. By means of weight changes, it is possible to determine evaporation processes, and, by temperature changes compared to the inert sample, melting processes can be measured. Based on the data on the conversion and melting energies, it is possible to determine the molten fraction in the sample.

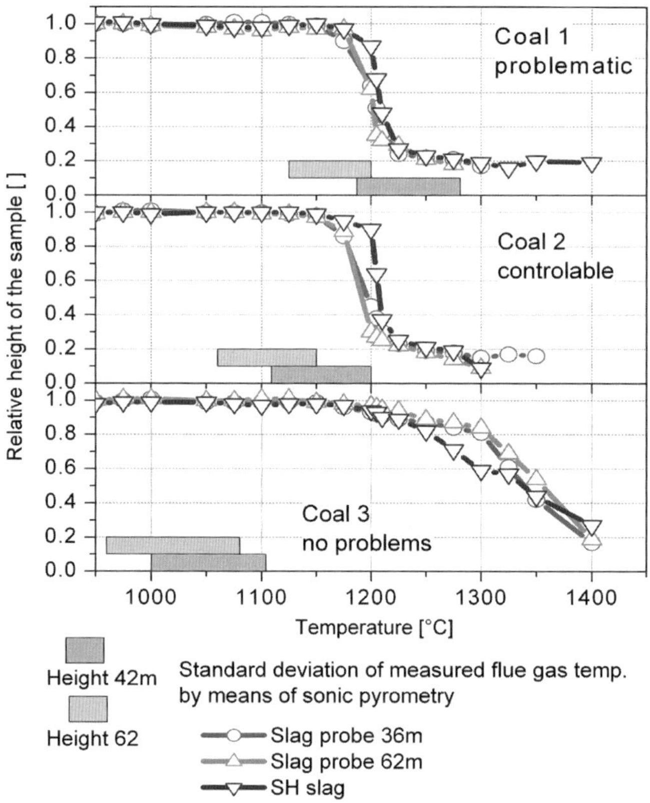

Fig. 5.83 Fusion behaviour of deposits and flue gas temperatures in the combustion of different brown coal types in a 325 MW_{el} pulverised fuel-fired furnace (Heinzel et al. 1997)

- Combustion tests and deposition analysis
  Fuels are burned in industrial plants or in experimental plants under realistic conditions, then deposits are sampled and examined by detailed chemical and mineralogical analyses in order to comprehend the conversion processes and be able to draw conclusions and make predictions about the formation and characteristics of deposits. Conversion models can be integrated into complex numerical models.

Because of the above, great importance is attributed to the methods of slag and ash analysis. Well-known methods are as follows:

- Scanning electron microscopy (SEM)
  Electron-microscopical examinations are a means to determine the structure of a sample by the size and the shape of particles, in order to make conclusions about the existence of molten particles and small condensates. The most frequently used technique is automatic image analysis.

- X-ray diffraction (XRD)
  The technique of X-ray diffraction serves to qualitatively verify mineral fractions in a sample. X-rays are diffracted at different degrees through the crystal lattices of mineral compounds. The diffraction angles are characteristic for the lattice distances. By measuring the intensity of the diffracted light as a function of the diffraction angle, it is possible to determine the crystalline structures and to assign the mineral types.
- Scanning electron microscopy with electron-dispersive X-ray analysis (SEM-EDX)
  In addition to the sample structure determined by the scanning electron microscope, X-ray analysis makes it possible to determine the elemental composition at the surface of the sample at selected locations. EDX is based on the excitement of single atoms by electron rays and the emission of X-radiation which is characteristic for the element.
- Computer-controlled scanning electron microscopy (CCSEM)
  An image received by a scanning electron microscope is numerically evaluated, contours are determined and particle sizes are measured. Afterwards, all particles and structures are analysed by EDX, determining their composition. The evaluation assigns size, composition and mass to the particles. This method can be used for mineral inclusions in the raw coal, but also for ashes and deposits from furnaces.

5.10.1.3 Impacts, Countermeasures and Remedial Actions

A general characteristic of coals with a slagging tendency is a low-ash deformation temperature. Design-wise, they require low furnace outlet temperatures and therefore low volumetric heat release rates. Another essential criterion in a design conceived to prevent slagging is to choose a low burner-belt heat release rate in order to decrease the temperatures in the burner zone.

Operational measures such as increasing the air ratio to decrease the temperature only have a limited range of application. The injection of additives to diminish the linkage forces between the deposits can be successful for iron-containing coals (Raask 1973).

By means of a homogeneous distribution of coal powder and air, sufficient burner distance to neighbouring and opposite walls and a restriction of the coarse particle fraction, it is possible to reduce the solid matter fraction that, in a molten or plastic state, may hit the wall. For coarse particles, it is in particular the residual carbon and the pyrite fraction that cause trouble. By secondary combustion of coarse particles on the walls, localised reducing zones are created which, according to the above-described correlations, are favourable for slagging.

Depending on the hardness and the degree of sticking of the slag deposits, removal by blowing is possible, using steam, water jet or air jet blowers. For cleaning in the region of the evaporator, water jet blowers are usually used, which are more effective against slag deposits than steam or air jet blowers. Figure 5.84 shows the cleaning mechanisms of a water jet and the arrangement of water cannons at

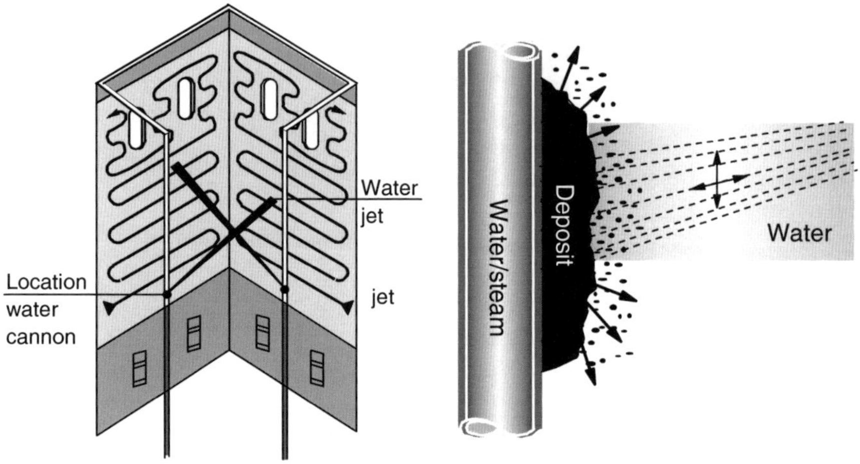

Fig. 5.84 Principle of slag cleaning by water cannons (Simon et al. 2006)

the furnace walls. In combination with a heat flux sensor-based diagnostic system, the technology enables effective online cleaning (Simon et al. 2006). Because of the high stress to the boiler material from thermal shock, water jet blowers are not used for superheaters or reheaters. Here, steam or air jet blowers are used, which are mounted on lances and can be moved along the heat exchangers inside the flue gas pass.

If slag deposits on furnace walls impede the take-up of heat even when wall blowers are used and if the furnace outlet temperature exceeds allowable values, it is necessary to reduce the boiler output. As a consequence of rising furnace temperatures, slag deposits can also form in the zone of the convective heat transfer surfaces (Stultz and Kitto 1992).

5.10.2 Fouling

The major part of the ash is carried out of the furnace as fly dust and flows through the area of the convective heat transfer surfaces. Small particles can stick to surfaces downstream of the furnace by adhesive power. During standard operation, the ash deformation temperature is not reached, so the dust deposits on the tubes are usually easily cleaned off.

In the combustion process in the furnace, part of the ash may vaporise, condensing afterwards during the cooling process in the region of the convective heat transfer surfaces. The alkalis sodium and potassium in particular become vaporous and are mostly released as chlorides, reacting with the sulphur oxide in the flue gas to form sulphates. In the temperature range of the convective heat transfer surfaces, the vapours condense either on the fly dust or on the tubes.

On the tubes, the condensates form a sticky basic layer in combination with the fly ash, in the process drawing more particles from the flue gas. Condensate and ash particles react with each other and the deposits sinter or melt totally. Sintering is the process of particles sticking together below the fusion temperature as a result of the localised stickiness of a particle surface. With time, the deposits solidify and eventually become rather difficult to clean off.

One method to determine the sintering behaviour involves taking fly ash from the furnace and pressing it into a cylindrical sample body. Then, in a laboratory furnace, the samples are subjected to various temperatures, up to about 1,000°C, for several hours. After cooling down, increasing pressure is exerted on the sample, determining the degree of compressive stress at which the sample breaks. This degree is a measure of the sintered state of the sample. For coals, the fouling behaviour can be correlated to the sodium content. The water-soluble sodium in particular, which is supposed to vaporise during combustion, has an impact on sintering. The influence of the potassium content, in contrast, is small. Potassium is mostly bound in the mineral phase in the coal and not vaporised during combustion (Stultz and Kitto 1992).

Low and homogeneous furnace temperatures reduce fouling in the convective section because fewer ash components vaporise. Fouling prevention measures are identical to a large extent to slagging reduction measures such as a low volumetric heat release, the use of flue gas recirculation, a high excess air ratio and a homogeneous distribution of the pulverised coal.

Furthermore, the design of the convective heat transfer surfaces has to be adapted to the coal type. For coals that have a slagging tendency, larger spacings are chosen. A careful arrangement of the tubes also helps to limit fouling. Since ash deposits cannot be avoided during operation, they have to be removed by regular soot blowing. The cleaning interval in this respect has to be set such that sintering of the deposits is prevented (Stultz and Kitto 1992; Hein 1984; Reidick and Schumacher 1985).

Due to the higher combustion temperatures in slag-tap furnaces, 5% of the ash may vaporise and condense on downstream heat transfer surfaces, contributing to fouling in these places (Kautz and Zelkowski 1985). At very high temperatures, the silicon of the ash may vaporise as well, forming stubborn deposits (Doležal 1990).

5.10.3 Erosion

In pulverised coal furnaces, the largest fraction of the ash, as fly ash, is carried out of the firing with the flue gas flow and through the tube banks of the convective heat transfer surfaces. Ash particles impinging on and rubbing along the tubes lead to material wear through erosion. Factors having a substantial effect on the degree of erosion are the ash content of the coal, the flow velocity and the properties of the fly dust. Fly dust is more abrasive than coal because the soft organic components of the coal are absent. Quartz fractions in the ash have a very abrasive effect. The shape of the ash particles, too, has an influence on erosion. If fly dust particles are smoothed by melting at high temperatures, their erosive effect diminishes.

Erosion occurs in areas of flow reversion. This is the reason why single-pass boilers are used for Rhenish brown coals, the fly ash of which contains high quartz fractions. Bridging of ash deposits in the area of the convective heat transfer surfaces may make the flow velocity rise locally, causing erosion (Couch 1994).

5.10.4 High-Temperature Corrosion

High-temperature corrosion is the term for chemical and physical–chemical wastage of the flue-gas-exposed tubes in solid fuel-fired steam generators. Tube wastage leads to tube ruptures and tube blower damage. According to an EPRI study, 50% of all breakdowns in fossil fuel-fired US power plants and a large majority of all tube damage can be put down to corrosion (Neumann and Kautz 1995). Fireside corrosion of steam generators accounts for a large part of the maintenance costs in waste incineration plants. The reasons for corrosion and potential to reduce it are known from comprehensive experience. Even so, it is not possible to control damage caused in solid fuels combustion to any degree of certainty (Born 2005).

High-temperature corrosion often occurs in the context of slagging and fouling, caused by the gaseous atmosphere or by interactions of slag and ash deposits. Conducive to corrosion are the contents of alkalis, chlorine and sulphur in the coal. Corrosion can occur both in the furnace section and in the section of the convective heat transfer surfaces. Corrosion occurring in low temperature regions shall not be a subject of the present discussion.

The following mechanisms of high-temperature corrosion are distinguishable and shall be discussed in more detail below:

- Gaseous corrosion by hydrogen chloride
- Chlorine-induced corrosion
- Molten salt corrosion

The chlorine to sulphur ratio in the fuel is the decisive parameter determining the dominating corrosion mechanism, and hence the corrosion rate. For low sulphur to chlorine ratios in the fuel, chlorine is mainly present as alkali chlorides, which may condense on heat exchangers to form deposits. The deposits can react with sulphur dioxide to release chlorine in the vicinity of the tube, inducing severe corrosion (i.e. chlorine-induced corrosion). In the case of high sulphur to chlorine ratios, alkali chloride is sulphated in the flue gas to form HCl and alkali sulphates. The gaseous corrosive attack by HCl is of much lower intensity than the attack by chlorine-induced corrosion.

5.10.4.1 Furnace Corrosion Through Hydrogen Chloride

Under oxidising conditions, metal tubes develop a protective oxide film of Fe_3O_4 (magnetite) and Fe_2O_3 (haematite). Under reducing conditions, the oxidic protective layer and the tube material are attacked by hydrogen chloride (HCl) and $FeCl_2$

forms. The corrosion rate in the furnace may even intensify if, through the reaction of SO_2 or SO_3 with deposited chlorine compounds, chlorine is released and also attacks the tube wall (chlorine-induced corrosion). In contrast, HCl reacts only slightly, if at all, with the protective layer on the tube wall in the presence of oxygen. Therefore, a sufficient oxygen-rich atmosphere can help to prevent corrosion (Neumann and Kautz 1995; Schirmer and Thömen 1984). In unfavourable conditions, wear rates of up to 600 nm/h have been observed, while the normal oxidation rates forming the protective film amount to 8 nm/h (Skorupska 1993).

In spite of a sufficiently high oxygen content in the flue gas, ferrous chloride formation may occur on near-white metal tubes at the initial start-up of a boiler. For this reason, metal wastage is higher during start-up than during standard operation. Besides the protection provided by an oxygen-rich atmosphere, chlorine-induced corrosion of the tube material still occurs, though only transitorily. This forming of ferrous chloride can be explained by the strong oxygen consumption of the protective oxide film.

When white metal tubes are repeatedly subject to strong soot blowing or erosion, the process is called erosion – corrosion (Neumann and Kautz 1995).

Evaporator wall corrosion is often correlated with unfavourable combustion conditions and deposits on the walls. Optimised combustion control parameters are a means to avoid reducing zones near the wall. A homogeneous distribution of the pulverised coal and air to the burners diminishes the forming of fuel-rich streaks, and a finer grinding accelerates the combustion and limits the extension of low-oxygen zones. Reducing zones may also form at the tube walls due to carbon-containing deposits which by reaction consume the oxygen, raising the tube wall temperature in the process. When air staging is applied in the furnace to reduce NO_x, care has to be taken (by ensuring an adequate air injection or by applying an air curtain) such that no reducing flue gas atmosphere exists in near-wall zones.

Corrosion in the furnace is largely dependent on the chlorine content of the fuel. Whereas in brown coal-fired furnaces, due to the low chlorine content, corrosion problems are extremely rare, problems have to be expected in furnaces fired with hard coals with a chlorine content above 0.15% (Kautz and Zelkowski 1985). Figure 5.85 shows the expected relative wear rates in the furnace as a function of the chlorine content for hard coals (Simon et al. 1997). Corrosion of the evaporator walls occurs to a higher degree in waste incineration plants (Neumann and Kautz 1995).

5.10.4.2 Corrosion of the Convective Heat Transfer Surfaces by Molten Salts

While under conventional steam conditions up to 540°C, high-temperature corrosion is negligible, this type of corrosion becomes appreciable at higher steam temperatures. The corrosion rate then mainly depends on the tube wall temperature and the gas temperature. Figure 5.86 shows the dependence of the corrosion rate on the tube wall temperature. Corrosion begins at temperatures somewhat above 600°C, reaching a maximum at about 700°C (Stultz and Kitto 1992).

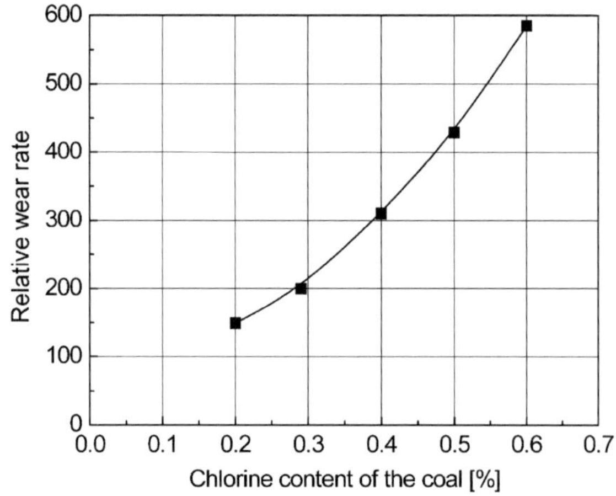

Fig. 5.85 Effect of the chlorine content on the corrosion rate in the furnace for hard coals (Simon et al. 1997)

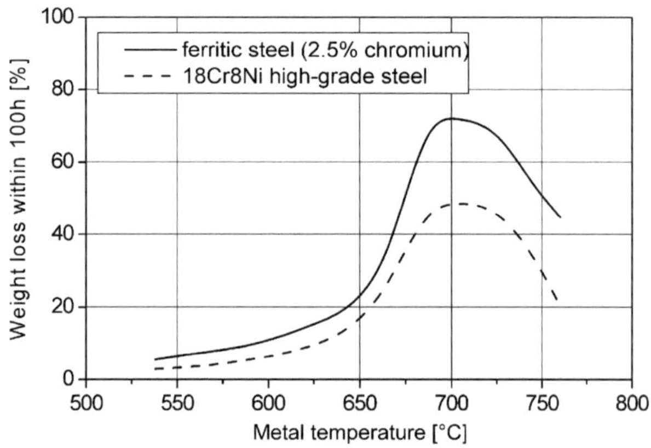

Fig. 5.86 Dependence of the corrosion rate on the tube wall temperature (Stultz and Kitto 1992)

High-temperature corrosion of the convection heating surfaces is due to complex alkali iron(III) sulphates and alkali aluminium(III) sulphates, which form in the deposits of the fly ash. These alkali sulphates and SO_2, which reacts in the deposits to become SO_3, diffuse through the fly ash and react with the iron or aluminium oxides of the fly ash:

$$3K_2SO_4 + Fe_2O_3 + 3\ SO_3 \rightarrow 2K_3Fe(SO_4)_3 \tag{5.29}$$

or

$$K_2SO_4 + Al_2O_3 + 3SO_3 \rightarrow 2\ Al(SO_4)_2 \tag{5.30}$$

Alkali iron/aluminium sulphates are molten in the temperature range of 550–700°C or so. In this state, they even attack high-alloy steels by corrosion. All types of hard coals contain enough alkalis and sulphur for the development of high-temperature corrosion (Stultz and Kitto 1992). Contents of alkaline earths, however, have a corrosion-inhibiting effect (Skorupska 1993). In the case of biomass and waste combustion, eutectic mixtures can melt at lower temperatures. Some chloride mixtures can show melting temperatures below 400°C (Born 2005).

A suitable arrangement of the heat transfer surfaces can help to limit high-temperature corrosion. For example, under unfavourable conditions around plate superheaters and plate reheaters, corrosion rates between 1 and 6 mm/year could be observed, which could be reduced to 0.1–0.5 mm/year at tube wall temperatures of 590–635°C (Stultz and Kitto 1992). Heating surfaces which are exposed to flame radiation, having the comparatively highest tube wall temperatures, should not be used. By transferring final-stage superheaters and final-stage reheaters to colder flue gas areas, or by flue gas recirculation, the tube wall temperatures of these superheaters can be decreased. If higher steam temperatures are applied, however, other measures have to be taken such as using corrosion-resistant materials or ceramic coatings or by mixing additives with the fuel (see Sect. 4.5.3).

5.10.4.3 Corrosion of the Convective Heat Transfer Surfaces by Chlorine-Induced High-Temperature Corrosion

Corrosion of the convective heating surfaces by chlorine is possible, too. If the alkalis are insufficiently sulphated, alkali chlorides form, which may condense in the regions of the superheaters or reheaters. They then react with the sulphur dioxide of the flue gas, releasing chlorine:

$$2\,NaCl + SO_2 + O_2 \rightarrow Na_2SO_4 + Cl_2 \tag{5.31}$$

$$2\,KCl + SO_2 + O_2 \rightarrow K_2SO_4 + Cl_2 \tag{5.32}$$

Close to the tube, chlorine attacks the steel via iron-chloride formation. Figure 5.87 shows the composition of the layers on the tube and possible mechanisms of chlorine-induced high-temperature corrosion (Schumacher 1996).

This form of corrosion occurs when the alkalis are not sulphated and leave the furnace as NaCl or KCl. If the alkalis are sulphated, Na_2SO_4 and K_2SO_4 are formed; chlorine then travels through the furnace as HCl. HCl in the flue gas causes only slight corrosion of the convection heating surfaces.

For fuels with low sulphur contents or a high sulphur retention in the fuel, alkalis are not sulphated and the above-described corrosion problems arise. Corrosion can also occur in the case of very low chlorine contents in the fuel. Chlorine corrosion was observed, for example, in the combustion of brown coals in fluidised bed furnaces.

A high degree of desulphurisation in the fluidised bed impedes the sulphation of chlorides, which then condense on the wall heating surfaces or heating surface banks, causing corrosion (Meyer et al. 1995). Straw, which contains a high content of chlorine and potassium, when co-combusted with coal in a circulating fluidised bed furnace, forms KCl, and hence imparts considerably higher corrosion

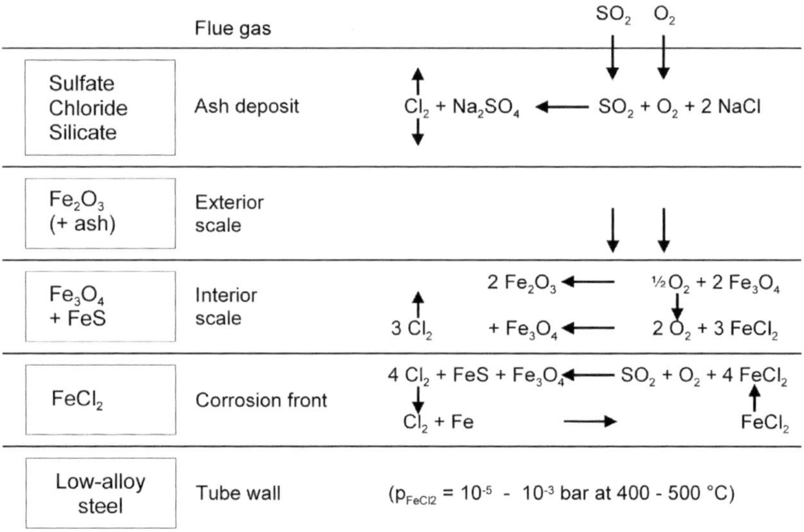

Fig. 5.87 Composition of layers on tubes and mechanisms of chlorine-induced high-temperature corrosion (Schumacher 1996)

rates than when co-combusted in a pulverised coal-fired furnace. Under the conditions of pulverised fuel firing, the formation of potassium sulphates and HCl is favoured, resulting in a lower corrosion rate (Henriksen et al. 1995). In addition, in the combustion of wood alone, alkali chlorides may cause corrosion in sulphur-poor conditions.

5.11 Residual Matter

5.11.1 Forming and Quantities

In the process of solid fuel combustion in coal-fired power plants, the production of mineral residues in the form of ash is inevitable. The type and the properties of the ashes depend on the fuel, the firing system and the location where the ash is removed.

When equipped with flue gas desulphurisation (FGD) systems, the coal firing process will involve additional residual matter (gypsum in most cases). The properties of the FGD residues are largely determined by the desulphurisation method and the fuel.

Figure 5.88 shows the quantities of the coal combustion residues fly ash, bottom ash, boiler slag and fluidised bed combustion residue, as well as the products from dry or wet flue gas desulphurisation, especially spray dry absorption products and flue gas desulphurisation (FGD) gypsum in the EU-15. The load of residues reflects the structure of the firing and the flue gas cleaning systems and their technological

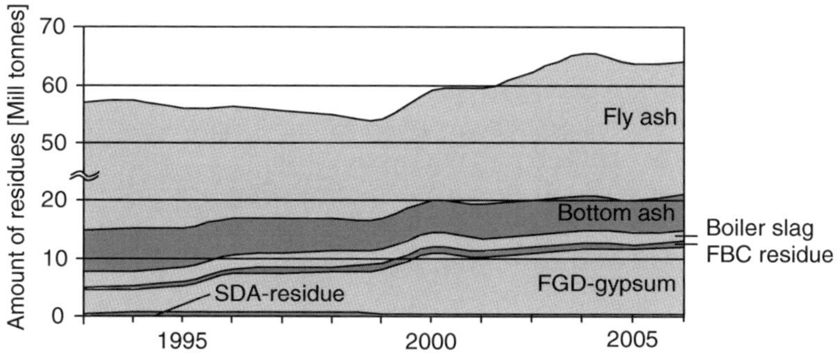

Fig. 5.88 Load of combustion and flue gas cleaning residues in the EU-15 from 1993 to 2005, data from (Ecoba 2006)

changes. Boiler slag or slag granulate from slag-tap firing systems is almost exclusively produced in Germany. The load of FGD gypsum can be put down to retrofits of desulphurisation installations since about 1980.

The total amount of coal combustion residues produced in European (EU 15) power plants was 64 million tonnes in 2005. Sixty-seven percent of the total residues were produced as fly ash. All combustion residues amounted to 81% and the FGD residues up to 19% by mass. The amount of residues in the EU 27, including the 12 new member states, is currently estimated at about 100 million tonnes per year (vom Berg and Feuerborn 2007). Worldwide coal combustion residues amount to 720 Mt, with 400 Mt being produced in India and China.

Residual matter can be categorised into ashes and slag granulates from pulverised hard coal combustion, ashes from brown coal combustion, ashes from fluidised bed furnaces and residues from flue gas desulphurisation units.

5.11.1.1 Ashes from Pulverised Hard Coal Combustion

Fly ash is the term for the mostly fine glass-like spherical residue which is separated by filters during flue gas dust collection after the combustion of hard coal. In dry-bottom firing systems, about 80–90% and in slag-tap firing systems, about 50% of the ash is removed as fly ash. Hard-coal fly ash consists of the oxides of silicon, aluminium, iron, alkaline earths and alkalis, as well as trace elements and unburned carbon. The composition of hard-coal fly ashes correspond to clays found in nature. Trace elements such as lead, nickel and arsenic are bound in the siliceous matrix, and so are neutral and hence not able to be mobilised. The chemical composition depends substantially on the origin of the burned hard coal. The extent to which the fly ash has a glass-like or amorphous structure indicates the amount of SiO_2 and Al_2O_3 present in the reactive state. The fly ash has pozzolanic properties, which means it can react with lime at room temperature and, similar to cement, form calcium silicate hydrates (Harder 1996).

Boiler slag forms from liquid slag, which, as a combustion residue, flows out of the slag-tap furnace in a molten state at combustion temperatures of 1,400–1,700°C.

The liquid slag flows into a water bath where it solidifies and, as a consequence of its residual stress, breaks into glass-like granules with a diameter of up to 10 mm. Removed fly ash can be returned into the slag-tap furnace and melted down.

Bottom ashes form in dry-bottom furnaces at combustion temperatures from 1,100 to 1,300°C. They clinker, forming larger ash lumps and are usually removed via a water basin. Boiler slag, bottom ash and fly ash from dry-bottom and slag-tap furnaces are alike in terms of chemical composition. Table 5.13 shows the range of variation in ash composition (Peters and vom Berg 1992).

5.11.1.2 Ashes from Pulverised Brown Coal Combustion

Rhenish brown coal has a mean ash content between 4 and 7%, while the lowest values found are below 2% and the maximum values may reach 20%. The cause of the high-ash contents is sandy or clayish matter. Brown coal is used predominantly in dry-bottom furnaces; its combustion produces 80% fly ash and 20% bottom ash on average. The ashes of brown coal, analogously to the variety of coal feedstocks, show very different element concentrations. In general, they have higher SO_3 and CaO contents than hard coal ashes (see Table 5.13).

5.11.1.3 Ashes from Fluidised Bed Combustion

In fluidised bed combustion plants, the residual matter is classified according to the location of its discharge from the process, i.e. as bed ash, cyclone ash or fly ash. If the residues are recombined in one storage bin, the resulting mixture is called mixed ash. The fluidised bed residues are a mixture of the coal ash, the products from the desulphurisation process and the unreacted absorbent – the composition is largely determined by the desulphurisation process. The furnace temperature between 800 and 900°C, which is low in comparison to pulverised fuel combustion systems, does not suffice to melt the mineral ash particles. The morphology of the clayish minerals is preserved (vom Berg 1993).

Table 5.13 Chemical composition of ashes [% by wt.] (Peters and vom Berg 1992)

	Pulverised fuel combustion		Fluidised bed combustion	
Oxide	Hard coal	Brown coal	Hard coal	Brown coal
SiO_2	40–55	20–70	7–55	3–17
Al_2O_3	23–35	1–15	2–26	1–6
Fe_2O_3	4–17	1.5–70	2–12	8–24
CaO	1–8	8–40	4–50	30–50
MgO	0.8–4.8	0.5–7	1–2.1	7–13
K_2O	1.5–5.5	0.1–1.5	0.4–4.3	0.1–0.7
Na_2O	0.1–3.5	0.1–2	0.1–0.9	0.1–0.9
SO_3	0.1–2	1.5–15	1–28	10–25
TiO_2	0.5–1.3	0.1–1	0.5–1.5	0.1–0.2

5.11.1.4 Residual Matter from Flue Gas Desulphurisation

Flue gas is usually desulphurised in a wet process using limestone ($CaCO_3$) or quicklime (CaO). The product of the desulphurisation process is initially a mixture of calcium sulphite and calcium sulphate, and, after oxidation with atmospheric oxygen, hydrous calcium sulphate ($CaSO_4 \times 2\ H_2O$), which is often called FGD gypsum. After dewatering the product is, as a rule, in the form of a powder with a content of free moisture of about 10%. The water bound by hydration to the calcium sulphate is about 20% by weight. FGD gypsum has a multitude of applications in the building materials industry. Its quality is equal to natural gypsum, and there are no restrictions on health grounds as to its use as building material (see Table 5.14).

In lime-spray drying for flue gas desulphurisation, the flue gas, after a dust collection step or in a dust-laden state, is injected with the lime slurry into a reaction vessel. The water evaporates and a dry desulphurisation product forms, which is removed from the flue gas in a downstream filter. Without prior dust collection, a mixture of additive (i.e. lime slurry), desulphurisation product and fly ash is produced; in the usual process with prior fly ash collection, the mixture consists only of the desulphurisation product and unreacted additive. The residual matter contains sulphite ($CaSO_3$), sulphate ($CaSO_4$), carbonate ($CaCO_3$), hydroxide ($Ca(OH)_2$), chloride (CaCl) and fluoride (CaF_2) (see Table 5.15).

In the so-called (in-furnace) dry sorbent desulphurisation process, absorbents such as ground limestone or hydrated lime are injected into the furnace to capture sulphur dioxide. This way, desulphurisation degrees of more than 50% can be achieved, which, however, are not sufficient for central power stations. The chemical composition of such an ash type is similar to the composition of fluidised bed ash.

Table 5.14 Chemical parameters of FGD and natural gypsum [% by wt.] (Peters and vom Berg 1992)

Parameter	Natural gypsum[a]	FGD gypsum[b]
pH	7.4	7.2
Water of hydration	16.5	20.3
CaO	35.8	31.7
SO_3	37.8	45.0
$CaSO_4 \cdot 2\ H_2O$	80	97
MgO, total	0.06	0.03
Na_2, water soluble	0.034	0.32
K_2O, water soluble	0.006	0.007
Fe_2O_3, total	0.19	0.12
HCl, insoluble	0.20	0.35
NH_4	0.003	0.003
SO_2	0.02	0.03
P_2O_5	0.003	0.0003
F	0.001	0.002

[a]Mean values of 12 samples, [b]Mean values of 15 samples.

Table 5.15 Composition of lime-spray drying products [% by wt.] (Peters and vom Berg 1992)

Component	Without prior dust collection	With dust collection
$CaSO_3$	9–47	17–685
$CaSO_4$	1.7–17	3.5–29
$CaCO_3$	4.5–13.7	5–13
$Ca(OH)_2$	1–15	0.5–15
CaCl	0.8–6.3	0.8–9.5
CaF_2	< 0.4	< 0.4
Fly ash	20–85	< 8

5.11.2 Commercial Exploitation

The following presents the potential for commercial exploitation of the residual material described in Sect. 5.11.1.

5.11.2.1 Ash from Combustion of Pulverised Hard Coal

Concrete and Concrete Products

The majority of the fly ash from hard coal-fired power stations in Europe is supplied to the concrete industry. As an example, about 75% of the fly ash from hard coal firing in Germany is used for concrete production. Fly ashes can be used in concrete or products of concrete as a direct additive or as an aggregate produced from fly ash, provided they meet specified quality standards. The utilisation of fly ash as a concrete additive is regulated by the European Standard EN 450 "Fly Ash for Concrete". The standard refers to siliceous fly ash from hard coal combustion only. Calcareous fly ash, commonly obtained from the combustion of lignite, cannot be utilised as a concrete additive according to EN 450 (vom Berg and Feuerborn 2005).

The first requirement of fly ashes used as concrete additives is that they be harmless, homogeneous and effective as desired. The guideline outlines limits for the loss of unburnt material (< 5% for category A), SO_3 (< 3%), Cl^- (< 0.1%), the content of free lime (CaO) (<2.5%), the content of alkalis (<5%), MgO (<4%), P_2O_5 (<100 mg/kg), the grain fraction (<0.045 mm) (>60%) and for the mechanical behaviour (solidification, volume constancy, compressive strength of the mortar and cement, carbonation behaviour, frost resistance). The revised EN 450 allows the co-combustion of secondary fuels such as wood and sewage sludge up to a certain mass fraction (Wiens 2005) (see also Sect. 6.5.3).

Processed granulate material and coarse ashes also have a use as a sand additive in concrete and in concrete products.

Cement

In cement production, fly ash can be added as a component of the basic material or to the milling process. The requirements are specified in the European Standard

EN 197: "Fly ash for cement". As a general rule the requirements for the fly ash quality are lower than for the use in concrete production. The guideline distinguishes between fly ash from lignite and hard coal power plants. For siliceous fly ash from hard coal, depending on the cement class, the allowable fly ash fraction is between 6 and 35%. The guideline outlines similar limits for the loss of unburnt material and the content of free lime as in EN 450 for concrete; however, limits on chlorine and sulphate, for example, are less stringent, as they are based on the concentrations in the cement and not the ash. Utilisation of the fly ash in cement production is generally less attractive, because revenues are lower than for utilisation in concrete production.

Road Construction and Earth and Landscaping Work

In the construction of roads and in earth and landscaping works, it is possible to use fly ashes that have a content of unburnt material of up to 15%. The most important field of application, in terms of quantity of use, for fly ash in road construction is the construction of base courses (i.e. the underlying layers of the road construction). Owing to their cement-like behaviour, hard-coal fly ashes help to cut down on the necessary quantity of binder. Fly ash can also be used as a filler in the construction of the road pavement (i.e. the surface layer of the road).

In earth and landscaping works, too, fly ash can be used as an additive to improve the granulometric composition of natural soil. In general, fly ashes meet all requirements with respect to soil statics and mechanics for stability. Considerable quantities of fly ash can be made use of, especially when there is an imbalance of the cut and fill in the construction of new roads. For embankments (i.e. earthworks), the use of fly ash may remove the need for any other hydraulic binder. When appropriate construction methods are used, fly ash has the same high quality of natural soil materials. The increasing frequency of the construction of vision and noise barriers offers a use for very large quantities of fly ash. The same is true for the backfilling of buildings, bridges and retaining walls.

Granulate matter from ash removal devices can be utilised with or without binder in all courses of the road construction. Granulates are impact-crushed before use if necessary. Granulates and coarse ashes are relatively equal in physical and chemical properties and can be used without restriction in earth and landscaping works. They also meet all requirements with respect to soil mechanics and are chemically inert.

Mining, Pneumatic Stowing and Backfilling of Cavities

In the production of mining mortar for applications in underground mining, fly ash is used as a filler or binder. If used as a filler, the mortar is produced by mixing fly ash with a binder and adding a coarse granulate. Alternatively, when acting as the binder, another material is used as the filler. Mining mortar can be used for the construction of gate side packs and for the sealing of rock. Fly ash or mixes of fly ash and FGD gypsum are used for backfilling cavities in underground mining.

Clay Bricks, Sandy Limestone Bricks and Aerated Concrete

Fly ash can be used in the manufacturing of clay bricks, sandy limestone bricks and aerated concrete. In brick fabrication, fly ash can be used to make clays leaner. Granulate material and coarse ashes are suitable for the production of sandy limestone bricks.

Mortar, Screed and Plaster

For mortar, screed and plaster production, fly ash can be used as an additive. The requirements for fly ash depend on the type of binder used. Granulates are suitable for use as additives without the need for any treatment. Coarse ash can, as a rule, be used for masonry mortar as well.

Other Purposes

Because of its low permeability to water, fly ash can be used with other sealing products in the construction of landfill sites. Boiler slag and bottom ash, because of their relatively high permeability and resistance to water, are suitable for use as filter layers and in bedding for drainage systems. Boiler slag material is used in the production of blasting media. It is also used as grit for road traction in winter.

5.11.2.2 Ash from Combustion of Pulverised Brown Coal

For brown coal ash, there is currently no potential for reuse in the conventional construction and building materials sector(s) or in underground mining. Due to the varying composition of German brown coal ash, no test marks exist as yet in Germany. However, brown coal ash, in the same way as hard coal ash, can serve as an additive to natural soils to improve the soil structure. Because of its mineral components, this ash can be used in agriculture and forestry, horticulture, earth and landscaping works as a plant nutrient and as a soil conditioning agent. The majority of brown coal ash is used for recultivation of landscapes in opencast brown coal mining.

5.11.2.3 Ash from Fluidised Bed Combustion

The fluidised bed combustion plants currently in service are operated mainly by communal utility companies. The plants are at a distance from each other and the local yield of residual matter is comparatively low. An additional fact is that the residues from different fluidised bed combustion plants, due to differing process technologies and fuel qualities, may have very dissimilar properties. These boundary conditions, unfavourable for any application, as well as the complex composition of the fluidised bed residues, make it more difficult to convert them to usable materials. Despite this, it is reported that much fluidised bed ash is reused. The predominant part is utilised in underground hard coal mining for pneumatic and hydraulic stowing and as a component for mining mortar. Repurchase by the coal supplier of the ash is usually a component of the contract between the supplier and the utility company.

5.11.2.4 Residual Matter from Flue Gas Desulphurisation

FGD gypsum from coal-fired power plants can be used as a fully adequate substitute for natural gypsum. Comparative investigations into natural gypsum and FGD gypsum have verified that FGD gypsum can be used for the fabrication of building materials without increased health risks. It is used in the gypsum and cement industries for the manufacturing of wallboards, gypsum blocks, projection plasters and floor screeds and for the production of cement.

Due to the varying properties of brown coal, FGD gypsum from brown coal-fired power plants in the past did not meet the requirements of the gypsum and cement industries for properties and purity. This gypsum was used instead to stabilise the backfilling material in opencast brown coal mining. The high demand for FGD gypsum has led to further development of the technology for multistage preparation of the gypsum, so that, today, gypsum from brown coal-fired power plants can also be used in the production of gypsum products and in cement production.

Today, all FGD gypsum from hard coal-fired and part of the FGD gypsum from brown coal-fired power plants is used as a raw material in the gypsum industry. The total amount used in 2004 was more than 7 million tonnes. As a comparison, the production of natural gypsum in Europe is about 25 million tonnes.

The reuse of desulphurisation products from lime-spray drying is more difficult and more complex technologically than those from wet flue gas desulphurisation processes. There is no potential for these products to replace a natural or industrial material. They have no outstanding technical properties to make them particularly useful in a technical application. So the development of a new product or an upgrading process is necessary in order to reuse these materials. Another disadvantage is the small quantity and varying quality produced by a plant. Additionally, the residual material contains corrosive chlorine. Despite these problems, in Germany 76% of the residual matter is utilised. Possible applications are as a raw material for mining mortar, as an additive for sandy limestone brick and aerated concrete and as a raw material for other processes, such as sulphuric acid production or anhydrite production. Another option is addition in a wet FGD process for the further oxidation of sulphite (Kolar 1995).

5.11.2.5 Heavy Metals and Leaching Behaviour of Residual Matter

In certain applications, in soils, for instance, it is necessary to know the concentrations of various elements in order to evaluate the general environmental impact of the reuse of residual matter. These elements are divided into potentially harmful substances and beneficial elements, with the boundaries being fluid depending on the concentrations.

Table 5.16 Heavy metal concentrations of power plant residues in comparison with maxima of the German Sewage Sludge Ordinance [mg/kg] (Peters and vom Berg 1992)

Heavy metal	Fly ash		Fluidised bed ash		Boiler ash[b]	FGD	Natural	Sewage	Standard value
	Hard coal[a]	Brown coal	Hard coal	Brown coal	Hard coal	gypsum	gypsum	sludge ord.	for soil
Lead	200	20	150	10	30	5	4	900	100
Cadmium	2	0.6	1.5	0.5	0.4	0.05	0.2	10/5	1.5/1
Chromium	120	40	130	20	25	3	7	900	100
Copper	160	15	130	15	30	3	5	800	60
Nickel	170	35	120	15	60	2	4	200	50
Mercury	0.1	0.5	0.5	0.7	< 0.1	0.5	0.03	8	1
Zinc	300	45	300	100	60	15	13	2,500/2,000	200/150

[a]Dry-bottom furnace; [b]Granulate matter and coarse ash.

Table 5.17 Eluate values of power plant products compared to the ordinance on drinking water and water for food processing companies [mg/l] (DIN 38414, EULAT 1:10) (Peters and vom Berg 1992)

	Fly ash[b]	Boiler ash[c]	Fluidised bed ash		Ord. drinking water
Parameter	Hard coal	Hard coal	Hard coal	Brown coal	
pH	8–12.5	6.5–11.5	10–12	10–12.5	6.5–9.5
El. Conductivity[a]	< 200	< 10	< 250	< 1,500	200
Arsenic	< 0.01	< 0.01	< 0.01	< 0.001	0.01
Lead	< 0.03	< 0.01	< 0.001	< 0.005	0.04
Cadmium	< 0.005	< 0.005	< 0.002	< 0.0002	0.005
Chromium	< 0.2	< 0.04	< 0.1	< 0.04	0.05
Copper	< 0.01	< 0.01	< 0.001	< 0.005	3[d]
Nickel	< 0.05	< 0.05	< 0.001	< 0.005	0.05
Mercury	< 0.001	< 0.0003	< 0.001	< 0.0002	0.001
Zinc	< 0.1	< 0.05	< 0.001	< 0.005	5[d]
Chloride	< 12	< 20	< 50	< 1,000	250
Sulphate	< 1,000	< 150	< 1,600	< 2,000	240

[a] El. conductivity in mS/m; [b] From dry-bottom furnace;
[c] Granulate or coarse ash; [d] Standard value of the *Ordinance on Drinking Water*.

The assessment of the environmental impact in Germany usually uses the maximum values laid down in the Sewage Sludge Ordinance. Table 5.16 presents the heavy metal concentrations of different kinds of power plant residues compared to the allowed maxima of the Sewage Sludge Ordinance. The application of sewage sludge is allowed only if the concentrations of the sludge and the soil it is added to fall below the values listed in the table. The concentrations of the different kinds of residual matter are significantly below the values allowed for sewage sludge application and partly below the maxima for soil. FGD gypsum and natural gypsum have comparable heavy metal contents.

A crucial feature besides the absolute concentration is the leaching behaviour of heavy metals. The comparison of eluate data of the power plant residues to the values of the Ordinance on Drinking Water in Table 5.17 shows that – with the exception of the values of pH, chromium and sulphate – the maximum values are seldom reached or exceeded (Peters and vom Berg 1992).

5.11.2.6 State of the Art of Reuse of Residual Materials

The rates of the utilisation and disposal of the different residues from coal combustion are shown in Fig. 5.89 for the EU 15 for the year 2005. Most of the coal combustion residues are used in the construction industry, in civil engineering and as construction materials in underground mining (52.8%) or for restoration of open cast mines, quarries and pits (36.3%). In 2004 about 8.7% were temporarily stockpiled for future utilisation and only 2.3% were disposed of. Utilisation is less practiced in the new EU member states. The legal definition of combustion residues, which are defined in the European Waste Directive as wastes, is still under discussion.

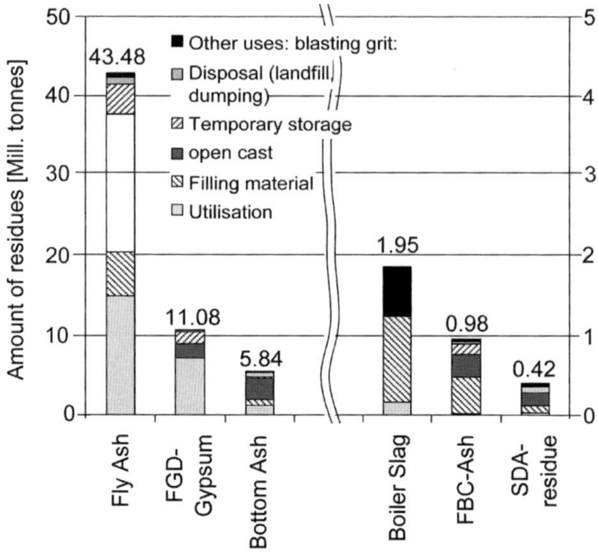

Fig. 5.89 Rates of residual matter utilisation and disposal in the EU 15 in 2005 (Ecoba 2006)

Whereas FGD residues are considered by-products, because they have undergone a further processing, the status of fly ash is still open (vom Berg and Feuerborn 2007).

Table 5.18 presents the state of commercial exploitation of the different kinds of residual matter from brown coal- and hard coal-fired power plants in Germany in

Table 5.18 Production and utilisation of by-products from coal-fired power plants in Germany in 2006 (VGB 2008)

	Hard coal		Brown coal		
Capacity in MW_{th}	68,605		64,311		
Burned coal in Mt	50		164		
By-product	Production in Mt	Utilisation (%)	Production in Mt	Utilisation	
				Opencast mining (%)	Other (%)
Boiler slag granulate	1.77	100	–	–	–
Bottom ash	0.62	97	1.94	86	14
Fly ash	4.40	98	8.90	96	4
Fluidised bed ash	0.35	100	0.36	90	10
FGD gypsum	1.92	100	5.57	10	90[a]
Lime-spray drying products	0.32	100	–	–	–
Total	9.53	99	16.52	59	41

[a]Of which 8% was temporarily stored.

2004 (VGB 2008). The main fields of application for hard coal fly ash are concrete and cement production, mining and road construction. Boiler slag and bottom ash are on the whole utilised in earth and road works, in products of sandy limestone, in concrete and concrete. Brown coal ash is almost entirely used for backfilling mined-out opencast mines and for recultivation of mine landscapes.

Although utilisation of fly ash has been practised in both North America and Europe since the 1950s, the rates of utilisation are still at a modest level in many countries. Whereas the utilisation of fly ash in Europe (EU 15) is nearly 100%, the rate of utilisation in the USA is only 35%. In China the rate of utilisation remained low in the 1980s but then grew rapidly during the 1990s. It reached 66% of an ash production rate of 150 Mt in 2002. The production of fly ash in China is predicted to increase to 350 Mt in 2010 and 600 Mt in 2020 (Smith 2005).

References

BImSchV (2004). 13. Verordnung zur Durchführung des Bundesimmissionsschutzgesetzes (Verordnung über Großfeuerungsanlagen, 13. BImSchV) vom 23.7.2004, BGBl. I Nr. 37, P.1717, from http://www.gesetze-im-internet.de/bimschv_13_2004/index.html.

BImSchV (2003). 17. Verordnung zur Durchführung des Bundesimmissionsschutzgesetzes (Verordnung über die Verbrennung und die Mitverbrennung von Abfällen. Bekanntmachung vom 14. August 2003 (BGBl. I P. 1633).

Adrian, F., Quittek, C. and Wittchow, E. (1986). Fossil beheizte Dampfkraftwerke. Handbuchreihe Energie, Band 6, Herausgeber T. Bohn. Technischer Verlag Resch, Verlag TÜV Rheinland

Albers, U., Holzapfel, T. and Wollner, G. (1987). Forschungsbericht Umweltschutztechnik Schadstoffanalytik. Bestimmung der Spurenelement-Verbindungen in Reingasstäuben aus steinkohle- und müllbefeuerten Kraftwerken. BMFT, VQ 223–1.

Albrecht, W. and Pollmann, S. (1980). Neuere Beiträge zum Verschlackungsmechanismus in trockenentaschten Steinkohlefeuerungen. *VGB Kraftwerkstechnik* **60**(2): 90–97.

Amdur, M. O. (1986). Coal combustion aerosols and SO_2 – an interdisciplinary analysis. *Environmental Science & Technology* **20**(2): 138–145.

Anthony, D. B. and Howard, J. B. (1976). Coal devolatilization and hydrogasification. *AIChE-Journal* **22**(4): 625–656.

Araoka, M., Iwanaga, A. and Sakai, M. (1987). Application of Mitsubishi "Advanced MACT" in furnace removal process. 1987 Joint Symposium on Stationary Combustion NOx Control. New Orleans.

Bals, M. (1997). Untersuchungen zum Verschmutzungsverhalten zukünftig anstehender Kohlen des Rheinischen Reviers. VGB-Fachtagung Feuerungen. Essen.

Baumbach, G. (1990). Luftreinhaltung. Berlin/Heidelberg, Springer.

Becker, J. (1986). Möglichkeiten der Stickoxidminderung durch SCR-Anlagen. *BWK* (Sonderdruck Fachreport Rauchgasreinigung): 19–26.

Becker, J. (1987). Aktiva und Passiva. Betriebserfahrungen zur NOx-Minderung der NOx-Emission. *Energie Spektrum* (6).

Beer, J. M. (1988). Stationary combustion – the environmental Leitmotiv. Proceedings of the 22nd Symposium (International) on Combustion, pp. 1–16. Pittsburgh, The Combustion Institute.

Benesh, W. A., Howe, G., Lümkemann (2001). 12 Jahre Betriebserfahrung mit der Brennstoffstufung im Kraftwerk Herne. VDI-Berichte 1629, pp. 25–30. Düsseldorf, VDI.

Bertacchi, S., Fioresi, S. and Pasini, S. (1997). Vado Ligure coal-over-coal reburn project. Proceedings of the 4th International Conference on Technologies and Combustion for a Clean Environment. Lisbon.

Bertram, J. (1986). Übersicht über Erfahrungen und Stand feuerungstechnischer NOx-Minderungsmaßnahmen bei Steinkohlenstaubfeuerungen mit flüssigem Ascheabzug. *VGB Kraftwerkstechnik* **66**(12): 1150–1159.
Biber, K. H. (1986). Erfolgreiche Primärmaßnahmen zur NOx-Reduzierung an einer Steinkohlen-Schmelzfeuerung. *VGB Kraftwerkstechnik* **66**(9): 835–843.
Blair, D. W., Wendt, J. O. L. and Bartok, W. (1978). Evolution of nitrogen and other species during controlled pyrolysis of coal. Proceedings of the 17th Symposium (International) on Combustion. Pittsburgh, The Combustion Institute.
Bonn, B. and Baumann, H. (1991). Kenntnisstand der N2O-Bildung in verschiedenen Feuerungsanlagen, VDI-Berichte 922, pp. 625–634.
Bonn, B., Baumann, H. and Pelz, G. (1993). Bildung und Abbau von N2O in Wirbelschichtfeuerungen. *VGB Forschung in der Kraftwerkstechnik*. Essen, VGB-TB 231.
Booth, R. C. and Borio, R. W. e. a. (1991). Reburn technology for NOx control on a cyclone-fired boiler. 1991 Joint Symposium on Stationary Combustion NOx Control. Washington D.C.
Born, M. (2005). Cause and risk evaluation for high-temperature chlorine corrosion. *VGB PowerTech* **4**(5): 107–111.
Brüggemann, H. (2008). Erfahrungen mit modernen Steinkohlefeuerungen. *VGB PowerTech* **88**(11): 36–42.
Brüggemann, H., Scheffknecht, G., Marx, F. J. and Koch, G. (1997). Erste Betriebserfahrungen nach dem Umbau des Kraftwerkes Bexbach I. *VGB-Fachtagung Feuerungen*. Essen.
Brüggemann, H., Stamatelopoulos, G., Krohmer, B. and Kessel, W. (2003). Die Entwicklung eines Rundbrennerkonzeptes – Simulation, Auslegung, Betriebserfahrung. VDI-Berichte 1750, VDI.
Bunthoff, D. and Meier, H. J. (1987). Umweltfreundliches Kraftwerk mit Druckwirbelschichtfeuerung, Planungskriterien für die Dampferzeuger. *VGB Kraftwerkstechnik* **67**(8): 751–757.
Chen, S. L., Clark, W. C., Heap, M. P. and Pershing, D. W. (1982a). NOx-reduction by reburning with gas and coal – bench scale studies. Joint Symposium on Stationary Combustion NOx Control. Palo Alto, CA.
Chen, S. L., Heap, M. P., and Pershing, D. W. (1982b). Influence of coal composition on the fate of volatile and char nitrogen during combustion. Proceedings of the 19th Symposium (International) on Combustion. Pittsburgh, The Combustion Institute.
Chugtai, M. Y. and Michelfelder, S. (1983). Schadstoffeinbindung durch Additiveinblasung um die Flamme. *BWK* **35**(3): 75–83.
Couch, G. (1994). Understanding slagging and fouling during PF combustion. London, IEA Coal Research. JEA CR72.
de Soete, G. (1981). Physikalisch-chemische Mechanismen bei der Stickoxidbildung in industriellen Flammen. *Gas Wärme International* **39**: 29–35.
Di Nola, G., de Jong, W., and Spliethoff, H. (2009). The fate of main gaseous and nitrogen species during fast heating rate devolatilization of coal and secondary fuels using a heated wire mesh reactor. *Fuel Processing Technology* **90**(3): 388–395.
Di Nola, G. (2007). Biomass fuel characterization for NO emissions in co-firing applications, TU Delft. Ph.D.
DoE (1999). Reburning technologies for the control of nitrogen oxides emissions from coal-fired boilers. Topical Report Number 14.
Doležal, R. (1954). Schmelzfeuerungen Theorie, Bau und Betrieb. Berlin, VEB Verlag Technik.
Doležal, R. (1961). Großkesselfeuerungen, Theorie, Bau und Regelung. Berlin, Göttingen, Heidelberg, Springer.
Doležal, R. (1990). Dampferzeugung: Verbrennung, Feuerung, Dampferzeuger. Berlin, Heidelberg, New York, Springer.
Eberius, H., Just, T. Kelm, S. and Schug, K. P. (1981). Messungen von NOx, HCN, NH3 und N2 im Abgas vorgemischter, CH3-NH2-dotierter Propanflammen. VDI Bericht Nr. 423, pp. 49–54.
Ecoba. (2006). EN10 Residues from combustion of coal for energy production, from http://themes.eea.europa.eu/Sectors_and_activities/energy/indicators/EN10,2007.04.

Effenberger, H. (2000). Dampferzeugung. Berlin, Heidelberg, Springer.

Epple, B., Brüggemann, H., and Kather, A. (1995). NOx-arme Steinkohletangentialfeuerung. *EVT Register* **54**: 37–49.

Farwick, H. and Rummenhohl, V. (1993). SCR-Katalysatoren. *VGB Kraftwerkstechnik* **73**(5): 437–441.

Farzan, H., Latham, C. E., Maringo, G. J., Hallstrom, J. E., Beard, C. T., Weed, G. E. and Prataps, J. (1995). Gas reburn retrofit on an industrial cyclone boiler. 1995 Joint Symposium on Stationary Combustion NOx Control. Kansas City.

Farzan, H., Kokkinos, A., Maringo, J. and Yagiela, T. (2004). B& W'S experience on the application of reburn technology on cyclone boilers. Conference on Reburning for NOx Control Reburning on Trial.

Fenimore, C. P. (1970). Formation of nitric oxide in premixed hydrocarbon flames. Proceedings of the 13th Symposium (International) on Combustion. Pittsburgh, The Combustion Institute.

Fenimore, C. P. (1976). Reactions of fuel-nitrogen in rich flame gases. *Combustion and Flame* **26**: 249–256.

Fenimore, C. P. (1978). Studies of fuel nitrogen species in rich flame gases. Proceedings of the 17th Symposium (International) on Combustion, pp. 661–671. Pittsburgh, The Combustion Institute.

Folsom, B., Sommer, T., Ritz, H., Pratapas, J., Bautista, P. and Facchiano, T. (1995). Three gas reburning field evaluations. Final Results and Long Term Performance. 1995 Joint Symposium on Stationary Combustion NOx Control. Kansas City.

Frank, M. J., Gutberlet, H. and Brandenstein, J. (2006). Betrieb von Rauchgas-Entstickungsanlagen mit DENOX-Katalysatoren. *VGB PowerTech* **86**(4): 72–77.

Gebel, K., Madlsperger, G., Hein, K. and Bökenbrink, D. (1989). Entwicklung eines Kombinations-DENOX-Verfahrens unter Verwendung feuerungstechnischer Minderungsmaßnahmen und Harnstoff-SNCR-Technik. *VGB Kraftwerkstechnik* **69**(12): 1200–1208.

Ghribelli, L., Pasini, S., Benedetto, D., De Santis, R. and Manini, G. (1999). Results from the application of a gas reburning technology on a 600 MW coal fired boiler. EPRI-DOE-EPA Mega Symposium, Atlanta 16–20 August 1999.

Glarborg, P., Jensen, A. D. and Johnsson, J. E. (2003). Fuel nitrogen conversion in solid fuel fired systems. *Progress in Energy and Combustion Science* **29**(2): 89–113.

Glass, J. W. and Wendt, J. O. L. (1982). Mechanisms governing the destruction of nitrogen species during the fuel rich combustion of pulverized coal. Proceedings of the 19th Symposium (International) on Combustion. Pittsburgh, The Combustion Institute.

Goidich, S. J., Fan, Z., Sippu, D. and Bose, A. C. (2006). Integration of ultra-supercritical OTU and CFB boiler technologies. Proceedings of the 19th International Fluidized Bed Combustion Conference, May 21–24. Vienna, Austria.

Görner, K. (1991). Technische Verbrennungssysteme. Grundlagen, Modellbildung, Simulation. Berlin, Heidelberg, New York, Springer.

Greul, U. (1997). Experimentelle Untersuchung feuerungstechnischer NOx-Minderungsverfahren bei der Kohlenstaubverbrennung. Dissertation Universtät Stuttgart.

Greul, U., Rüdiger, H., Spliethoff, H. and Hein, K.R.G. (1996a). NOx controlled combustion in a bench scale test facility. Proceedings of the 21st International Conference on Coal Utilization & Fuel Systems. Clearwater, FL.

Greul, U., Spliethoff, H., Magel, H. C., Schnell, U., Rüdiger, H., Hein, K. R. G., Li, C. Z., and Nelson, P. F. (1996b). Impact of temperature and fuel-n-content on fuel-staged combustion with coal pyrolysis gas. Proceedings of the 26th Symposium (International) on Combustion. Pittsburgh, The Combustion Institute.

Gumz, W. (1962). Kurzes Handbuch der Brennstoff- und Feuerungstechnik. Berlin, Göttingen, Heidelberg, Springer.

Günther, R. (1974). Verbrennung und Feuerungen. Berlin, Heidelberg, Springer.

Hannes, K., Schmahl, K.-H. and Weiler, H. (1987). Möglichkeiten zur NOx-Minderung für große Feuerungsanlagen. BWK, Fachreport NOx-Minderung in Rauchgasen, Oktober.

Harder, J. (1996). Moderne Entaschungstechnik und Ascheverwertung bei Steinkohlekraftwerken mit Trockenfeuerung. *VGB Kraftwerkstechnik* **76**(9): 791–798.

Haynes, B. S. (1977). The oxidation of hydrogen cyanide in fuel-rich flames. *Combustion and Flame* **28**: 113–121.

Hein, K. and Schiffers, A. (1979). Verbesserung der natürlichen Schwefeleinbindung bei der Verfeuerung Rheinischer Braunkohlen. Düsseldorf, VDI-Berichte Nr. 346, pp. 77–80.

Hein, K.R.G. (1984). Rauchgasseitige Probleme bei der Verfeuerung niederkalorischer Kohlen. *VGB-Konferenz Verschlackungen, Verschmutzungen und Korrosionen in Wärmekraftwerken*. Essen.

Hein, K.R.G. and Kallmeyer, D. (1989). Stand der NOx-Minderung bei braunkohlebefeuerten Großkesselanlagen. *VGB Kraftwerkstechnik* **69**(6): 591–596.

Heinzel, T. (2004). Ascheseitige Probleme bei der Mitverbrennung von Biomasse in Kohlenstaubfeuerungen. Dissertation Uni Stuttgart. Aachen, Shaker.

Heinzel, T., Maier, J., Spliethoff, H., Hein, K. R. G. and Cleve, W. (1997). Slagging tests on the suitability of alternative coals in a 325 MW_{el} PFC boiler. Engineering Foundation Conference "The Impact of Mineral Impurities in Solid Fuel Combustion". Kona, Hawaii.

Hemberger, H., Neckel, H. and Wolfrum, J. (1987). Lasermeßtechnik und mathematische Simulation von Sekundärmaßnahmen zur NOx-Minderung in Kraftwerken. 3. TECFLAM Seminar, Karlsruhe.

Henriksen, N., Larsen, O. H., Blum, R. and Inselmann, S. (1995). High-temperature corrosion when co-firing coal and straw in pulverized coal boilers and circulating fluidized bed boilers. VGB-Konferenz "Korrosion und Korrosionsschutz in der Kraftwerkstechnik", Essen.

Himes, R., Hubbard, D. and Stallings, J. (1995). Summary of SNCR applications to two coal-fired wet bottom boilers. 1995 Joint Symposium on Stationary Combustion NOx Control. Kansas City.

Hofmann, J. E., von Bergmann, J., Bökenbrink, D. and Hein, K. (1989). NOx control in a brown coal-fired utility boiler. 1989 Joint Symposium on Stationary Combustion NOx Control. San Francisco.

Hössle, H. (1985). Entschwefelung mit Trockenadditiven in Rostfeuerungen. *Energie* **37**(7): 32–36.

Hotta, A. and Venäläinen, I. (2006). Design and scale-up philosophy of once-through CFB boilers with supercritical parameters. *VGB PowerTech* **86**(4): 66–71.

Jacobs, J. and Hein, K. R. G. (1988). Bedeutung des N2O innerhalb der Stickoxidemissionen. *VGB-Kraftwerkstechnik* **68**(8): 841–843.

JBDT (1985). Jahrbuch der Dampferzeugertechnik, 5. Ausgabe, Essen Vulkan.

JBDT (1992). Jahrbuch der Dampferzeugertechnik, 7. Ausgabe, Essen Vulkan.

Jochum, B., and Reidick, A. (2005). Umrüstung der Feuerung des 680 MW Dampferzeugers Weiher III der SaarEnergie auf Babcock-DS-Brennertechnologie. *VGB PowerTech* **85**(3): 50–54.

Juniper, L. (1995). Applicability of ash slagging indices-revisited. *Combustion News*, published by Australian Combustion Technology Centre, February.

Just, T. and Kelm, S. (1986). Mechanismen der NOx-Entstehung und Minderung bei technischer Verbrennung. *Industriefeuerung* **38**: 96–102.

Kahlert, J., Wutzler, F. and Lange, P. (2008). Erfahrungen mit Korrosionsprozessen an den Dampferzeugern des Kraftwerkes Lippendorf. *VGB PowerTech* **88**(6): 37–42.

Kallmeyer, D. and König, J. (1987). Einführung von NOx-Minderungstechniken in mit rheinischer Braunkohle gefeuerten Kraftwerken. *BWK* **39**(5): 249–253.

Kather, A. (1995). Verfahrenstechnische und konstruktive Auslegung moderner Braunkohle-Dampferzeuger. *VGB Kraftwerkstechnik* **75**(9): 763–770.

Kauppinen, E. I. and Pakkanen, T. A. (1990). Coal combustion aerosols. A field study. *Environmental Science Technology* **24**: 1811–1818.

Kautz, K. and Zelkowski, J. (1985). Verschlackungs-, Verschmutzungs- und Korrosionsprobleme bei der Verfeuerung von Steinkohle weltweiter Herkunft. VGB-Konferenz *"Verschlackungen, Verschmutzungen und Korrosionen in Wärmekraftwerken"*. Essen.

Klingspor, J. S. and Vernon J. L. (1988). Particulate control for coal combustion. London, IEA Coal Research. JEA CR/05

Klose, E. (1992). Pyrolyse – Wissenschaftsgebiet und technische Anwendung. *Energieanwendung + Energietechnik* **41**(12): 488–496.

Kobayashi, H., Howard, J. B. and Sarofim, A. F. (1977). Coal devolatilization at high temperatures. Proceedings of the 16th Symposium (International) on Combustion. Pittsburgh, The Combustion Institute.

Kolar, J. (1995). Verwertungsmöglichkeiten für Reststoffe der Sprühabsorptionsverfahren. *VGB Kraftwerkstechnik* **75**(2): 167–173.

Kolb, T. (1990). Experimentelle und theoretische Untersuchungen zur Minderung der NOx-Emission technischer Feuerungen durch gestufte Verbrennungsführung. Dissertation Universität Karlsruhe.

König, J. (1996). N2O-Bildung bei der Verbrennung von rheinischer Braunkohle in Wirbelschichtfeuerungen. Düsseldorf, VDI-Fortschrittberichte Reihe 15, Nr. 147.

Koornneef, J. and Junginger, M. (2007). Development of fluidized bed combustion – An overview of trends, performance and cost. *Progress in Energy and Combustion Science* **33**(1): 19–55.

Laskus, L. and Lahmann, E. (1977). Korngrößenverteilung von Stäuben im Rauchgas von Kraftwerksstäuben und in atmosphärischer Luft. *Staub-Reinhaltung der Luft* **37**(4): 136–140.

Lehmann, H. (1990). Handbuch der Dampferzeugerpraxis. München, Resch.

Lehmann, H. (1996). Repetitorium Dampferzeuger. Herrsching, Energie & Management.

Leisse, A., Gräwe, H. and Streffing, M. (1993). Neuentwicklung NOx-armer Feuerungssysteme, VDI-Berichte 1029, pp. 165–181.

Leisse, A. and Lasthaus, D. (2008). New experience gained from operating DS (Swirl Stage) burners. *VGB PowerTech* **88**(11): 43–49.

Leuckel, W. (1985). Schadstoffbildung bei industriellen Verbrennungsanlagen und primäre Minderungsmaßnahmen. 1. TECFLAM Seminar, Heidelberg, pp. 35–50.

Löffler, F. (1988). Staubabscheiden. Stuttgart, Thieme Verlag.

Macphail, J., Rhine, J. M. and Hughes, M. (1997). The demonstration of gas reburning technology on a 600 MW boiler at Longannet, Scotland. Proceedings of the 4th International Conference on Technologies and Combustion for a Clean Environment. Lisbon.

Maier, H. (1998). Experimentelle Untersuchungen der Kohlenstaubverbrennung unter Berücksichtigung der Brennstoffaufbereitung. Dissertation Universität Stuttgart, Shaker, Aachen.

Maier, H., Bilger, H. and Mayer, G. (1992). Betriebserfahrungen mit der SCR-Stickoxidminderungstechnik. *VGB Kraftwerkstechnik* **72**(9): 796–800.

Maryamchik, M. and Wietzke, D. L. (2005). B&W IR-CFB operating experience and new development. Proceedings of the 18th International Fluidized Bed Combustion Conference, May 22–25. Toronto, Ontario, Canada.

McCarthy, J. M., Chen, S. L., Seeker, W. R. and Pershing, D. W. (1987). Pilot scale studies on the application of reburning for NOx control. 1987 Joint Symposium on Stationary Combustion NOx Control. New Orleans.

Mechenbier, R. (1989). Experimentelle Untersuchung der Stickoxid-Reduktion durch Brennstoffstufung mit Methan bei der Kohlenstaubverbrennung. Dissertation Universität Bochum.

Meyer, B., Willmes, O. and Röper, B. (1995). Mechanismen der chlorinduzierten Korrosion von Wirbelschicht-Heizflächen. *VGB Kraftwerkstechnik* **75**(12): 1043–1048.

Michel, W. (1992). Wirbelschichttechnik in der Energiewirtschaft. Leipzig, Deutscher Verlag für Grundstoffindustrie.

Morrison, F. (1986). Understanding Pulverized Coal Combustion. London, IEA Coal Research.

Murakami, N. (1985). Application of MACT in-furnace NOx removal process coupled with low-NOx SGR burner. 1985 Joint Symposium on Stationary Combustion NOx Control. Boston.

Myerson, A. L. (1974). The reduction of nitric oxide in simulated combustion, effluents by hydrocarbon-oxide mixtures. Proceedings of the 15th Symposium (International) on Combustion, pp. 1085–1092. Pittsburgh, The Combustion Institute.

Nalbandian, H. (2004). Air pollution control technologies and their interactions. London, IEA Clean Coal Centre, CCC/92.

Nalbandian, H. (2006). Economics of retrofit air pollution control technologies. London, IEA Clean Coal Centre, CCC/111.

Nalbandian, H. (2007). European legislation (Revised LCPD and EU ETS) and coal. London, IEA Clean Coal Centre. CCC/121.

Neumann, J. and Kautz, H. R. (1995). Auswertung der internationalen Literatur zur Hochtemperaturkorrosion in Kohle- und Müllkraftwerken. VGB-Konferenz Korrosion und Korrosionsschutz in der Kraftwerkstechnik. Essen.

Newell, R., Campbell, J., Wamsley, J., Gebhart, S., Yagiela, A. S., Maringo, G. J., Farzan, H. and Haggard, R. W. (1995). Coal reburning application on a cyclone boiler. 1993 Joint Symposium on Stationary Combustion NOx Control. Miami Beach.

Nolan, P. S. (1994). Flue-gas desulfurization in thermal power plants. *Thermal Engineering* **41**(6): 434–438.

Peters, F. and vom Berg, W. (1992). Entstehung und Eigenschaften von Kraftwerksreststoffen. *VGB Kraftwerkstechnik* **72**(4): 371–376.

Pohl, J. H. and Sarofim, A. F. (1976). *Devolatilization and oxidation of coal nitrogen.* Proceedings of the 16th Symposium (International) on Combustion, pp. 491–591. Pittsburgh, The Combustion Institute.

Pohl, J. H., Chen, S. L., Heap, M. P. and Pershing, D. W. (1982). Correlation of NOx-emissions with basic physical and chemical characteristics of coal. Joint Symposium on Stationary NOx Control. Palo Alto, CA.

Raask, E. (1973). Verschlackungen in steinkohlegefeuerten Kesseln – Verschlackungsmechanismus und Verhütungsmaßnahmen. *VGB Kraftwerkstechnik* **53**(4): 248–254.

Reidick, H. and Schumacher, D. (1985). Verschmutzungsverhalten bei der Verfeuerung von Braunkohlen. VGB-Konferenz "Verschlackungen, Verschmutzungen und Korrosionen in Wärmekraftwerken". Essen.

Reuter, F. D. and Hönig, J. (1988). Beeinflussung der Kesselauslegung durch Einbau einer SCR-Anlage. *VGB Kraftwerkstechnik* **68**(2): 118–124.

Rosenberg, H. S. and Oxley, J. H. (1993). Stand der Rauchgasentschwefelung in den Elektrizitätswerken der USA. *VGB Kraftwerkstechnik* **72**(3): 237–246.

Rüdiger, H. (1997). Pyrolyse von festen biogenen und fossilen Brennstoffen zur Erzeugung eines Zusatzbrennstoffes für Feuerungsanlagen, Universität Stuttgart, Shaker, Aachen. Dissertation.

Sarofim, A. F., Howard, J. B. and Padia, A. S. (1977). The physical transformation of mineral matter in pulverized coal under simulated combustion conditions. *Combustion Science and Technology* **16**: 187–204.

Sayre, A. N., Knill, K. J. and Smart, J. P. (1991). Coal characterization requirements for modelling pulverized coal flames. *IFRF Doc F/88/y/13*. Ijmuiden, NL.

Schirmer, U. and Thömen, H. J. (1984). Korrosionen und Verschmutzungen in Müllverbrennungsanlagen – Ursachen, Wirkungen, Gegenmaßnahmen. Internationale VGB-Konferenz "Verschlackungen, Verschmutzungen und Korrosionen in Wärmekraftwerken", 29.2.–2.3-1984, Essen.

Schroth, G. (1985). Rost-renaissance? *Energie* **37**(7): 27–30.

Schulz, W. (1985). Experimentelle Untersuchung der Bildung von Stickoxiden bei der Kohlestaubverbrennung. Dissertation Universität Bochum.

Schumacher, F. (1996). Korrosion in Abfallverbrennungsanlagen. Teil 1 Rauchgasseitige Korrosion. *BWK* **48**(11/12): 74–80.

Simon, E., Schönrok, M. and Henjes, F. (1997). Kohlekriterien in gestaffelten Bandbreiten und deren Einflüsse auf den Dampferzeugerbetrieb. *VGB Kraftwerkstechnik* **77**(8): 643–649.

Simon, S., Frach, M., Jochum, B. and Lang, A. (2006). Leistungssteigerung von Dampferzeugern beim Einsatz unterschiedlicher Kohlequalitäten durch neue intelligente Online-Reinigungstechnologie. *VGB PowerTech* **86**(5): 40–45.

Skorupska, N. M. (1993). Coal specifications – impact on power station performance. London, IEA Coal Research.

Sloss, L. (2003). Trends in emission standards, CCC/77. London, IEA Clean Coal Centre.

Smith, I. M. (2005). Cement and concrete - benefits and barriers in coal ash utilisation. IEA CCC/94. London, IEA Coal Research.

Smoot, L. D. and Smith, P. J. (1985). Coal combustion and gasification. New York, Plenum Press.

Solomon, P. R. and Colket, M. B. (1979). Coal devolatilization. Proceedings of the 17th International Symposium on Combustion, pp. 131–143. Pittsburg, The Combustion Institute.

Song, Y. H., Pohl, J. H., Beer, J. M. and Sarofim, A. F. (1982). Nitric oxide formation during pulverized coal combustion. *Combustion Science and Technology* **28**: 31–39.

Soud, H. N. (1995). Developments in particulate control for coal combustion. London, IEA Coal Research. JEA CR/78

Soud, H. N. and Fukasawa, K. (1996). Developments in NOx abatement and control. London, IEA Coal Research. JEA CR/89

Spliethoff, H. (1991). Large scale trials and development of fuel staging in a slag tap furnace. 1991 Joint Symposium on Stationary Combustion NOx Control. Washington, D.C.

Spliethoff, H. (1992). Großtechnische Untersuchung der Stickstoffoxidminderung mit dem Schwerpunkt Brennstoffstufung an einer Schmelzkammerfeuerung. VDI-Fortschrittberichte Reihe 6. Nr. 276. Düsseldorf, VDI.

Spliethoff, H. (1995). NOx-Minderung mit Reduktionsbrennstoffen aus der Steinkohle, VGB TW 207, Essen.

Spliethoff, H. (2000). Verbrennung fester Brennstoffe zur Strom- und Wärmeerzeugung: Verfahren und Stand der Technik – Wirkungsgrad, Betrieb, Emissionen und Reststoffe. VDI Fortschritt-Berichte, Reihe 6 Energietechnik, Nr. 443. Düsseldorf, VDI.

Spliethoff, H. and Haferkamp, A. (1991). Einsatz von Ammoniak bei der Brennstoffstufung an einer Schmelzkammerfeuerung - Projekt Brennstofftrennstufung. VGB Konferenz Kraftwerk und Umwelt. Essen.

Spliethoff, H. and Hein, K. R. G. (1997). Effect of fuel quality on low-NOx combustion. Proceedings of the 4th International Conference on Technologies and Combustion for a Clean Environment. Lisbon.

Spliethoff, H. and Hein, K.R.G (2002). Combustion engineering measures to reduce NOx in pulverised coal furnaces. *VGB PowerTech* **82**(4): 88–99.

Spliethoff, H., Gerhardt, T., Greul, U. and Hein, K. R. G. (1995a). Use of tar oil as reburning fuel – tests in a semi-industrial and large scale furnace. *TOTEM 11 "Liquid Fuels, Heavy Oils, Wastes and Slurries"*. Biarritz, IFRF.

Spliethoff, H., Greul, U., Maier, H. and Hein, K. R. G. (1995b). Low-NOx combustion for pulverized coal – a comparison of air staging and reburning. Proceedings of the 2nd International Conference Combustion and Emission Control. London.

Spliethoff, H., Greul, U., Rüdiger, H. and Hein, K. R. G. (1996). Basic effects on NOx emissions in air staging and reburning at a bench-scale facility. *Fuel* **75**(5): 560–564.

Stahlherm, D. (1973). Pyrolyse und Verbrennung einzelner Kohlekörner bei hohen Aufheizraten. Dissertation RWTH Aachen.

Stahlherm, D., Jüntgen, H. and Peters, W. (1974). Zündmechanismus und Abbrand von Kohlekörnern. *Erdöl und Kohle -Erdgas – Petrochemie* **27**(2): 64–70.

Stamatelopoulos, G.-N. and Weissinger, G. (2005a). *Die nächste Generation von Steinkohlekraftwerken*. VGB-Konferenz "Kraftwerke im Wettbewerb 2005, Ordnungspolitik, Markt und Umweltschutz", Potsdam.

Stamatelopoulos, G.-N. and Weissinger, G. (2005b). Die nächste Generation von Steinkohlekraftwerken. *VGB PowerTech* **85**(10): 49–53.

Staudt, J. E., Casill, R. P. and Tsai, T. S. A. (1995). Commercial Application of UREA SNCR for NOx RACT Compliance on a 112 MWe Pulverized Coal Boiler. 1995 Joint Symposium on Stationary Combustion NOx Control. Kansas City.

STEAG (1988). Strom aus Steinkohle: Stand der Kraftwerkstechnik. Berlin, Heidelberg, New York, London, Paris, Tokyo, Springer.

Strauß, K. (2006). Kraftwerkstechnik: zur Nutzung fossiler, nuklearer und regenerativer Energiequellen. Berlin [u.a.], Springer.

Strauß, K. and Thelen, F. (1989). Brennstoffaufbereitung als Beitrag zur NOx-Minderung. *VGB Kraftwerkstechnik* **69**(2): 207–211.

Stultz, S. C. and Kitto, J. B. (1992). Steam, its generation and use. Barberton, OH, The Babcock & Wilcox Company.

Takahashi, Y., Sakai, M. Kunimto, T., Ohme, S., Haneda, H. and Kawamura, T. (1982). Development of "MACT" in Furnace NOx removal process for steam generators. 1982 Joint Symposium on Stationary Combustion NOx Control.

Takeshita, M. (1994). Environmental performance of coal-fired FBC. London, IEA Coal Research. JEA CR/75.

Takeshita, M. (1995). Air pollution control costs for coal fired power stations. London, IEA Coal Research. ieaper/17.

Takeshita, M., Sloss, L.L. and Smith, I.M. (1993). N2O emissions from coal use, IEAPER/06. London, IEA Coal Research.

TALuft (2002). 1. Allg. Verwaltungsvorschrift zum Bundesimmissionsschutzgesetz (Technische Anleitung zur Reinhaltung der Luft) vom 27.2.1986, Gem. Ministerialblatt vom 30. Juli 2002 (GMBl. 2002, Heft 25–29, pp. 511–605).

Tigges, K. D., Leisse, A., Lasthaus, D. and Streffing, M. (1996). Entwicklung von schadstoffarmen Staubfeuerungssystemen. *VGB Kraftwerkstechnik* **76**(5): 388–393.

Tigges, K. D., Leisse, A. and Petzel, H.-K. (1997). Betriebserfahrungen mit dem Omega-Verfahren zur Primärentstickung im Heizkraftwerk Völklingen. *VGB Kraftwerkstechnik* **77**(5): 401–406.

van Heek, K. H. (1988). Die Pyrolyse als Schlüsselprozeß der Kohleumwandlung. *Glückauf* **124**(4–5): 259–265.

van Heek, K. H. and Mühlen, H. J. (1985). Heterogene Reaktionen bei der Verbrennung von Kohle. *BWK* **37**(1–2): 20–27.

Vernon, J. L. (1990). Pollution control for industrial coal-fired boilers. London, IEA Coal Research. ieacr/25.

Vernon, J. L. and Jones, T. (1993). Sulphur and coal. London, IEA Coal Research. ieacr/57.

VGB (1996). VGB-Handbuch "NOx-Bildung und NOx-Minderung bei Dampferzeugern für fossile Brennstoffe". Essen.

VGB (1997). VGB technical scientific report thermal power plants 212e: The future of fluidized bed combustion. Essen, VGB.

VGB (2007). VGB-Erfahrungsbericht: Primärmaßnahmen zur NOx-Minderung an Staubfeuerungen für Stein- und Braunkohle. VGB-TW 216. Essen, VGB PowerTech.

VGB. (2008). Aufkommen und Verwertung von Nebenprodukten aus kohlebefeuerten Kraftwerken in Deutschland im Jahr 2006. Retrieved 1.11.2008, from http://www.vgb.org/NAV.html.

vom Berg, W. (1993). Verwertung von Rückständen aus Wirbelschichtfeuerungsanlagen. *VDI-Berichte* **1081**: 265–283.

vom Berg, W. and Feuerborn, H.-J. (2005). Coal combustion products in Europe. *VGB PowerTech* **85**(12): 50–54.

vom Berg, W. and Feuerborn, H.-J. (2007). Coal ashes and FGD gypsum – byproduct or waste. *VGB PowerTech* **87**(12): 56–61.

Warnatz, J. (1985). Elementarreaktionen in Verbrennungsprozessen. *BWK* **37**(1–2): 11–19.

Wendt, J. O. L. (1980). Fundamental coal combustion mechanics and pollution formation in furnaces. *Progress Energy Combustion Science* **6**: 201–222.

Wendt, J. O. L. and Dannecker, K. M. (1985). Mechanics governing the destruction of nitrogen species during staged coal combustion. Joint Symposium on Stationary Combustion NOx Control. Boston.

Wickert, K. (1963). Versuche zur Entschwefelung vor und hinter dem Brenner zur Verringerung des SO2-Auswurfs. *Mitteilungen der VGB* **83**: 74–82.

Wiens, U. (2005). Neues aus den Regelwerken zur Verwendung von Flugasche in Beton. *VGB PowerTech* **85**(10): 73–79.

Wolfrum, J. (1985). Chemische Elementarprozesse bei der Bildung und Beseitigung von Schadstoffen in Verbrennungsvorgängen. 1. TECFLAM Seminar, Heidelberg, pp. 7–21.

Wu, Z. (2000). Prevention of particulate emissions, CCC/40. London, IEA Clean Coal Centre.

Wu, Z. (2002). NOx control for pulverised coal fired power stations. IEA CCC/69. London, IEA Coal Research.

Wu, Z. (2003). Understanding fluidised bed combustion. London, IEA Clean Coal Centre. ieaccc/76.

Wu, Z. (2005). Fundamentals of pulverised coal combustion, IEA CCC/95. London, IEA Clean Coal Centre.

Wu, Z. (2006). Developments in fluidised bed combustion technology. London, IEA Clean Coal Centre. ieaccc/110.

Zeldovich, J. (1946). The oxidation of nitrogen in combustion and explosions. *Acta Physiochimica URSS* **21**(4): 577–628.

Zelkowski, J. (2004). Kohlecharakterisierung und Kohleverbrennung. Essen, VGB PowerTech.

Zellinger, G. and Tauschitz, J. (1989). Betriebserfahrungen mit der nichtkatalytischen Stickoxidreduktion in den Dampfkraftwerken der Österreichischen Draukraftwerke AG. *VGB Kraftwerkstechnik* **69**(12): 1194–1200.

Zhu, Q. (2003). Developments in particulate control, CCC/73. London, IEA Clean Coal Centre.

Chapter 6
Power Generation from Biomass and Waste

The thermal use of biomass or waste is a cheaper and more technically complete option for reducing CO_2 emissions compared to other renewable energy sources. Possible biogenous fuels include forestry and agriculture residues such as wood and straw, and also specially cultivated reproducible raw materials such as *Miscanthus sinensis*, whole cereal plants, poplars or willows.

For the conversion of biomass into electric power and heat, a great number of processes are available. They can be classified according to the following:

- Fuel type: biomasses (such as wood or straw), organic residues, municipal solid waste, sewage sludge or refuse-derived fuel
- Fuel conversion: combustion or gasification (atmospheric or pressurised)
- Power production: combustion engine, gas turbine, steam turbine, Stirling motor or fuel cell
- Output capacity
- Products: electricity or heat or both combined
- Exploitation of biomass alone or combined with fossil fuels

6.1 Power Production Pathways

The pathways for heat or electric power production from solid biomass and waste can be subdivided into systems with combustion and systems with gasification. A systematic arrangement of the individual steps in each pathway is presented schematically in Fig. 6.1.

6.1.1 Techniques Involving Combustion

In combustion, the chemical energy of the fuel is converted into sensible heat which the flue gases carry, which is then transferred to a working medium in a heat exchanger. The working medium expands in a turbine or other such device to create mechanical energy for power production.

H. Spliethoff, *Power Generation from Solid Fuels,* Power Systems,
DOI 10.1007/978-3-642-02856-4_6, © Springer-Verlag Berlin Heidelberg 2010

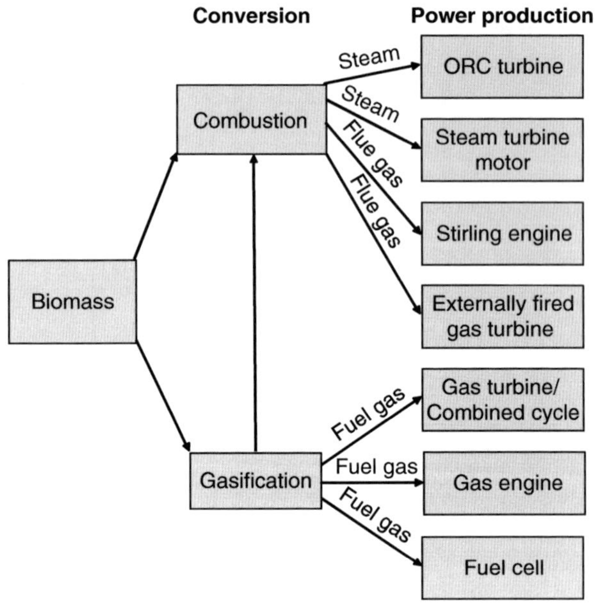

Fig. 6.1 Pathways for the production of power from biomass

Steam Turbines and Steam Engines: Production of power by steam turbine is the widest spread electrical power generation process in thermal power plants. The efficiency of the power generation is determined by the live steam conditions in front of the turbine and the exhaust steam conditions. For economic reasons, lower plant outputs mean lower achievable efficiencies. In the range of several megawatts, it is possible to achieve around a 20% plant efficiency and in the range of several tens of megawatts, about 30%. In comparison to this, the maximum efficiencies reach about 45% in the capacity range of several 100 MW. Steam turbines are on the market for capacities starting from 1 MW; for capacities less than 1 MW, steam engines are used. In the low capacity range, the high costs of the steam power process, which uses turbines, are a disadvantage.

Organic Rankine Cycle (ORC) Processes: The ORC process is based on the Rankine process, just as the conventional steam process is. Instead of water, however, the working fluid is of organic origin and has low boiling and condensation temperatures. Accordingly, the ORC process is suited to heat exploitation at a lower temperature. Due to the low temperature of the working fluid, the resulting efficiencies for biomass applications from a thermodynamic point of view must be lower than in a steam process, because the maximum steam temperature of the organic medium is in the range of 250°C. Electrical efficiencies for ORC plants in operation are in the range of up to 15%; however, it has to be considered that nearly all are operated as CHP plants (ADMONT 2001; Obernberger 2003). For the purposes of comparison, this would correspond to an electrical efficiency of about 20–24% for a pure power cycle. Improvements in the ORC power production process are currently being introduced by employing two parallel flows of the organic medium to improve the heat transfer (i.e. the split process) (Turboden 2009).

Externally Fired Combined Cycle (EFCC) Processes: This closed gas turbine process works with a high-temperature heat exchanger which can be operated using the exhaust gas from the firing. Here, the energy is transmitted to pre-compressed air which is then expanded in a turbine. Metallic materials may be used for the construction of smaller capacity gas turbines because of the lower turbine entry temperatures. Therefore, the EFCC process is ideally suited to new and existing biomass-fired furnaces for decentralised electric power and heat generation. The problems arising in gas cleaning in biomass gasification are avoided. In general, an efficiency in the order of magnitude of 25% seems to be achievable (Kaltschmitt et al. 2009).

Stirling Engines: The Stirling engine is a type of expansion engine. In contrast to combustion engines, the piston is moved not through the expansion of exhaust gases from internal combustion but through the expansion of a constant, closed gas volume, caused by heat transfer from an external heat source. This way, the generation of power is separate from the heat source, which means such engines can be fuelled by heat from biomass combustion plants. In the case of biomass-fired furnaces, there is a conflict concerning the design. On the one hand, the aim is to heat the smallest possible gas volume to high temperatures, on the other hand, for a high-efficiency and a low-heat flux, the heat exchanger should be designed at the largest possible scale. For natural gas-fired plants with capacities of between 10 and 40 kW, it is possible to achieve efficiencies of 20–25% by utilising the hot waste air as combustion air. Since this variant cannot be used in biomass-fuelled installations because of slagging, the efficiencies for biomass-fired engines range between 10 and 15%.

6.1.2 Techniques Involving Gasification

In gasification, the fuel is converted at air-deficient conditions into combustible gas by an addition of air or another gasification agent (such as water vapour). The combustible gases that are produced can be fed to a machine driven by internal combustion. Particular consideration is given to the gas engine and the gas turbine, each of which involve different efficiencies, costs and gas quality requirements.

Gas Engines: In gas engine applications, the product gas is mixed with air, conducted to the engine and burned in its cylinder according to either the spark ignition or the diesel principle. The mechanical work of the engine is converted into electrical energy by means of a coupled generator. Engines in conjunction with atmospheric fixed bed or fluidised bed gasifiers are suitable for capacities from about 50 kW_{el} to 10 MW_{el}. With gasification and subsequent motor-driven use of the combustible gas, it is possible even at small capacities to achieve high electrical efficiencies above those of combustion-based processes. For capacities around 500 kW_{el}, net efficiencies between 24 and 31% are indicated (Vogel 2007).

Gas Turbines: For outputs of 5 MW_{el} and over, it is reasonable to use gas turbines. Suitable gasifiers in this range are atmospheric or pressurised fluidised beds. Gas turbines have the potential of increasing the efficiency up to about 45% by installing a downstream heat recovery boiler with a steam turbine (output 20 MW_{el}).

Besides that, it is possible to conceive using gas turbines in innovative technologies presently still in development, such as fuel cells. In fuel cells (FCs), the chemically bound energy of the fuel is directly converted into electrical energy, with higher efficiencies than conventional technologies can achieve. Coupling of single FC technologies with downstream gas and steam turbine processes presents itself as a means for better fuel exploitation. FC technology, however, is still in different stages of development, depending on the FC type.

6.2 Biomass Combustion Systems

Biomass combustion systems can be classified by the same principle as used for coals (see Sect. 5.1) – that is, into fixed bed, fluidised bed and pulverised fuel systems. A lot of technologies have been derived from coal combustion systems; in the smaller capacity range, especially, a large variety exists (Nussbaumer 2003; Van Loo and Koppejan 2008; Eltrop et al. 2007; Kaltschmitt et al. 2009). In the course of this chapter only the most widespread systems are presented.

6.2.1 Capacities and Types

One essential criterion for the choice of the combustion system is the size of the plant that shall be built. In this respect, the classification distinguishes between small furnaces of up to a 15 kW thermal capacity, medium-scale plants of up to 1 MW and large-scale plants (Kaltschmitt 2001):

- Small furnaces are used in the household sector for hot water and room heating, with capacities of up to 15 kW_{th}. They will not be considered further in this text.
- Plants up to a thermal output of 1 MW_{th} are used in commerce and trade. The firing systems widely employed are shaft and underfeed firing furnaces. Investigations into the emission behaviour of plants in service in industry and trade show that, during changes of operating mode (i.e. start-up, shutdown, load change), higher emissions of dust, carbon monoxide and hydrocarbons are produced through incomplete combustion. These emissions are often caused not only by intermittent fuel feeding practices at start-up and shutdown but also from operation at part load, when the output is controlled by connecting and disconnecting the fuel supply and the fuel/air ratio is not set optimally. Newly developed underfeed firing systems for woodchips show that combustion and operation at low emission levels are possible even in this output range of up to 1 MW_{th}.
- Plants with capacities higher than 1 MW_{th} which are used for the production of heat, process steam and electricity are usually operated as combined heat and

power production plants (CHP plants). The upper capacity limit of plants fired exclusively with biomass is seen today as 50–100 MW_{th}, because transport and the logistics of the fuel supply become too complex and costly at higher capacities. In this capacity range, the predominant firing types are stoker-fired furnaces, which are suited to fuels that are moist, problematic and/or in lumps, and require little in terms of fuel preparation. Fluidised bed furnaces are attractive because of their lower emissions but have more complex constructions and are therefore economical only for output capacities above 10 MW_{th}. Furnaces designed for pulverised fuel are suitable for firing biomass, in particular if the fuel is available in a small-particle form. Pulverised fuel-fired furnaces are the dominating combustion technology for coal in large plants because of their high power density, good controllability and complete combustion. For fuels with fine and coarse fractions, it may also be practical to combine pulverised fuel and stoker firing. In Denmark, the so-called cigar burner proved to be a reliable technology for the combustion of straw bales.

- Co-combustion: Besides using biomass as the only feedstock, it is also possible to fire it combined with other, preferably solid, fuels. The respective firing and flue gas cleaning systems must be tested for their suitability in this case.

6.2.2 Impact of Load and Forms of Delivery of the Fuel Types

Besides considering the power plant capacity, the choice of the firing system has to take into account the shape (shavings, chaff, pellets, bales, etc.) in which the biomass is available. Figure 6.2 presents the field of application of firing systems as a function of the plant size and the biomass shape.

- Pieces of firewood (logs) can be used in shaft or pusher-type grate furnaces, depending on the thermal input.

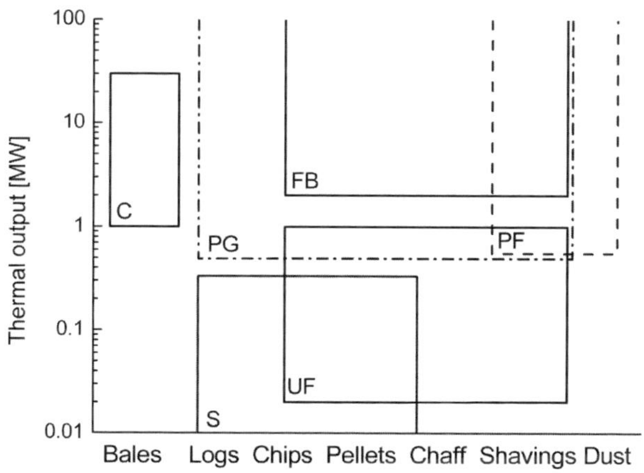

Fig. 6.2 Combustion systems as functions of plant size and biomass shape (PF pulverised fuel, S shaft furnace, UF underfeed firing, PG pusher-type grate, FB fluidised bed furnace, C cigar burner)

- Woodchips can be used in almost all firing systems (i.e. shaft, underfeed, stoker-fired and fluidised bed furnaces).
- Preparation into the form of woodchips is the standard technology for woody biomass types.
- Shavings of wood can be burned in underfeed firing up to a determined fraction. In pusher-type grate, pulverised fuel and fluidised bed firing systems, there are no such restrictions.
- Powdery biomass can be used only in pulverised fuel firing.
- Straw bales can be directly burned in cigar burners, which are used at capacities over 3 MW.
- Chaff from herbaceous biomass can be used in underfeed, in pusher-type and, to limited extent, in fluidised bed firing systems.
- Pellets are well suited for use in firing systems, with advantages similar to chopped material.

For herbaceous biomass types, there are essentially two competing methods of delivery. In the first, the biomass is mown in the field and pressed into bales. Bales are either fed as a whole to the cigar burner or are undone and cut into chaff before being fed to the combustion plant. The second delivery method involves the herbaceous biomass being pelletised in the field, thus increasing the density for transport. For underfeed, grate and fluidised bed firing systems, the fuel is delivered ready in the form most suitable for combustion.

In Denmark, the delivery method that has become generally accepted for the thermal utilisation of straw is the pressing into large-size Hesston bales, which can be performed at a relatively low cost using state-of-the-art technology (Kaltschmitt 2001; Nikolaisen 1992).

In biomass combustion, a high fuel storage capacity is required because of its low energy density. Storing on-site at the combustion facility over a long period is only possible for smaller facilities, so for plants of the megawatt order of magnitude or greater, biomass is usually stored on the producer's premises and delivered daily to the combustion facility. Often, these plants have an on-site fuel storage capacity of just 2–3 days fuel consumption.

6.2.3 Furnace Types

6.2.3.1 Shaft Furnaces

In the lower capacity range, from 20 kW up to about 250 kW, shaft furnaces are used for the combustion of both lump wood residues and woodchips. The fuel types that can be used in these plants usually do not require additional preparation (Eltrop et al. 2007; Spliethoff and Hein 1995a). Figure 6.3 shows an example of a shaft furnace with lateral burnout for thermal capacities between 50 and 150 kW.

The furnace design is based on the principle of bottom, or lateral burnout, which means that the flame extends from the side or the bottom into the combustion cham-

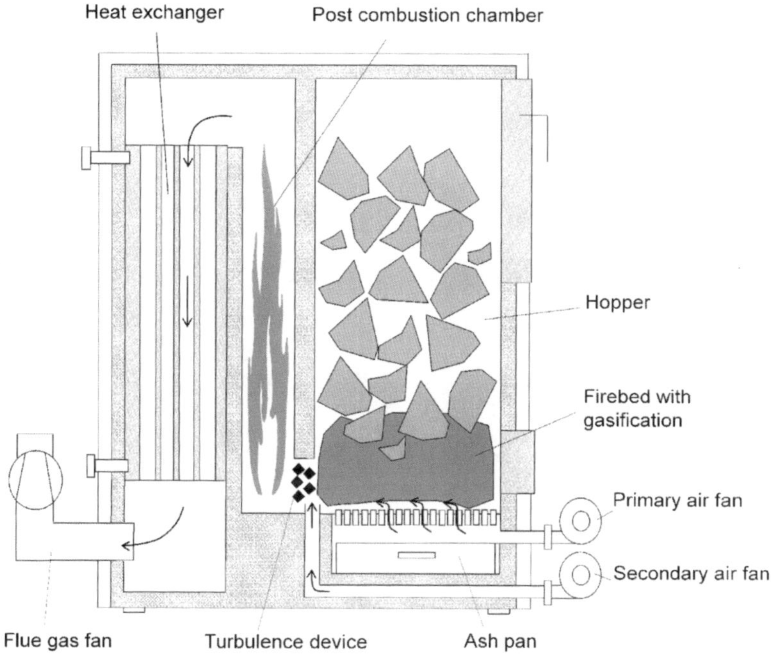

Fig. 6.3 A shaft furnace with lateral burnout (Kaltschmitt 2001)

ber. The air, supplied by natural draught or by fan, is divided into primary and secondary air in modern installations. The primary air is injected below and at the side of the grate, while the secondary air is directed into the combustion chamber. As a rule, these furnaces are fed manually and operated intermittently. The degree of automation and potential for control are smaller than for plants with continuous feeding. After the frequency and quantity of feeding, further control is generally limited to combustion air throttling. Modern firewood boilers with two-stage control allow adjustment of the firing to produce low emissions. The output in this case is controlled by the primary air flow; the secondary air, controlled by the exhaust gas composition, ensures the complete burnout.

The lack of the need for fuel preparation, the simple firing technology and the relatively low cost of purchase are the reasons for the widespread use of this firing type in the given capacity range.

6.2.3.2 Underfeed Firing

Underfeed firing systems, which are available on the market in a broad capacity range – from 20 kW_{th} to 6 MW_{th} – are suitable for the firing of chips, pellets, shavings and, to a certain extent, for pulverised wood residues as well – those with a moisture content between 5 and 40%. The maximum size of the fuel is limited by

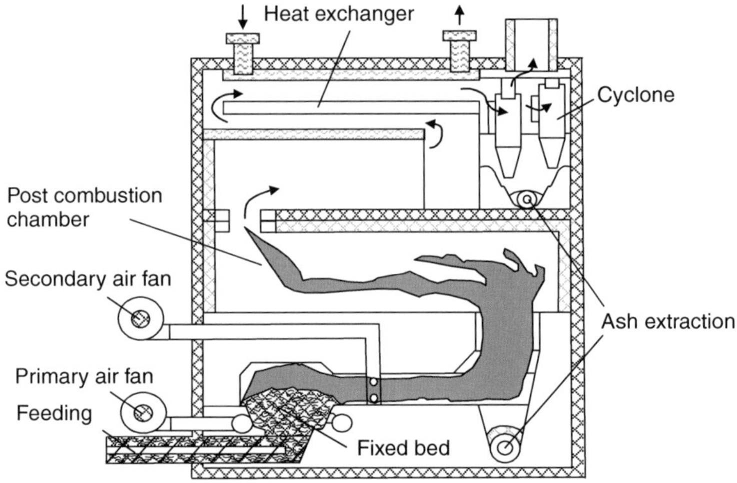

Fig. 6.4 Underfeed firing (Kaltschmitt et al. 2009)

the fuel feeding system. The fuel should be low in ash, finely grained and of a homogeneous structure.

The firing systems (Fig. 6.4) are automatically fed by screw conveyors directly from a silo. In a great number of the furnaces, the fuel is transported from below into a trough inside the combustion chamber, a so-called retort, then dried and devolatilised.

The pyrolysis gases and primary air enter the glowing bed layer from below, ignite and burn to completion following the addition of burnout air. Underfed fires are as a rule self-igniting and in principle controllable to an acceptable degree by co-coordinated fuel and air supply. Continuously working screw feeders and the small amounts of embers and fuel in the furnace result in operation with little emission of pollutants.

This combustion technology is widely employed for the thermal utilisation of residues from wood processing, because it

- works almost fully automatically,
- is built using simple technology and fewer components in comparison to other firing types such as pulverised fuel or stoker firing systems and
- is economically attractive, even if the storage, feeding and potential need for fuel preparation equipment is taken into account (Van Loo and Koppejan 2008; Spliethoff 2000).

6.2.3.3 Stokers

For capacities of around 1 MW_{th} and higher, stoker-type furnaces are the dominant technology for the combustion of biomass. Figure 6.5 shows a forward pusher-grate furnace, which is the dominant system used for woody biomass. Other grate

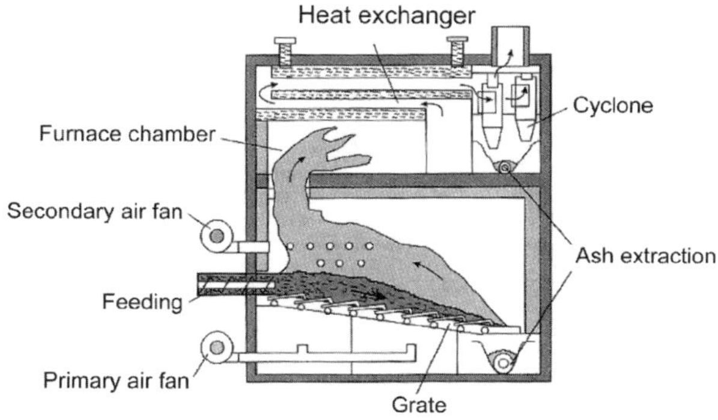

Fig. 6.5 A forward pusher-grate furnace (Kaltschmitt et al. 2009)

firing technologies, such as travelling or reverse reciprocating grates, can also be used; they are described in Sect. 5.5. Problematic fuels, for example, moist wood residues or high-ash bark residues can also be fired in these systems. In stoker firing, it is also possible to burn straw as the only fuel – as the practice in Denmark demonstrates.

The grate is either fed from the fuel storage via a hopper, by screw conveyor or hydraulic conveyor. The feedstock moves from the feeding point to the grate end at the grate speed. The fuel dries, pyrolyses and burns completely as it is pushed down the grate.

Primary air enters from below the grate and through the grate bars, and secondary air is injected above the grate and ahead of the partly firebrick-lined secondary combustion zones. In stokers, low fuel qualities can be combusted effectively because the residence times of the fuel and the combustion air flow can be adjusted to a wide range of fuel properties. Compared to underfeed-type furnaces, moving-grate stokers are harder to control and less suitable for fast load changes due to the greater fuel loads inside the furnace.

With their complicated plant mechanics, stokers are economical only from capacities of around 1 MW_{th} and above. For the combustion of wood and wood residues, these furnaces are partly operated in connection with a direct firing system for pulverised residues.

Stokers are also suited to the combustion of herbaceous biomasses such as straw, *Miscanthus* or grass. However, the ash fusion temperatures, lower than for wood, may result in fuel caking, which impedes the passage of air and thus combustion. In these cases, temperatures below the ash softening point can be set by lower bulk heights and grate heat release rates. Grate raking is not a reliable remedial action if it risks causing incomplete burnout (Biollaz and Nussbaumer 1996).

6.2.3.4 "Cigar Burner" for Herbaceous Biomass Bales

The so-called cigar burner of the Volund Company is in principle a grate firing system, as can be seen in Fig. 6.6. The bales, however, are not deposited on the grate in exactly the delivered form but first ignited at the front before being gradually pushed into the combustion chamber.

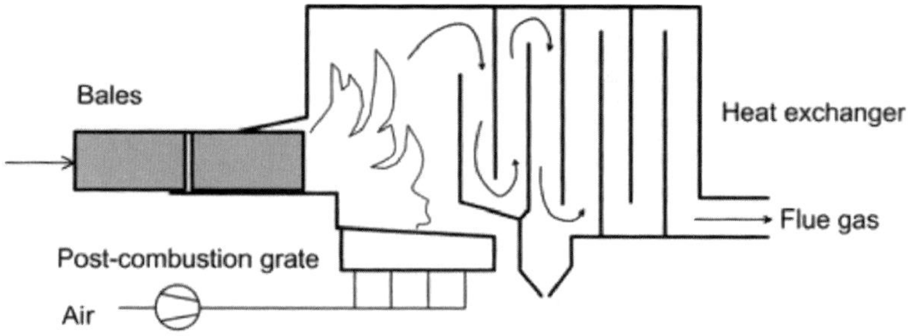

Fig. 6.6 A cigar burner

Unburned straw layers that break off fall into pieces onto the grate and burn to completion while moving along it. The ash gets discharged as a result of the forward movement of the grate. Water cooling and flashback dampers inside the bale charger prevent a burning back of the bales.

The advantages of this technology are the relatively minor preparation needed for the fuel, the continuous fuel supply and the relatively simple construction of the plant. The disadvantages are the narrow range of fuels and the restriction to one bale type. Burning straw layers falling onto the grate at intervals may cause increased CO emissions.

The bale shape and the minimum feed rate determine the minimum stable capacity of this firing type. Because of the complex automatic control technology of the storage and feeding installations, this firing type is justifiable from an economical point of view only for capacities of about 2–3 MW_{th} and above. A modified version of this type for smaller outputs involves a preparation step where the large bales are cut into several slices to be put onto the grate.

In Denmark, several plants of this type are in service, predominantly for straw bales, at capacities from 3 to more than 20 MW_{th}. The boilers can be used both for heat and for electricity production (Van Loo and Koppejan 2008; Nikolaisen 1992).

6.2.3.5 Fluidised Bed Combustion (FBC)

In fluidised bed combustion, the prepared fuel is burned at 800–900°C in a fluidised bed consisting of 95–98% inert material and only 2–5% combustible material. The process-related intensive mixing and combustion, the excellent heat transfer in the fluidised bed and the decoupling of the particle residence time and the flue gas velocity allow a wide range of fuels to be used. In particular, a wide range of moisture contents, compositions and fuel preparations can be exploited. Fluidised bed furnaces are particularly suited to the combustion of several, even very different, fuel types. In addition, fluidised bed combustion has advantages over grate combustion when sludge fuels are used.

Given the low combustion temperature, problems such as slagging and fouling are less severe in FBC furnaces. There is a risk of fluid bed sintering, but only when firing very high alkali fuels, such as straw. This risk may rule out the application of fluidised bed technology for such fuels if they are to be used in monofuel combustion. The characteristics of bubbling and circulating fluidised bed combustion are described in detail in Sect. 5.4.

Today, the design of modern, stationary fluidised bed firing systems for biomass clearly deviate from the design of a classic stationary fluidised bed for firing coal. When firing biomass, there is no need for an in-bed heat transfer surface because the bed is operated with a high deficiency of air and so only part of the fuel heat is released in the bed. By setting the air ratio in the fluidised bed, it is possible to reliably control bed temperatures between 650 and 800°C. The air ratio depends on the heating value of the fuel and the temperature required to assure the prevention of sintering of the bed. For wood chips the air ratio is typically in the range of 0.35--0.45 and the temperature is below 800°C (Bolhar-Nordenkampf et al. 2006). The temperature at which sintering of the bed can occur depends mainly on the ash composition of the fuel. When biomasses with high alkali contents are used, such as straw, sintering temperatures significantly below 800°C develop. The addition of the remaining air is performed in several stages in the freeboard of the furnace. In this process, for the sake of cleanliness, temperatures of about 1,050°C for clean wood should not be exceeded when air is added so that fouling in the freeboard is prevented. The temperatures can be controlled by adding the secondary air in several stages and by heat dissipation. The air-staged mode of operation leads to low emissions of NO_x. The principle of air-staged fluidised bed combustion is shown schematically in Fig. 6.7 and compared to the design of a stationary fluidised bed for coal.

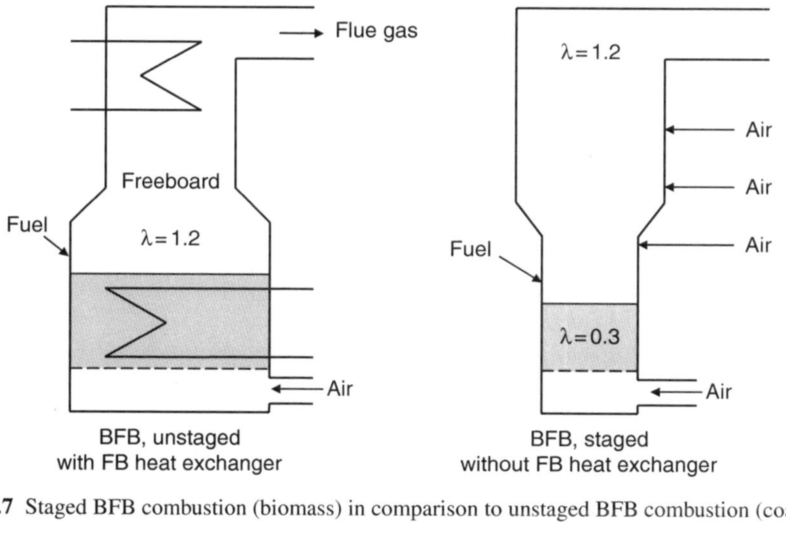

Fig. 6.7 Staged BFB combustion (biomass) in comparison to unstaged BFB combustion (coal)

The fuel preparation for a bubbling FBC furnace is a similar process to that used in a stoker-fired furnace. In both plants, wood cut into pieces smaller than 90 mm can be used. A circulating FBC furnace requires pieces smaller than 30–50 mm. Since fluidised bed combustion involves high capital costs, it can be operated economically only in larger units (bubbling FB greater than 10 MW_{th}, circulating FB greater than 30 MW_{th}). At Stadtwerke, Leipzig, a CFB with a capacity of 55 MW fuel input (clean wood) went into operation in 2005. The plant features an electrical efficiency of 37%, which is achieved with high steam conditions of 535°C, 130 bar and reheating (to 535°C) (Beckert and Schaarrschuch 2007). Such advanced conditions are the exception for a plant of this size.

Fluidised bed combustion, as a novel technology, competes with the predominant biomass combustion technology, i.e. the stoker. For monofuel combustion of wood, FBC is a proven technology, but not yet for herbaceous biomass. In the Scandinavian countries and increasingly in other countries, too, wood residues and sludge from paper and pulp production are mostly fired in fluidised bed furnaces that are an integrated part of the production process. Due to the economical output capacity, especially of circulating fluidised bed furnaces, biomass is used along with other fuels such as peat, coal and waste in such plants (Gockner and Rechberger 2008; Bolhar-Nordenkampf et al. 2006; McCann and Simons 1997; Seemann et al. 2008).

6.2.3.6 Pulverised Fuel Combustion (PFC)

The pulverised fuel furnace is a practical solution for biomass, in particular when the available fuel is already very fine. This is often the case in wood-processing enterprises, which usually employ high-speed machines, resulting in fine residues. For fuels with a certain fraction of coarser materials, a PFC furnace with a burnout grate may be reasonable. PF-fired furnaces stand out because of their high power densities, high furnace efficiencies and good controllability. They are offered on the market at capacities ranging between 500 kW_{th} and 50 MW_{th}.

Pulverised fuel furnaces for biomass combustion are usually designed as direct firing systems, i.e. by cyclone or through a muffle.

Figure 6.8 shows the principle of direct firing through a muffle for the combustion of pulverised wood. The fuel is injected tangentially into the combustion chamber together with the primary air. A whirling flow, similar to a vortex flow, develops. Larger particles are blown into and deposit in the rear of the combustion chamber, burning out after they have landed. Fine particles burn out while still airborne. Secondary air is injected at a constriction in the muffle. This constriction at the end of the muffle causes a rise in the velocity and thus a good mixing of the flue gases with the secondary air. Modern and bigger cyclone suspension furnaces are operated with continuous control – the fuel feeding is adjusted to the firing rate required at a given time and the combustion air supply is adjusted exactly to the fuel charge. These firing systems help to keep the emissions of unburned flue gas components

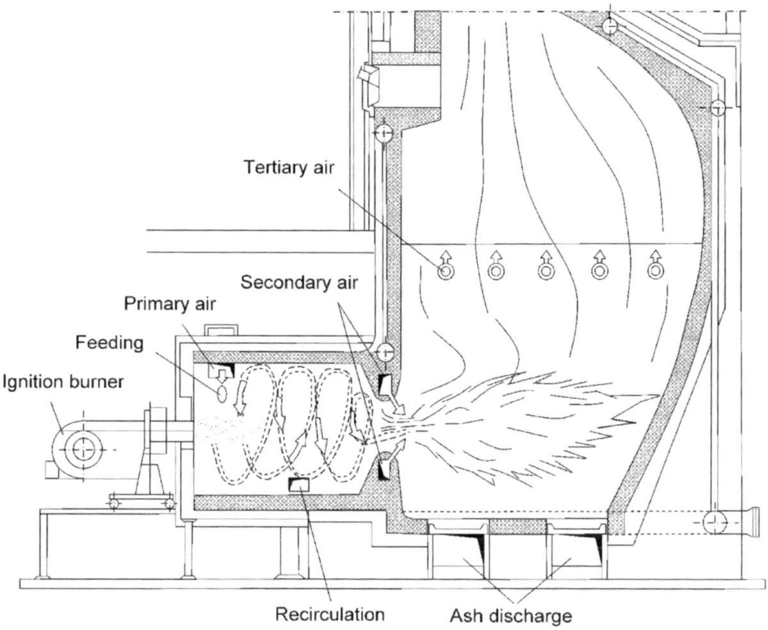

Fig. 6.8 A pulverised fuel muffle furnace (Kaltschmitt et al. 2009)

to a low level. The exact matching of fuel and combustion air makes it possible to apply effective primary measures of NO_x reduction.

6.2.4 Flue Gas Cleaning and Ash Disposal

The application of emission reduction techniques depends on the plant size, the fuel to be used and the emission limits to be complied with. Table 6.1 shows typical emission concentrations for the untreated gas (before flue gas cleaning) and after dust

Table 6.1 Typical flue gas emissions of woodchip combustion plants (Spliethoff 2000)

Emissions (mg/Nm^3)	Typical value	Range	Emission limit, Germany (TA Luft)[a] (mg/Nm^3)
SO_2	170	50–350	–
NO_x as NO_2	250	100–400	250
Particle (raw gas)	500	200–800	20–100[b]
Particle, multicyclone	300	200–400	
Particle, flue gas condensation	40	20–50	

[a] 1–50 MW_{th}

[b] Depending on capacity

collectors in wood chip combustion plants. These values, in practice, can strongly deviate beyond the given ranges in the table. They are compared to the emission limits of the *TA Luft* (the German Clean Air Code) for wood-firing appliances. It is obvious that, for larger biomass combustion plants, the design has to include flue gas particulate collectors. In biomass-fired plants, aside from dust collection, further flue gas cleaning units are not commonly installed.

6.2.4.1 Particulate Control

Dust collection for herbaceous fuels is more demanding than for woody biomass. Due to the greater lump sizes and the higher density of the fuel, the ash produced in wood combustion is coarser. With herbaceous biomass, the higher alkali, chlorine and sulphur contents trigger the development of salts (KCl, K_2SO_4), which form ultrafine fly ash particles that can be separated effectively only by a particulate filter such as an electrostatic precipitator (ESP) or a fabric filter.

Multicyclones used in wood combustion efficiently only remove particles larger thans 10 μm with tolerable pressure losses, which, in chip and bark combustion, results in residual flue gas dust contents of 120 to around 400 mg/Nm^3. The emission limits, dependent on the plant size, cannot be met in most instances, which means another dust collecting unit has to be installed downstream. For woody biomass, common practice is to use cyclones for coarse separation and ESPs or fabric filters for fine separation. This way it is possible to reduce dust contents to between 10 and 50 mg/Nm^3. ESPs are the preferred technology. For small plants the higher pressure loss of fabric filters exclude their application; for plants larger than 10 MW, ESPs are more economical (Kaltschmitt et al. 2009; Hasler and Nussbaumer 1996; Johnsen and Svendsen 1997; Obernberger 1997; Biollaz and Nussbaumer 1996).

ESPs can be problematic if used when firing herbaceous biomass, because the dust of dry straw ash has a higher resistance than coal ash. However, if ESPs are employed at lower flue gas temperatures, the fly ash will absorb water and the resistivity will decline. If temperatures become too low, the ash becomes sticky and will cause fouling problems. The result is a narrow temperature window around 115°C for the use of an ESP (Johnsen and Svendsen 1997). The separators well suited to these applications seem to be fabric filters, which, besides their high collection efficiency, have the advantage of a higher degree of capture of chlorine and sulphur in the fly ash, which is favoured by the intensive contact between flue gas and ash on the surface of the filter bags (Obernberger 1996).

For all the fuels mentioned, wet cleaning processes, in which the flue gas is conducted through a scrubber where spray water is injected, are also used. During this process, the dust particles are bound and discharged with the water. In addition, acidic pollutant gases, such as SO_2 and HCl, can be removed by wet cleaning. However, the wastewater creates an additional disposal problem (Nikolaisen 1992).

When moist fuels are used, flue gas condensation plants for heat recovery can double as particle collectors, if needed in combination with wet dust removal.

The flue gases are cleaned in a multicyclone and afterwards the water vapour is condensed for heat recovery. This way, it is possible to achieve dust contents of 40 mg/Nm^3. This method is applied only for very moist biomasses, having more than 30% moisture content – usually woodchips (Eltrop et al. 2007).

6.2.4.2 Nitrogen Oxides and Sulphur Oxide

The combustion of wood or straw does not require downstream DeNO$_x$ or desulphurisation installations, such as are state of the art in pulverised coal-fired power plants.

The sulphur contained in the fuel, though mostly released in a gaseous form during combustion, can be captured in the fly ash. Downstream of chip or bark combustion furnaces, for instance, the ash removed by multicyclone captures 40–70%, while during fine dust collection in a fabric filter, between 60 and 90% of the sulphur is captured. Only the remaining gaseous portion is released as SO_2 emissions. With straw or whole plants, the capture rates of the respective ashes are about 45–50%. The capture in the ash depends on the concentration of alkalis and alkaline earths (especially Ca), on the combustion temperature and the dust collection method employed. Due to the low sulphur contents of straw and wood, it is possible to do without desulphurisation in systems firing these fuels (Obernberger 1996).

The low fuel nitrogen content of wood results in only minor NO$_x$ emissions during combustion. For fuels with higher fuel nitrogen contents, such as straw or whole plants, it is possible to apply primary measures to meet the emission limits. Primary measures are effective because of the volatile matter fraction of the fuel nitrogen. Figure 6.9 shows NO$_x$ emissions measured in stokers in service (Nussbaumer 2003; Biollaz and Nussbaumer 1996).

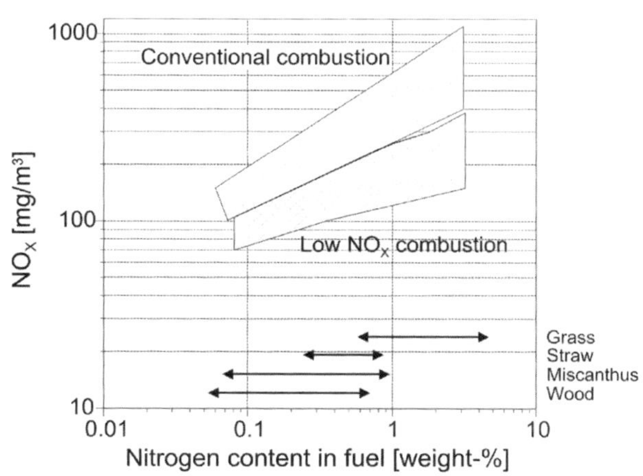

Fig. 6.9 NO$_x$ emissions from biomass-fired stokers (Biollaz and Nussbaumer 1996)

6.2.4.3 Chlorine

Chlorine can be captured in fly ash as well. At woodchip and bark combustion facilities, capture rates between 40 and 80% have been measured, with the rates for straw and whole plants ranging around 80–85% when applying a downstream fabric filter. Even with an 85% capture of chlorine, the German limit of 30 mg/m^3 for straw- and whole-plant firing systems is exceeded when there are fuel Cl concentrations over 0.15% by weight in the dry matter.

Higher chlorine concentrations in consequence require secondary measures such as a scrubber or ultrafine dust collector combined with dry sorption (Biollaz and Nussbaumer 1996). If fabric filters or ESPs are used, the flue gas is mixed with an alkaline solid sorbent such as calcium hydroxide ($Ca(OH_2)_2$) before entering the filter. In wet dust collection processes, the removal of acidic pollutant gases without additives is possible but is more effective when alkaline additives are used (Hasler and Nussbaumer 1996).

High chlorine contents may trigger dioxin formation. High combustion temperatures and a fast cooling down of the flue gas to the low-temperature area are a means to ensure low dioxin emissions. Measurements taken at various biomass-fired plants in service for different fuels (woodchips, bark, straw, whole plants) show that the applied firing technologies – underfeed, grate and cigar burner – in normal operation comply with the emission standards laid down in the *17*th *BImSchV* (Ordinance on Incinerators for Waste and Similar Combustible Materials) (Obernberger 1996).

6.2.4.4 Ash Utilisation

The residual matter from the combustion of natural wood can be used as a fertiliser in agriculture and forestry. While the dioxin and furan contents of these ashes are ecologically harmless, their heavy metal content must be assessed. The enrichment of heavy metals in the fine particulates of wood ash prohibits the utilisation in forests. Certain highly volatile heavy metals such as cadmium, lead and zinc concentrate in the finest ash fraction in the ESP or the fabric filter. Waste disposal strategies aim at enriching the heavy metals in the ultrafine filter ash (which is disposed of after removal) and utilising only the coarse ash from the cyclone together with the grate ash (Kaltschmitt et al. 2009; Obernberger and Biedermann 1996).

In Austria, the fertilisation of forests, fields and grassland with cyclone and grate ash from wood combustion is a widespread practice, controlled by the authorities. Filter ash and sludge from flue gas condensation are always sent to landfill (Kaltschmitt et al. 2009). In Denmark, woodchip ash is often used to fertilise the forest area from which the wood was harvested. The ash from straw-fired thermal power plants is mainly used as a fertiliser in agriculture (Nikolaisen 1992). In Germany, the utilisation of biomass ash as a fertiliser and soil conditioner in agriculture and forestry falls under the regulation for fertilisers. Furnace ash from monofuel combustion of untreated biomass can be used for the production of fertilisers and soil conditioners, whereas cyclone or ultrafine fly ashes cannot be used as a matter of principle. When using herbaceous biomass ash as a fertiliser or when using any ash on agricultural fields, however, the heavy metal concentration limits of the Biowaste Ordinance have to be complied with. These limits are significantly stricter than the limits of the sewage sludge regulation (Eltrop et al. 2007).

The combustion of waste wood, compared to natural wood, involves increasingly stringent requirements for flue gas cleaning, for instance, for HCl, which may require a secondary removal stage. With scrap or residual wood from chipboard fabrication, it may additionally be necessary to remove nitrogen. In waste wood combustion, heavy metals are removed for the most part together with the fly ash – which restricts the utilisation of the ash. Mercury, however, which can be emitted with the flue gas, is found only in small concentrations in waste wood and is therefore usually not a problem (Hasler and Nussbaumer 1996).

6.2.5 Operational Problems

The chemical components most critical for the smooth operation of biomass combustion plants are the chlorine and alkalis present in the ash. Causes and effects of slagging and corrosion are discussed in Sect. 5.10. Deposits of alkali chlorides are the reason for chlorine-induced corrosion, which limits the application of higher steam temperatures. The corrosion mechanism is similar to waste incineration (see Sect. 6.4). While there is little corrosion when woody biomass is the fuel, due to the low chlorine and alkali concentrations, serious corrosion problems may occur with herbaceous and petiolate biomass (straw, *Miscanthus* and whole plants).

As steam and tube wall temperatures rise, the corrosion rate increases. Figure 6.10 shows the correlation between corrosion rates and material temperatures measured at a straw-fired stoker. Superheater temperatures in monofuel straw combustion, according to the chart, should be below 500°C in order to keep corrosion at a tolerable level. At superheater temperatures of 450°C or lower, significant

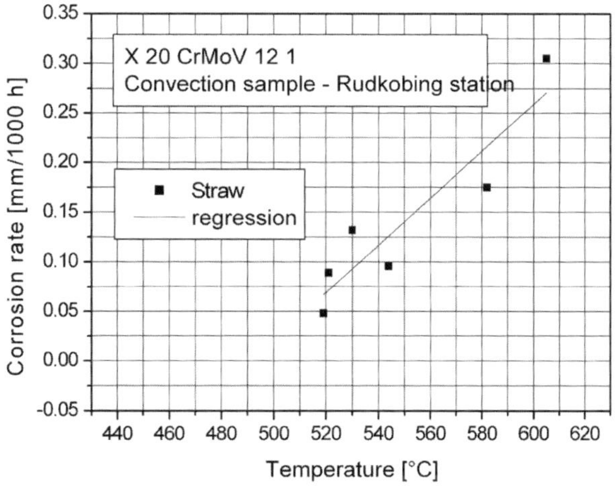

Fig. 6.10 Dependence of corrosion rate on material temperature (measured at a straw combustion plant by corrosion probe) (Clausen and Sorensen 1997)

corrosion does not occur in straw-fired stokers or cigar burner furnaces. If steam conditions above these temperatures are required, either a limited lifetime of the superheater has to be accepted or other fuels such as natural gas, wood or coal have to be used for superheating. A plant firing straw for the production of 470°C steam and firing wood for superheating to 540°C went into service in 1997 (Clausen and Sorensen 1997; Johnsen and Svendsen 1997). Severe corrosion can be avoided if austenitic steel (TP 347 FG, see Sect. 4.5) is used for the high-temperature section of the superheater. Inspections of corrosion at the Maribo Sakskobing plant (live steam temperature 540°C) indicated an expected lifetime of more than 25 years (Berg and Jensen 2008).

The important factors influencing the ash fusion behaviour are the concentrations of the alkaline earths Ca and Mg and the alkalis Na and K. Alkaline earths raise the fusion temperatures, while alkalis lower it. Chlorides, too, may cause the melting point to rise. Herbaceous biomass is the most likely to cause slagging on the grate, in the furnace and on the first superheater due to the high potassium content and the low ash softening temperature (Clausen and Sorensen 1997). The maximum furnace temperature should not be higher than 800–900°C (Obernberger 1996). Homogeneous fuel distribution, staged air injection to control the heat release, water-cooled grates, water- or steam-cooled walls and flue gas recirculation are means to limit the furnace temperatures and avoid slagging. In new straw-fired units precautions against slagging have been taken. A platen pendant superheater section is located above the furnace, operating as a slag condenser. The wet slag is designed to drip off the platen sections, and, because the spacing is so large, the slag deposits should not grow together to a massive slag formation (Berg and Jensen 2008).

In addition, fouling of the other heating surfaces is probable. Due to its smaller fly ash particles, herbaceous biomass poses a greater risk of fouling of the heating surfaces than woody biomass. Herbaceous biomass as a feedstock is finer and has a lower density and a higher alkali content than wood, which results in the smaller fly ash. In straw-fired plants in Denmark, the straw in some years had especially high chlorine and alkali contents, which caused various problems. Severe deposits were found throughout the entire furnace of both a grate and a cigar burner system, on radiant superheaters and air preheaters, in induced draught fans, ESPs and stacks. The problems could not be solved even by using soot blowers, so after 1–2 weeks of combustion operation with straw the plants had to be shut down and the heating surfaces cleaned (Obernberger 1996). However, satisfying boiler operating periods, inter-dispersed by manual cleaning intervals, are expected if pendant superheaters with large transverse pitches are employed.

In bubbling fluidised bed combustion, defluidisation of the bed resulting from the agglomeration of bed grains can be a major problem (Khan 2007; Khan et al. 2009). During biomass combustion in a fluidised bed, part of the ash and alkalis are released. Ash particles remaining in the bed can glue bed material grains together. This melt phase roughly matches the chemical composition of the ash. Another, even more severe type of agglomeration is the so-called coating-induced agglomeration (Visser et al. 2003). Sodium and potassium, released during combustion, form a very

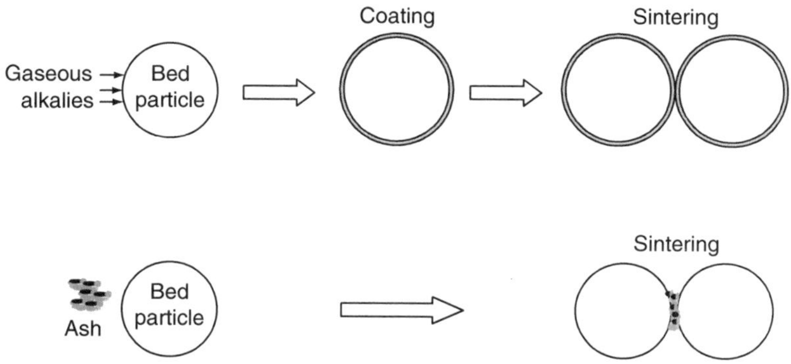

Fig. 6.11 Mechanisms of melt-induced and coating-induced agglomeration

thin and sticky coating of the bed grains. This type of agglomeration is believed to be the dominant process in commercial-scale installations. The mechanisms of melt-induced and coating-induced agglomeration are shown in Fig. 6.11.

High alkali fuels such as straw are known to provoke severe sintering of the bed, to an extent that the fluidisation can no longer be maintained, and operation must be halted. Fluidised bed technology is therefore limited in the fuels that it can fire alone – for instance, it is not suitable for the combustion of straw by itself. Investigations into straw combustion at a laboratory-scale furnace revealed that at a temperature of 800°C, defluidisation sets in after only 20 min. Tests with additives to prevent sintering were not very successful. The addition of coal was found to have a positive effect, though, suggesting a combined combustion of coal and straw in fluidised beds would be feasible (Lin et al. 1997; Bapat et al. 1997). In certain cases the use of quartz-free bed materials has also been successful in reducing sintering (Almark and Hiltunen 2005).

6.3 Biomass Gasification

The fundamentals of gasification are described in detail in Sect. 7.6 in the context of coal-based IGCC applications. Within this chapter, the focus is to discuss the special features of biomass gasification, though the principles of reactor design and gas cleaning are very much the same. The major differences in comparison to coal gasification are as follows:

- Due to the lower energy density and more limited availability of the feedstock, plant capacities only in the range of several megawatts up to 50–100 MW are considered. These sizes are much smaller than for coal gasification. The size of the plant is the dominating parameter for the power production costs and determines the choice of the gasification technology (the reactor design and the gasification medium). The preferred reactor technologies are fixed bed and

fluidised bed gasification, which, due to their lower gasification temperatures, give higher tar concentrations in comparison to the higher temperatures of entrained-flow gasification, the standard technology for coal gasification. For these capacities, gasifiers are usually air-blown.
– Biomass as a feedstock differs from coal. Biomass has a more inhomogeneous composition and particle size distribution, and generally fuel pre-treatment is required to produce a homogeneous feedstock suitable for the gasifier type. Hammer mills can only be used for relatively dry materials; to homogenise the size of wet materials, chippers equipped with blades can be used. Feeding must suit the higher bulk densities and the higher moisture contents that biomasses usually have, and therefore the feeding may require pre-drying. The higher volatile contents (in comparison to coal) are beneficial, requiring less residence time in the reactor for the conversion of the carbon, though the generally higher particle size may counteract this advantage. Alkali concentrations in the ash are higher than for coal.

6.3.1 Reactor Design Types

A large variety of gasification reactor designs are available for small and large scales. The designs can be classified in different ways (Kaltschmitt et al. 2009; Hofbauer 2007; Knoef and Ahrenfeldt 2005; Kaltschmitt and Bridgwater 1997; Spliethoff 2001; Kaltschmitt 2001; Higman and van der Burgt 2008; de Jong 2005), the most common being the following:

According to the gasification medium:

- Air-blown gasifiers
- Oxygen-blown gasifiers
- Steam gasifiers

According to the heat supply for the gasification:

- Autothermal or direct gasifiers: Heat is supplied by partial oxidation of the biomass, which results in a lower heating value of the product gas.
- Allothermal or indirect gasifiers: Heat is supplied from an external heat source, or for two-stage gasifiers, from the combustion zone to the gasification zone.

According to the pressure in the gasifier:

- Atmospheric
- Pressurised

According to the reactor design:

- Fixed bed
- Fluidised bed
- Entrained flow

The most common designs for biomass gasifiers are fixed bed and fluidised bed gasifiers. Fixed bed gasifiers are employed in the low capacity range of several MW_{th}; fluidised bed installations are typically larger than 5 MW_{th}, though this technology is undergoing further development for use at capacities down to below 1 MW. Fluidised beds can be subdivided into bubbling and circulating systems. Technologies such as pressurised fluidised bed gasification, entrained-flow gasification and oxygen gasification are only economical at larger scales. Figure 6.12 gives the typical fuel capacity ranges that different gasifier designs are used at. The gas quality depends on the fuel quality, the gasification agent (i.e. steam or air) and the reactor design. The choice of the gasification agent has the dominating effect on the heating value of the product gas. Table 6.2 shows the various component fractions and the heating value of the product gas using either air (autothermal gasification) or steam (allothermal gasification) as the gasification agent. The reactor design has the dominating effect on tar concentrations (see Table 6.3).

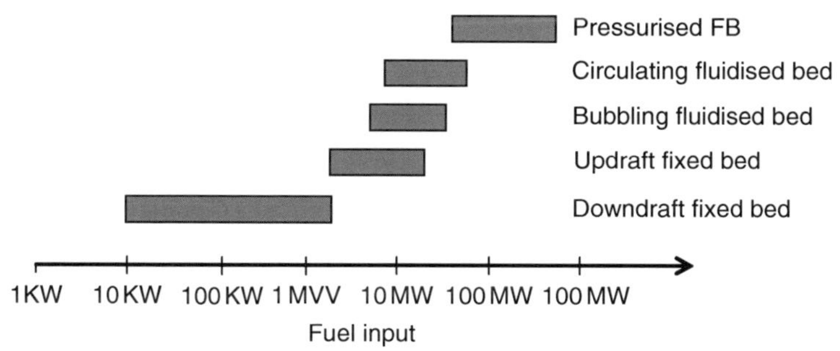

Fig. 6.12 Fuel capacity ranges for gasifier designs

Table 6.2 Heating value and product gas composition for air- and steam-blown gasification (Kaltschmitt 2001; FNR 2006; Knoef 2005)

	Gasification agent	
Gas [vol.-%]	Air	Steam
CO	10–20	25–47
H_2	9–22	35–50
CH_4	1–7	14–25
CO_2	10–15	9–15
N_2	40–55	2–3
LHV [MJ/Nm^3][a]	3.5–6.5	12–17

[a] Dry gas

Table 6.3 Tar and particle concentrations for different gasification systems (Kaltschmitt 2001)

	Fixed bed		Fluidised bed	
	Counter current	Co-current	BFB	CFB
		g/Nm^3		
Particle				
Range	0.1–3	0.2–8	1–100	8–100
Average	1	1	4	20
Tar				
Range	10–150	0.1–6	1–23	1–30
Average	50	0.5	12	8

6.3.1.1 Fixed Bed Gasifiers

In fixed bed gasifiers, gasification occurs in layers of the fuel bed, with different zones for the different gasification reactions that take place (pyrolysis, oxidation and reduction). A distinction (in respect to the flow) is made between counter-current and co-current gasifiers.

The most common type of counter-current gasifier is the vertical reactor, where the feedstock is fed from the top and the gasification agent added at the bottom. The directions of fuel flow and gas flow being opposed, separate reaction zones form in the reactor. The raw gas which is produced rises inside the reactor and leaves from the top, hence the common term "updraft gasification". Counter-current gasifiers have the advantage of not requiring any special fuel preparation, thus allowing the gasification of a wide range of biomass types with different particle sizes and moisture contents.

Through forced convection, the gas heated by oxidation in the bottom zone rises and transfers heat to the fuel. The gas leaves the gasifier with a relatively low temperature, which reflects the high gasification efficiency of this process. The drawback is that the volatile matter, gasified in the pyrolysis zone, becomes part of the rising gas stream. In consequence, the raw gas of counter-current gasifiers contains a considerable amount of tar compounds.

In a co-current gasifier, the fuel and the gasifying agent move together in the same direction (Fig. 6.13). The pelletised bio-fuel first dries and pyrolyses in a near absence of air in the upper zones, then enters the very hot oxidation zone, where it is transformed into char and ash, and finally falls into the reduction zone. The gases, for the most part produced in the pyrolysis zone, are heated to a temperature appreciably over 1,000°C in the oxidation zone. In this process, the overwhelming majority of the high-tar gaseous compounds entering the oxidation zone are converted into low-tar components, which then react with the char in the subsequent reduction zone, producing additional gas. The raw gas issues from the bottommost section of the reactor, hence the alternative term "downdraft gasification". In contrast to counter-current gasification, the heat transfer between the bio-fuel and the gasifying agent in co-current gasification is small, so the raw gas has a relatively high temperature and the gasification efficiency is lower than that of updraft gasification.

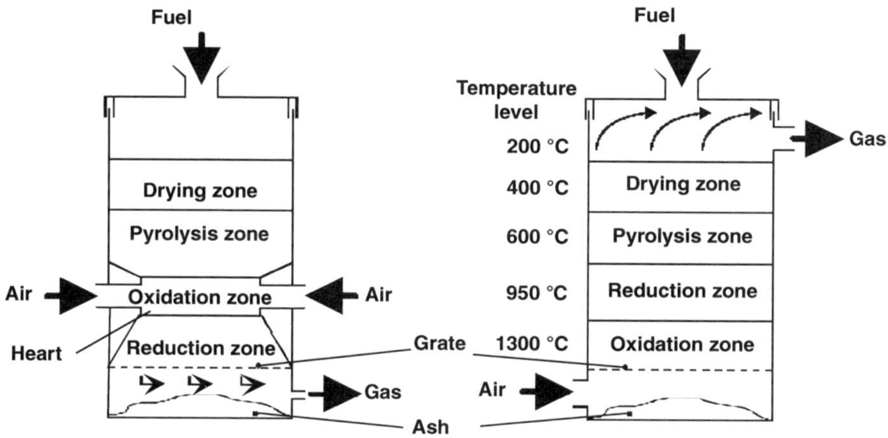

Fig. 6.13 Co-current gasifier (downdraft gasification, *left*) and counter-current gasifier (updraft gasification)

There is also a higher tendency for slag to form in co-current gasifiers than in counter-current gasifiers because of the high temperatures in the oxidation zone. A uniform temperature distribution within the individual reactor zones and a high permeability of the char to the gas are decisive factors for the gas quality. Co-current gasifiers therefore require a greater degree of fuel preparation to adjust the fuel particle size and the moisture content. The major advantage of co-current gasifiers is that the raw gas produced contains far less tar products than the gas from counter-current gasifiers.

Fixed bed gasifiers are generally atmospheric, air-blown (autothermal) and fuelled by wood, the latter of which results in a typical product gas heating value of between 4 and 6 MJ/m^3. The gas can be used for heating purposes or in gas engines. Co-current bed gasifiers are superior to counter-current installations because of their lower tar concentrations, which are in the range of several 100 mg/m^3. Quite a large number of fixed bed gasifiers have been tested, demonstrated or operated. However, most installations have had operational problems with fuel feeding, the gasifier itself, gas cleaning and prime movers. Current development concentrates on solving problems to do with automation, fuel feeding, the operation of the gasifier, gas cleaning and treatment of by-products. The tar concentrations are still a major concern.

A number of installations have been operated successfully in India for several years. These systems can be characterised by a near total absence of automation and by the use of wet gas cleaning with a sand bed filter for final tar removal. European standards, however, require fully automated installations and a gas cleaning system with disposable by-products.

Fluidised Bed Gasification

Fluidised bed gasification makes use of the advantageous mixing, reaction kinetics, gas – solid contact and heat transfer, as well as the ability to inject additives, of

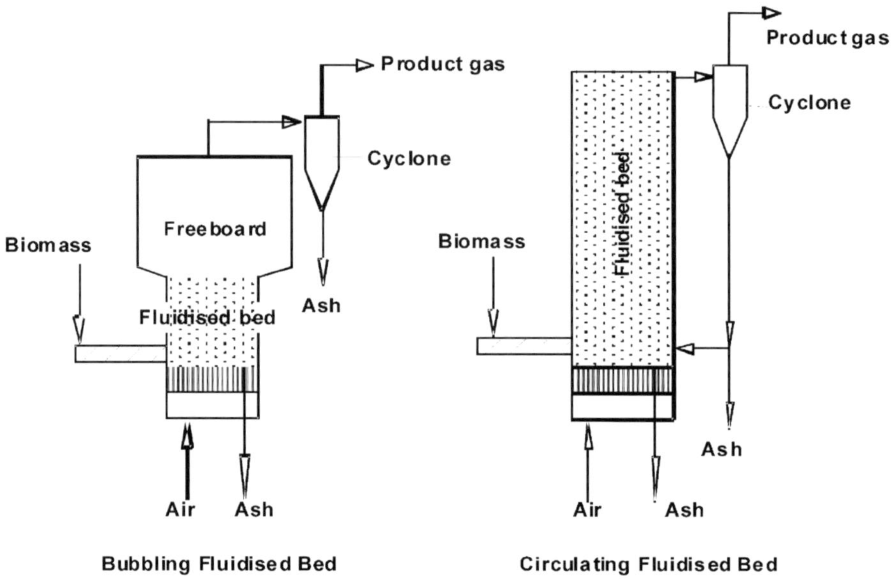

Fig. 6.14 Operating principles of fluidised bed gasifiers

fluidised beds. The bed material usually used is silica sand or, for high ash fuels, the ash of the fuel. The gasification temperature is typically around 800°C. The long residence time of the solid fuel and the intensive mixing are the reasons why very high gasification rates are achieved. The basic design types are bubbling (BFB) and circulating (CFB) fluidised beds, which can be seen in Fig. 6.14.

In a bubbling fluidised bed, the oxidant velocity is significantly lower than the terminal velocity of the bed material. The freeboard (the free space in the furnace above the bed) therefore has a gas flow which contains only small ash particles. In a circulating fluidised bed, in contrast, the oxidant approach velocity lies in the order of magnitude of the terminal velocity, and as a consequence the bed material is carried into the freeboard, thus forming a gas/solids flow through the entire reactor. By way of a cyclone, the transported bed material is separated from the gas stream and recirculated into the reactor. CFBs have significantly higher specific outputs. In addition, the gas/solids flow makes the mixing efficiency higher than in BFBs, which results in a better fuel conversion and lower tar contents. Drawbacks are the stricter fuel form requirements (the fuel needs to be in the form of grains) and the significantly higher pressure loss (meaning a higher process power consumption). What is more, controlling the bed material and the recirculation flow is more complex and the construction must be much taller than for a BFB. For lower outputs, a bubbling fluidised bed is therefore the better solution. With respect to the tar content, bubbling fluidised beds perform much worse (by about one order of magnitude) than co-current fixed bed gasifiers. Circulating fluidised beds are somewhat better, but do not reach the low concentrations of co-current fixed bed gasifiers.

In nearly all medium-to-large-scale electricity-producing biomass gasification demonstration plants to date, circulating fluidised beds (CFB) have been the preferred technology. The main reasons are that CFBs can handle a high throughput, are easy to scale up and accept a wide range of fuels. However, tar conversion or tar scrubbing is required for all cold gas applications, that is, where the gas is fed to a boiler or an engine at ambient temperatures (Knoef and Ahrenfeldt 2005). Table 6.4 gives some examples of fluidised bed gasifiers in operation, but does not intend to provide a complete list. All CFB gasifiers are air-blown and produce a product gas with an LHV of 4–6 MJ/Nm^3 (with wood as the fuel).

Pressurised gasification technology was successfully demonstrated from 1993 to 2000 in the world's first complete BIGCC (Biomass Integrated Gasification

Table 6.4 Medium-to-large-scale fluidised bed biomass gasification plants (Spliethoff 2001; Knoef 2005)

	Gasifier	Fuels	Use of product gas	Start of operation/ shutdown
Ruedersdorf, Germany	100 MW ACFB, Lurgi	wood, RDF, lignite, waste	Cement kiln	1996
Pietarsaari, Finland	35 MW ACFB, FW	Bark, Wood, waste	Lime kiln	1983
Norrsundet, Sweden	27 MW ACFB, FW	Bark, wood waste	Lime kiln	1985
Rodao Mill, Portugal	17 MW ACFB, FW	Bark, wood waste	Lime kiln	1986
Zeltweg, Austria	10 MW ACFB, AEE	Wood	PC co-firing	1997–2000
Lahti, Finland	40–70 MW ACFB, FW		PC/NG boiler	1997
Geertruidenberg, Netherlands	80 MW ACFB, Lurgi	Wood waste	PC co-firing	2000
Ruien, Belgium	86 MW ACFB		PC co-firing	2002
Greve-in-Chianti, Italy	2 × 15 MW ACFB, TPS	RDF	Steam cycle 7 MW$_e$	1993
Burlington, USA	Battelle Columbus interconnected CFBs		Initially steam cycle power plant	1997/shut down
Hawaii, USA	IGT Renugas Pressurised BFB		Gas cleanup testing	1995–1997
Värnamo, Sweden (Bioflow)	18 MW PCFB, FW		IGCC	1993/1999
Tampere, Finland Biocycle	7 MW$_e$ PBFB Carbona		IGCC	Unknown
Aire Valley, United Kingdom (ARBRE)	8 MW$_e$ ACFB, TPS	Willow, poplar	IGCC	2001
Güssing, Austria	8 MW FICFB	Woodchips	Gas engine	2001

ACFB: atmospheric circulating FB, PBFB: pressurised bubbling FB, PCFB: pressurised circulating FB, FICFB: fast internal circulating FB, FW: Foster Wheeler, AEE: Austrian Energy & Environment, TPS: Termiska Processor AB

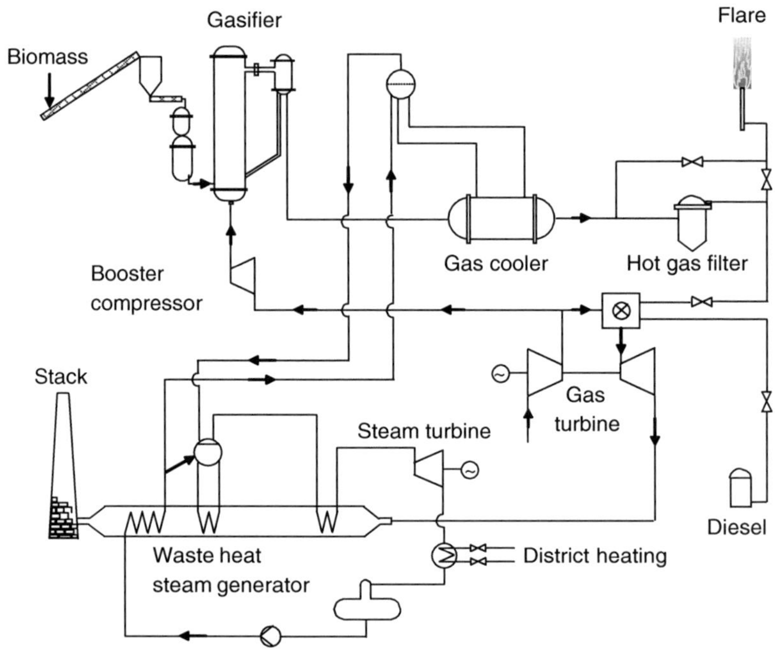

Fig. 6.15 Process flow diagram of the Värnamo plant (Kaltschmitt et al. 2009)

Combined Cycle) power plant in Värnamo, Sweden (Sydkraft 2001). The plant had a fuel input of 18 MW, an electrical output of 6 MW_{el} and a heat production of 9 MW_{th}. Figure 6.15 shows the process flow diagram of the Värnamo plant. Dried and crushed wood is pressurised in a lock hopper, then fed by screw feeders into the gasifier. The operating temperature of the fully refractory-lined air-blown CFB gasifier is 950–1,000°C and the pressure is about 18 bar. The gas produced in the gasifier is cooled to a temperature of about 350–400°C. After cooling, the gas enters a candle filter, where particles are removed. The clean product gas, with a heating value of 5–7 MJ/Nm^3, is fed to the turbine. By the end of the demonstration programme the plant had been operated in gasification mode for more than 85,000 h and the gas turbine had run on product gas for more than 3,600 h. It was shown that pressurised BIGCC technology works. Experiences from the demonstration programme are given in Knoef and Ahrenfeldt (2005).

Two-Stage Allothermal Fluidised Bed Gasifiers

Allothermal gasification offers the advantage of producing a product gas with a higher calorific value than that from autothermal gasification. The principles of autothermal and allothermal gasification are described in Sect. 7.6.3. Allothermal gasification requires an external heat source to supply the required energy for the gasification process. Steam is used as the gasification agent. In order to avoid the necessity of an external heat source for supplying the gasification heat, systems with two fluidised bed reactors have been developed. In one reactor, biomass is gasified, while biomass or char from the gasification reactor is combusted in the other. The transfer of heat from the combustion to the gasification reactor is accomplished by the transfer of sand, which acts as the energy carrier. Such a process has the advantage of producing a gas with a very low nitrogen content without the use of oxygen.

An example is the SilvaGas process, which has been demonstrated with a feed capacity of 44 MW in Burlington, Vermont, USA (see Fig. 6.16). The process consists of two CFBs. Biomass is fed to the gasification reactor, where it is mixed with hot sand and steam. Sand (the heat carrier) and the remaining char are separated in a cyclone and discharged to the combustor. The sand is reheated by burning the char with air in the combustor, then returned to the gasifier. The operating temperature of the gasifier is 815°C and the temperature of the combustor is 980°C (Knoef and Ahrenfeldt 2005).

The FICFB (fast internal circulating fluid bed) process developed by the University of Vienna is another example of such a process. It separates the steam gasification of the biomass from the combustion of the char, the latter of which is the heat source for the gasification. A commercial demonstration combined heat and power plant with a fuel power of 8 MW has been built in the town of Güssing, Austria. A schematic of the plant is given in Fig. 6.17. The synthesis gas is fired in a gas motor, generating 2 MW_{el} and 4.5 MW heat. The plant was put into operation in 2001. The gasifier operates as a bubbling fluidised bed with sand as the bed material. The sand and the ungasified char leave the bottom of the reactor and are transferred to the CFB combustor, where the char is burnt. The hot sand is separated from the flue gas in a cyclone and is returned to the gasifier, supplying the required gasification heat. The gasification reactor is operated at about 850–900°C, producing a gas with an LHV of 12 MJ/Nm^3 (dry).

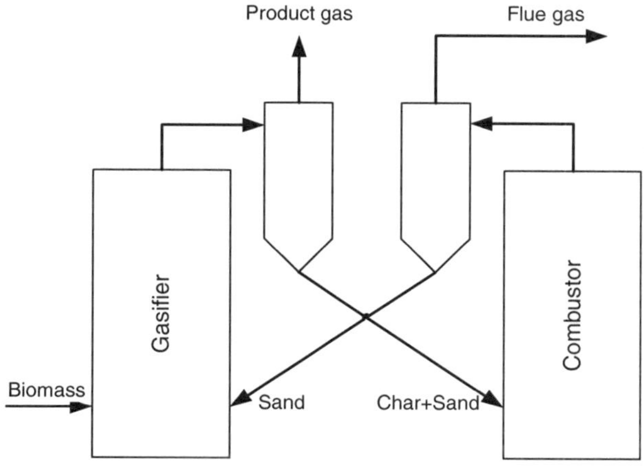

Fig. 6.16 Schematic of the SilvaGas (Batelle) gasifier

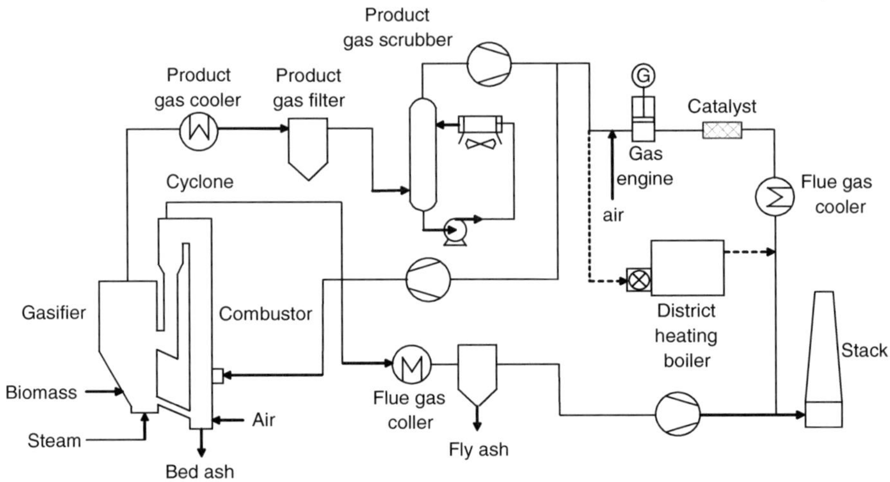

Fig. 6.17 Schematic of the Güssing plant (from Higman and van der Burgt 2008, © 2008, with permission of Elsevier)

Entrained-Flow Gasification

Entrained-flow gasification (described in more detail in Sect. 7.6) is the most common process for coal gasification. Despite the trend of using fluidised beds for biomass gasification, there are some cases of biomass gasification using entrained-flow technology.

One example is the Choren process, which is a combination of a low-temperature pyrolysis process with a two-stage entrained-flow gasifier, as shown in Fig. 6.18.

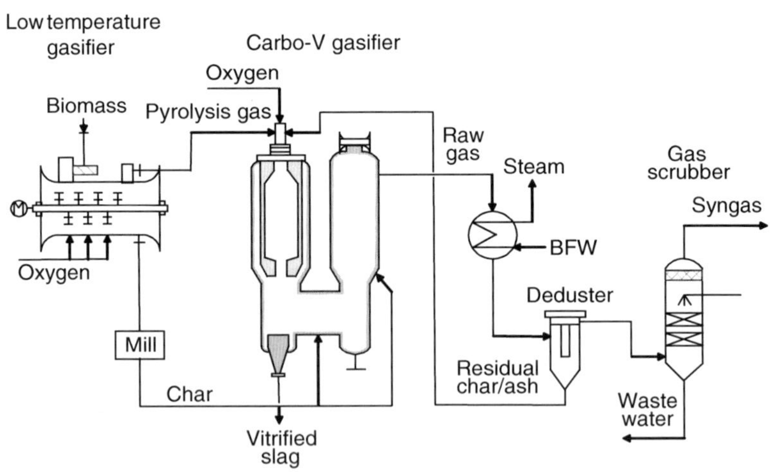

Fig. 6.18 Process flow diagram of the Choren process (from Higman and van der Burgt 2008, © 2008, with permission of Elsevier)

The biomass is fed into the stirred horizontal low-temperature gasifier and is pyrolysed in the presence of oxygen at a temperature of between 400 and 500°C. The pyrolysis gas and the char are extracted separately. The pyrolysis gas is then gasified in the first part of a high-temperature entrained-flow gasifier at temperatures of above 1,400°C, which effectively destroys the tars. Milled char from the pyrolyser is used as a chemical quench in the second part of the entrained-flow gasifier, where the hot gases from the first zone provide the energy for the gasification of the char. The raw gas leaves the reactor at a temperature of 800°C.

6.3.2 Gas Utilisation and Quality Requirements

The gas produced in the gasifier can be used in various ways for electricity production and for the production of process heat. These systems differ with respect to their efficiency, costs and gas quality requirements. The options for gas utilisation are shown in Fig. 6.19.

6.3.2.1 Gas Utilisation in Boilers and Cement Kilns

The most simple method of using gasification gas is to burn it in a steam generator or a cement kiln or to co-fire it in a coal power plant. This method requires only primary cleaning, such as by cyclones, of the gasification gas. If the gas is kept at temperatures above 500°C after it has been produced, tar conversion or tar scrubbing is not required. The majority of the commercially operated gasifiers supply gas for such thermal purposes. Among these, the plant at Rüdersdorf, Germany, with a thermal output of 100 MW, is the largest at present. The combustion of a gas instead of solid biomass simplifies the combustion process in the steam generator or the lime kiln (Rüdersdorf) and reduces ash-related restrictions. However there is no gain in efficiency over direct firing of biomass.

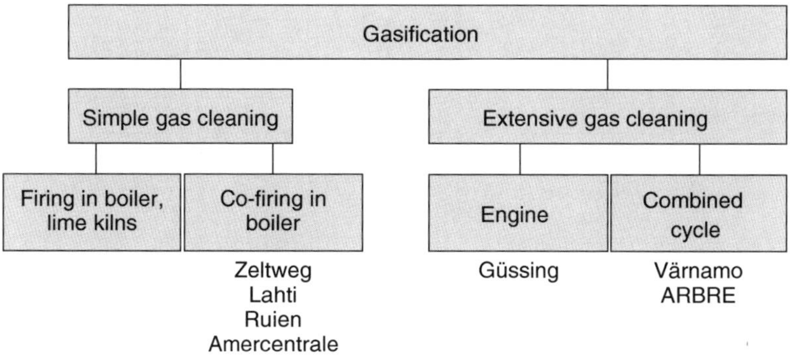

Fig. 6.19 Options for gas utilisation

Utilisation in Gas Engines and Gas Turbines

Gas utilisation in a turbine, motor or fuel cell for power production offers a higher efficiency than combustion in a steam generator.

The gas quality requirements for these applications are very high. In order to avoid fouling and deposits in the engine, the gas should be tar and dust free to a high degree. Typical concentration requirements for gas engines and gas turbines are listed in Table 6.5. Turbo-charged engines require an even higher gas quality for operation. It should be noted that the values given in published literature vary. Gasifiers available on the market today far exceed the indicated values when operated without gas cleaning. The removal of both tar and particles is therefore a prerequisite. The required concentrations are the result of a compromise between increased gas cleaning expenditure and a higher engine or turbine maintenance demand.

Engines are suitable for capacities between 50 kW_{el} and 10 MW_{el} and are used in connection with atmospheric fixed bed or fluidised bed gasifiers. With engines, the maximum electricity production efficiency is around 30%, although by including waste heat utilisation, the overall efficiency can be higher. Smaller plants yield lower efficiencies of up to about 25%. These efficiencies are somewhat above those that can be achieved by steam turbines in this capacity range.

From a capacity of about 5 MW_{el}, gas turbines are the better technology. The gasifiers suitable for use in connection with such turbines are atmospheric or pressurised fluidised bed reactors. With gas turbines, it is possible to increase the efficiency up to 45% by installing a tailing waste-heat boiler with a further steam turbine (capacities > 25 MW_{el}).

Only a few integrated gasification processes using gas turbines have been demonstrated, so experience with such plants is limited. In Vaernamo, Sweden, a pressurised fluidised bed furnace with an electrical output of 6 MW_{el} was in service from 1993 to 2000 (see Fig. 6.15). An atmospheric bubbling fluidised bed using the TPS (Termiska Processor AB) system, with an electric output of 8 MW_{el}, was put into service in 2000 (ARBRE Project, Great Britain). Both systems are now out of service due to economic reasons.

Table 6.5 Gas quality requirements for gas engines and gas turbines (FNR 2006; Spliethoff 2001; Kaltschmitt 2009)

	Gas engine	Gas turbine
LHV [MJ/Nm^3]	> 2	> 5
Tar [mg/Nm^3]	< 100	< 5
Particle [mg/Nm^3]	< 50	< 30
Particle size [μm]	< 3	< 5
Ammonia [mg/Nm^3]	< 30–55	
H_2S [mg/Nm^3]	< 1,150	< 1
Alkalis [mg/Nm^3]	< 50	< 3
Halogens [mg/Nm^3]	< 100	< 2

6.3.3 Gas Cleaning

Gas cleaning for biomass gasification remains an area of uncertainty, in need of development. Gas cleaning depends on the cycle arrangement and the gas quality requirements of the gas engine or gas turbine.

Atmospheric gasification systems require the removal of both particulates and tars. In atmospheric systems in which the product gas is fed to an engine or a compressor, cooling of the product gas down to about 50°C is required. In such cases, the most simple method for cleaning is to perform the separation of particles and tars in one step at the temperature that the gas is used at. Separation uses physical methods which require the cooling of the gas prior to cleaning, such as wet scrubbing or wet electrostatic precipitation.

The separate collection of tars and particles is a strategy applied in dry, high-temperature gas cleaning systems, using hot gas particle filtration and a thermal or catalytic cracker for tar reduction. Separate removal processes can also be advantageous when the tar separation unit, for example, a scrubber with an organic solvent at low temperatures, requires a low product gas particle content. Disposal restrictions, for instance, the forbidding of a mix of tars, particles and water to be discharged, may also necessitate separate removal processes.

Pressurised systems with gas utilisation in a turbine do not require a tar separation step if the gas flow can be kept above the dew point of the tars. In this case the chemical energy of the tars can be exploited in the turbine. Solids are removed in hot gas filters at temperatures of 400–500°C; typically, ceramic or metal candle filters with back-pulsing are employed. The temperature of the filtration step has to be below 500°C so that alkalis can be removed by condensation on the particles. Temperatures below 400°C have to be avoided to prevent tar condensation. Lower temperatures reduce the efficiency of the cycle because of losses of sensible heat and chemical energy in the tars (FNR 2006; Knoef and Ahrenfeldt 2005; Kaltschmitt 2001; Spliethoff 2001).

6.3.3.1 Tar Formation in Gasification

Tars are organic compounds (hydrocarbons), which condense at room temperature. The tar species formed in gasification are aromatic, heterocyclic aromatic or polyaromatic hydrocarbons. The species that are commonly found in fractions above 5% are toluene, naphthalene and, when process temperatures are below 800°C, phenol. A great number of compounds occur only as trace elements, but taken as a group they can also constitute a considerable fraction of the tar quantity.

While the literature does not provide a uniform definition of the term "tar species", efforts have been undertaken in European projects to standardise tar classification and tar measurements. Based on the so-called tar guideline (or "tar protocol"), a CEN (Comité Européen de Normalisation) standard (CEN Technical Specification) has been developed (Neeft et al. 2002; Good et al. 2005; Coda et al. 2004; DIN 2006). The intention is to make tar measurements in gases produced from biomass comparable.

Class	Name	Example
1	GC undetectable tars. Heavy tars that condense at high temperature even at very low concentrations.	Gravimetric tars
2	Heterocyclic aromatic components. These components have a high water solubility.	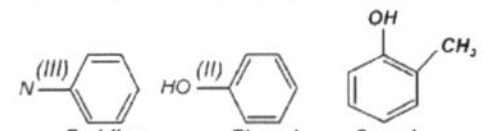 Pyridine, Phenol, Cresol
3	Aromatic (1-ring) components. Light hydrocarbons that are not important in condensation and water solubility issues.	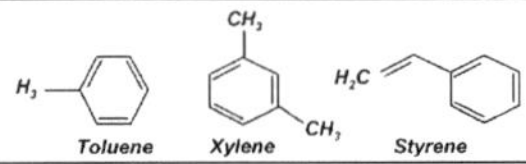 Toluene, Xylene, Styrene
4	Light polyaromatic hydrocarbons (2-3 rings PAH`s). They condense at relatively high concentrations and intermediate temperatures.	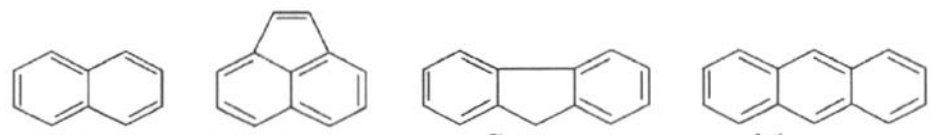 Naphtalene, Acenaphtylene, Fluorene, Anthracene
5	Heavy polyaromatic hydrocarbons (4-5 rings PAH`s). They condense at relatively high temperature already at low concentrations.	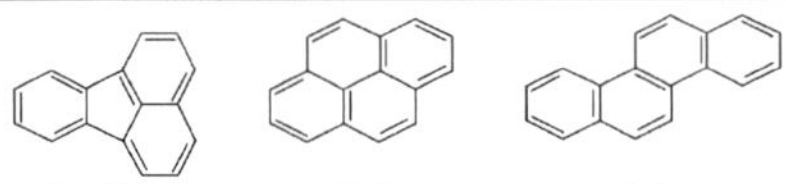 Fluoranthene, Pyrene, Chrysene
Not considered as tar	Light hydrocarbons, that do not condense at ambient temperatures in normal concentrations	 Benzene

Fig. 6.20 Tar classification and chemical structure of selected tars. GC = gas chromatograph

The "tar guideline" defines tars as the group of all organic compounds excluding gaseous hydrocarbons (C1–C6) and benzene. Benzene is not included because it does not condense at room temperatures at the concentrations typical of biomass gasification. The classification system in the "tar guideline" reports five different classes of individual tar compounds, as shown in Fig. 6.20 along with chemical structures of some typical tar components.

Class 1 consists of heavy poly-aromatic hydrocarbons (PAH) which cannot be detected with a gas chromatograph (GC) and is determined by subtracting the GC-detectable tar fraction from the total gravimetric tar. Class 2 tars are aromatic compounds with hetero atoms (pyridine, phenol) while class 3 tars are light compounds with one aromatic ring (xylene, styrene, toluene). The last two categories, classes 4 and 5, consist of light polycyclic aromatic hydrocarbons (PAH) with two or three aromatic rings (naphthalene, fluorene, anthracene) and heavy PAH with four to five aromatic rings (pyrene) (Neeft et al. 2002).

Whether a component condenses when the temperature falls below the boiling point depends on the steam pressure and on the concentration. Figure 6.21 shows the saturation concentration of some typical tar components and benzene in nitrogen. Benzene at 25°C, for instance, reveals a saturation concentration of more than $300\,g/m^3$. Since this value is higher than the typical concentrations in product gases by orders of magnitude, the condensation of benzene is not to be expected and benzene is not considered a tar. On the other hand, components such as fluorene have saturation concentrations of only a few milligrams per cubic metre at 25°C, so an almost complete condensation has to be taken into account.

The measurement of the dew point gives the highest temperature at which the first tar molecules can condense, information which is required to prevent a

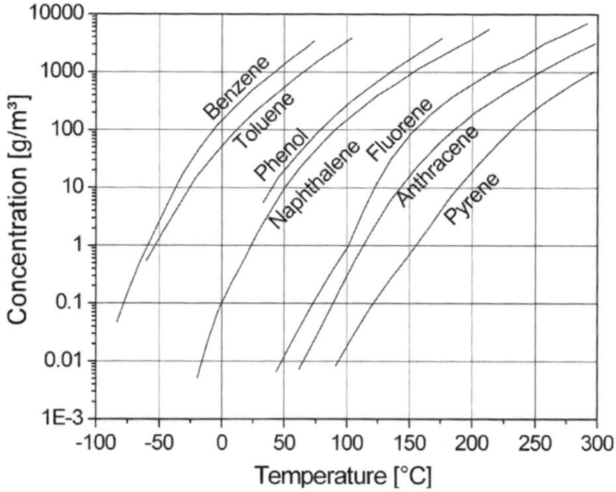

Fig. 6.21 Saturation concentrations of some tar components in nitrogen (Spliethoff et al. 1998)

trouble-free operation of the gas cleaning train as well as trouble-free gas utilisation. The dew point is primarily dependent on the molecular mass of the compound and secondarily on the concentration of the compounds. An online tar dew point analyser has been developed based on the principle of tar condensation on an optical surface (van Paasen et al. 2005).

The tars grouped in each class of the "tar guideline" show a similar condensation behaviour. Class 3 tars have a very low dew point, much below ambient temperatures, so they will not condense in gasification systems. Class 5 tars have a high dew point (120°C at 0.1 mg/m^3), so they will always condense.

Different methods, using different principles, exist for the measurement of the total tar concentration and the individual concentrations of tar compounds. The most widely used method is the tar guideline, which uses a modular sampling train in which the gas from biomass gasification flows through a series of impinger bottles filled with an organic solvent. Since the tars are diluted and collected in the bottles, they are analysed gravimetrically. Individual organic compounds can be determined by GC and high-performance liquid chromatography (HPLC). Another method traps tar vapours in a polypropene cartridge using a solid amino-phase adsorbent. The analysis of the fractions is then made by means of a GC-flame ionisation detector (FID) technique (Braage et al. 1997).

A quasi-online tar measurement method has been developed based on continuous GC-FID measurements. Total hydrocarbons are measured both before and after removal of tars by condensation, the difference corresponding to the mass of the tars (Mörsch 2000). The application of laser spectroscopy for quantitative tar measurements is under investigation (Mitsakis et al. 2008).

The tar quantity and composition at the outlet of a gasifier highly depend on its construction and the operating parameters. Co-current fixed bed gasifiers produce a gas with a relatively low tar content, whereas the gas from counter-current gasifiers contains a high level of tar. In a co-current gasifier, the pyrolysis gases flow through

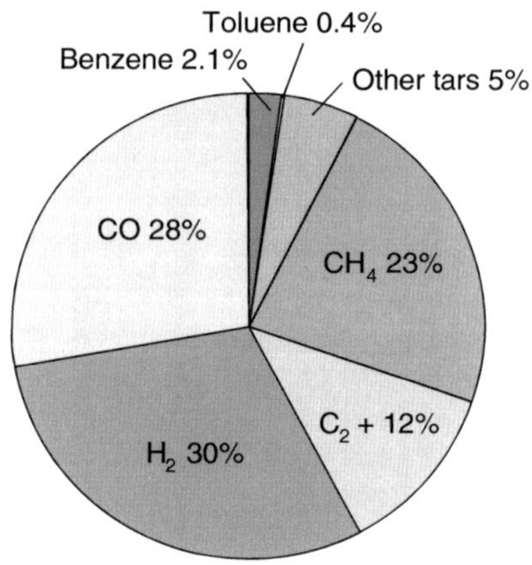

Fig. 6.22 Contribution of each gas component to the chemical energy of the product gas (beach wood, 800°C, $\lambda = 0.25$) (Mörsch 2000; Spliethoff et al. 1998)

the hot reaction zone, which results in the tar species getting cracked. The tar content of the product gas from fluidised bed gasification lies between these two values.

Investigations carried out in a bench-scale fluidised bed gasifier have revealed the impact of process conditions on tar formation. The reference case is the standard test with beech wood as the fuel, sand as the bed material, a gasification temperature of 800°C and an air ratio of $\lambda = 0.25$. The tar concentration in this case was $8.2\,g/m^3$. Figure 6.22 shows the contribution of each of the gas components to the total heating value of the product gas of $5.2\,MJ/m^3$.

Figure 6.23 gives the tar content of the product gas under variations of the operating and fuel parameters. The graphic shows for each case the tar concentrations after

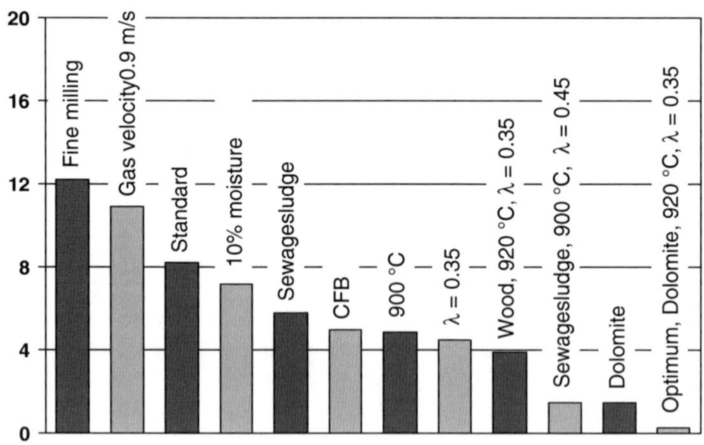

Fig. 6.23 Influence on the tar content of the tested operating parameters compared to the standard test case for a bench-scale fluidised bed (Mörsch 2000; Spliethoff et al. 1998)

a modification of a parameter in comparison to the reference case. With increasing process temperatures and air ratios, the tar content drops. At higher temperatures, there is also a shift in the tar composition towards lighter components. It should be noted that an electrically heated facility such as this bench-scale fluidised bed gasifier enables the air ratio and the temperature to be varied independently, whereas in a real autothermal gasifier, the air ratio and the gasification temperature are coupled. High temperatures in the bed can be limited by bed agglomeration.

The most effective measure to reduce tar concentrations established in the tests was the use of a catalytic material – in particular, dolomite. When employing this material, tar concentrations suitable for the direct utilisation of the gas are reached at temperatures above 900°C. The lowest tar concentrations using a dolomite bed, of just 250 mg/m^3, were produced at 920°C and an air ratio of $\lambda = 0.35$. The resulting calorific value was 5.2 MJ/m^3 at a cold gas efficiency of 85%. The main problem in using this technique is the low abrasion resistance of burned dolomite, meaning the catalyst gets entrained in the gas flow along with some of the bed material. To make up for this loss, it is necessary to constantly add a certain amount of bed material.

6.3.3.2 Secondary Tar Reduction

Tar cleaning is required if the product gas has to be cooled prior to use (for example, when used in gas engines) or if it has to be compressed (as in a combined cycle process with an atmospheric gasifier). The amount of tars present in the product gas, as described previously, depends on the gasification temperature, the reactor design and the biomass type. In most cases primary measures for tar reduction are not sufficient to meet the requirements for gas utilisation, so secondary measures for tar removal have to be implemented. These measures can take the form of a physical separation or a chemical conversion of the tars.

- In most cases physical methods are used for tar reduction. The two most common methods are scrubbing and electrostatic precipitation. Both methods remove the tars in a condensed form, which requires the gas to be cooled prior to the cleaning step. If the separated tars are not recycled to the gasification reactor, their chemical energy is lost from the process, reducing the cycle efficiency. Physical separation of tars and particles can be performed in a single step in one device at low temperatures.
- Chemical conversion of tars into lighter gas components can be achieved by thermal or catalytic tar cracking. After such processes, the chemical energy of the tars can be used to increase the heating value of the product gas.

Scrubbers

Wet scrubbing is the most common method for tar removal. If water is used as the scrubbing medium, the tar separation efficiency is limited and multistage cleaning may be required. In order to be able to separate tars in a wet scrubber, the tars have to be condensed so that the aerosols and droplets collide with the water and increase their particle size. For this reason the product gas has to be cooled and saturated with water before the cleaning step. Because tars are hydrophobic and have a low solubility in water, tars which remain in the vapour phase cannot be removed.

Using a wet water scrubber to remove tar from the product gas requires a gas temperature of 30–60°C. Different washer types are in operation, for example, tower, rotating tower and Venturi washers. To achieve a higher tar removal efficiency, washers tend to be multistage, meaning a higher pressure drop. Clean gas concentrations of 20–40 mg/m^3 can be achieved.

By using lipophilic liquids, which can act as solvents and so are used as a scrubbing medium, gas phase tars can also be removed. At the Güssing plant in Austria, a wet scrubber is used with oil (RME) as the scrubbing liquid. Used oil, saturated with tars and condensate, is vaporised and recycled to the gasifier. Another scrubbing technology (OLGA) has been developed and patented by ECN. The removal of tars is accomplished by scrubbing the tar-loaded product gas with a specially developed liquid oil in an absorption column. Tar removal efficiencies of 99% of the heterocyclic tars and almost all the heavy and light tars have been measured. The dew point of tars is as low as -17°C. The current design of the OLGA technology requires a dust-free gas.

Fixed Bed Filters

Fixed bed filters employing sand, saw dust or other materials as the filter medium are usually used in smaller units due to their simple design. Sand filters have a high removal efficiency of up to 95%, but use of them is problematic because of the need to dispose of the contaminated bed sand.

Wet Electrostatic Precipitators

A wet ESP, which is operated at temperatures of about 60°C, is a more attractive solution than a wet scrubber due to a higher removal efficiency, a lower pressure drop and a lower quantity of waste water being produced. The gas is cooled and saturated with water prior to the cleaning, then the liquid droplets and particles are separated by electrostatic precipitation. The removal of the condensed tars is improved by the use of a small water stream to flush them away. The ESP has the advantage of high particle and tar removal rates. The water – tar – dust mixture can be fed back into the gasifier.

Catalytic Tar Reduction

Higher tar reduction rates than those of physical gas cleaning can be achieved with catalytic hot gas cleaning at temperatures between 800 and 900°C. In this process, the gasifier is backed by a "tar cracker" – a fixed bed, a fluidised bed or a honeycomb structure filled with catalytically active material. Materials that have revealed themselves to be extremely effective in reducing tar are limestone and dolomite. Nickel catalysts are also known for giving a very high reduction efficiency, being used in particular in commercial applications for steam reforming. The advantages of these systems are that theoretically, no waste matter is produced, and the chemical energy of the tar species remains in the gas, thus having an efficiency-enhancing effect. Catalytic hot gas cleaning is a feasible option for tar reduction if high-temperature fuel cells are used to exploit the product gas. In some investigations, dolomite has brought about tar reduction rates of more than 99.5%; nickel compounds achieved rates up to as high as 99.99%. The drawback of secondary tar crackers is the relatively high cost of the additional equipment. What is more, to raise the temperature to the optimum process level, a certain quantity of air often needs to be added to partially combust the gas, a factor unfavourably affecting the efficiency.

As well as using a catalyst in a secondary tar cracker behind the gasifier, it is possible to use one directly as part of the bed material in a fluidised bed gasifier. The choice of possible catalyst materials for this purpose is limited to limestone, dolomite and nickel compounds. Nickel catalysts, however, have a number of disadvantages if used as a bed material or additive. First, nickel dust is toxic – a particular concern for the ash, which gets contaminated by nickel. Second, the catalysts very quickly lose their effectiveness in the presence of carbon, as carbon deposits form on the surface of the nickel compounds. In consequence, nickel compounds are only suitable for application after a dust removal stage. The effectiveness of calcium-based sorbents for in situ tar cracking in a fluidised bed gasifier is shown in Fig. 6.23. In order to reduce the consumption of the catalytic bed material, the catalyst should have a minor attrition rate. A positive side effect of catalytic tar cracking is the reduction of ammonia to N_2, with conversion rates of 70–80% (see Table 6.6).

Table 6.6 Removal efficiencies of different tar cleaning devices (Kaltschmitt 2001)

	Temperature [°C]	Particle reduction [%]	Tar reduction [%]	NH_3 reduction [%]	H_2S reduction [%]	HCI reduction [%]
Sand filter	10–20	70–99	50–97	> 95	80–95	90
Washing tower/rotating	50–60	60–80	10–25			
Venturi washer			50–90			
Rotating sprayer	< 100	95–99			> 95	90
Wet ESP	40–50	> 99	0–60			
Bag filter	130	70–90	0–50			
Rotating filter	130	85–90	30–70			
Catalytic cracker	900		> 95	70–80		

Thermal Tar Reduction

Thermal tar reduction (cracking) offers an effective method for removal of tars and is typically performed at temperatures of about 1,200°C. The application of thermal tar crackers to fluidised bed gasifiers therefore would require a temperature increase from 800 to 1,200°C, which would have to be accomplished by partial oxidation. As partial oxidation would reduce the heating value of the product gas and would reduce the cycle efficiency, it is not used in biomass gasification.

Table 6.6 gives an overview of particle and tar removal efficiencies for different devices.

6.3.3.3 Particle Cleaning

The principles of particle cleaning are discussed in detail in Sect. 5.8 in the context of dust removal from combustion systems and in Sect. 7.4 in the context of pressurised fluidised bed combustion.

Low-temperature gas cleaning devices such as wet scrubbers or wet electrostatic precipitators have been described in Sect. 6.3.3.2. The systems operate at temperatures of about 50°C and are capable of removing both particles and tars. The particle removal efficiency is generally higher than that for tars.

Barrier filters are the common choice for dry particulate removal. Barrier filters with a rigid, porous, metal or ceramic candle can be operated at temperatures up to 500°C. However due to the higher costs and lower reliability, barrier filters such as bag house filters, which can be operated at temperatures of up to 350°C, are usually preferred. Depending on the dew point of the tars, condensation of tars in the filter can occur, blocking the filter. At the Güssing plant the fabric filter is operated at a temperature of about 160–180°C. Since some tar condensation on the filter does not inhibit the operation of the process as a whole, the filter is coated with a material that is cleaned from the filter together with the particulates when it is back-flushed. The cleaned material and particulates are fed back to the combustion section of the gasifier.

6.3.4 Power Production Processes

The focus of the following comparison of potential processes is power production at the medium to large scale. Within this range, circulating fluidised beds are considered the most promising technology, and therefore only these will be considered in this comparison, though the concepts can be transferred to other gasification systems. The processes under consideration are shown in Fig. 6.24.

A. *Pressurised gasification with a combined gas–steam cycle*: The concept of pressurised fluidised bed gasification has been put into practice at the Värnamo Biomass Integrated Gasification Combined Cycle (BIGCC). Pressurisation

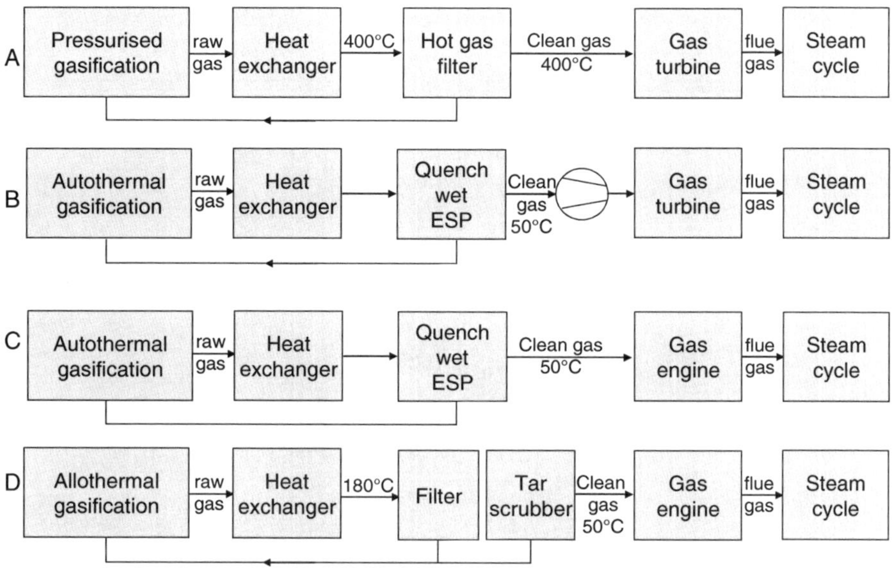

Fig. 6.24 Power production processes (Knoef and Ahrenfeldt 2005)

allows the use of the product gas in the gas turbine after gas cleaning without further compression. Therefore tar removal is not required and the heating value of the tars can be used in the gas turbine. The gas conditioning and cleaning consists of a heat exchanger to cool down the product gas to about 400°C and then dust removal in a hot gas filter. The cleaned gas is combusted in the gas turbine, which also supplies the gasification reactor with pressurised air. The waste heat of the gas turbine is transferred to a waste heat steam generator.

B. *Autothermal atmospheric gasification with a combined gas–steam cycle*: The product gas from the autothermal atmospheric gasifier is cooled in a heat exchanger down to a temperature of about 400°C. Additional cooling is achieved by quenching until a suitable temperature is reached for dust and tars to be removed in a wet electrostatic precipitator. The removal of tars is required for the compression of the product gas to the pressure of the gas turbine.

C. *Autothermal atmospheric gasification with a gas engine and waste heat utilisation*: The gasification and gas cleaning is the same as in case B, but instead of a gas turbine, a gas engine is used for power production. The flue gas from the gas engine has to be treated catalytically to achieve the CO emission limits. The waste heat is used in a steam cycle.

D. *Allothermal gasification with a gas engine*: This concept is based on the design of the Güssing plant. The allothermal fluidised bed gasifier produces a gas with a low nitrogen content, which is cooled, de-dusted in a bag house filter and cleaned from tars using a solvent scrubber.

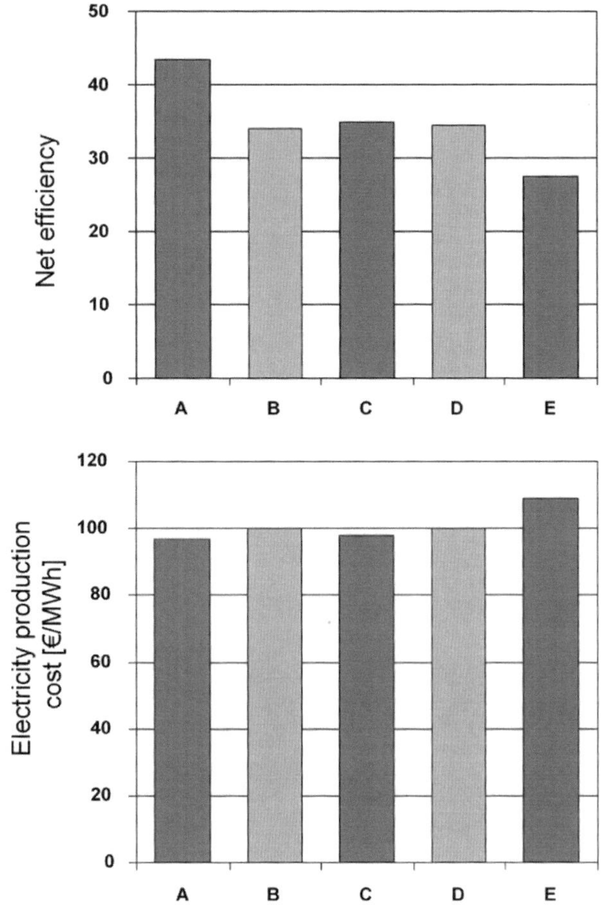

Fig. 6.25 Net electrical efficiency and production costs for biomass CFB processes (Knoef and Ahrenfeldt 2005)

The results of thermodynamic and economic assessments of the above process concepts are given in Fig. 6.25. The different gasification concepts are compared to a combustion-based conventional steam cycle (case E). The assessment has been carried out for an electrical output of 20 MW_{el}; details can be found in Knoef and Ahrenfeldt (2005). Pressurised fluidised bed gasification yields the highest efficiency of 44%, whereas the other cases achieve efficiencies of about 35%. The lower efficiency of atmospheric gasification with a gas turbine is due to the power demand for the compression stage. All gasification processes achieve much higher efficiencies than the 28% of the reference steam cycle.

The importance of the scale is demonstrated in Fig. 6.26, which shows that the pressurised process has lower power production costs only above a certain scale (6 MW_{el}), because of the higher capital costs of the pressurised system.

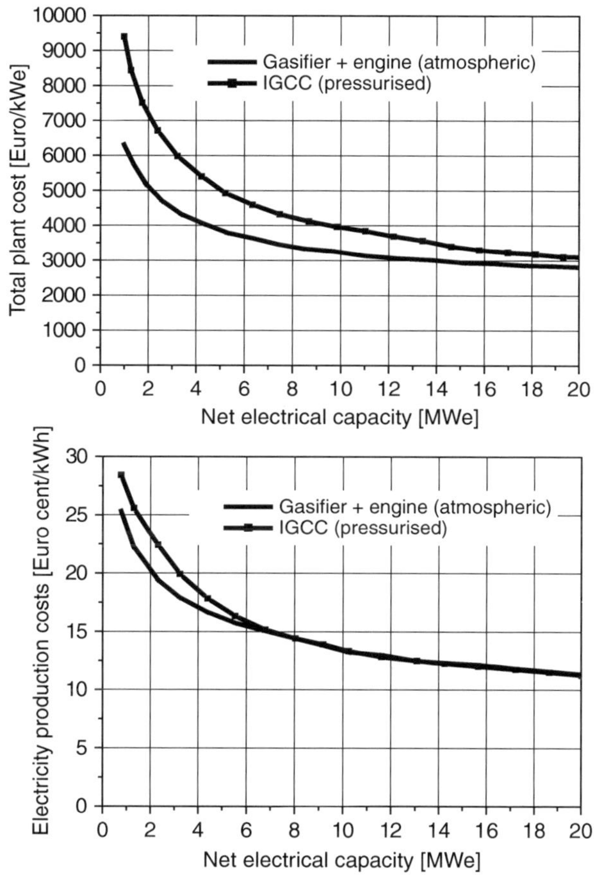

Fig. 6.26 Capital and electricity production costs as a function of the capacity for biomass CFB processes (Knoef and Ahrenfeldt 2005)

6.4 Thermal Utilisation of Waste (Energy from Waste)

Waste is an unwanted or undesired material or substance. The European Union, under the Waste Framework Directive (EU 2008), more precisely defines waste as an object the holder discards, intends to discard or is required to discard. Thermal waste treatment is an important element in modern waste management.

Table 6.7 lists the amount of various wastes fired in thermal waste treatment plants in Germany in 2006. The rise of thermal waste treatment plants was largely driven by the regulation restricting the disposal of MSW (AbfAblV 2000; TASi 1993). Today thermal waste treatment fulfils several purposes:

- To destroy, convert, separate, concentrate or immobilise harmful or hazardous components of the wastes
- To reduce the volume and amount of waste as far as possible

Table 6.7 Thermal treatment of waste in Germany in 2006 (Statistisches_Bundesamt 2008)

	Amount treated in 2006 (millions of tonnes)
Municipal solid waste (MSW)	18,142
Sewage sludge	1,669
Hazardous waste	1,056
Waste used as a fuel in power/CHP plants and others (wood residues, RDF, paper)	12,330

- To transform the remaining residues into usable substances or to put them into a disposable form
- To use the thermal energy released in the process to the greatest extent possible

Until 2008, the distinction between thermal treatment being "disposal" or "recovery" was made in Germany using the heating value of the waste, according to the German waste law "Kreislaufwirtschaftsgestetz" (KrWG 1994). The incineration of waste with heating values above 11 MJ/kg was considered "recovery", whereas below this value, the process was considered "disposal". The distinction followed a similar pattern in Italy, where the utilisation of so-called CDR (combustibles from waste, "Combustibile derivato dai rifiuti"), with heating values of more than 14 MJ/kg, was considered recovery. This kind of division seems to be rather arbitrary or politically motivated and is not consistent from the engineering point of view. The efficiency of an energy from waste (EfW) system depends on numerous factors, such as the process technology, the energy consumption of pre-treatment and transportation and generally speaking whether heat or electricity is used in the process. The heating value in itself only plays a minor role.

The European Waste Framework Directive (EU 2008) of 2008 introduced a new methodology for distinguishing between "disposal" and "recovery" with respect to thermal treatment:

- "Recovery" means "any operation the principal result of which is waste, serving a useful purpose by replacing other materials which would otherwise have been used to fulfil a particular function, or waste being prepared to fulfil that function, in the plant or in the wider economy." In plainer English, recovery is a process in which waste replaces another material or where it is prepared so that it can replace another material. The recovery criterion (R1-recovery) based on energy efficiency, described in Annex II of the directive, is explained below.
- "Disposal" means any operation which is not recovery, even where the operation has as a secondary consequence the reclamation of substances or energy. The disposal criterion (D10-treatment) is described in Annex I of the directive.

In practice, the distinction between "disposal" and "recovery" is made depending on the energy efficiency of the process. Plants gaining approval after 2008 need to have an energy efficiency corresponding to an $R1$ criterion above 0.65. The $R1$ criterion is defined in the footnote of Annex II of the Waste Framework Directive (EU 2008) as follows:

$$R1 \text{ energy efficiency} = (E_p - (E_f + E_i))/(0.97 \times (E_w + E_f))$$

In which

- E_p is the annual energy produced as heat or electricity. It is calculated by multiplying the electrical energy by 2.6 and the heat produced for commercial use by 1.1 (GJ/year)
- E_f is the annual energy input into the system from additional (i.e. fossil) fuels contributing to the production of steam (GJ/year)
- E_w is the annual energy contained in the treated waste calculated using the net calorific value of the waste (GJ/year)
- E_i is the annual energy imported into the process excluding E_w and E_f (GJ/year) (i.e. the auxiliary power consumption)

In a simplified way an $R1$ of 0.65 corresponds to a gross power generation efficiency of 24%, if only electricity is produced and no fossil fuel is used. This also assumes a perfect availability of the power generation components over the year (i.e. no unscheduled shutdowns and operation at full capacity) as well as an electricity in-plant consumption of 3% (referring to the gross heat input).

EfW plants in Switzerland and Amsterdam do not use fossil fuels, as assumed above. In most other plants in Europe, Japan or the USA, fossil fuels are used for start-up, shutdown and during unstable combustion conditions (for instance, as soon as the furnace temperature drops below 850°C at the 2 s-level of control). The use of, for example, 2.5% fossil fuel (by heat input) would decrease the $R1$ by 0.04. For this reason, it is estimated that in some cases an efficiency at the design point of up to 28% is necessary for reaching, over the year, an $R1$ value of 0.65 in practice.

The topic of energy efficiency in EfW systems is further detailed in Sect. 6.4.6.

One of the particularities of using waste as a fuel in combustion systems for energy generation is the need to meet the special emission requirements for both the flue gas and ash. In Europe, these emissions are subject to strict limits laid down in the Waste Incineration Directive (EU 2000), which shall be replaced in 2009 by the Industrial Emissions Directive (IED). The limits of the Waste Incineration Directive can be found in Table 5.7.

There are numerous kinds of waste materials, each of which needs to be treated in an environmentally sustainable manner. The focus in this chapter will be on municipal solid waste (MSW), refuse-derived fuel (RDF) and sewage sludge. More information on these waste streams can be found in Sects. 2.2.1.2, 2.21.3 and 2.2.1.4. They are produced in significant quantities and have properties such that it is reasonable to use them for energy recovery. Because half of the energy contained in MSW is biogenic, EfW is considered an important additional renewable energy source.

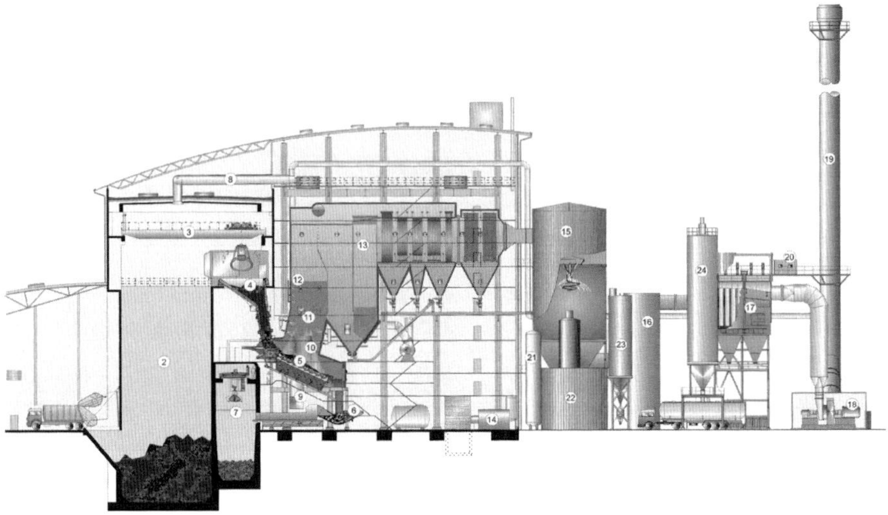

Fig. 6.27 Classical EfW system suitable for MSW, RDF and the co-combustion of sewage sludge (*Source*: Martin)

Municipal Solid Waste (MSW): In Germany, the thermal treatment of MSW and the resulting energy generation in classical EfW plants uses the most waste, with about 18 million tonnes being treated in 70 plants. An example of a typical plant employing classical technology is shown in Fig. 6.27. The key components are a waste bunker, a waste feeding process, a grate-based combustion system, a boiler, a turbine and a flue gas cleaning process. Typical boiler steam conditions are 380–420°C and 40 bar. The flue gas cleaning system is based on the injection of lime and activated carbon to remove pollutants and a downstream baghouse filter for removing the fly ash and the products of reaction from the pollutant removal. NO_x is reduced by ammonia injection in the furnace.

This type of EfW system is typically used for untreated municipal solid waste (also called residual waste), which can be quite heterogeneous in composition. In addition, residues from mechanical – biological waste treatment plants, as well as industrial wastes, are also fed to classical EfW plants. On top of that, co-combustion of between 5 and 15% sewage sludge (by mass input) is quite common.

Refuse-Derived Fuel (RDF): RDF (also called solid recovered fuel (SRF) or substitute fuel) is produced from municipal solid waste by mechanical – biological treatment, in most cases including commercial waste as a feedstock. RDF usually has a higher heating value and a narrower particle range than untreated MSW, which makes it possible to use it in fluidised bed combustion systems, unlike untreated MSW. Another option is co-combustion in coal power plants or in cement plants. For RDF processed from MSW it is most common to use grate-based combustion systems.

Sewage Sludge: Sewage sludge is a type of waste which arises in a relatively homogeneous form, largely because of its high water content. Typical sewage sludge, even after digestion and mechanical drying, has a water content of between 60 and 80%. It can be co-combusted in this state in waste incineration plants and brown coal-fired power stations or used as the only fuel in fluidised bed systems. For co-combustion in hard coal-fired power stations, thermal drying is usually necessary. In 2004, more than one-third of the 2.2 million tonnes of dry sewage sludge matter produced was combusted in Germany: 0.4 million tonnes in 21 monofuel combustion plants and 0.3 million tonnes in 25 coal power stations, with 0.05 million tonnes being co-combusted in EfW plants (Quicker et al. 2005; Schmelz 2006; Hermann 2004).

The following subchapters cover

- the historical development of EfW systems (with a focus on MSW, being the most used waste source for energy generation),
- grate-based combustion systems (most common for MSW and also used for RDF and the co-combustion of sewage sludge),
- pyrolysis and gasification systems (used for MSW and RDF),
- RDF systems (combustion systems being designed specifically for RDF using suspension or fluidised bed combustion),
- sewage sludge systems (mono-combustion),
- boilers for MSW, RDF and sewage sludge as a group and
- flue gas treatment for MSW, RDF and sewage sludge as a group.

The co-combustion of RDF and sewage sludge in coal power plants is described in Sect. 6.5.

6.4.1 Historical Development of Energy from Waste Systems (EfW)

The incineration of MSW began around the end of the 19th and the beginning of 20th centuries, when the first plants were constructed in England, the USA and Germany. The main purpose of these plants was to promote public health, as the practice of distributing waste on fields had been instrumental in the spread of cholera epidemics. The technology was very simple: brick-lined cell ovens with a fixed metal grate over an ash pit below, with one opening in the top or side of the oven for loading and another opening in the side for removing the solid residues, clinker or ash. Since that time, the technology has made huge advances in terms of emission control and energy efficiency (Gohlke and Spliethoff 2007).

An efficient energy from waste industry developed in Europe in the 1960/1970s and in the USA in the 1980/1990s. Typical EfW plants constructed in the USA during this period used steam parameters of 60 bar/443°C and, generally speaking, were more energy efficient than the European 40 bar boilers or the Japanese 20 bar boilers of this time. Waste incineration or energy from waste plants is often described in the USA as waste-to-energy plants (WTE), which can be considered as essentially the same.

In the 1980s, waste incineration plants became *the* symbol of environmental contamination: citizens were opposed to the throw-away society and "dioxin spouting" on the outskirts of cities. One of the reasons for the protests against waste incineration was the discovery of polychlorinated dibenzo-dioxins and polychlorinated dibenzo furans (PCDD/F), often simply called "dioxins", in the flue gas of waste incineration plants. Even though these dioxin concentrations in the flue gas were comparatively low, a connection was made to the dioxin accident at the chemical plant in Seveso, Italy, in 1976.

The protest led, in Germany, for example, to important developments: by 2005 more than half of all household waste (55%) was recycled as bio-waste, waste paper, waste glass or packaging waste. Since June 1, 2005, untreated waste has no longer been sent to landfill. Furthermore, because of stringent emission regulations, waste incineration or EfW plants are no longer significant in terms of emissions of dioxins, dust or heavy metals. This is the case even though total waste incineration capacity has almost doubled since 1985 (UBA 2005b). Table 6.8 shows the historical development of waste treatment capacity in classical EfW plants in Germany.

Regardless of the European or international increase in total EfW capacity, cost reduction remains one of the main targets for the development of EfW systems in Europe, because it is necessary to compete with the very cheap alternative of landfill. The inexpensiveness of landfill will obviously only remain as long as the long-term environmental burden is not considered. In some of the more advanced European countries, landfill of MSW is restricted and efforts are being concentrated on further improving the energy efficiency of EfW plants (Bonomo 1998; Van der Linde 2003; Wandschneider 2005; Fischer 2005; Seguin 2004). The driving force behind the implementation of high-efficiency systems is usually a premium for renewable electricity, for example, up to 170 €/MWh under the CHP6 program in Italy (which has now ended) (Pfeiffer 2003). In countries with sophisticated waste management systems like Sweden, Denmark, the Netherlands, Belgium, Germany and Switzerland, more than 170 kg of MSW per person per year is used for energy generation. Consequently, less than 40 kg per person per year is sent to landfill in these countries.

Table 6.8 Historical development of total waste treatment capacity in classical EfW plants in Germany (UBA 2005b)

Year	Number	Capacity, in 1,000 tonnes per year
1965	7	0.718
1970	24	2.829
1975	33	4.582
1980	42	6.343
1985	46	7.877
1990	48	9.2
1995	52	10.87
2000	61	13.999
2005	66	16.9
2007	72	17.8

The development in the USA was different: in 1994, the US Supreme Court rejected the practice of waste flow control (US_Supreme_Court 1994). Waste flow controls are legal provisions that allow state and local governments to designate the places where municipal solid waste is treated or disposed. This rejection then led to an abrupt halting of the development of energy from waste schemes. Instead, waste was transported by truck to mega-landfills which had been developed at low cost in more remote areas. This resulted in the very negative environmental impacts of landfilling: land consumption, transportation by truck over long distances, methane emissions, leaching of pollutants to soil and to groundwater. The development of more sustainable waste management systems was no longer possible. However, after landfill taxes and premiums for renewable electricity were introduced and fuel prices for transport by truck increased, there has been an increasing trend towards the development of improved EfW processes since the middle of the first decade of the 2000s.

Waste incineration is very widespread in Japan, as landfill has traditionally not been an option due to the lack of space. In the 20th century most waste was incinerated in numerous small plants. In the year 2000, there were still more than 10,000 plants with a capacity of less than 2 t/h, while there were more than 2000 plants with a capacity of more than 2 t/h (Konda 2000). The UNEP (United Nations Environment Programme) published a study in 1995 which showed that 4 kg of the global 10 kg TEQ (toxic equivalent units, a measure of all toxic dioxins, furans and PCBs in terms of the most toxic form of dioxin, 2,3,7,8-TCDD) of dioxin emissions per year came from Japan. This resulted in a major policy change and a new dioxin regulation was put into effect in 2003. The dioxin emissions for existing plants larger than 4 t/h were limited to 1 ng TEQ/Nm^3 and for new plants to 0.1 ng TEQ/Nm^3. Up until the end of 2002, a dioxin emission of 80 ng TEQ/Nm^3 was still acceptable. As a consequence, most small incineration plants had to shut down, because retrofitting dioxin reduction technologies was not economically feasible. The formation and destruction of dioxins is discussed in Sect. 6.4.8.

These new regulations and the general policy from 1995 to 2005 in Japan were drivers for the development and installation of new gasification and ash melting processes. Today, there are more than 90 gasification and over 90 ash melting processes in operation or under construction in Japan. The main purpose of these processes is to improve ash quality (by reducing the extent of leaching of the contaminants from the ash) and to reduce the total dioxin output per tonne of waste. This development peaked in 2000, when a total capacity of 3 million tonnes per year of plants were ordered. Half of this capacity was realised through gasification plants and the other half through stoker-based systems combined with ash melting (Vehlow 2006). However, both these processes are considered to be somewhat inadequate in terms of energy/cost efficiency and availability (Martin et al. 2005). The same experience was had in Europe in the 1990s with gasification systems, which have now all been shut down. For this reason, a policy change was announced in Japan in 2005. The new policy states that gasification or ash melting would no longer be required if plants were situated in remote areas or if it could be proved that the plants had at least 15 years of landfill capacity for ash or other ash recycling options. As a consequence, development has once again focused on grate-based processes with an improved ash quality (for example, by oxygen enrichment, bottom ash washing, reduction of dioxin from the fly ash and the use of bottom ash in the cement industry).

6.4.2 Grate-Based Combustion Systems

The classical EfW system, a grate-based system, is shown in Fig. 6.27. They are widely used for MSW, RDF and the co-combustion of sewage sludge. In the USA, this kind of grate-based combustion systems for MSW are also called "mass-burn" combustion. Fluidised bed or suspension firing plants, which are specially designed for RDF, are described in Sect. 6.4.4. Section 6.4.5 deals with systems for the mono-combustion of sewage sludge.

Grate-based EfW systems are offered by a great number of manufacturers, though the principal processes differ only slightly from one another. Such a plant consists of the following functional sections:

- Waste unloading, storage and pre-treatment
- Feeding and combustion
- Waste heat recovery
- Bottom ash (also called slag) removal and residual material treatment
- Flue gas cleaning (Bilitewski et al. 2000)

Figure 6.28 shows in more detail the schematic of the combustion system, which is the core component of EfW systems. The delivered waste is stored in the refuse pit at first. There, in order to ensure an even fuel quality, it gets mixed and homogenised by means of cranes. Via a feed chute and special feeding and charging devices, the

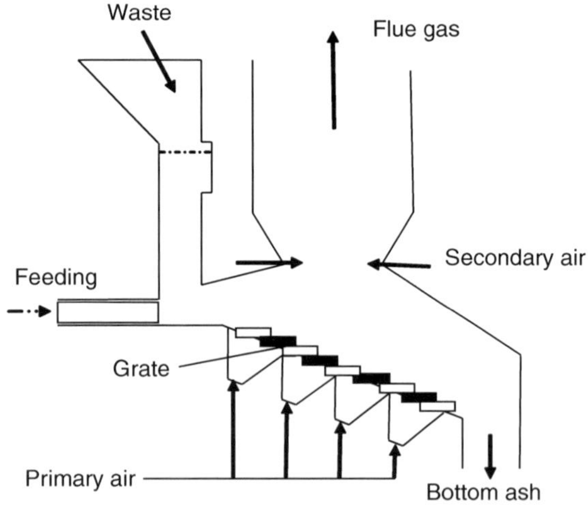

Fig. 6.28 Schematic drawing of a grate-based combustion system for MSW

waste is loaded onto the stoker. The grate transports the fuel through the various zones of the combustion process while providing a good mixing. On the whole, the waste remains on the grate for about an hour until all processes are completely finished. The remaining ash is discharged into a basin filled with water, called the ash discharger. Subsequently, the cooled bottom ash can be subjected to various treatments, such as the separation of metals, crushing and sieving, to make it more suitable for use as a filler material.

The hot flue gases are cooled in a steam boiler, creating steam, which is normally used for electricity production. Additionally, it can be used to deliver heat for district heating. In EfW plants, the design of the steam generator has to be given special attention, because the flue gases from the combustion of waste are very aggressive, reducing the heat transfer through fouling and slagging and potentially corroding the heat transfer surfaces. For this reason, the allowable steam conditions in EfW plants are markedly lower than the conditions in large steam power plants – a fact contributing to the lower power generation efficiency of a waste treatment plant. The design, the restrictions on it and the possible means to increase the efficiency are discussed in Sect. 6.4.7.

The cooled flue gases are treated in a flue gas cleaning section, where the remaining fly ash, including components such as heavy metals and organic toxic components, e.g. dioxins, is removed. Additionally, gaseous pollutants such as hydrochloric acid, hydrofluoric acid, sulphur dioxide and nitrogen oxides are removed. A large variety of technologies and systems are available for this purpose and are discussed in Sect. 6.4.9.

For the combustion of untreated municipal solid wastes, the grate furnace is the most widespread technology employed in the large EfW markets of Europe, the USA and Southeast Asia. The essential advantage of the grate furnace is its capacity to process lump-sized waste and accommodate fluctuating fuel properties (Bilitewski et al. 2000; Scholz et al. 2001; Doležal 1990; UBA 2005a; Effenberger 2000; VDI 2006, 2007; EIPPCB 2005). In a grate furnace, the combustion process can be divided into different phases, though these overlap to a large degree:

- *Drying and devolatilisation*: In the upper area of the grate, the waste is heated, dried and devolatilised by heat radiation from the refractory-lined walls or ceilings or by thermal convection of hot flue gases. The devolatilisation process usually takes place in a reducing atmosphere.
- *Gasification and combustion*: In these processes, two kinds of reactions run concurrently – heterogeneous fuel reactions and homogeneous reactions of the released volatile components in the combustion zone above the grate.
- *Burnout zone*: In the final area of the grate, complete burnout should be achieved. The burnout of bottom ash in modern waste incinerators is characterised by a loss of unburnt carbon of less than 2%.
- *Secondary combustion*: For the complete burnout of CO and fly ash particles, there is always a secondary combustion zone. Here, air or recirculated flue gas is added to complete the combustion.

The primary air injected below the grate provides the combustion air for the combustion reactions on the grate; the remaining combustion air is added as secondary air above the grate for the post-combustion of the unburnt components. In order to be able to adapt the primary air injection to the requirements of the individual combustion zones, the primary air ducts are divided into compartments. Large grate furnaces can have compartments not only in terms of length but also in terms of width. The amount of air per compartment can be controlled. With the aim of a complete burnout at low NO_x emissions in mind, the combustion of the bulk is mostly set at near-stoichiometric to air-deficient conditions. To give an example, at a total air ratio of about 1.6, 60% of the air is, as primary air, divided up and distributed to the drying zone (10%), to the combustion zone (40%) and to the burnout zone (10%). The rest of the total air is secondary air. The grate bars are protected and cooled by the primary air and the ash layer. The combustion air can be preheated for low waste LHVs. For high LHVs, e.g. above 12 MJ/kg, no preheating is applied and for LHVs of 6 MJ/kg, the preheating temperature can be up to 150°C. Forward pushing grate systems may need water cooling in case of high LHVs.

At the end of the grate, the bottom ash falls into an ash discharger, which is effectively a water basin, and is cooled down for further treatment. The amount of bottom ash is approximately 25% by weight or 10% of the volume of the original waste. A small portion of the fine ash may fall through the grate. These riddlings (also called grate siftings) are collected and transported pneumatically or via mechanical conveyers to the ash discharger.

The maximum combustion temperatures along the grate should be set so that the mineral components on the grate become sintered while avoiding the formation of large agglomerates, the latter because they would hamper the passage of air and thus the combustion process. Such an incidence is counteracted regardless, however, by the stoking effect of the grate firing system. Generally speaking, waste bed temperatures between 800 and 1,250°C are desired. The waste bed temperature can be monitored by infrared cameras in the furnace roof. A few plants have been constructed with oxygen enrichment of the primary air in order to achieve an improvement in the ash quality through the resulting higher waste bed temperatures (Martin et al. 2005). This enriching of primary air with oxygen also improves the general effectiveness of the combustion. Other positive impacts are the decreases of the flue gas flowrate and the emissions and the increase in thermal efficiency brought about by less flue gas loss. However, the consumption of electricity required for oxygen production cancels out the increase in the thermal efficiency.

6.4.2.1 Design of the Grate Firing System

The decisive parameter for the design of grate furnaces is the heating value. For residual waste, the heating value determination includes the ash and water contents. Theoretically, according to Fig. 6.29, MSW can be burned without back-up firing if the low heating value lies above 3,500 kJ/kg. However, the state-of-the-art combustion conditions required by today's environmental regulations for emissions and

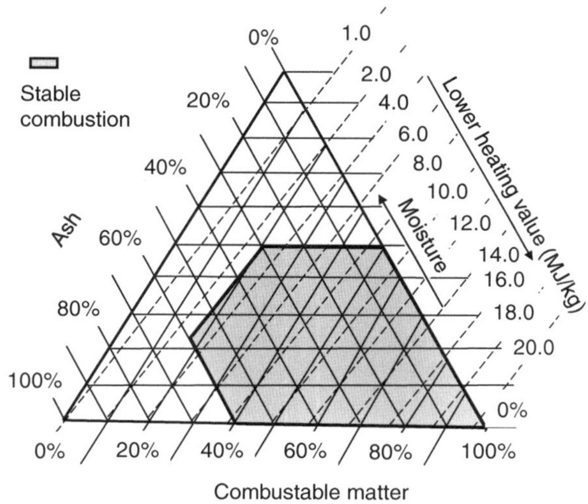

Fig. 6.29 Heating value, moisture and ash content triangle (Bilitewski et al. 2000)

ash quality make it necessary to have a heating value of above 6 MJ/kg (without any additional fossil fuel). For the combustion of untreated MSW in Germany, an average LHV of 9–10 MJ/kg, with fluctuations between 8 and 12 MJ/kg, is taken as a basis.

Grate incineration is very flexible in regard to the heating value and the waste characteristics. Figure 6.30 provides an example of a waste incineration diagram, indicating the relationship between waste throughput, LHV and thermal output. Waste incinerators are designed for large variations of the average LHV, normally between 6 and 12 MJ/kg. Wastes with low LHVs require air preheating. Because of variations in the waste composition, spare steam generator capacity is required for peaks in the heat input.

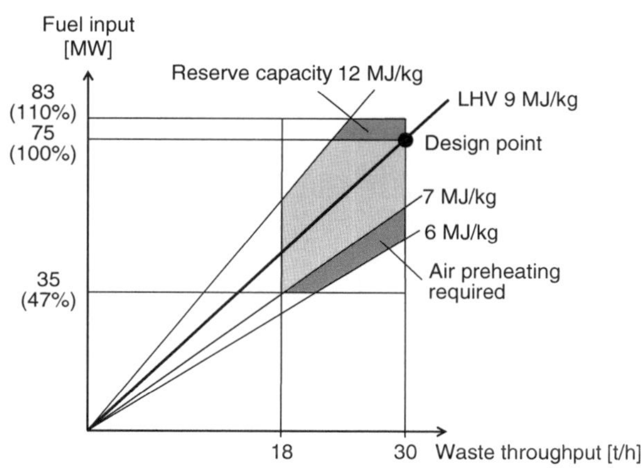

Fig. 6.30 Thermal power and throughput diagram

Typical grates are between 7 and 10 m long. The width of the grate determines the incineration capacity and can vary from 1 to 14 m for the largest units, which have a capacity of up to 40 t/h. Typical cross-sectional heat releases from the grate are in the range of 0.7–1.1 MW of fuel input per square metre (Effenberger 2000). The heat release for EFW grates is often characterised by the thermal width load, which relates the fuel input to the width of the grate. Typical width loads are 25–35 GJ/m h (related to the heat of the fuel) or 2.5–3.5 t/m h (related to the mass).

The specification of the adiabatic temperature is an important criterion in designing a waste incineration plant. Since the heating value of the waste is given, the adiabatic combustion temperature can be controlled by the air ratio or by recirculating flue gas. A low air ratio leads to high adiabatic combustion temperatures. This diminishes the flue gas volume, with the benefit that the following flue gas passes and the flue gas cleanup train can be built more compactly. A negative side effect of the higher flue gas temperatures is the shortening of the refractory lifetime. It is sometimes also suspected that a negative fouling and corrosion impact may occur (Foster et al. 2007).

6.4.2.2 Grate Variants

Modern grate furnaces use various grate systems, as shown in Fig. 6.31.

Pusher-type grate: The grate rods of pusher-type grates overlap and alternate between being fixed and being moveable, meaning there is a fixed rod overlapping a moving rod which overlaps a fixed rod and so on. Lifting movements of the movable rods set in motion the waste charge towards the bottom ash removal and also rake the fuel. Since the transport of the waste is brought about by the movement of the grate, it is possible to install the grate horizontally. A slight incline, though, helps the transport of the charge.

Reciprocating grates: The reciprocating grate, like the pusher-type grate, is comprised of movable and fixed grate rods, though they are arranged in a manner such that the lifting movement is directed towards the fuel inlet. The advantage of the reciprocating grate is an intensive thorough mixing of the waste charge as the rod movement constantly pushes embers from the main combustion zone back to the upper end of the grate. For this reason, the reciprocating grate is also suited to moist low calorific waste, which needs to be agitated more intensively during combustion. In contrast to the pusher-type grate which, due to the less intensive mixing, develops a prolonged drying, devolatilisation and ignition zone, the reciprocating grate features a long burnout zone. Since the inclination of the grate has to ensure the transport of the fuel towards the discharge chute, for reciprocating grates a steeper incline is required.

Roller-type grates: The roller grate is composed of rollers roughly 1.5 m in diameter arranged one after the other. The roller grate is inclined in the direction of the discharge at about 30°. The rollers are driven electrically and separately, which makes it possible to adapt the rate of feeding to the combustibility of the waste. Some agitating of the waste bulk is necessary in the gussets, the gaps between two rollers. The air is distributed via slots cut all over the surface of the rollers

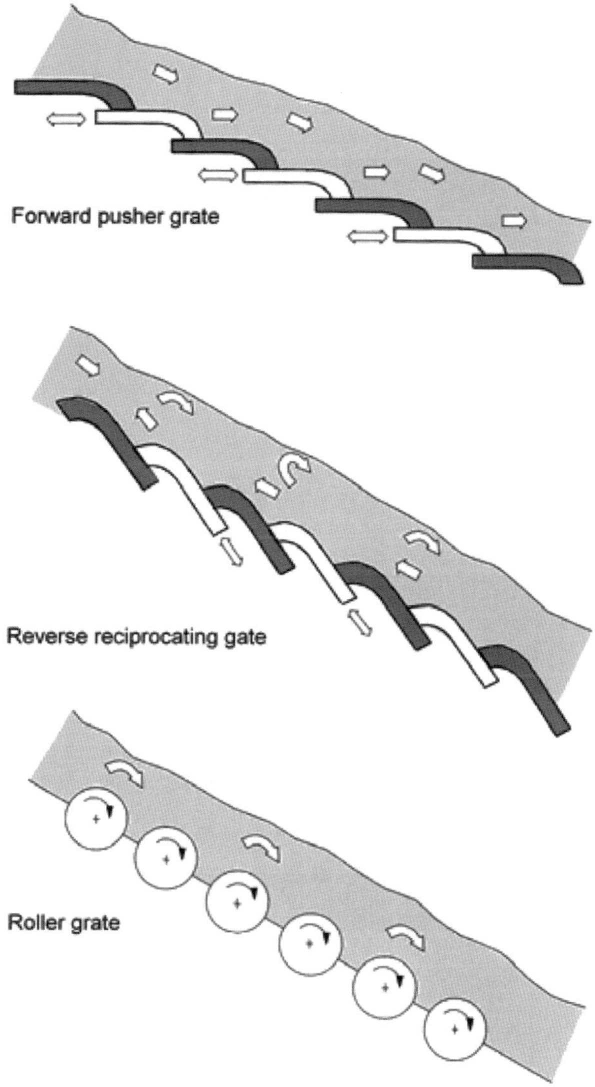

Fig. 6.31 Different grate types

and through the gussets. One advantage of the roller grate is the slight amount of wear and tear, a consequence of the grate surfaces only being exposed to the high furnace temperatures briefly during the turning movement and being cooled by the primary air the rest of the time. The disadvantages are the great accumulation of "riddlings" and the fact that there is only slight grate agitation, meaning a lesser ability to intensify the combustion process.

In normal grate furnaces, the lifetime of the grate material is limited, especially the material in the incineration zone. Higher heating values further reduce the lifetime, even if specific advanced alloys are used for the grates. In consequence, water-cooled grates have been developed. By water cooling, the temperatures can be kept significantly lower and the air distribution can be made more flexibly. However, water cooling is costly and has a negative impact on the efficiency, because at the low temperature of the cooling water the heat cannot be used in the thermal process. Approximately up to 5% of the thermal capacity is lost and is therefore not available for steam and electricity production. This corresponds to a decrease in the electrical efficiency of about 1%. With air-cooled grates, the grate design also has an impact on the grate temperatures. The reciprocating grate, owing to the intensive mixing of the grate bed and reliable covering of the grate with waste and ash, is able to manage even higher heating values without water cooling. The use of water-cooled grates, on the whole, is rare. Excessively high grate rod temperatures can be expected when the grate is not covered with fuel. This is avoided by appropriate operation.

6.4.2.3 Furnace and Boiler

The waste heat of the hot flue gases is recovered in a steam generator. EfW steam generators differ significantly from coal-fuelled steam generators, for example, by having comparatively large combustion chambers which ensure the necessary burnout and allow the substantial cooling of the flue gases. Different sections of the steam generator are distinguished:

- The first flue gas pass is the combustion chamber (or furnace) above the grate, including the primary and secondary combustion zones as well as the remaining first radiation pass. The combustion chamber size is limited by water/steam-cooled membrane wall heating surfaces, which are refractory-lined or protected by SiC tiles in the lower part. This serves to protect the metal heating surfaces from corrosion, but it also aids in reducing the heat transfer so that the stipulated flue gas temperature of 850°C can be met. In this refractory-protected area, one or several fossil fuel-fired burners are installed in the side walls. These burners are used for start-up and shutdown of the plant in order to maintain the minimum temperature conditions at all times. The membrane wall of the upper part of the furnace, from 8 to 12 m above the grate, is generally unprotected or equipped with overlay-welded nickel-based alloys (cladding). In the first part of the 2000s, a few plants with combustion chambers where the membrane walls are completely cladded down to the grate have been constructed.
- Subsequently, in the second flue gas pass, there are metal, not refractory-lined, membrane-wall heating surfaces. In many EfW plants, up to three empty flue gas passes (including the combustion chamber) are built in to cool the flue gases down to low temperatures by radiant heat transfer.
- Convective heating surfaces for superheating and, in the cooler area, for feed-water heating (the economiser) are arranged, preferably in the area of flue gas temperatures below 650°C.

The design of the combustion chamber above the grate (the incineration chamber) has to be suited to the grate and depends on the chosen flue gas routing above it.

In the combustion of waste, due to the high volatile matter content, only a small part of the combustion process takes place on the grate, the larger part occurring in the area above. Therefore, it is necessary to have a sufficient residence time at high temperatures in that upper zone. The furnace geometry has to be chosen such that a high and even flue gas velocity above the grate will develop. The combustion chamber walls and the ceiling in the part above the grate are refractory-lined in order to limit the heat transfer and to protect the walls from corrosion. The refractory lining of the combustion chamber ceiling favours the heating of the fuel by radiant heat from the hot walls.

The design of the grate and combustion chamber differs according to the suppliers and their specific experiences. In general, three basic designs can be distinguished. The nomenclature comes from the flow direction of the flue gases in relation to the waste flow: co-current, counter-current and medium current, as shown in Fig. 6.32.

Co-current or parallel flow furnace: In a co-current combustion arrangement, primary combustion air and waste are guided together in a co-current flow through the combustion chamber. Accordingly, the flue gas outlet is located at the end of the grate. Only a comparatively low amount of energy is exchanged between the combustion gases and the waste on the grate. The advantage of co-current designs is that the hot flue gas passes above the ash bed and improves the burnout of the ash, which is why it is most commonly used for roller grate systems. To facilitate ignition for wastes with very low heating values, the primary air must be preheated.

Counter-current furnace: In this case, the primary combustion air and the waste are guided in opposing directions in a counter-current flow arrangement through the combustion chamber and the flue gas outlet is located at the front end of the grate. The hot flue gases facilitate drying and ignition of the waste. Special attention must be paid to avoid the slip of unburned gas streams.

Medium current furnace: The medium current design is a compromise system suitable for a wide range of fuel properties. In this case, the flue gas outlet is located in the middle of the grate. Mixture-promoting chamber contours and/or secondary air injections can be used to promote a thorough mixing of the various flue gas

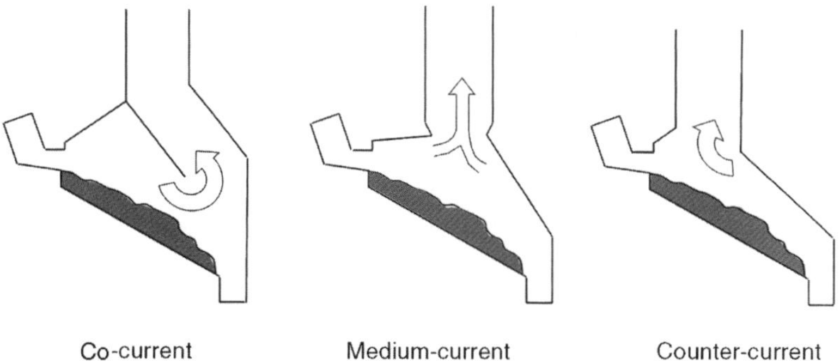

Fig. 6.32 Furnace and grate arrangements for EfW systems

streams. Nearly all modern EfW plants in Europe, the USA and Japan built in the 2000s have been designed using characteristics lying in between medium-current and counter-current furnace designs.

The primary combustion zone is followed by the secondary combustion zone, where the secondary air (or overfire air) is added for the combustion of the unburned gaseous components. The following secondary combustion conditions are favourable for a complete burnout of the flue gases:

- Availability of sufficient oxygen. An oxygen concentration of approximately 6% was considered as a minimum in the German regulation “17. Bundesimmissionsschutzverordnung (17. BImSchV – the 17th Amendment to the Federal German Emissions Control Act)” of 1993. In later revisions of the 17. BImSchV (17.BimSchV 2009) this requirement was removed in order to be consistent with the EU Waste Incineration Directive (EU 2000).
- Sufficient residence time at a high temperature. According to the European Waste Incineration Directive (EU 2000), a minimum residence time of 2 s with temperatures above 850°C is required to ensure the destruction of dioxins (see Sect. 6.4.8)
- Complete mixing of secondary air and homogeneity of flue gas flow. This condition is included in the combustion engineering principle 3T, meaning that time, temperature and turbulence are required for good quality combustion.

In some cases, the primary and secondary combustion chambers are separated by a constriction. Adding the remaining secondary air prior to but near the constriction provides an intensive mixing of the flue gases, which is followed by a homogeneous flow profile on the other side of the constriction.

6.4.2.4 Ash Deposition

Even though the dust load in the flue gases of grate-based EfW plants is much lower than in fluidised bed systems, special care must be taken to avoid dust-related slagging, fouling and corrosion. Reduced underfire air flows and flue gas velocities in the furnace generally favour less carryover of fly ash to the boiler. MSW ashes contain various components such as relatively inert aluminosilicates, lime-related compounds (in the form of oxide carbonates or sulphates) and salts. The majority of the salts are alkali and earth alkali chlorides, but significant amounts of heavy metal chlorides also occur. Alkali chlorides, and especially the lead and zinc chlorides, form eutectic mixtures with low melting temperatures. Therefore, the above-mentioned flue gas temperatures of 850°C in the first flue gas path should not be markedly exceeded, because at flue gas temperatures of 1,000°C or so, sticky particles may lead to severe fouling and slagging in this section. Because it is nearly impossible to completely avoid the formation of deposits in the flue gas path, adequate measures have to be taken to control the growth of deposits and to avoid unscheduled shutdowns of the power plant for the manual cleaning of deposits. Several techniques have been developed to reduce the formation of deposits during operation. Cleaning can be accomplished with water cannons or shower cleaning, soot blowers using steam or air, mechanical rappers or explosives.

6.4.2.5 Corrosion

High-temperature corrosion is the biggest operational problem in the thermal treatment of waste, and it limits the efficiency of the conversion into electrical power. Corrosion is worst when there is, at the same time, high chlorine and low sulphur contents in the MSW and a high content of heavy metal chlorides in the fly ash. The membrane wall heating surfaces and the superheater surfaces can be severely corroded. The corrosion mechanisms are discussed in Sect. 5.10.4.

In the area of the membrane wall, corrosion is caused by molten salts. In order to diminish the effects of corrosion in the furnace above the refractory-protected area, coatings made of nickel–chromium alloys are applied by deposition welding (also called cladding) to protect the steel of the membrane wall tubes. This cladding may cover the entire upper furnace (first pass) and the first part of the second pass and is considered state of the art for the most common 40–60 bar boilers used in Europe, Japan and the USA. It can be applied either in the workshop during the construction of the plant or retrofitted if corrosion problems occur. Consequently, membrane wall corrosion no longer presents a frequent problem in EfW plants in service today.

On the other hand, however, high-temperature corrosion of superheater tubes is still an unsolved problem and limits the allowable live steam temperatures to values of 400–450°C. In this corrosion mechanism, sulphation of alkali chlorides in boiler tube deposits sets a chlorine cycle going which involves very strong wear and tear. In the design of EfW plants, this is the reason the allowable working medium temperatures are chosen as a function of additional parameters such as the flue gas temperature or the flue gas velocity, i.e. in order to avoid superheater corrosion. This design practice is represented in Fig. 6.33 by the so-called Flingern corrosion diagram.

The investigation of corrosion processes and the development of corrective measures are the subject of extensive research (Born 2007; Born 2005; VDI 2006, 2007). Corrective measures for reducing superheater corrosion take two forms: developments of new materials and combustion engineering measures. Making use of the

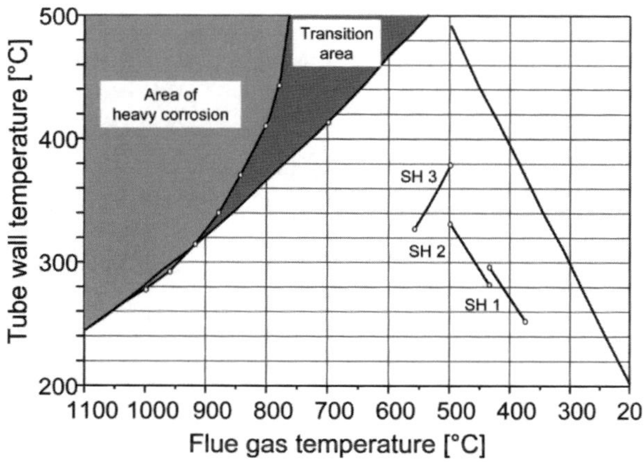

Fig. 6.33 Corrosion diagram

most highly alloyed materials and using cladding in the area of the superheater have not yet gained acceptance due to technical and economic reasons. Combustion engineering measures aim at preventing ash deposits on the superheaters or altering their consistency. For example, the sulphation of the chlorine in the fly ash before hitting the deposits can be favoured by injecting sulphur-containing additives. Another measure employed for this purpose is the use of intensive mixing and low flue gas flow velocities, which provide more time for natural sulphation. Furthermore, some authors state that by avoiding or removing deposits it is possible to diminish the corrosion potential (Warnecke 2007; Warnecke 2006; VDI 2006, 2007).

6.4.3 Pyrolysis and Gasification Systems

In the USA and Europe, interest in gasification technologies for waste processing was kindled during the 1970s, when the oil shocks struck. However, the development of the technology was unsatisfactory at that time. The European market only began to seriously reconsider the technology during the early 1990s, and this was driven by the political desire to avoid the use of incineration and to maximize recycling and resource recovery in a more sustainable way. A number of high profile companies developed processes for waste treatment that combined pyrolysis, gasification and combustion in various configurations. Most of the processes required extensive pre-treatment of the MSW (making RDF or SRF (secondary recovered fuel)).

In Japan, gasification technologies for MSW reached the highest degree of development in the late 1990s. As of 2007, more than 90 gasification plants were in operation or under construction. The main purpose of these processes is to improve ash quality and to reduce the total dioxin output per tonne of waste. Table 6.9 shows the installed capacity of gasification systems in Japan in 2008.

6.4.3.1 Pyrolysis in Rotary Kilns

The pyrolysis or gasification system which reached the highest degree of development in Europe was the Siemens SBA process ("Schwel-Brenn Anlage") – a schematic of which is shown in Fig. 6.34. Development started in 1988 at a small-

Table 6.9 Installed capacity (in 2008) of the processes for the pyrolysis or gasification of waste realised in Japan in the 2000s (Themelis 2007)

Gasification system	Installed capacity (t/day)
JFE – thermoselect	1,980
JFE – fluidised bed	1,300
Nippon steel – shaft furnace	6,200
Ebara – fluidised bed	1,700
Other fluidised bed processes	3,200

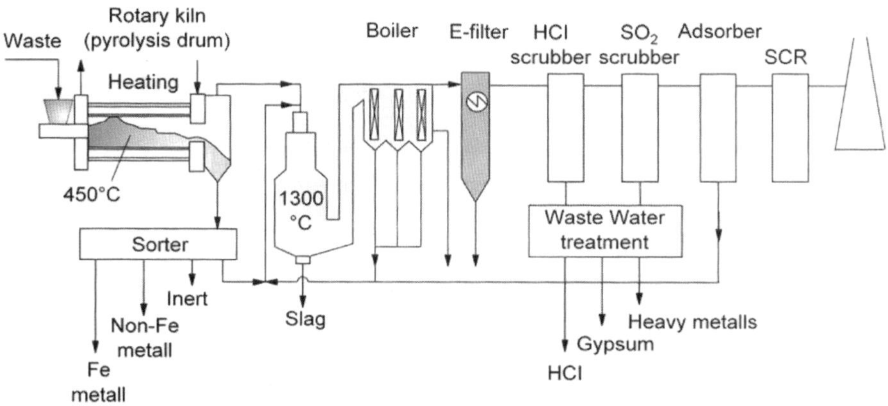

Fig. 6.34 Siemens SBA gasification of MSW (pyrolysis in rotary kiln followed by slag-tap combustion)

scale plant in Ulm, consisting of a rotary pyrolysis reactor followed by a mechanical sorting system of the residues, and then the combustion of the separated coke with the pyrolysis gases in a melting furnace. The start-up of the first large-scale plant was in 1997 in Fürth (Germany), where power generation was via a classical steam cycle. This plant was designed for a 100,000 t/year (two lines each with a 5 t/h capacity) waste throughput and had a capital cost of approximately 150 million euros. The gasification system consisted of two lines each with a 5 t/h capacity (after pre-sorting of the MSW). The SBA process had severe problems with the waste pre-processing, clogging of the rotary kiln-type pyrolysis drums and instabilities in the post-combustion melting chamber. The final decision for dismantling the plant was taken after a minor incident involving some entangled metal on 12 August 1998, which caused pyrolysis gas to escape through a damaged gliding ring seal (Schwarzmann 1999). Since then, no further projects with this technology were seriously considered in Europe, but licenses were acquired in Japan by Mitsui and Takuma. These companies constructed several commercial size plants.

6.4.3.2 Gasification with Pure Oxygen and Integrated Melting

Another gasification technology with quite a high degree of development is the thermoselect gasification and melting process, developed in Switzerland between 1985 and 1992. A demonstration facility with a capacity of 110 t/day was built in Fondotoche, northern Italy, and used to validate the technology; the facility operated under a commercial licence from 1992 until 1999 (Themelis 2007).

A larger commercial facility with a design capacity of 792 t/day (or 2,250,000 t/year) was built in Karlsruhe in Germany and commenced operation in 1999. The plant suffered technical and commercial problems and none of the lines operated at full design capacity for sustained periods (Whiting and Schwager 2006). It was finally shut down in 2004 by the owner EnBW, a large German power utility (partly

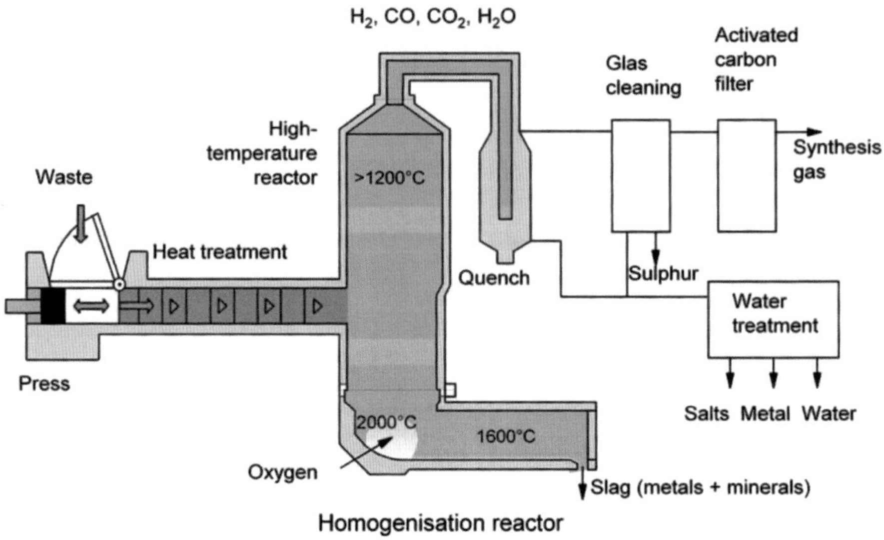

Fig. 6.35 Thermoselect gasification of MSW (gasification with pure oxygen and integrated melting of the ash as well as post combustion in a boiler)

owned by EDF – Electricité de France). Since then it has been "mothballed" pending the outcome of litigation between the supplier Thermoselect S.A. and the owner EnBW. The third European thermoselect plant, which was constructed in Ansbach, never went in operation. The schematic of the thermoselect process is shown in Fig. 6.35.

In Japan, similar plants have been built by JFE (Japan Steel Engineering), licensing the thermoselect technology. The first plant was completed in 1999 at a steel mill in Chiba with the synthesis gas produced being used in the mill. A further six JFE plants using the thermoselect technology had begun operation by 2006.

6.4.3.3 Fluidised Bed Gasification

In Europe and the USA, no highly developed fluidised bed gasification systems have been realised. In contrast, by the beginning of the 2000s in Japan, 13 companies were engaged in the development of fluidised bed gasification systems for MSW. The most utilised system in Japan is the Twin Internally Circulating Fluidised Bed Gasifier (TIFG), developed by the Ebara Corporation (Fujimura et al. 2001). It gasifies the wastes first and uses the heat content of the gases to raise the temperature in the following slag combustion furnace stage.

For all the gasification technologies that have been developed to a significant extent and have some market impact, the pyrolysis and gasification steps are only a pre-treatment. These steps are directly followed by combustion of the gases and, in the case of pyrolysis, of the pyrolysis char as well. The only exception is the fixed bed gasification technology "Sekundärrohstoff-Verwertungszentrum

Schwarze Pumpe" (SVZ) that was developed in Schwarze Pumpe/Spremberg (Brandenburg/Germany), based on German Democratic Republic (GDR – i.e. the former East Germany) coal gasification technology and used at the end of 1990s and beginning of the 2000s for MSW. Insolvency in 2004 meant the final withdrawal of this technology from the waste business in 2007 (Mielke 2007). The example of SVZ-Schwarze Pumpe seems to show that it makes little sense to use MSW, with its difficult properties, as a raw material for complex biomass to liquid (BtL) or methanol production processes, as long as coal is available.

The main driving forces for the development of gasification processes have been a desire to improve ash qualities and reduce dioxin outputs. However, the more complex gasification processes result in lower efficiencies, lower process availabilities and higher costs. Although in principle, gasification offers a better efficiency, the real plant efficiencies are even lower than conventional EfW plants. Gasification with oxygen and integrated melting has produced electrical efficiencies well below 10%, for example. Because of this, gasification technologies are no longer developed and demonstrated in Europe, while in Japan, the focus of development has been readjusted to grate-based processes producing improved ash qualities (e.g. by using oxygen enrichment, bottom ash washing, reduction of dioxin from the fly ash and the use of bottom ash in the cement industry).

6.4.4 Refuse-Derived Fuel (RDF)

The basic idea of the combustion of refuse-derived fuel by itself is that the EfW plant can be designed especially for the quality of this more homogeneous fuel. In doing so, there are design and operational advantages, and in addition it becomes possible to decentralise the plants. RDF contains chlorine, as MSW does, which means that the steam parameters are restricted in a similar way to about 60 bar and 450°C. In this respect, then, there is no advantage over waste incineration plants. However, the construction and the design of the flue gas cleaning facilities is often simpler than for EfW plants using unprocessed MSW (as a rule reduced to an SNCR process followed by calcium hydrate and activated carbon addition upstream of a fabric filter), because there are fewer pollutants. Designing the flue gas cleaning process for a narrower fuel range, though, can give rise to problems if the composition of the RDF fluctuates, especially if the chlorine and sulphur contents change (Neukirchen 2008).

The majority of RDF plants use grate furnaces. The technology is rather similar to conventional EfW plants, which have been thoroughly discussed in Sect. 6.4.2. Attention has to be paid to the higher LHV, for example, by using water-cooled grates. Fluidised beds are a reasonable technology for RDF combustion, owing to the more homogeneous fuel quality, but they are less common than grate combustion.

Four fluidised bed combustion plants were built in France between 1995 and 1997 in Mulhouse, Giens, Guerville and Monthyon (ActuEnvironment 2005). After that, no other fluidised bed projects were constructed in France. A similar scenario occurred in Japan, where 44 fluidised bed plants for MSW or RDF were operating in 1999, with a further 11 being installed between 1999 and 2003. After 2003, there is no report of any other fluidised bed combustion plants for MSW or RDF being constructed, but various types of fluidised bed gasification processes were developed (see Sect. 6.4.3). The reasons for the low acceptance seems to be the higher pre-processing requirement in comparison to grate combustion and some negative experiences with increased dioxin formation rates in the past. The latter, however, can be solved by an appropriate design of the flue gas train, as discussed in Sect. 6.4.8.

In Germany, some circulating fluidised bed (CFB) plants were built in the 2000s to treat the increasing amounts of coarse fractions from mechanical–biological treatment (MBT) plants. One relatively well-documented example is the CFB plant in Premnitz, operated by the large German utility EON-Energy from waste. It was designed for steam parameters of 97 bar and 500°C, with superheating realised in an external bed heat exchanger. Superheating steam in the external fluidised bed heat exchanger of the recirculation loop in principle provides the opportunity to superheat steam to higher temperatures than superheaters situated directly in the flue gas path, because of a less corrosive environment. However, erosion and corrosion problems still seem to occur, resulting in a reduction of the steam temperature and pressure to 75 bar and 450°C. The CO and dioxin formation rates are much lower than in the Japanese and French examples of fluidised bed combustion and are well below the emission limits (Borghart 2008).

In most cases, RDF plants in the USA do the pre-processing of the waste on-site and utilise suspension combustion, as shown in Fig. 6.36 (Themelis 2007). RDF is

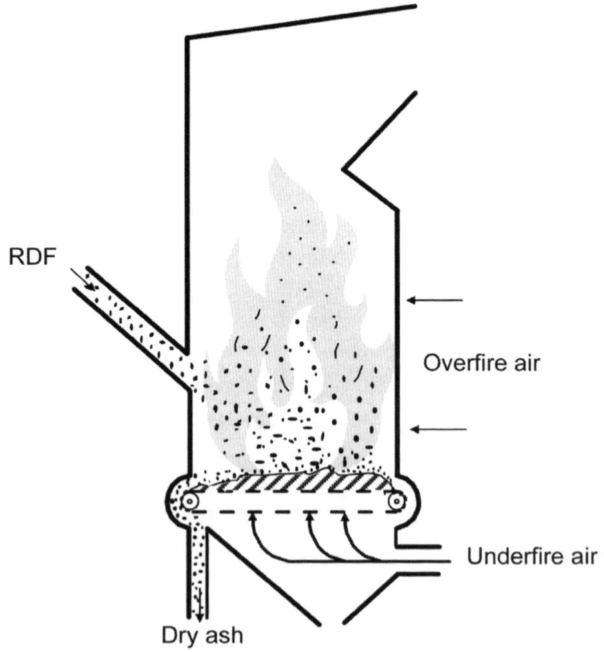

Fig. 6.36 A suspension combustion system for RDF in the USA

fed through multiple feeders into the furnace and burned part in suspension and part on a travelling grate stoker.

6.4.5 Sewage Sludge

The exclusive purpose of the monofuel combustion of sewage sludge is volume reduction for the subsequent disposal of the sludge. Power, in general, is not produced for more than in-plant use.

To ensure economic operation, the combustion of sewage sludge should develop and continue self-sufficiently without employing a support fuel. This is possible only when the heating value is larger than the power necessary to vaporise the water, heat the combustion air to the combustion temperature and compensate for the heat losses. For self-sufficient combustion, a heating value around 4 MJ/kg is required, which corresponds to a total solid (TS) content of about 40% in digested sewage sludge.

Mechanical dewatering of the sewage sludge before combustion serves to raise the heating value and to reduce the sludge volume. The degree of dewatering depends on the dewatering method and can be increased by adding conditioning agents. The total solid content after dewatering ranges between 30 and 45% TS. This is usually not enough for self-sufficient combustion. However, using a thermal drying system, it is possible to achieve TS contents of up to 95%, as Sects. 2.2.1.4 and 2.2.3.4 explain.

According to data from the German Federal Environmental Agency (*Umweltbundesamt*) in Berlin, 15 public monofuel combustion plants for sewage sludge were in service in Germany in 2004. In addition, there are six more in-house combustion plants in the chemical industry. The public and the industrial plants together had an installed capacity of about 780,000 t TS/year. The dominating technology in this context is the stationary fluidised bed. Furthermore, only three multistage grates and one single-stage swirler were in service in 2004 (Hermann 2004).

Figure 6.37 shows a sectional view of the layout of a stationary fluidised bed furnace used for the combustion of sewage sludge with a TS content of 50% (Treiber and Schroth 1992). The fluidised bed temperature has to be kept below the ash deformation temperature in order to prevent a sintering or fusion of the bed ash. The minimum temperature stipulated by law for the combustion of wastes (850°C according to *17th BImSchV*) confines the allowable operating temperatures to a lower range. The operating temperature can be set by controlling the excess air and, if needed, by heat extraction from the fluidised bed.

The good heat and mass transfer in the fluidised bed allows nearly complete combustion at a low temperature with an even temperature distribution. CO and NO_x emissions are a function of the temperature, with CO emissions decreasing and NO_x emissions increasing with a rising temperature. Temperature peaks, which are crucial for NO_x emissions, can be avoided in fluidised beds. Emission limits can be complied with in most cases.

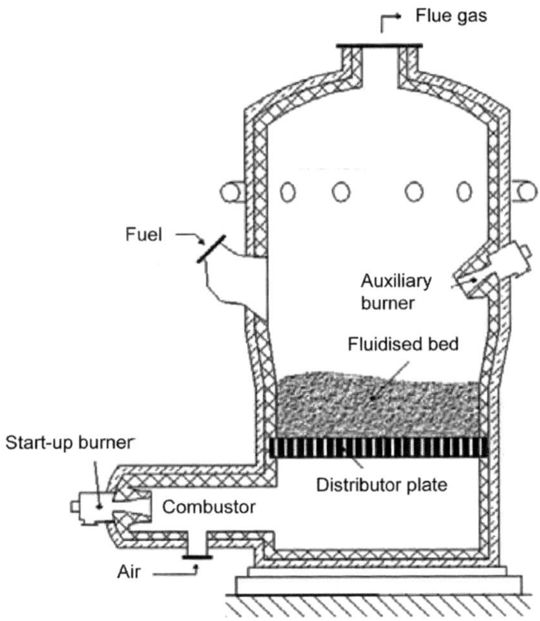

Fig. 6.37 Bubbling fluidised bed for sewage sludge combustion (Treiber and Schroth 1992)

Another advantage of fluidised bed combustion is the possibility of capturing SO_2 in the fluidised bed. Owing to the alkaline earth matter contained in the sewage sludge, part of the sulphur dioxide is bound without any additive being supplied. For a higher degree of capture, calcium-based additives such as limestone are usually used. The capture capacity shows a maximum at temperatures around 900°C, a fact which can be used if an operational mode is sought to optimise the capture of SO_2. However, the option of running waste incineration processes without any secondary desulphurisation measures cannot be inferred from this capture, because low SO_2 emission limits cannot be met with these operating parameters alone.

6.4.6 Steam Boilers

The heat released in the combustion of waste is used in a boiler (steam generator) to produce steam which in turn works in a steam turbine to produce electrical power. The flue gases, initially at about 1,000–1,200°C, are cooled in the boiler to temperatures typically in the range of 140–300°C. This temperature range is usually required for the subsequent flue gas cleaning process. Boilers for waste need a design suited to the often particular and rather difficult composition of the flue gas, with its corrosion and fouling effects.

The envelopment of the furnace, the following empty passes and the passes where evaporator and superheater tube bundles are located are generally designed as water-cooled membrane walls. The first pass generally needs to be empty, as hot gases are too corrosive and particulate matter is too sticky for convective heat exchangers.

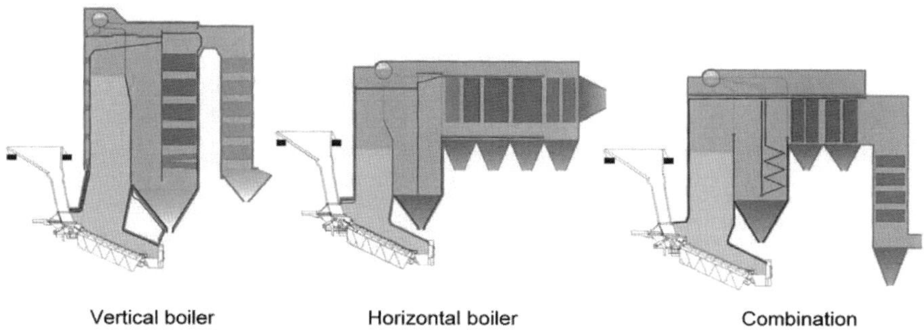

Fig. 6.38 Boiler arrangements for waste combustion (Source: Martin)

Convective heat exchangers are typically arranged in the third and following boiler passes at temperatures below 650°C. Different boiler configurations can be used in waste incineration plants. The arrangement of the heat exchangers is shown in Fig. 6.38 for a vertical boiler, a horizontal boiler (also called a "tail-end" boiler) and a combination of both. Other widely used boiler arrangements are two-drum boilers with platen superheaters in the second pass and boilers with two empty passes and a low horizontal "tail-end" section (e.g. Amsterdam).

The design of the boiler mainly depends on the flue gas characteristics (the corrosion, erosion and fouling potentials), which are themselves highly dependent upon the waste content. Hazardous wastes, for example, tend to have very wide variations in composition and, at times, very high concentrations of corrosive substances (e.g. chlorides) in the raw gas. This has a significant impact on the possible energy recovery techniques that may be employed. In particular, the boiler can suffer significant corrosion and steam pressures may need to be reduced with such wastes.

A compromise is required when setting steam parameters for waste-fired boilers. Higher steam parameters can lead to significantly increased corrosion problems, especially for the superheater surfaces and the evaporator. In EfW plants, it is common to use 40 bar and 400°C if electricity is produced, although higher values are used, especially where incentives for renewable power production are in place. In these cases, values of 60 bar and 520°C are often employed, with special measures to prevent corrosion. Because of the rather low steam parameters (low compared to coal-fired power stations), natural circulation steam boilers are selected almost exclusively. For heat production, steam at lower conditions may be produced.

6.4.7 Efficiency Increases in EfW Plants

In this section, possible improvements to waste incineration plants in service today (and how to implement them technically) will be discussed. As a reference case, a power plant with the following characteristics, produced by today's widely implemented state-of-the-art technology, will be used:

- Steam parameters of 40 bar/380°C
- A boiler outlet flue gas temperature of 209°C
- An excess air ratio of 1.75 (flue gas O_2 content of 8.4%, dry)
- Condenser pressure of 150 mbar
- An in-plant power consumption of 2.1 MW (0.1 MWh/t of waste)

The power plant, with its live steam conditions of 40 bar and 380°C, has a net electrical power production efficiency of 20.6%. In comparison, the average efficiency of plants in service in Europe lies at about 13%.

Within the framework of a study, the potential of technically feasible solutions to improve the efficiency has been investigated using this reference power plant. The results are compiled in Table 6.10 (Gohlke and Spliethoff 2007; Spliethoff et al. 2008).

Excess air ratio: Since the development of EfW plants in the 1960s, the excess air rates have remained relatively high, between 1.8 and 2.2. This was necessary in order to compensate for fluctuating combustion conditions and to avoid the added wear from refractory and membrane wall corrosion. Today, some EfW plants have reduced the excess air ratio in conjunction with special precautions.

Mixed municipal waste is relatively homogenous and therefore well suited to decreasing the excess air ratio. Should pre-processed waste fractions or commercial waste be used, it is preferable to use higher excess air ratios (leading to lower efficiencies). The decrease of the air ratio from 1.75 for the reference case to 1.4 results in a net increase of the efficiency of 0.6% (Fig. 6.39).

Boiler exit temperature: Typically, the flue gas is cooled down in the economisers to approximately 200°C at the outlet of the boiler (209°C in the example of the reference case). The remaining energy is lost by quenching the flue gas with water to typical scrubber temperatures of 150°C or less. Alternatively, cooling can be realised by additional heat exchangers used for condensate preheating (in this case where no direct use of heat is considered). As temperatures fall below the dew points, advanced materials are needed for these heat exchangers. The reduction of the flue gas temperature to 135.5°C leads to an efficiency increase of 0.7–21.3% (Fig. 6.40).

Condensation pressure: The condenser temperature has a strong influence on the plant efficiency. The reference plant employs air condensers with a condensation temperature of 54°C, corresponding to a pressure of 150 mbar. The efficiency can be increased by 2.8 percentage points to 23.4% if a water-cooled condenser with a temperature of 23°C and 30 mbar pressure is used. However, this option for increasing the efficiency is limited in the "real world" because cooling water is rarely available (Fig. 6.41).

Steam pressure and temperature: An increase in the pressure and temperature of the steam results in an increase in the efficiency of the thermal cycle. The gain in efficiency by increasing the live steam pressure, the live steam temperature and reheating can be evaluated from a temperature/entropy ($T - s$) diagram (see Fig. 6.42). Basically, the efficiency of the thermal cycle is $\eta = 1 - (T_{out}/T_{in})$, where T_{in} and T_{out} are the average medium temperatures of heat addition and heat extraction. Steam parameters can be increased in convective superheaters up to around 73 bar

Table 6.10 Overview of measures to increase efficiencies of electricity generation ($R1$ criterion of European Draft Waste Framework Directive is 0.6 and 0.65 after 2009) (Gohlke and Spliethoff 2007). D = Germany, I = Italy, NL = Netherlands, E = Spain

Measure of energy increase	Net efficiency in % (power)	Simplified $R1$ criteria (only waste and power generation)	Advantages	Disadvantages	Existing plant
Typical EfW excess air 1.75 40 bar/380°C, p_c = 150 mbar, $T_{\text{flue gas}}$ = 209°C	20.6	0.64	High availability, moderate cost	Limited efficiency	Zella Mehlis (D)
Reduced excess air rate to 1.4	21.3	0.66	Reduced flue gas losses. Smaller equipment	Increased risk of corrosion and refractory wear in furnace. Risk of CO peaks	Brescia (I)
Wall superheater with 60 bar/460°C	23.2	0.71	Superheaters behind rear-ventilated tiles are protected by sealing air	Tiles must be removed for maintenance of superheater tubes	No large-scale reference yet
Reduced condensate pressure to 30 mbar	23.4	0.72	Reduced losses of thermal cycle. Reduced in-plant power consumption	No water cooling available at most sites. Water content in steam after turbine. Increased steam temperature or reheat necessary	Amsterdam (NL)

Table 6.10 (continued)

Measure of energy increase	Net efficiency in % (power)	Simplified $R1$ criteria (only waste and power generation)	Advantages	Disadvantages	Existing plant
Increased steam parameters to 74 bar/480°C	24.0	0.73	Reduced losses of thermal cycle	Increased risk of superheater and membrane wall corrosion. Limited options for improved materials	Brescia (I)
130 bar/440°C with intermediate reheat (AEB Amsterdam energy concept)	28.1	0.84	Efficiency gain without increase of superheating temperature	Additional heat exchanger necessary (saturated steam/steam from first stage of turbine)	Amsterdam (NL)
100 bar/540°C with external superheating in combined cycle power plant	42 (referred to waste and natural gas)	–	Efficiency gain with limited corrosion risk of superheaters	Increased risk of membrane wall corrosion. High capital cost. Consumption of natural gas	Bilbao (E)

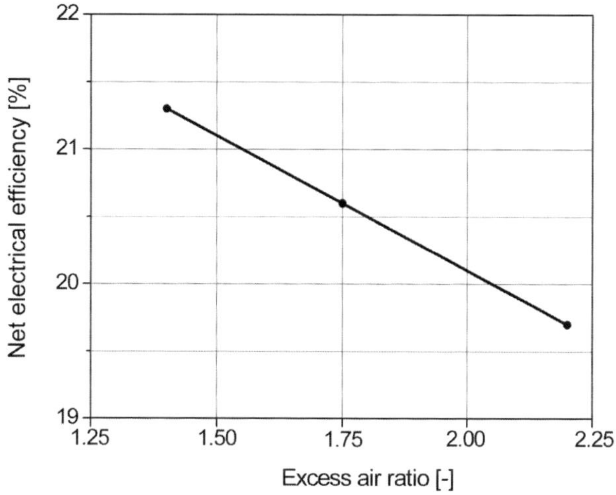

Fig. 6.39 Influence of the excess air rate on efficiency (Gohlke and Spliethoff 2007)

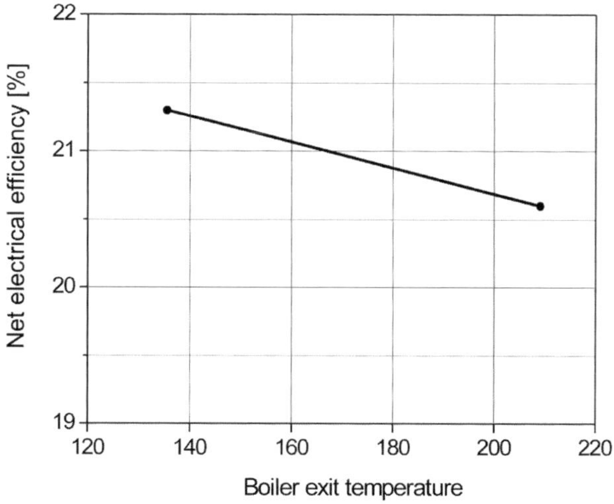

Fig. 6.40 Influence of boiler exit temperature on net electrical efficiency (Gohlke and Spliethoff 2007)

and 480°C with a classical boiler design, resulting in an efficiency increase of 3.4 percentage points to 24%. However, this increase is associated with a significantly higher risk of corrosion in comparison to a typical process which uses 40 bar and 380°C.

Another innovative approach to increase steam parameters is to use wall superheaters, where the most critical superheating tubes are placed behind rear-ventilated furnace tiles. In this way, the superheating temperature can be increased without additional corrosion problems. The critical tubes are protected by the sealing air of the rear-ventilated tiles. It is possible to obtain an efficiency increase of 2.6 percentage points to 23.2% with wall superheaters and steam parameters of 460°C and 60 bar.

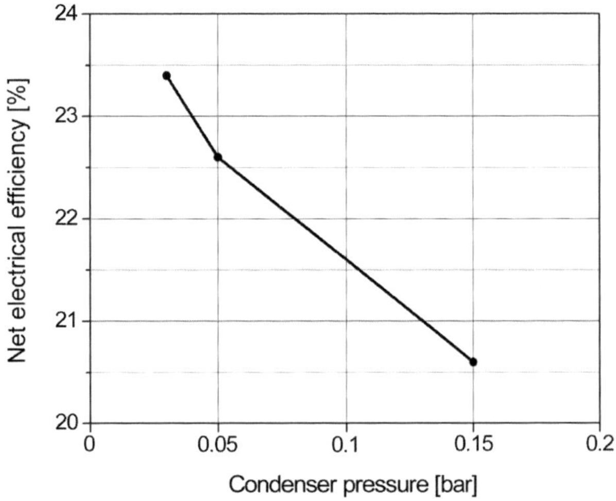

Fig. 6.41 Influence of condensation pressure on net electrical efficiency (Gohlke and Spliethoff 2007)

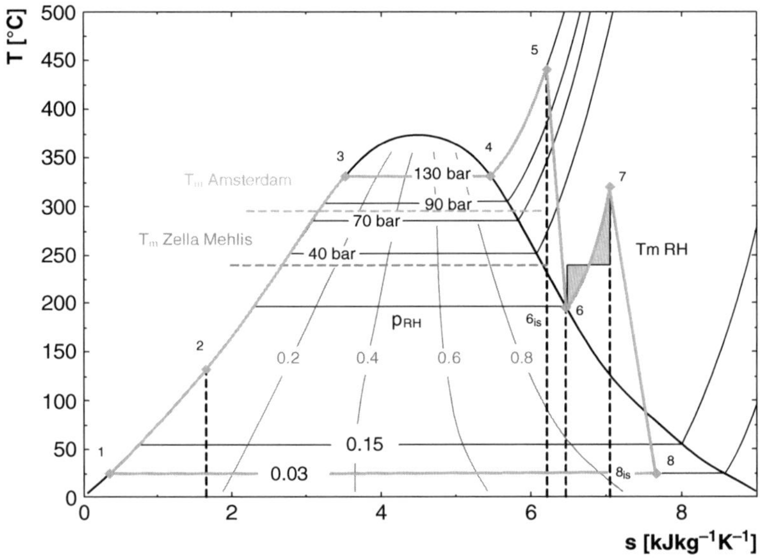

Fig. 6.42 Medium temperature of heat addition of the reference plant and of a plant with reheating (Gohlke and Spliethoff 2007)

Intermediate reheating: To achieve a further increase in efficiency, it is necessary to use intermediate steam reheating. This process has been developed by *Afval Energie Bedrijf (AEB) Amsterdam* ("the City of Amsterdam Waste and Energy Company") and is now used in the new Amsterdam plant, which began operation in spring 2007 (Van Berlo 2006). The system is run with a steam pressure of 130 bar, a superheating temperature of 440°C and reheating of the steam, using saturated drum steam, to 320°C after the first stage of the turbine (see Fig. 6.43). Compared to the reference EfW plant mentioned above, the configuration provides an efficiency increase of 7.5 percentage points to 28.1%. In Amsterdam, an even higher efficiency

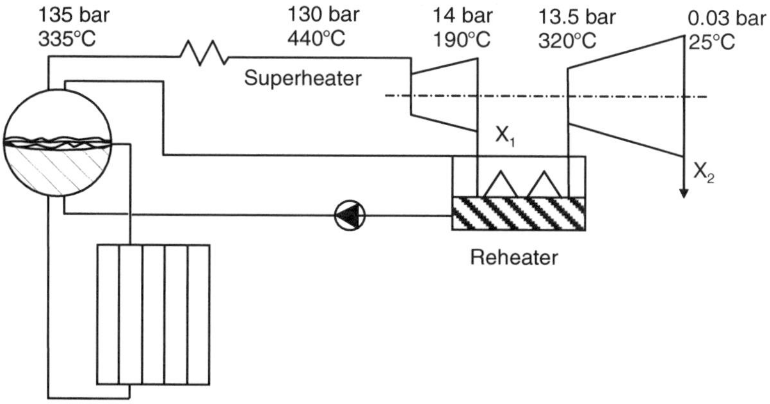

Fig. 6.43 Water-steam schematic diagram of a 130 bar/440°C system with intermediate reheating (Gohlke and Spliethoff 2007)

of over 30% is obtained, because other measures, such as reduced excess air and a lower condensate pressure, are employed as well. Due to the low reheat steam conditions at the Amsterdam plant, the higher efficiency is not the result of intermediate reheating – it is caused by the high live steam pressure; reheating is required to limit the exhaust moisture in the turbine.

External superheating: Even higher efficiencies can be obtained if superheating of the steam is performed in external fossil-fired boilers which do not have the corrosion limitations of waste-fired boilers. An example is the EfW plant in Mainz, where steam at 40 bar and 400°C is used in the intermediate reheater of an adjacent combined cycle natural gas power plant. Another example is the EfW plant in Bilbao. This waste boiler is operated with 100 bar pressure; superheating to 540°C takes place in the boiler of an integrated combined cycle power plant. In this way, the overall plant efficiency is increased to 42% (Seguin 2004).

Criterion for energy efficiency: The draft of the European Waste Framework Directive defines the efficiency criterion $R1$ as needing to be above 0.65 (from 2009) for an EfW process in order to be considered recovery (EU 2008). In the simplified case, where an EfW process does not use any additional (fossil) fuels or energy and produces only electricity, the calculation is

$$\boxed{R1 = \frac{E_p \times 2.6}{0.97 \times E_w}}$$

E_p: the energy annually produced and used in the form of electricity (GJ/a)
E_w: the energy supplied annually by the waste (GJ/a)
0.97: a factor taking the inevitable energy losses through radiation and bottom ash into account

The $R1$ criterion refers to the electricity produced (gross).

A typical EfW process as described above was calculated to have an efficiency $\eta_{net} = 20.6\%$ and $\eta_{gross} = 24\%$, which results in $R1 = 0.64$. In real cases, additional fossil fuels used for start-up and shutdown, as well as limited availability of the power generating components (i.e. failures, etc.), will reduce the $R1$. On the other hand, the use of heat will increase it significantly. Generally speaking, it can be estimated that EfW plants with a design similar to the reference EfW process will in fact reach the $R1$ efficiency criterion currently set at 0.6 for plants in operation and permitted before 2009 (EU 2008).

Conclusions: electrical efficiency: The main objectives of typical EfW processes are the transformation of waste into ash (incineration), the destruction of pollutants and the conversion of energy (to heat and power). A typical new EfW plant in Germany with steam parameters of 40 bar and 380°C was calculated to have a net electrical efficiency of 20.6%. This corresponds to an $R1$ of 0.64. Additional measures would therefore be necessary to meet the criterion of the European Draft Waste Framework Directive of $R1 > 0.65$ for plants in operation and permitted from 2009.

Major increases of the energy efficiency can be obtained by the following measures:

- Increasing the steam parameters (the pressure and temperature of the superheated steam)
- Reducing flue gas heat losses (via the temperature at the boiler outlet and the excess air ratio)
- Improving the steam condensation conditions (using water instead of air condensers)
- Optimising the thermal cycles (by using intermediate superheating or external superheating)
- Reducing in-plant power consumption (by using SNCR instead of SCR or a smaller excess air ratio)

Examples of recent innovative EfW plants applying these measures to increase the efficiency of electricity generation can be found in Brescia, Amsterdam, Mainz and Bilbao (Bonomo 1998; Van Berlo and de Waart 2008; Fischer 2005; Seguin 2004). The average net efficiency of electrical power generation by EfW processes is 13% in Europe (EIPPCB 2005). This could be increased in the Brescia plant to more than 25% through increased steam parameters, reduced flue gas losses and minimised auxiliary power consumption. The new plant in Amsterdam has achieved 30% with additional reheating and water condensers. Any further increase in energy efficiency is then only possible by external superheating with natural gas in combined cycle plants, as in Bilbao.

EfW is an important additional source for renewable energy, as half of the energy contained in municipal waste is biogenic. Over 50 TWh per year of renewable electricity could be generated in the EU, which is more than 10% of today's total renewable electricity generation (in the EU). To achieve this potential, it will be necessary to avoid disposal of municipal waste by imposing landfill taxes and to put in place incentives for increasing the efficiency of EfW systems.

Combined use of heat and power: In addition to the above-mentioned measures for increasing the electrical efficiency, the combined use of heat and power should be considered. This is particularly true for the use of energy from MSW, as this fuel is usually generated close to the centres of heat consumption in the densely populated and industrial areas of the world (big cities in western Europe, the east and west coasts of the USA, Japan and China, for instance).

For plants producing both electricity and heat, it is an ongoing discussion how to weight the two products. Within the EU Waste Framework Directive, power is weighted by the factor 2.6 and heat by the factor 1.1. In Fig. 6.44, the *energy efficiency performance indicators* of the EU Waste Framework Directive and the Swiss Electricity Generation Directive (EU 2008; Bundesamt 2008) are plotted as a function of the gross electrical efficiency and the heat recovery rate. The line labelled "EU Directive $R1 = 0.65$" marks the minimum requirement a plant must fulfil to get the recovery status. The line labelled Switzerland marks the energy efficiency threshold for plants in Switzerland.

Figure 6.44 also includes data points derived from applying the energy efficiency criteria to different modern plants which reach particularly high efficiencies of power generation or combined heat and power generation, as well as two instances of the reference plant with combined heat and power generation.

– The example of the reference case (40 bar, 380°C) with an electrical efficiency of 20.6% achieves an $R1$ of 0.64 (if only electricity is produced) which is close to the threshold. Extracting heat from this plant reduces the electrical power production, but increases the $R1$ efficiency criterion. This is shown for heat production at temperatures of 130 and 95°C. The slope of these lines is determined by the power loss coefficient, giving the ratio of lost electricity production to heat pro-

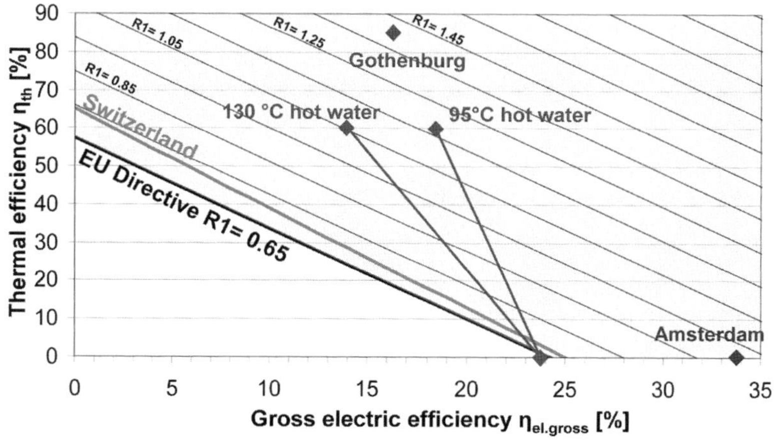

Fig. 6.44 Gross electric efficiency–heat recovery rate diagram (Gohlke and Murer 2009)

duction. The exergy efficiency, which includes the exergy of power (which is 1) and the exergy of the heat according to the temperature, is for all cases about 20.6%.
- A sophisticated heat production system in Gothenburg results in an $R1$ value of 1.42, which exceeds the required $R1$ value of 0.65 for new plants by far, whereas the exergy efficiency is only 29.7%.
- The new plant in Amsterdam focuses on the production of electricity and reaches an exergy efficiency of 30.6%, whereas the resulting $R1$ value is 0.91, which is quite low compared to Gothenburg, but still fulfils the $R1$ requirement easily.

Amsterdam and Gothenburg are two extraordinary examples with special economic and geographical constraints. However, with a reasonable combined heat and power production the $R1$ criterion can be fulfilled by state-of-the-art plants with 40 bar/380°C boilers, though for locations without a demand for heat energy, the fulfilment of the $R1$ criterion requires measures such as those described earlier to increase the electrical efficiency. The comparison of the energy performance indicators for the different plants shows that heat production is overrated if the thermodynamic value (exergy content) of heat is considered. For the author, the exergetic efficiency seems to be a more appropriate performance indicator.

6.4.8 Dioxins

The discovery of polychlorinated dibenzo-dioxins and polychlorinated dibenzo furans (PCDD/F), often simply called "dioxins", in the flue gas of waste incineration plants had a major influence on the technical development of EfW plants. In order to point out the possibilities of dioxin reduction, the formation and destruction of dioxins will be briefly discussed in this section. The most toxic dioxin, 2,3,7,8-tetrachlorodibenzo-*p*-dioxin (TCDD), became well known as a result of the accident at the chemical plant in Seveso, Italy, in 1976. The toxic equivalent unit (TEQ) measures all toxic dioxins and furans in terms of 2,3,7,8-TCDD.

Dioxin emission from EfW plants principally results from two sources:

- dioxins may exist in the waste or they
- can be newly formed (de novo) when cooling down the flue gas.

Dioxins fed with the waste into the EfW plant can be effectively destroyed at high temperatures and sufficient residence time. Accordingly, a residence time of 2 s at 850°C is required in the flue gas path of an EfW plant.

The de novo formation of dioxins is a heterogeneous gas– solid reaction, in which the fly ash or solid carbon provides the surface for the reaction. De novo dioxin formation requires the presence of chlorine, oxygen and aromatic species. Gaseous HCl in the flue gas can be converted to molecular chlorine by the Deacon reaction:

$$4\,HCl + O_2 \rightarrow 2\,H_2O + 2\,Cl_2 \tag{6.1}$$

Molecular chlorine reacts with aromatic species and soot in the flue gas to form PCDDs and PCDFs, the amount and particular species depending on the temperature and the boiler design. Copper in the fuel or the ashes may act as catalyst for formation. It has been shown that the de novo reaction takes place in a temperature window of 180–450°C, with a maximum formation at about 300°C, and is dependent on the residence time of the gas and in particular the fly ash in that temperature range.

Effective primary measures to reduce the de novo formation of dioxins are as follows:

- A complete burnout, which reduces the potential of dioxide formation by destroying the aromatic compounds and soot. The residence of 2 s at 850°C, which serves to destroy existing dioxins also promotes a complete burnout.
- A low residence time of the flue gas and fly ash in the temperature range of 180–450°C. This can be achieved by rapid cooling or quenching of the flue gases. Particle filters should be installed at lower temperatures, preferably below 180°C.

Dioxin emissions can be effectively reduced by primary measures to destroy existing dioxins in the fuel and to prevent de novo formation by appropriate boiler design and flue gas handling. Dioxin emissions are influenced by the burnout behaviour and depend on the temperature course of the flue gas train. High dioxin emissions from EfW plants in the past were mainly caused by de novo formation in particle filters installed at excessively high temperatures in combination with an incomplete burnout. Additionally, dioxins can be reduced by secondary flue gas cleaning. Secondary measures can be catalytic dioxin reduction or adsorption on char coal, which then has to be treated thermally. As a result of emission control requirements, incineration in developed countries is now a very minor contributor to dioxin emissions (Baumbach 1990; Nussbaumer 2004).

6.4.9 Flue Gas Cleaning

To comply with the stipulated emission limits of the European Waste Incineration Directive, a great number of methods for each individual pollutant are available. In the flue gas cleaning process, the following pollutants have to be removed from the flue gas:

- Fly ash (removed by ESP or bag filter)
- Sour gases like HCl, HF and SO_2 (removed by wet scrubbing, spray dry absorption, etc.)
- Heavy metals like Hg, Pb, Zn (removed by activated carbon filter or filsorption)
- Organic substances like PCDD/F (removed by activated carbon filter or filsorption)
- Nitrogen oxides (removed by SNCR or SCR)

Since the different methods of flue gas treatment have largely been discussed in Chap. 5, they shall be examined in the following only in regard to their use in waste incineration plants (Nethe 2008; Vehlow 2006).

Dust separation: Both fabric filters (also called baghouse filters) and electrostatic precipitators (ESPs) are suitable for removing particulate pollutants. To combine dust separation with wet absorption of acid gases (HCl, SO_2), ESPs are usually employed. They are cheaper and associated with a lower pressure loss than baghouse filters. The removal of dust and pollutants is lower with ESPs, but sufficient if associated with wet scrubbers for the additional removal of pollutants. In combination with spray dryer processes, the fabric filter is advantageous, because the sorptive effect on the filter linings can be exploited. The operating temperatures of baghouse filters are usually between 140 and 200°C. ESPs are also used at temperatures up to 280°C, upstream of the last economiser (last in relation to the flue gas stream; first in relation to the condensate water stream) in combination with SCR catalysts.

Separation of acid gases: For the separation of acid flue gas components such as HCl, HF, SO_2 and SO_3, wet and dry processes rival each other. Wet processes use scrubbing slurries, putting them into close contact with the flue gas to be cleaned. The separation process usually runs in two steps: in the first step, HF, HCl and Hg compounds are scrubbed with water and in the second step, SO_2 and SO_3 are separated by the addition of a lime slurry or sodium hydroxide. Wet scrubbing methods, with their good mass transfer between the gas and the liquid, are very effective and work under nearly stoichiometric conditions, so the consumption of absorbents is low. A drawback is the wastewater that is produced, which is of a type requiring wastewater treatment or concentration by evaporation. Wet processes are preferred in countries where it is authorised to discharge effluent to the river, such as in Switzerland and Austria or to the sea like in the Netherlands.

Dry or spray drying processes, in contrast, do not produce wastewater. In dry absorption or in an entrained-flow absorber, solid absorbents like calcium hydrate or sodium carbonate are fed to the reactor to separate the acid components, whereas in spray drying, an aqueous lime slurry is finely atomised and completely evaporated. The good mass transfer between the gas and the liquid in spray draying is again advantageous. The salt particles formed as a consequence of evaporation of the water and chemisorption are removed from the gas flow in a filtering separator. In this case, fabric filters offer the advantage of further removal via the solid layer of matter on the filter. Spray drying processes typically run at temperatures of 150–170°C.

A further reduction of organic pollutants or heavy metals can be effected by adding surface-active adsorbents such as activated carbon or activated charcoal. Dry or spray drying flue gas cleaning systems are preferred in countries where it is difficult to get authorisation for effluent discharge to rivers or seas. This is the case in most western and southern European countries as well as in the USA and Japan. In Germany, dry or spray drying processes are favoured because of the option of disposing the fly ash and flue gas cleaning residues in abandoned salt mines at relatively low cost. In France and Japan, these residues are solidified with cement or concrete, which is costly, and there is contention over the long-term stability of such procedures.

Toxic heavy metals (e.g. Hg, Cd, Pb, Zn) and organic substances (e.g. PCDD/F): The technologies used for these pollutants are entrained-flow and fixed bed adsorbers. The methods to separate heavy metals and organic matter are based on the adsorption of the pollutants by carbonaceous surfactants such as activated carbon or lignite coke and not on absorption as in the removal of acid gases. Entrained-flow processes are designed for the separation of heavy metals and dioxins by adsorption onto reactants which are injected to the flue gas stream. In a fixed bed adsorber, the separation process occurs as the flue gas flows through a packed bed of carbonaceous adsorbents. Single-chamber systems with fillings of activated carbon/lignite coke and multi-chamber systems with various adsorbents are used.

Both entrained-flow and fixed bed adsorbers can be used as safety or "police" filters at the end of the flue gas cleaning train. Remaining heavy metals or dioxins are removed by adsorption onto carbonaceous material and remaining acid components can be absorbed by the addition of calcium hydroxide.

NO_x reduction: Nitrogen oxides can be reduced by primary measures, by selective non-catalytic reduction (SNCR) or by selective catalytic reduction (SCR). SCR technology not only gives the highest reduction rates but also involves the highest cost. In the low-dust SCR configurations which are most common in Europe and Japan, the catalyst is arranged after the scrubber to prevent deactivation. This configuration has the disadvantage of the need to reheat the flue gases to the operating temperature of the catalyst (above 240°C). In the USA it is common to use SNCR (and not SCR) in EfW plants.

Flue gas cleaning configurations: All modern waste incineration plants are equipped with an efficient flue gas cleaning system which guarantees a reliable compliance with emission limits. Figure 6.45 shows possible process variants. Configurations with wet gas cleaning (configuration a) are clearly more complex than dry processes. In German plants with wet gas cleaning, concentration of the blowdown solutions by evaporation is required, which can be performed by an external evaporator or by installing a spray dryer and a fabric filter into the hot flue gas stream (configuration b). In contrast, the dry variant in combination with NO_x control by SNCR is clearly less complex (configuration c). By means of the spray dryer, both the acid gases and mercury and dioxins can be separated by adding activated carbon. In the following fabric filter, the products from flue gas cleaning are removed together with the fly ash. In some plants, the fly ash is removed separately beforehand.

The dry process has the disadvantage that it has to be operated with a higher sorbent stoichiometry and so produces considerably higher amounts of residues, which, because of their solubility in water, are more difficult to dispose of (Vehlow 2006). Police filters were used in Germany and Austria in the 1990s but are not common in newer plants because of the general advancement of flue gas cleaning technologies.

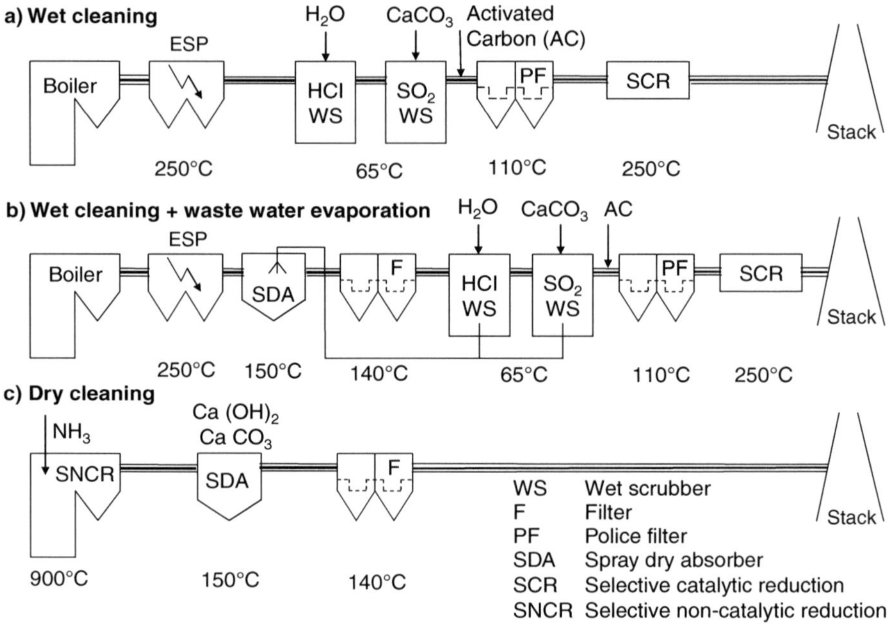

Fig. 6.45 Configurations for flue gas cleaning

6.5 Co-combustion in Coal-Fired Power Plants

The combined thermal utilisation of biomass and fossil fuels provides a cost-efficient option for the short-term exploitation of the biomass that is currently available. Given that biomass is a solid fuel, it is reasonable from the technical point of view to use it in combustion or gasification plants in combination with other solid fuels. Existing coal firing power plants, with outputs up to a maximum of 2,500 MW_{th}, are almost all combustion plants and, because of their firing and flue gas cleaning installations, are ideally suited to the co-combustion of biomass (Spliethoff 2000; Spliethoff et al. 2001).

The share of the biomass in the total thermal output of a co-fired power plant is limited by the biomass flow that can be used without inhibiting the operation of the plant and also by the quantity of biomass that can be supplied to a given plant site. This second constraint sets a limit of around 50–100 MW of bio-fuel energy input. Higher capacities strongly increase goods traffic, which becomes problematic when attempting to gain the approval of authorities. A fraction of 10–30% of biomass in the total thermal output from a pulverised hard coal plant should not be exceeded, so that adaptation measures in the plant can be kept to a minimum. In brown coal-fired and fluidised bed furnaces, it is possible to use greater fractions. The exact values have to be determined for each plant individually.

Power generation by co-firing biomass in existing coal-fired power plants has a number of advantages over generation in small plants fired with biomass only:

- The large total existing power plant capacity – even with a relatively small biomass fraction of the total fuel input – allows biomass utilisation at a large scale without a delay for construction of biomass-specific plants. This holds true despite the fact that not all large power plants are generally suitable for the co-combustion of biomass.
- The power generation efficiencies of large power plants are high compared to the efficiencies of small plants fired with biomass only.
- In case of seasonal non-availability or of shortfalls in bio-fuel supply due to weather conditions, the generation of power can still be guaranteed based on coal (i.e. a high security of supply combined with high fuel flexibility).
- The additional capital costs needed to co-fire biomass in existing coal-fired power plants are relatively low compared to new dedicated biomass combustion systems. While the capital costs for the new construction of a local biomass combustion plant amount to between 2,500 and 3,000 €/kW_{el} of installed capacity, retrofitting an existing power plant for co-combustion requires capital costs of about 300 €/kW_{el} of biomass input. In the main, these additional costs are allotted to fuel preparation (VGB 2008).

Besides biomasses such as wood and straw, other biomasses that are produced in adequate and homogeneous quantities can, in general, also be used in co-combustion, for example, sewage sludge. Co-combustion is also suitable for biomass types that are rather problematic from the combustion engineering point of view. In Germany, if the co-fired fuels derived from waste are within the legally defined range of between 0 and 25% of the thermal capacity, the plant can still be licensed and operated according to the German Ordinance on Large Combustion Plants (*13*th *BImSchV*). Only part of the statutory regulations for waste incinerators (re *17*th *BImSchV*) apply to co-combustion (see also Sect. 5.6).

Co-firing technologies are presently being routinely commercially practiced in the USA, Finland, Denmark, Germany, Belgium, the Netherlands, Austria, Spain, Australia, Britain and a number of other countries. An inventory of the application of co-firing worldwide in 2004 indicated that more than 150 coal-fired power plants had experience with co-firing of biomass or waste, at least on a trial basis. A large range of biomass materials including herbaceous and woody materials, wastes and energy crops are co-fired today (Van Loo and Koppejan 2008; Baxter 2005; Fernando 2005; Fernando 2007).

The focus of co-combustion activities varies from country to country and depends on the availability of biomass types, but mainly on legal issues and national incentives. Many countries have initiated incentives to encourage the use of biomass for electricity production. Some examples are as follows:

- According to a political agreement in Denmark from 1993, power stations in Denmark have to use 1.4 million tonnes of biomass, of which at least 1.0 million tonnes must be straw, every year beyond 2000 (Berg and Jensen 2008).
- In 2002, the Dutch government agreed a coal covenant with the six major utilities in which they were obliged to reduce CO_2 emissions from coal by 5.8 Mt/year in the period 2008–2012. More than half of this target, namely 3.2 Mt, is to be

achieved by the substitution of biomass for coal. This equates to an installed biomass capacity of 508 MW. To achieve these targets, the incentives for electricity production from biomass of a subsidy of 4.8 € cents/kWh and a tax redemption of 2.9 €cents/kWh were put in place. In the Netherlands, all eight coal-fired power plants co-fired biomass, industrial waste, RDF or sewage sludge up to 2005 (Fernando 2005).

- In Germany, the Renewable Energy Act promotes the utilisation of biomass only in units up to 20 MW_{el}. Electricity generated from clean biomass is subsidised at the rate of 7.7–10.6 € cents/kWh (from 2009: 5.8–10.2 € cents/kWh). The lower numbers are for larger plants in the range extending to 20 MW_{el}. Due to this reason, clean biomass is hardly used for co-combustion. The focus of co-combustion is on waste fuels, which are not subsidised. In Germany, nearly 20 plants co-fired sewage sludge and another 6 co-fired industrial waste or RDF up to 2007 (Fernando 2007).

6.5.1 Co-combustion Design Concepts

The different co-combustion design concepts are shown in Fig. 6.46. Co-firing can be direct, with biomass and coal being fed into the same boiler, or indirect, where a pre-treatment like gasification is carried out prior to the combustion in the main unit. In parallel combustion, biomass and coal combustion are separate and the boilers are connected only on the steam side.

Direct co-firing of biomass in furnaces is a simple method of combined biomass and fossil fuel utilisation. The firing technologies to be considered for the co-combustion of solid biomass and waste matter are those used for coal – grate, fluidised bed and pulverised fuel firing. In direct co-firing, the additional fuel is simply added to a boiler designed for the base fuel, usually coal. This is the most convenient method, able to be used in connection with both fluidised bed and pulverised fuel boilers. Fluidised bed combustion is quite suitable for co-combustion because of its

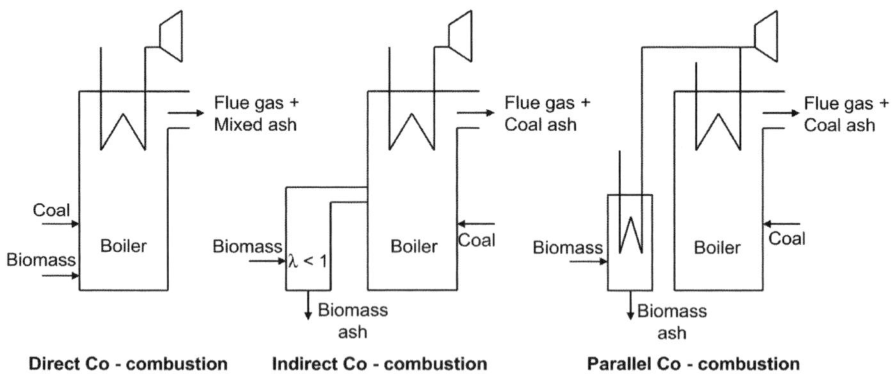

Fig. 6.46 Co-combustion arrangement options

fuel flexibility, whereas the pulverised coal combustor requires a well-defined fuel size distribution. Experience shows that only minor quantities of additional fuels (a few per cent of the fuel power) can be prepared together with the coal in the existing coal mills of a PC plant. If larger quantities are to be used, special mills and burners are desirable. There are no such limitations in FBC. In fluidised beds, larger quantities and larger fuel particle sizes can be used (Leckner 2007; Spliethoff 2000).

Because there are only a small number of grate firing systems used for coal, this technology shall not be considered further in the context of co-combustion. Grates are used for mono-combustion of biomass and predominantly for waste incineration, since their capacities are adequate for the quantities of waste produced at many locations and the technology is suitable for a wide range of fuel types.

Direct co-combustion of biomass in coal-fired plants, however, can have negative effects on operation and the quality of the residual matter. It can impair the plant performance through fouling, slagging, or corrosion, or reduce the potential for use of the fly ash. An additional pre-treatment step for the biomass such as pyrolysis, gasification or washing can solve these problems (see Fig. 6.47).

In contrast to direct co-combustion, it is possible to separately remove and utilise the ashes of coal and biomass by topping with a pyrolysis or gasification step. Combustible, low-calorific-value gas is injected and burned in the main combustor, whereas most of the ashes remain in the gasifier, from where they are removed,

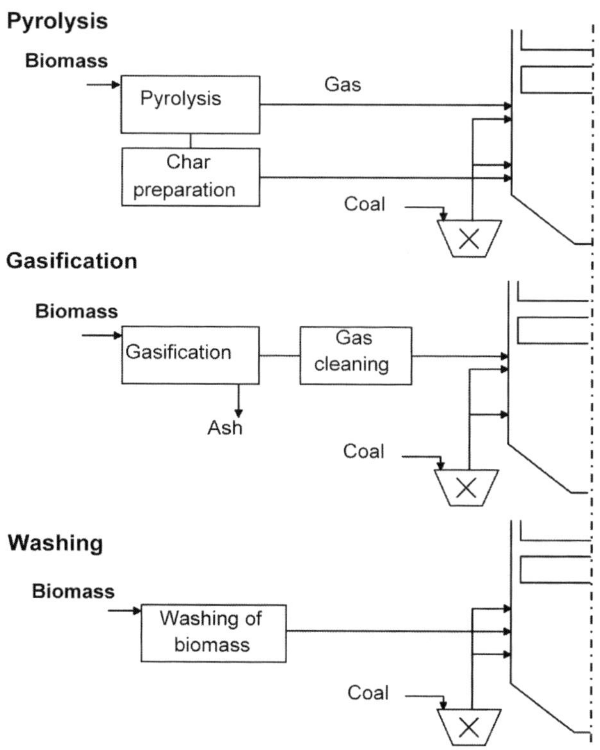

Fig. 6.47 Indirect co-combustion configurations

separately to the ash removal from the main boiler. However, gaseous components such as volatile alkali compounds or trace elements from the additional fuel will enter the main combustor if no adequate gas cleaning is provided for. Such systems have been built in the Netherlands, Finland and Austria. Used as a reburn fuel, the pyrolysis gases are able to reduce the nitrogen oxides formed in the combustion of the pulverised fuel. Fouling or erosive effects caused by biomass ashes are avoided or reduced, and the commercial value of the fly ash from the coal is not impaired (Leckner 2007; Fernando 2002).

In parallel co-combustion, biomass is fired in an entirely separate combustion system which is connected to the main boiler only on the steam side. Steam produced in the biomass combustor can be either fed directly to the joint turbine or to the coal boiler to be superheated. In this way, the additional furnace can provide heat to the steam cycle at a suitable steam temperature. Another advantage of connecting only on the steam side is that any harmful substance released in the additional boiler will not affect the main combustor, and both ashes and flue gases can be treated separately. This promotes the utilisation of both the coal and the biomass ash. An example of such a plant is the 600 MW_{el} multi-fuel power plant at Avedore, Denmark. Straw of a thermal input of 100 MW is burnt in a grate combustor, with the steam produced having the same steam conditions as the main boiler – 300 bar and 580°C. The multi-fuel concept enables efficiencies of 48% for the main unit and 45% for the straw unit (Noppenau 2003; Fernando 2002).

Direct co-firing is straightforward but can lead to several technical problems. Indirect and parallel co-firing incurs greater costs and is in particular suitable for biomass containing troublesome or harmful components or when the quality of the ash is of importance. It is necessary to weigh up the costs and the other pros and cons of the available technologies. Indirect and parallel co-firing is less common than direct co-firing.

6.5.2 Biomass Preparation and Feeding

The required preparation for bio-fuel for direct co-combustion in existing power plants depends on the type of biomass and on the firing technology. For wood, woodchips can be regarded as the state of the art. Experience in Denmark suggests that providing herbaceous biomass in the form of bales is suitable.

Different methods for supplying the fuel to the furnace can be distinguished. Preparation for and feeding of PC boilers usually require more steps, for example, milling or even separate burners for the additional fuels. The methods are shown in Fig. 6.48.

The simplest option is to blend the bio-fuel with the coal, introducing the mixture through the existing fuel-handling system and through the existing pulverised coal burners. The main restriction is the different milling behaviours of coal and biomass. Most coal mill designs base the pulverisation of the coal on its brittleness. This milling principle can be used in the case of sewage sludge, but it is not

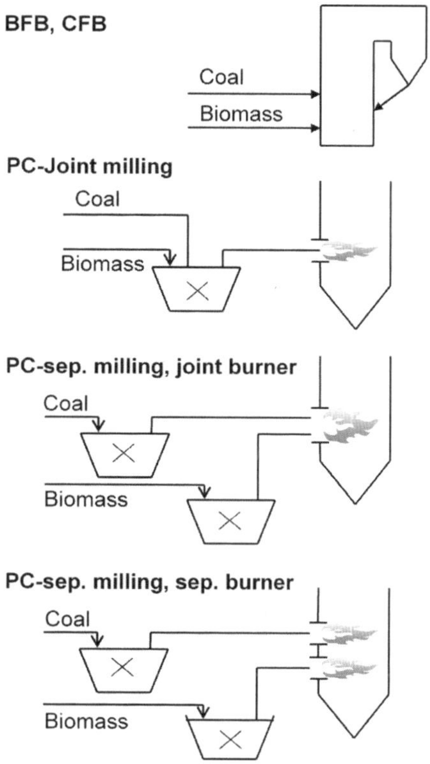

Fig. 6.48 Fuel supply arrangements for PF and FB co-firing

suitable for woody or herbaceous biomass. For these fuels, the technologies usually applied are cutting milling or hammer milling. If the biomass has a considerable moisture content, the heat balance of the furnace limits the addition of the fuel to a few per cent of the total fuel power. For straw and wood co-firing, separate milling in cutting or hammer mills, as well as separate feeding to the burner, is the standard technology. The pulverised fuel can be injected into the furnace together with coal at the same burner or to a separate biomass burner. For fluidised beds the system is less complicated as there are no burners or grinding devices.

Wood in the form of woodchips can be used in fluidised bed furnaces without further milling. Maximum chip sizes are about 50 mm in circulating fluidised bed furnaces and about 90 mm in stationary FB furnaces. Bales of herbaceous biomass have to be undone and the biomass cut into chaff of lengths between 10 and 30 cm. So in regard to fuel particle sizes, there is practically no difference between co-combustion and combustion with biomass only. Overall, fluidised bed firing requires very little preparation of the fuel and is in addition insensitive to hydrous fuel types with low calorific values.

To achieve complete combustion, pulverised fuel firing requires the biomass (e.g. woodchips or bales of herbaceous biomass) to be milled to a high degree, adequate devices for which are cutting and hammer mills. In order to determine the

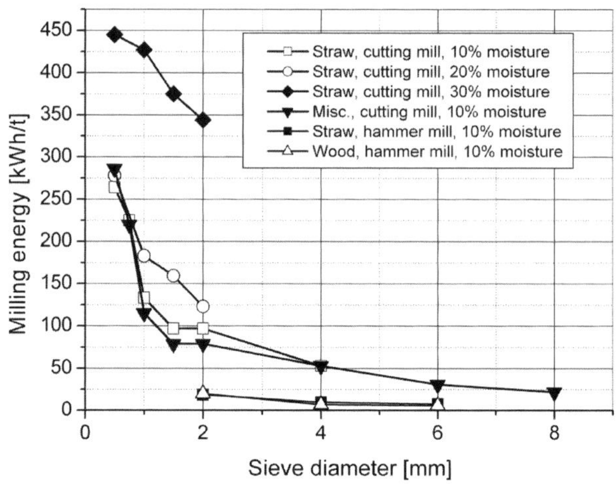

Fig. 6.49 Milling energy required for cutting and hammer mills of different sieve insert diameters (Siegle 2000; Spliethoff 2000)

energy requirement for milling, tests were carried out using a cutting mill and a hammer mill. In both systems, the biomass fibres were repeatedly reduced in size until they fell through the holes of a sieve insert. The electrical energy needed for this size reduction increased as the particle size (i.e. the sieve insert hole size) decreased, ranging between 0.8 and 2% of the calorific value of the biomass if a cutting mill was used with sieve holes between 2 and 6 mm (see Fig. 6.49). With a hammer mill, the energy required was only 0.5–1% of the calorific value (Siegle 2000). As a rule, the biomass milling energy requirement increases with the moisture content of the biomass – with moisture contents of more than 10–20%, the energy demand rises especially high. When straw with a moisture content of 30% and a sieve hole of 2 mm was chosen, the energy demand was more than 8% of the calorific value. A cutting mill was used, however, which is less efficient for such a task.

Wood particles produced in cutting mills have a rather cubic form, whereas straw and *Miscanthus* particles are small, elongated, rectangular plates. Figure 6.50 shows the average particle diameters determined by sieving for various sieve inserts, milling methods and biomass types. It should be observed that fibres longer than the diameter of the sieve holes can fall through the sieve. For these reasons, the values in Fig. 6.50 are only approximate.

The bio-fuel has to be appropriately milled to guarantee complete combustion during pulverised fuel firing. In a 0.5 MW experimental plant, to give an example, the maximum particle size to ensure ignition and complete combustion was a diameter of about 6 mm for straw and about 4 mm for *Miscanthus* (due to its wood-like structure). Wood co-fired with pulverised coal required a milling degree of 2–4 mm (Kicherer 1996; Spliethoff and Hein 1996). In co-combustion in plants with higher thermal capacities of up to several thousand megawatts, coarser particles can also be fired because of the longer residence times. In one industrial plant, for instance, straw with stem lengths up to 10 cm could be burned completely (Bemtgen et al. 1995).

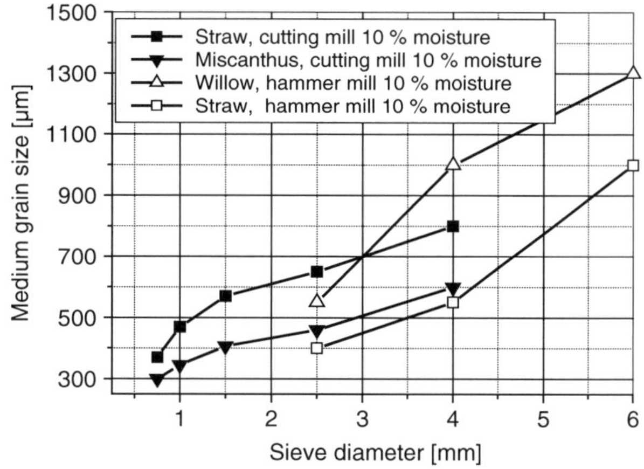

Fig. 6.50 Medium particle size as a function of sieve diameter (Siegle 2000; Spliethoff 2000)

Preparation and feeding of mechanically dewatered or (additionally) thermally dried sewage sludge can usually be achieved using the existing installations for coal handling and preparation. For this purpose the sewage sludge is added before the milling process, so that it is completely dried, milled and injected into the firing together with the coal. If, however, only mechanically dewatered sewage sludge is fed to the mill (i.e. not thermally dried sludge) along with the coal, the amount of sewage sludge for co-firing is limited by the mill's capacity to evaporate the moisture. The process of milling and drying the fuels together ensures that coal and sewage sludge are well mixed.

Another possible requirement might be to dry the biomass. While it is usually not necessary to dry wood and straw for the actual combustion, there may be an energy efficiency advantage in doing so, in particular when previously unused waste heat can be utilised. Sewage sludge, in contrast, does as a rule require at least partial drying. Mechanical dewatering by centrifuge or compartment-type filter press can raise the dry matter fraction of the sewage sludge from about 3–5% to between 20 and 45%. This process, though, is already carried out at the sewage treatment works. Combustion or thermal drying can effect a further decrease of the water content. While mechanical sludge dewatering is in all cases used at sewage treatment plants for volume reduction, the application of thermal drying is debatable. The issue, however, is not whether adequate process technology makes thermal drying superfluous, but where the drying should be carried out (external to or inside the combustion plant), how the dewatered sewage sludge should be integrated into the thermal process, what the costs are for the individual solutions and what the outcomes are of the respective energy balances.

For an external drying process, using a low-energy drying medium can optimise the heat utilisation. Sewage sludge drying at a power plant offers the opportunity of employing either low-temperature steam, flue gas or combustion air. The less favourable option is drying at the sewage sludge works either by using a dryer fired with gas collected from the treatment works (mainly methane) and/or natural gas or directly in a firing system.

6.5.3 Co-combustion in Pulverised Fuel Firing

The issues of prime importance when co-firing biomass in existing pulverised fuel firing plants are the mass and volumetric flowrates and their rates of change. They must be compared to those of coal firing alone. The existing fuel transport arrangements and the fuel preparation process must be suitable for the new volumetric fuel flow and have adequate capacity. Likewise, the change in the moist volumetric flue gas flowrate has to be determined, because of its substantial influence on the heat transfer and the residence time in the steam generator and on the functioning of the downstream flue gas cleaning equipment.

When biomass is co-combusted in existing coal-fired power plants, attention must be paid to the impacts on all the units occurring as a result of the bio-fuel properties and the deviation from the design fuel. Figure 6.51 identifies the possible impacts on the components of a pulverised coal-fired power plant. These impacts will be discussed in the following detailed descriptions of wood and herbaceous biomass co-firing. In addition, sewage sludge is taken into consideration as a supplementary fuel because, so far, it is industrially the most significant waste fuel.

6.5.3.1 Volumetric and Mass Fuel Flowrates and Flue Gas Flowrate

The volumetric flow of the fuel increases considerably as a result of the admixture of ligneous (wood and wood waste) and herbaceous biomass to the coal. This is due to the far lower calorific value and the greater fuel volume of the organic material com-

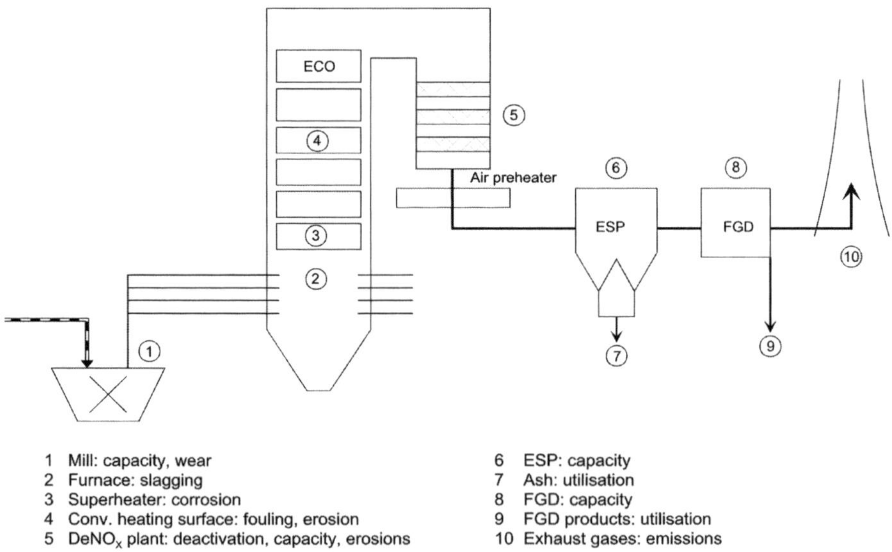

Fig. 6.51 Possible impacts of co-combustion (Spliethoff 2000)

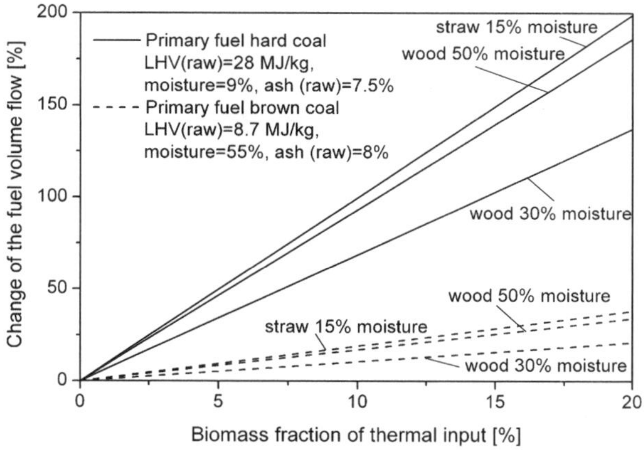

Fig. 6.52 Increase in the volumetric as-received fuel mass flow in biomass co-combustion (bulk density of coal = 870 kg/m^3, brown coal 740 kg/m^3, chopped material (30% moisture content) = 250 kg/m^3, straw bales (15% moisture content) = 150 kg/m^3)

pared to coal (Fig. 6.52). To give an example, the total volumetric fuel flow doubles when straw is co-fired in pulverised hard coal firing when straw has a 10% fraction of the thermal input. For this reason, both milling and transport to the furnace should be separate. Often there is no choice, as the coal milling installations in most cases cannot be used for biomass because of the difference in material structures. Biomass therefore requires specific milling installations.

The change in the volumetric wet flue gas flowrate depends on the moisture content of the primary fuel and the biomass. Co-firing of straw with a moisture content of 15% with hard coal as the primary fuel results in only a slight increase, of 5%, when the biomass input is 20% of the total thermal input. In contrast, very moist biomass (e.g. bark or fresh cuttings) may have a stronger effect and the resulting increase in the flue gas volume may limit the substitutable fuel quantity. If brown coal with a high moisture content is being fired as the primary fuel, the flue gas flow decreases slightly (see Fig. 6.53).

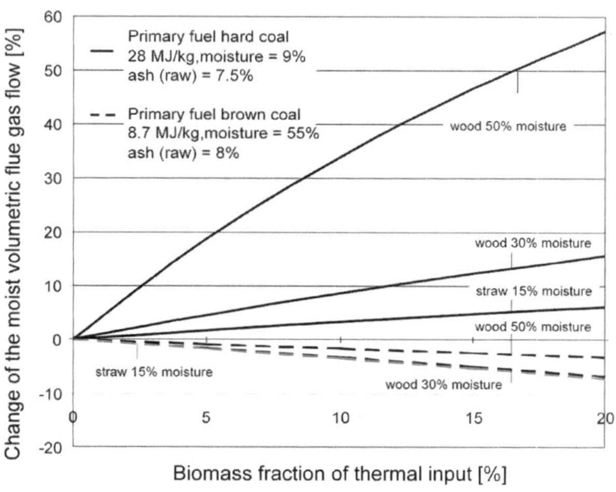

Fig. 6.53 Change of moist flue gas volume in biomass co-combustion

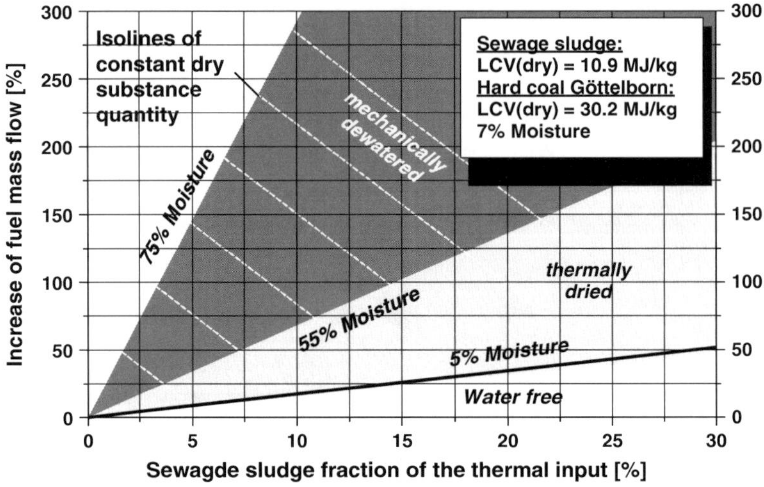

Fig. 6.54 Influence of co-combustion of sewage sludge on the fuel mass flow (Gerhardt et al. 1997)

The influence of the moisture content of sewage sludge in co-combustion with hard coal is described in Fig. 6.54 for the fuel mass flow and in Fig. 6.55 for the volumetric flue gas flow (Gerhardt et al. 1996, 1998). Both diagrams are presented as a function of the sewage sludge fraction of the fuel heat input. When reading these charts it must be remembered that the same fraction of sewage sludge with a higher moisture content would need more dry substance to be fired with it for the evaporation of its moisture.

For this reason, the diagrams in Figs. 6.54 and 6.55 also feature isolines of equal dryness and constant population equivalents (the population equivalent gives the

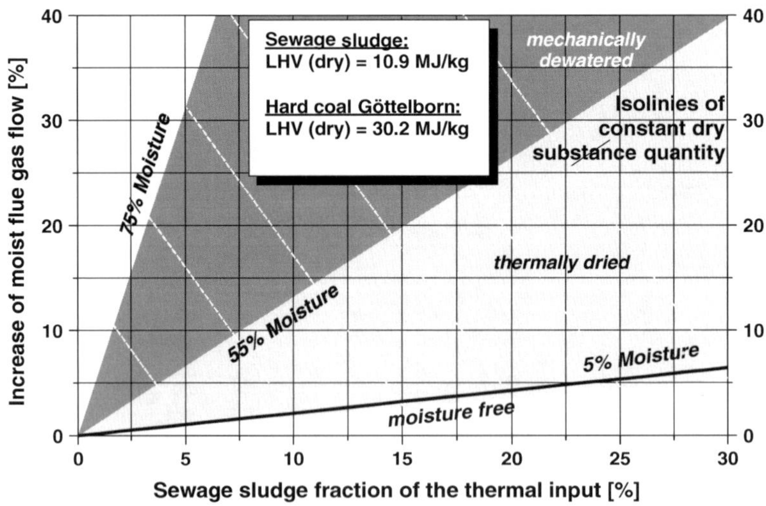

Fig. 6.55 Influence of sewage sludge co-combustion on the moist flue gas flow (Gerhardt 1997)

average amount of sewage sludge to be disposed of per inhabitant). While the fuel mass flow amounts to 18% for a thermally dried sewage sludge with 95% dryness by weight, contributing a 10% fraction to the thermal capacity, the fuel mass flow rises to about 70% for mechanically dewatered sludge at 45% dryness by weight if the same thermal fraction is to be contributed. In this latter case (mechanically dewatered sewage sludge), a higher dry substance mass flow is fed to the firing, which means that a greater sewage sludge quantity, given in population equivalents, is disposed of. If the same population equivalent is taken as a basis both for the mechanically dewatered and for the thermally dried sewage sludge, the result is a 50% larger fuel mass flow for the mechanically dewatered sludge, as seen in Fig. 6.54. The thermal input fraction is around 7% compared to 10% for thermally dried sewage sludge.

There is a strong case for the mechanical handling equipment, the mills and the fuel feeding via the burners to be modified if the sewage sludge will bring excessively high moisture into the installed equipment of the power plant. The same is true, more or less, for the flue gas system. The higher flow velocity leads to changed heat transfer rates and also impairs the separating performance of the flue gas cleaning train.

If the primary and additional fuels have nearly the same moisture contents, only minor modifications are necessary. In practice, therefore, either mechanically dewatered sewage sludge is combined with raw brown coal (moisture about 50%) or thermally dried sewage sludge is combined with hard coal (moisture about 7%). If there is only a small fraction of mechanically dewatered sewage sludge, however, it is also possible to use it in hard coal firing.

6.5.3.2 Combustion Process

Due to the high volatile matter content of biomass in comparison to coal, much coarser biomass particles can be used in pulverised fuel firing. Herbaceous biomass types are more reactive than wood in this respect and therefore require less fine milling.

The significantly coarser milling degree of the biomass particles has an impact on the combustion process. Figure 6.56 shows the course of combustion for three fuel combinations: pure coal, 20% milled straw/80% coal and 25% sewage sludge/75% coal. The gradients of the mean oxygen concentration and the temperatures *(not shown)* of the combustion courses illustrate how the ignition of the coarse straw particles is delayed. After ignition, though, oxygen is consumed faster and the combustion proceeds more rapidly to completion in comparison to firing coal only. For thermally dried sewage sludge with a similar particle size distribution to coal, combustion proceeds faster.

6.5.3.3 Slagging, Fouling, Erosion

The process of deposit formation is described in detail in Sect. 5.10. There are two principal mechanisms through which co-combustion can affect slagging and

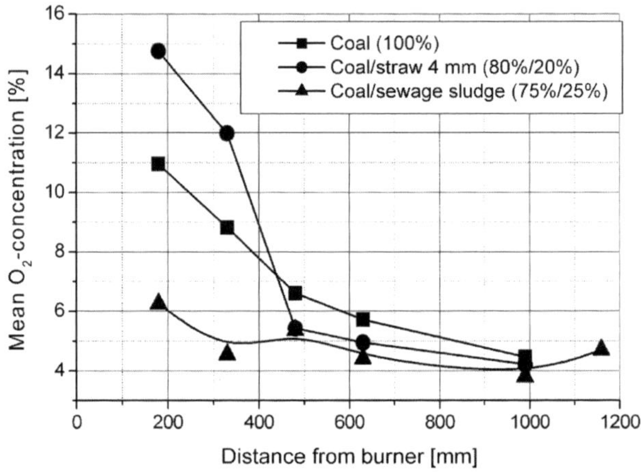

Fig. 6.56 Course of the combustion process of a mixed biomass/coal firing

fouling. The first mechanism is dependent on the melting of the bulk ash and is mainly related to the formation of molten deposits (Tortosa Masiá 2006). When biomass is co-fired with coal in low to medium ratios, the behaviour of the resulting ash will be dominated by the coal ash. However, the presence of significant levels of alkali and alkaline earth compounds in the mixed ash can change the behaviour of the coal ash, reducing ash fusion temperatures by 100–200°C and promoting formation of molten deposits. This effect is more pronounced for coal ashes with high fusion temperatures. In these cases co-firing at even modest ratios can have a major impact on the ash fusion behaviour. The effect is less dramatic with coal ashes originally having lower fusion temperatures and already having significant slag formation propensities. The effects of lower ash fusion temperatures differ between dry-bottom and slag-tap furnaces. While low fusion temperatures may be rather welcome in slag-tap furnaces, in dry-bottom firing they can lead to slagging in the combustion chamber, especially around the burner, thus impairing the firing process.

When co-firing woody or herbaceous biomass fuels (which have moderate ash contents), the volatilisation and condensation of alkali metals is the major mechanism for the initiation and growth of fouling deposits. Alkalis vaporise at flame temperatures, undergo chemical transformations and settle on surfaces that have suitable temperatures for condensation. The sticky layer of condensed alkalis acts as a glue for other solid ash constituents, thus initiating deposit formation. Most types of biomass or wastes are high-fouling fuels, and in almost all cases co-firing with coal increases the likelihood of fouling in comparison to coal alone. On the other hand, deposit formation is lower than in pure biomass combustion (Fernando 2007; Leckner 2007).

Although fouling and slagging problems on convective heating surfaces are increased by the addition of low-melting biomass ash, investigations at a 0.5 MW plant showed that the fouling rate in straw co-combustion is only slightly higher than in the combustion of a low fouling tendency coal. Dust layers were easily removed as well. If the biomass fraction is not very high, the coal ash characteristics dominate. However, when the fraction is larger, incompletely burned straw particles form deposits, resulting in slagging because of the low fusion temperatures of straw ash (Heinzel 2004).

The large-scale experience in Europe suggests that slagging and fouling are unlikely to be a problem for co-firing ratios less than 10%. At one plant using a 20% co-combustion ratio of straw, the boiler performance was still satisfactory, although there was need for additional soot blowing, and some slagging occurred (Fernando 2005).

With sewage sludge, lower fusion temperatures and a higher ash content than the standard fuel may be the causes of slagging inside the furnace. The ash deformation temperatures of investigated sewage sludges have been found to be between 1,200 and 1,300°C, 100°C lower than the temperature of hard coal ash (Gerhardt 1998). In slag-tap firing, lower fusion temperatures may be favourable, but they can cause slagging in a dry-bottom furnace. The high ash content may cause erosion of the convective heating surfaces, a risk that also occurs with poultry litter, which contains a high ash content as well. However, the impacts of the much higher ash fraction of sewage sludge should still be controllable by soot blowing.

6.5.3.4 Corrosion

Herbaceous biomasses such as straw or cereals have significantly higher chlorine contents than coal and most other bio-fuels, which may increase high-temperature corrosion of the heat exchange surfaces. The high-duty surfaces of the superheater, where there are high steam and flue gas temperatures, are most likely to be affected. In a Danish 130 MW_{el} pulverised coal furnace, corrosion tests were carried out for two superheater materials. Using a 10% thermal input fraction of straw, the corrosion rates were moderate and only slightly higher than in coal mono-combustion (Fig. 6.57) (Bemtgen et al. 1995; Spliethoff and Hein 1995b). The results were con-

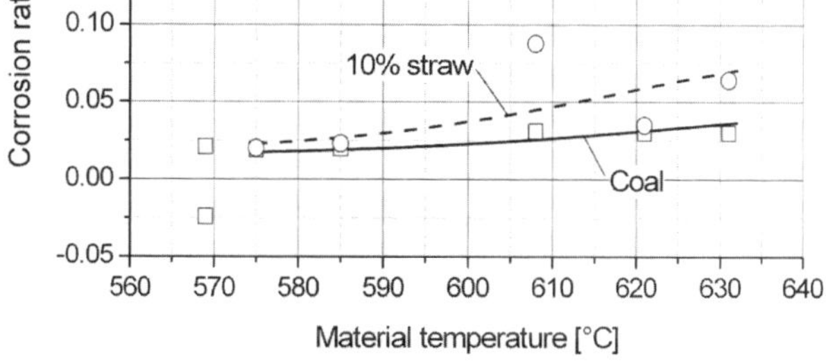

Fig. 6.57 Corrosion rates of straw co-combustion in a 130 MW_{el} pulverised fuel firing system (Spliethoff and Hein 1995; Bemtgen et al. 1995)

firmed in a 2-year investigation into co-firing with straw in a 150 MW_{el} coal-fired boiler. Up to steam temperatures of 580°C, the corrosion risks seemed tolerable. The introduction of a 20% straw fraction increased the corrosion rates by a factor of 1.5–3 at steam temperatures up to 540°C. Nevertheless, the corrosion rates are the same as those of low to medium corrosive coals (Wieck-Hansen et al. 2000). During these tests the fly ash and deposits that were formed mainly consisted of potassium aluminosilicates and potassium sulphate. Most of the chlorine left the boiler as less corrosive HCl, and there was little KCl in the deposits (Andersen et al. 2000). This example demonstrates the beneficial effect of the minerals and the sulphur in the coal on the transformation of the potassium in the straw (Leckner 2007).

Waste fuels containing high levels of alkalis or chlorine, such as refuse-derived fuel (RDF) and poultry manure, can increase corrosion rates by factors comparable to straw. RDF, if it has a high plastic content, can contain chlorine levels five to six times higher than average coals. Any efforts to increase the efficiency of the unit by increasing steam parameters will promote high-temperature corrosion. Sewage sludge has chlorine levels similar to coal and hence is less likely to cause high temperature corrosion.

In pulverised coal co-combustion, the extent of corrosion by high alkali and chlorine fuels is reduced by the presence of sulphur dioxide in the flue gas, an effect not occurring in monofuel combustion. SO_2 reacts with alkali and alkaline earth chlorides to form less corrosive sulphates. These sulphate compounds, however, are only stable under oxidising conditions, hence corrosion may still occur in overfire air systems (Fernando 2007).

6.5.3.5 Emissions

One result of co-firing ligneous and herbaceous biomass is a reduction of the major pollutants in the uncleaned flue gas – sulphur dioxide (SO_2) and nitrogen oxides (NO_x). CO emissions do not rise if the biomass has an adequately high milling degree. Supporting evidence came from investigations within the framework of an EU project, which produced more detailed information about emissions from different plant types co-firing herbaceous and woody bio-fuels (Bemtgen et al. 1995).

NO_x emissions: The high volatile matter content is the reason why biomass is especially suitable for the application of NO_x reduction measures like air and fuel staging (Spliethoff and Hein 1998). Although the fuel nitrogen in straw in relation to the calorific value is in the same order of magnitude as the content in coal, the result of the higher release of pyrolysis products and volatile nitrogen compounds from straw is less NO_x formation. Figure 6.58 shows the NO_x emissions measured at an experimental plant co-combusting different biomass types with hard coal, the biomass comprising 25% of the thermal input. The results reveal roughly the same emissions of NO_x, regardless of the fuel nitrogen content. The suitability of biomass to nitrogen-reducing combustion engineering measures is indicated by the decreasing NO_x emissions at diminishing primary air ratios. Co-combustion of biomass in pulverised coal firing therefore does not depend on the biomass nitrogen content. Even higher concentrations in sewage sludge or poultry litter can be controlled, to

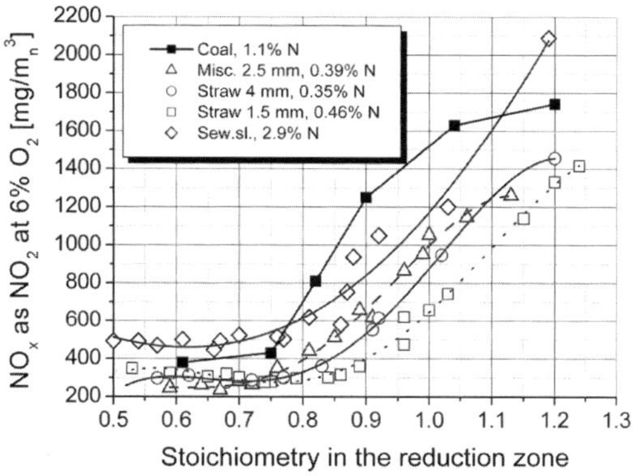

Fig. 6.58 NO_x emissions with air staging for different biomass types, biomass fraction: 25% (Kicherer 1996; Spliethoff and Hein 1996)

a certain extent, by combustion engineering measures. Biomass fuels promote the formation of ammonia instead of HCN in the primary combustion zone (DiNola 2007; Di Nola et al. 2009). Detailed investigations into the emission behaviour of biomass-coal blends in different combustion processes are described in Kicherer (1996). Due to its high volatile content, biomass is also suitable for reburning to reduce NO_x emissions. As shown in Fig. 5.53, pulverised *Miscanthus* as a reburn fuel is nearly as effective as natural gas. A pyrolysis gas from biomass can even be better as a reburn fuel than natural gas (Rüdiger 1997).

Despite the much higher fuel nitrogen content of sewage sludge, the NO_x concentrations after the fuel-burning system are not correspondingly higher, as they are limited by a lower conversion rate. However, within the typical range of use, up to 25% of the thermal input, an increase in emissions when incorporating sewage sludge co-firing must be expected if there are no additional methods to reduce NO_x. By applying the investigated in-furnace reduction methods of air and fuel staging, NO_x emissions from sewage sludge co-combustion were comparable to the combustion of coal only. Still, the high fuel nitrogen content should be taken into consideration when designing and constructing the furnace and considering the distribution of air in the combustion chamber.

SO_2 emissions: In pulverised coal combustion, the sulphur contained in the fuel is almost completely oxidised into sulphur dioxide (SO_2), so that SO_2 emissions can be directly correlated with the sulphur input. Figure 6.59 plots SO_2 emissions as a function of the fraction of biomass or sewage sludge, respectively, in investigations carried out at a 0.5 MW furnace. Increasing the share of biomass (i.e. wood, straw or *Miscanthus*) caused a decrease in SO_2 emissions. This reduction can be attributed to the low sulphur input. However, it is observed that sulphur is additionally captured in the biomass ash. Starting from about 90% in pure coal combustion, the conversion rate drops linearly to reach a value of 50% in pure biomass combustion. The coarse milling of the biomass and its moisture content delay combustion, thus favouring, because of the lower temperatures, the capture of SO_2 in the biomass ash (Spliethoff and Hein 1998).

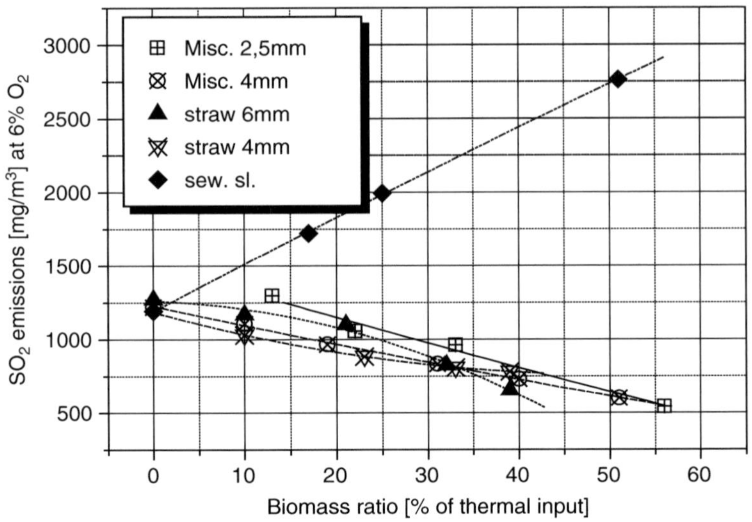

Fig. 6.59 SO_2 emissions as a function of the biomass ratio for different blends. (Kicherer 1996; Spliethoff and Hein 1996)

In co-combustion of thermally dried sewage sludge with hard coal, the concentrations of SO_2 and NO_x after the firing depend on the input fraction of the sewage sludge, the fraction of sulphur or nitrogen in relation to the calorific value and the conversion rate. The nitrogen contained in the sludge, in relation to the calorific value, is about —six to eight times higher than that contained in the coal. For fuel sulphur, the ratio is —three to four times higher than coal. The sulphur-to-SO_2 conversion rate of about 90% is not affected by the sewage sludge fraction, so the result is a rise in SO_2 emissions in proportion to that fraction. The high CaO content of the sewage sludge ash did not have any reducing effect on SO_2 emissions in the tests. This inactiveness of CaO can be explained by surface sintering as a result of the high combustion temperatures in pulverised fuel firing, as sewage sludge is milled to a similar degree to coal.

HCl emissions: The chlorine content of herbaceous biomass fuels such as *Miscanthus*, grass and straw and also of waste fuels such as RDF and municipal solid waste (MSW) can be considerably higher than coal. Straw can contain chlorine concentrations of about 1%, which is about 10 times greater than typical bituminous coals. The reactions of chlorine have been discussed in the context of deposit formation. In pulverised coal co-combustion, the availability of sulphur will lead to the sulphation of alkalis and formation of HCl, which is beneficial with respect to corrosion. HCl is completely removed by FGD scrubbers.

Dioxins: Fuels containing chlorine are suspected of producing harmful polychlorinated dibenzo-*p*-dioxins (PCDDs) and polychlorinated dibenzofurans (PCDFs), especially if the ashes of the fuels contain copper or other catalysts for dioxins. HCl in the flue gas can be converted to molecular chlorine by the Deacon reaction:

$$4HCl + O_2 \rightarrow 2H_2O + 2Cl_2 \quad (6.2)$$

Molecular chlorine reacts with aromatic species in the fuel to form PCDDs and PCDFs, depending on the temperature and the boiler design. When chlorine-containing biomass and waste fuels are co-fired with coal, the formation of PCDDs and PCDFs is inhibited. The possible mechanisms for PCDD/F inhibition include the depletion of molecular chlorine concentrations by the reaction with SO_2:

$$Cl_2 + SO_2 + H_2O \rightarrow 2HCl + SO_3 \quad (6.3)$$

Results from laboratory and large-scale investigations show that in co-combustion, due to the SO_2 in the flue gas, the emissions of PCDD/F are as low as for coal-fired plants (Fernando 2007; Leckner 2007).

6.5.3.6 Effects on Residual Matter

The effects of biomass co-combustion in coal-fired power plants on residual matter should be split into two: the consequences of a greater ash load and the effect on the commercial exploitability of the ash.

Ash load: The low ash content of wood and straw bio-fuels reduces the workload of the dust removal equipment. All in all, less ash will be produced in comparison to coal firing alone when co-firing ligneous and herbaceous or petiolate bio-fuels. In contrast, the typically high ash contents of sewage sludge dry matter can lead to higher workloads for the electrostatic precipitator when larger sludge fractions are used.

Commercial exploitability of fly ash: The composition of the fly and bottom ashes in coal firing determines their possible uses. For the utilisation of fly ash in the cement and concrete industries, the critical parameters are the concentrations of alkalis, SO_3, Cl, CaO and unburned carbon.

Since fly ash utilisation was one of the major obstacles for the broad application of co-combustion in hard coal-fired power plants, a review of EN 450 was initiated in 1999 and, since 2005, a new European Standard "fly ash for concrete" has replaced the earlier one. Fly ash from co-combustion of specific secondary fuels such as woodchips, straw, olive shells, cultivated biomass, municipal sewage sludge and paper sludge can now be used for concrete if the percentage of secondary fuel does not exceed 20% by mass of the total fuel, the derived amount of ash from the co-combustion material is not greater than 10% of the total ash and the requirements of the fly ash quality can be met (Wiens 2005). The maximum allowed contents of total alkalis, Cl and residual carbon for fly ash to be used in concrete production are 5% by wt., 0.1% by wt. and 5% by wt., respectively (see Sect. 5.11).

The commercial operation of straw co-firing at the 350 MW_{el} pulverised coal-fired Studstrup Unit 4 in Denmark began by burning straw with a maximum share of only about 10% on an energy basis. The fly ash is presently used for cement production (Zheng et al. 2007).

In the Netherlands, a programme was carried out to determine whether the fly ash produced from co-firing met the quality requirements for its utilisation as a filler material or as a cement replacement and for the production of concrete. Tests were undertaken at several power plants using up to 10% by mass of the secondary fuel. The fuels used included several types of pellets, sewage sludge, pet cokes, wood chips, poultry manure, MBM (meat-and-bone meal) and liquid hydrocarbons. The resulting fly ashes demonstrated that even high biomass co-firing percentages can produce fly ashes that meet European standards (Fernando 2007).

The ash from brown coal firing is commonly used as a filler material in opencast mines, and the regulations extend to this use, with the most critical parameter being the leachability of the ash components.

Waste fuels such as sewage sludge or RDF may contain higher concentrations of heavy metals. The evaluation of the ash properties, therefore, and in contrast to coal ash, above all has to take into account these heavy metals, which display an accumulation behaviour in the biosphere. Heavy metal concentrations in waste fuels are highly dependent on the origin of the fuel. As far as municipal sewage sludges are concerned, surveys certify that despite the given variations only a few sewage sludges have higher pollutant concentrations than is permitted for use in farming.

Apart from substances such as mercury, selenium and arsenic, which escape in elemental or compound form in appreciable fractions in the flue gas flow because of their low boiling points, most of the trace elements from sewage sludges are found in the solid residues from the furnace or the flue gas cleaning processes. Figure 6.60 shows a direct comparison between heavy metal ash concentrations of sewage sludge ash and of typical hard coal ash (Gerhardt et al. 1996; BMU 1996; Fahlke 1994). According to this comparison, the trace element concentrations, taking into account the different ash contents, approximate each other. The co-combustion of sewage sludge and hard coal therefore does not result in a serious change of pollutant concentrations in the ash.

For heavy metals that are partly carried out in the flue gas flow from the plant, it is necessary to check the removal efficiency in the cleaning sections downstream of

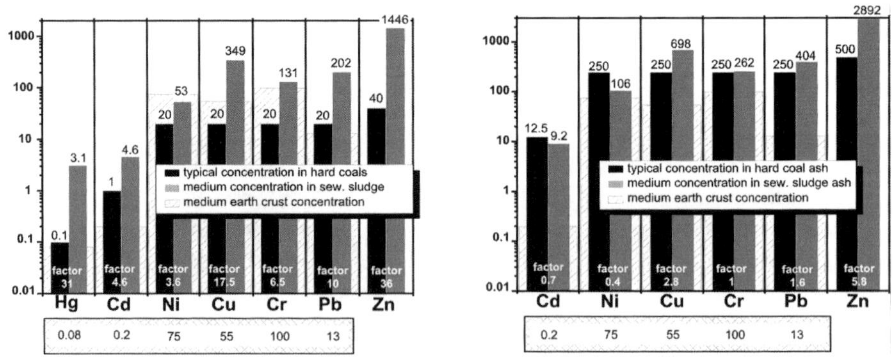

Fig. 6.60 Concentration of trace metals in dry fuels and ashes (Gerhardt et al. 1996; BMU 1996; Fahlke 1994)

the furnace. In the wet flue gas desulphurisation units common in power plants, only some of these heavy metals are removed, so the remaining part is emitted as part of the exhaust gas flow (Tauber et al. 1996). For mercury, for instance, a removal efficiency of about 50% is given (Fahlke 1994); the rest gets emitted. This circumstance can make it necessary to improve the common flue gas desulphurisation systems in power plants, for instance, by the addition of specially adapted precipitants or by the downstream installation of an additional filtering stage (e.g. an activated charcoal filter).

6.5.3.7 NO_x Control Equipment

When considering the impacts of co-firing on NO_x control, two scenarios have to be considered: plants with high-dust removal configurations and those with low-dust configurations. There is a smaller impact on low-dust configurations, because NO_x cleaning is preceding by the gas cleaning steps of ESP and FGD, which mean low-dust configurations are better suited to co-firing than high-dust ones.

Common practice in hard coal-fired dry-bottom furnaces is to install NO_x control in high-dust configurations. This puts the catalyst at risk, especially if straw is used, when employing biomass co-combustion. Various mechanisms may work to deactivate the catalyst. One of these is based on reactions of the catalyst with potassium and sodium. Accordingly, catalyst manufacturers set limits on the alkali fraction ($K_2O + Na_2O < 4\%$ by weight of ash). Depending on the coal type used, this amount may be reached even with small straw fractions. Another mechanism is triggered by alkalis and alkaline earths blocking up pores of the active catalyst cells. Arsenic and phosphorus, too, can poison the catalyst. Catalyst deactivation can be limited by installing the catalyst after the flue gas desulphurisation unit – that is, by employing a low-dust configuration. In brown coal firing, where sulphur is removed by combustion engineering techniques, this problem does not occur.

Testing of SCR catalyst elements in the slip stream of a power plant resulted in high deactivations when co-firing with 20% straw. After 3,000 h of operation the catalyst activity was reduced by 35% with a high-dust configuration, whereas with a low-dust configuration employing dry flue gas desulphurisation, the loss was between 10 and 15%. Due to the set-up of the facility, the test conditions are considered to be the worst-case scenario. More than 7,500 h can be expected before the activity is reduced to 50% (in the high-dust configuration), which is still considered a high level of activity (Wieck-Hansen et al. 2000). Results of 2 years of 7% co-firing of straw at Studstrup Unit 4 showed that there was no decrease in the removal efficiency of the high-dust SCR (Fernando 2005).

The high ash content of sewage sludge can cause fouling and erosion in high-dust configurations. Because of the high nitrogen content of sewage sludge, the NO_x concentrations after the furnace may rise to a level such that they have to be reduced by NO_x control.

In co-combustion processes with sewage sludge and meat-and-bone meal as secondary fuels in hard coal-fired power plants, an increased deactivation of the SCR-$DeNO_x$ catalysts for flue gas denitrification was observed. Investigations revealed a correlation between the phosphorus content in the fuel and the degree of catalyst deactivation. In combustion, the phosphorus is released from the fuel, leading to increased concentrations of both particulate and gaseous phosphorus compounds. Gaseous phosphates in particular penetrate the catalyst surface, effecting severe deactivation. Particulate phosphorus, too, contributes to the deactivation by obstructing the catalyst pores and reacting with sulphuric acid. A calcium addition can abate the deactivation by phosphorus (Beck 2007).

6.5.3.8 Flue Gas Desulphurisation (FGD) Equipment

The low sulphur content of biomass reduces the load on the flue gas desulphurisation plant. However, the increase in other flue gas components as a result of the use of the biomass may impair the function of FGD or necessitate additional FGD capacity. These potential consequences set a limit on the biomass fraction, particularly in regard to the chlorine input into the FGD.

Besides sulphur, the FGD unit also removes a number of other flue gas components. The volatile ash components leaving with the flue gas, such as mercury, arsenic, lead and other heavy metals, are partly removed together with the FGD residual matter, the quality of which has to be checked if it is to be commercially used. However, the concentrations of these substances in the biomasses that have been investigated can be neglected in comparison to coal. This is not the case when municipal sewage sludge is co-fired – in this case, the quality of the residues can sometimes be negatively affected.

Sulphur in sewage sludge exceeds the usual sulphur content of coal considerably. The FGD unit must have sufficient capacity to deal with this additional load. With a 25% thermal input fraction of sludge, the SO_2 to be dealt with rises to 1.6 times the quantity from coal alone.

6.5.4 Co-combustion in Fluidised Bed Furnaces

Fluidised bed furnaces are suited to a wide range of fuels, including biomasses such as wood or straw and wastes. Biomass can be co-fired both in bubbling and in circulating fluidised bed furnaces. Co-combustion in a fluidised bed is uncomplicated and in most cases limited only by the heat balance of the bed. When co-firing with herbaceous biomass or waste fuels, steam conditions can be limited by the need to avoid deposition and corrosion (Leckner 2007).

Given its higher volatile content, biomass tends to have post-combustion reactions in the freeboard volume of the furnace, in particular in bubbling fluidised bed furnaces. In these furnaces, lightweight particles such as straw can easily be carried away from the fluidised bed, which raises the temperature in the freeboard if they post-combust. The well-mixedness of a circulating fluidised bed creates an even distribution of furnace temperatures. Nevertheless, an upward temperature shift can also be observed.

As fluidised bed firing is especially suited to high-ash and high-moisture fuel types, it seems a good technology for mechanically dewatered sewage sludge. While the higher moisture content of the sludge increases the volumetric flue gas flow when co-firing with hard coal, there is little effect when co-firing with brown coal.

The impact of biomass co-combustion on gaseous emissions was investigated at experimental and industrial plants of various thermal capacities within the framework of a research project funded by the European Union (Bemtgen et al. 1995). Except for HCl emissions in straw co-combustion, the results showed that biomass addition has a positive effect. In all plants, the observed result was a reduction in SO_2 emissions with an increasing biomass fraction of the thermal input. This effect can be put down on the one hand to the low sulphur contents of the biomasses and on the other hand to the fact that SO_2 is captured in the biomass ash.

The correlations between NO_x emissions and co-combustion were diverse. With a low biomass fraction, the emissions of NO_x changed very little in some of the plants. In other plants, the emissions were reduced, in particular by co-firing wood (the reduction increasing with the wood fractions).

While in co-combustion of woody bio-fuels, additional operational problems are not expected, herbaceous fuels may cause severe corrosion, slagging and fouling, with the potassium chloride contained in such bio-fuels playing a major part (Binderup Hansen et al. 1997).

6.5.4.1 Co-combustion of Coal and Straw in an 88 MW_{th} CFBC

Coal and straw have been co-fired for over 10 years at the CHP plant in Grenaa, Denmark. The 88 MW_{th} circulating fluidised bed furnace is designed to fire up to 60% straw and up to 100% coal on an energy basis. During initial operation, co-firing of straw and a coal type with a high sulphur content of 3% (each with a 50% thermal input fraction) resulted in severe slagging in the furnace, in the cyclone and in the superheater area, so that the operational parameters could not be maintained even shortly after start-up. In consequence, coals with a sulphur content below 1% were later used exclusively, and deposit formation occurred only to a minor extent. The only remaining problem was in the superheater area, where the narrowly designed spacing (tube pitch) of 37 mm was favourable for the build-up of deposits. The measures used to prevent deposits in this area are to employ hanging superheaters with a spacing of 50 mm (tube pitch) (thus avoiding bridging) and to lower the flue gas temperature at the superheater. At flue gas temperatures below the melting point of potassium chloride (770°C), solid deposits do not pose any problem. In comparison to combustion of coal only, the fouling rate from using a straw fraction of 50% quintupled, though the deposits could be removed easily (Binderup Hansen et al. 1997; Clausen and Sorensen 1997). Special attention is paid to the bed inventory in order to limit the enrichment of potassium in the bed and to prevent bed agglomeration. Instead of sand with a high silica content, ash from a stoker-fired furnace with a high alumina concentration is used (Wieck-Hansen and Sander 2003).

Severe corrosion was observed at the final convective superheater, which had to be replaced after only 1 year of operation. Corrosion studies were carried out in the fluidised bed furnace by testing several different materials. Using a different

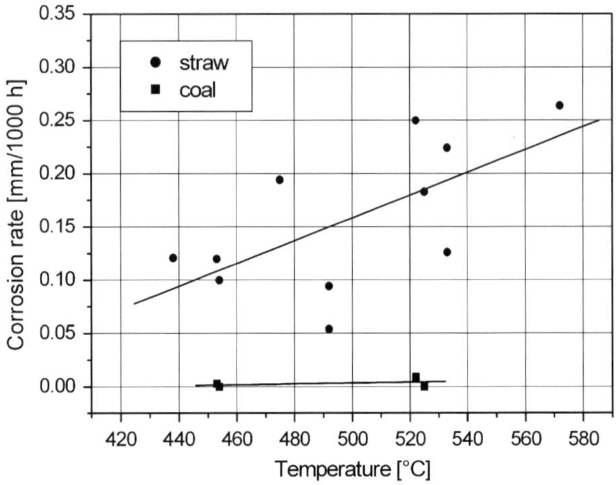

Fig. 6.61 Corrosion rate during co-combustion as a function of the steam temperature when using a 50% straw fraction in a circulating fluidised bed furnace (Binderup Hansen et al. 1997)

high-alloy steel did not result in any substantial improvement. The results, although uncertain considering the short test periods of 500–1,000 h, revealed that considerable corrosion problems occurred on convective superheater surfaces with straw co-combustion in the circulating fluidised bed furnace. For martensitic steel, type X 20 CrMoV 12 1, the corrosion rate was about one order of magnitude higher in straw co-combustion than in coal mono-combustion, and also considerably higher than in pulverised fuel firing with the same straw fraction (see Fig. 6.61). The cause of these high corrosion rates is assumed to lie in the in situ desulphurisation in the fluidised bed, which favours the formation of potassium chloride. The potassium chloride condenses on the superheater tubes, where it forms potassium sulphate, releasing the corrosive chlorine in the process. In contrast, pulverised fuel firing has less potassium sulphate and HCl, and hence lower corrosion rates (Henriksen et al. 1995). The principles of corrosion are discussed in detail in Sect. 5.10.4.

Various measures were implemented to reduce the rate of corrosion of the convective superheater. The main measure was the reduction of the bed temperature by about 60°C, so that it is now 860°C or lower. The changes were successful to the extent that the superheater was still in service after 7 years of operation (Wieck-Hansen and Sander 2003).

New techniques have made the superheater an in-bed heat transfer surface in the fluidised bed rather than placing it in the flue gas path. However, perhaps because there may still have been unburned straw particles in the cyclone return pipe, which form KCl as they burn, corrosion probes inserted in the return pipe measured comparably high wear rates. In order to avoid corrosion, it became necessary to arrange an uncooled section upstream of the fluidised bed superheater to ensure complete combustion. This design turned out to be successful, though erosion occurred after several years (Wieck-Hansen and Sander 2003).

Uses for the mixed ashes of this coal/straw CFBC have not yet been found, and so they have been disposed of in landfills to date (Clausen and Sorensen 1997). There was a significant reduction of N_2O emissions from the 88 MW_{th} circulating fluidised bed furnace in comparison to coal firing alone, a fact explained by the higher temperatures in the upper part of the furnace and in the cyclones. The chlorine input into the process using straw at a thermal input fraction of 60% was 20 times higher than in the combustion of coal only. The inputted chlorine was found almost entirely in the flue gas.

6.5.4.2 Co-combustion of Sewage Sludge in a CFBC

Tests in a 230 MW_{th} brown coal-fired circulating fluidised bed for sewage sludge co-combustion have shown that emissions of SO_2, NO_x, CO and dust are within the normal operational range. The mechanically dewatered sewage sludge has a moisture content of 70%. This experience has demonstrated that the performance of circulating fluidised bed furnaces does not deteriorate through sewage sludge co-combustion. Depending on the calorific value of the sludge input, there is only a drop in the thermal output and thus in the steam production. The percentage regulation limits imposed by the *17*th *BImSchV* (17th Amendment to the Federal German Pollution Control Act) were not reached. For continuous operation, an additional flue gas cleaning stage, consisting of an entrained-flow absorber using lignite-derived coke, was installed downstream of the ESP in order to ensure compliance with the mercury limits (Bierbaum et al. 1996). The plant has been in continuous operation since 1995 and co-fires approximately 200,000 t of sewage sludge a year (Roper and Kipshagen 2003).

Given that sewage sludge has an ash content of about 15% (raw), the ash load increases considerably. The pollutants in the sludge are captured inertly in the ash, except for mercury, which is transported in the flue gas. The ash is, as before, utilised for regeneration of opencast brown coal mines. It meets the requirements for landfill grade 1 of the German Technical Instructions on Municipal Solid Waste (*TA Siedlungabfall*).

References

17.BImSchV (2009). 17. Verordnung zur Durchführung des Bundesimmissionsschutzgesetzes (Verordnung über die Verbrennung und die Mitverbrennung von Abfällen. Bekanntmachung vom 14. August 2003 (BGBl. I S. 1633), geändert durch Artikel 2 der Verordnung vom 27. January 2009 (BGBl. I P. 129). §5 Absatz 5 mit Jahresmittelwert 100 mg/Nm^3.

AbfAblV (2000). Verordnung über die umweltgerechte Ablagerung von Siedlungsabfällen und über biologische Abfallbehandlungsanlagen [in Deutschland]. Bundesgesetzblatt Part 1 No. 10, pp. 305–324 (Abfallablagerungsverordnung AbfAblV nachfolgend der TASi-TA Siedlungsabfall).

ActuEnvironment (2005). L'incinérateur de Gien plombe le bilan annuel des incinérateurs de tout le territoire. Actu Environment 1.6.2005, from http://www.actu-environnement.com/ae/news/1153.php4.

ADMONT. (2001). Biomass fired CHP Plant Based on a ORC cycle- Project. ORCSITA-Admont. Final publishable report, 1.6.1998–31.12.2000, from http://www.bios-bioenergy.at/uploads/media/Report-ORCAdmont-Thermie-2001-03-26.pdf.

Almark, M. and Hiltunen, M. (2005). Alternative bed materials for high alkali fuels. Proceedings of the 18th International Fluidized Bed Combustion Conference, May 22–25, 2005, Toronto, Canada.

Andersen, K. H., Frandsen, F. J., Hansen, P. F. B., Wieck-Hansen, K., Rasmussen, I., Overgaard, P. and Dam-Johansen, K. (2000). Deposit formation in a 150 MWe utility PF-boiler during co-combustion of coal and straw. *Energy Fuels* **14**(4): 765–780.

Bapat, D. W., Kulkarni, S. V. and Bhandarkar, V. P. (1997). Design and operating experience on fluidized bed boiler during burning biomass fuels with high Alkali Ash. Proceedings of the 14th International Conference on Fluidized Bed Combustion, Vancouver, Canada.

Baumbach, G. (1990). Luftreinhaltung. Berlin/Heidelberg, Springer-Verlag.

Baxter, L. (2005). Biomass-coal co-combustion: opportunity for affordable renewable energy. *Fuel* **84**(10): 1295–1302.

Beck, J. (2007). The behaviour of phosphorus compounds and their effect on DeNOx-catalysts in coal fired power plants, Universität Stuttgart. Dissertation.

Beckert, F. and Schaarrschuch, H. (2007). Das erste Biomassekraftwerk der Stadtwerke Leipzig mit zirkulierender Wirbelschichtfeuerung und Zwischenüberhitzung. *VGB Powertech* **87**(4): 50–54.

Bemtgen, J. M., Hein, K. R. G. and Minchener, A. (1995). Combined combustion of biomass/sewage sludge and coals volume II: Final Reports, APAS Clean Coal Technology Programme, ISBN -3-928123-16-5.

Berg, M. E. and Jensen, J. P. (2008). Biomass and secondary fuels in Denmark, history and perspective. *VGB PowerTech* **88**(1/2): 62–64.

Bierbaum, K., Lambertz, J. and Thomas, G. (1996). Klärschlamm-Mitverbrennung in einem Industriekraftwerk. Bamberg, VDI-Seminar Klärschlammentsorgung II.

Bilitewski, B., Härdtle, G. W. and Marek, K. A. (2000). Abfallwirtschaft : Handbuch für Praxis und Lehre. Berlin [u.a.], Springer.

Binderup Hansen, P. F., Lin, W. and Dam-Johansen, K. (1997). Chemical reaction conditions in a Danish 80 MWth CFB-boiler co-firing straw and coal. Proceedings of the 14th International Conference on Fluidized Bed Combustion, Vancouver, Canada, ASME.

Biollaz, S. and Nussbaumer, T. (1996). Einsatz von Rostfeuerungen für Holz- und Halmgüter. 4. Holzenergiesymposium "Feuerungstechnik, Ascheverwertung und Wärme-Kraft-Kopplung" Zürich.

BMU (1996). Bericht gemäß Artikel 17 der EG-Richtlinie 86/278/EWG über die Klärschlammverwertung in der Bundesrepublik Deutschland, Berichtszeitraum 1991–1994. N. u. R. Bundesministerium für Umwelt, Referat WAII 5(L).

Bolhar-Nordenkampf, M., Gartnar, F., Tschanun, I. and Kaiser, S. (2006). Combustion of biomass in bubbling fluidised beds – operation experience. Proceedings of the 19th International Fluidized Bed Combustion Conference, May 21–24. Vienna, Austria.

Bonomo, A. (1998). Combining energy efficiency, renewable energy, environmental protection, cost efficiency. First operational results from the Brescia WtE plant. ISWA 4th International Management Conference. Amsterdam, 24.11.1998.

Borghart, R. (2008). Einsatz von EBS in industriellen Kraftwerken – Erfahrungen und Weiterentwicklungen. 13. Fachtagung thermische Abfallbehandlung [11. Bis 12. März 2008 in München]. Kassel University Press, 2008. ISBN 3899583841.

Born, M. (2005). *Dampferzeugerkorrosion : [Teil 1: Stand der Kenntnisse, Teil 2: Vorträge zum 2. Diskussionsforum "Rauchgasseitige Dampferzeugerkorrosion", Freiberg, 27. und 28. Oktober 2005].* Freiberg, SAXONIA Standortentwicklungs- u. -verwaltungsges.

Born, M. (2007). Dampferzeugerkorrosion 2007. Freiberg, Saxonia Standortentwicklungs- und Verwaltungsges.

Braage, C., Yu, Q., Chen, G. and Sjöström, K. (1997). Use of amino phase adsorbent for biomass tar sampling and separation. *Fuel* **76**: 137–142.

Clausen, C. and Sorensen, M. (1997). Anlagen- und Betriebserfahrungen mit strohgefeuerten Kesseln in Heizkraftwerken. *VGB Kraftwerkstechnik* **77**(10): 802–806.

Coda, B., Zielke, U., Suomalainen, M., Knoef, H., Good, J., Liliedahl, T., Unger, C., Ventress, L., Neeft, J., v.d. Hoek, H. and Kiel, J. (2004). Tar measurement standard: a joint effort for the standardisation of a method for measurement of tars and particulates in biomass producer gases. Proceedings of the 2nd World Conference on Biomass for Energy, Industry and Climate Protection, Rome, Italy.

de Jong, W. (2005). Nitrogen compounds in pressurised fluidised bed gasification of biomass and fossil fuels, TU Delft. Ph.D.

Di Nola, G., de Jong, W. and Spliethoff, H. (2009). The fate of main gaseous and nitrogen species during fast heating rate devolatilization of coal and secondary fuels using a heated wire mesh reactor. *Fuel Processing Technology* **90**(3): 388–395.

DIN (2006). Biomass gasification-Tar and particles in product gases – Sampling and analysis. German Version DIN CEN/TS 15439, DIN Deutsches Institut für Normung e.V. Berlin.

DiNola, G. (2007). Biomass fuel characterization for NO emissions in Co-firing applications, TU Delft. Ph.D.

Doležal, R. (1990). Dampferzeugung: Verbrennung, Feuerung, Dampferzeuger. Berlin, Heidelberg, New York, Springer.

Effenberger, H. (2000). Dampferzeugung. Berlin, Heidelberg, Springer.

EIPPCB (2005). Integrated pollution prevention and control. Reference document on the best available techniques for waste incineration (BREF) E. I. P. P. a. C. Bureau.

Eltrop, L., Raab, K., Hartmann, H., Schneider, S., Schröder, G., Kaltschmitt, M., Jahraus, B. and Heinrich, P. (2007). *Leitfaden Bioenergie: Planung, Betrieb und Wirtschaftlichkeit von Bioenergieanlagen*. Gülzow b. Güstrow, Fachagentur Nachwachsende Rohstoffe.

EU (2000). EU commission: Directive 2000/76/EC of the European Parliament and of the Council of 4 December 2000 on the incineration of waste.

EU (2008). EU Commission: Directive 2008/98/EC of the European Parliament and of the Council of 19 November 2008 on waste and repealing certain directives (Waste Framework Directive).

Fahlke, J. (1994). *Spurenelementbilanzierungen bei Steinkohlefeuerungen am Beispiel einer Trocken- und einer Schmelzkammerfeuerung unter Berücksichtigung der Rauchgasreinigungsanlagen*. Düsseldorf, VDI, VDI Fortschrittsberichte Reihe 15, Umwelttechnik Nr. 120.

Fernando, R. (2002). Experience of indirect cofiring of biomass and coal. London, IEA Clean Coal Centre, CCC/64.

Fernando, R. (2005). Fuels for biomass cofiring. London, IEA Clean Coal Centre, CCC/102.

Fernando, R. (2007). Cofiring of coal with waste fuels. London, IEA Clean Coal Centre, CCC/126.

Fischer, B. (2005). Energieeffizienz des Müllheizkraftwerkes Mainz. In Thomé-Kozmiensky, K. J. (ed.), Optimierung der Abfallverbrennung 2. pp. 565–578 Berlin, TK.

FNR (2006). Analyse und Evaluierung der thermo-chemischen Vergasung von Biomasse. Münster, Landwirtschaftsverlag.

Foster, N. A., Dräger, R., Daublebsky von Eichhain, C. and Warnecke, R. (2007). Wärmetechnische Auslegung von Kesseln für Verbrennung von Reststoffen – Grundlagen und Korrosionsdiagramm. Seminar des VDI-Wissensforums "Beläge und Korrosion, Verfahrenstechnik und Konstruktion in Großfeuerungsanlagen", Frankfurt/Main, VDI.

Fujimura, H., Oshita, T. and Naruse, K. (2001). Fluidised bed gasification and slagging combustion systems. IT3 Conference May 14–18th 2001, Philadelphia, PA.

Gerhardt, T. (1998). Thermische Behandlung von kommunalen Klärschlämmen in Kohlenstaubfeuerungen. Essen, VGB-Kraftwerkstechnik, Verl. Techn.-Wiss. Schriften.

Gerhardt, T., Spliethoff, H. and Hein, K. R. G. (1996). Thermische Nutzung von Klärschlämmen in Kraftwerksfeuerungen. Untersuchungen an einer Staubfeuerung im Pilotmaßstab. *Entsorgungspraxis* **3**:50–58.

Gockner, L. and Rechberger, N. (2008). Biomassekraftwerke. *VGB PowerTech* **88**(1/2): 5661.

Gohlke, O. and Murer, M. J. (2009). Anwendung von aktuellen Kennzahlen der Energieeffizienz auf moderne europäische Beispielanlagen. 14. Fachtagung Thermische Abfallbehandlung. München, Kassel University Press.

Gohlke, O. and Spliethoff, H. (2007). Innovative approaches to increase efficiency in EfW plants – potential and limitations. Proceedings of the ISWA conference. Amsterdam.

Good, J., Ventress, L., Knoef, L., Zielke, U., Lyck Hansen, P., van de Kamp, W., de Wild, B., Coda, B., van Paasen, S., Kiel, J., Sjöstrom, K., Unger, C., Neeft, J., Suomalainen, M. and Simell, P. (2005). Sampling and analysis of tar and particles in biomass producer gases, Technical Report, http://www.tarweb.net/

Hasler, P. and Nussbaumer, T. (1996). Brennstoffeigenschaften und Anforderungen an Feuerungstechnik und Abgasreinigung für die Nutzung von Industrie- und Altholz. Tagung "Biomasse als Festbrennstoff – Anforderungen, Einflussmöglichkeiten, Normung", Schriftenreihe Nachwachsende Rohstoffe, Band 6, Landwirtschaftsverlag, Münster.

Heinzel, T. (2004). Ascheseitige Probleme bei der Mitverbrennung von Biomasse in Kohlenstaubfeuerungen. Dissertation Uni Stuttgart. Aachen, Shaker.

Henriksen, N., Larsen, O. H., Blum, R. and Inselmann, S. (1995). High-temperature corrosion when co-firing coal and straw in pulverized coal boilers and circulating fluidized bed boilers. Essen, VGB-Konferenz "Korrosion und Korrosionsschutz in der Kraftwerkstechnik".

Hermann, T. (2004). *Daten zur Anlagentechnik und zu den Standorten der thermischen Klärschlammentsorgung in der Bundesrepublik Deutschland.* Berlin, Umweltbundesamt.

Higman, C. and van der Burgt, M. (2008). Gasification. Amsterdam [u.a.], Elsevier/GPP, Gulf Professional Publishing.

Hofbauer, H. (2007). Conversion technologies: gasification overview. Proceedings of the 17th European Biomass Conference & Exhibition. Berlin, Germany.

Johnsen, J. B. and Svendsen, J. J. (1997). Straw-fired Benson boiler of 120 t/h steam production. Essen, VGB, VGB-Konferenz Feuerungen 1997.

Kaltschmitt, M. (2001). Energie aus Biomasse: Grundlagen, Techniken und Verfahren. Berlin [u.a.], Springer.

Kaltschmitt, M. and Bridgewater, A. V. (1997). Biomass Gasification and Pyrolysis: State of the Art and Future Prospects. Newbury, CPL Press.

Kaltschmitt, M., Hartmann, H. and Hofbauer, H. (2009). Energie aus Biomasse: Grundlagen, Techniken und Verfahren. Dordrecht, Heidelberg, London, New York, Springer.

Khan, A. A. (2007). Combustion and co-combustion of biomass in a bubbling fluidized bed boiler, TU Delft. Ph.D.

Khan, A. A., de Jong, W., Jansens, P. J. and Spliethoff, H. (2009). Biomass combustion in fluidized bed boilers: potential problems and remedies. *Fuel Processing Technology* **90**(1): 21–50.

Kicherer, A. (1996). Biomasseverbrennung in Staubfeuerungen – Technische Möglichkeiten und Schadstoffemissionen. Düsseldorf, VDI.

Knoef, H. and Ahrenfeldt, J. (2005). Handbook biomass gasification. Enschede, BTG Biomass Technology Group.

Konda, C. (2000). Measures against dioxins in waste management. In: German/Japanese Consultations on Environmental Policies 2000, Part I, Workshop on dioxin policies, 18–19 September, Berlin.

KrWG (1994). Gesetz zur Förderung der Kreislaufwirtschaft und Sicherung der umweltverträglichen Beseitigung von Abfällen (Kreislaufwirtschafts- und Abfallgesetz – KrW-/AbfG (German Waste Law).

Leckner, B. (2007). Co-combustion – a summary of technology. *Thermal Science* **11**(4): 5–40.

Lin, W., Krulsholm, G., Dam-Johansen, K., Musahi, E. and Bank, L. (1997). Agglomeration phenomena in fluidized bed combustion of straw. Proceedings of the 14th International Conference on Fluidized Bed Combustion. Vancouver, Canada.

Martin, J., Gohlke, O., Tabaries, F., Praud, A., Yanagisawa, Y. and Takuma, M. (2005). Defining inert. *Waste Management World Sept./Okt*(70–73).

McCann, D. and Simons, H. A. (1997). Design review of biomass bubbling fluidized bed boilers. Proceedings of the 14th International Conference on Fluidized Bed Combustion. Vancouver, Canada, ASME.

Mielke, T. (2007). Sustec hält an Spreewald fest. Lausitzer Rundschau vom 7.11.2007, from http://www.lr-online.de/regionen/Weisswasser;art13826,1832164.

Mitsakis, P., Karellas, S. and Spliethoff, H. (2008). Application of laser spectroscopy for the quantitative analysis of biomass gasification tars. Proceedings of the 16th European Biomass Conference & Exhibition. p. 652–658. Valencia, Spain.

Mörsch, O. (2000). Entwicklung einer online Methode zur Bestimmung des Teergehalts im Gas aus der Vergasung von Biomasse. Düsseldorf, VDI.

Neeft, J. P. A., van Paasen, S. V. B., Knoef, H. A. M., Buffinga, G. J., Zielke, U., Sjöström, K., Brage, C., Hasler, P., Simell, P. A., Suomalainen, M., Dorrington, M. A. and Thomas, L. (2002). Tar guideline – A standard method for measurement of tars and particles in biomass producer gases. Proceedings of the 12th European Conference and Technology Exhibition on Biomass for Energy, Industry and Climate Protection. pp. 469–572 Amsterdam, ETA Florence and WIP-Munich.

Nethe, L.-P. (2008). *Abfallverbrennung – Emissionsminderung durch Sekundärmaßnahmen.* 4. Fachtagung "Verfahren und Werkstoffe für die Energietechnik": Biomasse & Abfall – Regionale Brennstoffe richtig nutzen, Sulzbach-Rosenberg, Dorner.

Neukirchen, B. (2008). Erfahrungen in Monoanlagen. 13. Fachtagung Thermische Abfallbehandlung, München, Kassel University Press.

Nikolaisen, L. (1992). Stroh als Energieträger. Technik-Umwelt-Ökonomie. Arhus, Center für Biomasse-Technologie.

Noppenau, H. (2003). Concept and first operating experience with Avedore 2. *VGB PowerTech* **83**(5): 88–91.

Nussbaumer, T. (2003). Combustion and co-combustion of biomass: fundamentals, technologies, and primary measures for emission reduction. *Energy & Fuels* **17**(6): 1510–1521.

Nussbaumer, T. (2004). Dioxin- und PAK-Emissionen der privaten Abfallverbrennung, Literaturstudie und Situationsanalyse. http://www.bafu.admin.ch/publikationen/publikation/00265/index.html?lang=de

Obernberger, I. (1996). Erforderliche Brennstoffeigenschaften holz- und halmgutartiger Biomasse für den Einsatz in Großanlagen (1 bis 20 MW thermische Leistung), Tagung "Biomasse als Festbrennstoff – Anforderungen, Einflussmöglichkeiten, Normung", Schriftenreihe Nachwachsende Rohstoffe, Band 6, Landwirtschaftsverlag, Münster.

Obernberger, I. (1997). Nutzung fester Biomasse in Verbrennungsanlagen unter besonderer Berücksichtigung des Verhaltens aschebildender Elemente. Graz, dbv.

Obernberger, I. (2003). Biomasse-KWK auf Basis des ORC-Prozesses, from http://www.bios-bioenergy.at/uploads/media/Presentation-Obernberger-ORC-LienzAdmont-2003-11-26.pdf

Obernberger, I. and Biedermann, F. (1996). Anlagenkonzept zur fraktionierten Ascheabscheidung bei Holzfeuerungen. 4. Holzenergiesymposium Feuerungstechnik, Ascheverwertung und Wärme-Kraft-Kopplung, Zürich.

Pfeiffer, A. E. (2003). Governmental support to waste-to-energy. ISWA beacon conference, Malmö.

Quicker, P., Mocker, M. and Faulstich, M. (2005). Energie aus Klärschlamm. 1. Fachtagung "Verfahren und Werkstoffe für die Energietechnik": Energie aus Biomasse und Abfall, Sulzbach-Rosenberg, Förster.

Roper, B. and Kipshagen, F. J. (2003). Operational experience with sewage sludge and waste wood co-combustion in the Berrenrath power plant. *VGB Power Tech* **83**(6): 60–64.

Rüdiger, H. (1997). Pyrolyse von festen biogenen und fossilen Brennstoffen zur Erzeugung eines Zusatzbrennstoffes für Feuerungsanlagen, Universität Stuttgart, Shaker, Aachen. Dissertation.

Schmelz, K.-G. (2006). Klärschlammmengen und Entsorgungskosten im Vergleich zum europäischen Ausland. Bonn, Perspektiven der Klärschlammverwertung.

Scholz, R., Beckmann, M. and Schulenburg, F. (2001). Abfallbehandlung in thermischen Verfahren: Verbrennung, Vergasung, Pyrolyse, Verfahrens- und Anlagenkonzepte. Stuttgart [u.a.], Teubner.

Schwarzmann, O. (1999). Schwelbrennanlage Fürth (SBA Fürth): Plant concept, commissioning, cancellation of the contracts. VDI – Seminar Waste Disposal at a Turning Point, Reuse/Co-Incineration/Secondary fuel. Munich: Paper No. 11.

Seemann, B., Steer, T. and Hein, D. (2008). Consistent further development of the BFB-technology – application of the BioCOM for problematic residues of the wood and fibre-board industry. Proceedings of the 9th International Conference on circulating Fluidized Beds, Hamburg.

Seguin, P. (2004). Zabalgarbi Bilbao plant: integration of a waste to energy unit into a combined cycle with a high energy efficiency. Karlsruhe, GVC-DECHEMA Jahrestagung.

Siegle, V. (2000). Biogene Brennstoffe in Aufbereitung und Verbrennung. Dissertation Universität Stuttgart. Aachen, Shaker.

Spliethoff, H. (2000). Verbrennung fester Brennstoffe zur Strom- und Wärmeerzeugung: Verfahren und Stand der Technik – Wirkungsgrad, Betrieb, Emissionen und Reststoffe. VDI Fortschritt-Berichte, Reihe 6 Energietechnik, Nr. 443. Düsseldorf, VDI.

Spliethoff, H. (2001). Status of biomass gasification for power production. *IFRF – Combustion Journal*: 1–25.

Spliethoff, H. and Hein, K. R. G. (1998). Effect of co-combustion of biomass on emissions in pulverized fuel furnaces. *Fuel Processing Technology* **54**(1–3): 189–205.

Spliethoff, H. and Hein, K. R. G. (1995a). Nutzung und Chancen von Ligno-Cellulose-Pflanzen als Energiequelle. *Landinfo* **3**: 21–27.

Spliethoff, H. and Hein, K. R. G. (1995b). Eignung von Kohlenstaubfeuerungen zur Mitverbrennung von Biomasse und Klärschlamm. Düsseldorf, VDI-Berichte 1193, 1995, pp. 125–133.

Spliethoff, H. and Hein, K. R. G. (1996). Effect of co-combustion of biomass on emissions in pulverized fuel furnaces. Snowbird, UT Biomass Usage for Utility and Industrial Power.

Spliethoff, H., Gohlke, O. and Seitz, A. (2008). Effiziente Nutzung von Energie aus Abfall_40. Kraftwerkstechnisches Kolloquium, Dresden.

Spliethoff, H., Moersch, O. and Hein, K. R. G. (1998). Gasqualitäten von Biomassevergasern – Stand und Perspektiven. Gülzower Fachgespräch "Energetische Nutzung von Biomasse mit Brennstoffzellenverfahren", 9–10. Dezember 1998, Gülzow.

Spliethoff, H., Unterberger, S. and Hein, K. R. G. (2001). Status of co-combustion of coal and biomass in Europe. Proceedings of the 6th International Conference on Technologies and Combustion for a Clean Environment, Porto.

Statistisches_Bundesamt (2008). Fachserie 19 Reihe 1: Abfallentsorgung 2006. Wiesbaden.

Sydkraft (2001). Värnamo Demonstration Plant. Trelleborg.

TASi (1993). TA Siedlungsabfall: Technische Anleitung zur Verwertung, Behandlung und sonstigen Entsorgung von Siedlungsabfällen (Dritte Allgemeine Verwaltungsvorschrift zum Abfallgesetz) vom 14. Mai 1993 (BAnz. Nr. 99a vom 29.05.1993).

Tauber, C., Klemm, J. and Schönrok, M. (1996). Mitverbrennung kommunaler Klärschlämme in Steinkohlekraftwerken – Erfahrungen der Preussen Elektra aus einem Langzeitversuch. Bamberg, VDI-Seminar Klärschlammentsorgung II.

Themelis, N. (2007). Thermal treatment review. *Waste Management World* **8**(4): 37–44.

Tortosa Masiá, A. and Spliethoff, H. (2006). *Experimental study on deposition behaviour during co-firing of coal with biomass fuels*. Snowbird, UT Impacts of fuel quality on power production.

Treiber, H. and Schroth, H. (1992). Die Klärschlamm-Wirbelschichtverbrennungsanlage II im Hauptklärwerk Stuttgart-Mühlhausen. *BWK/TÜ/UT-Spezial* **10**: M38–M46.

Turboden (2009). HRS Einheiten mit hohem elektrischem Wirkungsgrad. Retrieved 2009, from http://www.turboden.eu/de/public/downloads/Turboden_HRS_brochure_deutsch.pdf.

UBA (2005a). Integrierte Vermeidung und Verminderung der Umweltverschmutzung (IVU) BVT-Merkblatt über beste verfügbare Techniken der Abfallverbrennung Juli 2005, from http://www.bvt.umweltbundesamt.de.

UBA (2005b). Waste incineration a potential danger: bidding farewell to dioxin spouting, Report by Germany's Ministry of the Environment, from http://www.seas.columbia.edu/earth/wtert/sofos/Waste_Incineration_A_Potential_Danger.pdf

US_Supreme_Court (1994). Petition declaring flow control unconstitutional: C & A Carbone, Inc. ET AL., Petitioners v. Town of Clarkstown, New York, Certiorari to the appellate division, Supreme Court of New York, Second Judicial Department No. 92-1402, Argued December 7, 1993, Decided May 16, 1994.

Van Berlo, M. (2006). Waste is innovation: Amsterdam's high efficiency waste fired power plant. ISWA World Congress 2006, Copenhagen.

Van Berlo, M. and de Waart, H. (2008). Unleashing the power in waste. Proceedings of the 16th Annual North American Waste to Energy Conference. Philadelphia.

Van der Linde, K. D. (2003). 4th-generation waste incineration. ECOTECH Conference. Amsterdam.

Van Loo, S. and Koppejan, J. (2008). The handbook of biomass combustion and co-firing. London, Earthscan.

van Paasen, S., Boerrigter, H., Kuipers, J., Struijk, F. and Stokes, A. (2005). On-line tar dewpoint measurements. Proceedings of the 14th European Biomass Conference, Paris, France.

VDI (2006). Seminar des VDI-Wissensforums "Beläge und Korrosion in Großfeuerungsanlagen", Würzburg.

VDI (2007). Seminar des VDI-Wissensforums "Beläge und Korrosion, Verfahrenstechnik und Konstruktion in Großfeuerungsanlagen", Frankfurt/Main, VDI.

Vehlow, J. (2006). Internationale Entwicklungen der Thermischen Abfallbehandlung. 11. Fachtagung Thermische Abfallbehandlung, München, 14. bis 15. März, Kassel University Press.

VGB (2008). Advantages and Limitations Biomass Co-combustion in Fossil Fired Power Plants. From http://www.vgb.org/en/news_biomass_co_combustion_march2008.html

Visser, H. J. M., van Lith, S. C. and Kiel, J. H. A. (2003). Biomass ash – bed material interactions leading to agglomeration in FBC. Proceedings of the 17th International Fluidized Bed Combustion Conference, Jacksonville, FL.

Vogel, A. (2007). Dezentrale Strom- und Wärmeerzeugung aus biogenen Festbrennstoffen: eine technische und ökonomische Bewertung der Vergasung im Vergleich zur Verbrennung. Leipzig, Inst. für Energetik und Umwelt.

Wandschneider, J. (2005). Development of technology and basic engineering for two high-efficiency incineration lines at the HR-AVI Amsterdam, from http://www.wg-ing.de/en/dok/hr-avi.pdf (July 15, 2006).

Warnecke, R. (2006). *Stand der Forschungs- und Praxisergebbnisse für die HT-Chlor-Korrosion*. Seminar des VDI-Wissensforums "Beläge und Korrosion in Großfeuerungsanlagen", Würzburg.

Warnecke, R. (2007). Beläge und Korrosion, Verfahrenstechnik und Konstruktion in Chlorbelasteten thermischen Anlagen. Seminar des VDI-Wissensforums "Beläge und Korrosion, Verfahrenstechnik und Konstruktion in Großfeuerungsanlagen", Frankfurt/Main, VDI.

Whiting, K., Schwager, J. (2006). Why are novel technologies such as gasification for MSW processing struggling to make an impact in Europe? Proceedings of the 4th i-CIPEC Conference, September 26–29, 2006, Kyoto, Juniper Consultancy Services Ltd., Sheppards Mill, South Street, Uley (GB).

Wieck-Hansen, K., Overgaard, P. and Larsen, O. H. (2000). Cofiring coal and straw in a 150 MWe power boiler experiences. *Biomass and Bioenergy* **19**(6): 395–409.

Wieck-Hansen, K. and Sander, B. (2003). 10 years experience with co-firing straw and coal as main fuels with different types of biomasses in a CFB boiler in Grena, Denmark. *VGB PowerTech* **83**(10): 64–67.

Wiens, U. (2005). Neues aus den Regelwerken zur Verwendung von Flugasche in Beton. *VGB PowerTech* **85**(10): 73–79.

Zheng, Y., Jensen, P. A., Jensen, A. D., Sander, B. and Junker, H. (2007). Ash transformation during co-firing coal and straw. *Fuel* **86**(7–8): 1008–1020.

Chapter 7
Coal-Fuelled Combined Cycle Power Plants

Combined cycle power plant, when used as a generic term, refers to a plant which converts heat into mechanical energy in a combined gas and steam turbine process. Combined cycle processes with coal gasification or coal combustion turn solid fuels into a fuel gas or a hot pressurised gas which is then used in the gas and steam turbine processes. Coal-fuelled combined cycle plants will be discussed in detail in the following sections. A start will be made by describing the basic technical features and the characteristic data of combined cycle power plants fuelled by natural gas for the purposes of comparison to coal.

7.1 Natural Gas Fuelled Combined Cycle Processes

The combined cycle process offers a number of advantages over the simpler steam – water only process. These are

- highly efficient generation of electrical power,
- a straightforward process,
- low investment costs and
- a smaller environmental impact.

The only requirement is a fuel gas which is suitable for gas turbines, for example natural gas.

The high efficiency results from combining the high-temperature gas turbine process with the low-temperature steam process. The fuel is fed to the process only via the gas turbine combustion chamber. While it is being combusted with compressed air, hot flue gas is produced under pressure in the combustion chamber. The gas turbine then converts the energy from the pressurised hot flue gas into mechanical energy. This causes the gas to expand, having lost most of its pressure, at low temperature at the turbine outlet. The residual heat from the flue gas has a temperature of 500–600°C and is transferred to the downstream steam process.

In a natural gas fired combined cycle, about two thirds of the electrical power is produced in the gas turbine and one third in the steam turbine. The gas turbine

H. Spliethoff, *Power Generation from Solid Fuels*, Power Systems,
DOI 10.1007/978-3-642-02856-4_7,

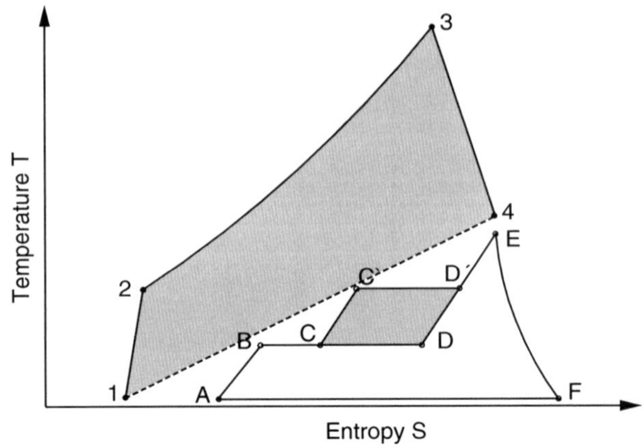

Fig. 7.1 Combined cycle process in a T–S diagram with a gas turbine process (1-2-3-4) and a single pressure (A-B-C-D) or dual-pressure steam process (A-B-C-C′-D′-E-F)

process is normally called the topping process and the steam process the bottom process. A T–S diagram of the two cycle processes is given in Fig. 7.1.

Figure 7.2 shows a natural gas fuelled combined cycle process. The gas turbine installation consists of a compressor, a combustion chamber and a gas turbine. The steam plant consists of a heat recovery steam generator (HRSG), a steam turbine and the subsequent steam – water cycle.

Gas turbines are available commercially as standard products consisting of an integrated gas turbine, compressor and combustor. They feature a high power density and fixed design data, to which only minor changes can be made. Figure 7.3 shows a sectional view of a modern gas turbine with a capacity of 292 MW_{el}. The largest gas turbines today have a capacity of around 340 MW_{el} (Schöler 2007; Ratcliff et al. 2007).

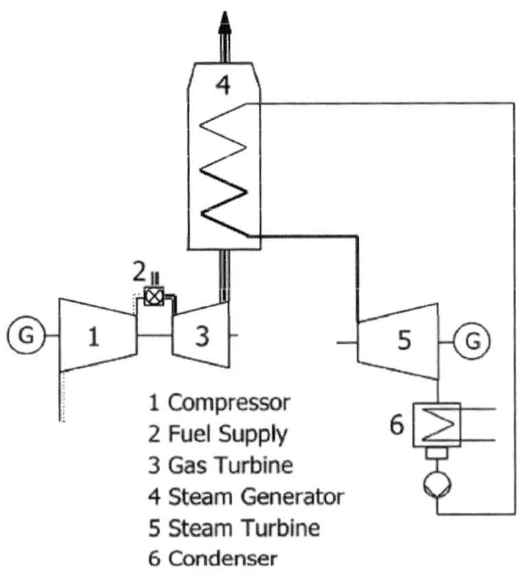

Fig. 7.2 Diagram of the combined cycle process

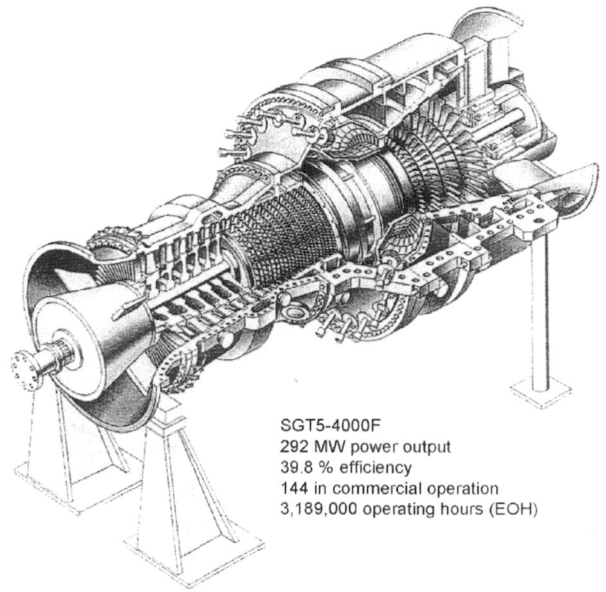

Fig. 7.3 State-of-the-art gas turbine (*Source*: Siemens)

Natural gas has an adiabatic combustion temperature of about 2,200°C. For this reason, the furnace of the gas turbine is operated at a high excess air level in order to reduce the flue gas temperature to the permissible inlet temperature of the gas turbine itself. The flue gases in the newest gas turbines enter the turbine at a temperature of around 1,400–1,500°C. The turbine blades closest to the entry point, in particular, are therefore subjected to high thermal and mechanical stresses, as well as chemical stress from oxidation and corrosion.

Since the metallic turbine materials currently in use can only be exposed to surface temperatures of up to around 950°C, the first stages of a gas turbine are cooled by air from the compressor. Over the last decades cooling has evolved from convection over impingement to film cooling. The blades are made from a composite material; the base material provides the mechanical properties, and one or more surface-protective coatings ensure corrosion and oxidation resistance. Thermal barrier coatings can also offer thermal insulation and allow higher gas turbine inlet temperatures (Bohn 2007).

There is no standard definition of the turbine inlet temperature (TIT). The term can, in fact, refer to different temperatures. The three definitions of relevance are as follows:

(1) The temperature at the outlet from the combustor (i.e. at the inlet to the first turbine stator).
(2) The temperature at the outlet from the first turbine stator (i.e. at the inlet to the first rotor). At this point, the temperature is typically 40–70°C lower than in definition 1 due to the cooling air or steam for the first stator. This definition of the TIT is used by General Electric.

(3) The calculated temperature (not physical) that would result from mixing all the cooling air with the combustor exhaust gas. This temperature is typically 70–110°C lower than definition 2. Siemens uses this definition, which is also known as the "ISO TIT".

It should be added that all three definitions would be identical for a gas turbine without cooling (Maurstad 2005). In this book, definition 1 (TIT) and definition 3 (ISO TIT) will be used.

Given a TIT of roughly 1,400°C at the outlet from the combustor, the ISO temperature calculated for a modern gas turbine is around 1,200°C. The highest ISO gas turbine inlet temperature is around 1,320°C in current practice.

Figure 7.4 shows how the efficiency is influenced by the design parameters (e.g. the ISO turbine inlet temperature and pressure ratio of the gas turbine), both for the gas turbine itself and for the turbine in the gas/steam combination of a combined cycle power plant.

When the gas turbine is operated by itself, the turbine's pressure influences the efficiency, while the inlet temperature influences the output. With a constant gas turbine inlet temperature, the efficiency rises as the pressure increases. This is because the flue gas outlet temperature behind the turbine and hence the losses decrease. If the pressure rises above the optimum level, the energy required for compression increases more quickly than the additional gain in the gas turbine's power output.

The gas turbine inlet temperature has only a minor impact when the gas turbine is being operated alone. This is because the higher temperature of the heat input is partly compensated for by the higher temperature of heat dissipation. In combined cycle operation, the inlet temperature of the gas turbine has a discernible effect on efficiency. The gain in efficiency obtained by the higher temperature of the heat input is conserved. This is because the waste heat of the high flue gas temperatures is used in the heat recovery process.

The optimum pressure of a gas turbine for the combined cycle process is lower than for a gas turbine operated by itself, because the steam process only achieves a high efficiency at flue gas temperatures above 600–650°C, which generates

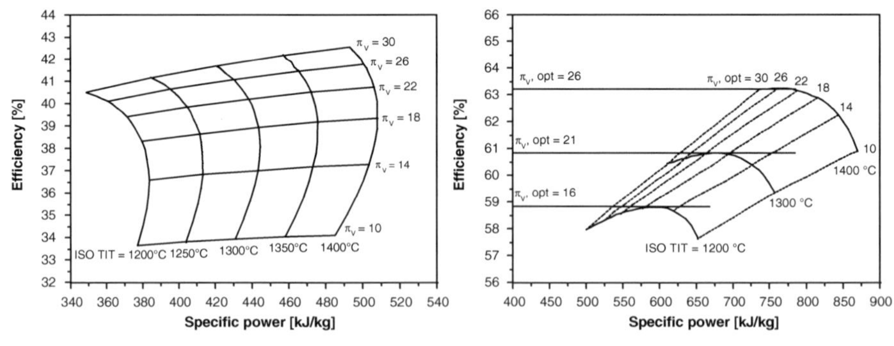

Fig. 7.4 Impact of pressure and the gas turbine inlet temperature (ISO) on the efficiency and output of a gas turbine and a combined cycle process (Kloster 1999)

state-of-the-art high live steam temperatures of 550–600°C. As a result, a gas turbine designed to be operated by itself and which features a high efficiency, a high pressure and a low gas turbine outlet temperature does not necessarily result in a high efficiency when used in a combined cycle. Flue gas temperatures below 600°C lead to a decrease in the efficiency of the steam turbine process higher than the efficiency increase in the gas turbine. On the other hand, flue gas temperatures higher than the level required for optimum efficiency lead to exergy losses in the heat recovery process, because the steam temperature is limited by the materials. The outlet temperatures of modern gas turbines, however, are close to those temperatures which make high-output steam production possible (Kail and Rukes 1995; Kloster 1999).

The number of gas turbine types available from manufacturers is limited. This places restrictions on how combined cycle power plants can be designed. Unlike steam turbines (which are designed for a given steam cycle on a case-by-case basis), gas turbines on the market feature specific capacities only. Modifying the gas turbine is expensive, because significant design work and manufacturing modifications are involved, in particular for the compressor.

Once a gas turbine has been selected, the main design parameters for the downstream heat recovery process (the gas mass flow and the gas turbine outlet temperature) are specified. A high total plant efficiency is produced when the heat is transferred from the gas turbine flue gas to the steam – water cycle with minimum energy losses. In practice, this can be achieved by generating the steam at various pressures, so that the difference in temperature between the flue gas to be cooled and the heat-receiving medium (and hence the energy losses) can be decreased. This is shown in Fig. 7.5 (Riedle et al. 1990). Today, large plants optimised for high electrical efficiency use a triple-pressure heat recovery process. In addition to the number of pressure stages, the heating surfaces of the heat recovery steam generator (HRSG) allow the energy losses of the waste heat transfer to be diminished. Selecting the number of pressure stages and the design of the heat exchanger are subject to cost-effectiveness considerations: the additional costs are pitted against the savings resulting from the higher efficiency (Warner and Nielsen 1993).

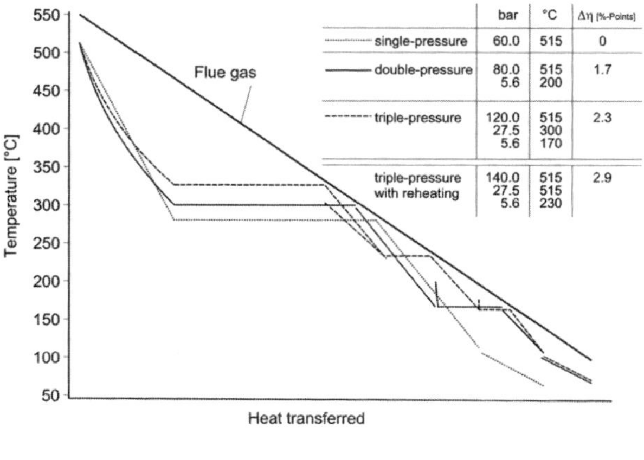

Fig. 7.5 Temperature course in a waste heat boiler (Riedle et al. 1990)

Table 7.1 Possible development of combined cycle processes (Bohn 2005)

Parameter	Reference	Phase 1	Phase 2	Phase 3
Efficiency	57.4%	61.7%	63.3%	65.0%
Combustion exit temperature	1,500°C	1,500°C	1,520°C	1,520°C
Cooling GT	21.46%	11.7%	12.4%	9.9%
ISO turbine inlet temperature	1,172°C	1,432°C	1,439°C	1,473°C
Max. substrate temp. (GT)	850°C	900°C	950°C	990°C
Max. material temp. (ST)	560°C	580°C	595°C	650°C
Required cooling steam	–	2.2%	1.0%	0.3%

State-of-the-art gas turbine technology, designed for gas turbine only plants, achieves efficiencies of up to 38%. In a combined cycle plant with an optimised heat recovery process, efficiencies rise to 58% (Jopp 2005). A new gas turbine is currently being tested with an efficiency of around 40% as a stand-alone gas turbine and above 60% as part of a combined cycle plant (Schöler 2007). Gas turbine technology has undergone continuous, high-level development in recent years. A further rise in the ISO gas turbine inlet temperature, and hence in the gas turbine efficiency, is expected in the future. A higher ISO TIT results in higher gas turbine outlet temperatures and promotes the use of advanced live steam conditions. Table 7.1 shows potential developments in the coming years (Bohn 2005; Bohn 2007).

The high efficiencies of the combined cycles shown above are only possible with a clean fuel such as natural gas. Otherwise, the purity requirements of the gas turbine for the working medium (pressurised hot flue gas) cannot be met.

7.2 Overview of Combined Processes with Coal Combustion

7.2.1 Introduction

For coal-based combined cycle processes, the solid fuels must first be converted into a fuel or a hot gas suitable for gas turbines. Impurities in the fuel prevent it from being used directly in today's gas turbines. Compared to the natural gas fired combined cycle process, more steps are required (COORETEC 2003; JBDT 1992; Rukes 1993; Wittchow and Müller 1993; Jahraus and Dieckmann 1989; Böhm 1994). The processes shown in Fig. 7.6 have been investigated, tested and partially implemented on an industrial scale in order to put a coal-based combined cycle into practice. They are

- integrated gasification combined cycle (IGCC),
- combined cycle with pressurised fluidised bed combustion (PFBC),
- combined cycle with pressurised pulverised coal combustion (PPCC) and
- externally fired combined cycle (EFCC).

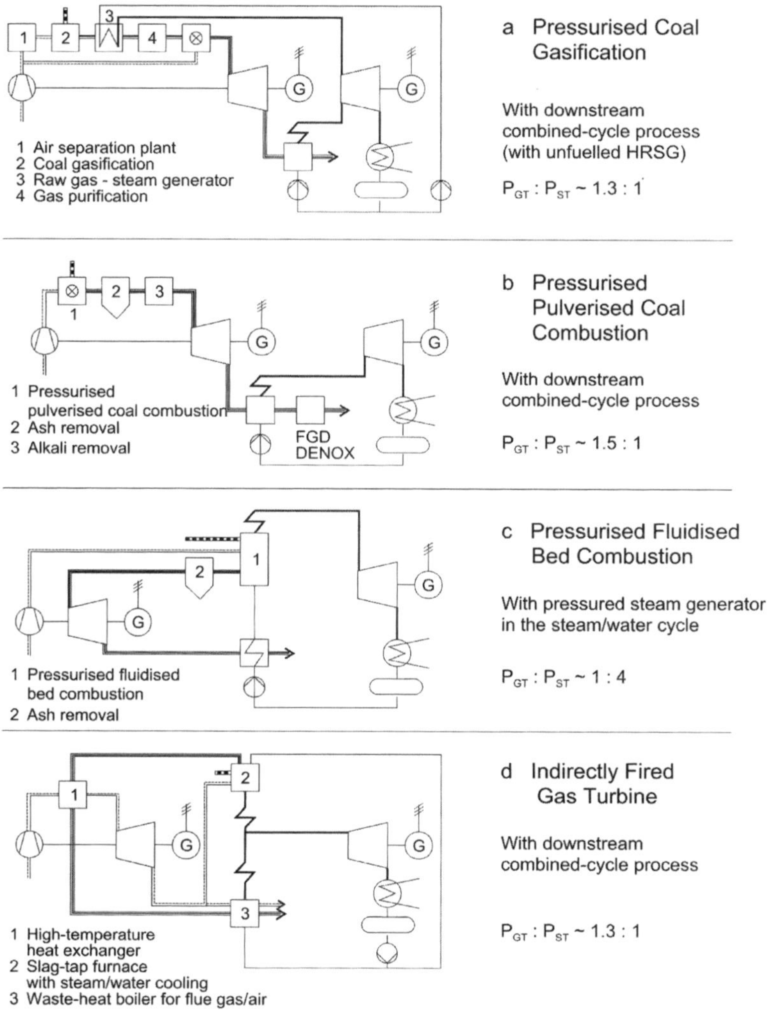

Fig. 7.6 Coal-based combined cycle processes (Böhm 1994)

In the integrated gasification combined cycle, the process starts by transforming the solid fuel into a fuel gas at high temperatures in a gasifier. The gas then has to be purified. Following this, the fuel gas is burned in the gas turbine. The fuel gas produced is cooled by water/steam for the flue gas cleaning step. This results in a low ratio (about 1.3:1) of the gas to the steam turbine power outputs, and hence to a lower efficiency compared to the natural gas fired combined cycle.

In the combustion-based combined cycles of pressurised fluidised bed combustion (PFBC) and pressurised pulverised coal combustion (PPCC), the fuel is completely combusted under pressure. Before entering the gas turbine, the hot flue gas has to be cleaned so that it meets the required gas purity standard. Whereas in

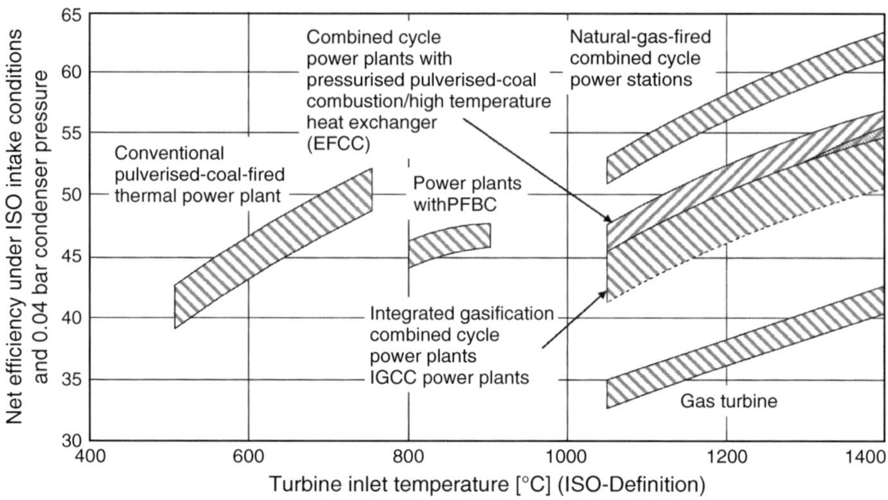

Fig. 7.7 Efficiency of combined cycle processes depending on the gas turbine inlet temperature

the PPCC process a 1.5:1 ratio of the output of the gas and steam turbines can be achieved, the PFBC process, with its steam – water cooling, yields a mere 1:5 ratio of the gas and steam turbine outputs. The ratio of the gas to the steam turbine power outputs correlates directly with the efficiency of the overall cycle; the higher the ratio, the higher the efficiency.

There is an essential difference between the combined cycles with gasification and with combustion: the volumetric flow to be cleaned is 10 times higher for combustion. In a power plant with an integrated gasification combined cycle, the fuel gas

Table 7.2 Comparison of power plant processes

	Steam power plant	Natural gas fired combined cycle	Coal gasification combined cycle	Pressurised fluidised bed combustion	Pressurised pulverised coal combustion	Externally fired combined cycle
Stage of development	State of the art	State of the art	Can be demonstrated	Being launched on the market	In development	In development
Relative investment costs	100%	30%	110–120%	100%	?	?
Relative availability	> 90%	> 90%	Lower than steam power plant	> 90%	?	?
Efficiency (1,250° C ISO)	46%	58%	51%	45% (51%)[a]	53	53

[a] Hybrid process

flow can be cleaned at a lower temperature than that of gasification. By contrast, in the combustion-based combined cycles with pressurised fluidised bed combustion and pressurised pulverised coal combustion, the flue gas has to be cleaned at temperatures above the gas turbine inlet temperature. Otherwise, there would be a high loss in efficiency.

In the externally fired combined cycle (EFCC) process, a high-temperature heat exchanger is used to avoid the problems of hot gas cleaning. The energy yielded by combustion is transferred to a clean working medium in a heat exchanger. This working medium, which is suitable for use in a gas turbine, charges the turbine.

These different methods are at differing stages of development. While both PFBC and IGCC are already being used in industry, the EFCC and PPCC processes have not been implemented at industrial scale.

In Fig. 7.7, the efficiencies of coal and natural gas based combined cycles are compared with each other and with the conventional steam power plant at a specific turbine inlet temperature. Table 7.2 shows a comparative evaluation.

7.2.2 Hot Gas Purity Requirements

The purity of the working medium for the gas turbine required by gas turbine manufacturers is a parameter which determines how the conversion process is designed and which purification steps are selected. When evaluating impurities in the fuel gas or hot flue gas, a distinction must be made between components that lead to

- high-temperature corrosion, erosion and deposits in the gas turbine,
- corrosion at the cold end of the heat recovery steam generator or
- undesirable emissions.

The damage to the flue gas charged gas turbine caused by corrosion, erosion and deposits shall be discussed below, because monitoring these problems is decisive in ensuring the success of coal-fired combined cycles.

Deposits on the turbine blades lead to a decline in the turbine efficiency. They can be removed by scrubbing the gas turbine. Such a process uses water jets installed in the combustion chamber. Desalted water is sprayed into the chamber under pressure and reaches the turbine via the air flow. This water washes away the water-soluble compounds and penetrates the pores and crevices of water-insoluble compounds. When the gas turbine heats up, the water in the pores and crevices evaporates. The resulting steam pressure causes the deposits to spall (JBDT 1992).

Erosion and corrosion wear the turbine down and lead to a reduction in the turbine's lifetime and efficiency. Solid as well as liquid particles reach the turbine blades in a number of ways (Thambimuthu 1993):

- Inertial impaction:
  Large particles do not follow the gas flow due to their inertia. The particles which hit the blades cause erosion but can also add to deposits. Impinging particles can

also carry deposits away again under certain conditions. Particles smaller than 5 μm or so follow the gas flow and therefore do not cause erosion.

- Turbulent and Brown's diffusions:
  Inertia is not the only factor which can cause particles to deviate from their course. Turbulent and Brown's diffusions can have a similar effect. In the case of turbulent diffusion, the particles are caught by the eddies of the turbulent boundary layer flow. In Brown's diffusion, the kinetic energy of the gas molecules is transferred to small particles. Both turbulent and Brown's diffusion phenomena catch even the smallest particles, which are deposited over the whole inner surface of the gas turbine.
- Thermophoresis:
  Thermophoresis refers to transportation by thermal diffusion as a result of thermal gradients between the gas and the surfaces. While thermophoresis can be ignored in the case of low gas turbine inlet temperatures, it can play a significant role where blade cooling (required for higher tube inlet temperatures) is involved.

Of particular importance with regard to contamination and corrosion are sodium and potassium. These gaseous alkalis are released during combustion, and even small traces of them can shorten the lifetime of the gas turbines. Gaseous alkalis condense in the turbine while it is cooling down. They then form molten alkali sulphates (in compounds using SO_2 from the flue gas, for example) which are deposited on the turbine blades or on ash particles. Alkalis cause the ash fusion temperatures to fall, and this in turn causes some or all of the ash particles to melt, either before or after they reach the blade. In the latter case, they form deposits on the blades. Melting or sintering of the ash causes deposits to form on the blades that are difficult to remove (Thambimuthu 1993).

The molten alkali compounds attack the gas turbine blades severely. By means of high-temperature corrosion, they destroy the oxidic protective layers of the parts they come in contact with. This causes an intense corrosive attack on the unprotected base material, resulting in a drastic shortening of the lifetime of the blades. The rate of corrosion depends on the chemical composition of the deposits; the greater the alkali content, the faster the rate of corrosion. The available information on the corrosive impact of alkalis is based on experience of the combustion of oil distillation residues and heavy fuel oil in gas turbines. Literature on this issue often uses the limit of 0.024 mg/kg fuel, set by General Electric, for the combustion of oil distillation residues, though it varies depending on the fuel used. Even if it seems impossible to apply this data directly to the conditions of coal-fuelled firing, they do form a basis for specifying the alkali requirements to be met by the hot gas.

Certain metals such as vanadium, lead and zinc can cause the same destruction of protective layers as do alkalis. Chlorine, fluorine and their acids can wear away the protective layers by forming gaseous chloride and fluorides (JBDT 1992). Alkaline earth metals can lead to hard deposits on the turbine blades (Hannes et al. 1989).

The limiting values for turbines depend on the gas turbine inlet temperature. Lower permissible limits are reported for higher temperatures. Much less stringent requirements are expected after PFB firing due to the gas turbine inlet temperatures, which are lower in comparison to pressurised pulverised coal firing.

Table 7.3 Permissible guideline concentrations for dusts and trace elements in the hot gas for gas turbine V94.3 (now SGT5-4000F) (data from Jansson 1996; Mitchell 1997)

Dust	Total	[mg/kg]	1
Dust distribution	> 10 μm	% by wt	0
	2–10 μm	% by wt	7.5
	0–2 μm	% by wt	92.5
Trace elements	Ca	[mg/kg]	0.4
	V + Pb	[mg/kg]	0.01
	Na + K	[mg/kg]	0.01

Table 7.3 gives the permissible flue gas concentrations in front of the turbine for the Siemens V94.3 (now SGT5-4000F) gas turbine at a gas turbine inlet temperature of 1,120°C (ISO) which are required after combustion of natural or coal gas. The table shows the limits for the total particulate matter content, the maximum size of particles and the concentrations of heavy metals – lead (Pb) and vanadium (V) – the alkaline earth metal calcium (Ca) and the alkali metals sodium (Na) and potassium (K) in the flue gas (Jansson 1996). For combined cycles with pressurised coal firing, limits comparable to natural gas fuelled gas turbines are used as a basis.

The fact that pollutants can also be sucked in together with the combustion air must be taken into consideration as well. At coastal locations, for instance, the marine salt contained in the fresh air can make a major contribution to the alkali load in the process.

Although gas turbines have been operated with hot gas generated by PFB combustion for several years now, no detailed data on erosion, deposits and corrosion, or relevant limits for prevention of such damage, is available. The values achieved with the hot gas filters currently used for PFB furnaces, i.e. 250–650 mg/Nm^3, reveal much higher concentrations of particulates in the flue gas than the values given in Table 7.3. All particles are smaller than 10 μm and the mean diameter is 2–3 μm (Jansson 1995b). The concept developed by ABB (now Alstom) is based on a modified robust gas turbine, designed to minimise erosion.

The technical data currently available is provided by experimental plants with gas turbine cascades. In order to limit the erosion caused by hot gases in fluidised bed combustion (FBC) furnaces, limits are suggested for concentrations of particles larger than 4 μm as such (Stringer 1989):

> 20 μm: 1 mg/kg
10–20 μm: 1 mg/kg
4–10 μm: 10 mg/kg

Particles smaller than 4 μm are likely to appear but they will not cause erosion, contamination or corrosion. At lower gas turbine inlet temperatures in PFB furnaces, concentrations of particulates up to 100 mg/Nm^3 are considered tolerable as long as all particles are smaller than 5 μm (Emsperger and Brückner 1986).

The alkali emissions from PFB furnaces can be one or more orders of magnitude higher than the 0.024 mg/kg fuel indicated as a limit for turbine corrosion for combustion of oil distillation residues. Although the installed PFB furnaces are not equipped with an alkali remover, the reported corrosion is mild.

Table 7.4 Required flue gas purity for pressurised pulverised coal combustion

Dust content	[mg/Nm^3]	3
Maximum particle diameter	[μm]	<3
Gaseous alkalis	[mg/Nm^3]	0.01

The values in Table 7.4 for the required purity of the flue gas at the turbine inlet temperature for the pressurised pulverised coal firing concept are somewhat higher than the values given in Table 7.3 (Hannes 1986; Förster et al. 2005).

7.2.3 Overview of the Hot Gas Cleaning System for Coal Combustion Combined Cycles

In a combined cycle with coal gasification, cold gas cleaning of the fuel gas is the standard procedure. By contrast, hot flue gas cleaning, an efficiency-improving option, is essential for combustion-based combined cycles. FBC systems can operate with no more than dust removal, whereas pressurised pulverised coal firing also requires alkali removal. The removal of pollutant gases (sulphur dioxide or nitrogen oxide, for example) which the gas turbine can tolerate can take place either before or after the gas turbine. While sulphur dioxide removal can take the form of an in situ process in fluidised bed firing, pressurised pulverised coal firing requires desulphurisation after the heat has been recovered in the steam generator.

The choice of the dust removal technique depends on the fusion behaviour of the fuel ash. Below the ash deformation temperature, fly ash is separated in solid form; at temperatures above ash fluid temperatures, it is separated as liquid slag. In the temperature range of roughly between 900 and 1,300°C, hot gas cleaning of ash is possible in theory. However, difficulties will arise when cleaning the filter components, because the particles stick to the equipment.

Table 7.5 offers an overview of the possible dust removal techniques (Weber and Pavone 1990; Weber et al. 1993; Pruschek et al. 1990). They can be classified as follows:

- Mass force separators
- Wet scrubbers
- Filter separators
- Electrostatic precipitators (ESPs)

The class of filters used for *separation by mass force* comprises all separators that use only mass forces to clean the gas, i.e. gravity, inertia or centrifugal force. The names of the various separators are therefore derived from the effective force in the case in question. The majority of mass force separators are straightforward in design and pose few engineering problems. Compared to other dust-collecting devices, they are cost-effective and easy to service and operate. However, they are not suitable for the removal of soft or sticky particles. They are well suited to the removal of liquid slag at temperatures above 1,500°C, though material-related problems may arise.

Table 7.5 Summary of temperature windows for use of particulate matter collection technologies

Flue gas temperatures	Mass force separator	Wet scrubber	Filter separators	ESP
< 900°C (solid)	Possible	Possible with suitable scrubbing liquids	Possible with ceramic filter elements	Possible
850–1,300°C (melting range)	Ash removal not possible because particles are sticky	No suitable scrubbing agent known	Not possible because particles are sticky	Not possible because particles are sticky
Above 1,300°C (liquid)	Possible; material problems	Possible; material problems	Not applicable for liquids	Not possible

Wet separators can, in principle, be used for high-temperature cleaning if a suitable, thermally stable scrubbing liquid is used at low vapour pressures. The wet scrubbing process is complicated by the treatment of the scrubbing liquid and the material used. At temperatures up to 850°C (and possibly higher), this method is more complicated and expensive than other dust-collecting methods. The scrubbing medium used for the pressurised pulverised coal fired furnace is liquid slag. When injected into a venturi scrubber, it helps the liquid slag droplets to agglomerate, thus enhancing the removal efficiency in a downstream cyclone. For more information, see Sect. 7.4.2.

In their various technical designs, *filter separators* generally form suitable high-temperature particle collectors if suitable thermally stable filter materials are available and the particles to be separated are solid. If the particles are sticky, removing the particles from the filter material becomes a problem. For temperatures above 1,250 °C, the principle is not applicable because of the liquid consistency of the ash. In contrast to the other particulate collection technologies, not only do the electrostatic precipitators in the high-temperature range present material and construction problems, but the physical conditions for collection must also remain stable at extreme gas temperatures. Investigations at temperatures up to 1,000 °C show that the production of charge carriers (required for particle removal) is possible. At temperatures of 1,300 °C or above, the ESP principle cannot be applied due to the conductivity of the flue gas.

The individual methods for removing solid particles are discussed in Sect. 7.3.2 in relation to pressurised fluidised beds; the removal of molten ash particles is discussed in Sect. 7.4.2 in relation to pressurised pulverised coal firing.

7.2.4 Effect of Pressure on Combustion

Combustion of the residual char is a major factor in determining how long coal combustion will take. This is described in detail in Sect. 5.2. The effect of pressure on the burning speed of the char depends on whether chemical reactions or transport processes during combustion of the solid char influence the reaction velocity. At low temperatures, chemical reactions slow down the reaction velocity. Given that the chemical reaction velocity is proportional to the oxygen concentration, an increase in pressure will accelerate char combustion. In contrast, pressure exerts only a minor influence on the diffusion-controlled reaction.

A clear improvement of the combustion is therefore expected in the temperature range of the fluidised bed combustor. With its higher temperatures, however, the conditions under which pulverised fuel combustion occurs leave little room for a substantial acceleration of combustion. In this case pressure cannot raise the speed of combustion, but it does influence the dimensions of the furnace. Figure 7.8 shows the effect of pressure, as calculated for different temperatures (Gockel 1994).

If the furnace has the same geometry as in the process at ambient pressure, the air mass flow (which enters the furnace at constant inlet speeds) will increase proportionally to the pressure. The fuel mass flow will then change accordingly if the carrier gas has a constant fuel load. Since the pressure does not cause any change in speed, the residence time required for the purpose of combustion remains constant. From a combustion perspective, the output can be increased proportionally to the pressure as long as the furnace volume remains the same. Alternatively, the volume can be decreased proportionally to the pressure with no change in output and without taking the pressure effect on the speed of combustion into account.

If the furnace has to dissipate heat in addition to releasing it (as in the case of stationary fluidised bed), the impact of pressure on the heat transfer also plays a role in determining the furnace's dimensions.

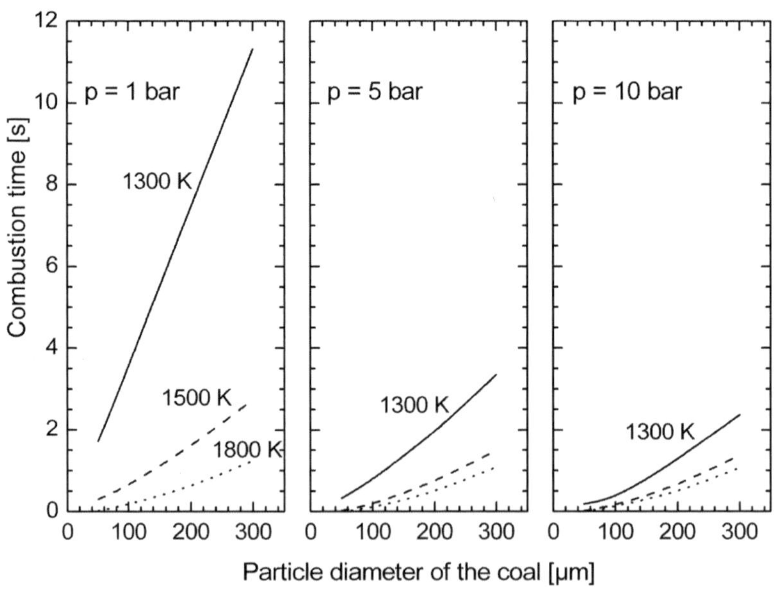

Fig. 7.8 Effect of pressure on combustion (Gockel 1994)

7.3 Pressurised Fluidised Bed Combustion (PFBC)

7.3.1 Overview

The pressurised fluidised bed is one method of using coal as a fuel in a combined cycle process. The solid fuel is burned in a bubbling or circulating fluidised combustion bed at temperatures between 850 and 950°C and pressures of up to 16 bar. Following cleaning, the hot flue gas is brought to the gas turbine.

The optimum temperature in the fluidised bed is determined by the requirement that it must remain below the ash deformation temperature at all times, in order to prevent agglomerations from forming in the fluidised bed and in order to achieve optimal desulphurisation. Therefore it is necessary to cool the fluidised bed, as the adiabatic combustion temperature of coal is above 2,000 °C. Various methods can be used for the purpose of cooling: operation at a high excess air level, air cooling or vapour – water cooling (Emsperger and Brückner 1986). In natural gas fired combined cycles, by comparison, the gas turbine inlet temperature (which is much higher than the fluidised bed though) is set by operating at 100% excess air. Figure 7.9 shows the different cooling process variants, which are described below:

- *Adiabatic pressurised fluidised bed*:
  An adiabatic fluidised bed with no in-bed heating surface needs an excess air level of 300% (an air ratio of 4) in order to limit the fluidised bed temperatures. This results in a furnace with large dimensions and particulate removal downstream. The ratio of the gas turbine output to the steam turbine output is about 2:1. The adiabatic fluidised bed has been studied conceptually, but as yet a plant has been neither planned nor constructed.
- *Pressurised fluidised bed with an air-cooled heating surface*:
  In this configuration, compressed air is heated to about 700°C in the in-bed heat exchanger and mixed with the cleaned flue gas before the gas turbine. Due to the lower air temperature, the gas turbine inlet temperature drops below the temperature of the fluidised bed, resulting in a lower efficiency. The ratio of the gas turbine output to the steam turbine output is about 2:1. This variant presents some major drawbacks, such as the poorer heat transfer of air as a cooling medium (compared to steam and water) and the higher material temperatures.
- *Pressurised fluidised bed with steam – water cooling*:
  Steam – water cooling is the method used in today's PFBC furnaces due to the more compact, and hence more cost-effective, design of these furnaces. Therefore, only this variant shall be discussed in more detail. A disadvantage is the fact that the gas turbine makes no more than a minor contribution (one fifth to one quarter) to the total power output. The efficiency depends on the steam conditions used. In the case of a high-quality steam process, the efficiency is higher than that of the adiabatic fluidised bed; in the case of low-quality steam conditions, the efficiency is lower.

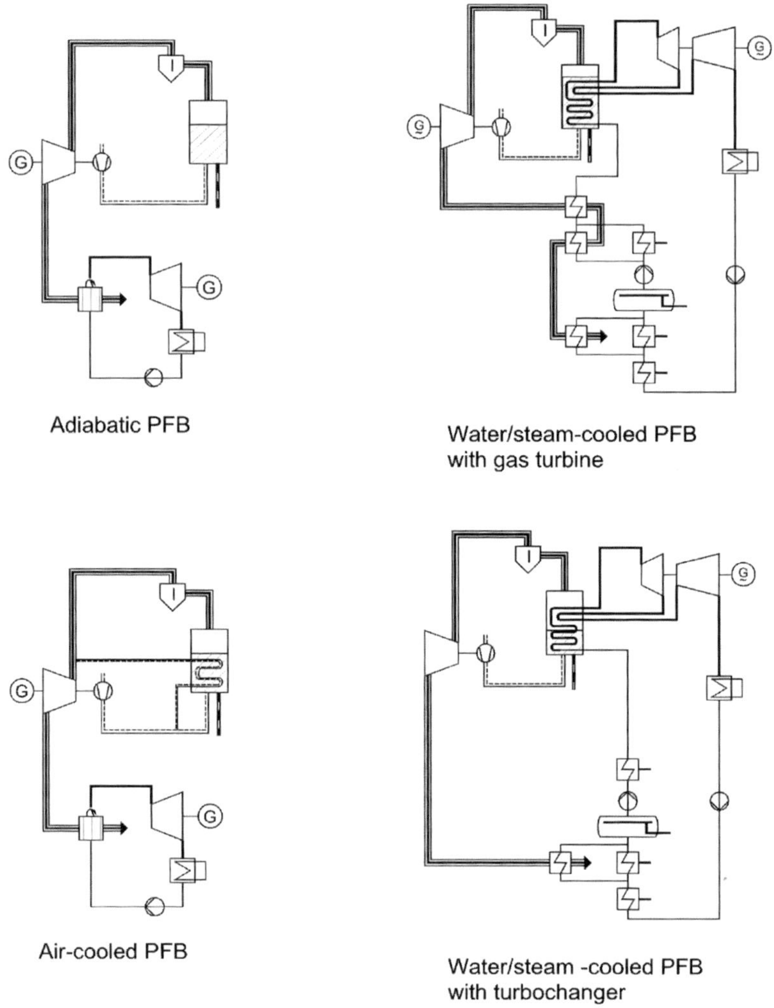

Fig. 7.9 Cooling of PFBC furnaces (Emsperger and Brückner 1986) and amendments

- *Pressurised fluidised bed with steam – water cooling and turbocharger*:
  In this configuration, the flue gases are cooled down by convective heat exchange surfaces to about 400–500°C before particulates are collected. This avoids the problem of hot flue gas cleaning at temperatures of around 850°C. At these low gas turbine inlet temperatures, however, the gas turbine only provides the power necessary to drive the compressor. The advantage of this configuration is limited to its compact design. Its efficiency depends solely on the steam process (Emsperger and Brückner 1986; Dibelius and Pitt 1989). Such a plant, with a thermal capacity of 40 MW and a maximum pressure of 4 bar, was used between 1984 and 1989 in the thermal power station operated by the Aachen University of

Technology. Its purpose was to prepare design data and test components as part of a research project (Terhaag et al. 1995; Thelen 1993; Emsperger and Brückner 1986).

Figure 7.10 shows various flue gas cleaning flow diagrams for PFBC furnaces (Thambimuthu 1993). Flue gas cleaning works at temperatures between 800 and 900°C. Its purpose is to prevent fouling and to protect the gas turbine from erosion and corrosion. Commercial plants mainly use cyclone separators. Ceramic hot gas filters have been tested in the Tidd, Wakamatsu and Escatrón plants. However, this technology is not considered to be ready for commercial PFBC plants. Alkali removal is not used in commercial PFBC plants (McMullan 2004; Sasatu et al. 2001; Wright et al. 2003; Wu 2006).

The upper limit of FBC temperatures in order to avoid bed agglomeration is about 950°C. This sets a limit to the efficiency. Because the steam cycle provides approximately 80% of the power output, the PFBC efficiency depends mainly on the steam conditions used. Plants currently in service achieve efficiencies of up to 42% (Wu 2006). Other projects aim to achieve higher efficiencies by combusting the clean gaseous fuel in addition to the hot cleaned flue gas from the fluidised bed (supplementary combustion). The purpose of this is to raise the flue gas temperature to standard gas turbine inlet temperatures. For this purpose, either natural gas or a fuel gas produced by coal gasification can be used. The latter is shown in Fig. 7.10.

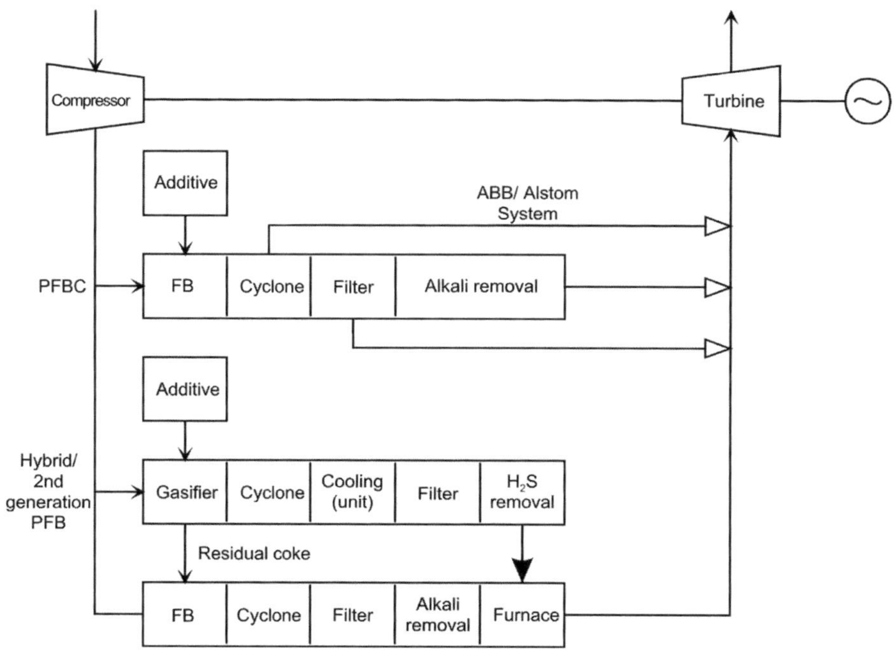

Fig. 7.10 Configurations of PFBC furnaces (Thambimuthu 1993)

For a hybrid or second-generation PFBC process of this nature, efficiencies of up to 52% are reported (when advanced gas turbine inlet temperatures and an advanced steam-production process are used). However, the higher gas turbine inlet temperatures generated by supplementary combustion place higher requirements on the gas cleaning stage after the PFB furnace. This in turn requires advanced filter systems and an alkali removal stage (Robertson et al. 2005).

A PFBC furnace is designed in such a way that the steam generator is installed inside a pressure vessel, because the compressive forces generated by the furnace pressure cannot be absorbed by the furnace's heat-absorbing enclosing walls. Because of the steam generator size and geometry, the vessels considered are cylindrical or spherical in shape.

The pressurised design facilitates a much more compact furnace compared to atmospheric plants. The advantage of operating under pressure lies in the fact that the oxygen partial pressure of the combustion air increases as the operating pressure increases. As a result, the fuel throughput can be considerably higher for a given steam generator size.

The cross-sectional heat release rate $\dot{q}_{\mathrm{f}}/a_{\mathrm{bed}}$ specifies the specific thermal output in relation to the bed surface. It is therefore a factor in determining the size of the fluidised bed combustor. Assuming equal fluidising velocities, the cross-sectional area heat release rate increases in proportion to the pressure (Bunthoff and Meier 1987).

In Fig. 7.11, the bubbling (stationary) and the circulating FBC types are compared with each other, with and without pressure. The parameters used for comparison are the superficial velocity and the cross-sectional heat release rate.

Atmospheric bubbling firing has fluidising velocities between 1 and 2.5 m/s. The speed cannot be further increased as it could cause erosion of the in-bed heat exchange surfaces. Due to this erosion hazard and the higher level of emissions of CO and NO_x in bubbling FBC furnaces, circulating FBC has become the usual technology for plants operated at atmospheric pressure. Here, the fluidising speed can be raised to 8 m/s, because there are no heat transfer surfaces inside the fluidised bed. As a result, the cross-sectional heat release rate in atmospheric combustion plants rises from 1.5 $\mathrm{MW/m^2}$ in stationary operation to 5–7 $\mathrm{MW/m^2}$ in circulating operation.

Pressurisation allows the plant capacity to be increased considerably. In order to prevent erosion of the heating surfaces, the fluidising speed is limited to 1 m/s for bubbling pressurised fluidised beds. Despite this, the cross-sectional area heat release rate is between 10 and 17 $\mathrm{MW/m^2}$ because of the increase in pressure. If this combustion technology is changed to a circulating type of pressurised fluidised bed, cross-sectional heat release rates of up to 50 $\mathrm{MW/m^2}$ can be achieved. One outcome of this is a relatively slim pressure vessel design. The bubbling and the circulating FBC furnaces thus differ in geometry or, more precisely, in the proportion of the bed cross-section to the boiler height. Circulating FBC furnaces have a slimmer body and also promise lower emission levels. However, a circulating type features a more complicated plant design (JBDT 1992).

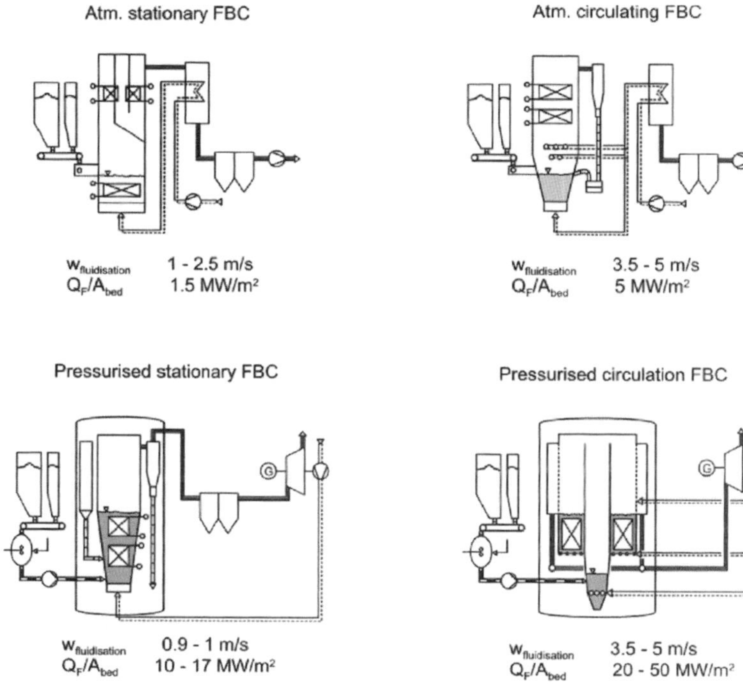

Fig. 7.11 Comparison of bubbling (stationary) and circulating fluidised beds with and without pressure (JBDT 1992)

Although pressurised circulating FBC furnaces offer a range of advantages, only stationary combustion types are used commercially. Pressurised circulating FBC furnaces have been developed in Germany and Finland (Renz 1994). A pilot pressurised circulating fluidised bed furnace has been tested by Foster Wheeler as one component in a second-generation PFBC (Wheeldon et al. 2001). Figure 7.12 shows the commercial plants which have been built to date and their electrical capacities. Most of the pressurised FBC furnaces which have been planned and built are in Japan (McMullan 2004; Schemenau and van den Bergh 1993; Wu 2006).

In designing a fluidised bed, complete combustion must be ensured either through a sufficient residence time in the furnace or by ash recirculation. The particles emitted from the bed of a stationary FBC are completely combusted by means of higher combustion reaction velocities at a higher pressure. The size of the emitted particles depends on the fluidising speed and the density of the particles. Under the conditions offered by a bubbling fluidised bed with a fluidising speed of about 0.9 m/s, the emitted particles are smaller than 250 μm. Since these particles are completely combusted in the freeboard volume of the furnace, they do not have to be recirculated under pressurised fluidised bed con ditions. As a result of the higher fluidising velocities of circulating combustion, coarser particles are also emitted. This means

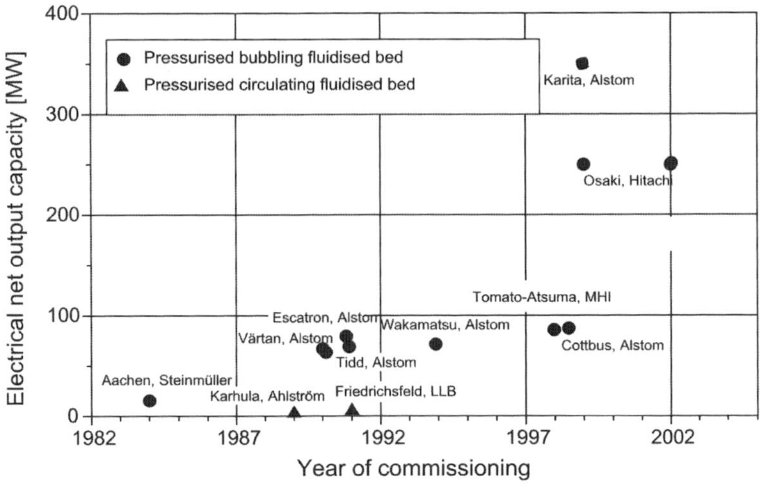

Fig. 7.12 Commercial pressurised FBC furnaces (data from Wu 2006; Schemenau 1993)

that the particle residence time in the freeboard stays below the necessary burnout time, and hence complete burnout is only achieved after several recirculations.

In order to reduce sulphur dioxide emissions, limestone is added to the combustion. It binds the developing sulphur dioxide and forms calcium sulphate. Atmospheric fluidised beds require ash recirculation in order to make better use of the limestone and to improve the sulphur capture. Recirculation is not needed in pressurised FBCs because of the better reaction kinetics of desulphurisation under pressure (Bunthoff et al. 1990).

In bubbling atmospheric FBCs, the bed height, generally around 1 m, is limited by the bed's pressure loss. An excessive loss should not be chosen for economic reasons. In comparison, the high thermal load and the large heat extraction surface of the bubbling PFBC furnace require higher beds of between 3 and 6 m. It is possible to use these high beds because the relative pressure loss, in relation to the process pressure, is low (Bunthoff and Meier 1987).

Pressure has a positive effect on the heat transfer. The heat transfer coefficients depend on the grains of the bed material, the fluidising velocity and the tube wall temperatures. The relationship between the pressure and the heat transfer coefficient is shown in Fig. 7.13. This figure shows that, compared to atmospheric operation, an increase in pressure to 16 bar (through the increase in gas density and the dynamic viscosity) improves the heat transfer of a bubbling fluidised bed by about 80%. Consequently, smaller heat transfer surfaces can be built for the same thermal output (Bunthoff and Meier 1987).

Oscillations of the grains of the fuel influence the heat absorbed by the in-bed heat transfer surfaces of the bubbling fluidised bed. An in-bed heat transfer surface which is large enough for one fuel type may be too large or too small for another fuel and may affect both the steam generation and the gas turbine inlet temperature. The

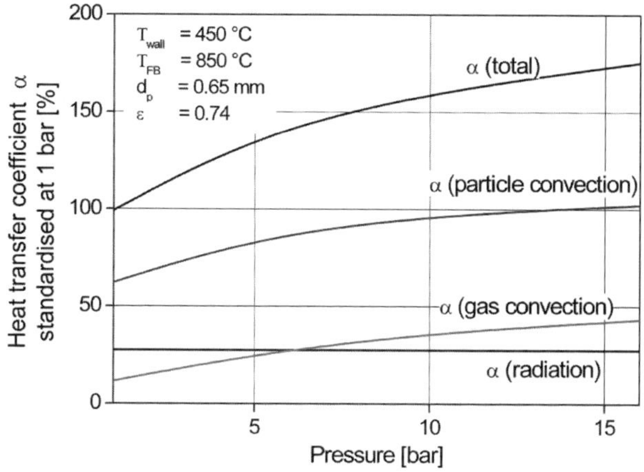

Fig. 7.13 Effect of pressure on heat transfer in a pressurised fluidised bed (Bunthoff and Meier 1987)

division between coal combustion and heat dissipation in a circulating FBC process avoids these problems (JBDT 1992).

The design of a PFBC plant is determined by the gas turbine chosen. The compressor supplies the air mass flow required for fluidisation, combustion and cooling of the pressure vessel. It thus determines the boiler's capacity. The operating regime of the gas turbine at part load is determined by the need to set as large as possible a constant air excess level for the fluidised bed over the part-load range. This causes the pressure and the air mass flow to the fluidised bed to diminish in proportion to the thermal output (Stuhlmüller et al. 1995).

When used as a fuel, natural gas has a ratio of flue gas to air mass flow of about 1.02. In contrast, wet coal paste feeding gives between 1.07 and 1.11 and can rise to 1.3 with run-of-mine brown coal. Accordingly, the compressor of a gas turbine designed for natural gas would be too big or the turbine too small for wet coal paste FBC.

Large, stationary gas turbines are usually single shaft. The compressor, gas turbine and generator have a common shaft, and the revolutions per minute are determined by the power frequency. As a result, they cannot be used for load control (of the mass flow). For natural gas, it is possible to vary the mass flow by controlling the inlet vane for loads slightly above 50%; for smaller loads, the turbine inlet temperature is decreased. If inlet vane control is used in pressurised FBC to adjust the smaller compressor mass flow, the possible inlet vane control range is limited during part load. Below an inlet vane benchmark (which lies at about 70% of the rated useful heat output for hard coal and at about 80% for brown coal), single-shaft turbines make it necessary to bypass the fluidised bed in order to keep the excess air in the combustion at a constant level. However, the mixing of hot flue gas with the cold compressor air results in a decreased efficiency.

In comparison to single-shaft turbines, dual-shaft turbines allow better adaptation of the mass flow and the compressor pressure to the conditions of the pressurised fluidised bed, because the compressor's revolutions can be set independent of the grid's frequency. With two exceptions, all existing plants with bubbling PFBC implement dual-shaft turbines of the same design. However, modern stationary gas turbines from all manufacturers, without exception, are currently single-shaft turbines (Stuhlmüller and Schauenburg 2001).

Combined cycle processes with pressurised fluidised bed combustion also differ from the natural gas fuelled combined cycle due to the much higher pressure losses in the fluidised bed boiler, filtering system and pipework – about 1 bar in total as opposed to 0.3 bar with natural gas. The pressure losses diminish the turbine's output. Since the impact on the efficiency is limited, the loss in efficiency is regarded as not too serious, and the fact that adapting the compressor blading is highly complicated further discourages addressing this drawback.

The gas turbine inlet has a low temperature of around 850°C. This is significantly lower than the temperatures of gas turbines in use today. With these lower temperatures, blade cooling can be simplified or dispensed with, meaning less cooling air is required.

7.3.2 Hot Gas Cleaning After the Pressurised Fluidised Bed

Hot gas cleaning after fluidised bed combustion concentrates on particulate matter removal (fly ash), although in some plants alkali removal may be taken into consideration as well (see Sect. 7.4.3). Sulphur is captured in the fluidised bed, while nitrogen oxide formation can be limited by the conditions of the combustion.

FB combustion has a temperature of up to 900°C, which is well below the ash deformation temperature. As a result, the separation process has to remove solid particulate matter from the hot gas flow. Various technologies for hot gas fly ash removal in PFBC furnaces are discussed below. They are

- cyclone separators,
- electrostatic precipitators (ESPs) and
- particulate collectors such as candles, tubulars, bags or cross-flow filters and packed-bed filters.

7.3.2.1 Cyclone Separators

Cyclone separators use the principle of separation by centrifugation. The flue gas, laden with fly ash, enters the cyclone tangentially, creating a downward rotating flow, changing direction as it reaches the bottom of the cyclone. It then leaves the cyclone at the top, free from fly ash (see also Fig. 5.76). The ash particles are pushed outwards by centrifugal force, hit the walls and fall by gravity to the bottom of the cyclone, where they are removed.

Cyclone separators are simple in design and feature high throughputs and removal rates. The removal rate depends on several factors. The higher the inlet speed, particle size and density and the smaller the cyclone diameters, the greater the centrifugal

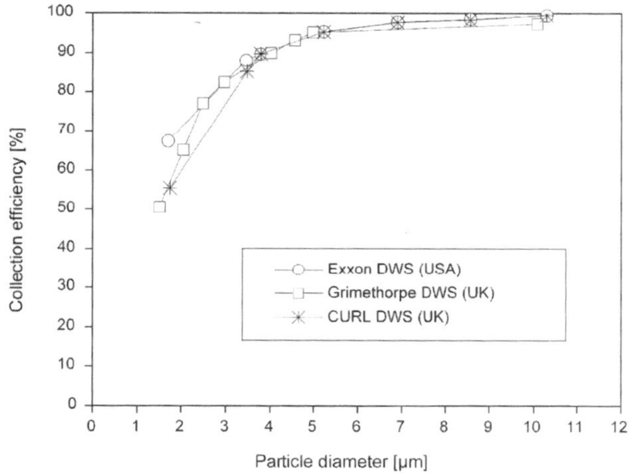

Fig. 7.14 Cyclone collection efficiency as a function of particle diameter (Thambimuthu 1993)

force and hence the removal rate of a cyclone. The removal rate diminishes at higher temperatures due to the increasing gas viscosity (Thambimuthu 1993). Reported total removal rates under pressurised fluidised bed conditions range from 83 to 98% with one-stage and from 98 to 99.6% with multistage cyclone separators (Schiffer 1989).

A cyclone's efficiency also depends on the particle size. It decreases considerably for particles between 5 and 10 μm. Figure 7.14 shows the relationship between the removal rate and particle size as determined by measurement after FBC furnaces in experimental facilities (Thambimuthu 1993). Increasing the removal rate for small particles by means of changing design parameters and operating conditions has only limited effectiveness, because acceleration of the flue gas speed results in increased pressure losses and erosion. A pressure loss of 0.8 bar is assumed for a two-stage cyclone separator after an industrial-scale pressurised FBC furnace (Mustonen et al. 1991).

Single-stage cyclones are used for pre-separation; two or three series-connected cyclones are used for final fly ash separation. State-of-the-art ash separation in the pressurised FBC furnaces currently in use is performed by two-stage cyclone separators. Given a removal rate of 99%, the flue gas shows a cleaned gas particulate concentration of 200–500 mg/kg before entering the turbine, with a maximum particle size of 10 μm (Jansson 1995b). In this case, the gas turbine features a modified design in which the turbine parts subject to erosion have a high wear resistance to the remaining fly ash in the gas.

Cyclone separators are always designed for a defined volumetric flow. If the flow, and hence the inlet speed, decreases, the removal rate also falls. The volumetric flue gas flow should therefore be kept constant in order to ensure a sufficient removal rate, even under partial-load conditions.

7.3.2.2 Electrostatic Precipitators

An electrostatic precipitator (ESP), used to remove particulate matter, is a well-established piece of technology in coal-fuelled power plants. The principles of ESPs have been discussed in Sect. 5.9.2 in the context of pulverised coal combustion. For pulverised coal combustion, the ESP is located at the cold end of the flue gas path in the temperature range of about 150°C. In the temperature and pressure range of pressurised fluidised bed combustion, the removal behaviour of ESPs has only been investigated in laboratories. These investigations revealed a remarkably high ESP energy demand, which has to be attributed to the lower electrical resistance of fly ash at higher gas temperatures.

Although negative results, which would exclude the principle of the use of an ESP under pressurised FBC conditions, have not been reported, there is no information available about R&D projects which continue the work in this field (Weber et al. 1993; Takahashi et al. 1995; Renz 1993).

7.3.2.3 Filtration Separators

Filtering separators use filter media such as granular bulk materials, sintered ceramic material, tissue, felt, non-woven material or shaped pieces of fibre. Particulates deposit on the surface of these and are cleaned off periodically. For fly ash collection in pressurised fluidised beds, packed filter beds made from temperature-resistant granular material, ceramic filter media in the form of bags, and rigid elements in the form of plates or cylinders are the types of equipment investigated today (Wu 2006; Jansson et al. 1996; Sasatu et al. 2001; Toriyama et al. 1999; Newby et al. 2001; Newby et al. 1999; Weitzel and McDonald 1999; Wheeldon et al. 2001; Wright et al. 2003).

- *Packed-bed or granular bed filters*

  In a packed-bed filter, the gas containing the particulates flows through a bulk bed made of granular material, where it deposits the particles. Figure 7.15 shows such a packed-bed filter. The untreated gas containing the particulates from combustion flows into the bed through a concentric duct from above. It is deflected inside the bulk bed and extracted in the form of cleaned gas at the top. As the particulate deposits accumulate in the bulk bed, the bed's flow resistance increases. The polluted bulk material is continuously discharged at the bottom while fresh bulk material is supplied from above.

  Today, packed-bed filters are used to collect particulate matter contained in hot waste gases from industrial furnaces containing abrasive, chemically aggressive and/or sticky dusts. Investigations into the sorption of gaseous pollutants show that packed-bed filters can also be used to remove pollutant gases such as HCl, SO_2 and alkalis (Schiffer 1989).

  Packed-bed filters are efficient at removing particulates from hot gases and satisfy the purity requirements of gas turbines. During investigations under FBC conditions, removal rates of 97–99% and a particulate concentration of

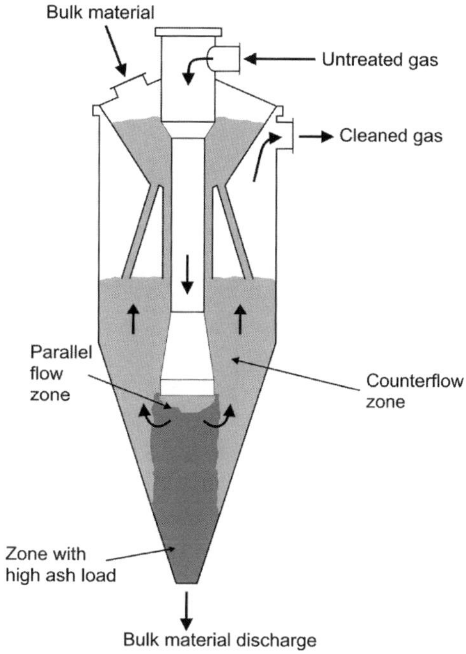

Fig. 7.15 Schematic drawing of a packed-bed filter (Thambimuthu 1993)

4–6 mg/m^3 in the cleaned gas were achieved, with pressure losses of 50–80 mbar. The flow velocities range from 0.1 to 0.3 m/s (Renz 1993). Compared to cyclone separators, it is clear that less pressure is lost when packed-bed filters are used. Even if a cyclone is installed upstream, the total pressure loss reported is about 200 mbar. This is also lower than the rate of the two-stage cyclone (Mustonen et al. 1991).

A packed-bed filter installation needs considerably more equipment, however. This includes vessels for cleaning and for the polluted bulk material, the appropriate lock-hopper system for pressurised operation and a regeneration system for the polluted bulk material (Mustonen et al. 1991).

- *Candle filters*

  Candle filters are named after their shape, which resembles a hose with a closed end. Figure 7.16 shows the installation of a candle filter in a pressure vessel, while Fig. 7.18 shows the design for a 150 MW$_{el}$ PFBC plant. The untreated flue gas containing the particulates enters the lower part of the vessel, flows through the filter candles and is extracted in the upper part of the vessel. The filter candles are held in place both vertically and horizontally by a perforated plate. The untreated gas and cleaned gas zones are separated from each other by the candles and plate. The filter medium is porous and only allows very small fly ash particles to pass through. The maximum size depends on the pore diameter of the filter material. A filter cake forms on the outside of the filter candle and plays a major part in fly ash collection and also increases the pressure loss. The filter cake can be removed by

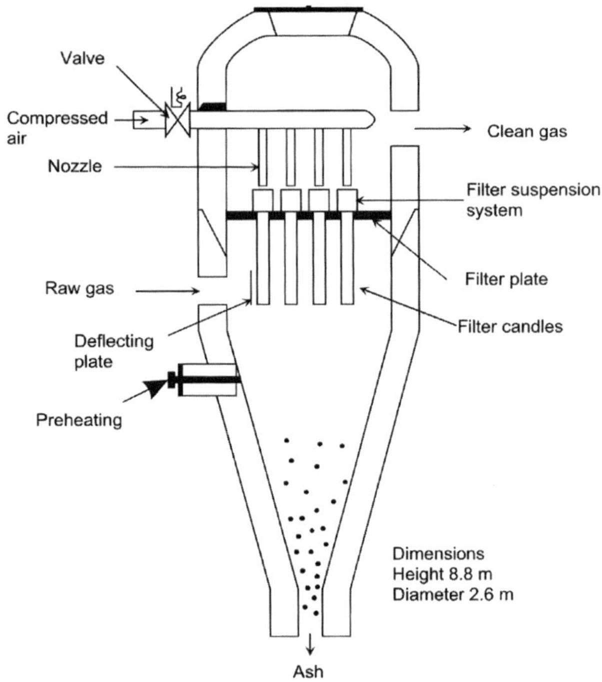

Fig. 7.16 Schematic drawing of a candle filter (Thambimuthu 1993)

means of compressed air pulses from the side of the cleaned gas. The particulates are extracted from the bottom of the vessel.

The candle tubes, closed at their bottom end, are made of a coarsely porous sintered ceramic material. Typical filter elements have an outer diameter of 60 mm, a wall thickness of 10–20 mm and a length of 1–1.5 m. If the material is silicon carbide, the pore diameter ranges between 30 and 100 μm. The filter efficiency can be improved by applying a finely porous surface layer of ceramic fibres to the side of the ceramic material facing the untreated flue gas. These fibres are firmly joined by sintering with the main body. When a two-component system such as this is used, the pore sizes can be as small as 3 μm (Thambimuthu 1993).

The filter shown in Fig. 7.16 was installed after a cyclone separator for long-term tests at an experimental plant. Given an initial particle concentration of 1,000–4,000 mg/kg in the untreated gas, with an average particle diameter of 3–7 μm, removal rates of 99% were reached. This meant that particulate concentrations below 10 mg/kg were achieved in the cleaned gas. The approach speed at the filter candles ranged between 0.03 and 0.07 m/s (Thambimuthu 1993). At an approach speed of 0.07 m/s, the pressure loss was 230 mbar. The basic disadvantage of the configuration shown in Fig. 7.16 is that filter suspension and the cleaning system cannot be scaled up for larger filter units (Renz 1993).

The only practical method for cleaning the filter candles while ensuring a constant flue gas mass flow to the gas turbine is pulse jet cleaning. A method involving constant switching of single filter modules would not be compatible with the operating regime of the gas turbine. What is more, there are no sufficiently reliable hot gas valves available for this purpose (Stuhlmüller et al. 1995).

Even though ceramic candle filters are highly developed, their suitability for use in industry has not yet been proven. At temperatures above 775°C, it gradually becomes more difficult to remove the filter cake as ash properties begin to change and as the operating pressure loss increases (Jansson and Svensson 1997). In different PFBC demonstration projects, ceramic filters have been tested using a partial flow of hot flue gas. At the Tidd plant, ceramic candle filters were installed after one of the primary cyclones. The main problems encountered in the test included ash bridging between candles, difficulties in cleaning the candles and difficulties in draining the filter vessel. The bridging reduced the effective filtration area and, more seriously, led to mechanical failure of the elements under tensile stress. At the Escatrón plant, ceramic candle filters were tested at 750–820°C. The filters were also subjected to ash bridging and mechanical failure (Wu 2006; Wright et al. 2003).

- *Bag filters*
  In contrast to the self-supporting candle filters, bag filters need a support cage to maintain the form of a hose because of their fabric filter. Ceramic or metallic materials are the fabrics considered for the application of bag filters in the temperature range of FBC furnaces. A survey of filter materials which have already been investigated can be found in Thambimuthu (1993).
- *Tube filters*
  Tube filters work according to the same principle as candle filters. The untreated gas, however, approaches from inside the tubes. The design of a tube filter is shown in Fig. 7.17. The gas flow containing the particulates enters the vessel and ceramic tubes from above and exits as clean gas. The particulates are removed as the flue gas radially passes through filter tube walls. The particulate matter collects inside the tube, is removed by compressed air, pushed downwards, collected at the vessel bottom and discharged. The tubes have large internal diameters of 140 mm in order to prevent clogging of the filter media. The tube is up to 6 m in length, which gives the advantage of a compact design (Thambimuthu 1993).

This configuration was used in a circulating PFBC test furnace of 10 MW_{th} and in a stationary PFBC test furnace of 15 MW_{th}. An outlet particulate load of around 3 mg/Nm^3 was achieved in the cleaned gas. At the 71 MW_{el} stationary PFBC furnace in Wakamatsu, Japan, the tube filter technique was tested during demonstrative operation (Sasatu et al. 2001). Following a coarse separation stage in cyclones, the whole flue gas flow was cleaned in two tube filters. Each tube filter was 3.2 m in diameter and 16 m in height. Under trouble-free operation conditions, the design parameters of about 2 mg/Nm^3 for the particulate matter content and 100 mbar for the pressure loss were achieved. Tests were carried out in two phases, with a total of 11,500 h of operation. The longest surviving filter had a lifetime of approximately 8,000 h (Wu 2006). On the whole, however, a series of problems such as filter failure and ash clogging occurred.

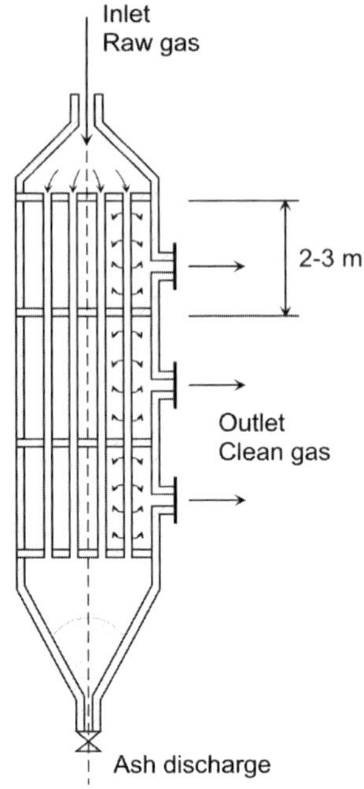

Fig. 7.17 Schematic drawing of a tube filter by Asahi Glass, Japan (Thambimuthu 1993)

7.3.2.4 Comparison of Methods and Techniques

In combined processes with pressurised FBC, the removal of particulate matter such as fly ash from the hot gas is an essential component in the production of hot gas suitable for a gas turbine. In addition to cyclone separators currently in use, other methods are being developed and demonstrated.

Cyclones for fly ash removal are characterised by both high velocities and high pressure losses (Thambimuthu 1993). It is possible to achieve removal rates of 99% using two cyclones, though the requirements of modern gas turbines are not met by doing so. However, cyclone separators are nevertheless used in PFBC furnaces because proof of the operating reliability of the other techniques has not yet been established. Ceramic candle filters work with low approach velocities of less than 0.04 m/s, involving low pressure losses of about 200 mbar. These filters are the most advanced of the various filtration separation methods described above.

Based on several years of investigations, the candle filter technique has been demonstrated at various industrial-scale plants. Tube filters are almost as highly developed as ceramic candle filters. Although they exhibit an acceptable filtration performance in demonstration plants, ceramic candle and tube filters are not yet sufficiently reliable for the advanced power generation demonstration plants. Their unreliability makes operation uneconomic. Thus the continuing development of hot gas filter technology is focussed on improving reliability (Newby et al. 2001). Given the high mechanical and thermal stresses, to attain greater reliability, the filter elements and the material and better constructions and configurations have been highlighted as potential areas for development. Ceramic-based bag filters are clearly not as highly developed as candle filters.

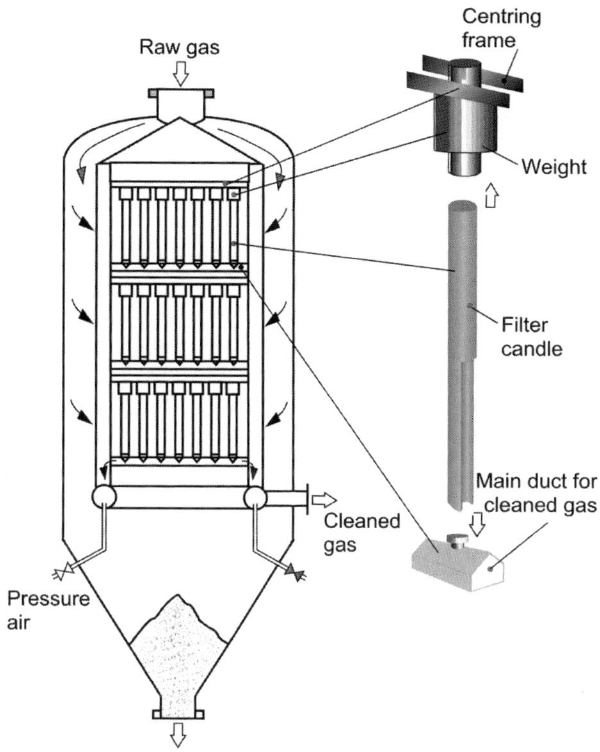

Fig. 7.18 Candle filter of a 150 MW_{el} power plant with circulating PFBC furnace (Bauer et al. 1994; Rehwinkel et al. 1992)

Figure 7.18 shows the design of a ceramic candle filter for a planned 150 MW_{el} power plant with a circulating PFBC furnace (Bauer et al. 1994; Rehwinkel et al. 1992). The flue gas containing the particulates is fed into the filter vessel and candles from above. An impact plate acts as a flow deflector, ensuring a steady flow of solid flue gas across the entire cross-section of the filter.

In the filter unit shown, 1,800 ceramic filter candles are arranged upright on supporting headers on three levels. This vertical mounting has several advantages compared to the suspended mounting. First, the weight load on the ceramic material takes the form of compressive stress and not tensile strain, which is the preferable design for ceramics. Furthermore, the filter candles, with an increasing degree of pollution, are pressed more firmly into their sealed fit. The concentration of outlet particulates in the cleaned gas is less than 5 mg/Nm^3 (Bauer et al. 1994).

Packed-bed filters work at higher velocities (0.2 m/s) and have lower pressure losses of less than 200 mbar. In addition, they allow the removal of other substances which are likely to damage the gas turbine. Although packed-bed filters are not yet as highly developed as candle filters, they are expected to be just as suitable as them in general.

7.3.3 Pressurised Bubbling Fluidised Bed Combustion (PBFBC)

7.3.3.1 State of Development

The technology used in bubbling pressurised fluidised bed combustion is highly advanced. A total of eight PBFBC plants have been built around the world, giving a cumulative installed capacity of 1,125 MW_{el} to date (McMullan 2004; Wright et al. 2003; Wu 2006). Some of the plants initially functioned as demonstration units; today, however, the majority are operated on a commercial basis. The company known as ABB Carbon, now part of Alstom Power, has supplied most of the installations; two units have been built in Japan with Japanese PFBC technology. Overall, the uptake of PBFBC technology has been progressing slowly. Table 7.6 summarises the details of existing PBFBC plants.

Four demonstration plants were built around 1990. They were based on ABB Carbon's P200 module and had a thermal output of about 200 MW_{th}. Two of them are operated by Stockholm Energi in Värtan, Stockholm, Sweden, and one by ENDESA in Escatrón, Zaragoza, Spain. The Tidd plant near Brilliant, Ohio, operated by American Electric Power, was taken out of service in 1995 after operating as a demonstration plant for several years. The Tidd and Escatrón plants were designed for an electrical output of 70 and 76 MW_{el}, respectively, whereas the Värtan plant (two P200 modules) is operated as a CHP power station with an electrical output of 135 MW_{el} and a thermal heat extraction rate of 225 MW. In Escatrón, hot gas filtration using a partial flue gas flow was investigated (Jansson 1995a; Jansson et al. 1996; Jansson 1995b; Martinez Crespo 1995).

In 1993, another plant entered service in Kyushu, Japan. This plant is also equipped with a filtration separator for the whole flue gas flow (Goto 1995; Sasatu et al. 2001). In 1999, a pressurised fluidised bed based on the P200 module and fuelled by brown coal entered service in Cottbus, Brandenburg, Germany (Walter et al. 1997). Based on the experience with the P200 module, an FBC furnace was designed and manufactured for a plant in Karita, Japan. This furnace featured a thermal capacity of 800 MW_{th}. The plant began demonstration operation in 1999 and entered commercial operation in 2001. It features advanced supercritical steam conditions (24.1 MPa/566°C/593°C) and a net thermal efficiency rate of nearly 42% (HHV), corresponding to about 44% (LHV) (Koike et al. 2003; Asai et al. 2004). This represents state-of-the-art PFBC technology (Jansson and Anderson 1999).

Other PBFBC plants have been built by Japanese companies in Japan: the 85 MW_{el} Tomatouatsuma Unit No. 3 built by Mitsubishi Heavy Industries (MHI), which uses a ceramic filter, and the 250 MW_{el} Osaki Plant by Hitachi (Shimuzu and Itoh 2001; Hokari et al. 2001).

The apparent lack of market penetration in Europe and North America is believed to be a result of PBFBC's perceived higher costs and complexity compared to competing systems, coupled with an increasing focus on natural gas fired combined cycle gas turbine plants. As Japan does not have access to substantial gas reserves and is a net importer of LNG, the future market potential is seen as being mainly in Japan. Due to rationalisation of Alstom resources, PBFBC is no longer actively

Table 7.6 Summary data for PBFBC plants currently in service (data from Wu 2006 and additions)

Project/typical feature	Värtan	Escatrón	Tidd	Wakamatsu	Cottbus	Tomatouatsuma	Osaki	Karita
Start-up date	1989	1991	1991	1994	1998	1999	1999	1999
Type	ABB 2xP200 CHP	ABB P200 Cond.	ABB P200 Cond.	ABB P200 Cond.	ABB P200 CHP	MHI	Hitachi	ABB P800 Cond.
P_{el}, net	135.0 MW	79.5 MW	70.5 MW	71.0 MW	74 MW	85 MW	250 MW	360 MW
P_{th}	224.0 MW	–	–	–	120 MW			
Gas turbine	2xGT35P	GT35P	GT35P	GT35P	GT35P	MW-151P	GE F7EA	GT140P
Boiler pressure	12 bar	12 bar	12 bar	12 bar	12 bar		10 bar	12 bar
Bed temp.	860°C	860°C	860°C	860°C	840°C	870°C	865°C	870°C
Gas clean-up	Cyclones 2 stages	Cyclones 2 stages	Cyclones 2 stages	Cyclones + ceramic filter	Cyclones 2 stages	Cyclones + ceramic filter	Cyclones 2 stages	Cyclones 2 stages
Coal	Hard coal, Poland	Brown coal	Hard coal, Ohio	Hard coal	Brown coal, dried		Hard coal	Hard coal
Sulphur[a]	0.1–1.5%	2.9–9.0%	3.4–4%	0.3–1.2%	< 0.8%			0.3–1.2%
Ash[a]	8–21%	23–47%	12–20%	2–18%	5–6%			2–18%
Moisture[a]	6–15%	14–20%	5–15%	8–26%	18–20%			8–30%
Coal feeding	Paste	Dry	Paste	Paste	Dry	Dry	Paste	Paste
Sorbent feeding	Paste with coal	Dry	Dry	Paste with coal	Dry		Paste with coal	Paste with coal
Sorbent	Dolomite	Limestone	Dolomite	Limestone	Limestone		Limestone	Limestone
Live steam pressure	137 bar	94 bar	90 bar	103 bar	142 bar	166 bar	167 bar	241 bar
Live steam temp.	530°C	513°C	496°C	593°C/ 593°C	537°C/ 537°C	566°C/ 538°C	571°C/ 596°C	570°C/ 595°C
Bed height	3.5 m	3.5 m	3.5 m	3.5 m			4 m	3.5 m
Net efficiency,%		36 (HHV)	35 (HHV)	38 (HHV)		41 (HHV)	42 (HHV)	44 (LHV)

[a] As-mined coal

marketed, although support is still provided for existing installations. It seems that the PBFBC technology will only flourish if its use is promoted by the Japanese licensees and their local competitors or if it is utilised in some form of a hybrid cycle (McMullan 2004).

7.3.3.2 Industrial-Scale Configurations

The design of ABB Carbon's P200 module will be described in more detail below, as it is the most widely used PFBC technology and information about its design and operation has been published. Figure 7.19 shows the plant in Cottbus, which is based on pre-dried brown coal (Walter et al. 1997; Jansson and Anderson 1999). It can be divided into three sections – a gas turbine cycle, a steam – water cycle and a fluidised bed.

The combustion air is taken in from the ambient air, pre-compressed in a low-pressure compressor (LP compressor), intercooled and then further compressed to reach the required working pressure for combustion. Intercooling is chosen so that low inlet temperatures of around 300°C can be set in order to cool the pressure vessel. Afterwards, the air is injected into the furnace via the distributor plate.

The flue gas produced by combustion is cleaned in a two-stage cyclone separator. It is then directed, in the form of hot gas, to the high-pressure gas turbine via a coaxial duct, inside which the hot gas flows, while the compressed air flows on the outside. In the high-pressure gas turbine, the hot flue gas is pre-expanded, then further expanded in the low-pressure gas turbine, in order to drive the low-pressure compressor.

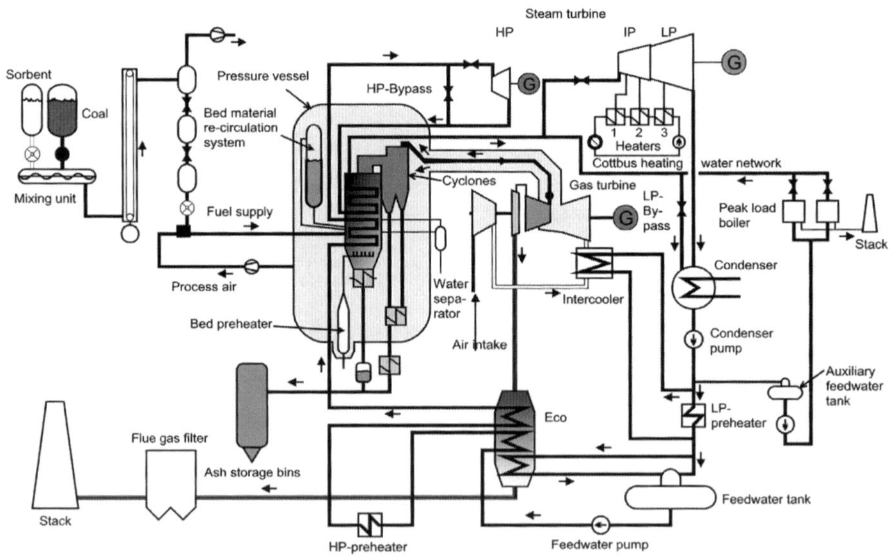

Fig. 7.19 Diagram of the PBFBC power plant in Cottbus (Walter et al. 1997)

The low-pressure cycle, featuring an LP gas turbine and a pre-compressor, is driven at variable speed; the high-pressure cycle, featuring a HP turbine, a HP compressor and a generator, is driven at constant speed. Power output is generated exclusively by the HP gas turbine, which drives the generator via gears. After passing through the two cyclone separators, the particulate matter typically comes to between 200 and 500 mg/kg (Jansson 1995b), with a maximum particle size of about 10 μm. The gas turbine used has been modified and adapted in order to suit the cleaned hot gas from the PFB – to limit the impact of erosion, corrosion and fouling and to achieve a longer lifetime.

A standard gas turbine was modified for low flow velocities by increasing the number of stages (in order to reduce the deflections) and by reinforcing the blade profiles of the parts at risk. This made it more suitable for operation after the fluidised bed combustion furnace. With its pressure conditions and mass flow, the gas turbine chosen determines the design of the pressurised FBC furnace. Since it is too expensive to develop a gas turbine especially for pressurised FBC furnaces, recourse is made to suitably modified existing gas turbines. The compressor mass flow then sets the key values for the design of such a furnace.

The waste heat from the gas turbine can only be used for feed water heating at relatively low gas turbine outlet temperatures in the steam – water cycle. This heat accounts for 10% of the steam produced. The far greater portion of heat in the steam produced in fluidised bed combustion – about 90% – is transferred directly to in-bed heating surfaces and furnace walls and used for vaporisation and superheating. The gas turbine accounts for only a small share of the total power output. This is because the steam – water cooling of the fluidised bed (by means of in-bed heating surfaces, in order to maintain the FBC temperatures) has to stay below 850°C and because this heat is not used in the gas turbine. The output ratio of the gas turbine to the steam turbine is about 1–5.

The pressure vessel in the P200 module in the installed plants is 20 m high and has a diameter of 13 m. In these plants, the cyclones are located to the side of the fluidised bed component. In the Wakamatsu and Cottbus plants, the cyclones are located above the fluidised bed which, with a diameter of 11 m and a height of 32 m, makes the body of the pressure vessel higher and more slender.

The vessel interior is cooled by the combustion air coming from the compressor before it is used for fluidisation and combustion in the fluidised bed. Because of the low temperature, low-alloy steel types can be used for the pressure vessel.

In order to be used in FBC, the coal has to be ground to a grain size of less than 5 mm. Any further preparation depends on the calorific value and the sulphur content of the coal. In the case of the Värtan power station, coal and limestone are mixed with water and pumped into the firing facility by means of a slurry pump. The water content in this coal – water suspension amounts to about 25%. From a process-engineering perspective, wet feeding is less complicated than dry feeding using pneumatic conveying and a lock-hopper system. Coals with higher calorific values, ash contents below 25% and lower sulphur contents require less additives for desulphurisation. If such coals are used, losses in efficiency due to the necessary evaporation heat and the increased volumetric flue gas flow are compensated for by the lower auxiliary power requirement of wet feeding. Wet feeding works without coal drying and pressurisation of the lock-hopper system.

Dry feeding is used for coals with a low calorific value and a high sulphur content, as in the Escatrón plant, because if wet feeding were used the high fuel and additive mass flows in this plant would lead to a reduced efficiency. In the Tidd plant, the fuel is fed in wet, while the additive for desulphurisation is dry.

A high fluidised bed of 4 m and a relatively low flow speed result in a sufficient residence time of about 4 s for the gas in the fluidised bed; the residence time in the freeboard above the bed is also around 4 s. The relatively long residence times, as well as the pressure conditions, favour burnout, resulting in a low carbon level of about 1% in the fluidised bed. This is the same as the level of unburned matter in the extracted ash. It is not necessary to recirculate the ash in order to raise the burnout in PFBC (Schemenau 1993).

7.3.3.3 Control

At partial load, the operating pressure is reduced to the level of the fluidisation speed and the inlet velocities to that of the cyclones, and the excess air has to be kept constant. The operating pressure and volumetric air flow are set by the variable-speed LP compressor in the cross-compound gas turbine, and the HP compressor unit is driven at a constant speed. The speed of the LP compressor is controlled by adjustable inlet guide vanes in the LP turbine (Keppel 1995). The power output of the gas turbine diminishes with a diminishing pressure.

Partial loads in the steam generator and steam turbine are set by means of the fluidised bed height. For this, the heating surfaces are only partly submerged into the fluidised bed, because heat generated by the gas is not transferred as effectively as that generated by the fluidised solid matter in the bed. As a result, heat absorption by the in-bed heating surfaces diminishes and steam production decreases. The flue gas cooling by the free-lying, in-bed heating surfaces (i.e. those portions of the heating surfaces that are not emerged in the bed) causes the gas turbine inlet temperatures to drop, resulting in a decreased efficiency.

So that the ash balance within the system is maintained during partial-load operation, the pressure vessel contains a buffer bed ash storage unit. As the load decreases, this unit removes bed ash from the fluidised bed and, in return, provides stored ash to refill the fluidised bed according to the (pre-determined) rated loads of ash. A pneumatic transport system stores the ash and discharges it. In the Cottbus plant, fuel oil is supplied to the firing unit at partial load in order to keep the freeboard temperature or the gas turbine inlet temperature constant across the entire load range. This also helps to reduce nitrogen oxide loads by means of ammonia injection before the cyclones (SNCR) (Jansson and Anderson 1999).

Experience at the Karita plant has shown that the minimum stable capacity is 40% of the nominal power output, and the load change rates are about 3%/min between a 40 and 90% load and 2%/min between a 90 and 100% load. A cold start from ignition to full load takes 11 h, a warm start takes 4 h and a hot start takes 3 h. When the plant is starting up, the compressor is first of all run by the generator, which is operated as a motor, whereupon the fluidised bed is heated up by natural gas or fuel oil (Asai et al. 2004).

7.3.3.4 Emissions

Table 7.7 shows the emissions for selected PBFBC demonstration plants. Thanks to the low combustion temperature of the fluidised bed, NO_x emissions are generally lower than 200 ppm, but vary widely from plant to plant. As these plants use similar bed temperatures of around 860–870°C, this variation is probably due to the differences in the fuel-N content, fuel-volatile content and other parameters. The Värtan plant burns a bituminous coal with 1.3 wt% nitrogen and 27% volatile matter and has relatively high emissions of 165–191 ppm. At the Osaki plant, however, the uncontrolled NO_x emissions are as low as 14.4 ppm (Wu 2006). As the power output decreases, the NO_x emissions increase, because the heterogeneous reduction of the nitrogen oxides by the fluidised bed ash diminishes due to the lower bed height.

Compliance with emission limits requires further measures. In addition to injecting ammonia in order to control nitrogen oxide (selective non-catalytic reduction), manufacturers suggest a "freeboard" firing stage in order to increase the temperature and stage the fuel. Both methods have been applied in the Cottbus plant (Walter et al. 1997; Almhem 1996). In the case of high uncontrolled emissions, selective catalytic reduction (SCR) is the most effective, but also the most expensive, NO_x reduction technology. For example, the Karita plant has achieved NO_x emissions of 35–42 ppm by using SCR.

The plants can achieve desulphurisation of more than 90% by adding limestone or dolomite, depending on the Ca/S molar ratio. A sulphur retention of 90% is typically achieved at Ca/S molar ratios of 1.8–2.0. As shown in Table 7.7, the SO_2 emissions for these plants vary widely from 5 to 350 ppm. The Värtan plant has particularly low SO_2 emissions due to the low-sulphur (0.1–0.6%) coal burned and a high sulphur retention (96–98%). The Escatrón plant burns a high-sulphur (3–9%) coal but with a moderate sulphur retention (90–95%). This results in relatively high

Table 7.7 Emissions from PBFBC plants in operation (Wu 2006)

Plant	Värtan	Tidd	Escatrón	Osaki	Karita
NO_x without control, ppm	165–191	86–102	120–170	14.4	
NO_x with control, ppm	20–33 (SNCR)				35–42 (SCR)
Sulphur retention, %	96–98	93	90–95	97.7	
Ca/S molar ratio	3.3	2.0–2.2	1.7–2.0		
Ca/S ratio at 90% sulphur retention	2.0	1.8–1.9	1.8		
SO_2 emissions, ppm	5–9		350	7.1	7–36
Particulates, mg/m^3	< 30	18	76	≤ 3.5 (2-stage cyclones + bag filters)	5–15 (2-stage cyclones + ESP)

SO_2 emissions. The Värtan plant meets the German standard emission limits of 400 mg SO_2/Nm^3 at 6% O_2, whereas both Tidd and Escatrón exceed these standard limits by far. This is due to the high sulphur content of the coal used in these plants (Jansson et al. 1996).

Particulate emissions for the PBFBC plants in operation are also relatively low, between 3.5 and 76 mg/m^3. The levels depend on the initial gas cleaning before the gas turbine (cyclones or ceramic filters) as well as final cleaning (ESP or bag filters) prior to discharge from the stack. For example, the Osaki plant employs two-stage cyclones for initial cleaning and bag filters for final cleaning. This plant has achieved particulate emissions as low as 3.5 mg/m^3 (Wu 2006).

7.3.3.5 Residual Material

In order to operate a power plant, it is essential to use, rather than dispose of, any residual material produced. In a PFBC furnace, residual material is produced in the cyclone filters (20–50%), the bag filter after the turbine (2%) and as a result of ash removal from the bed material (45–75%). According to manufacturers' data, the residual material mainly contains calcium sulphate (Ca_2SO_4), additive ($CaCO_3$) and small quantities of quicklime (CaO), sulphites ($CaSO_3$) and sulphides (CaS), in addition to coal ash. After the addition of water, the residual material is self-capturing, and eluted trace elements are scarce. Using the ash presents the same opportunities and problems as ash from atmospheric FBC furnaces (see Sect. 5.11). Examples of options for PFBC ash include using it as filling material in underground constructions or for road substructures (Rogbeck 1996).

7.3.3.6 Operating Expertise

The operational expertise gained at the different experimental plants is compiled and summarised in Stringer (1989). The major problems are erosion and corrosion of in-bed heating surfaces and gas turbines. Extensive investigations at the test plants revealed the need to improve the design of the in-bed heating surfaces, to choose lower flow velocities and to limit the grain size in order to minimise erosion of the surfaces. The corrosion phenomena of the in-bed heating surfaces can be compared to those in atmospheric FBC furnaces. Corrosion can be limited by the choice of material and by preventing air-deficient zones. An efficient particle removal technique is essential in preventing erosion of the turbine. The use of two-stage cyclone separators should help to meet these requirements, even though there is little leeway when there are disturbances. At FB temperatures below 850°C, problems are not expected to arise from corrosion of the gas turbine, though they are likely to occur at higher temperatures.

An overview of the operational expertise gained through use of the demonstration plants is summarised in Wright et al. (2003) and Wu (2006). A series of operational problems arose in all plants, as is quite common when new technologies are introduced for the first time. However, most of the problems were not associated with pressurised fluidised bed combustion. The specific problems of PBFBC furnaces reported are discussed below.

Erosion of In-Bed and Membrane Wall Heating Surfaces

The furnace walls feature a water-cooled membrane design and are lined with various refractories in order to prevent the bed materials from causing erosion. Heat transfer surfaces may be located within the furnace in the form of evaporator tubes, superheaters and reheaters. The evaporator tubes are typically made of low-grade steel with a 1 mm thick flame-sprayed coating (Metco-2 with a sintered bond coating). The high-temperature superheater or reheater tubes are made of high-chromium steel or austenitic stainless steel. They are left uncoated because the intrinsically formed protective oxide layer on the surfaces has a high resistance to wastage (Wu 2006).

In the early plants (Värtan, Tidd and Escatrón), the furnace membrane walls were left uncoated and subsequently experienced various degrees of wastage. In Värtan, those parts of the membrane wall which were exposed (e.g. tubular offsets for inspection openings) showed erosion damage caused by the return flow of the bed material along the wall (Schemenau and Anderson 1992). However, the membrane walls performed satisfactorily following the application of suitable refractory coatings (Wright et al. 2003).

The material performance of the evaporator tubes depended on the steam conditions and the corrosive environment. At the Värtan plant, one of the two units experienced a failure of the Metco-2 coating on the evaporator tubes. The tube bundle was replaced with a new one, incorporating modifications to the bed flow pattern in order to minimise erosion, and alternative coatings were tested. The Escatrón plant also experienced flaking of the coatings in some local areas. However, the Metco-2 coating performed well at the Tidd plant. There was only minor tube erosion, apparently resulting from local flow disturbances near the bottom of the tube bundle. This might be attributed to the lower steam pressure at Tidd, and thus a lower tube surface temperature than at the Värtan and Escatrón plants (Wright et al. 2003).

Overall, the uncoated superheater tubes performed satisfactorily. The Värtan and Tidd plants experienced some wastage in locations exposed to increased particle flow. This can be treated by taking suitable measures such as heat exchanger redesign or coating (Bauer and Marocco 1995). Only minor erosion damage was detected in the Escatrón plant. The significantly stronger erosion in the Värtan plant can be attributed to the different ash behaviour within it (Jansson 1995a).

Sintering

Sintering inside the fluidised bed and in the cyclones causes severe problems. In higher temperature zones in the fluidised bed, bed material sinters and forms agglomerates. These are no longer fluidised and stick to heating surfaces and walls. This interferes with the heat transfer of the fluidised bed and impairs steam production. In extreme cases, fluidisation of the bed cannot be maintained. Improvements to the fuel injection in order to set a homogeneous temperature distribution counteract the sintering processes. Furthermore, uncooled tubes should be avoided in the fluidised bed because these "hot spots" can cause sintered deposits. Such deposits will then accumulate in precisely these "hot spots". Sintering phenomena also occur in the cyclones: deposits form inside the cyclone, then break off during load changes, clogging the ash discharge from the cyclones (Martinez Crespo 1995).

In Escatrón, both plant components – the fluidised bed and the cyclone – are affected by sintering. As a remedial action, the fluidised bed is operated at a lower temperature and power output. Low fluidised bed temperatures prevent clogging due to sintered pieces in the cyclones. At the plants in Tidd and Värtan, problems also arose with limestone, but not with dolomite. No sintering phenomena were detected at the Wakamatsu plant (Käser 1996).

Sintering depends on the composition of the coal ash. The alkalis in the fuel lower the deformation point of the ash particles, which stick together and sinter. The additives used for desulphurisation have a similar effect. Limestone intensifies sintering more than dolomite. However, the reason behind this has not yet been clarified. One possible reason is that lower deformation and sintering temperatures develop when the mixture features limestone with other ash components, such as alkali compounds, than with dolomite (Käser 1996). Another cause may be the different comminution behaviour during calcination. It is assumed that the pressure, with limestone, suppresses calcination so that larger and heavier particles form. These fail to mix properly in the fluidised bed and are therefore more likely to sinter (Weitzel et al. 1996). The SO_2 concentration of the flue gas also appears to have an effect. At high SO_2 emissions and a low desulphurisation degree, the problems which have arisen have been more severe.

The findings collected at the industrial-scale plants in service make it possible to correlate the problems encountered with the alkali contents of the fuel, the SO_2 content of the flue gas and the additive used. As a result, the suitability of fuels can be assessed in advance (Käser 1996).

Gas Turbine Erosion

At several plants, erosive wear of the turbine blades was detected (despite the use of erosion/corrosion-resistant coatings) as well as of the inlet guide vanes. This occurred mainly as a result of plugging of the cyclone ash discharge legs. This, in turn, caused an excessive loading of dust and more damaging coarse particulates to be transferred to the turbine. However, when the cyclone ash system performed without problems, the erosive wear was generally limited. At the Escatrón plant, there was lower erosion because of the relatively soft ash which formed from the black lignite used (Wright et al. 2003). Even though the erosion damage does not cause a premature shutdown, it shortens the service life of the machines and raises the maintenance costs. Filtering separators are absolutely essential in order to prevent lifetime-curtailing erosion and fouling of the gas turbines. On the basis of investigations at the Tidd plant, featuring trouble-free operation of the cyclones, a blade lifetime of more than 20,000 h can be expected (Bauer and Marocco 1995).

The lifetimes of turbine blades and in-bed heating surfaces are estimated at 20,000–30,000 h (Renz 1994; Schemenau and van den Bergh 1993).

7.3.3.7 Comparison of Bubbling Pressurised Fluidised Beds to Conventional Pulverised Coal Firing

Pressurisation of the FBC furnace results in a very compact structure, and hence to reduced expenditure on material. A comparison of material requirements for a 330 MW_{el} power plant with a pressurised FBC furnace to those with a pulverised fuel fired furnace gives a weight difference of more than 40%. Whereas the equipment requirements for coal feeding and ash removal, mechanical equipment and auxiliary systems in both power plant types are more or less equal, the mass of material for the boiler installation alone is cut by more than 60% for PFBC. Although the weight comparison is not the only tool for comparing the expected costs, it explains the reason behind the development of PFBC furnaces (Rehwinkel 1989).

Based on published data on the Cottbus PFBC plant, the specific costs for a 100 MW_{el} plant in condensing operation were about €2,050/kW_{el} in 1999. This is comparable to the cost of other power plant technologies for coal. In the lower output range, the PFBC furnace has an efficiency of about 42% and is thus more advantageous than conventional steam cycles, whose efficiency typically falls below 40%. Cost comparisons for higher outputs show that the capital costs for 350 MW and 800 MW_{el} PFBC power plants are equal to, or lower than, pulverised coal fuelled power plants with conventional steam conditions (Weitzel et al. 1996; Wauschkuhn 1994).

7.3.4 Pressurised Circulating Fluidised Bed Combustion (PCFBC)

Thanks to its advantages, circulating fluidised bed combustion (CFBC) has become accepted as the preferred option to stationary fluidised bed combustion for atmospheric operation (see Sect. 5.4.2). These advantages also hold true in operation under pressure. In spite of this, pressurised circulating fluidised bed combustion (PCFBC) has not yet reached a similar stage of development. No large-scale plants have been built to date, even though extensive investigations have been carried out and complete plans for industrial power plants with capacities of up to 150 MW_{el} have been prepared (Bauer et al. 1994; DOE 2003a). As a result, no operating expertise has been gained in this area.

One essential characteristic of a CFBC furnace is the decoupling of heat release and dissipation (see Fig. 7.11). All the heat is released via the entire furnace. There are no heating surfaces inside the furnace beyond the cooled furnace walls. The heat is dissipated via the furnace walls and an ash fluid bed heat exchanger. The latter cools the ash that was removed in the cyclone.

The necessary temperatures in the fluidised bed are adjusted by recirculating the cooled ash. Depending on the ash recirculation flow, the temperature of the cooled ash lies between 400 and 600°C. The output of the circulating PFBC is controlled by the fuel flow; the ash recirculation flow controls the temperature in the fluidised bed and the flue gas temperature. Increasing the output also causes an increase in the ash recirculation flowrate.

In bubbling PFBC furnaces, by contrast, heat release and heat dissipation take place in the fluidised bed. Although this makes for compact plant structures, it couples combustion, pollutant formation and heat transfer. Separating the combustion and heat transfer functions from the steam – water cycle in circulating FB combustion offers several advantages (Rehwinkel et al. 1993):

- Without having to take the in-bed heating surfaces into account, the flow velocity can be increased to about 5 m/s in order to generate a high cross-sectional heat release rate and a compact construction body. By contrast, the speed of a stationary FBC is limited to 1 m/s in order to prevent erosion on the in-bed heating surfaces.
- In the ash fluid bed heat exchanger, the fluidisation speed can be chosen independently of the combustion air quantity. Given the small grain size of the circulation ash, low velocities of 0.3 m/s are sufficient for fluidisation. As a result, erosion problems do not arise.
- The flue gas temperature at the outlet of the CPFB furnace can be kept constant over the whole load. This is because the output is controlled by the fuel mass flow and the temperature by the ash recirculation flow. In the freeboard, it is possible to set flue gas temperatures of 900–1,000°C, even at low loads and low ash recirculation flows, without impairing the desulphurisation process. As a result, the hot gas temperature always lies at 850°C or above. In stationary fluidised bed combustion, the temperature of the fluidised bed is set by changing the bed height. Because of the low fluidisation speed and the lower mass flow, it is not possible to balance the temperature between the bed and freeboard. The in-bed heating surfaces projecting from the bed at part load cause a further decrease in the freeboard temperature of the stationary PFB. At half load, the temperature ranges around 550°C, compared to 850°C at full load. At an inlet temperature of 550°C, the useful output of the gas turbine falls to zero.
- Any influence of the fuel's grain size on the heat transfer in circulating PFBC can be counteracted by modifying the ash recirculation flow. Depending on the reaction kinetic conditions, the plant can be operated with fine or coarse grains. By contrast, a change in grain size influences the heat transfer in stationary FBC. It is impossible to counteract this effect. This results in changes to the steam production from the in-bed heating surfaces and to the flue gas temperature.
- Air staging is one possible method of reducing NO_x emissions. Using this method, only part of the combustion air is injected into the fluidised bed via the distributor plate. The remaining combustion air is injected further up in the furnace at a later point in time. In stationary PFBC furnaces, in-bed heating surfaces would be damaged not only by erosion but also by increased corrosion in air-deficient zones.

7.3.4.1 Bubbling and Circulating Pressurised Fluidised Bed Combustion: Comparison in a Pilot-Scale Plant

A 15 MW_{el} pilot-scale power plant, the original design of which included hot gas filters based on ceramic filter candles, was run in stationary operation from 1989 to 1991. Afterwards, it was rebuilt for circulating operation. The tests in circulating operation were performed up until 1992 (Rehwinkel et al. 1993).

The bubbling PFB boiler is shown in Fig. 7.20. The plant comprised wet feeding for coal and limestone and a multistage compressor for supplying the system with combustion air. The PFB firing, the ash removal system and the hot gas filters were located in the pressure vessel. Designed as a bubbling PFB, its equipment included in-bed heating surfaces with a 4 m high bed and a freeboard above it, also 4 m in height. Operated as bubbling FB combustion, the height of the bed had to be lowered at part load in order to reduce the heat transfer to the in-bed heating surfaces. The excess ash was buffered in a bed ash buffer storage facility.

In 1991, the plant was retrofitted for a circulating process (see Fig. 7.21). The pressure vessel then contained the circulating PFB firing with a cyclone downstream

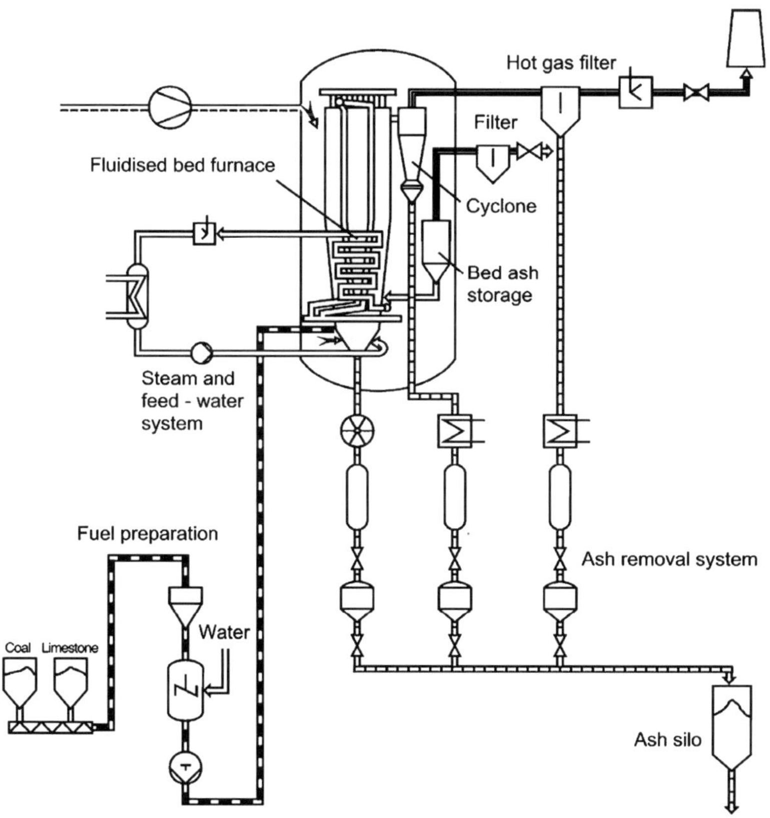

Fig. 7.20 15 MW_{th} test plant with bubbling PFB combustion (Rehwinkel et al. 1993)

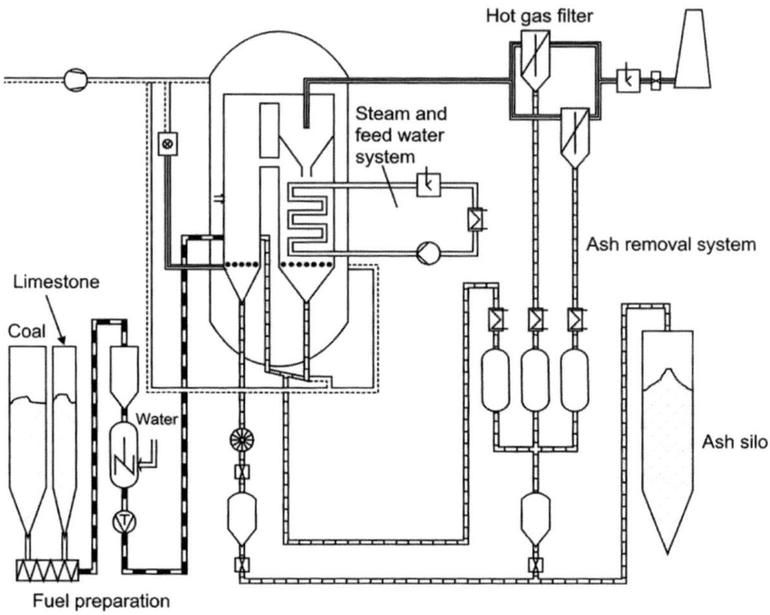

Fig. 7.21 15 MW_{th} test plant with circulating PFB combustion (Rehwinkel et al. 1993)

for ash recirculation and a fluid bed heat exchanger to cool the recirculated ash. The bed ash buffer storage facility, which was required for stationary PFB combustion, was removed. The peripheral systems for coal feeding, ash removal, steam cycle and hot gas filtration were left more or less unchanged. The circulating PFB combustion worked with a stationary fluidised bed in the lower part of the combustion reactor to facilitate the admixture of the coal – water suspension. The primary air entered via the distributor plate. Secondary air was injected above the fluidised bed.

The following cyclone removed the ash. In the ash fluid bed heat exchanger, the sensible heat was transferred to the steam – water cycle and directed towards the first pass at 400–600°C. The ash recirculation rate, i.e. the ratio of the ash recirculation flow to the mass flow of fuel and additive fed to the firing, was about 10–20. Lower ash recirculation rates caused the temperature in the freeboard to rise to levels above that of the bed temperatures. By contrast, an increase in the ash recirculation rate caused the excess temperature in the freeboard to fall.

Figure 7.22 shows the freeboard temperatures, measured in tests, as determined by the load. While in bubbling PFB combustion, the in-bed heating surfaces projecting from the fluidised bed at part load caused the freeboard temperatures to fall. These temperatures remained constant in circulating PFB combustion. At full load, the freeboard temperatures in circulating PFB combustion were about 50°C higher than those in the stationary process.

The CO emissions more or less depended on the temperature. The bubbling FBC had low CO emission levels at full load. However, it exhibited a strong increase at part load due to the corresponding decrease in freeboard temperatures. The

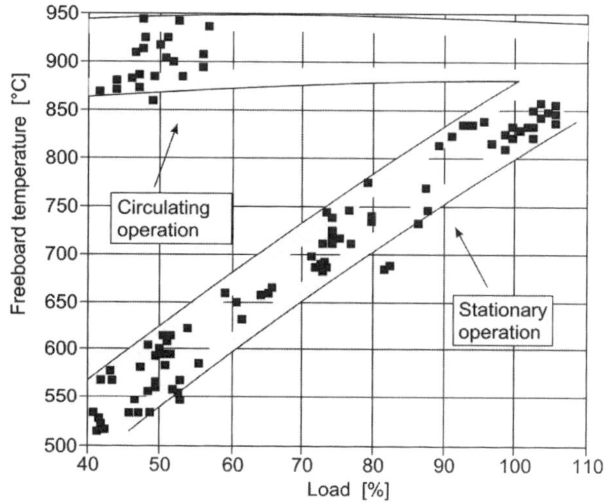

Fig. 7.22 Freeboard temperature as a function of load (Rehwinkel et al. 1993)

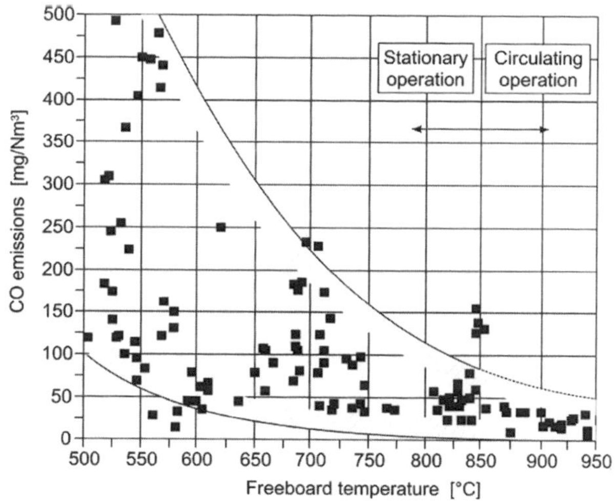

Fig. 7.23 CO emissions as determined by the freeboard temperature (Rehwinkel et al. 1993)

circulating PFBC, at all load ranges, worked at freeboard temperatures equal to, or higher than, the bed temperature. This resulted in low CO emissions. Figure 7.23 shows the CO emissions as determined by the range of freeboard temperatures: below 850°C in stationary and between 850 and 950°C in circulating PFB combustion.

Figure 7.24 shows the NO_x emissions as determined by the air ratio for the stationary process. The crucial influence of the excess air ratio is obvious. Because of the potential corrosion of heating surfaces, an excess air ratio below 1.2 is impossible for bubbling PFB combustion. This results in an increase in NO_x emissions, which range between 200 and 400 mg/m^3. In accordance with German emission limits, this makes secondary reduction measures necessary.

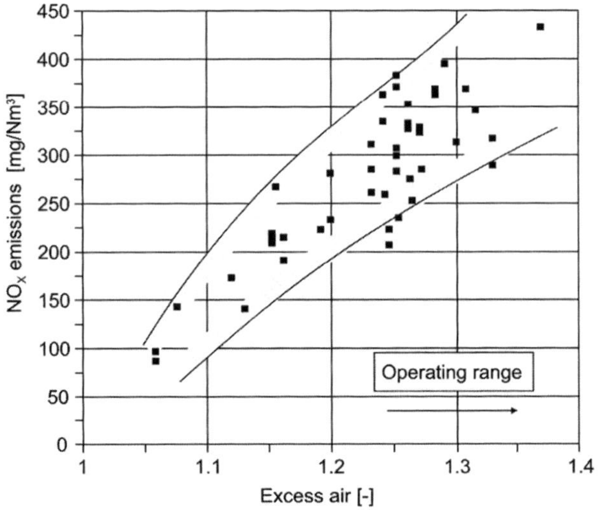

Fig. 7.24 NO_x emissions as a function of excess air, bubbling PFBC (Rehwinkel et al. 1993)

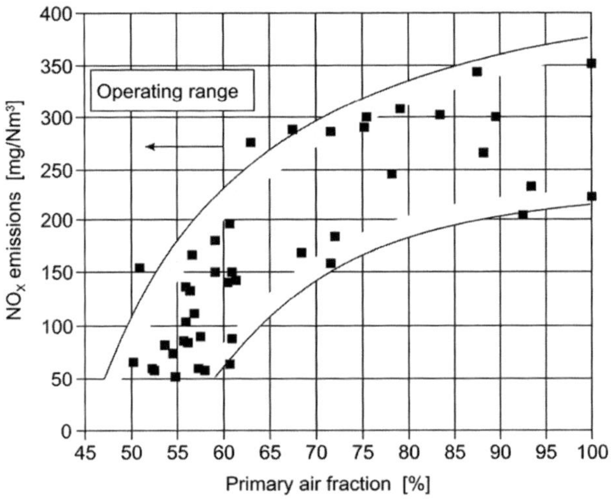

Fig. 7.25 NO_x emissions as determined by the primary air fraction, circulating PFBC (Rehwinkel et al. 1993)

By applying the air staging technique in circulating PFBC – impossible in bubbling PFBC because of the risk of corrosion to in-bed heating surfaces – it is possible to limit NO_x emissions without further NO_x measures. Figure 7.25 plots the NO_x emissions in circulating PFB combustion as determined by the primary air fraction. The operating range lies below a 60% primary air fraction. This produces NO_x emissions between 100 and 200 mg/m^3.

At combustion temperatures below 900°C, typical for bubbling PFB firing, another type of nitrogen oxide develops – N_2O (nitrous oxide, or laughing gas). Tests at a stationary PFBC furnace showed that N_2O emissions are influenced almost exclusively by the temperature. Below 900°C, they rise dramatically as the

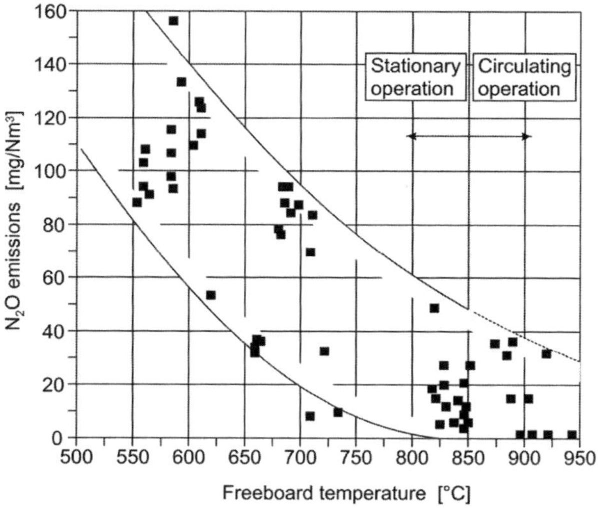

Fig. 7.26 N_2O emissions as determined by the freeboard temperature (Rehwinkel et al. 1993)

temperature decreases. At freeboard temperatures of 850–950°C in circulating PFB combustion, N_2O emissions can be disregarded. Figure 7.26 shows the N_2O emissions of the bubbling and circulating PFB combustion as determined by the temperature for both cases.

The test results obtained at the pilot-scale plant show how much more advantageous circulating PFB combustion is than bubbling (Rehwinkel et al. 1993). These advantages include

- lower emissions,
- higher part-load efficiency and
- a wider range of fuels.

Studies and plans for circulating PFBC furnaces have been undertaken based on expertise gained with the test plants.

Four companies – VEAG, LLB, Steinmueller and Siemens (Bauer et al. 1994) – planned a power plant with pressurised circulating fluidised bed combustion and a capacity of 150 MW_{el}. Raw brown coal was used as the fuel. The planned efficiency of the plant was 45%, with a gas turbine output fraction of 46 MW and a steam turbine power output of 112 MW. The gas turbine inlet temperature was 880°C, the pressure 16 bar. The steam parameters were 190 bar/580°C/585°C (see Fig. 7.27). By awarding the contract for the thermal power plant to Cottbus, the bubbling FBC furnace was chosen over the PCFBC concept.

It was planned to build a 137 MW_{el} PCFBC plant in Lakeland, FL, USA, under the US DOE Clean Coal Technology Demonstration Program. The plant would have used Foster Wheeler's pressurised circulating fluidised bed (PCFB) technology and Siemens Westinghouse's hot gas filtration system. The projected net efficiency was 36% (HHV basis). However, this project was cancelled because technical and economic issues could not be resolved. A subsequent plan to convert the installation

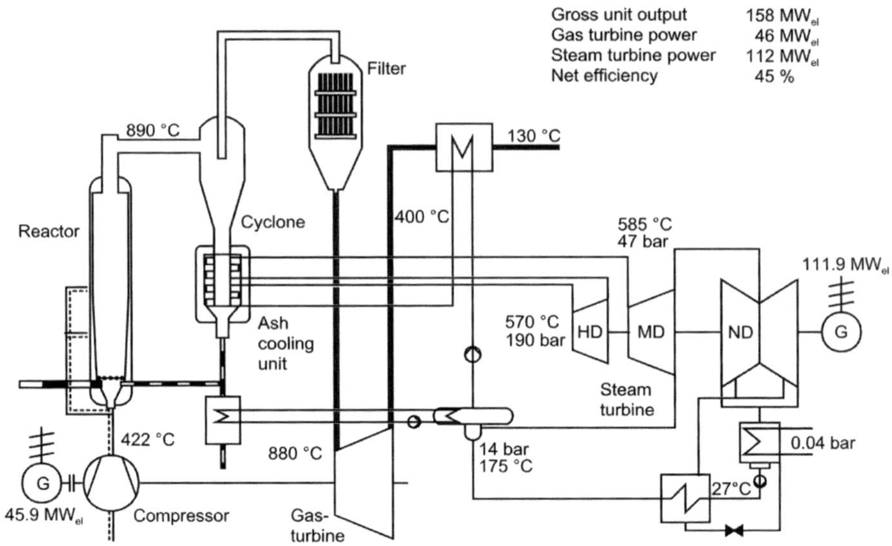

Fig. 7.27 Projected 150 MW pressurised CFBC furnace (Bauer et al. 1994)

into a second-generation PCFBC by adding a carboniser was developed. This is discussed in Sect. 7.3.5 (DOE 2003a; Wu 2006).

7.3.5 Second-Generation Fluidised Bed Firing Systems (Hybrid Process)

The disadvantage of pressurised fluidised bed firing systems is the limited process temperature of the hot gas before it enters the gas turbine (a maximum of about 900°C). Modern gas turbines are capable of operating at higher temperatures, a potential unable to be exploited by the PFB furnace alone.

This potential for an increased efficiency can only be realised when the temperature of the flue gas after the fluidised bed firing and after gas cleaning can be raised to the common gas turbine inlet temperature by firing a "gas turbine compatible" fuel. The usable gaseous fuel can be either an ash-free fuel such as natural gas or a fuel gas produced by coal gasification (see Fig. 7.28). Such procedures to increase the gas turbine inlet temperature are called hybrid or second-generation pressurised fluidised bed combustion processes ("2nd PFBC").

It should be noted that higher gas turbine inlet temperatures also present higher purity requirements for the hot gas resulting from PFB combustion. The particulates collection technique currently used is that of two-stage cyclones. This is insufficient for second-generation PFBC processes, as an increase in the temperature would cause the remaining ash particles to soften, resulting in fouling deposits in the turbine. The increased fraction of fines with its enrichment of alkalis causes the

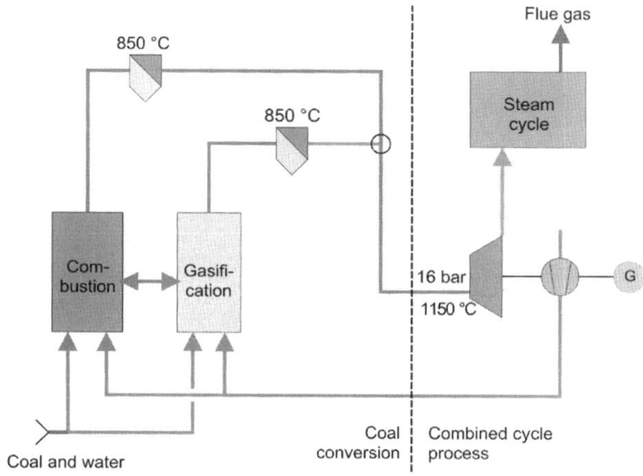

Fig. 7.28 Schematic of a second-generation PFBC

deformation temperature to fall to about 1,000–1,050°C, which is much lower than the deformation temperature of the original coal ash of between 1,150 and 1,250°C. As a result, a hybrid scheme requires a more effective particulate collection system, as well as a hot gas cleaning stage which extracts gaseous alkalis by means of getters. Otherwise, the purity requirements for the gas turbine cannot be met (Jansson 1996).

In the PFBC furnaces currently in use, the output of the gas turbine is one fifth of the overall output. This low figure is due to the cooling of the furnace by steam and water. As a result, an increase in the gas turbine inlet temperature does not cause a corresponding increase in efficiency. A higher efficiency is achieved if the entire fuel is fed to the gas turbine and steam is produced only in the heat recovery boiler. In this case, either the fluidised bed needs to be cooled by a higher excess of air or the heating surfaces have to be air-cooled. The additional gaseous fuels react with the excess oxygen of the hot gas in a special gas turbine combustion chamber. This causes the hot gas temperature to increase to the permissible gas turbine inlet temperature. In this configuration, the output of the gas turbine is about 50–60% of the total output (Rehwinkel et al. 1993). However, cooling the fluidised bed by means of excess air or air-cooled heat exchangers is not state of the art. Those second-generation PFBC systems featuring steam cooling of the fluidised bed combustor which are currently under consideration have a lower share from the gas turbine of the total output. Therefore, the potential for efficiency increases is smaller than for the competing coal-based combined cycles (IGCC, PPCC, EFCC).

Figure 7.29 shows second-generation pressurised fluidised bed combustion as proposed and developed by Foster Wheeler. Coal is fed into an air-blown pressurised bubbling fluidised bed partial gasifier, called a carboniser. The air is supplied in sub-stoichiometric quantities, and the carboniser operates at 900–950°C and 19 bar to produce a low calorific, coal-derived syngas and a char-sorbent residue. After passing through a cyclone and a ceramic barrier filter to remove particulates and alkali vapours, the hot syngas is burned in a specially designed gas turbine

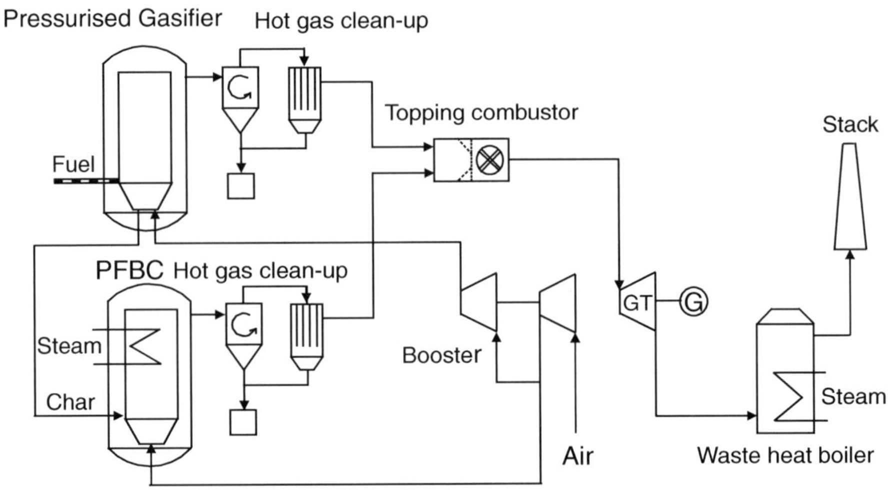

Fig. 7.29 Foster Wheeler's second-generation PFBC concept (Nagel 2002)

combustor called a topping combustor. The gas turbine drives a generator, and its compressor feeds air into the carboniser and the PCFB boiler. The carboniser char is burned in the PCFB boiler using a high amount of excess air to produce superheat and reheat steam for the steam turbine. The hot, oxygen-rich PCFB boiler flue gas passes through its own cyclones and ceramic barrier filters to facilitate the removal of particulate and alkali vapours and the combustion of the syngas in the topping combustor. The efficiency of the design case is 46% (HHV basis); both filters have a temperature of 870°C, the gas turbine firing has a temperature of 1,480°C and the steam is 160 bar and 565°C/565°C. Reducing the temperature of the syngas filter to 540°C and the PFBC filter to 650°C decreases the efficiency to 45.8%. Using double reheating and advanced steam conditions of 275 bar/595°C/595°C/595°C increases the efficiency to 50.5% (Robertson et al. 2005; Robertson et al. 2001).

A second-generation fluidised bed power plant was developed by Foster Wheeler at a pilot-scale facility to test the carboniser, the PCFB boiler, candle filters and the topping combustor. The plant had a coal feed rate of 2,500 kg/h. The tests showed that the original 870°C design temperature of the filter behind the PCFB was not feasible. The limit was determined more by the properties of the ash than by those of the candles. The operating limit was 760°C for bituminous coal, but was about 540°C for petroleum and sub-bituminous coal. Operating at lower temperatures would also meant that an alkali getter would not be required (Wheeldon et al. 2001).

A large-scale demonstration of Foster Wheeler's advanced PCFBC was originally planned in Lakeland, FL, USA, under the US DOE's Clean Coal Technology Demonstration Programme. The plant would have had an output of 240 MW_{el}, a gas-to-steam turbine output ratio of 28% and a net efficiency of 40.6% (HHV basis). A net efficiency of 45% (HHV basis) was indicated for greenfield operation. The design temperature of the gasifier was 950°C, while that of the syngas filter was 650°C. The gas turbine firing temperature was 1,290°C. However, it became clear

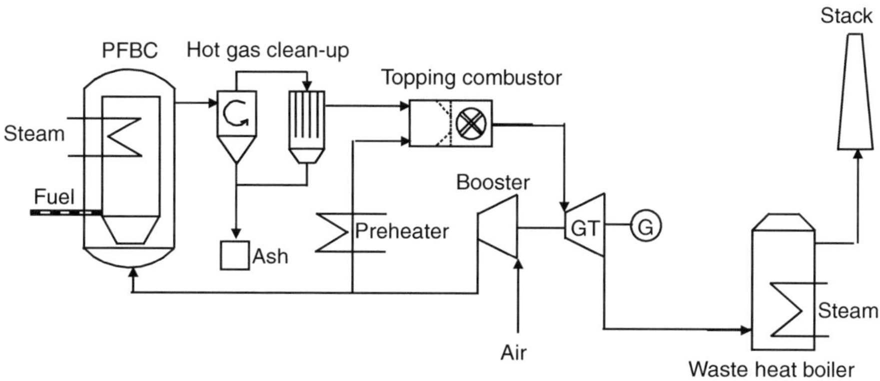

Fig. 7.30 Schematic of a pressurised fluidised bed with staged combustion (Nagel 2002)

that no suitable gas turbine was available. As a result, a decision was taken towards late 2003 to terminate this project (DOE 2003b).

Another possible method of implementing a hybrid scheme is air-deficient operation of the fluidised bed firing followed by flue gas cleaning at the fluidised bed temperature. The addition of air causes the conversion of the unburned gas components and an increase in temperature (Moersch et al. 1999; Nagel et al. 1998).

Figure 7.30 shows a pressurised fluidised bed with staged combustion. Combustion takes place at sub-stoichiometric conditions with air ratios of 0.6–0.8 and temperatures of 800–900°C. Before the particulates are removed by the cyclone and ceramic candle filters, the temperature is lowered to below 700°C. The LCV gas (low calorific value gas with an LHV much less than $10\,MJ/m^3$) produced in the first stage is then combusted in a separate combustion chamber to increase the temperature to 1,200°C prior to expansion in the gas turbine. The major advantage of this process over the configuration shown in Fig. 7.29 is the considerably lower capital cost, since only one reactor is required (Nagel 2002). The process has been investigated in greater detail at a pilot scale (Chalupnik et al. 2001).

7.3.6 Summary

There is little market potential for PBFBC. While the technology has been demonstrated in several parts of the world, hot gas clean-up remains a key issue. PBFBC also appears to offer less scope for efficiency increases and environmental performance improvements than does IGCC. In addition, it is more complex than supercritical PCC and CFBC, which can offer comparable efficiencies. All these factors may have resulted in the current market difficulties. Alstom, the original developer of PBFBC, is no longer actively marketing the technology, although it does maintain support for existing installations. PBFBC is thus "stalled" at present, and its future development and deployment have become uncertain. If this technology is to be taken forward, it is more likely to be in the form of advanced PBFBC/PCFBC, which offers greater scope for efficiency increases (Wu 2006).

7.4 Pressurised Pulverised Coal Combustion (PPCC)

7.4.1 Overview

The aim of pressurised pulverised coal combustion (PPCC) is to produce a hot gas at a high temperature and pressure by the direct combustion of pulverised coal and then to convert its energy into electrical energy in a gas turbine. A downstream heat recovery steam generator (HRSG) and steam turbine extract more electrical energy from the hot gas. The principle of PPC combustion is shown in Fig. 7.31. In the furnace, pulverised coal is burned, producing temperatures of 1,400–1,600°C (which is above the ash fluid point) and a pressure of around 18 bar. Upstream of the gas turbine, it is necessary to remove the molten slag and the gaseous alkali compounds from the hot gas. The heat recovery process, after the hot gas has been cleaned, corresponds to a combined cycle process which is fuelled by natural gas or crude oil. Additionally, though, a $DeNO_x$ and a desulphurisation stage for the flue gases is necessary at the cold end in order to comply with allowed emission limits.

Pressurised pulverised coal combustion, compared to other advanced methods of power generation by coal, has the potential for the highest efficiency. The PPCC process reaches efficiencies of 53% at ISO gas turbine entry temperatures of 1,200°C (see Sect. 7.1 for a definition of the ISO temperature in relation to gas turbines). Further development of gas turbines towards higher turbine entry temperatures are expected to increase the PPCC process efficiency up to 55% (Schuknecht 2003). In

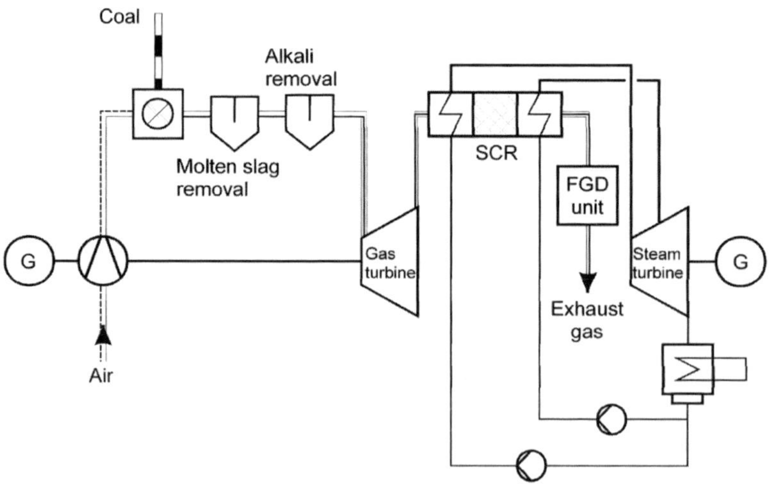

Fig. 7.31 Schematic diagram of a pressurised pulverised coal firing system (Förster et al. 2001)

contrast to technologies that are already demonstrated at an industrial scale (thermal power plants with advanced steam conditions, pressurised fluidised bed combustion and coal gasification), PPCC has not yet reached the demonstration stage. Investigations to date have focussed on slag and alkali removal after the furnace process. Research aiming at implementing the concept of "directly pulverised coal fuelled gas turbines" has been carried out at pilot-scale plants in the USA (Parsons and Byam 1989) and in Germany (Hannes 1996; Hannes 2002; Förster et al. 2005; Förster et al. 2001). Development was discontinued in the USA after approximately 1990, while development in Germany ended in 2005.

All PPCC designs feature in common combustion occurring above the ash fluid temperature, which means part of the ash can be removed early, in its molten state in the furnace. Because the hot flue gas is supposed to serve directly as the working medium of the gas turbine, cleaning of slag particles from the gas is necessary to meet "turbine-compatible" particulates concentration limits. For the conditions of PPCC, a maximum particulate content of 3 mg/Nm^3, a maximum particle diameter of 3 μm and a maximum alkali content of 0.01 mg/Nm^3 STP are desired (Förster et al. 2005).

The designs differ in the gas turbine entry temperature (see Fig. 7.32). The research carried out in the USA focussed on gas turbine entry temperatures of about 1,000°C (Parsons and Byam 1989). The reduction of nitrogen oxide emissions was achieved by running the combustion in two stages. Cleaning of the flue gas was

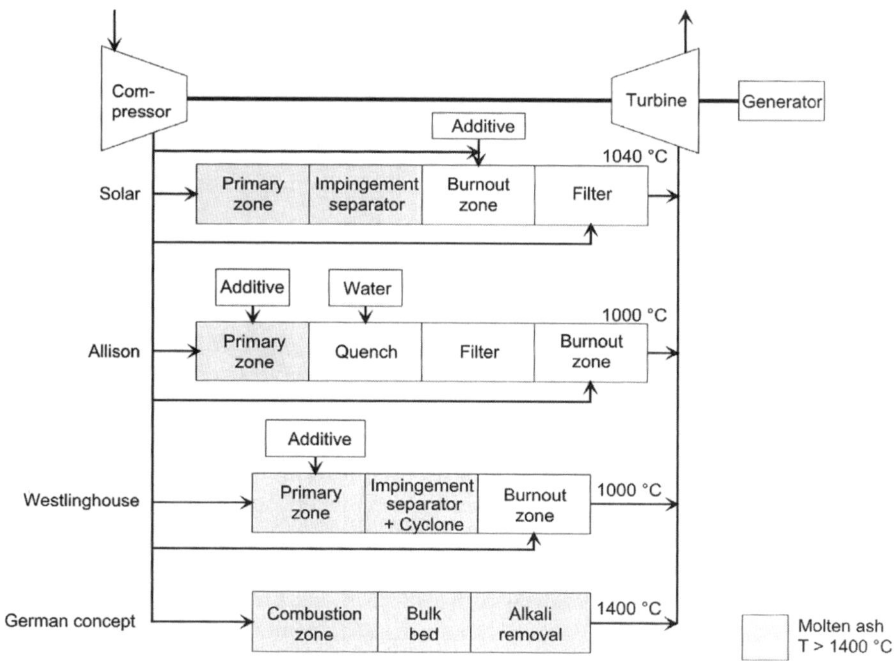

Fig. 7.32 PPCC concepts (Thambimuthu 1993)

carried out either in or immediately after the furnace at high temperatures, or in a range somewhat above the gas turbine entry temperatures, after cooling the flue gas by air or water vapour. The concept favoured in Germany featured a one-stage combustion process and flue gas cleaning at temperatures of 1,400–1,500°C with molten ash removal. The flue gas was conducted to the gas turbine without further cooling (Hannes 1996).

For all PPCC designs, particulates removal at high temperatures and the impacts of gaseous substances forming in the process of combustion at high temperatures are problematic factors. Essentially, these gaseous substances are the alkalis in the flue gas. The alkalis condense in the gas turbine and, in the relevant temperature range, lead to deposition on the turbine blades and to corrosion.

In both the US and the German PPCC designs, hot gas cleaning is at least partly carried out at high temperatures above the ash fluid point. While the German concept uses only high-temperature cleaning, the US concept, with temperatures of about 1,000°C, has the option of cleaning the hot gas both above the ash fluid and below the ash deformation temperature.

Hot gas cleaning, besides slag components, has to remove gaseous gas turbine incompatible components as well. In the following, only the techniques of molten ash removal will be discussed, as the dry ash removal techniques have already been covered in the context of pressurised fluidised bed combustion.

7.4.2 Molten Slag Removal

Given the high temperatures, molten slag removal is the only option considered for PPCC with high gas turbine entry temperatures. Therefore the gas cleaning temperature must be above the ash fluid temperature, which depends on the coal-specific composition of the ash. The fluid temperatures of hard coal ashes range between 1,350 and 1,500°C. Since thin fluid ash with a low viscosity can be removed more easily, the operating temperature of the gas cleaning should be about 100°C above the fluid temperature. Compared to separation techniques for solid ash particles, the removal of molten ash has only had limited development. Investigations have been carried out at pilot-scale plants in Germany and the USA within the framework of developments of PPCC furnaces (Thambimuthu 1993; Weber and Pavone 1990; Weber et al. 1993; Hannes 1996; Hübner et al. 1988; Förster et al. 2001; Hannes 2002; Förster et al. 2005).

Filtration separators using ceramic filter candles have a high efficiency in the temperature range of that after a fluidised bed combustion zone. For the higher gas temperatures and the conditions of PPCC, suitable filter materials are not yet available. Ceramic filters which have as a base aluminium silicate, aluminium oxide and silicon carbide are attacked and destroyed by molten slag (Hübner et al. 1988). Furthermore, it is unclear whether the removal mechanism of ceramic filter candles, based on surface filtration, can also be applied to liquid matter or molten ash. On the one hand, the risk is that molten ash may penetrate through the pores; on the other, molten ash may hinder the gas penetration when small pore cross-sections are used. The filtering effect that the layer of dust deposition has in the lower temperatures of FBC processes does not occur for PPCC (Weber et al. 1993).

Separation by ESP in the temperature range of PPCC is not possible because the conductivity at such temperatures means a corona cannot form on the spray electrodes (Förster et al. 2001).

Investigations into molten ash removal focus almost exclusively on mass force separators. Among these, inertia and centrifugation separators are the most suitable technologies, since they are much more efficient than gravitation separators. Mass force separators only play a minor role nowadays in conventional dust collection technology, removing solid particles from flue gases under moderate temperature and pressure conditions. Their use has declined because they do not collect small particles and so yield only modest removal rates. The case can be different when molten particles have to be removed. They are more likely to merge before or during the removal process into larger, heavier drops and also stick to the wall of the collector, so that the secondary flows interfering with the gas cleaning process are of minor importance (Weber et al. 1993).

Different variants of slag removal were investigated by the project partners of the combined project "Pressurised Pulverised Coal Combustion". For the most part, results were obtained from tests at atmospheric plants and later verified in part in pressurised operation (Weber et al. 1993).

By means of a centrifugation separator (a two-stage cyclone) it was possible to achieve outlet particulate loadings of the cleaned gas as low as $20\,mg/Nm^3$. Figure 7.33 shows the fraction collection efficiencies of the cyclone separator measured under atmospheric conditions. According to the diagram, the particle sizes are very small. In contrast to solid ash particle removal, molten ash removal is much more efficient despite the higher temperatures and the higher gas viscosity. This can be put down to the collision and agglomeration of small droplets, as mentioned above, and the drops deforming and sticking when they impinge upon the wall.

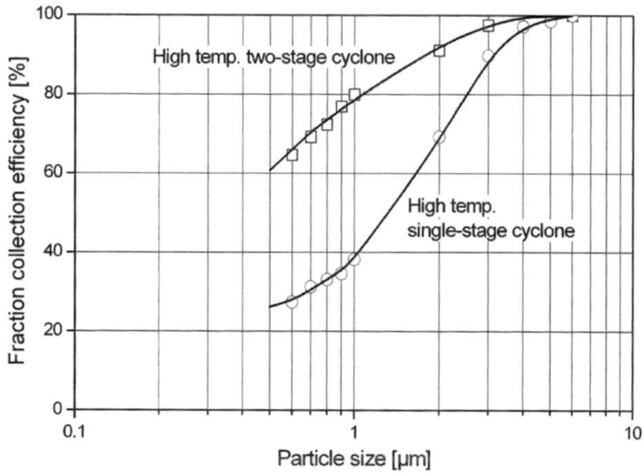

Fig. 7.33 Cyclone removal rate in PPCC as a function of particle size (Weber et al. 1993)

Unlike solid particles, they lose impingement energy through deformation. Also, dispersion of the removed particles is impossible because of the liquid consistency.

From the point of view of material engineering, fewer problems are expected from the use of cyclones because they can be refractory-lined. However, industrial-scale plants with pressurised pulverised coal firing would need a great number of cyclones of a small diameter to achieve high removal efficiency rates – which means the configuration and homogeneous flue gas charging of the cyclones could pose a problem (Weber and Pavone 1990).

Another concept investigated involves a venturi scrubber connected to a cyclone separator. In the venturi scrubber, molten slag is added vertically to the speeding flue gas flow (of a velocity of about 100 m/s) in order to agglomerate the molten ash particles. Therefore a coarser particle spectrum is fed into the cyclone, which results in a higher removal efficiency. Tests carried out discontinuously with slag under atmospheric conditions at temperatures up to 1,600°C showed a reduction in the amount of particulates after the cyclone to 7 mg/Nm^3, as opposed to 32 mg/Nm^3 without a venturi scrubber. The addition of molten slag is assumed to also favour a higher alkali capture in the slag (Hoberg and Gudenau 1998). Problems in applying this concept may arise with liquid ash recirculation, necessary for a continuous process, and the materials' ability to resist corrosion by molten ash.

Although packed-bed or granular bed filters for solid dust particle removal are classified as filtration separators, the removal of liquid matter by such means must be classed as separation by inertia and mass force. Divided into many sub-streams, the untreated gas flows through the packed bed. Each sub-stream is deflected by bulk material, and droplets, due to their inertia, impinge upon it. The packed bed offers sufficiently many flow paths for the slag to drain out. Outlet particulates loadings of the cleaned gas between 50 and 200 mg/Nm^3, depending on the filter velocity, are reported for atmospheric tests (Hannes 1996).

The developments at the PPCC test facility at Dorsten in its last years focussed on molten slag removal in a packed-bed filter (Förster et al. 2005; Hannes 2002; Förster et al. 2001). As a result of the long years' tests, it was concluded that inertia separators, such as the packed-bed filter, are not capable of achieving concentrations lower than about 350 mg/Nm^3 or removing particles with a diameter smaller than 3 μm. The removal rates of the pressurised 1 MW facility were considerably higher than the previously reported atmospheric small-scale test results. Tests with higher velocities, increased bed heights and smaller bed material did not improve the removal efficiency.

This is explained by the electrical behaviour of the flue gas, which can be considered to be a cold plasma at the existing process temperatures, meaning that the flue gas behaves neutrally macroscopically, but a separation due to electrical charges occurs. This fact can be verified by the conductivity of the flue gas, as well as by the charges of the particles. The positive charges of the small particles cause an electrostatic repulsion, so that small particles can no longer agglomerate and therefore cannot be separated.

This knowledge provided the basis for developing a novel concept for fine particulates separation by exploiting the electrical properties. By disassociating the charges from the particles, the resulting non-charged particles can agglomerate and be separated by adhesion on the bed material. Employing a suitable electrical field can enhance the separation of charged particles. Disassociating the charges requires conductive ceramic materials such as those containing ZrO_2. Investigations were carried out to design a local electrical gradient by combining suitable ceramic materials and employing SiC (silicon carbide) electrodes behind the packed bed. As a result of this development, particulates concentrations below 1 mg/Nm^3 with all particles smaller than 1 μm were claimed to be achieved, which would fulfil the requirements of advanced gas turbines (Förster et al. 2005).

7.4.3 Alkali Release and Capture

7.4.3.1 Fundamentals

Alkali Release

Coal ashes contain the alkalis sodium (Na) and potassium (K) in very differing concentrations depending on the coal type. Hard coal shows typical concentrations in the order of magnitude of 0.5–6, with a maximum of 15% by weight of the coal ash. Alkalis are organically bound in the coal matrix or exist in the mineral phase either as simple salts or as complex aluminosilicates. The type of bond has a major impact on the behaviour of alkalis during the combustion process.

Alkalis are released from combustion as gaseous components mainly in the form of NaCl and KCl or else retained firmly bound in the coal ash. Alkalis released during combustion are termed "active alkalis" (see Table 7.8). Active alkalis consist primarily of simple, inorganic salts and organically bound alkalis. Those alkalis bound in clay minerals as aluminosilicates stay almost inert during combustion. While for hard coals, alkalis are more frequently bound in the mineral components, younger coals, such as brown coal, have a higher alkali fraction in the organic components (Singer 1991).

The main parameter of the combustion process influencing the degree of gaseous alkali release is the temperature. The alkali release of a fuel is often correlated with the chlorine content of the coal. This correlation is valid for coal types which contain alkalis in the form of simple salts, such as German hard coals. Although not generally applicable to fuels with organically bound alkalis, it has been observed for German brown coals as well (Oleschko and Muller 2007).

Another possible method to determine the active alkali content is by solution in water or weak acids. Depending on the type of solvent used, this method either

Table 7.8 Classification of alkalis in coal

Active	Inactive
• Simple inorganic salts	Clay and schist minerals:
∘ NaCl, Na_2SO_4, Na_2CO_3	Complex aluminium silicates
∘ KCl, K_2SO_4, K_2CO_3	$Na_2Al_2Si_6O_{16}$
• Organically bonded alkalis	$K_2Al_2Si_6O_{16}$

detects only the simple salts dissociated in the coal moisture or, depending on the strength of the acid, organic compounds as well.

Active alkalis contribute substantially to fouling and slagging in conventional firing systems. In combined cycle processes with coal combustion, gaseous alkali compounds condense during expansion in the gas turbine and can, like aerosols, deposit on the blades, inducing fouling and corrosion.

Alkali Compounds in the Flue Gas and Their State of Aggregation

For a combined cycle process with coal combustion, the type of alkali compounds in the flue gas and their state of aggregation are also of interest, not just the quantity of released alkalis. Alkalis, after release and conversion, are found mainly as low molecular weight compounds such as sulphates (Na_2SO_4, K_2SO_4), hydroxides (NaOH, KOH) or chlorides (NaCl, KCl).

To determine the states of aggregation of the compounds, one way is to employ the vapour pressure, equilibrium states of which are shown in Fig. 7.34 as a function

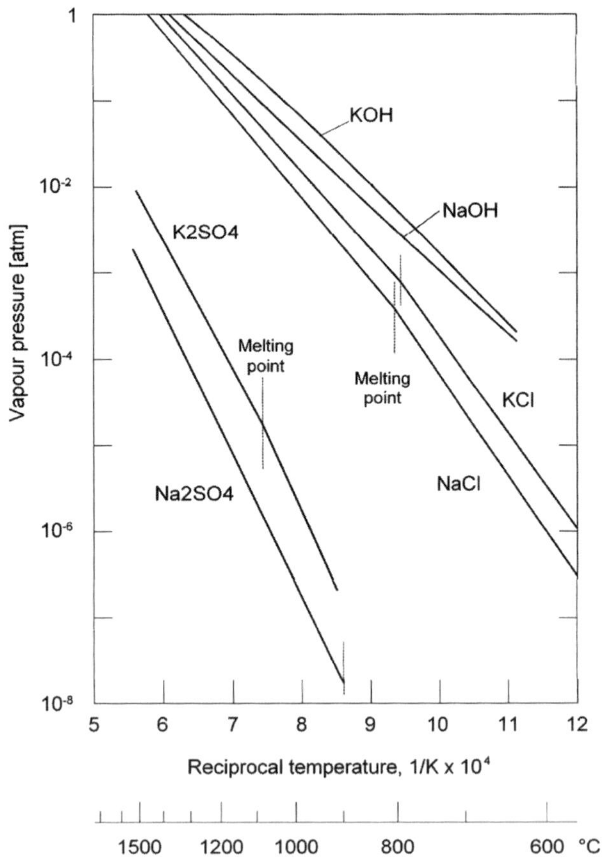

Fig. 7.34 Vapour pressures of the chlorides, hydroxides and sulphates of sodium and potassium (Scandrett and Clift 1984)

Table 7.9 Saturation-phase pressures and concentrations of alkali compounds at 1,173 K (Scandrett and Clift 1984)

		Concentration [ppmw]	
Component	Saturated-state pressures [bar]	Salt	Na or K
Na_2SO_4	2.94×10^{-8}	0.014	0.004
K_2SO_4	2.33×10^{-7}	0.13	0.06
NaCl	2.12×10^{-3}	410	160
KCl	4.86×10^{-3}	1,200	620
NaOH	13.68×10^{-3}	1,800	1,000
KOH	23.3×10^{-3}	4,300	3,000

of the temperature. Table 7.9 gives the corresponding saturation concentrations for the saturated-state pressures at a temperature of 900°C.

For chlorides and hydroxides, the saturation concentrations are on a level far above those usually measured in flue gas; for sulphates, they are below. Thus it can be concluded that alkali compounds in the flue gas present as chlorides and hydroxides are found in the form of vapours, and alkali compounds in the flue gas present as sulphates are found in the form of condensates (Scandrett and Clift 1984).

It is assumed that, in the process of the release of alkalis, chlorides and hydroxides form initially and can then be converted by gas-phase reactions (depending on the composition of the flue gas). At the same time, gaseous alkalis can undergo heterogeneous reactions with solids or liquids, which offers the possibility of capture of the gaseous alkalis.

Of particular importance to the alkali speciation is the reaction with sulphur oxides (SO_x) to form sulphates, according to the following mechanism:

$$2MCl_{(g)} + H_2O_{(g)} + SO_{(g)} + 0.5\ O_{2(g)} \leftrightarrow M_2SO_{4(C)} + 2HCl_{(g)} \tag{7.1}$$

(g) = gas phase, (c) = condensate.

At the high temperatures of combustion processes, sulphur oxides exist as sulphur dioxides, SO_2. The equilibrium of the alkali reactions with SO_2 depends on the temperature, pressure and flue gas composition. Figure 7.35 gives the results of thermodynamic equilibrium calculations for peat based on the minimisation of the Gibbs free enthalpy (Mojtahedi and Backman 1989).

Given fluidised bed temperatures, oxidising conditions and low HCl concentrations, a large fraction of the alkali chlorides react with sulphur oxides to form sulphates, which either condense on particles or in turn form aerosols. The alkalis condensed on particles can be removed by particulate filters. Removal of the condensed sulphate aerosols is more difficult due to their small size (smaller than 1 μm). Filter separators with a deep-layer effect, such as packed-bed filters, seem to be most suitable for this purpose.

The influence of pressure on the gaseous alkalis for fluidised bed temperatures is shown in Fig. 7.36. Higher pressures favour sulphate formation and impede the formation of gaseous compounds. Increasing concentrations of HCl in the flue gas

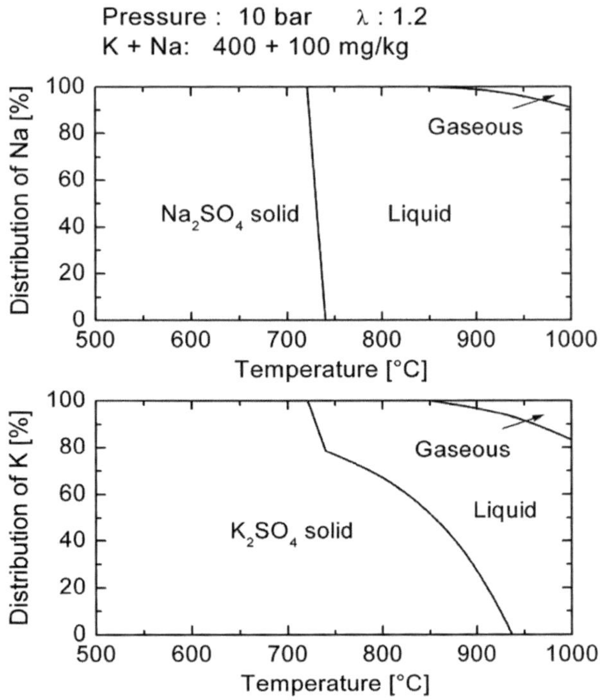

Fig. 7.35 States of aggregation of sodium (Na) and potassium (K) compounds under pressurised fluidised bed conditions (Mojtahedi and Backman 1989)

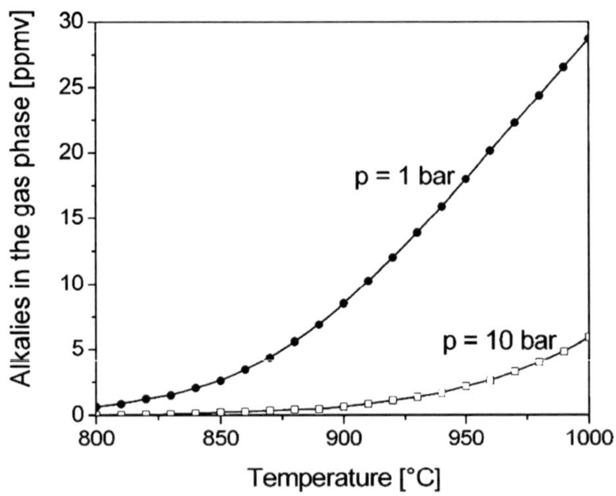

Fig. 7.36 Effect of pressure on alkalis in the gas phase, data from Mojtahedi and Backman (1989)

shift the equilibrium of the reaction towards the gaseous compounds and impede the formation of sulphates (see Fig. 7.37). The alkali content of the fuel itself only plays a minor role in determining the concentration of gaseous compounds (Mojtahedi and Backman 1989).

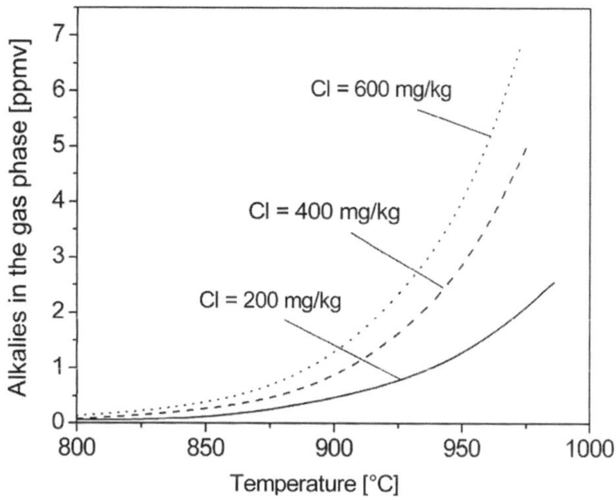

Fig. 7.37 Effect of chlorine content on concentrations of gaseous alkalis, data from Mojtahedi and Backman (1989)

At the higher temperatures of pressurised pulverised coal firing, gaseous chlorides and hydroxides predominate. Elemental gaseous alkalis occur only in trace quantities across the entire temperature range (Hannes 1986).

In conditions without SO_2, for example in gasification, no sulphates but only gaseous alkali compounds form, such as chlorides and hydroxides. Two or three orders of magnitude higher levels of gaseous alkali compounds are thus reported for pressurised fluidised bed gasification in contrast to pressurised fluidised bed combustion.

Alkali Removal

Alkalis can be separated by chemical reaction or removed by physical adsorption on solid or liquid sorbents (known as getters). In the case of chemical sorption, the getters become exhausted, while in the case of physical adsorption, they can be regenerated. In the temperature range of fluidised bed combustion, the basic procedure is physical adsorption, possibly followed by chemical capture. Because of its negative temperature gradient, adsorption is not suitable for higher temperatures, which means chemical capture has to be applied.

Suitable alkali getter materials are quartz (SiO_2), aluminosilicates ($Al_2O_3 \cdot SiO_2$), clay (Al_2O_3) and naturally occurring substances such as bauxite, kaolinite, emathlite or diatomaceous earth, essentially composed of aluminium oxide and silicon oxide (Thambimuthu 1993; Punjak et al. 1989). Table 7.10 shows their composition by weight.

Thermodynamic equilibrium calculations of systems with different getter materials can be used to evaluate their potential for the chemical capture of alkalis. Even though these evaluations are valid only for the state of equilibrium, they are effective to some extent in predicting the practicability of various getter materials.

In principle, a differentiation has to be made between the type of bonding process of the sulphates and the chlorides. In the temperature range of fluidised beds, the

Table 7.10 Composition by weight of additives for alkali capture (Punjak et al. 1989)

Component	Bauxite	Diatomaceous earth	Kaolinite	Emathlite
SiO_2	0.11	0.92	0.521	0.734
Al_2O_3	0.842	0.05	0.449	0.139
Fe_2O_3	0.048		0.008	0.034
TiO_2	–		0.022	0.004
CaO	–		–	0.05
MgO	–		–	0.026
K_2O	–		–	0.012
Na_2O	–		–	0.001

alkali compounds are bound mainly as sulphates, but also as chlorides; at the high temperatures of pressurised pulverised coal firing, binding as chlorides dominates:

(a) Reactions of the alkali sulphates (Scandrett and Clift 1984):
The chemical bonding of the alkali sulphates can occur according to the following reaction mechanisms:

(1) Silica sand:

$$Na_2SO_{4(c)} + SiO_{2(s)} \leftrightarrow Na_2O \cdot SiO_{2(c)} + SO_{2(g)} + 0.5O_{2(g)} \tag{7.2}$$

$$K_2SO_{4(c)} + SiO_{2(s)} \leftrightarrow K_2O \cdot SiO_{2(C)} + SO_{2(g)} + 0.5O_{2(g)} \tag{7.3}$$

(2) Aluminium silicate:

$$Na_2SO_{4(c)} + Al_2O_3 \cdot SiO_{2(s)} + 5SiO_{2(s)} \leftrightarrow 2NaAlSi_3O_{8(s)} + SO_{2(g)} + 0.5O_{2(g)} \tag{7.4}$$

$$K_2SO_{4(c)} + Al_2O_3 \cdot SiO_{2(s)} + 5SiO_{2(s)} \leftrightarrow 2KAlSi_3O_{8(s)} + SO_{2(g)} + 0.5O_{2(g)} \tag{7.5}$$

(g) = gas phase, (c) = condensate, (s) = solid matter.
All bonding reactions of alkali sulphates are strongly endothermic reactions and so require a certain temperature to proceed. Reactions with silica sand need a temperature of above 1,300 K to achieve bonding of the sulphates. For a chemical capture process following fluidised bed combustion, reactions with silica sand therefore do not seem suitable at these temperatures. Aluminium silicates react and bond with alkalis at temperatures above 1,000 K.

(b) Reactions of the alkali chlorides (Scandrett and Clift 1984):

(1) Silica sand:

$$2NaCl_{(g)} + SiO_{2(s)} + H_2O_{(g)} \leftrightarrow Na_2O \cdot SiO_{2(c)} + 2HCl_{(g)} \tag{7.6}$$

$$2KCl_{(g)} + SiO_{2(s)} + H_2O_{(g)} \leftrightarrow K_2O \cdot SiO_{2(c)} + 2HCl_{(g)} \tag{7.7}$$

$$2NaCl_{(g)} + 2SiO_{2(s)} + H_2O_{(g)} \leftrightarrow Na_2O \cdot 2SiO_{2(c)} + 2HCl_{(g)} \tag{7.8}$$

(2) Aluminium silicates:

$$2NaCl_{(g)} + Al_2O_3 \cdot SiO_{2(s)} + 5SiO_{2(s)} + H_2O_{(g)} \leftrightarrow 2NaAlSi_3O_{8(s)} + 2HCl \tag{7.9}$$

$$2KCl_{(g)} + Al_2O_3 \cdot SiO_{2(s)} + 5SiO_{2(s)} + H_2O_{(g)} \leftrightarrow 2KAlSi_3O_{8(s)} + 2HCl \tag{7.10}$$

(3) Alumina:

$$2NaCl_{(g)} + Al_2O_{3(s)} + H_2O_{(g)} \leftrightarrow 2NaAlO_{2(s)} + 2HCl_{(g)} \tag{7.11}$$

(g) = gas phase, (c) = condensate, (s) = solid matter.

All bonding reactions of alkali chlorides are strongly exothermic reactions, so the capture rates of all of the getters decrease with rising temperatures. Silica sand getters yield very low capture rates. With pure alumina as a getter, substantial conversion takes place only at temperatures below 800 K (530°C). The use of aluminium silicates, however, makes it possible to achieve a capture rate of 99% at 1,000 K (730°C) and above 95% at 1,200 K (930°C). From this it is evident that aluminium silicates are the only practical material for getters for gaseous alkalis (see Fig. 7.38). For the calculations with pressures of 1 and 10 bar, it was assumed that HCl develops exclusively by reaction and that the water vapour content was 3% by volume. Higher water vapour contents favour, while higher HCl concentrations in the flue gas impede the chemical conversion (Scandrett and Clift 1984).

Thermodynamic calculations substantiate that capturing is also possible at a high-temperature range of around 1,400 °C. At such temperatures, the aluminium silicates mentioned before have an optimal capture rate at an Al_2O_3-to-SiO_2 ratio of 1:8. The alkalis are captured in the liquid phase (Willenborg et al. 2006).

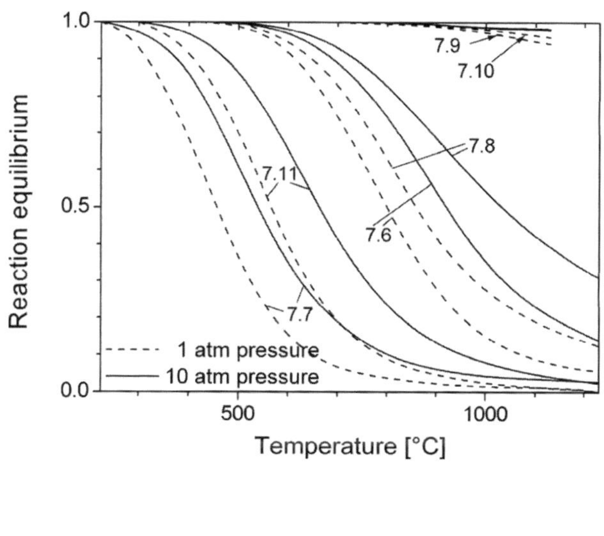

Fig. 7.38 Equilibrium of alkali capture reactions (Scandrett and Clift 1984)

Measurements of Gaseous Alkali Compounds

When comparing concentrations of gaseous alkalis, attention must be paid to the measurement method as well as to the measurement location. The ability of each of the measurement methods to detect condensed aerosol particles as well as gaseous components, or only gaseous components, must be known. If the measurement method detects the sum of all the gaseous, liquid and fine particles portion of the solid alkalis, the filter system has the task of separating the liquid and solid alkalis to exclusively feed the sampling setup with gaseous alkalis. While the filtration of condensed and particle-bound sulphates is possible, it is more difficult to remove aerosols, and hence errors of measurement can occur. In the worst case, when the filtration is insufficient, the solid alkali compounds present in small particles are also measured.

Measurements of the gaseous alkalis under fluidised bed conditions are usually carried out behind filters, so that most of the sulphates have already been separated and only gaseous compounds are detected. At higher temperatures, gaseous chlorides and hydroxides predominate instead of liquid sulphates. Measurements obtained at combustion plants are influenced by the alkali release and gas-phase and heterogeneous capture reactions with bed, fly and liquid ash.

Concentrations are given either per unit mass in mg alkalis/kg flue gas or per unit volume in ppmv.[1,2] Wet-chemical measurement methods detect the ions in the solvent, while on-line measurement methods detect alkali atoms or alkali ions in the gas (Bonn 1996). Results of on-line alkali concentration measurements using simultaneous surface ionisation (SI) and excimer laser-induced fragmentation fluorescence (ELIF) have been compared for a PFBC (Monkhouse et al. 2003). Both methods measure in real time but are complementary in that ELIF can discriminate towards gas-phase alkali species, whereas SI detects alkali both in the gas phase and on aerosol particles (Monkhouse 2002).

Mass concentrations can be given either as pure alkali metals or as compounds. For a comparison and the conversion of concentrations per unit mass and per unit volume, it is necessary to know the compound used as a basis as well as the flue gas composition. Usually, however, the relevant data cannot be found in the literature.

By way of example: for alkali chlorides at an air ratio of 1.3, the following conversion holds true:

$$\begin{aligned} &1\ \text{ppmv Na} = 1\ \text{ppmv NaCl} = {}^1\!/\!_2\text{Na}_2\text{SO}_4\ \text{ppmv} \\ &\sim 0.75\ \text{mg Na/kg}(0.75\ \text{ppmw}) \sim 1.9\ \text{mg NaCl/kg} \sim 4.6\ \text{mg Na}_2\text{SO}_4/\text{kg} \end{aligned} \tag{7.12}$$

$$\begin{aligned} &1\text{ppmv K} = 1\ \text{ppmv KCl} = {}^1\!/\!_2\text{K}_2\text{SO}_4\ \text{ppmv} \\ &\sim 1.27\ \text{mg K/kg}(= 1.27\ \text{ppmw}) \sim 2.42\ \text{mg KCl/kg} \sim 5.65\ \text{mg K}_2\text{SO}_4/\text{kg} \end{aligned} \tag{7.13}$$

---

[1] ppmv = volume parts per million

[2] ppmw = mass parts per million

7.4.3.2 Alkali Emissions from Combustion

Most published research into alkali emissions from pressurised fluidised bed combustion has been carried out at a temperature range of about 800–900°C. Only a limited number of measurements have been reported for the conditions of high-temperature pressurised pulverised coal combustion.

Measurements taken after a filter unit under fluidised bed conditions reveal a great scattering of values, from several μg/kg up to some mg/kg. These differences can be put down to the different fuels and test parameters, in particular the temperature. The measurement technology for gaseous alkalis also contributes significantly to this degree of scattering.

To give an example: for peat, a young fuel type, reports give flue gas sodium concentrations between 0.2 and 1.5 mg/kg and potassium concentrations between 0.07 and 0.4 mg/kg. These concentrations increase with rising temperatures and decrease with pressure. They are also dependent on the additive for sulphur retention, with dolomite showing lower concentrations than limestone. These concentrations mean conversion rates of the fuel alkalis of 1–4% for sodium and 0.2–0.5% for potassium. In the case of a hard coal, the flue gas sodium concentrations ranged from 0.01 to 0.06 mg/kg and the potassium concentrations from 0.02 to 0.07 mg/kg – even though the sodium and potassium contents of this hard coal were equal to or higher than those of peat. Thus the resulting conversion rates are 0.01–0.05% for the sodium in the fuel and 0.01–0.02% for the fuel potassium (Hippinen et al. 1991).

The consideration of the thermodynamics of the reactions of alkali chlorides and hydroxides in forming sulphates in Sect. 7.4.3.1 (Eq. 7.1) suggests that removal in the temperature conditions of pressurised fluidised bed firing should be complete. Tests, however, did not substantiate this. The assumed cause was the presence of HCl, which impedes the formation of sulphates but favours the formation of gaseous chlorides which cannot be removed in the filter (Thambimuthu 1993).

Measurements by optical laser, which only detects gaseous alkali chlorides and hydroxides, show big differences for different coal types. Measured after a cyclone filter for a brown coal type, the concentrations of sodium in the gas phase amounted to 4–5 ppmv and for potassium to 3.5–4 ppmv, whereas for hard coals, the concentrations of sodium and potassium were 30–70 and 10–30 ppbv, respectively. This variability corresponds to the range of dispersion of other measurements. The addition of kaolin reduced the flue gas concentrations by a factor of 2–3 due to capture by the increased amount of aluminosilicate. The use of the hard coal ash as the bed material for lignite combustion resulted in a decrease in concentrations by one order of magnitude. For both coals, potassium was more effectively captured than sodium (Gottwald et al. 2001).

Research into the in situ capture of alkalis in fluidised bed combustion has been carried out at various experimental plants. At one pressurised fluidised bed test plant, combustion tests using a brown coal determined the influence of different bed materials on the alkali concentrations after a two-stage cyclone. With alumina, which consists of 99% aluminium oxide, a reduction ratio of nearly 90% was achieved. The capturing mechanism is based on reactions of alumina with the silicon compounds in the coal, converting vaporous alkali chlorides and alkali sulphates into high melting point alkali aluminium silicates. Using silica sand, the capture rate was lower. However, the test results show that, even with alumina as an additive, the limit of 0.024 mg/kg was usually exceeded by one order of magnitude (Radhakrishnan et al. 1986). Further investigations with additions of bauxite and kaolinite at another PFBC plant revealed that both materials are suitable for the capture of alkalis. Although tests with a high-alkali young brown coal and with an older brown coal with sodium added to it achieved capture rates of up to 90%, the concentrations of gaseous sodium compounds were in the order of just below 1 mg/kg – thus significantly higher than the previously mentioned limit (Mann and Ludlow 1997).

At the pressurised fluidised bed demonstration plant in Wakamatsu, alkali concentrations of approximately 0.03 mg/kg were detected in the flue gas after a ceramic filter unit, using different hard coals at fluidised bed temperatures of 800°C (Daijou et al. 1997).

Aho (Aho et al. 1995) carried out systematic tests at an electrically heated entrained-flow reactor at particle temperatures between 1,000 and 2,400°C. The gaseous release of sodium was higher than that of potassium for two hard coals and one lignite. In these tests, the conversion of the sodium was several percent at temperatures somewhat above 1,000°C and between 25 and 50% at higher temperatures (see Fig. 7.39). This data shows that the alkali concentrations, even at low temperatures of around 1,000°C, are at least two orders of magnitude higher than those allowed for gas turbines.

The alkali release was studied during comprehensive investigations by Reichelt at a pressurised entrained-flow reactor (Reichelt 2001). As in the investigations by Aho, the reactor was operated differentially according to the entrained-flow principle. This means that the fuel mass flow is very small in comparison to the volumetric gas flow. The volumetric gas flow, with an oxygen concentration of 6%, is preheated to reactor temperature; the slight loading of the fuel, of $6\,g/m^3$, causes minor changes in the concentration in the reactor (i.e. a differential mode of operation). Because of the small fuel load and the corresponding low ash concentration, it is assumed that, in contrast to industrial plants, capturing of alkalis in the solid phase is suppressed to a great extent. For two hard and two brown coals, the gaseous chlorides and hydroxides were detected by means of optical lasers. The results reveal rising alkali emissions with increasing temperatures and falling emissions with increasing pressure. The alkali concentrations presented for a hard coal, at a reactor temperature of 1,400°C, and atmospheric pressure, correlate with a sodium release of about 50% and potassium release of only 20%.

Investigations by other authors indicate a release of sodium of about 50% and of potassium of about 20% for a hard coal at 1,600°C, whereas at temperatures of 1,800–1,900°C, the studies identified an almost complete release (Wen et al. 1992).

Reichelt (Reichelt 2001) also investigated an atmospherically operated combustion reactor. This reactor was operated integrally, at an air ratio of 1.2, as it is customary for industrial firing systems operating atmospherically. In contrast to the differential tests presented above, reactions of alkali capture in the ash and slag

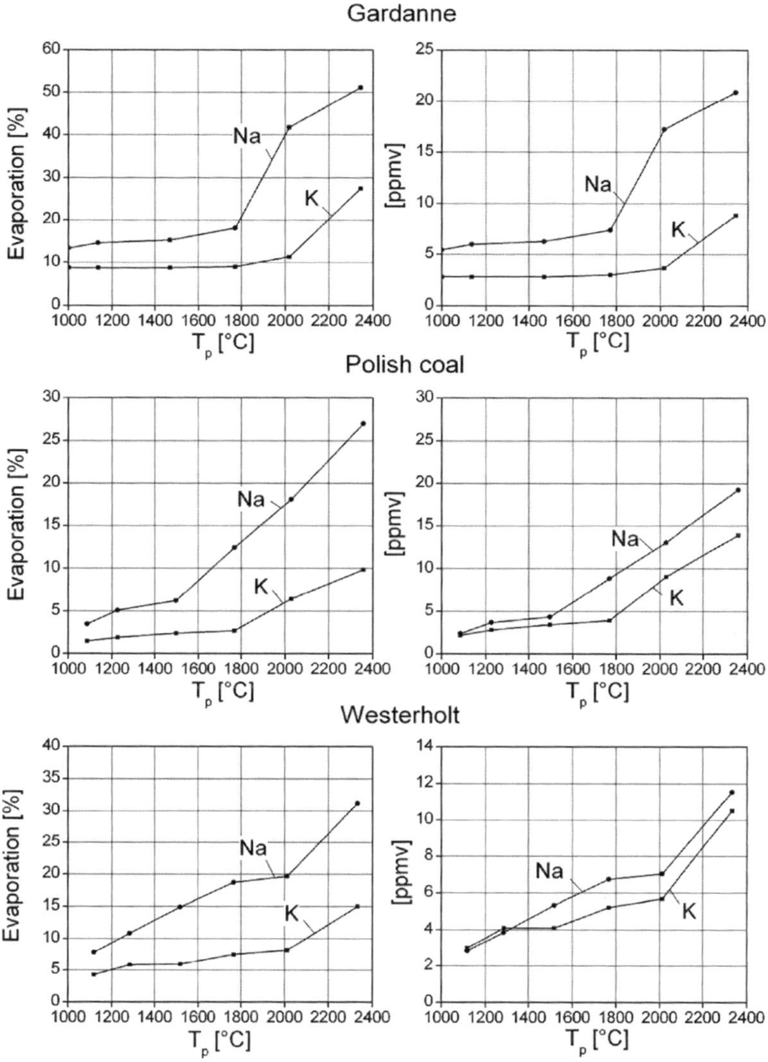

Fig. 7.39 Evaporation of sodium and potassium for different coal types and concentrations in the gas phase as a function of the particle temperature (Aho et al. 1995)

could be observed. This means that a clear decrease of alkali concentrations occurs as the residence time in the reactor increases. Figure 7.40 shows the measured concentrations for different hard and brown coals. The maximum rate of release in the tests for the Ensdorf hard coal, at 1,400°C and a short residence time, was about 10% for sodium and about 5% for potassium. With longer residence times, a high capture in the ash was observed, entailing a reduction of the gaseous alkali concentrations of more than one order of magnitude. The capturing effect in the ash was significantly weaker with brown coal due to the lack of aluminium silicates. By

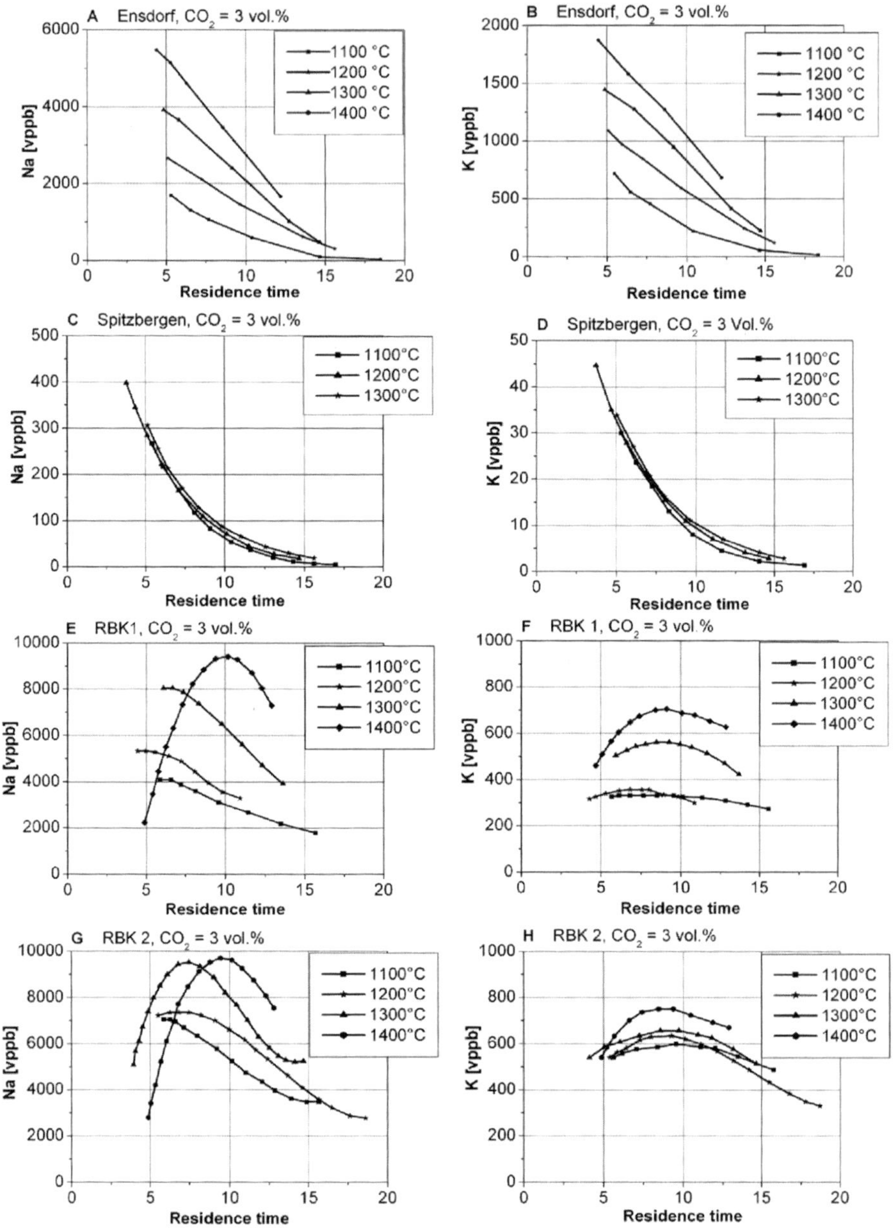

Fig. 7.40 Gas-phase sodium and potassium concentrations for combustion of different coal types (Reichelt 2001)

injecting additives such as alumina and silica in the combustion of the Ensdorf coal, reductions up to one order of magnitude were achieved (Schürmann et al. 2001).

7.4.3.3 Secondary Alkali Removal

For pressurised pulverised coal firing, the removal of alkali compounds is a prerequisite to preventing damage of the gas turbines. The low temperatures of pressurised fluidised bed combustion involve lower concentrations of volatile alkali compounds, and what is more, the turbine is less sensitive to alkali loading because of the low gas turbine entry temperature. The PFBC plants currently in service operate without alkali removal. For hybrid pressurised fluidised bed processes, however, the higher gas turbine entry temperatures entail similar requirements to pressurised pulverised coal firing.

The abatement of alkali concentrations hazardous to the turbine can be achieved by a combination of methods, proposed as follows:

- Leaching of the coal to reduce the contents of alkali chlorides.
- In situ absorption in the furnace by injection of additives and selection of process parameters to raise the rate of capture.
- Absorption in a filter downstream of the combustion. A packed-bed filter, for instance, besides having the function of particle separation can in addition be used for alkali removal (see Sect. 7.3.2).

The most preferable method of alkali removal is to pass the combustion gases through a fixed bed of non-volatile inorganic solid sorbents. Potential sorbents should show a high temperature stability, fast sorption kinetics and a high loading capacity. Kaolin, bauxite and emathlite have been identified as suitable sorbents for gas-phase alkalis. The capturing mechanism can be based upon the principle of physical absorption and/or upon a chemical reaction.

In the temperature range of fluidised bed firing systems, the most effective removal process is physical absorption of condensed alkali components followed by chemical bonding. In contrast, given the high temperatures, the chemical reactions are most effective in pressurised pulverised coal firing. A chemical reaction consumes the getters, while exclusively physical absorption allows their regeneration. As physical absorption is a reversible reaction and alkalis could be released by pressure and/or temperature changes, chemical reactions are the preferred option.

Studies into the suitability of potential getter materials have for the most part used synthetic gases at temperatures in the range of fluidised beds (Punjak et al. 1989; Uberoi 1990; Turn et al. 1998). Only recently have results for higher temperatures become available (Oleschko and Muller 2007; Willenborg et al. 2006; Escobar and Muller 2007; Escobar et al. 2008).

In another investigation, using a portion of the flue gas flow from a PFBC plant, different sorbents in a granular bed filter were tested for alkali removal. The purpose of these tests was to develop an alkali measurement technology employing absorption of the alkalis in a bulk bed, with subsequent analysis of the absorbed alkalis. Measurements at bulk bed temperatures of 800–850°C and a pressure of 9 bar showed an 88% removal of the gaseous sodium using activated bauxite and an 86% removal with diatomaceous earth (Thambimuthu 1993; Lee and Carls 1990; Lee and Swift 1991).

At a laboratory-scale PFB furnace with particulate and alkali removal, data was obtained in practical tests with a packed-bed filter (Zakkay et al. 1989). During the tests, the plant was operated at a temperature of between 809 and 871°C and a pressure of between 7 and 9.3 bar. The particle- and alkali-laden flue gas from the PFB furnace flowed into a granular bed filter after an initial cleaning step in a cyclone. The filter medium was of alumina granulates. The measurements showed an average dust removal degree of 97.8–99% and an alkali removal degree of 90–96%. The alkali emissions at the outlet of the granular bed filter ranged between 0.0019 and 0.041 mg/kg for sodium and between 0.001 and 0.031 mg/kg for potassium, which meant most of the measurements were below the limit of 0.024 mg/kg given by gas turbine manufacturers.

Documentation of industrial-scale experience with alkali capture by such a filter arrangement is not available.

Studies into alkali capture at temperatures above 1,300°C are being carried out at a 1 MW_{th} experimental plant within the framework of the joint pressurised pulverised coal combustion project. The subject of the investigations is the capture of gaseous alkalis by the molten coal slag either in the furnace or in the downstream packed-bed filter for molten ash removal. Given that the getter materials silicon and aluminium oxide are the main components of coal ash, it seems reasonable to make use of this getter potential. The main influencing parameters in alkali removal by slag are the temperature and, to a great extent, the residence time (Hannes 1996).

In tests at the 1 MW PPCC furnace, considerable reductions of the alkali concentrations were achieved, supporting the theory of the molten ash removal process. Starting out from an initial 18 mg/m^3, the alkali concentrations were reduced to about 8 mg/m^3 by using inertia separators and reduced further to less than 6 mg/m^3 by fine particulates removal. The downstream alkali removal by non-optimised aluminosilicate sorbents affected another clear reduction to less than 2 mg/m^3. Because vaporous alkalis are present in an ionised state at high temperatures, the alkali capture by means of slag or getters can be improved by the implementation of electrical fields. This has been proven in laboratory-scale investigations at high temperatures of 1,300°C (Hübner et al. 2003). By also using electrical fields, the alkali concentrations in the 1 MW PPCC furnace could be reduced to 0.2 mg/Nm^3 (Förster et al. 2005; Müller et al. 2008).

Accompanying fundamental investigations at lab scale were carried out to determine the alkali retention potential of the coal ash present in the combustion chamber and the liquid slag separator. For this reason, alkali partial pressures over coal ash slags with and without additives were determined by Knudsen effusion mass spectrometry (KEMS). Although alkali vapour reductions of one order of magnitude could be achieved by the addition of SiO_2 or TiO, the vapour pressures are still about 2 orders of magnitude higher than those required by gas turbine manufacturers. Therefore, a separate alkali removal unit is necessary to achieve alkali concentrations in the hot flue gas compatible with these specifications. In further investigations, model sorbents were exposed to sodium-containing gases. Aluminium silicates having an Al_2O_3/SiO_2 ratio of about 1:8 showed the highest sorption capabilities (Willenborg et al. 2006). In laboratory-scale flow channel experiments at 1,400°C, kaolin and silica-enriched bauxite have shown the best potential for sufficiently removing the alkalis. The alkalis are bound in a glass melt state formed during alkali sorption. The total NaCl concentration can be reduced to less than 30 ppbv using kaoline or silica-enriched bauxite. The sorption reaction of potassium is favoured over that of sodium. However, with sufficient water, both reactions take place to equal extents (Escobar and Muller 2007).

Thermodynamic calculations based on the experimental results were used for a first estimation of the risk of hot corrosion to the turbine blading of a PPCC (Escobar et al. 2008; Müller 2008). By expanding and cooling the flue gas in the gas turbine, alkali components can condense and cause corrosion. For corrosion to occur, two conditions are necessary:

- The components have to condense, which means that the blade temperatures must be below the dew point or condensation temperature.
- The dew point has to be higher than the melting temperature, because corrosion is caused by liquids.

Hot gas corrosion (type I) is caused by the formation of liquid Na_2SO_4 above its melting point (884°C). Another type of hot gas corrosion (type II) is caused by the formation of a eutectic melt of $NiSO_4$ and Na_2SO_4 above 671°C. $NiSO_4$ is formed by the reaction of the oxide scale of Ni based alloys with SO_3 according to the SO_3 partial pressure in the hot flue gas. Therefore, the coexistence of Na_2SO_4 and $NiSO_4$ or the formation of liquid Na_2SO_4 is taken as the criterion for the risk of hot gas corrosion.

The thermodynamic stability of the sulphates and other species in the gas turbine was calculated using a three-staged reactor model consisting of a combustion chamber, hot gas cleaning and a gas turbine. The thermodynamic equilibrium was calculated for all stages. In the hot gas cleaning stage, the alkali content of the gas was set to 2 ppbv at 16 bar according to the experimentally obtained values of about 24 ppbv at 1 bar. The ultimate thermodynamic stability of the sulphates depended on the equilibrium calculation of the gas turbine stage.

The results of the thermodynamic calculations are shown in Fig. 7.41. Na_2SO_4 is the first alkali species that condenses – coal contains high amounts of sulphur and Na_2SO_4 is the least volatile sodium species in the flue gas. The condensation temperatures of Na_2SO_4 range from 650°C at 1 bar to 720°C at 16 bar. These temperatures are below the melting temperature. Furthermore, there should be no condensation of Na_2SO_4 on the gas turbine blades at all, because the calculated dew point is lower than the temperature of each blade. Therefore, no hot gas corrosion (type I) should take place. The dew point temperature of $K_3Na(SO_4)_2$, the thermodynamically most stable potassium sulphate, is 20 K lower again, so the corrosion risk is even lower.

The shadowed area marks the region in which both Na_2SO_4 and $NiSO_4$ are stable above the eutectic temperature of 671°C, where hot gas corrosion type II may occur.

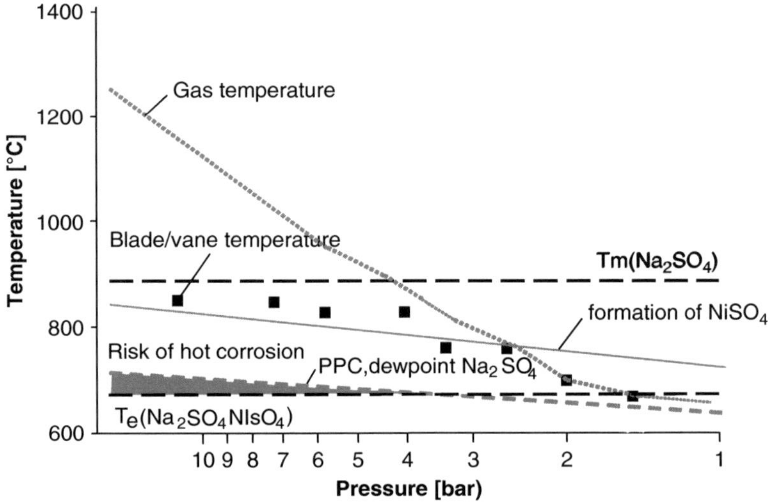

Fig. 7.41 Results of thermodynamic calculations for the estimation of hot corrosion risks (from Escobar et al. 2008, © 2008, with permission of Elsevier)

However, operation of the blades does not occur at these critical conditions (Escobar et al. 2008; Müller 2008).

7.4.4 State of Development

7.4.4.1 Germany, Pressurised Pulverised Coal Combustion Project

Following preliminary investigations into cleaning hot flue gases from a small-scale pressurised pulverised coal combustion (PPCC) furnace at the *Institut für Energie- und Umweltverfahrenstechnik der Universität Duisburg-Essen (IUVT)*, "Institute for Energy and Environmental Protection Technologies, Duisburg-Essen University", the process of PPCC was investigated and developed by several German industry partners from 1989 to 2005. The objective of the joint project was the production of hot gases from a PPCC furnace suitable for gas turbines. The various planned and accomplished steps of development are compiled in Table 7.11. The project was discontinued after 2005, and a pilot-scale plant was not constructed.

Within the framework of the joint project, a test plant using PPCC with a thermal capacity of 1 MW was erected and operated in Dorsten, Germany. The test plant comprised of the furnace, a molten slag and alkali removal unit and a testing section for turbine blade materials. The schematic of the test plant is shown in Fig. 7.42, while Fig. 7.43 shows the combustion chamber with the fly ash and alkali removal stages. The furnace was designed for a temperature of 1,700°C and a pressure of up to 20 bar. The construction was an uncooled multiple-leaf refractory-lined upright chamber with a roof burner and bottom outletting of slag. Slag flowing down the

Table 7.11 PPCC Development Programme (Förster et al. 2005)

Phase	Planned development	Period
1	Planning, construction and putting into service of a 1 MW_{th} test-scale plant in Dorsten, North Rhine-Westphalia	1989–1992
2	Selection and development of suitable materials resistant to molten ash for the construction of the furnace and the molten ash separator	1993–1995
3	Development of the molten ash and alkali separator	Since 1996
3a	Investigation of the various inertia separators	1996–2002
3b	Development of the fine particulates and alkali separation technologies	2003–2005
4	Planning, construction and operation of a pilot-scale plant (10 MW_{th} capacity); operation of the demonstration plant	–

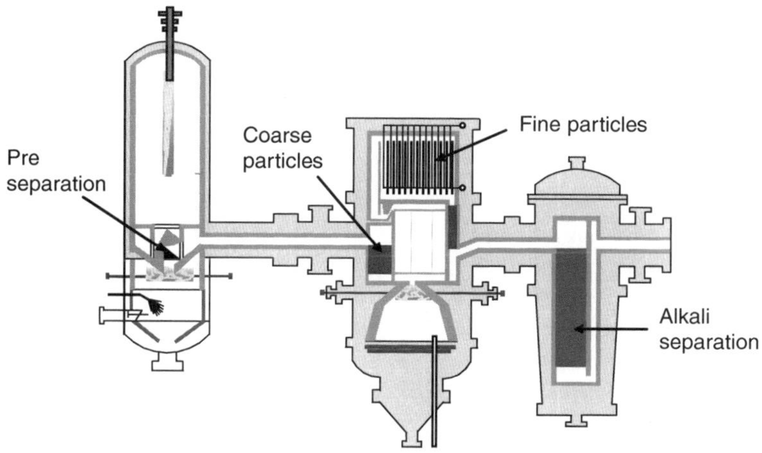

Fig. 7.42 Schematic drawing of the 1 MW PPCC facility (Förster et al. 2005)

chamber walls and over the floor granulated in a water bath. The slag-tap furnace was followed by a flue gas cleaning unit with molten slag and alkali removal. Integrated into the flue gas duct, a testing segment served to investigate the behaviour of different turbine blade materials in the flue gas (Hannes 1986; Preußer and Spindler 1988; Förster et al. 2001; Hannes 2002; Förster et al. 2005).

The results of the investigations into particle and alkali removal have been presented in Sects. 7.4.2 and 7.4.3.3. As a result of many years of development, it became possible, after separating coarse and removing fine particles, to ascertain particle concentrations below 1 mg/m^3 for particles smaller than 3 μm. By doing so, the strict requirements of modern gas turbines could be complied with (Förster et al. 2005).

As far as alkali removal is concerned, it is assumed that, by the primary capture of alkalis in the molten ash inside the combustion chamber and by a secondary cleaning

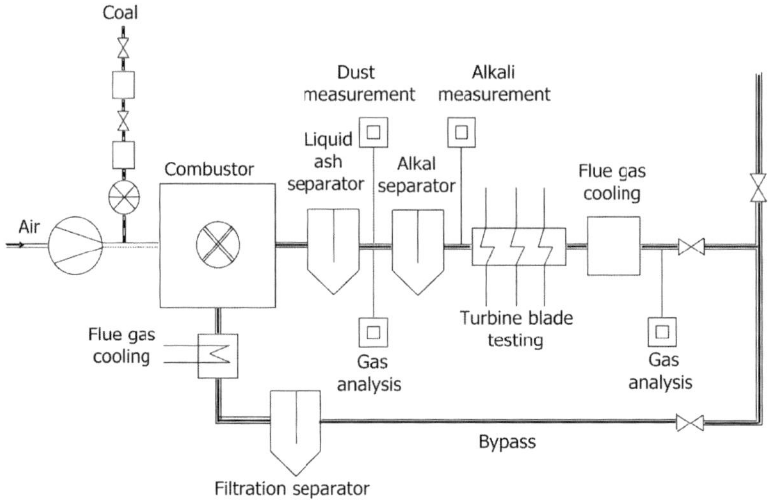

Fig. 7.43 1 MW PPC combustion chamber and hot gas cleaning (Förster et al. 2005)

step, it will be possible to comply with the required limit values of 0.01 mg/m^3 (Förster et al. 2005).

Considerable work within the PPCC project was also directed towards testing and developing materials suitable for the high-temperature zones of the combustion chamber, the molten ash separator and the alkali removal unit. Such materials should have a high temperature stability, resistance and density to withstand the molten slag. Suitable materials are isostatically pressed or fusion-cast ceramics. They are temperature and corrosion resistant and have a high density and low porosity, which prevent the slag from penetrating the refractory lining (Weber et al. 1993). A drawback, however, is the high chromium content, which evaporates at the temperatures used, forming, in particular, compounds with alkalis, which is likely to mean deposits in the gas turbine. Potential solutions to the problem are coatings to reduce the evaporation of the chromium or the use of hafnium oxide based ceramics, which are free from chromium (Förster et al. 2001; Müller 2008).

7.4.4.2 Efficiency Potential and Design of PPCC Furnaces

The joint PPCC project was the framework for the planning of industrial-scale plants with capacities of 150 and 300 MW_{el} (as of 1992). The configuration of the 150 MW_{el} PPCC furnace was a two-line concept, each line with a combustion chamber and a gas turbine. The type of gas turbine taken as a basis was an existing gas turbine, V64.3 by Siemens, with a capacity of 53 MW_{el} at an ISO gas turbine entry temperature of 1,150°C. Each combustion chamber was designed for a thermal output of 159 MW. The combustion chamber had an inner diameter of 3 m and a length of 6 m. The waste heat recovery process was shared by the two lines and was executed as a dual-pressure (steam) cycle, with moderate steam conditions of 70 bar and 535°C. The efficiency of the concept plant was 48%. An electrical output of 300 MW_{el} could be produced by doubling the number of modules. Depending on the cogeneration process quality, it is possible to achieve efficiencies of up to 51% (Leufert 1993).

For the industrial-scale design of a combustion chamber, two variants are distinguished, as in the case of pressurised fluidised bed furnaces. The combustion chamber can be designed either with water – steam cooling, like in conventional atmospheric furnaces, or without cooling for adiabatic operation. From the thermodynamic point of view, the adiabatic combustion chamber has advantages because the total fuel heat can be fed to the gas turbine by this technology. With an air ratio of about 1.7, temperatures are limited to around 1,600°C in the combustion chamber, and gas turbine entry temperatures are achieved by adding additional air. Using adiabatic hot gas production, the resulting gas-to-steam turbine output ratio is 65:35, roughly similar to a natural gas fired cogeneration process. The advantage of water – steam cooling is the thermal design flexibility and gas tightness of the wall. Using this method, cooling of the flue gases is achieved by heat dissipation to the membrane walls, not by setting a high excess air level, as in adiabatic firing. The ratio of the gas-to-steam turbine output then diminishes to 55:45, while the efficiency decreases from 48 to 44%. In practice, though, it will be advisable to opt for a middle course where the wall thicknesses of the refractory linings are reduced by the use of a certain amount of cooling. This way, the plant has a better thermal flexibility; it stores less heat and has shorter heat-up and cooling-down times (Preußer and Spindler 1988; Reichert et al. 1988).

Recent calculations of PPCC systems give overall efficiencies between 53.3 and 55%, using higher gas turbine entry temperatures of 1,200–1,400°C and pressure conditions optimised to these temperatures. Extensive studies using various parameters were carried out with a view to analysing pressure and heat losses from the hot gas cleaning process and the combustion chamber cooling. The results are compiled in Table 7.12 (Schuknecht 2003).

The efficiency of the PPCC process is about 6% lower than that of the natural gas fuelled combined cycle process at the same gas turbine inlet and waste heat

Table 7.12 PPCC cycle calculations (Schuknecht 2003)

Turbine inlet temperature (ISO)	°C	1,200	1,300	1,400
Optimal gas turbine pressure ratio	–	16	17	30
GT exit temperature	°C	583	637	622
GT flue gas mass flow	kg/s	628	632	635
Live steam temperature	°C	565	619	604
Live steam pressure	MPa	18	18	18
Boiler exit temperature	°C	140	140	140
Fuel input (LHV)	MW	632.1	753	855.2
Power – gas turbine	MW	220.7	259	299
Power – steam turbine	MW	131.3	172	194.3
Auxiliary power	MW	13.7	16.1	17.1
Electrical net power	MW	338.4	414.9	476.2
Efficiency (LHV)	%	53.5	55.1	55.7

steam generator conditions. The efficiency loss in comparison to a natural gas fired combined cycle is due to several reasons:

- In the combined process with PPCC, the auxiliary power demand, of about 4% of the gross electric power generation, is about double that of the natural gas fuelled process, due to the power needed for coal preparation, flue gas desulphurisation and $DeNO_x$. This causes an efficiency decrease of 2% in comparison to a gas-fired combined cycle.
- If the heat loss of the combustion chamber and the hot gas cleaning unit is 0.5%, accordingly the efficiency decrease is then about 0.5%.
- The pressure loss in the combustion chamber and the gas cleaning unit decreases the efficiency by approximately an additional 0.5%.
- Water/steam cooling of the combustion chamber further decreases the efficiency. The larger the heat flow dissipated to the steam turbine using combustion chamber cooling, the higher the efficiency loss.
- Another reason for the efficiency difference is the higher flue gas exit temperature needed when using coal as the fuel (Preußer and Spindler 1988; Schuknecht 2003).

7.4.4.3 USA

In the USA, the Westinghouse, Solar Turbines and Allison Gas Turbines companies developed processes with coal-fuelled gas turbines supported by the Department of Energy (DoE). While Solar and Allison wanted to provide plants with a capacity of 20 MW_{el} for industry, Westinghouse focussed on the power plant sector, with capacities of 200 MW_{el}.

The principal configurations of these variants were shown previously, in Fig. 7.32. The gas turbine entry temperature of all of the process variants was only around 1,000°C, which limits the efficiency. The reasons for developing these processes was cost reduction (in comparison to conventional power plants) and the ability to construct small-capacity units using coal as fuel (Parsons and Byam 1989).

The processes were investigated at pilot scale up until 1993; publications on newer investigations do not seem to be in the public domain. It seems reasonable to assume that, due to operational problems (especially with hot gas cleaning), PPCC development has been discontinued in the USA. Publications in the USA to do with new coal-firing concepts do not involve PPCC, which supports this assumption.

7.4.4.4 Westinghouse

The Westinghouse Company carried out investigations at a furnace with a capacity of 3.5 MW_{th}. The schematic diagram of the furnace is shown in Fig. 7.44. The process concept is based on two-stage combustion, molten slag removal and low gas turbine entry temperatures of 1,000°C, controlled by a high excess air level in the second combustion stage.

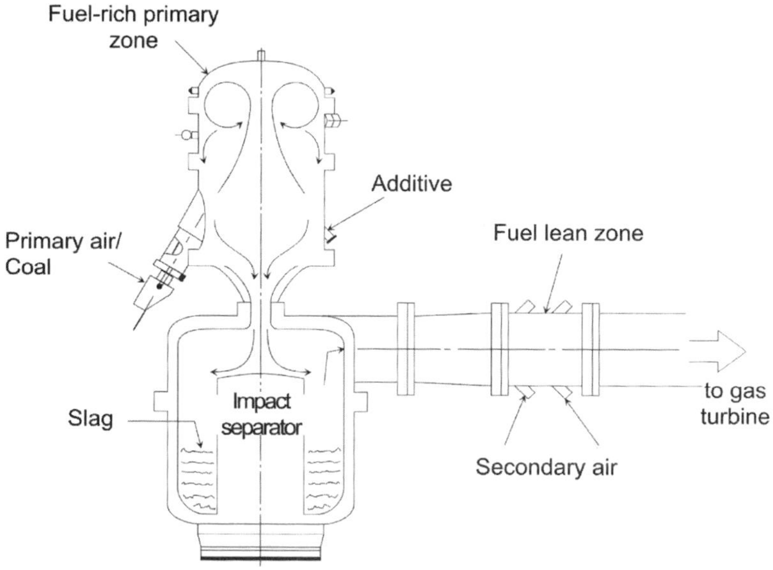

Fig. 7.44 Westinghouse's PPCC facility (Pillsbury et al. 1989)

The slag-tap furnace was operated under a pressure of 6 bar, at a temperature of 1,600°C and an air ratio of 0.7. This way 98–99% of the carbon was converted into the gas phase, i.e. into CO and CO_2. Four burners directed upwards at an angle combined to produce a single flame shape, in the process forming a swirl, deflected downwards from the top of the furnace. This flow pattern ensured both stabilisation of the flame and a high ash retention.

During firing, a slag layer would form on the chamber walls, flowing by gravity towards the bottom of the chamber, where lime or limestone was added for capture of sulphur oxides. An impingement separator followed after the bottom of the chamber. Here, the flue gases of the primary combustion chamber were sped up by an outlet nozzle so that ash particles, while accelerating, impinged and stuck to the separator. Its geometrical design acted in such a way that particles greater than 5 μm were separated. Molten slag flowed down the walls of the separator. In test operation, a total ash removal of more than 90% could be achieved.

By adding secondary air, the burnout, especially of the gas phase, was attained in the secondary combustion chamber. The total air ratio was set at 2.5 to achieve a cooling to the desired gas turbine entry temperatures.

Using two-stage combustion, it was possible to achieve NO_x emissions below 250 mg/Nm^3 (Bannister et al. 1990). The stated sulphur retentions ranged from 30 to 40% at a Ca/S ratio between 2 and 4. Analyses of the slag showed that about 80% of the alkalis were captured in the slag (Pillsbury et al. 1989). After the secondary combustion chamber, about 1–2% of the alkalis from the coal were still found in the gas phase. Later, an additional cyclone was installed in the test plant between the impingement separator and the secondary combustion chamber. This way, the total ash removal was increased to 99% (Bannister et al. 1992).

Studies discussed drafts for a 200 MW_{el} plant. Featuring a low gas turbine entry temperature of 1,010°C and a corresponding low turbine exit temperature of 370°C, a steam turbine output of 50 MW_{el} and a 160 MW_{el} output from the two gas turbines, the stated total efficiency was 37.5%. The concept was meant to give 20% cheaper electric power production compared to a conventional coal-fuelled power plant with flue gas desulphurisation. A precondition to that estimate, though, was that additional hot gas cleaning was not necessary (Pillsbury et al. 1989). Difficulties with the flue gas cleaning in the slag-tap furnace and in the secondary combustion zone and, more obviously, the efficiency of 37.5% not being commensurate with the set target led to a change in the concept. In a later publication, a modified concept was presented (Bannister et al. 1992) which differed from the PPCC concept, including gasification process elements as well as externally fired gas turbine process features.

7.4.4.5 Solar Turbines

After several preliminary investigations, Solar Turbines developed a pulverised coal furnace for a 3.8 MW_{el} gas turbine (Cowell and LeCren 1992; Cowell et al. 1992b). The schematic drawing of the pilot-scale plant is shown in Fig. 7.45. It consisted of a first combustion stage, an uncooled, adiabatic slag-tap furnace, a particle separator and a second combustion stage. The fuel was a coal – water suspension. The first combustion stage was operated at air deficiency (an air ratio of 0.7) and high temperatures of 1,600°C. The combustion pressure was 9.6 bar. The furnace was followed by a molten slag separation unit. Additional air injection completed the combustion in a second stage, with the flue gases being cooled to 1,040°C in the process. The total air ratio ranged around 2.5. Before the cleaned hot gas entered the gas turbine,

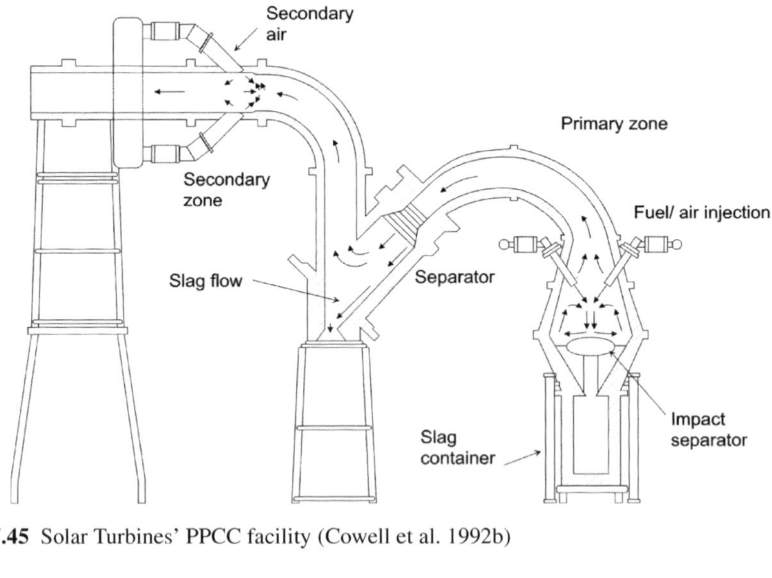

Fig. 7.45 Solar Turbines' PPCC facility (Cowell et al. 1992b)

the plant had another dust removal unit, using ceramic filters, at the secondary combustion temperature (Cowell et al. 1992a).

The separation unit installed between the combustion chamber and secondary air injection consisted of seven staggered rows of ceramic rods with a diameter of 3.2 cm, arranged with a clearance of 1.9 cm. The flue gas duct cross-section was widened to reduce the flow velocity, thus preventing the ash removed by the rods from being re-entrained (Cowell et al. 1992b).

During the investigations at the test plant, an ash removal of 61% in the combustion chamber and of 98% after the ceramic rods between the primary and secondary zones could be achieved. The particle content was thus about 40 mg/kg flue gas. The staged combustion made it possible to achieve low NO_x emissions of 150 mg/Nm^3 at 6% O_2 (calculated as NO_2) along with equally low CO emissions of 30 mg/Nm^3 at 6% O_2. There are no results available from the operation of the pilot-scale plant, including for the ceramic filters and the gas turbine.

7.4.4.6 Allison

The concept of Allison differed from the concepts of Westinghouse and Solar by not including slag removal. Only dry fly ash was separated, at temperatures around 1,000°C. The coal – water suspension was burned in a dry bottom furnace at temperatures of 1,200°C, at air deficiency. The flue gases were cooled to below 1,000°C by a water quench; the fly ash was removed in a cyclone. After a second combustion stage, with injection of the remaining combustion air, and another cyclone stage for fly ash removal, the hot gases were fed to the gas turbine (Parsons and Byam 1989).

Allison carried out tests with a 3.5 MW_{el} gas turbine firing coal at part load so that an entry temperature of 816°C was achieved. After each 4 h test, the gas turbine was unmounted and the blading inspected. A mass balance revealed that about 0.2% of the fuel ash had been deposited in the turbine. An analysis of the deposits indicated an enrichment of alkalis and sulphur. The deposits could be easily removed from the blades by washing, but with longer operation, it would be expected that adhesive, stable deposits of ash would form. In addition, corrosion would occur. As a result, alkali separation (from the flue gas) would have to be introduced into the process for any practical applications. The authors wanted to achieve this by a further lowering of the quench stage temperature.

7.4.5 Summary and Conclusions

PPCC offers the potential for a high efficiency in comparison to the competing coal-based combined cycle processes. To exploit this potential, high-temperature particle and alkali removal at temperatures of 1,400–1,600°C is a prerequisite. Although substantial progress in both fields has been achieved, the PPCC development in Germany and, it seems, in the USA, has been discontinued.

7.5 Externally Fired Gas Turbine Processes

Combined cycles with integrated coal gasification or pressurised pulverised coal firing use a gas cleaning stage to produce a clean fuel (i.e. a hot gas) for the gas turbine. The externally fired gas turbine process, in contrast, uses a heat exchanger to heat a gas already clean enough to meet the gas turbine related requirements. This indirectly fuelled or externally fired combined cycle (EFCC) presents an alternative to the combined cycle processes that integrate coal gasification or operate using pressurised firing (Baum 2001; Spliethoff 2000; Spliethoff and Baum 2002a, b; Benson 2000).

7.5.1 Structure, Configurations, Efficiency

In an indirectly fired combined cycle, the heat from the fuel is released by combustion in a firing system. A heat exchanger follows the furnace, the hot flue gases giving heat to a clean, pressurised turbine working fluid. Hot gas cleaning is not needed for the indirectly fired process, as the flue gases and turbine fluid are segregated by the heat-exchanging walls. Therefore, any necessary flue gas cleaning to meet emission limits can be performed at low temperatures, as in conventional coal-fired power plants.

There are various configurations possible for indirectly fired gas turbine combined cycles. Open gas turbine processes use air or cleaned flue gas as the turbine working fluid, while closed gas turbine processes work with gases such as helium or carbon dioxide. These gases have better thermodynamic properties than air or hot flue gas but can only be used in closed gas turbine processes.

Figure 7.46 shows a schematic diagram of an open process using air as the working fluid for the gas turbine. After compression, the air is heated to the gas turbine inlet temperature in the high-temperature heat exchanger. Part of the gas turbine exhaust air is fed to the firing as combustion air. The heat of the remaining exhaust air is used in a heat recovery process together with the remaining hot flue gas heat which was not transferred in the high-temperature heat exchanger.

The configuration of an open EFCC process using flue gas as the working fluid for the gas turbine shown in Fig. 7.47 employs pressurised slag-tap firing. The hot pressurised flue gas transfers heat in a high-temperature heat exchanger and in a heat recovery process. It is cleaned at a low temperature before being reheated in the high-temperature heat exchanger, then enters the gas turbine. In contrast to the open variant using air, this process has the advantage of not subjecting the heat exchanger to compressive stresses except through its own pressure drop and that of the gas cleaning process. Equipment costs, however, are significantly higher than using air as the working fluid for the gas turbine.

Figure 7.48 shows a schematic diagram of a closed EFCC process. The working fluid of the gas turbine is conducted in a closed circuit, while the firing and the flue gas path remain unmodified. The basic advantage is the potential to use gases with better thermodynamic properties for the heat transfer. The high compressor inlet temperature, however, has a negative effect on the overall efficiency.

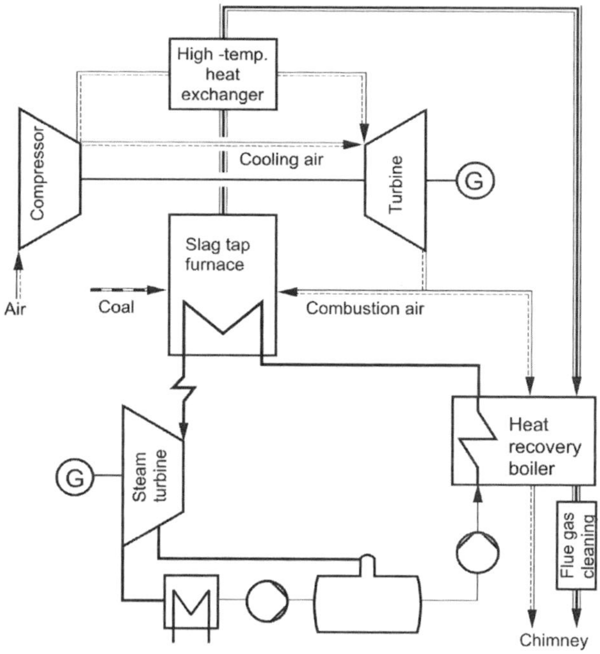

Fig. 7.46 An open EFFCC process using air (atmospheric slag-tap furnace) (Spliethoff and Baum 2002)

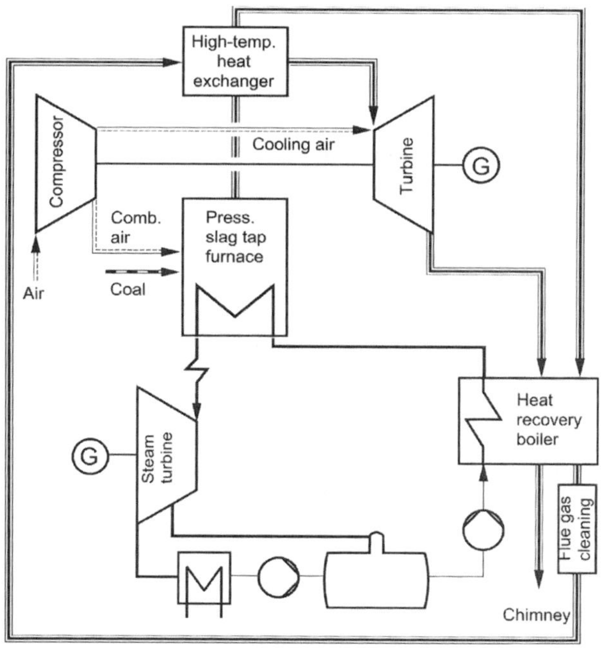

Fig. 7.47 An open EFCC process using flue gas (pressurised slag-tap furnace) (Spliethoff and Baum 2002)

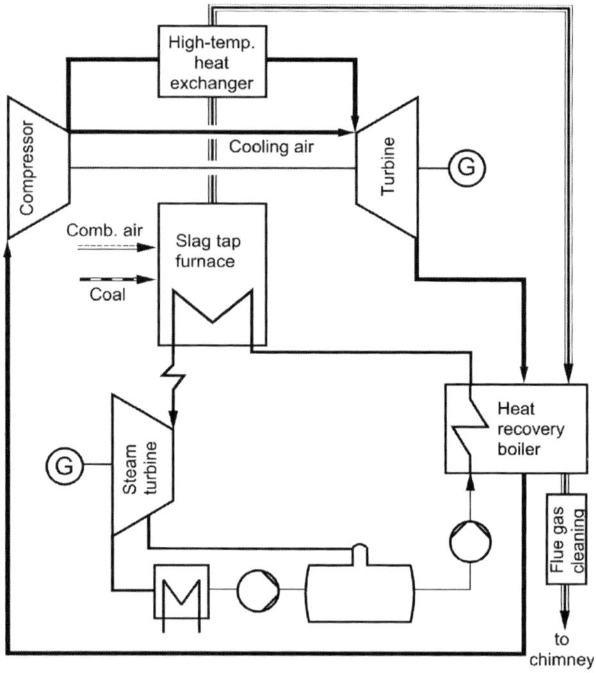

Fig. 7.48 A closed EFCC process (atmospheric slag-tap furnace) (Spliethoff and Baum 2002)

The EFCC process with auxiliary natural gas firing shown in Fig. 7.49 is analogous to a natural gas fired combined cycle with combustion air preheating by coal flue gas in a high-temperature heat exchanger. In this process, this exchanger is not subjected to the high thermal stress as in the variants described so far. The coal firing need not necessarily use a slag-tap furnace. The type of furnace depends on the temperature the heat exchanger operates at. The natural gas demand decreases with an increasing temperature.

The following material will refer to open gas turbine processes using air as the working fluid because this configuration is rather uncomplicated while having potential for a high efficiency. Most of the data presented, however, is applicable to the other EFCC configurations as well.

The efficiency of the EFCC process essentially depends on the following parameters:

- The heat exchanger outlet and the gas turbine inlet temperatures
- The cooling of the furnace
- The cycle efficiency of the steam generation process
- The flue gas losses

The rest of this section (Sect. 7.5.1) shall illustrate, using calculations over the cycle, the effect on the efficiency of the major parameters. The baseline case of these calculations is shown in Fig. 7.50, a power plant with an electrical capacity of 350 MW (Baum 2001).

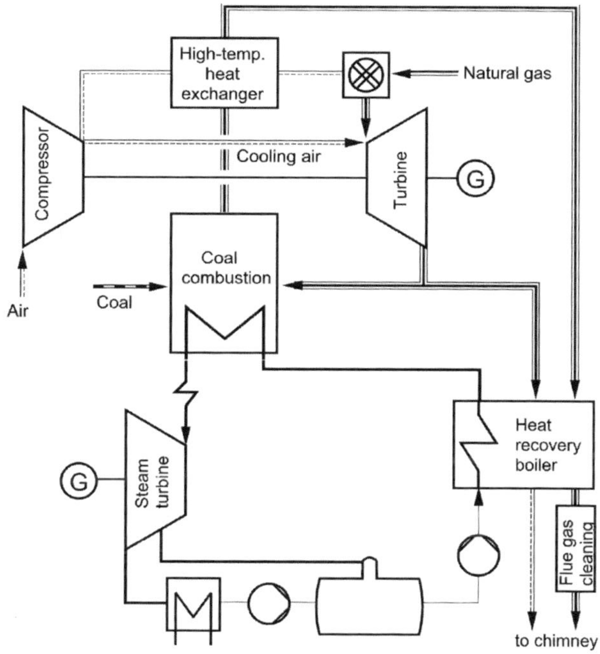

Fig. 7.49 An EFCC process with additional natural gas firing (Spliethoff and Baum 2002)

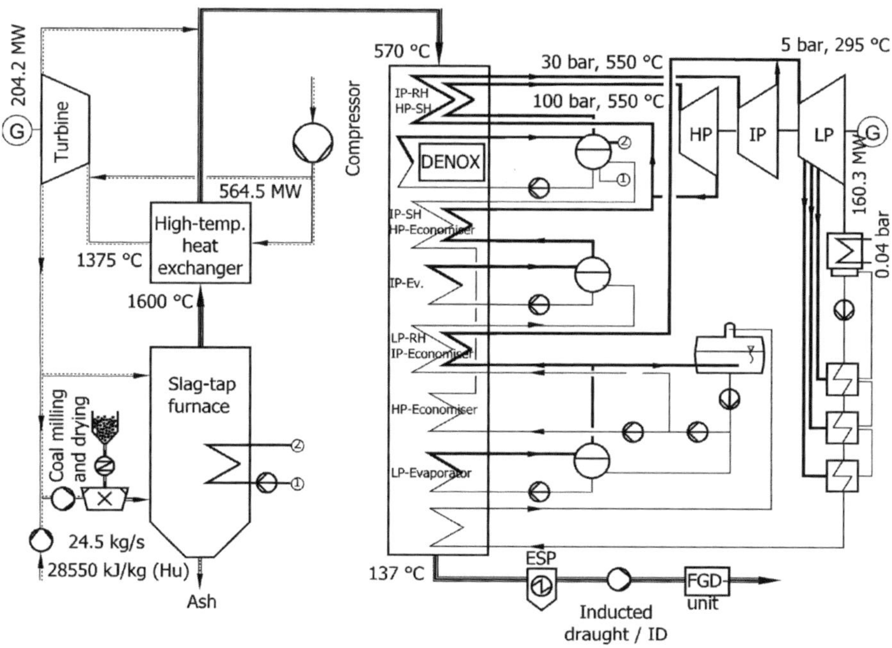

Fig. 7.50 Cycle diagram with design data of a 350 MW_{el} EFCC process (Spliethoff and Baum 2002; Baum 2001)

The input of heat from the fuel (coal) to the slag-tap furnace is 700 MW in the model case. The furnace uses steam – water cooling, and about 7% of the total fuel heat input is directly transferred to the steam – water cycle. Further cooling to the flue gas outlet temperature of 1,600°C is achieved by operating at an air ratio of 2.1. In the high-temperature heat exchanger which follows, the flue gas transfers its heat to the compressed air, heating it to 1,375°C.

In order to cool the gas turbine, part of the compressor air is conducted directly to it. The result is an ISO gas turbine inlet temperature of 1,184°C for this baseline case. That part of the gas turbine exhaust which is not used as secondary air is either used for the drying-milling process of the coal, and for its transport, or is mixed with the flue gas from the combustion of the coal and used for waste heat steam generation.

Taking into account the auxiliary power demand for coal preparation, pumps and the induced-draught, and other losses, totalling about 6% of the gross electrical output, the gross efficiency of 52% is reduced to a net efficiency of 49% for the baseline case. It should be mentioned that, given the conservative assumptions about the machine efficiency, the terminal temperature difference in the waste heat boiler and so forth, the process discussed here is not efficiency optimised. The efficiency potential of the EFCC process is comparable to the IGCC process. With higher ISO turbine inlet temperatures of 1,250°C and an improved steam cycle it should be possible to achieve a net efficiency of about 52%, comparable to the IGCC 98 case (see Sect. 7.6) (Edelmann and Stuhlmüller 1997).

The influence of the heat exchanger outlet temperature (which is to say the real gas turbine inlet temperature) on the efficiency of an EFCC process is shown in Fig. 7.51. The heat exchanger outlet temperature and the thermodynamic efficiency increase with higher mean heat input temperatures. The gas turbine share of the total process output increases as well. At a constant mass flow, higher gas turbine inlet temperatures make larger heat transfer surfaces necessary. As the high process temperatures make it impossible to use metal, it is only possible to use ceramic

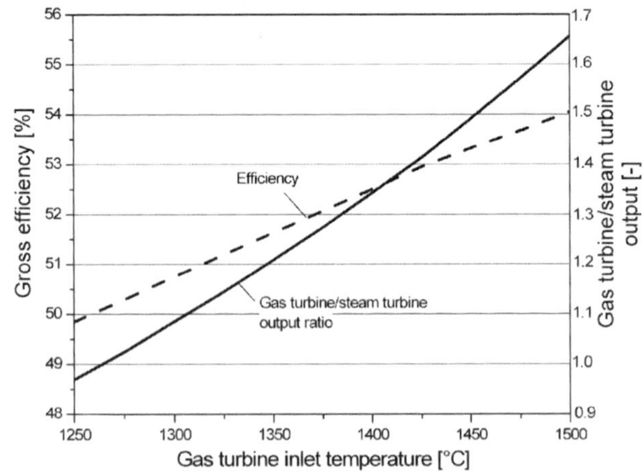

Fig. 7.51 Efficiency and the gas turbine/steam turbine output ratio as a function of the real gas turbine inlet temperature (Spliethoff and Baum 2002; Baum 2001)

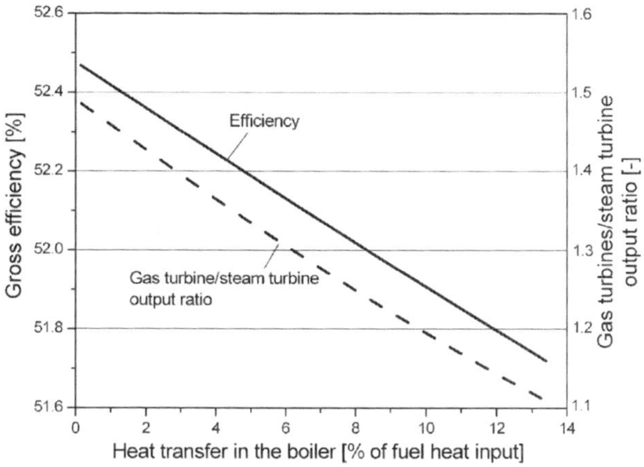

Fig. 7.52 Influence of furnace cooling on the efficiency and the gas turbine/steam turbine output ratio (Baum 2001)

materials for these surfaces. The ceramic heat exchanger is the only component in the indirectly fired combined cycle which is not yet state of the art. It is discussed in detail in Sect. 7.5.2.

The cooling technique for the furnace has a substantial influence on the EFCC process, in a similar way as in pressurised pulverised coal firing (see Fig. 7.52). The efficiency is highest if the heat of the furnace and heat exchanger is transferred to the working fluid of the gas turbine and the steam-generating process is heated only by the waste heat from the gas turbine.

This can be achieved by an adiabatic furnace without cooling or by an air-cooled furnace. Water/steam cooling, as a third option, is simpler to construct. It was therefore chosen as the basis for calculations.

The cycle efficiency of the steam process is another important parameter influencing the total efficiency of the EFCC process. Since the steam process contributes 40 – 50% of the total output, depending on the type of furnace cooling and the gas turbine inlet temperature, an improvement of the steam process by 1% means an increase of the total efficiency of 0.4 – 0.5%.

7.5.2 High-Temperature Heat Exchanger

The concept of an indirectly fired gas turbine is an alternative to combined cycle processes with integrated coal gasification or pressurised firing which have enjoyed the majority of research efforts. A critical aspect of the indirect firing concept if it is to be put into practice is the high-temperature heat exchanger where the heat of the flue gas is transferred to the working fluid of the gas turbine.

7.5.2.1 Requirements

The requirements in process-engineering terms for such a heat exchanger are the following:

- **Process temperatures of up to a maximum of 1,600°C**
  Modern gas turbines work at ISO gas turbine inlet temperatures of up to 1,300°C, which correspond to real gas turbine inlet temperatures of up to 1,500°C. To achieve these temperatures on the side of the cleaned gas of the high-temperature heat exchanger, it is necessary to be able to set flue gas temperatures of up to 1,600°C on the side of the untreated gas.
- **Compressive stress of the heat exchanger**
  In the case of the open EFCC process using air, the heat-transferring walls are subjected to stress by the pressure difference between the turbine working fluid and the flue gas. This difference is the result of the compression ratio of the gas turbine, which is around 16–18 for gas turbines designed for natural gas with gas turbine inlet temperatures of 1,200–1,300°C. It must be observed in this respect that ceramic material should not be subject to tensile stress.
- **Tightness of the heat exchanger**
  The heat exchanger should be leak-tight because losses by leakage leads to lower efficiencies. Problems in this respect are to be expected at tube/tube or tube/tube-sheet joints.
- **Pulverised coal firing durability**
  As the heat exchanger is charged with fly ash laden flue gas, it is susceptible to the hazards of corrosion and fouling.

The heat exchanger material must be resistant to flue gas and slag at high temperatures. Fouling of the heat exchanger has to be either prevented by appropriate upstream removal or minimised by deposit removal during operation. The decisive factors are the temperatures at which the heat exchanger is operated and the state of the arising ash or slag. In the temperature range above 1,400°C, i.e. higher than the ash fluid temperature, slag is removed in a molten state. At temperatures below the ash deformation point, the ability of soot-blowing to remove fouling deposits should be tested. At temperatures above the ash deformation point and below the ash fluid point, it may be necessary to heat up and melt the deposits.

Last but not least, there has to be safety during all operating states – start-up and shutdown and continuous operation, and cases of outages. None of these cases must lead to damage of the heat exchanger.

7.5.2.2 Selection of the Material

The material properties of the heat exchanger decide the upper limit of the process temperatures and hence the efficiency of the EFCC process. Figure 7.53 highlights the temperature-dependent stability of selected ceramic and metal materials. It is clear from the figure that, for the operation of the EFCC process in the high-temperature range, only ceramic materials can be considered (Kainer and Willmann 1987).

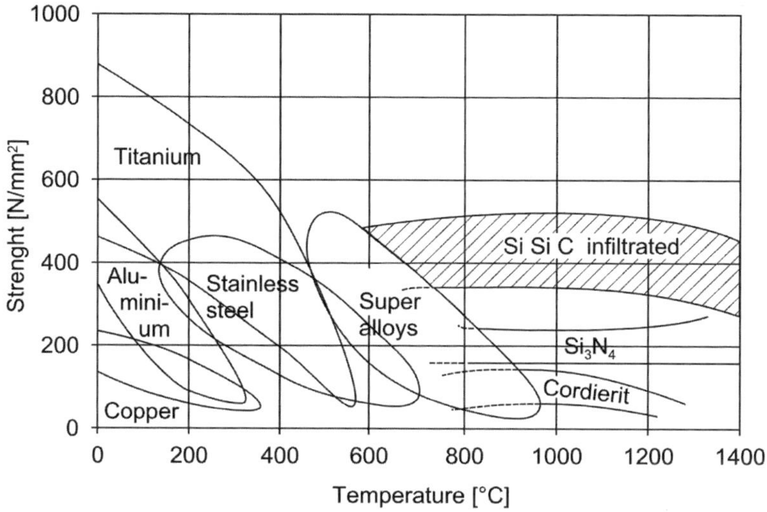

Fig. 7.53 Strength of metallic and ceramic materials (Kainer and Willmann 1987)

Steels

Low-alloy steel types (represented typically by 15 Mo 3, 13 CrMo 4 4, 10 CrMo 9 10) can be used at temperatures of up to about 500°C. Higher-alloy ferritic steels – such as the type used for the final superheater stage in steam power plants, X 20 CrMoV 12 1 – or martensitic steels can be used up to temperatures of around 600°C. Austenitic steels can be used up to 750°C.

Nickel-Based Alloys

Nickel-based alloys, where nickel is the main element of the alloy, have a higher temperature resistance and strength than austenitic steels. The temperature limit is about 850°C. Due to the composition, these materials are relatively expensive and rather difficult to handle.

ODS Superalloys

ODS (oxide dispersion strengthened) alloys are powder metallurgical (PM) manufactured superalloys based on nickel or iron. They feature a high temperature resistance along with comparably good strength properties. Typical are Inconel MA 754 (Ni base) and PM 2000 (Fe base). They show highly promising properties such as high temperature and corrosion resistances up to temperatures of 1,150°C (Aquaro and Pieve 2007; Hurley et al. 2003).

Ceramic Materials

Ceramic materials, with respect to temperature resistance paired with strength, are largely superior to metallic materials. Of disadvantage are the limited thermal shock resistance and the limited tensile and bending strengths. Additional limitations to the use of ceramics arise due to their brittleness; these materials are not able to relax stress peaks by deformation. They are not at all ductile – on reaching the elastic limit fracture occurs immediately, without any deformation. Designs incorporating ceramics are suitable if they avoid high loads that are uncontrollable. Simple construction forms such as tubes and constructions that only put pressure stresses (rather than tensile) on the ceramic material are advantageous in this respect (Landfermann and Hausner 1988).

In principle, suitable materials for the high temperatures of EFCC heat exchangers are aluminium-based oxide ceramics and non-oxide ceramics such as Si_3N_4 and SiC. The properties determining their suitability as construction materials for high-temperature heat exchangers are assessed in Table 7.13. According to this assessment, silicon carbide is the most suitable material meeting the requirements of the EFCC process. Table 7.14 draws a comparison of the properties of ceramic and other materials for the use in high-temperature heat exchangers. Fibre-strengthened ceramic materials are still under development. While they should be considered for use at high temperatures, problems persist with respect to their protection against long-term oxidation.

The corrosion resistance of ceramics is the major bottleneck. Only some types of ceramics, and only to a maximum of 1,300°C, meet the resistance requirements. This knowledge is the result of investigations into materials carried out in two coal-fired semi-industrial test plants within the framework of a research project (Kuhnle

Table 7.13 Suitability of ceramic materials as construction materials for high-temperature heat exchangers (Baum 2001; Kuhnle et al. 1997; Fichtner 1992)

	Aluminium oxide	Silicon nitride Si_3N_4		Silicon carbide SiC	
	Al_2O_3	RBSN	SSN H(i)PSN	SSiC H(i)PSiC	SiSiC
Strength at high temperatures	+	(–)	+	+	(–) max 1,400°C
Thermal conductivity	(–)	(–)	(–)	+	+
Fatigue resistance	–	+	+	+	+
Resistance to oxidising atmospheres at high temperatures	+	–	(–)	+	+
Porosity/gas tightness	–	–	+	+	+
Resistance to molten coal slag			No resistance		

+ suitable, – not suitable, (–) only suitable to a limited extent *Silicon carbide SiC* (HPSiC hot pressed silicon carbide, HiPSiC hot isostatically pressed silicon carbide, SSiC sintered silicon carbide, SiSiC silicon-infiltrated silicon carbide)
Silicon nitride Si_3N_4 (HPSN hot pressed silicon nitride, HiPSN hot isostatically pressed silicon nitride, SSN sintered silicon nitride, RBSN reaction-bonded silicon nitride)

Table 7.14 Data for ceramic materials compared to other recuperator materials (Kainer 1988)

Material	Bulk density [g/cm^3]	Bending strength (20°C) [MPa]	Tensile strength [MPa]	Modulus of elasticity (20°C) [GPa]	Thermal expansion coefficient (20–1,000°C) 10^{-6} K^{-1}	Coefficient of thermal conductivity [W/mK]	Open porosity [%]	Gas permeability [nPm]	Temp. limit [°C]
Alumina-rich fireclay	2.5	< 10	–	15	5	1–2	17	$\sim 1 \cdot 10^{-2}$	1,700
Cordierite	2.1	10–100	–	60	2	1–3	0–10	n.d.	1,200
Plasma-coated Al_2O_3	3.2–3.4	~ 20	–	22	3–8	1–3.5	10–12	$\sim 5 \cdot 10^{-3}$	1,200–1,500**
Silicon nitride	2.4	200	–	130	3	13	20	n.d.	1,350
Si-infiltrated SiC (SiSiC)	3.1	300	150	300	4.5	125–140	0	0	1,400
Sintered SiC (SSiC)	3.1	460	–	410	4.5	125–140	0	0	1,650
Stainless steel 25 CrMo	7.9	–	900	210	11*	40	0	0	400
Tantalum	16.6	–	280	190	6.5*	55	0	0	700
Graphite	1.9	60	30	30	8*	120	0	0	200 (ox.) 2,500 (red)
Glass	2.23	70	–	68	3*	1	0	0	300

* between 20–500°C

** depending on posttreatment

et al. 1997) and confirmed by experiments where materials were placed in an electrical furnace with synthetic flue gas and ash. Up to 1,300°C, the monolithic ceramic types – aluminium oxide, silicon nitride and silicon-infiltrated and sintered silicon carbide – showed promising corrosion behaviour for the hard coal used. In contrast, carbon fibre strengthened silicon carbide, with mechanical properties better suited to high-temperature heat transfer, showed poor oxidation behaviour despite the coating (Baum 2001).

When the materials in the preceding paragraph were tested in a cyclone slag-tap furnace at high flue gas temperatures of 1,500 °C, all showed severe damage even after less than 30 h, which excludes their application in high-temperature heat transfer components. The corrosive attack comes especially from the aggressive molten slag. The use of ceramic materials under these conditions needs further development. Apart from finding suitable ceramic materials and construction types for them, development of how to make adequate connections and joins of ceramic constructions is necessary. For temperatures below 1,150 °C, the oxide dispersion strengthened (ODS) superalloy PM 2000 seems to be suitable for use. This metallic material, in comparison to monolithic ceramics, has great mechanical property and manufacturability advantages.

7.5.2.3 Classification of Heat Exchangers

There are primarily two techniques of heat transfer available: recuperative heat exchangers (or recuperators) and regenerators. While recuperators transfer heat from one heat-carrying medium to the other via separating walls, regenerators exchange heat after a time lag via intermediary media. Figure 7.54 gives an overview of heat exchanger technologies (Kainer and Willmann 1987). Recuperators work continuously in transferring heat, while regenerators, by alternating between heat storage and heat dissipation, are discontinuous processes.

Only regenerators with many single-storage elements, which, as they circulate, alternate the processes of heat storage and heat dissipation, make quasi-continuous operation possible. Regenerators with temperature-resistant, thermally conductive

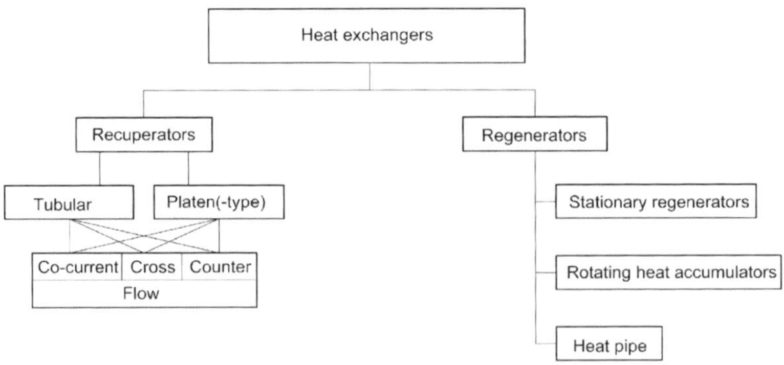

Fig. 7.54 Heat exchanger systems (Kainer 1988)

storage elements such as ceramics, as well as temperature-resistant supports, allow high working temperatures. Disadvantages of regenerators are gas exchanges or gas losses associated with every switching and the extra cost of construction and control engineering (Kainer 1988).

A well-known example of regenerators is hot blast stove for air preheating in blast furnace processes (see Fig. 7.55). Part of the hot blast stove is a static heat accumulator which is built of single ceramic hollow blocks placed upon each other and penetrated by the air flow. The heat accumulator is placed on a metal grid connected to a furnace which, in the case pictured, is mounted externally. The heat accumulator and furnace are surrounded by thermal insulation and steel plating. A

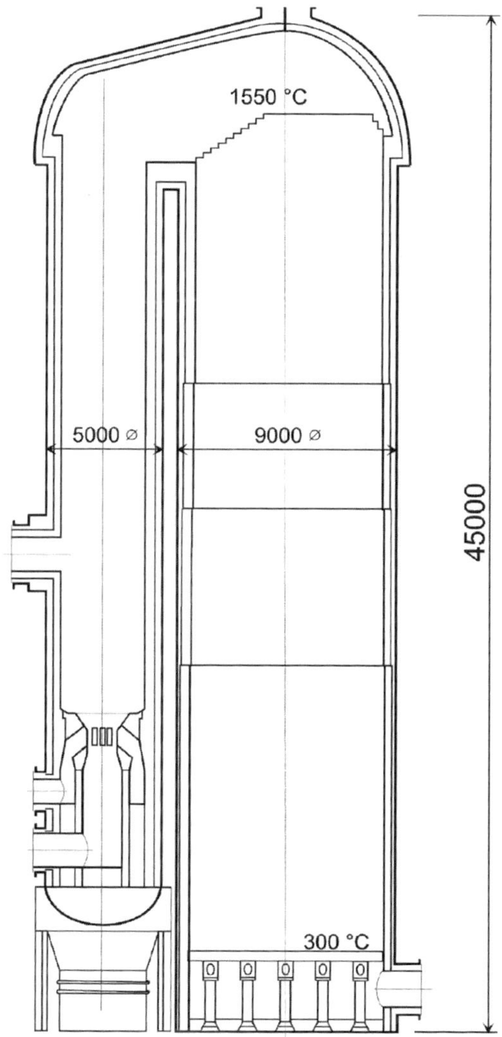

Fig. 7.55 A typical regenerator – a hot blast stove with an external furnace for blast furnace operation (Kainer 1988)

complete air preheating plant for a blast furnace needs at least two, often up to four, regenerators because the single regenerator can only be operated discontinuously. By means of such plants, it is possible to heat air flowrates of up to 450, 000 Nm^3/h to temperature maxima of 1,400°C at pressures of up to 6 bar (Kainer 1988).

Another heat exchanger type considered for EFCC processes is the regenerative Ljungstroem heat exchanger, featuring a rotating heat accumulator (Wilson et al. 1991; Wilson 1993a). However, further development of sealing engineering is necessary to reduce the leakage loss that results from rotation. One possibility is to rotate the cylindrical heat accumulator only in short intervals. The gaskets are then released and lifted only during rotation, otherwise pressed against the heat accumulator. Another technology considered is regenerator where rectangular heat-accumulating blocks circulate between the hot and the cold gas flow. This way the lengths to be sealed are shorter than for a rotating heat exchanger (Wilson 1993b).

The biggest problem with all regenerators, however, is fouling. Because the medium which temporarily stores the heat is in contact with both the polluted flue gas and the clean working fluid of the gas turbine, valves and gaskets are subject to fouling. This way, fly ash or alkalis can also pollute the turbine working fluid.

Yet another technology considered for high-temperature processes is heat pipe, the principle of which is shown in Fig. 7.56 (Bliem 1985). It is a closed pipe, one

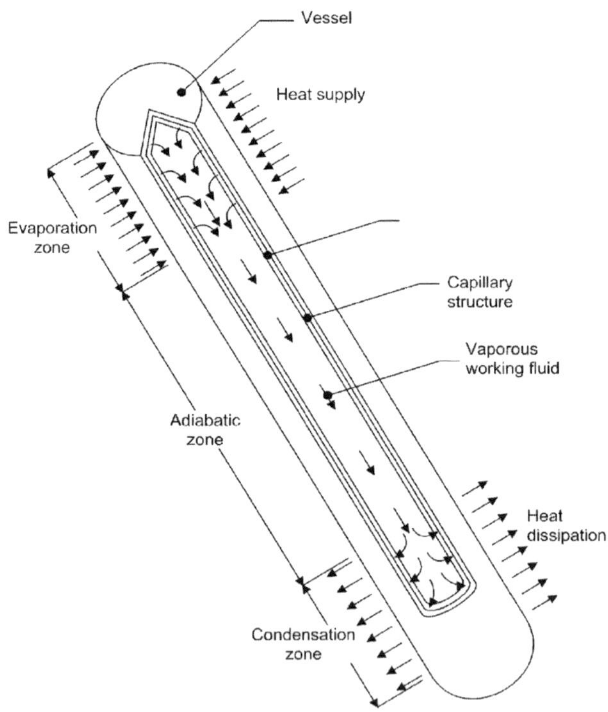

Fig. 7.56 Schematic drawing of a heat pipe (from Bliem 1985, © 1985, with permission from Noyes Publications)

half of which projects into the untreated flue gas, the other into the cleaned gas. The tube, the inside of which is in a vacuum, has a capillary structure saturated with a liquid. When the half of the tube in the flue gas is heated, the working fluid vaporises and flows to the other end of the tube where, dissipating heat, it condenses internally. The capillary action transports the formed condensate back to the heating zone.

For use in an EFCC process, the heat carrier and its supporting structure have to be adjusted to high temperatures. For temperatures up to 1,600°C, lithium is considered the most suitable heat carrier and SiC the most suitable structural material (Groll 1980). These heat pipes, however, are not state of the art. As with other types of heat exchangers, heat tubes require a solution to the problem of corrosion. Heat pipes with sodium as the working fluid, transferring heat between two small-scale fluidised beds at 800°C, have been successfully tested (Kuhn 2007).

Recuperators that are used at high temperatures have to be made of ceramic materials. Up until now, there have been two different types of designs developed: the modular or block system and the shell-and-tube construction.

Figure 7.57 shows the block of a module-type heat exchanger. By series connection of the single modules it is possible to conduct mass in a cross-counterflow regime. The modules consist of extruded plates which are stacked together and joined by a special substance during baking. Their susceptibility to fouling and clogging and their limited accessibility for cleaning suggest that such module-type heat exchangers are less suitable for a coal-fired EFCC process.

Figure 7.58 shows the working principle of a ceramic cross-flow recuperator. The straight ceramic tubes are flexibly mounted at both ends so that the tubes can extend without stress. Plugging gaskets of ceramic fibres serve to minimise leakages. Another possibility is to clamp the tubes at one end – preferably at the colder

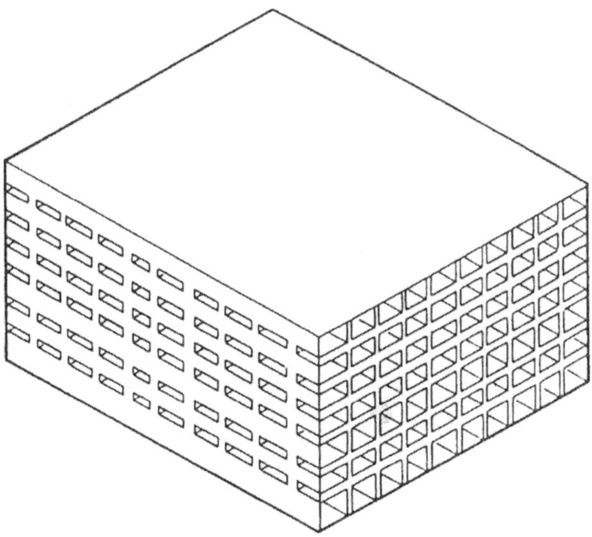

Fig. 7.57 Unit of a module-type heat exchanger (from Bliem 1985, © 1985, with permission from Noyes Publications)

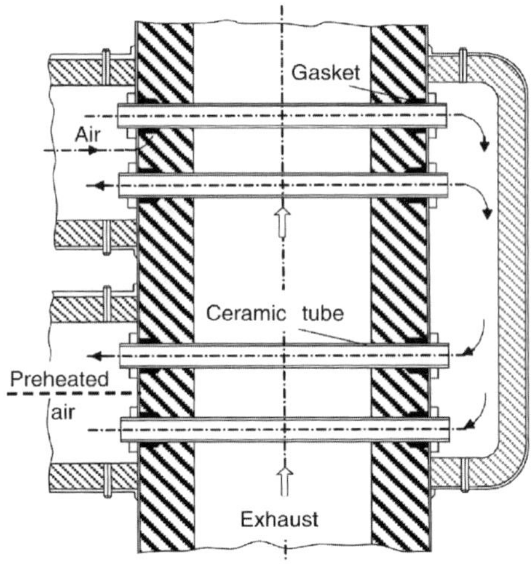

Fig. 7.58 Working principle of a ceramic recuperator (Kainer and Willmann 1987)

one – so as to mount them flexibly while allowing for extension at one end (Kainer and Willmann 1987).

The principle of clamping at one end is also applied in the construction of the double-tube recuperator shown in Fig. 7.59. Both tubes are clamped at one end and can freely extend in the other direction. The tubes are suspended vertically so gravity affects the gasket between the housing and the tube flange. The outer tube has a diameter of about 95 mm, a wall thickness of 6 mm and a length of nearly 2,000 mm. The tube is made of SiC (Bliem 1985; Harkins and Ward 1989).

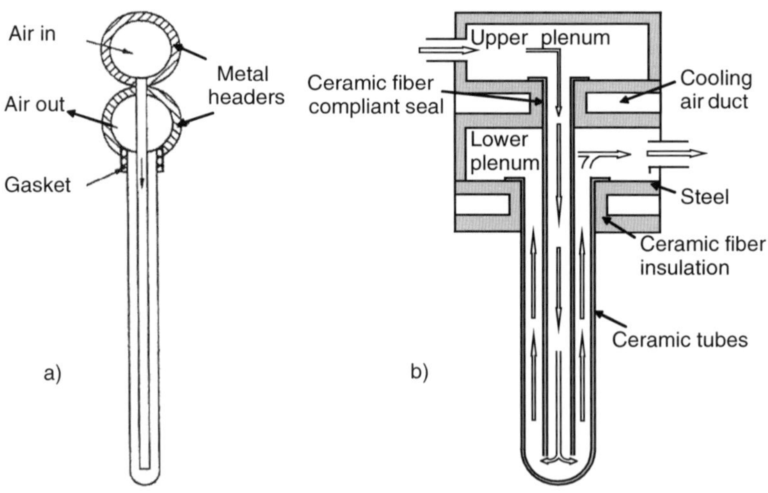

Fig. 7.59 Tube-in-tube recuperators (**b** from Bliem 1985), © 1985, with permission of Noyes Publications)

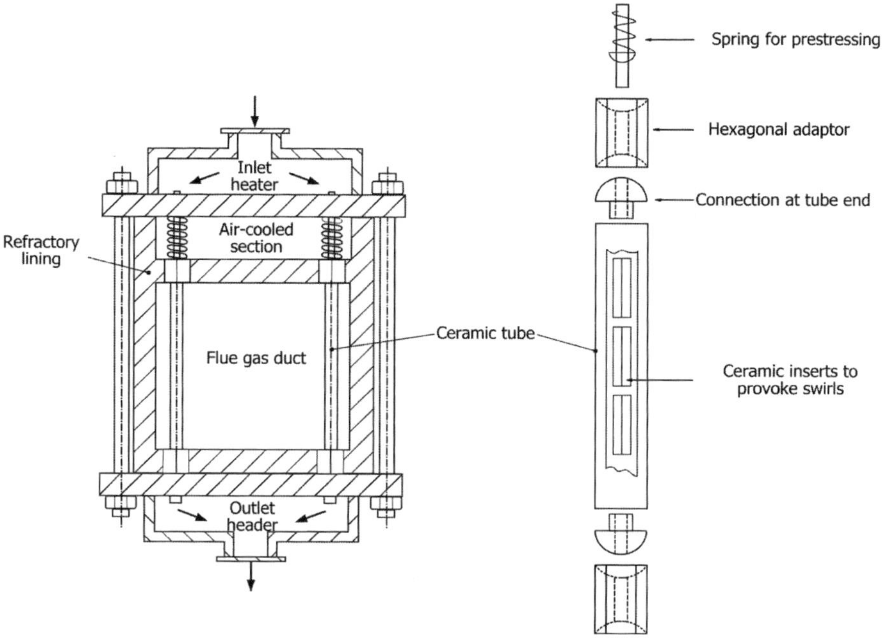

Fig. 7.60 Recuperator by Hague International (LaHaye 1989, 1986)

When ceramic heat exchangers are used in an open process with air as the working fluid of the gas turbine, attention must be paid to the pressure difference between the air and the flue gas. Given that ceramic materials, with their minimal ductility, should preferably be loaded with compressive stress, it is logical to conduct the pressure-free flue gas inside and the working fluid outside the tubes of a shell-and-tube heat exchanger. Conducting the slag- or ash-carrying flue gas inside the tubes is a valid counterargument, though.

The heat exchanger concept for EFCC pursued by Hague International is to conduct the clean working fluid of the gas turbine inside the tubes and the hot flue gas outside, as shown in Fig. 7.60. The tubes are pre-stressed by means of a system of springs in order to be able to load the ceramic material with compressive stress and to seal the tubes. Tube circumferences typically measure between 75 and 100 mm and have a length of about 1,200 mm. Semi-circular fittings and adaptors guarantee the absorption of shearing forces (lateral extensions) (LaHaye and Feldmann 1986; LaHaye and Zabolotny 1989; Vandervort 1991). A heat exchanger designed for a demonstration power plant is described in Sect. 7.5.3.2.

7.5.3 State of Development

Although the EFCC process has been well known for a long time, only the further development of ceramic materials during the past 10–15 years has given it a new impetus. Ceramic materials for high gas turbine inlet temperatures are indispensable if a higher efficiency of the EFCC process is to be achieved compared to other gas and steam turbine processes or processes with a steam turbine alone. At lower temperatures, metallic materials are more advantageous, because of their more suitable mechanical and thermal properties.

7.5.3.1 EFCC Processes with Metallic Heat Exchangers

Development of EFCC processes using metal as the base construction material began more than 70 years ago. In 1939, an experimental plant with a capacity of 2 MW_{el} was put into service by *Escher Wyss*, Switzerland (Keller 1946). After 1945, test operations in Scotland, using coal and peat, followed. In 1956, the first commercially operated plant, with an electrical capacity of 2.5 MW, went on-line in Ravensburg, Baden-Wüerttemberg (Germany). Further plants followed in Germany up until 1960 – Coburg, Bavaria, at *Haus Aden/Monopol*, and in Oberhausen and Gelsenkirchen, North Rhine-Westphalia – with electrical capacities between 6.4 and 17 MW, and one plant in Moscow, Russia, with 10 MW_{el} manufactured by a consortium composed of *Escher Wyss*, Switzerland, and *Gutehoffnungshütte* and *Kohlescheidungsgesellschaft*, both German (Keller and Gaehler 1961).

The corresponding cycle diagram of the closed gas turbine process using air is shown in Fig. 7.61 (Bammert 1986). The air was compressed to a pressure of 41 bar in two steps by an intercooler and subsequently heated up to 450°C by the hot gas turbine exhaust in a recuperator. In the furnace, the air was heated to a gas turbine inlet temperature of 710°C. The efficiency of the largest plant reached 31%.

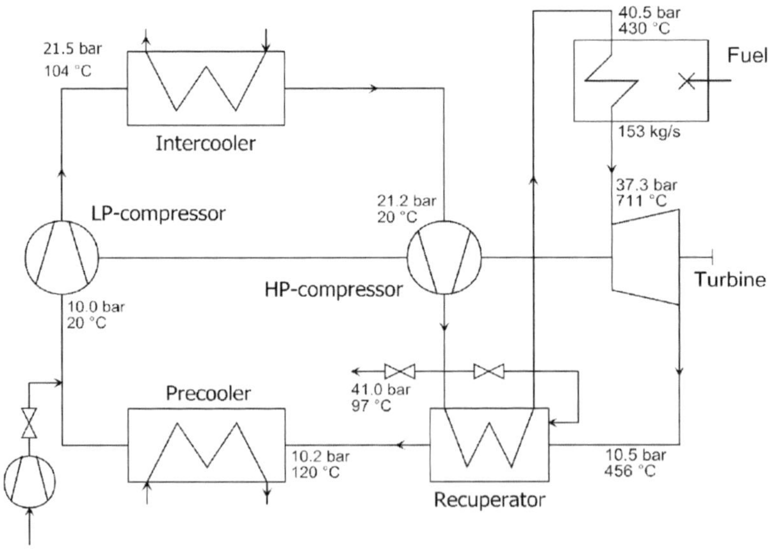

Fig. 7.61 Cycle diagram of the EFCC plant, which has a metal heat exchanger, in Gelsenkirchen (Bammert 1986)

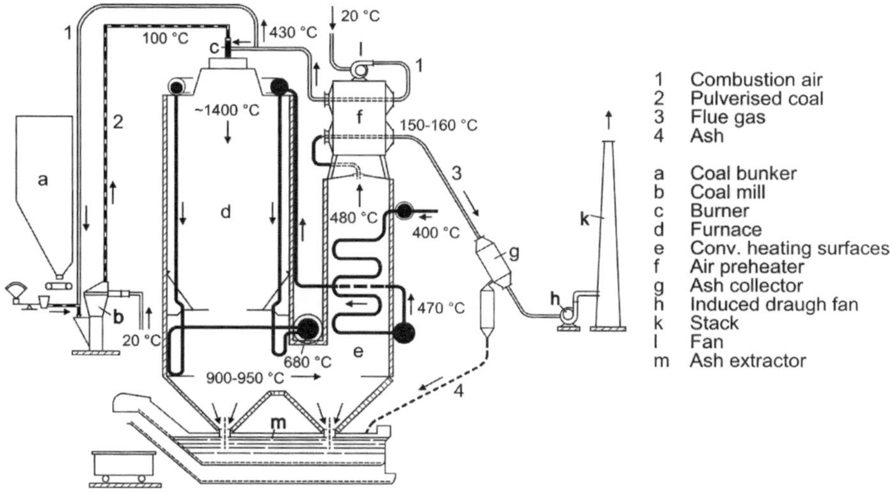

Fig. 7.62 Schematic diagram of the EFCC plant in Ravensburg, Baden-Württemberg (Keller and Gaehler 1961)

The schematic diagram of the combustion plant is shown in Fig. 7.62. The furnace was refractory-lined and designed as a two-pass construction. The pulverised coal burner was mounted at the furnace top. The air, preheated to 450°C in the recuperator, was heated by convective heating surfaces arranged in the second pass and then by radiant wall heating surfaces. The heating surfaces were designed as in conventional steam generators, with inlet headers, single tubes and outlet headers. The tube dimensions were an inside diameter of 32 mm and a wall thickness of 3 mm; the number of single tubes of the 6.6 MW_{el} plant in Coburg was 320 (Keller and Gaehler 1961). At a gas turbine inlet temperature of 710°C, the highest tube wall temperatures were between 770 and 790°C, which austenitic materials could still cope with. The highest alloyed material used was austenite of the 16 Cr13Ni type. All plants built in Germany reached 120,000 h of operation; only the plant in Coburg was in service for more than 160,000 h (by 1986) (Bammert 1986).

Another type of EFCC process using metal as a construction material was put into practice in a model power plant in Völklingen, Saarland (Germany). In it, the in-bed heat transfer surfaces of two fluidised bed modules were used to preheat the air to about 700°C. The gas turbine inlet temperature could be raised to 820°C by additionally firing a gaseous fuel. The combustion gases of both fluidised beds were conducted into the pulverised coal firing of the steam generator, while the gas turbine exhaust was made use of as a fluidising medium and oxygen carrier in both fluidised bed modules and as secondary air for the pulverised coal firing. Topping the gas turbine made the efficiency rise to 2% higher than the steam process (Stoll and Bleif 1986).

The programme "Combustion 2000", funded by the US Department of Energy (DoE) for the development of efficient technologies for the generation of power from coal, also pursued the EFCC process as a concept. Contracts were issued to two independent consortia, led by the Foster Wheeler Development Company (FWDC) and United Technologies Research Center (UTRC), respectively. The development of the overall process, called "High Performance Power Systems" (HIPPS), however, only relied on currently available technology. The most recent feasible concept design involved a metallic heat exchanger which is used to preheat air, with natural gas additionally being fed to raise the temperature.

Besides the high-temperature heat exchanger, the Foster Wheeler concept design included a pyrolyser and a char-fired combustion system, with heat being transferred to preheat the air and to produce steam. Air is preheated up to 760°C using tube banks constructed of alloyed steel; in a topping combustor that is fired with fuel gas from the pyrolyser, air is further heated to a gas turbine inlet temperature of 1,288°C.

The UTRC HIPPS concept was based on a turbine working fluid heat exchanger outlet temperature of 1,000°C and a gas turbine inlet temperature of 1,260°C. With natural gas contributing about one third of the thermal power output, an efficiency of 50.7% (LHV) was given (Klara 1994a, 1994b; Ruth 1997, 2001). However, Phase III of the HIPPS programme, which would have involved construction of a demonstration plant, was terminated and a demonstration plant will not be built (Benson 2000).

7.5.3.2 EFCC Processes with Ceramic Heat Exchangers

In the 1970s and 1980s, several US companies carried out investigations into the use of ceramic heat exchangers in EFCC processes. The results, however, have not been translated into practice in commercial EFCC plants.

In 1977, Solar Turbines Inc., supported by the US Department of Energy and the Electric Power Research Institute (EPRI), started a project investigating a ceramic high-temperature heat exchanger (SolarTurbines 1980). In the course of this research project, material tests were carried out in different atmospheres and at different temperatures, followed by a strength test. Furthermore, investigations into joining techniques of ceramic heat exchanger components were a subject of the research. In the end, a vertical shell-and-tube heat exchanger with a counterflow configuration made of silicon carbide tubes was built.

In a subsequent project, this heat exchanger was successfully subjected to 1,370°C hot flue gas from the firing of oil (Ward et al. 1983). There are no subsequent publications on tests using coal flue gas, which suggests that operational problems arose.

In the area of indirectly fired gas turbine processes, Solar Turbines further planned to use a ceramic high-temperature heat exchanger in waste incineration. The research project was to conclude with field tests using a "high-pressure ceramic heat exchange system" (HiPHES) with a thermal capacity of 10 MW in a waste incineration plant in Houston, Texas (Harkins and Ward 1989). Again, there are no current publications on their experiences.

From the mid to the end of the 1970s, investigations into high-temperature heat exchangers in coal flue gas atmospheres were carried out by AiResearch with the support of the Electric Power Research Institute. For instance, a small recuperator model was tested at a flue gas temperature of 1,260°C. The material temperature in this test was 1,093°C (Pietsch 1978). In a following project, the fundamentals of the manufacturing process and the design of the heat exchanger were adapted to a closed gas turbine process using argon as the working fluid of the gas turbine. Use was also made of the previous material characterisations during this adaptation (Coombs et al. 1979).

Probably the most advanced development programme of an EFCC process incorporating a ceramic heat exchanger was carried out in the USA by Hague International and other industrial partners and supported by the US Department of Energy. Within the framework of the first phase, from 1987 to 1989, suitable materials were tested; high-pressure and high-temperature experiments were carried out with different heat exchanger tubes; the effect on fouling and corrosion by the coal flue gases was determined; and methods of cleaning were trialled (Vandervort and Orozco 1992; LaHaye et al. 1990). According to the published data, these investigations show that it is possible to control the material stress, fouling and slagging at material temperatures of 1,200°C.

In Phase II of the "Combustion 2000" programme (1990–1994), an experimental plant with a thermal capacity of 7.4 MW was built, equipped with a ceramic heat exchanger of 2 MW. The execution of the project was halted in 1996, however, because the operation of the plant was impossible beyond a period of 50 h (DoE 1997). Phase III, which was designed to retrofit an existing coal-fired power plant to be an EFCC process, and had already begun in 1994, was stopped as well.

The design of the Phase III demonstration retrofitted power plant, Warren, had an efficiency of 37%, but was stopped before it was completed. The gross gas turbine output was 22 MW_{el} and the gross steam turbine output reached 48 MW_{el} – so the resulting total output, taking into account an auxiliary power requirement of 4 MW_{el}, amounted to 66 MW_{el} (LaHaye and Bary 1994). The planned outlet temperatures of the high-temperature heat exchanger were at roughly 900°C at tube wall temperatures of about 1,150°C. The layout seems to be such that tube wall temperatures should be kept well below 1,200°C. Since the ceramic tubes used are supposed to be suitable for temperatures up to 1,480°C, this limitation to 1,200°C could be an indication of fouling and corrosion problems. The low efficiency of the planned retrofit is to be put down to the low gas turbine inlet temperature and the applied steam cycle (470°C, 60 bar) being not of high enough quality because of the retrofitting.

A schematic diagram of the furnace and heat exchanger of the 7.4 MW_{th} EFCC test plant is shown in Fig. 7.63. The planned design of the Warren power plant described hereafter follows the layout of the test plant. The combustion takes place in a slag-tap furnace, the operation of which is air-staged for NO_x reduction. The furnace was designed as a membrane wall construction with water/steam cooling, refractory-lined to minimise the heat absorption by the walls.

Particle-laden gases exit the combustor and enter the slag screen to remove particles above a certain size. This should prevent ash deposition in the ceramic heat exchanger. The slag screen was to consist of single rods and to be designed such that particles larger than 12 μm would not follow the flue gas as it deflected but collect on the rods.

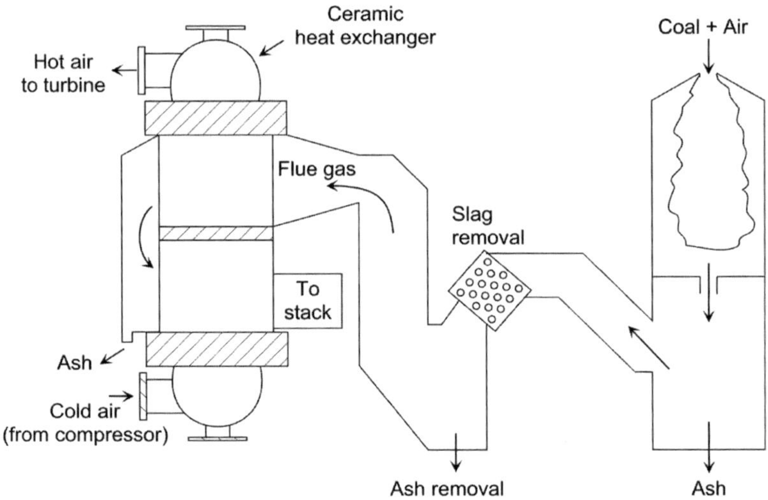

Fig. 7.63 Schematic diagram of a 7.4 MW_{th} EFCC test plant (Vandervort 1991, Vandervort and Orozco 1992)

The dimensions of the ceramic heat exchanger of the Warren power plant were to be 26.8 × 8.2 × 2.4 m, comprising 700 tubes in total. The flue gas was to flow on the outside of the tubes through four heat exchanger passes altogether, the compressed air through the vertical tubes. The ceramic tubes of one pass were to have a length of 4.9 m and an outside diameter of 10 cm. The heat exchanger material to be used was ceramic on an aluminium/silica carbide base (LaHaye and Bary 1994). The pressure loss on the side of the flue gas given for the dust collector and the heat exchanger was below 50 mbar, and on the side of the air in the heat exchanger, lower than 0.2 bar. The total leakage loss was assumed to be less than 0.5% (LaHaye and Bary 1994).

In Germany, in the context of the development of pressurised pulverised coal fired furnaces, the use of a high-temperature heat exchanger was also considered (Hannes 1986). In contrast to the variant shown in Fig. 7.47, the concept design included a low-temperature heat exchanger as well as a high-temperature one. After the transfer of the flue gas heat to the cleaned gas, the flue gas, at a temperature still above the ash fluid temperature, was to be cooled below the ash deformation temperature by adding warm air at 300°C in order to prevent uncontrollable sticky deposits in the following convective heat exchanger. The air guidance was also chosen to prevent sticking in the multi-fuel combustion chamber by producing an air curtain (Neumann et al. 1996).

The firing system is arranged as a cyclone furnace boiler, where part of the ash is removed in a molten state. In the radiation section which follows, heat is transferred by a high-temperature heat exchanger. The construction of the heat exchanger can be seen in Fig. 7.64. It consists of two concentric ceramic tubes set in an air-cooled pressure vessel. The flue gas transfers its heat by radiation to the ceramic tube which forms the furnace wall. Within the cross-section of the two ceramic tubes, the

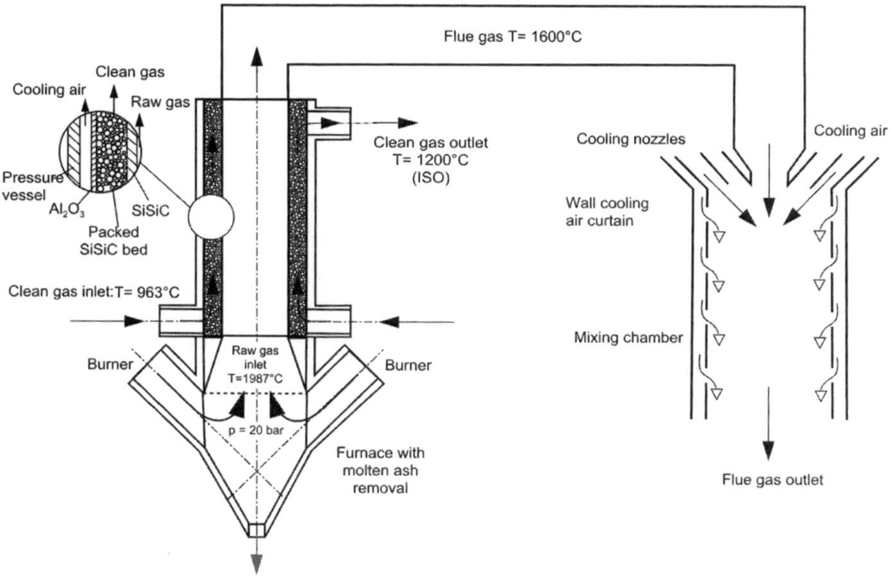

Fig. 7.64 An EFCC process with a furnace, heat exchanger and multi-fuel combustion chamber (Neumann et al. 1996)

working fluid of the gas turbine absorbs the dissipated heat, mostly by convection. The annular clearance contains ceramic packing to improve the heat transfer. The exterior pressure vessel is cooled by conducting air between the exterior ceramic tube and the vessel wall, so that the ceramic heat exchanger is not loaded with pressure (because both the flue gas and the working fluid of the turbine are pressurised).

A 600 MW_{el} plant needs eight parallel trains each with a furnace and a high-temperature heat exchanger, where the high-temperature heat exchanger has a height of 12.3 m and the pressure vessel a diameter of 3.5 m. The efficiency depends on the temperature of the gas cleaning stage. With flue gas cleaning at the cold end, the efficiency is 48%; with hot gas cleaning at 850°C, the calculated efficiency amounts to 51% (Neumann et al. 1996).

The Italian company Ansaldo Ricerche, in collaboration with its European partners, is undertaking a programme which aims to realise a ceramic heat exchanger module. Initial programmes concentrated on the development of production techniques and material development for reliable and cost-effective ceramic tubes for use in aggressive and high-temperature environments. Research has mainly concentrated on coated carbon fibre reinforced silicon carbide and one monolithic SiC. Continuation of this research focussed on the design of a high-temperature heat exchanger module.

The design of the heat exchanger and its modules is shown in Fig. 7.65. The heat exchanger is designed to operate with the flue gas at 1,395°C and 0.1 MPa and the compressed air at 1.7 MPa with a maximum temperature of 1,300°C. The flue gas and compressed air flow in opposite directions through a series of modules.

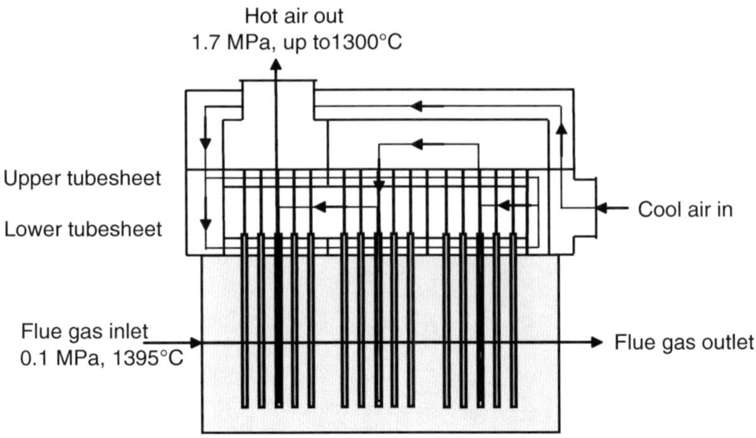

Fig. 7.65 Ceramic heat exchanger module (Benson 2000)

Those modules which are only exposed to temperatures below 850–900°C can be constructed from heat exchange tubes made from ODS alloys. The tubes in the high-temperature modules are ceramic. The structural elements of the heat exchanger are made from metallic materials and cooled by air. After preheating, the air is conducted through the ceramic tubes, which are suspended in the flue gas, where the air is heated to the TIT of 1,300°C. The flue gas flows directly through the module, transferring heat to the compressed air.

A ceramic heat exchanger is being tested at a 5 MW multi-fuel boiler burning coal to generate the flue gas. In the proposed design, the heat exchanger consists of four modules: two modules containing 42 ceramic U-tube arrangements and two modules containing 40 ODS alloy U-tubes. The outer tube dimensions are 1.55 m in length by 7.5 cm outer diameter. The modules are spaced 50 cm apart, with soot-blowing equipment incorporated between them. It is planned to initially use monolithic SiC tubes in the ceramic heat exchanger modules and then swap to CMC (ceramic matrix composite) tubes based on C/C-SiC when a suitably developed material becomes available (Benson 2000).

7.5.4 Conclusions

The EFCC process offers the advantage of high efficiency; however, the problems associated with the high temperatures are not yet solved. The major bottleneck seems to be the corrosion attack by liquid slag when the ceramic heat exchangers are exposed to temperatures of 1,400–1,600°C. At temperatures up to 1,300°C, oxide dispersion strengthened alloys are preferable as they are easier to handle. Whereas for large-scale coal-fired power stations the technology is not yet applicable, the technology could be suitable for smaller biomass-based systems with lower gas turbine inlet temperatures.

7.6 Integrated Gasification Combined Cycle (IGCC)

7.6.1 History of Coal Gasification

Gasification, in the broadest sense, means the transformation of a solid carbonaceous fuel into a gas with a useful calorific value. The wider variety of applications and the advantages of a gaseous fuel in contrast to solid coal spurred the development of coal gasification. This development began in the 19th century with the powering of street lighting by coal gas (1812 saw the first commercial gasworks for illuminating gas production in London), while in later years gasification for heating purposes became the dominant motivator. This gas, called town gas, was produced in a discontinuous pyrolysis process similar to a coking process. From about 1880, this method was superseded by the water gas shift reaction, which was then used for town gas production until the middle of the 20th century. In this process, with the supply of heat and steam, the solid carbon is transferred into a mixture of carbon monoxide and hydrogen, as expressed in Eq. (7.37). The rising availability of cheap natural gas put an end to the consumption of town gas in Europe from about 1970, and coal gasification became limited to niche applications (Higman and van der Burgt 2008).

A step of great importance in the development of gasification was the commercialisation of cryogenic air separation by Carl von Linde in the 1920s, which then made it possible to operate a continuous, oxygen-blown gasification process for the production of synthesis gas (commonly known as syngas) and hydrogen. Developments took place around that period that were the precursors of today's gasification technologies: the Winkler process (a fluidised bed gasifier process) in 1926, the Lurgi process (a fixed bed coal gasification process) in 1931 and the Koppers-Totzek process (an entrained-flow process for the gasification of pulverised coal) in 1940. After these processes had been established, there were only minor advances in the following 40 years, save for the rise of a petrochemical industry based on coal gasification and the Fischer – Tropsch synthesis process by the Sasol Company in South Africa.

In the wake of the oil crisis in the 1970s and the feared shortage of natural gas and oil, extensive efforts were undertaken then and in the 1980s to further develop and demonstrate gasification technologies for the production of syngas, liquid or gaseous energy sources, or electric power. By way of example, Lurgi and British Gas developed a molten bath gasifier; Koppers and Shell together worked on a pressurised version of the Koppers-Totzek gasifier; and Rheinbraun developed the high-temperature Winkler process. As oil prices sunk, however, the interest in coal gasification or liquefaction decreased at the end of the 1980s, restricting further development.

Coal gasification technology for electric power production was demonstrated in Europe in several large-scale plants (Lünen, 170 MW, 1972; Buggenum, 250 MW, 1992; Puertollano, 335 MW, 1997) with success. The motivation for pursuing this technology in the future is the potential for better environmental performance at a lower marginal cost. A broad introduction into the electricity generation market of this technology, however, has not yet taken place due to its cost disadvantages and low availability.

Nonetheless, there are a great number of gasification plants worldwide at the moment. The cumulative gasification capacity reported for 2008 ranged around 70 GW_{th} (Higman and van der Burgt 2008). These plants are mostly designed and built to produce syngas for ammonia, hydrogen or transportation fuel production. The feedstocks in many cases are difficult fuels, such as refinery waste or petrol coke, which are hard to exploit with other technologies. Here, the advantages of the low environmental impact of gasification come fully to fruition.

A strong increase over the past 10 years in installed gasification capacities is noticeable. The various advantages of gasification technology give reason to believe that gasification will see a revival:

- Gasifiers are suitable for a wide range of fuels. In an entrained-flow gasifier for instance, coal can be used together with biomass and residual material. Besides solid fuels, liquid fuels can also be used.
- The emissions of all gaseous pollutants and trace components, given that gas cleaning has to be installed for process-engineering reasons, are significantly lower than from conventional power plants.
- Using two additional common process steps (a CO shift reactor and CO_2 capture), it is possible to separate CO_2. The required additional effort is lower, owing to the process pressure and the small volumetric flow, than in downstream CO_2 scrubbing following a steam power cycle. Considering this, the advantage in efficiency that gasification already has would become still greater. The present disadvantage of higher capital costs could be balanced out.
- Gasification technology offers the highest product flexibility. Gasification together with synthesis gas generation form the first process step to produce liquid fuels (Fischer – Tropsch), gaseous secondary energy sources such as hydrogen or synthetic natural gas, and methanol or ammonia for the chemical industry (Higman and van der Burgt 2008).

7.6.2 Applications of Gasification Technology

7.6.2.1 Generation of Secondary Energy Sources

Gasification technologies are suited to providing a high-energy gas for the production of basic products for the chemical industry or of secondary energy sources. Figure 7.66 presents an overview of the various possible uses of product gases from gasification processes.

The synthesis process for the generation of the final product determines the entire cycle and defines the requirements for the gasification process and the gas treatment. The major gas components and allowable gaseous pollutants have to be taken into consideration. The requirements differ from process to process but, in general, the level to be met is high. The sulphur content by volume, for example, has to be

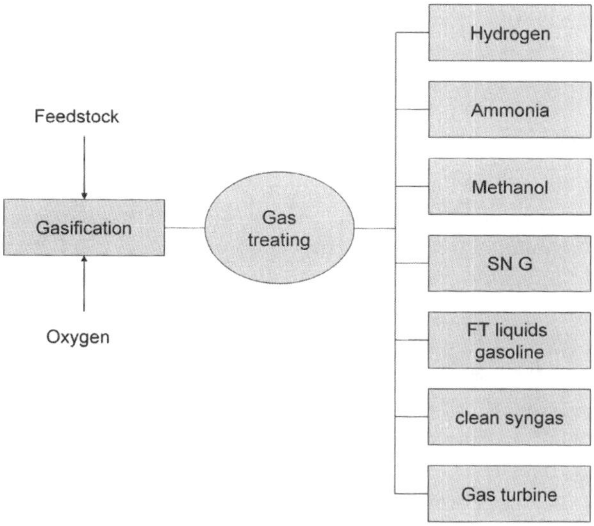

Fig. 7.66 Production possibilities with gasification

below 1.0 ppm or even below 0.1 ppm. A summary of the gas purity requirements of different synthesis processes can be found in Radtke et al. (2006), Ogriseck and Meyer (2005) and Higman and van der Burgt (2008).

It is plausible to say that for all processes, nitrogen is an unwelcome component in the final gasification gas. Given that removing nitrogen from the gas produced is more complex than the fractionation of air, almost all industrial-scale gasification processes are operated with oxygen. In principle, alternative methods to an oxygen-blown gasification process exist, such as allothermal processes, where heat is transferred from external sources, thereby generating a nitrogen-free gas product. These processes, which are considered for medium-scale biomass applications (see Sect. 6.3), are significantly more costly for larger-scale units and are therefore not today's generally accepted practice. Instead, high-temperature entrained-flow gasification is the technology considered feasible for the production of basic chemical substances or secondary energy media. In this process syngas is produced, which consists almost exclusively of H_2 and CO. Further processing of the gas, such as the CO shift process, is then required so that it conforms to compositional requirements as a product. The removal of trace elements such as sulphur, phosphorus and chlorine is required so that deactivation of the catalyst of the syngas is minimised.

7.6.2.2 IGCC With and Without CO_2 Capture

The integrated gasification combined cycle (IGCC) produces electricity from a solid or a liquid fuel. First, the fuel is converted to a syngas, which is then converted to electricity in a combined cycle power plant consisting of a gas turbine process and a steam turbine process with a heat recovery steam generator (HRSG). The combined cycle technology is similar to the technology used in modern natural gas fired power plants.

IGCC Without CO_2 Capture

Figure 7.67 shows the principle units of a coal-based IGCC plant without CO_2 capture. The coal is supplied to the gasifier, where it is partially oxidised under pressure (30–80 bar). The plant uses oxygen as the oxidant and therefore has an air separation unit (ASU). In the gasifier, which is normally of the entrained-flow slagging type, the temperature may exceed 1,500°C. In addition to its chemical energy (heating value), the hot raw syngas contains sensible heat, which may be recovered in heat exchangers to produce steam for the steam turbine. It would be desirable to clean the raw syngas without cooling and to deliver the hot syngas to the gas turbine, which would result in a higher efficiency. However, hot gas cleaning (discussed in Sect. 7.6.5.6) is not state of the art, and proven technologies operate at near ambient temperatures. In the gas clean-up process, particles, sulphur and other impurities are removed.

For the sake of efficiency, it is practical to integrate the air separation as completely as possible into the IGCC process. It is talked of as 100% integration if the air flow compressed in the compressor of the gas turbine is fed entirely to the air separation unit and the nitrogen stream separated under pressure is completely

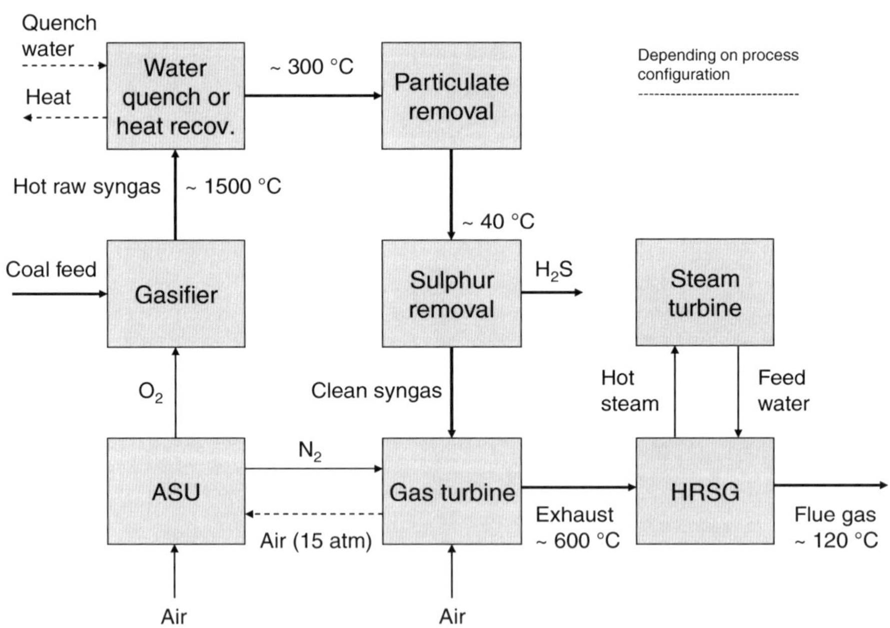

Fig. 7.67 An IGCC process without CO_2 capture (Maurstad 2005)

expanded in the gas turbine. Full integration yields the highest efficiency; partial integration can result in a higher output and a higher operational flexibility, for example in the start-up process.

IGCC with CO_2 Capture

The IGCC power plant offers favourable conditions for capturing CO_2, because the separation of CO_2 under pressure involves less energy loss than downstream removal from the atmospheric flue gas. The capture process of CO_2 in an IGCC power plant needs two additional components, as shown in Fig. 7.68:

- a so-called shift reactor, which converts the CO of the syngas into CO_2 and hydrogen using water vapour, and
- a unit for CO_2 removal by chemical or physical absorption, which removes the CO_2 from the hydrogen/carbon dioxide mixture.

The gas turbine is thus fed with hydrogen or with a fuel rich in hydrogen.

The two additionally necessary process steps are discussed in more detail in Sects. 7.6.5.4 and 7.6.5.5.

In an IGCC process with CO_2 capture the efficiency is lower because, due to the exothermic shift reaction, some of the chemical energy of the fuel is converted into

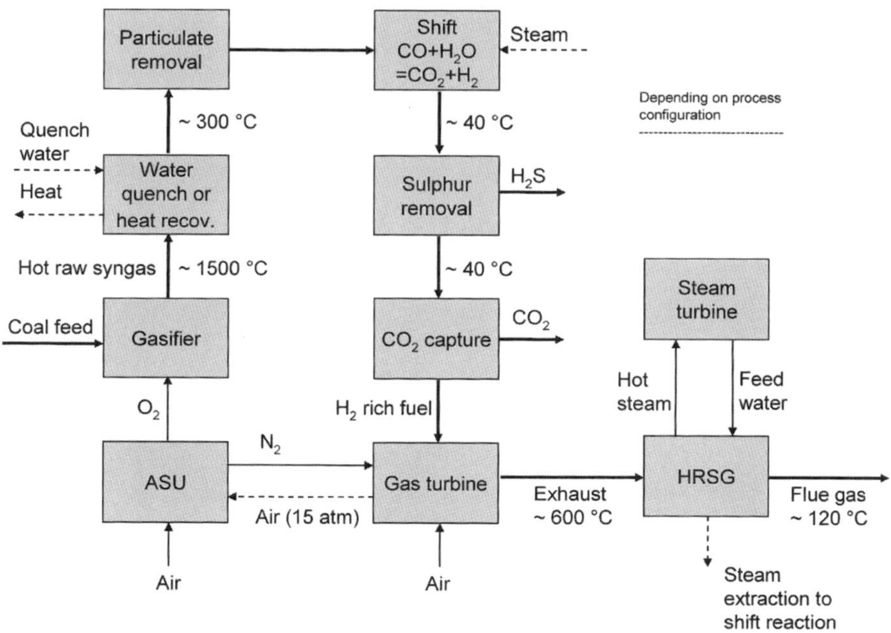

Fig. 7.68 IGCC process with CO_2 capture (Maurstad 2005)

heat at a low temperature, and this heat energy can only be converted to electrical power with a lower efficiency than chemical energy. In most gasification processes, the steam – carbon ratio is too low for the shift reaction, so high-quality steam has to be added. In addition, energy is needed for compressing the CO_2. On top of that, energy is also need for the separation process, for example for regenerating a solvent.

7.6.2.3 Factors Affecting the Efficiency of an IGCC

Figure 7.69 shows the energy flows of a simplified IGCC system. In the gasification stage, the chemically bound energy of the fuel $\dot{Q}_F$ is converted into the energy $\dot{Q}_{Gas}$ of the syngas in the gasification island (comprised of the gasifier, syngas cooling and gas cleaning stages). $\dot{Q}_{Gas}$ includes both the sensible heat and the chemically bound energy.

The (warm gas) gasification efficiency, η_{Gas}, is defined as

$$\eta_{Gas} = \frac{\dot{Q}_{Gas,Chem.} + \dot{Q}_{Gas,sensible}}{Q_F} \tag{7.14}$$

The gasification efficiency of the gasification island is dependent on the heat losses $\dot{Q}_{Loss}$ and the steam production $\dot{Q}_{St,Gas}$, which mainly arises from syngas cooling. The dimensionless parameters α and β relate the losses and the steam production to the fuel input:

$$\alpha = \frac{\dot{Q}_{Loss}}{\dot{Q}_F} \quad \beta = \frac{\dot{Q}_{St,Gas}}{\dot{Q}_F} \tag{7.15}$$

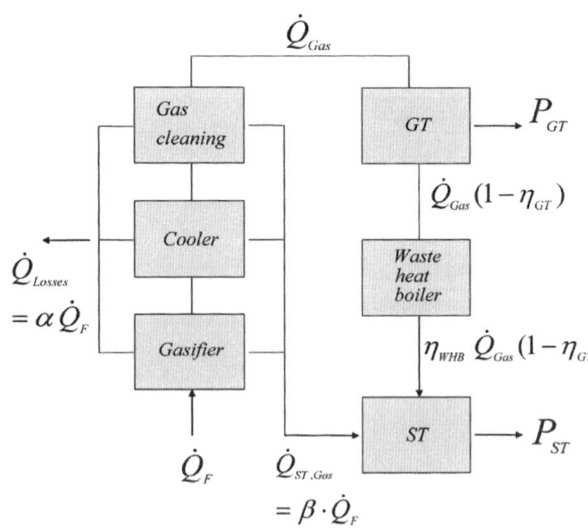

Fig. 7.69 A simplified IGCC process for efficiency calculations

whereas the cold gas efficiency, $\eta_{\text{Cold Gas}}$

$$\eta_{\text{Cold Gas}} = \frac{\dot{Q}_{\text{Gas,Chem}}}{\dot{Q}_{\text{F}}} \tag{7.16}$$

only considers the chemically bound energy of the product gas.

In an IGCC system, power is produced in gas and steam turbines. The efficiency η_{IGCC} is defined as

$$\eta_{\text{IGCC}} = \frac{P_{\text{ST}} + P_{\text{GT}}}{\dot{Q}_{\text{F}}} \tag{7.17}$$

In the gas turbine, the energy $\dot{Q}_{\text{Gas}}$ is converted into power P_{GT} with the efficiency η_{GT}:

$$P_{\text{GT}} = \eta_{\text{GT}} \dot{Q}_{\text{Gas}} = \eta_{\text{GT}} \cdot \eta_{\text{Gas}} \cdot \dot{Q}_{\text{F}} \tag{7.18}$$

Steam is produced both in the gasification island $\dot{Q}_{\text{St,Gas}}$ and in the waste heat boiler $\dot{Q}_{\text{St,WHB}}$.

The heat which is not converted into power in the gas turbine is fed to the waste heat boiler and converted with the efficiency of the waste heat boiler η_{WHB} into steam:

$$\dot{Q}_{\text{St,WHB}} = \eta_{\text{WHB}} \cdot \dot{Q}_{\text{Gas}} \cdot (1 - \eta_{\text{GT}}) = \eta_{\text{WHB}} \cdot \eta_{\text{Gas}}(1 - \eta_{\text{GT}}) \cdot \dot{Q}_{\text{F}} \tag{7.19}$$

Both the steam of the waste heat steam generator and the steam from the gasification island are converted into power with the efficiency η_{ST} via steam in the steam process:

$$P_{\text{ST}} = \eta_{\text{St}}(\dot{Q}_{\text{ST,Gas}} + \dot{Q}_{\text{ST,WHB}}) = \eta_{\text{St}}(\beta \cdot \dot{Q}_{\text{F}} + \eta_{\text{WHB}} \cdot \eta_{\text{Gas}} \cdot (1 - \eta_{\text{GT}}) \cdot Q_{\text{F}}) \tag{7.20}$$

The efficiency of the IGCC process as a whole is then given as

$$\eta_{\text{IGCC}} = \eta_{\text{Gas}}(\eta_{\text{GT}} + \eta_{\text{ST}}\eta_{\text{WHB}}(1 - \eta_{\text{GT}})) + \eta_{\text{ST}} \cdot \beta \tag{7.21}$$

Using Eq. (7.21), it is possible to discuss the influence of the gasification efficiency, η_{Gas}, on the total efficiency, η_{IGCC}. If the gasification efficiency η_{Gas} deteriorates, less output is produced in the gas and steam turbines and the efficiency of the IGCC process decreases. Usually an effort is made to utilise the losses occurring in the gasification island for steam production, the measure of this being the parameter β. This steam, however, is converted into electrical power only in the steam turbine, not in the more efficient gas and steam turbine process.

The efficiency of the gasification stage in an IGCC power plant means that compared to a natural gas fired combined cycle process, the overall efficiency is lower. State of the art in gas cleaning are processes at low temperatures, so that the gas turbine is not supplied with sensible heat but only with chemical energy of the syngas. The gasification efficiency, η_{Gas}, corresponds to the cold gas efficiency, $\eta_{Cold\ Gas}$. Using a hot gas cleaning process it would be possible to increase the efficiency by providing the combined cycle process with additional sensible heat. An additional aim should be to convert the sensible heat into chemically bound energy. That is why the evaluation of the energy conversion in gasification processes generally brings into play cold gas efficiencies. These are typically at about 80%, with high-temperature gasifiers having lower and fluidised bed gasifiers having higher cold gas efficiencies.

The cold gas efficiency includes the carbon conversion, which is also used for drawing comparisons between gasifiers:

$$\eta_{carbon} = 1 - \frac{c_{Res}[\mathrm{kmol/h}]}{c_{Fuel}[\mathrm{kmol/h}]} \tag{7.22}$$

where c_{Res} is the carbon in the gasification residue in kmol/h and c_{Fuel} is the carbon in the fuel in kmol/h. The carbon conversion in high-temperature entrained-flow gasifiers reaches about 99%; in fluidised bed gasifiers, however, the conversion of carbon is lower and is a function of the reactivity and the volatile matter content of the fuel.

A further reduction of the efficiency of IGCC power plants comes about as a result of the energy used by the air separation process. By integrating the air separation unit into the process as a whole, the input energy can be reduced or energy can be recovered.

For natural gas fuelled IGCC processes, efficiencies of 50–60% are the current state of the art. This is the result of Eq. (7.21) if the efficiency η_{Gas} is set at 1 and the steam fraction β is set at 0. The gas turbine efficiency, η_{GT}, ranges around 40%; the efficiency of the steam process, η_{ST}, is about 35% and the efficiency of the heat recovery steam generator, η_{WHB}, about 90%. So the efficiency of the combined cycle, η_{GuD}, reaches 59% as a result. If the coal gasification efficiency is around 80%, the IGCC efficiency falls to 47% if no steam is produced. Assuming all losses are used for steam production ($\beta = 0.2$), the efficiency could be increased by a maximum of 7%. In practice though, the extra efficiency is significantly lower, being partly used up by the air separation process. These considerations clarify the need to achieve a cold gas efficiency as high as possible.

7.6.3 Gasification Systems and Chemical Reactions

7.6.3.1 Allothermal and Autothermal Gasification

Gasification systems can be categorised according to the heat input into autothermal and allothermal gasification processes (Jüntgen and van Heek 1981).

The principle of autothermal gasification using steam is depicted in a simplified form in Fig. 7.70. The fuel, depending on the reactor type, is converted at temperatures between 800 and 1,800°C by means of oxygen and water vapour into a gas which, besides CO_2 and perhaps also CH_4, mainly contains CO and H_2. The mixture of the latter two gases is termed synthesis gas because it is used in the chemical industry for the synthesis of methanol, ammonia and hydrocarbons (the latter using the Fischer – Tropsch process). The term autothermal comes from the fact that the process runs without external heat supply and because the heat consumption of the endothermic reaction of the coal with steam and the heat generation of the exothermic reaction with oxygen compensate each other. If air is used instead of oxygen, the gas produced becomes diluted by nitrogen and is usually only suited to be burned on the spot for heating purposes. One disadvantage of autothermal gasification is the lower gasification efficiency due to burning part of the coal. Furthermore, the production of a highly calorific gas needs oxygen as a feedstock and thus a more complex air separation plant.

The principle of allothermal gasification is also shown schematically in Fig. 7.70. Heat in this case is externally supplied to meet the heat requirement of the endothermic gasification reaction of the coal with water vapour. The heat can be fed to the gasification medium by recuperative tube-bundle heat exchangers or regeneratively via a circulating heat-carrying medium. The heat transfers from heating medium to tube wall and from tube wall to the fuel (recuperative heat exchanger) or

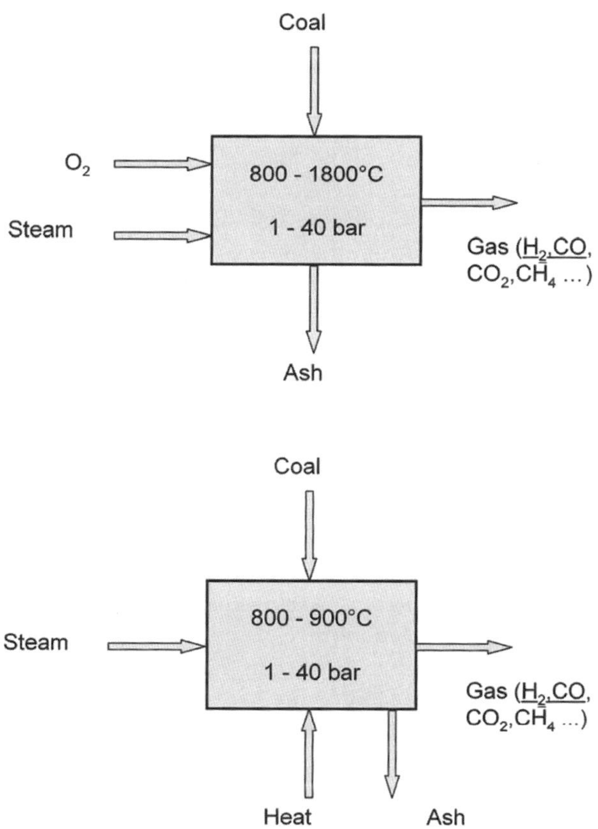

Fig. 7.70 Principle of autothermal (*above*) and allothermal gasification (*below*)

to and from the heat-carrying medium (regenerative heat exchanger) are parameters determining the size and economy of the process. The fact that fluidised beds are characterised by a good heat transfer is the reason why they are taken into particular consideration for allothermal gasification. For greater gasification capacities, however, allothermal gasification is yet to be accepted because of the heat transfer problems. The method is practical for smaller capacities because the additional work and expense of heat transfer is compensated by not having an air separation step.

7.6.3.2 Basic Chemical Reactions

In gasification, similar to combustion (see Sect. 5.1), different phases are distinguished:

- Heating-up of the fuel
- Drying of the fuel
- Pyrolysis reactions
- Conversion of the solid carbon by gasification in the presence of oxygen, water vapour or CO_2

In the same way as in combustion processes, heat has to be supplied for the partial processes of heating-up, drying and pyrolysis. The dominant factor of the gasification process is the heat demand of the gasification reactions. For this purpose, a corresponding amount of heat has to be fed from outside the process or released by partial oxidation.

In the entire process of coal gasification, numerous reactions take part. On the one hand, there are heterogeneous reactions in which the gasifying medium and also the product gases react with the solid matter. On the other hand, homogeneous reactions take place in the gaseous phase where the primary gaseous products go through conversion reactions involving both each other and the gasifying agent. The final composition of the gas produced depends on the interaction of all these reactions.

Despite the very complex molecular structure of coal, it is reasonable when discussing the reactions of the coal gasification to consider carbon alone as a first approximation, and only then, as a second approximation, the pyrolysis which precedes the gasification process. In the following, the most important heterogeneous and homogeneous reactions of carbon and of the pyrolysis are compiled. These can be used as basic reactions for describing the complex gasification reactions.

The principle possible pathways for gasifying carbonaceous fuel or fuel-containing hydrocarbon(s) are the partial oxidation reaction according to Eq. (7.23), the Boudouard reaction (Eq. 7.26), the heterogeneous water gas reaction (Eq. 7.27) and the methanation reaction (Eq. 7.28), all compiled in Table 7.15. The Boudouard reaction is of secondary importance in gas production from coal but it is significant for the blast furnace process. In this process, the carbon dioxide formed from the combustion reacts when flowing through the layers of coke above to form carbon monoxide, which for its part becomes oxygenated to form CO_2 by reducing the iron

Table 7.15 Gasification reactions (Higman and van der Burgt 2008), (Jüntgen and van Heek 1981)

Combustion reactions		
	Δh	
$C + {}^1/_2\, O_2 \leftrightarrow CO$	-111 MJ/kmol	(7.23)
$CO + {}^1/_2\, O_2 \leftrightarrow CO_2$	-283 MJ/kmol	(7.24)
$H_2 + {}^1/_2\, O_2 \leftrightarrow H_2O$	-242 MJ/kmol	(7.25)
Heterogeneous gasification reactions		
Boudouard reaction		
$C + CO_2 \leftrightarrow 2CO$	$+172$ MJ/kmol	(7.26)
Water gas reaction		
$C + H_2O \leftrightarrow CO + H_2$	$+131$ MJ/kmol	(7.27)
Methanation		
$C + 2\,H_2 \leftrightarrow CH_4$	-75 MJ/kmol	(7.28)
Homogeneous gasification reactions		
Homogeneous water gas reaction		
$CO + H_2O \leftrightarrow CO_2 + H_2$	-41 MJ/kmol	(7.29)
Steam reforming		
$CH_4 + H_2O \leftrightarrow CO + 3H_2$	$+206$ MJ/kmol	(7.30)
Pyrolysis reactions		
$C_1H_xO_y \rightarrow (1-y)\,C + y\,CO + x/2\,H_2$	$+17.4$ kJ/mol[1)]	(7.31)
$C_1H_xO_y \rightarrow (1-y-x/8)\,C + y\,CO + x/4\,H_2 + x/8\,CH_4$	$+8.1$ kJ/mol[1)]	(7.32)

[1)] For gas coal: $x = 0.874,\ y = 0.0794$

oxides. The methanation reaction is of great importance in hydrogasification, i.e. the gasification of coal by means of hydrogen. The objective of hydrogasification is to produce methane directly from coal as a substitute for natural gas. In allothermal steam gasification, the dominant process is the heterogeneous and strongly endothermic water gas reaction. In order to procure the reaction heat for 1 kg of carbon for steam gasification, about 0.29 kg of carbon has to be burned into CO_2. For most industrial gasifiers heated autothermally, the gasification process can be described by means of the partial oxidation and heterogeneous water gas reactions.

The principle reaction of the gas phase is the homogeneous water gas reaction (Eq. 7.29), where the carbon monoxide (formed previously) and water vapour are converted into hydrogen and carbon dioxide in an exothermic process. This reaction is used downstream in other industrial processes, not only in gasifiers, to convert CO completely or partially into hydrogen and to create a suitable feed gas for syntheses or other kinds of use. This step is also termed conversion or the CO shift reaction.

Another important reaction is the endothermic steam-reforming reaction (Eq. 7.30), which serves to convert methane into a synthesis gas. Running in the reverse direction, it is a strongly exothermic methanation reaction, which is used in downstream catalytic processes to produce methane.

The combustion reactions with oxygen (Eqs. 7.23, 7.24 and 7.25) mostly run to completion under gasification conditions, so they need not be considered for determining the equilibrium. The three heterogeneous reactions (Eqs. 7.26, 7.27 and 7.28) suffice for this purpose. Assuming a complete conversion of carbon as in the case of entrained-flow gasifiers, it is possible to transform the reaction equations (7.26), (7.27) and (7.28) in order to obtain the two reaction equations (7.29) (subtraction of reaction 7.26 from reaction 7.27) and (7.30) (subtraction of reaction 7.28 from reaction 7.27). In entrained-flow gasifiers, the temperatures are commonly so high that in terms of thermodynamics, and in practice, hydrocarbons are no longer able to be found in noticeable concentrations.

So far the first reaction phase in the process of heating the coal up to the reaction temperature – the so-called pyrolysis reaction – has been ignored. In this phase, the coal is decomposed to coke and liquid and gaseous substances, which, depending on the reaction conditions, continue to react with the gasifying agent to form the gasification products. In general, pyrolysis can be described as follows:

$$\text{Coal} \rightarrow CH_4,\ C_mH_n,\ CO,\ CO_2,\ H_2,\ H_2O,\ \text{tars},\ \text{char} \tag{7.33}$$

If the pyrolysis products in the gasification reactor convert into the successor products of gasification, the following simple reaction scheme is sufficient:

$$\text{Coal} \rightarrow C + CH_4 + CO + CO_2 + H_2O \tag{7.34}$$

or for high-temperature processes:

$$\text{Coal} \rightarrow C + CO + H_2 \tag{7.35}$$

Table 7.15 also includes the pyrolysis reactions, showing the possible processes of conversion into carbon and the gasification products CO and H_2 (Eq. 7.41) or CO, H_2 and CH_4 (Eq. 7.42). It should be noted that the analyses in Eqs. 7.41 and 7.42, in the same way as the previous reactions, are made on a molar basis. For a bituminous coal with a weight composition of 85% C, 6% H_2 and 9% O_2, for instance, the result is a molar elemental formula of $C_1H_{0.847}O_{0.0794}$ and a molar weight of 14.12 kg/kmol. The pyrolysis reactions are slightly endothermic.

Besides the components of carbon, hydrogen and oxygen, most fuels also contain smaller fractions of sulphur and nitrogen. In the gasification process, sulphur is mainly converted into hydrogen sulphide; COS, CS_2 and other sulphur-containing molecules form only in small quantities. Fuel nitrogen is converted into molecular nitrogen, NH_3 and HCN. These compounds have little influence on the principle gas composition of the synthesis gas. Further information is found in Sect. 7.6.5.1.

7.6.3.3 Considerations of the Thermodynamic Equilibrium

Analyses of thermodynamic equilibrium help to calculate the reaction progress achieved therein, i.e. the greatest possible stable reaction progress of the reactions involved. For the reversible reactions in Table 7.15, chemical equilibrium, where forward and reverse reactions run at the same rate, is a function of temperature and pressure. Time, however, is not considered as a factor in such analyses, so information on how fast these states develop is not generated.

Quantitative information on the conversion and the composition of products is obtained by applying the law of mass action for equilibrium reactions. Each chemical equilibrium reaction can be formulated according to the general form

$$\sum_{i=1}^{N} \nu_i \, A_i = 0 \tag{7.36}$$

where ν_i is the molar ratio, or stoichiometric coefficient, and A_i is the substance involved in the reaction.

The state of the chemical equilibrium is described by the law of mass action, which expresses the correlation between the parameters depending on the concentration (measured, for instance, as the partial pressure or the volume fraction). Using the partial pressures p_{Ai} of the substances involved, the result for the equilibrium constant k_p is

$$k_p = \prod_{i=1}^{N} p_{Ai}{}^{\nu_i} \tag{7.37}$$

The law of mass action can also be applied to heterogeneous reactions. However, since the steam concentration is independent of the concentration of solids, and depends only on the temperature, this fixed value is included in the equilibrium constants, so the concentration of the solids need not be taken into account.

For the temperature-dependent equilibrium constant k_p, the gas constant R, the temperature and the reaction enthalpy ΔH, the following relation applies:

$$d\frac{\ln k_P}{dT} = \frac{\Delta H}{RT^2} \tag{7.38}$$

This means that, for endothermic reactions ($\Delta H > 0$), k_p rises with rising temperatures, i.e. a temperature increase pushes the reaction in the direction of the products. Inversely, k_p, for exothermic reactions ($\Delta H < 0$), rises with falling temperatures, so in this case a decrease in temperature is favourable for the formation of the products.

As the composition of the product gases is of interest in gasification, it makes sense to represent the law of mass action as a function of the volume fractions x_i. The equilibrium constant k_x describes the correlation of the volume fractions x_i:

$$k_x = \prod_{i=1}^{N} x_i{}^{\nu_i} \tag{7.39}$$

Dalton's law gives the relation between the volume fraction x_i and the partial pressure p_{Ai}:

$$x_i = \frac{p_{\acute{A}i}}{p} \tag{7.40}$$

Consequently, the following relation between k_x and k_p applies:

$$k_x = \prod_{i=1}^{N} x_i{}^{\nu_i} = \frac{k_p}{p\Sigma\nu_i} \tag{7.41}$$

This means that for reactions where the molar ratio $\Sigma\nu_i$ is not changed, the equilibrium constants k_x and k_p are equal. An example of a consequence of this can be shown for the homogeneous water gas reaction with $\Sigma\nu_i = 0$:

$$k_P = \frac{p_{CO_2} \cdot p_{H_2}}{p_{CO_2} \cdot p_{H_2O}} = k_x = \frac{x_{CO_2} \cdot x_{H_2}}{x_{CO} \cdot x_{H_2O}} \tag{7.42}$$

In reactions with a changing molar ratio, the total pressure p has an additional influence on the equilibrium composition. For the heterogeneous water gas reaction with a change in the molar ratio $\Sigma\nu_i$, for instance, the expression is

$$k_x = \frac{x_{CO} \cdot x_{H_2}}{x_{H_2O}} = \frac{k_P}{p} = \frac{p_{CO} p_{H_2}}{p_{H_2O}} \frac{1}{p} \tag{7.43}$$

and for the steam-reforming reaction with a change in the molar ratio of 2, the expression becomes

$$k_x = \frac{x_{CO} \cdot x_{H_2}^3}{x_{CH_4} \cdot x_{H_2O}} = \frac{k_P}{p^2} = \frac{p_{CO} p^3{}_{H_2}}{p_{CH_4}\, p_{H_2O}} = \frac{1}{p^2} \tag{7.44}$$

In both reactions, the equilibrium shifts towards the reactants with increasing pressure; with a temperature increase, it shifts in favour of the products because of the endothermic nature of the reactions. This corresponds to the principle of the minimal constraint developed by Le Chatelier, which says that a rise in pressure shifts the equilibrium in the direction of a volume decrease and a lowering of the pressure shifts the equilibrium towards a volume increase.

The equilibrium constants serve to calculate the conversion of the reactant material or the composition of the product gases. The equilibrium constants can be calculated either by experiment, using measured equilibrium concentrations, or using thermodynamic variables such as enthalpies and entropies.

Another, often applied, calculation method is the minimisation of the free enthalpy, also called the Gibbs free energy:

$$dG = dH - T\,dS \tag{7.45}$$

The method is based on the fact that in the state of equilibrium, a minimum of the free enthalpy arises. For the calculation of the free enthalpy of the mixtures, the thermodynamic data of the enthalpy and the entropy has to be known. Software and databases are commercially available for this purpose.

Thermodynamic analyses only describe the reality in gasification processes if the reactions run very quickly in comparison to the residence times. This does apply in some cases, for instance for the homogeneous reactions in entrained-flow reactors, due to the high temperatures of 1,500–1,600°C involved. Assuming an almost complete conversion of carbon, it is possible to calculate the gas-phase composition using Eqs. (7.39) and (7.40). In fluidised bed reactors, with low temperatures of 800–900°C, the gas composition cannot be determined by means of equilibrium analyses, though they may help to determine qualitative effects. The reactions involving the solid matter, in particular the gasification reactions of the solid carbon with steam, are the slowest reactions in the gasification process, so in this case kinetic analyses are required.

Effect of Pressure and Temperature on Gas Composition and Gasifier Efficiency

High pressures in gasification processes bear many advantages. This is why almost all industrial-scale gasifiers are operated at pressures of between 10 and 100 bar. High pressures reduce the energy demand of the total process, because the compressive work needed for the mass flows fed to the gasifier (oxygen, steam, fuel) is less than the work required for compressing the obtained synthesis gas. Furthermore, high pressures involve more compact and thus less expensive components. The pressure of the gasifier has to be adjusted to align with the rest of the process as a whole. For the exploitation of the syngas in combined cycle processes, pressures up to 30 bar are reasonable. For methanol synthesis, pressures from 130 to 180 bar would be advantageous, but such high pressures are not feasible in gasifiers.

Figure 7.71 shows the effect of the pressure on the gas composition from coal gasification at a temperature of 1,000°C based on thermodynamic equilibrium calculations. With rising pressure, the fractions of methane, carbon dioxide and steam increase; conversely, the contents of CO and H_2 diminish. This can be explained by the pressure effect, described above, on the steam-reforming reaction:

$$CH_4 + H_2O \leftrightarrow CO + 3H_2 \tag{7.30}$$

and on the reaction of methane with carbon dioxide:

$$CH_4 + CO_2 \leftrightarrow 2CO + 2H_2 \tag{7.46}$$

because the pressure works towards volume reduction. At higher temperatures, the effect of pressure is smaller, because in endothermic reactions higher temperatures shift the equilibrium in the direction of the products.

Figure 7.72 shows the effect of the temperature on the equilibrium composition at a pressure of 30 bar. As low methane contents are desired in the production of synthesis gas, temperatures higher than 1,300°C are required. For an IGCC process, these high temperatures are not necessary in this respect, because methane is rather advantageous. However, entrained-flow gasifiers are nevertheless designed for high

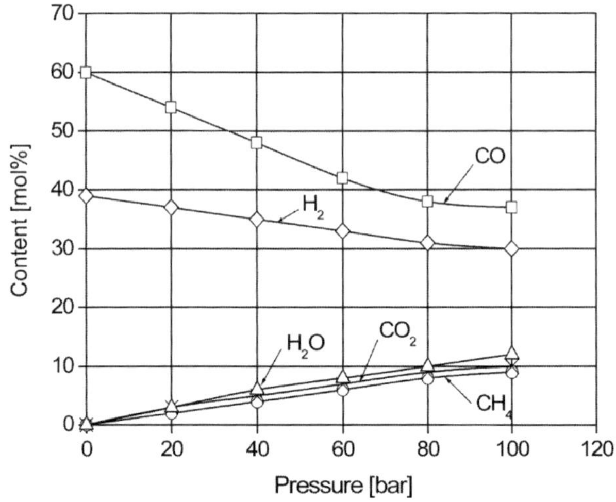

Fig. 7.71 Variation of syngas compositions with pressure at a temperature of 1,000°C (from Higman and van der Burgt 2008, © 2008, with permission from Elsevier)

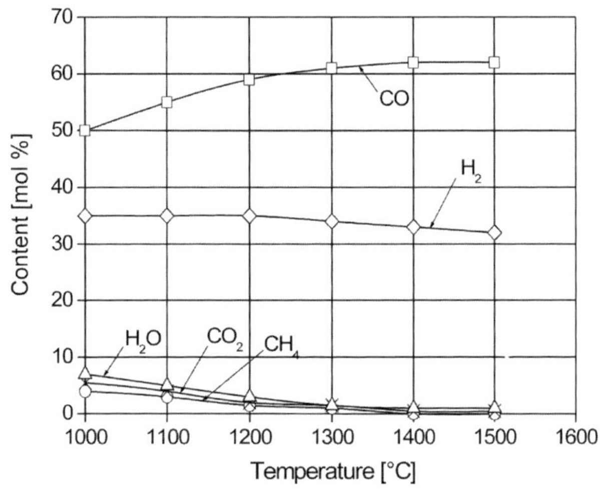

Fig. 7.72 Variation of syngas compositions due to temperature at a pressure of 30 bar (from Higman and van der Burgt 2008, © 2008, with permission from Elsevier)

temperatures, because the fuel's ash fluid temperature has to be exceeded for safe operation with molten slag removal.

Knowing the composition of the generated gas product, the gasifier efficiencies can also be determined (assuming a complete conversion of the carbon). Figure 7.73 depicts the isolines of a complete carbon conversion for constant gasifier temperatures as a function of the necessary amounts of steam and oxygen. The higher the temperatures in the gasifier, the more the oxygen needed. Higher temperatures mean that the cold gas efficiency decreases, so in order to achieve an efficiency as high as possible for IGCC power plants, the temperature should be the lowest possible. For the entrained-flow gasifier, however, temperatures below 1,400–1,500°C, depending on the coal type, are impossible. The operating mode has to be adjusted so that, with

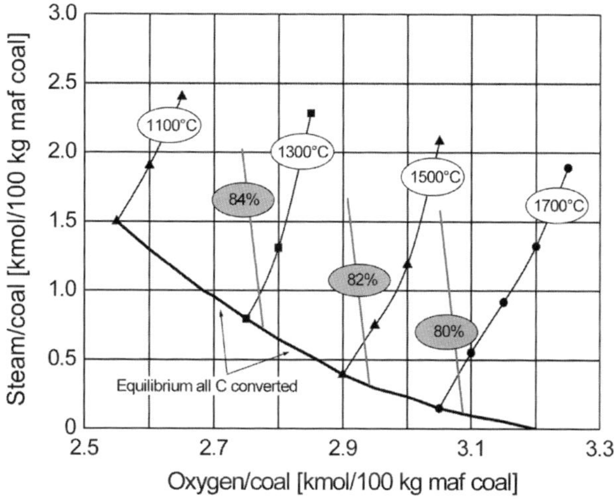

Fig. 7.73 Cold gas efficiencies (from Higman and van der Burgt 2008, © 2008, with permission from Elsevier)

as little oxygen and steam as possible, the carbon is completely converted and the temperature can be reliably maintained during operation.

7.6.4 Classification of Coal Gasifiers

For coal gasification, a great number of methods have been developed which, in terms of process engineering, can be traced back to three basic principles. Analogous to combustion methods (see Chap. 5 and Fig. 5.1), the gas/solids reactor types employed can be categorised according to the state of motion of the solid matter:

- Fixed bed gasification
- Fluidised bed gasification
- Entrained-flow gasification (Strauß 2006; Higman and van der Burgt 2008; Jüntgen and van Heek 1981)

Table 7.16 gives an overview of the gasification methods and their characteristics.

7.6.4.1 Fixed Bed Gasifiers

In a fixed bed reactor, the fuel is at rest. The gas flow can be conducted through the bed in co-, counter- or cross-flow. For better heat transfer, the flow of gas is usually conducted in a counterflow arrangement, as depicted in Fig. 7.74. In the direction of the fuel flow, three zones thus form in the fixed bed: a drying zone, a gasification zone and a combustion zone. In the direction of the gas flow, first the combustion of the solid carbon takes place; the rising hot gases then supply the gasification zone (which lies just above the combustion zone) with heat for the conversion of the

Table 7.16 Characteristics of different gasification processes

Category	Moving-Bed		Fluid-Bed		Entrained-Flow	
Ash conditions	Dry ash	Slagging	Dry ash	Agglomerating	Slagging	Slagging
Feeding	screw	screw	screw	screw	Pneumatic	slurry
Typical process	Lurgi	BGL	Winkler, HTW, CFB	KRW, U-Gas	Shell, Siemens E-Gas	GE
Feed characteristics						
Size	6–50 mm	6–50 mm	6–10 mm	6–10 mm	< 100 μm	< 100 μm
Acceptability of fines	limited	better than dry ash	good	better	unlimited	unlimited
Acceptability of caking coal	yes (with stirrer)	yes	possibly	yes	yes	yes
Preferred coal rank	any	high	low	any	any	any
Operating characteristics						
Reactor temperature	800–1200°C	800–1200°C	900–1000°C	900–1000°C	> 1400°C	> 1400°C
Outlet gas temperature	425–650°C	425–650°C	900–1000°C	900–1000°C	1250–1600°C	1250–1600°C
Residence time	10–30 min	10–30 min	1–10 min	1–10 min	< 1s	< 1s
CO_2 vol. % (dry)	5		20		2	20
CO vol. % (dry)	56		40		66	42
H_2 vol. % (dry)	28		35		28	36
N_2 + Ar vol. % (dry)	3		2		4	2
CH_4 vol. % (dry)	7		3		0	0
C_nHm^2 vol. % (dry)	1		0		0	0
Oxidant demand	low	low	moderate	moderate	high	high
Steam demand	high	low	moderate	moderate	low	low
Cold gas efficiency	very high	very high	high	high	low (82%)	low (72%)
Carbon conversion	high	high	low	low	very high	very high

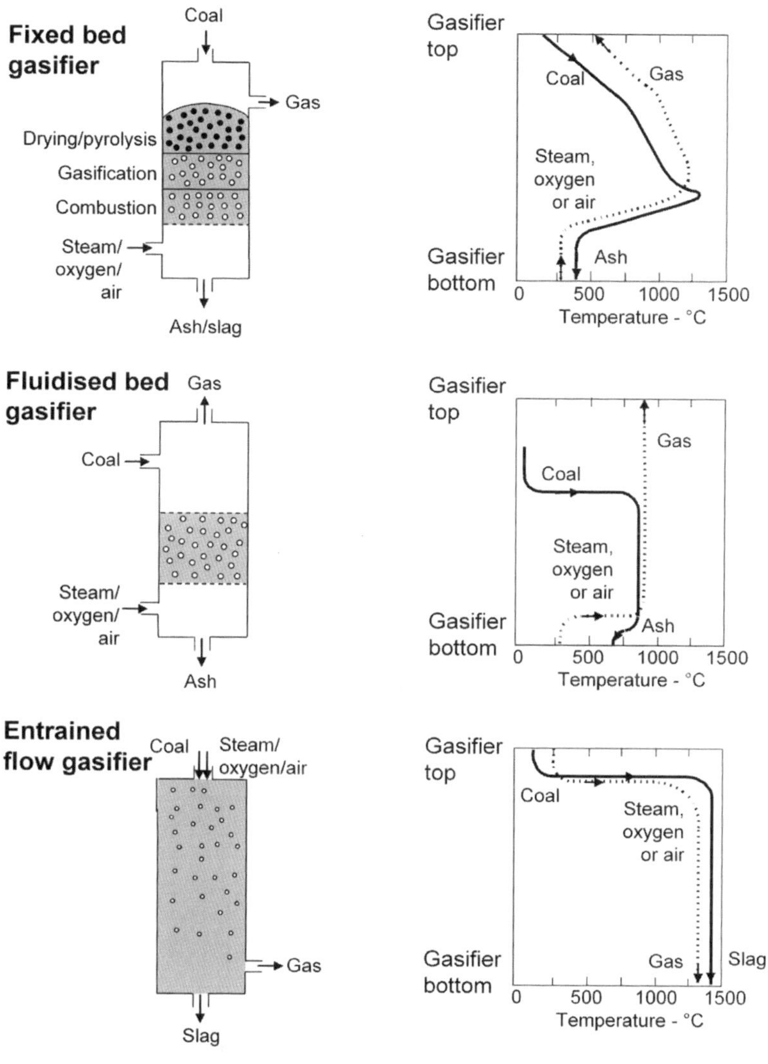

Fig. 7.74 Major types of gasifiers

solid carbon. Subsequently, the residual heat of the rising gases is used to pyrolyse the entering fuel and to dry it.

The advantage of the high level of heat recovery is countered by the disadvantage of the high tar content of the product gas – as the product gas cools down, the tar condenses. The fraction of the tar components can amount to 25% with respect to the supplied fuel energy. The gas produced exits the gasifier at a temperature of 400–500°C, with a content of CH_4 as high as between 10 and 15%. If the energy content of the tar is used, the cold gas efficiency may be around 90%, but such waste heat utilisation in a raw gas cooler is problematic because of the very fact of the high tar fractions.

Owing to the long residence time in the fixed bed, the carbon is almost completely converted, while consuming little oxygen. The coal types preferred are non-caking ones with a particle size ranging from 5 to 30 mm; the proportion of solid fuel fines has to be limited to ensure that there is a continuous flow through the fixed bed.

The method developed by Lurgi, called the Lurgi Dry Ash Process, goes back to a patent from 1927 and, for years, was the only pressurised gasification system. In Sasol, South Africa, 97 of a total of 152 fixed bed gasifiers worldwide are being operated at present, being used for the production of a synthesis gas for the Fischer – Tropsch process. A modified process is the British Gas/Lurgi slagging gasifier, which has been further developed for molten ash removal.

7.6.4.2 Fluidised Bed Gasifiers

The operation of a fluidised bed is described in principle in Sects. 5.1 and 5.4, and illustrated in Fig. 7.74. The solid particles are either made to float by the upward-directed flow (stationary fluidised bed) or carried out and circulated by the flow (circulating fluidised bed). In the fluidised bed reactor, the intensive solids mixing means the temperature is nearly constant. The temperature in the fluidised bed is kept below the deformation temperature of the ash so that the ash does not stick together and can be removed in a dry state. In consequence, the fluidised bed temperature is specified as a function of the fuel and determines the carbon conversion. The typical temperatures for coals range between 950 and 1,100°C, while for biomass, they are between 800 and 950°C. Fine coal particles are carried out of the fluidised bed by the gas flow and thus diminish the carbon conversion. Typical carbon conversion rates lie between 95 and 97% and so are significantly lower than the rates in entrained-flow reactors. The advantages of fluidised beds compared to fixed bed gasifiers are the small contents of condensable by-products and the potential for use of coals with high ash contents. In contrast to the entrained-flow gasifier, milling is not needed, as it is sufficient to crush the coal to particle sizes smaller than 10 mm. The fluidised bed process is particularly suited to reactive coal types like lignite and for biomass; the cold gas efficiency ranges around 85%.

The atmospheric Winkler process was the first modern continuous gasification process with oxygen as the gasifying medium. The process was patented in 1922, and since then about 70 plants of that type have been constructed worldwide. Today, however, the process is only of historical interest, because almost all plants are no longer operated because of economic reasons. The gasifiers were operated at temperatures between 950 and 1,050°C and a fluidising velocity of 5 m/s. After a radiant cooler, the ash flow was removed along with a considerable amount of carbon, in the order of 20% of the fuel flow. Rheinbraun, in the 1970s, developed the high-temperature Winkler process (HTW), the essential characteristic of which was the upgrading of the fluidised bed process to handle pressures of up to 30 bar. A test plant of 600 t/day and a pressure of 10 bar was operated for more than 12 years with availabilities of 84%, the gas being used for methanol synthesis. Today, the HTW process is considered a possibility for IGCC power plants using lignite.

Circulating fluidised beds, with a more intensive gas – solids mixing and the inherent solids recirculation, have a number of advantages over stationary fluidised beds. There is, for instance, the higher carbon conversion rate and a performance insensitive to the particle size and form. The high gas velocities of 5–8 m/s ensure that most of the bigger particles are carried out and separated by the cyclone, then recirculated via a seal pot. Circulating fluidised beds have been developed by Foster Wheeler and Lurgi and are used today almost exclusively for biomass. In this process, air is used as the gasifying medium; the biomass has to be reduced to a size of 25–30 mm.

Allothermal fluidised bed gasifiers have the advantage of being able to produce a highly calorific product gas even without complex air separation. Developments of allothermal coal gasification processes, such as the *Mannesmann-Bergbauforschungsverfahren (MBG)* ("Mannesmann Mining Research Method") in the 1970s and 1980s for example, have aimed at utilising high-temperature heat from high-temperature gas-cooled reactors (HTGR). In later concepts, heat produced from fossil fuels was incorporated into the fluidised bed by helium as a heat-carrying medium. With its necessarily large heat exchanger surfaces, however, the process failed to gain popularity. Two-stage allothermal processes, such as those sometimes used successfully for biomass, have not often been developed for or applied to coal gasification. The characteristic of the two-stage processes are separate reactors for combustion and gasification. In the combustion reactor, the residual coal from gasification is burned and the inert material of the fluidised bed is heated. The sensible heat of the sand, which serves as the heat-carrying medium, is used to gasify the fuel in a second fluidised bed. Two-stage gasifiers for biomass are described in Sect. 6.3.1.

7.6.4.3 Entrained-Flow Gasifier

In an entrained-flow process, finely milled coal with a particle size of less than 0.1 mm is fed directly into the reactor and converted in parallel flow with the gas at high temperatures and pressures, as shown in Fig. 7.74. The temperatures in the reactor are above the ash fluid point; depending on the calorific value of the coal type, they rise to between 1,400 and 1,600°C. The pressures of entrained-flow gasifiers in operation are between 20 and 70 bar. Given that the heat is provided through the combustion of part of the coal (i.e. an autothermal process), the cold gas efficiency lies below 85%. On account of the high temperatures, the inside walls of the gasifier are lined with a fireproof material; for high-ash fuels it is possible to do without such lining if the formation of a protecting slag layer on a cooled membrane wall can be ensured. The residence time of the coal particles in the reactor is about 1 s. In consequence, significantly higher volumetric power densities result for the entrained-flow reactor than for the other two reactor types. The carbon gets converted almost completely, the conversion rates amounting to more than 99%. Owing to the high reaction temperature, the entrained-flow process is also suitable for less reactive fuels. However, expenditures of energy for the very fine milling have to be taken into account when considering the viability of this process for such fuels. High-moisture fuels, too, can be used, but the oxygen demand will rise and the cold gas efficiency will decrease in this case. About 60% of the ash is removed from the reactor in a liquid state, the rest being removed as fly ash from the product gas. Because of the high temperature, the product gas consists mainly of hydrogen and carbon monoxide (synthesis gas) and contains no tars.

Process Variants

The advantages of the autothermal entrained-flow process – the high power density, the production of a tar-free synthesis gas and the fuel flexibility – have meant that the vast majority of gasifiers installed worldwide are autothermal entrained-flow types. The disadvantages of the high oxygen demand and lower cold gas efficiency can be limited through the selection of an appropriate process variant. The processes applied today are distinguished by the fuel feeding method (dry or wet), the cooling method for the reaction vessel (membrane wall or refractory lining) and the way the sensible heat of the hot product gas is used (a gas quench, water quench or a synthesis gas cooler). Furthermore, a distinction is made between one-stage and two-stage gasifiers. The design has a considerable influence on the cold gas efficiency and the overall efficiency, in which the effects can differ for IGCC and chemical applications. In the following, attention shall be given to IGCC applications with respect to the following:

- Wet or dry coal feeding: Wet coal feeding or using very moist fuels diminishes the cold gas efficiency, because water has to be vaporised and heated and because the oxygen demand increases. The more a fuel gets oxidised, the lower the cold gas efficiency will be. Wet coal feeding has the advantages of being simpler and having the ability to achieve very high pressures of up to 200 bar. Table 7.17 shows a comparison (Radtke et al. 2005; Uhde 2008).
- Product gas cooling: The product gas heat can be utilised by means of a radiant cooler, which comes in the form of a water- or gas-quench cooler or a chemical quench. A radiant cooler produces steam which is used in the steam process of the combined cycle power plant. As this heat is not fed to the gas turbine, the conversion process runs at the lower efficiency of the Rankine process, so the efficiency is lower than for a natural gas fired combined cycle process. A gas quench in the following radiant cooler lowers the temperature of the waste heat utilisation process, for instance by quenching with recirculated product gas. As long as the remaining heat contained in the product gas is sufficient to achieve high steam temperatures, there are no negative effects from the radiant cooler. For the water quench, the same observations as for the gas quench hold true, but an additional drawback arises through the vaporisation of the water. The energy supplied for the vaporisation loses usefulness as it drops to the temperature of the condensation heat. Most satisfactory from the energetic point of view is a chemical quench, where the residual heat propels an endothermic reaction, and

Table 7.17 Gas quality of dry and wet feeding (Radtke et al. 2005), (Uhde 2008)

	Dry Feed EF Gasifier	Slurry Type EF Gasifier
Coal/Petcoke feed	Dry pneumatic	Water slurry
Gasifier	Membrane wall	Refractory lining
Syngas cooling	Gas quench and convective cooler for IGCC applications	Water quench or radiant and convective cooler. (Radiant cooler is a very large piece of equipment.)
Solids removal	Almost completely dry	Wet scrubbing
Water treatment	Simple	Water quench is a potential source of corrosion, causes formation of formic acid, high water treatment costs.
Cold gas efficiency	82%	72%
Oxygen Consumption	0.93 kg/kg coal	1.07 kg/kg coal
Typical gas analysis vol. %		
CO_2	2–3	15
CO	62	32
H_2	27	27
$H_2S + COS$	0.7	0.9
Inerts	5	1
H_2O	2	23

the chemical energy of the product gases, or the cold gas efficiency, increases. This corresponds to concept designs that use two-stage gasifiers.

- Two-stage gasifiers: The purpose is to cool the product gas exiting the gasifier from temperatures between 1,500 and 1,600°C down to temperatures of about 1,000–1,100°C. While cooling the product gas, the heat extracted is used to propel the endothermic reactions of gasification, meaning that coal can be fed without oxygen to the second stage of the gasifier, bypassing the first. The lower temperatures mean that longer residence times for gasification are necessary and there is a risk that pyrolysis products do not react completely. For this reason there is a lower limit to the temperature. Carbon that has not reacted can be removed and recirculated after the product gas cooler. In another variant, incomplete gasification takes place in the first stage followed by an addition of steam for the complete gasification of the carbon in the second stage. Two-stage gasifiers have higher fractions of CH_4 and CO_2. The decisive factor for the cold gas efficiency of two-stage gasifiers is the outlet temperature of the gasifier. A two-stage gasifier with an outlet temperature of 1,100°C has an efficiency corresponding to the cold gas efficiency of a one-stage gasifier with the same outlet temperature.
- Cooling of the gasifier: Gasifier cooling has the same effect as product gas cooling by steam. The heat removal is generally very low, though.

Technologies

The development of entrained-flow gasification began with the development of the atmospherically operated Koppers-Totzek process in the 1950s. Commercial plants were operated in a great number of countries, mainly for ammonia synthesis. In recent years, no additional plants that follow this method have been built. Based on the Koppers-Totzek process, Shell and Koppers developed pressurised versions, first in co-operation and later separately. Both the Shell Coal Gasification Process (SCGP) and the Prenflo Process by Krupp-Koppers (later: Krupp-Uhde, today: Uhde) are one-stage entrained-flow gasifiers with dry coal feeding. Both designs feed coal in a dense flow to four burners using an inert gas. During the upward flow through the bed, where 90% of the particles are smaller than 90 μm, the gasification of the fine coal particles takes place, with the molten ash flowing down into a water bath where it granulates. The temperatures in the gasifier are typically around 1,500°C and the pressure at about 30–40 bar. The reactor wall is a membrane wall construction that is studded and covered with a castable refractory mix in order to protect the metal wall from direct radiation and from the liquid slag. The tubes of the membrane wall are steam-cooled. The heat losses depend on the size of the reactor and on the quality and amount of slag and typically lie in the order of magnitude of 2–4% of the fuel heat. The hot product gas gets cooled down to 900°C by recirculated cold product gas at 280°C before exiting the gasifier. By means of this fast cooling down, cutting through the temperature range between the ash deformation and ash fluid temperatures, an attempt is made to prevent the formation of agglomerations. In the steam-cooled raw gas cooler, the product gas gets cooled from 900 to 280°C, afterwards being conducted to a particulate removal unit via ceramic filters. About half of the product gas is conducted to the quench cooler by use of a compressor, while the other half is led to a wet scrubber, becoming the net product. The Shell Coal Gasification Process is shown in Fig. 7.75.

In 1994, a 2,000 t/day Shell gasification unit was built for Demkolec (now NUON) in Buggenum in the Netherlands, using coal as feedstock. In 1997, Krupp-Koppers (now Uhde) built a 3,000 t/day unit for Elcogas in Puertollano, Spain, using a blend of high-ash coal and petcoke as the feedstock.

The Siemens gasification process is also a one-stage gasification method using a dry fuel supply. It can be traced back to developments of lignite gasification by the *Deutsches Brennstoffinstitut* in Freiberg in the 1970s. The entire fuel flow, steam and oxygen are fed through a burner. The reactor has a downward flow, with both molten slag and the hot product gas being discharged at the bottom (see Fig. 7.76). The discharge of slag and product gas together avoids a blocking of the slag outflow. Depending on the application, partial quenching or full quenching may be employed. For high-ash fuels, a cooling screen lined with SiC is used, becoming covered by a protective slag layer during operation. For low-ash fuels, the design incorporates refractory lining.

The GE (formerly Texaco) process for coal gasification uses a slurry-feed downward-flowing entrained-flow gasifier. The coal is wet milled to a particle size of about 100 μm and slurried in conventional equipment, then charged to the reactor with a membrane pump. The reactor pressure is typically about 30 bar for IGCC

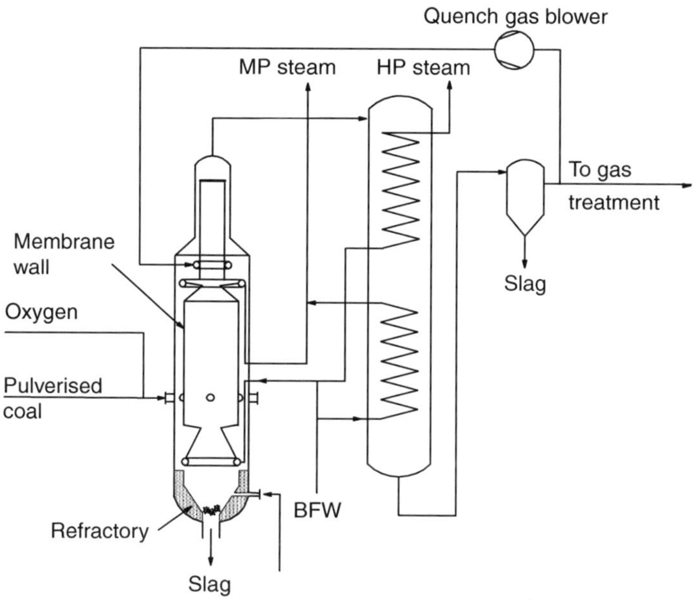

Fig. 7.75 The Shell Coal Gasification Process (from Higman and van der Burgt 2008, © 2008, with permission from Elsevier)

applications; for chemical applications it may be as much as 70–80 bar. The reactor shell is an uncooled refractory-lined vessel.

Syngas cooling can be performed by a radiant boiler or via a total water quench. In the quench configuration, the hot syngas leaves the reactor at the bottom together with the liquid ash and enters the quench chamber. The gas leaves the quench chamber fully saturated and at a temperature of between 200 and 300°C, which are suitable conditions for direct CO shift conversion. In the radiant cooler configuration, which is part of the Cool Water and Polk IGCC plants, full use is made of heat recovery for maximum efficiency. The GE gasifier is the most inexpensive gasifier on the market, but is maintenance-intensive. To achieve high availabilities of production, a standby reactor is required.

The E-Gas gasifier is currently the only two-stage process with an operating commercial-scale demonstration plant. A sub-bituminous coal – water slurry is injected into the hot gases coming from the first slagging stage, resulting in a temperature drop from 1,400 to 1,040°C. Unreacted char is separated by metallic candle filters and reinjected into the first slagging stage.

7.6.5 Gas Treatment

The composition of the raw gases produced by gasification differ considerably depending on the production method. For example, from gasification in a fluidised bed or an entrained-flow reactor, a raw gas has a high particulate content, whereas

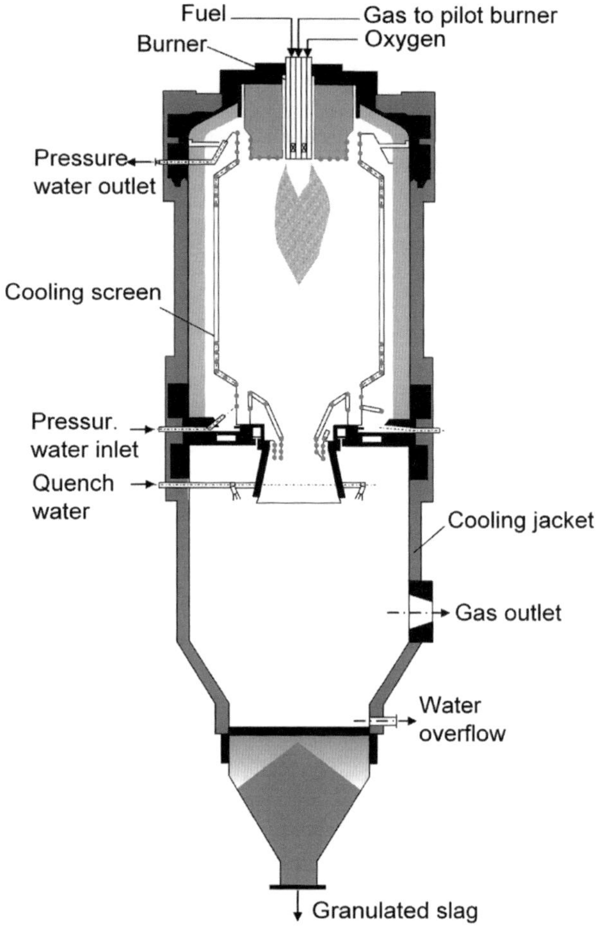

Fig. 7.76 Siemens gasifier with cooling screen (*Source*: Siemens Fuel Gasification)

from gasification in a fixed bed, the gas contains less particulates but a considerable amount of condensable pyrolysis products. Trace concentrations of pollutant gases depend on the fuel used and the gasification technology. The aim is to bring dry gas cleaning processes, using hot particulate removal, desulphurisation and dehalogenation, to a level considered state of the art. Such methods, however, have not yet been developed for application at an industrial scale. Wet gas cleaning, already tried and tested, is complex to implement and has a negative effect on the efficiency, because during the necessary cooling of the raw gas, about 15–20% of the supplied energy passes into the steam – water cycle (Maurstad 2005; Higman and van der Burgt 2008).

Besides depending on the gasification process and on the fuel, the necessary gas treatment steps depend on the purity requirements of the total process or on the syngas purity requirements. Correspondingly, the necessary gas treatment procedures differ, but in general comprise the following steps:

- Raw gas cooling of the syngas
- Gas cleaning of pollutant components such as particulates, sulphur, chlorine, nitrogen
- Gas preparation to adjust the composition of the product gas to that desired for downstream use

7.6.5.1 Impurities in the Gas

Sulphur compounds: Sulphur compounds in raw synthesis gas act as a catalyst poison for most chemical applications and for the low-temperature shift reaction. Used in IGCC, if untreated, these compounds are oxidised in the gas turbine and emitted as SO_2.

In high-temperature gasification, over 90% of the sulphur components in the feedstock are converted to hydrogen sulphide (H_2S) and the rest to carbonyl sulphide (COS). Compounds such as SO_x and CS_2 are essentially absent in the syngas. This is not the case in low-temperature gasification, such as in the fixed bed process, where tars and other species are not completely cracked.

The relationship between H_2S and COS contents of a raw gas is determined by the hydrogenation reaction:

$$H_2 + COS \leftrightarrow H_2S + CO + 7\,\mathrm{MJ/kmol} \tag{7.47}$$

and the hydrolysis reaction:

$$COS + H_2O \leftrightarrow H_2S + CO_2 - 34\,\mathrm{MJ/kmol} \tag{7.48}$$

the rest being CO_2.

Up to 99.8% of the sulphur can be removed in the acid gas removal process. As COS is not easily removed, a catalytic hydrolysis unit of COS to H_2S prior to the acid gas removal is required.

Nitrogen compounds: Nitrogen enters the gasifier both as molecular nitrogen (supplied with the coal or oxygen flow) and as organic nitrogen in the fuel. The bulk of the nitrogen in the syngas is present as molecular nitrogen. Hydrogen cyanide (HCN) and ammonia (NH_3) are produced in gasification in small amounts; due to the reducing conditions NO_x is negligible. Most of the HCN and NH_3 originate from fuel-bound nitrogen, not from the molecular nitrogen, which has strong chemical bonds. The proportions of HCN and NH_3 are dependent on the coal characteristics. For the formation of NH_3 and HCN in a gasifier, the same fundamentals of NO_x formation as discussed in Sect. 5.2.1 apply; however, due to the reducing conditions the oxidation is inhibited. Typical concentrations of HCN and NH_3 are in the low ppm range, up to several tens of ppm.

If untreated, both ammonia and hydrogen cyanide in the raw syngas would result in NO_x emissions from IGCC processes. HCN can react with the amines used in the acid gas removal (AGR) unit and degrade them, while also being a poison for some catalytic processes (e.g. the Fischer – Tropsch synthesis). Both HCN and NH_3

have very high solubilities in water and may therefore be easily removed by water scrubbing.

Chlorine compounds: Chlorine released from the coal is converted to hydrogen chloride (HCl) gas, which in contact with metals can form chlorides, such as sodium chloride (NaCl), with melting points in the range of 350–800°C. These metal chlorides pose a fouling risk in heat exchangers. Chlorine compounds from the coal will also react with ammonia to form ammonium chloride (NH_4Cl) which is in the vapour phase at higher temperatures and becomes a solid at around 280°C. In addition to fouling problems, chlorine is a catalyst poison for the low-temperature shift reaction. Many of the chlorides may be removed in a water scrubber.

Solid carbon and ash: Ash and a small amount of remaining char will always be entrained in the exit flow of the gasifier. The quench or syngas cooler has to ensure that these particles will be non-sticky so that fouling problems are prevented. After capture in a filter or scrubber, char can be recycled to the gasifier to increase the carbon conversion efficiency.

Other trace components: Besides the major components, a variety of trace elements such as lead (Pb), mercury (Hg) and arsenic (As) are present in coals. Metal carbonyls such as nickel carbonyl ($Ni(CO)_4$) and iron carbonyl ($Fe(CO)_5$) can be formed.

7.6.5.2 Raw Gas Cooling

Raw gases exit gasifiers at high temperatures, ranging from 550°C from a fixed bed gasifier to 1,600°C from an entrained-flow gasifier. As the cleaning processes for the removal of these pollutants run at lower temperatures, the syngas needs to be cooled down. This cooling is required even for hot gas (or more correctly warm gas) cleaning. Figure 7.77 shows configurations for raw gas cooling and particulate removal units for different gasification systems.

Particular attention has to be paid to the cooling from gasification temperatures to temperatures below 900°C, because the ash is liquid in this temperature range and may cause build-ups of slag. For the sake of the continued availability of the process, the raw gases should therefore be cooled as fast as possible down to around 900°C, where the ash is present in solid form. For cooling in this range, the following process variants are used:

- Radiant cooling
- Water quenching
- Gas quenching
- Chemical quenching

In a radiant cooler, the heat of the syngas is transferred to the water – steam-cooled walls by radiation. Due to the high heat transfer rate, saturated steam is produced exclusively. The radiant cooler is an expensive component and susceptible to fouling and slagging.

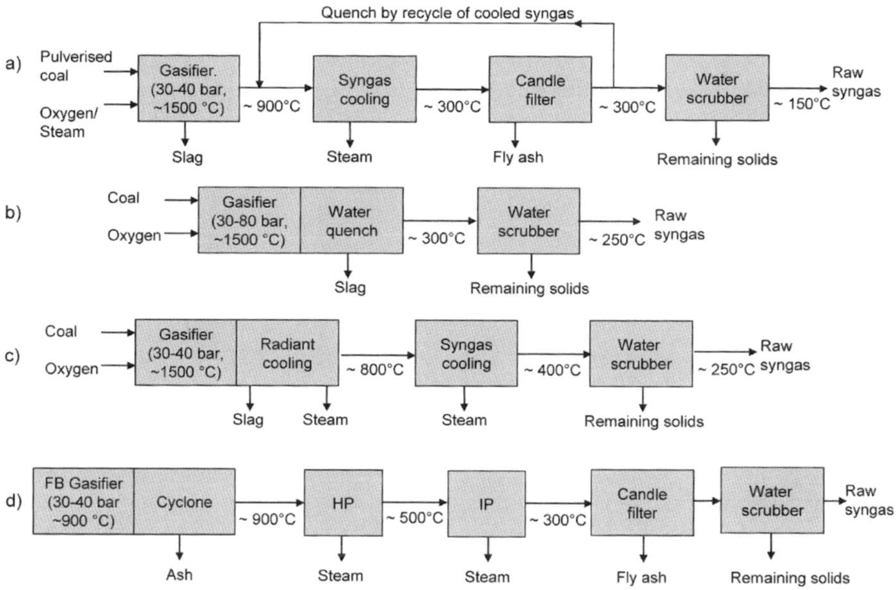

Fig. 7.77 Process flow diagram for different gasification processes (Maurstad 2005) and additions (**a**: EF + gas quench, **b**: EF+ water quench, **c**: EF + radiant cooling, **d**: fluidised bed)

A water quench uses the sensible heat of the syngas to vaporise the injected water. With a partial quench, the syngas is cooled down to 900°C and the sensible heat (but not the latent heat) can be utilised for steam production. With a full quench, no high-pressure steam is generated and the syngas is saturated with steam. Water quenches, and the full quench in particular, are disadvantageous for IGCC processes without CO_2 separation, because the heat of evaporation cannot reasonably be used at the low temperature of condensation. For an IGCC power plant with CO_2 separation, the addition of water shifts the equilibrium of the water gas reactions in the direction of higher H_2/CO contents.

One example of the use of a gas quench is in the Shell gasifier, where syngas that has already been cooled is mixed with the 1,500°C (hot) untreated synthesis gas, thus cooling to 900°C. This way, the heat is used within an unproblematic temperature range. Apart from the increased power consumption of the recirculation process, no other negative efficiency effects arise for an IGCC power plant because only sensible heat, not condensation heat, is produced.

Chemical quenching is advantageous because the gasifier exit temperature is lowered and thus the cold gas efficiency increased (see also Sect. "Process Variants").

For the process of further cooling of the flue gas by convective heat exchangers, the factors that should be considered are particulate removal and the behaviour of condensing components. The latter can be in the form of tars in biomass gasification, ammonium chloride in coal gasification or, simply, water. Typically, this cooling stage lowers the temperature from 900°C to about 300°C. At 900°C, the ash particles are no longer sticky; at 300°C, deposits of NH_4Cl do not occur. A distinction is made between water-tube and fire-tube boilers, both of which are in successful service. Fire-tube boilers conduct the synthesis gases inside the tubes, with the water flowing on the outside of them; in water-tube boilers, the high-pressure steam is produced inside the tubes. Typical steam pressures range from 100 to 150 bar for both types, but water-tube boilers can also be designed for even higher pressures. Fouling has to be taken into consideration for both types, so adequate cleaning facilities have to be included in the design. Superheating of the steam is possible. In order to limit corrosion, high-alloy materials are used and the material temperatures are restricted to values of 500–600°C.

7.6.5.3 Particulate Removal

A dry particulate removal process should run at temperatures ranging between 300 and 500°C. Only at temperatures below 500°C do the alkali compounds achieve the maximum possible condensation on the fly ash, so that they are then removed along with the particulates. Below 300°C, the filters may be clogged by deposits of ammonium chloride. Cartridge filters are employed. By using topping cyclones prior to the filter, the filter load can be reduced.

In most existing plants the (remaining) solids are washed out in venturi scrubbers or wash towers. The scrubbing takes place below the dew point of the gas, so that the finest solid particles can act as nuclei for condensation, thus ensuring that all solids are removed effectively. In wet scrubbing the water-soluble gaseous components such as NH_3, HCN, HCl and HF are also separated. The disadvantage of wet scrubbing is that the ash contains many regulated substances such as lead, zinc and cadmium, so the removed components must be disposed of carefully and according to relevant directives.

7.6.5.4 CO Shift

In an IGCC configuration with CO_2 removal or in a hydrogen production plant, the water gas shift reaction is used to push the chemical composition of the syngas towards a maximum H_2 yield:

$$CO + H_2O \leftrightarrow H_2 + CO_2 - 41.2\,\mathrm{MJ/kmol} \tag{7.29}$$

According to the reaction, one mole of hydrogen can be produced from every mole of CO. The heating value per mole is less for H_2 (241.8 MJ/kmol) than for CO (283.0 MJ/kmol), which means that chemical energy is converted to heat (exothermic reaction). The reaction itself is equimolar and therefore largely independent of pressure. The reaction is normally carried out in two stages, a high-temperature shift and a low-temperature shift. The high-temperature stage has the advantage of high reaction rates, while the low-temperature stage favours an equilibrium for maximum hydrogen production. Typical operating temperatures of the two stages are between 200 and 500°C, depending on the catalyst. The types of catalysts are distinguished by their operating temperature range and the maximum sulphur content

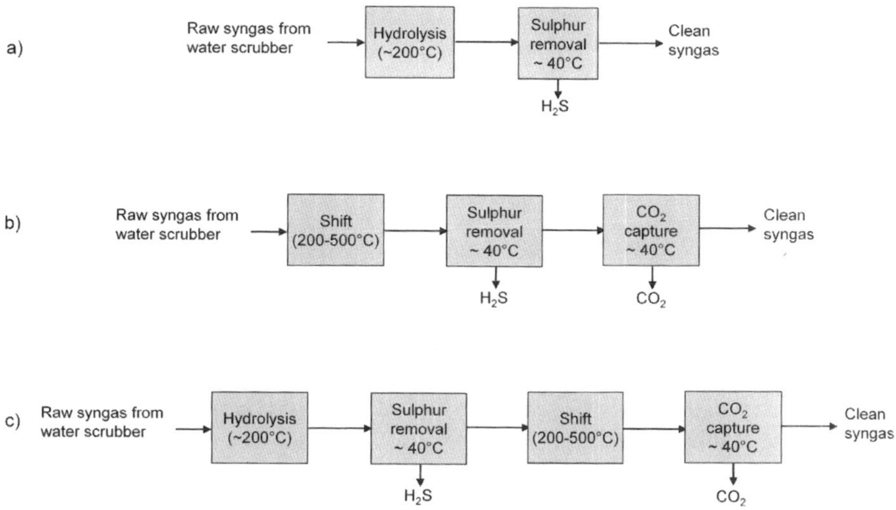

Fig. 7.78 Process flow diagrams of gas cleaning (**a**) without shift conversion, (**b**) sour shift conversion, (**c**) clean shift conversion (Maurstad 2005)

of the syngas to be treated. The minimum molar H_2O/CO ratio is around 2. If there is not sufficient steam present in the syngas for the reaction, steam is extracted from the steam cycle.

Figure 7.78 shows the principle gas clean-up steps for processes with and without CO_2 capture. If CO_2 is not captured and the syngas is used to feed a turbine, then a shift is not required. In this case, a separate hydrolysis reactor is needed to convert COS to H_2S for easier sulphur removal. If there is a shift reaction, this conversion takes place simultaneously and no additional reactor is needed. For CO_2 capture there are two alternative processes for the shift reaction:

- Sour shift (or raw shift)
- Clean shift

The sour shift is the preferred process when considering costs and efficiency. Gasifiers with a water quench are not suited to the clean shift, as a lot of valuable steam in the syngas would have to be condensed before sulphur removal and then, before the shift, a lot of steam would have to be added again. For a gasifier with dry gas quenching, the clean shift has some advantages, such as a cheaper catalyst and easier sulphur removal, as less CO_2 is present. However, the more complex clean shift, with more heating and cooling, is less attractive from a capital cost and efficiency point of view for an IGCC plant (Maurstad 2005).

7.6.5.5 Acid Gas Removal (H_2S, COS, CO_2)

The term acid gas removal is often used as a synonym for desulphurisation, but strictly speaking, in the context of gasification, it also includes the acid gas CO_2. A large number of different processes can be used for acid gas removal, which can be categorised according to the following principles used:

- Absorption or adsorption by a liquid solvent with a subsequent desorption step (chemical or physical washing)
- Absorption or adsorption on a solid material
- Diffusion through a permeable membrane

Acid gases such as H_2S and CO_2 cannot be removed in a water wash process due to their low water solubilities. For acid gas removal, chemical (absorption) or physical (adsorption) washes with liquid solvents are normally used. The different principles for removal in gasification environments are illustrated in Fig. 7.79. The loading capacity of a physical solvent primarily depends on Henry's law and is therefore proportional to the partial pressure of the component to be removed. In contrast, the loading capacity of a chemical wash is limited by the quantity of the active component of the solvent. Generally, the solvent can be regenerated by flashing, stripping or reboiling or a combination of these. Both stripping and flashing reduce the partial pressure of the acid component and are used for physical solvents. In physical washes, reboiling raises the temperature and thus reduces the acid gas solubility. In chemical washes, the increased temperature breaks the chemical bonds and releases the components in the same chemical form in which they were absorbed.

Criteria for the selection of the appropriate process are as follows:

Gas purity: The demands of the syngas purity vary extremely with the application. For an IGCC power plant with a limit of 5 ppm SO_2 in the flue gas, about 40 ppm H_2S at the outlet of the AGR is sufficient. For chemical applications such as ammonia, methanol or SNG production, 100 ppb may be required.

Raw gas composition: The washing solution must cope with the impurities in the raw gas. HCN in the raw gas, for instance, can react with amines, causing solution degradation.

Selectivity: The selectivity of a gas treatment process is the ability to remove H_2S while leaving CO_2 in the synthesis gas.

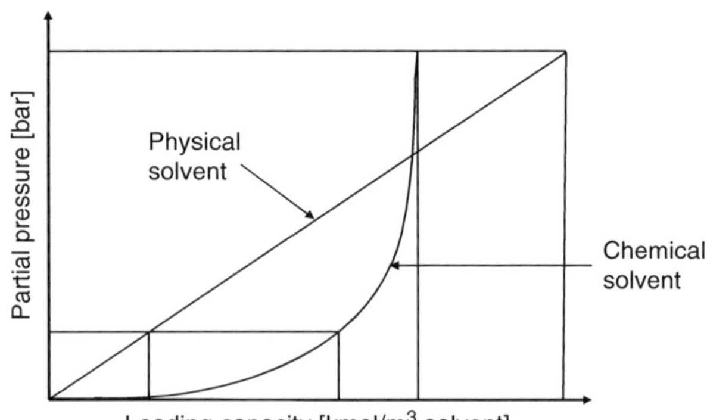

Fig. 7.79 Loading capacity of physical and chemical solvents (from Higman and van der Burgt 2008, © 2008, with permission from Elsevier)

Solutions of amines in water are commonly used for chemical washes; examples of amines are mono- and diethanolamine (MEA and DEA) and methyldiethanolamine (MDEA). MDEA is the most widely used amine today. Examples of physical washes are the Rectisol, Selexol and Purisol processes. The Rectisol process, which uses methanol as a solvent, operates between temperatures of -30 and $-60°C$. The process can achieve very high gas purities and is used for chemical applications where synthesis catalysts require sulphur removal to less than 0.1 ppmv. The Selexol process, which uses dimethyl ethers of polyethylene glycol (DMPEG), is operated at 0–40°C, reducing refrigeration requirements, and can achieve gas purities of 1 ppm H_2S and COS, respectively. The Purisol process shows similar characteristics, but has a higher H_2S/CO_2 selectivity.

Physical – chemical washes make use of the principles of both physical and chemical removal. They generally use an amine together with an organic physical solvent. An example of this is Shell's Sulfinol solvent; the modified m-Sulfinol solvent uses MDEA as the chemical component and is applied at the Buggenum plant.

Sulphur Removal

The sulphur removal process consists of three process steps:

- Acid gas removal (AGR)
- Sulphur recovery (SR)
- Tail-gas treating (TGT)

The AGR process removes the H_2S from the syngas. In present-day IGCC plants, the two preferred processes are chemical washing, based on aqueous methyldiethanolamine (MDEA), and the Selexol process, based on a physical solvent. Both methods can reduce the total sulphur ($H_2S + COS$) to levels below 20 ppmv in the cleaned syngas. For deep sulphur removal, required for chemical applications, the more expensive Rectisol process, using a physical solvent, may be applied. For CO_2 capture a second-stage AGR has to be added to remove the CO_2 from the sulphur-free syngas.

The purpose of the sulphur recovery unit (SRU) is to convert the H_2S into a chemical product which can be reused. The most common method for SRU is the Claus process which produces elemental sulphur by sub-stoichiometric combustion with air or oxygen. Different versions of this process are available. The sulphur may be fixed as elemental sulphur in liquid or solid form or as sulphuric acid. In order to achieve high enough degrees of sulphur recovery, the thermodynamics of the Claus process requires some treating of the tail gas, which usually contains mostly H_2S and SO_2, but also small amounts of COS, CS_2 and elemental sulphur vapours. In the TGT process, the sulphur species are converted to H_2S, which can then be absorbed in a liquid solvent.

CO_2 Removal

CO_2 capture requires a second stage to the AGR process for the treatment of the sulphur-free syngas. A two-stage Selexol process is the preferred option for selective removal of sulphur and CO_2. If combined capture of H_2S and CO_2 is acceptable for a downstream storage or an enhanced oil recovery (EOR) project, significant cost reductions are possible because of a simpler AGR process and elimination of the SRU and TGT units in the sulphur removal process.

7.6.5.6 Hot Gas Cleaning

Hot gas cleaning units (HGCUremove particulates, sulphur compounds and other pollutants at higher temperatures than traditional processes such as water scrubbers and acid gas removal systems. HGCUs provide several advantages in comparison to state-of-the art cold gas cleaning units (CGCU) operating at ambient temperatures or below:

- The total process efficiency increases, because syngas cooling is not required and water does not have to be removed from the syngas.
- Sour water treatment requirements are eliminated. Sour water is produced in CGCUs when syngas is cooled below the dew point of the water.
- Troublesome ash-char water mixtures produced in water quenching or wet scrubbing of particulates from the syngas can be avoided.
- There is the potential to reduce capital and operating costs (Korens et al. 2002; Holt 2003).

The development of hot gas clean-up systems has been pursued in the USA, Europe and Japan since the 1970s. The development has focussed primarily on syngas from air-blown gasification, because air-blown gasification systems produce over twice the volume of oxygen-blown systems due to dilution by nitrogen. Conventional cold gas cleaning for air-blown gasification incurs a costly and substantial efficiency loss, making it uneconomic. Therefore the success of air-blown gasification depends on the development of HGCUs. However, hot gas cleaning is also applicable to oxygen-blown gasification and improves its process efficiency in comparison to cold gas cleaning.

The temperature at which the product gas is used determines the temperature of the gas cleaning train. As most HGCU development programs have focussed on hot gas cleaning for IGCC applications, the temperature has been the highest possible at which the gas turbine fuel control and delivery systems can be designed. The requirement for very low alkali contents in the flue gas to prevent alkali corrosion of hot gas turbine components, and the desire to avoid expensive materials and unreliable refractory-lined pipes, sets this level at about 500–550°C. At this temperature the alkali vapour condenses on particles in the hot syngas which are then removed in the barrier filters. Since large-scale gasifiers operate at 1,400–1,600°C, significant cooling, the extent of which depends on the gasifier design and the feedstock, is still

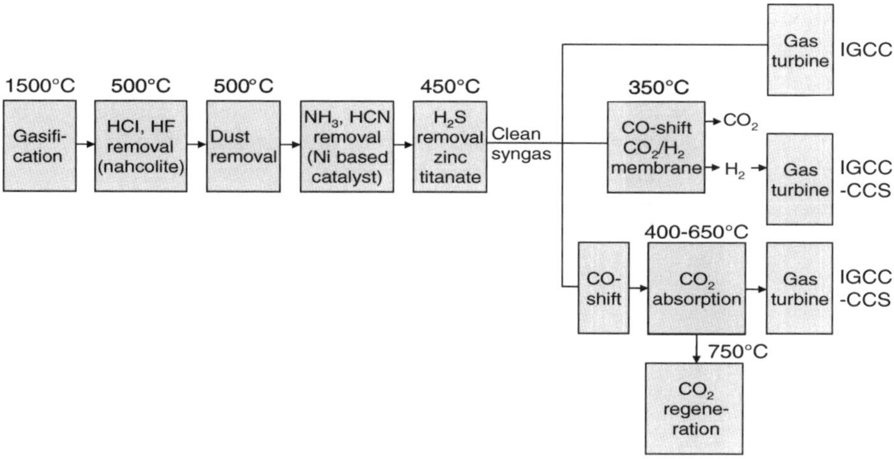

Fig. 7.80 Schematic diagram of a hot gas cleaning process

required. Most of the hot gas cleaning demonstration units have operated between 400 and 500°C, so that the term hot gas cleaning is misleading. Instead the term warm gas cleaning describes the actual temperature more appropriately.

Development of hot gas cleaning has focussed mainly on particle separation and removal of chloride, alkalis and sulphur components from syngas for gas turbine applications. In order to fulfil emission requirements, additional components such as HCl, HCN, NH_3 and mercury also have to be considered.

Figure 7.80 shows a schematic diagram of a possible hot gas cleaning process for an IGCC process with and without CO_2 removal. In the following, the principle process steps of hot gas cleaning are discussed and the state of development is described.

Hot Gas Filtration

Barrier filters are the only currently commercially available HGCU technology and have been successfully demonstrated in gasification projects. Candle filters – ceramic or metal tubes mounted in bundles, themselves within a filter vessel – are being used for final particulate removal for large syngas flows (refer to Fig. 7.18). The syngas flows from the outside through the porous tube walls, into the ceramic or metal tubes, and flows out of the vessel through the inside of the tubes. Back-pulsing the filtered gas dislodges the deposited ash from the outside of the candles (the ceramic or metal tubes), and the ash is discharged from the bottom of the vessel. The details are described in Sect. 7.3 in the context of pressurised fluidised bed combustion.

As described previously, the operating temperature of barrier filters is chosen to be below 550°C to promote the condensation of alkalis on particulates. The actual operation temperatures of the hot gas filters in the IGCC demonstration plants are even lower. The temperatures of the hot gas filters are ~ 350°C at Wabash (metallic elements) and 250°C at Buggenum (ceramic elements) and Puertollano. Solids bridging, candle degradation and breakage and fouling and corrosion of metallic components have been the major problems. Further improvements are still needed to increase the filter element lifetime and to reduce filter installation, operating and maintenance costs. Upstream cyclones are important components of a hot particle removal system – minimising the load on the hot filters. Overall, hot gas filtration offers definite advantages to IGCC over water scrubbing (Korens et al. 2002; Holt 2003).

Alkali Cleaning

There are two main methods employed for cleaning vapour-phase alkali compounds:

- Cleaning at low temperatures. When the gas temperature is lowered below 550–600°C, alkali vapours condense and can be removed by particle removal systems.
- Syngases with alkali compounds can also be cleaned by passing through alkali getters, such as activated bauxite or activated alumina, at higher temperatures. Alkali compounds are then physisorbed or chemisorbed on the getter surface, with chemisorption suggested to be the dominant pathway when moisture is present. More details can be found in Sect. 7.4.3.

Hot Gas Desulphurisation

Metal oxide sorbents, which come as regenerable or disposable types, are able to capture H_2S at elevated temperatures. Disposable sorbents, such as limestone or dolomite, are typically calcium based and injected into the gasifier for in situ desulphurisation. H_2S reacts with these materials at 950–1,050°C and pressures over 20 bar to form CaS. Due to thermodynamic limitations, only 90% sulphur removal can be achieved, which means a typical outlet H_2S concentration of 300–500 ppmv. These sorbents can only be used once, which increases the amount of solid waste to be continuously removed from the process. Because CaS is not environmentally stable, it has to be converted to $CaSO_4$, which requires a separate oxidation stage (Atimay 2001).

Regenerable sorbents are usually used in a separate fixed or movable bed reactor after the gasifier. Having the unit separate makes it easier to regenerate the sorbent. Zinc oxide sorbents give the best results for H_2S cleaning at elevated temperatures (in the range 350–750°C). The desulphurisation reaction is

$$ZnO + H_2S \rightarrow ZnS + H_2O \quad (7.50)$$

and the regeneration reaction:

$$ZnS + 1.5O_2 \rightarrow ZnO + SO_2 \tag{7.51}$$

A great disadvantage of ZnO is the fact that it quickly reduces in reducing fuel gas atmospheres at high temperatures, so that the syngas is polluted with vaporised zinc. Zinc titanate can achieve the same residual H_2S level, but is more stable and shows a better attrition resistance. In general, zinc titanate (Zn_2TiO_4) is considered the best option for H_2S removal at high temperatures of up to 850°C (Aravind 2007).

The only two large-scale hot gas desulphurisation systems have been installed in the USA. They have never been demonstrated, however. Both systems were based on the reaction of H_2S with zinc oxide/nickel oxide solid sorbents followed by regeneration of the sorbent by contact with air. The regenerator off-gas contained SO_2, which had to be converted to elemental sulphur or sulphuric acid in a final recovery operation. At the 260 MW coal-fired IGCC in Tampa, the HGCU system was designed to treat 10% of the syngas flow. The HGCU was a moving bed absorption process designed for temperatures of 480°C. One of the reasons for cancelling the demonstration was the sorbent attrition behaviour, which led to extremely high annual sorbent costs. Interest in HGCU processes such as hot desulphurisation has been decreasing of late, partly because of disappointing results in finding solid sorbents with the necessary attrition resistance (Korens et al. 2002; Tampa_Electric 1996).

HCl Removal

Hydrogen chloride has to be removed from the syngas to prevent corrosion in the gas cleaning train and in the gas turbine. Sodium and potassium compounds are effective for dry removal of HCl and HF from syngas down to the ppm level at temperatures of 600°C. Higher temperatures result in an increase of gaseous alkalis (Aravind 2007). Nahcolite (naturally occurring sodium bicarbonate, $NaHCO_3$) and sodium carbonate/bicarbonate mixtures are considered for use as sorbents. In the case of sodium carbonate, the following absorption reaction takes place:

$$Na_2CO_3 + 2HCl \rightarrow 2NaCl + CO_2 + H_2O \tag{7.52}$$

If the sorbent is injected before the particle filter, the solid salt can be separated together with the dust.

Mercury Removal

The prospect of stringent mercury emissions standards for coal conversion plants seriously dampens the outlook for hot or warm gas clean-up. It is believed that mercury removal becomes more difficult as the syngas temperature increases. If it is necessary to cool the syngas for mercury removal, then the motivation for hot or warm gas desulphurisation disappears unless related economic benefits can be demonstrated (Korens et al. 2002).

Hot Gas Cleaning at Temperatures Above 1,400°C

For the process of pressurised pulverised coal combustion, comprehensive investigations have been carried out to develop a gas cleaning system at a temperature above the fluid temperature (1,400–1,600°C). Results of these developments have been described in Sect. 7.4. It was proven that it is possible to remove liquid slag and gaseous alkalis down to very low concentrations in the gas. It is assumed that it will also be possible to apply this technology to gasification systems, to remove liquid ash, alkalis and possibly heavy metals. If such cleaning is needed, the removal of sulphur and chlorine would require temperatures of about 600°C as described above. The higher temperature gas cleaning process would be beneficial for gasification systems combined with a membrane shift reactor. The requirements for such a reactor are not yet known. Investigations are currently being carried out in the framework of a research project (Müller et al. 2009; Spliethoff et al. 2009).

7.6.5.7 CO_2 Separation at High Temperatures

Figure 7.80 shows a process configuration with hot gas cleaning and a CO_2 separation step at the end of the gas cleaning train, similar to corresponding arrangements for cold gas cleaning. For the removal of CO_2, a water gas CO shift reactor is required prior to the separation to produce a H_2/CO_2 mixture. There are two options for H_2/CO_2 separation:

- CO/O_2 separation by means of high-temperature membranes. The membranes separate the syngas, already shifted to H_2 and CO_2, into nearly pure streams of the two gases. In this process, however, the H_2 is produced at near-atmospheric

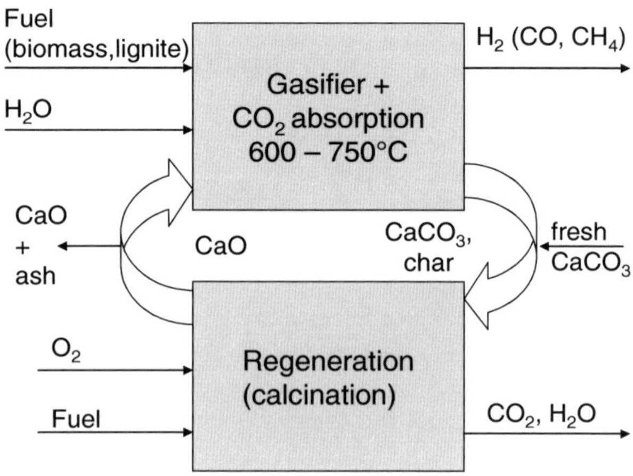

Fig. 7.81 Sorption-enhanced reforming

pressure, requiring compression for use in IGCC or refinery processes. Most probably, any hot gas membrane would require the prior removal of all particulate material and other trace components that may be corrosive or plug the membrane pores, though the gas cleaning requirements depend on the membrane material and are not yet known. It has to be pointed out that this technology is far from mature and that further development is required before it is applied, mainly due to the high cost of ceramic membranes. Conceptual designs try to combine the functions of the CO shift and membrane separation processes in a water gas shift membrane reactor (WGSMR).

- Absorption of CO_2 by using solid sorbents at temperatures between 400 and 650°C and regeneration at 750°C. This process can be applied as a post-combustion capture technology (termed carbonate looping) and will be discussed in Sect. 8.5.2.

CO_2 absorption by CaO and regeneration of $CaCO_3$ can be combined with steam gasification (Weimer et al. 2008; Florin and Harris 2008). This process is known as lime-enhanced gasification of solid fuels (LEGS), sorption-enhanced reforming (SER) or absorption-enhanced reforming (AER). Figure 7.81 shows a schematic diagram of the process, which consists of a steam gasification reactor as well as a regeneration reactor. For both reactors, fluidised beds are proposed because of their excellent gas/particle contact and the fact that the convenient temperatures are appropriate for fluidised beds.

- In the gasifier, CaO absorbs CO_2 and increases the H_2 concentration of the product gas. In the case of biomass, H_2 concentrations of 40–50%, which are reported for steam gasification without CO_2 capture, increase to about 80% after CO_2 absorption. The process conditions of the gasification reactor have to be selected according to the carbon conversion and CO_2 absorption requirements. Carbon conversion is favoured at high temperatures, whereas the exothermic CO_2 absorption is favoured at lower temperatures. To achieve a carbon capture of 80% it is necessary to increase the pressure as the temperature increases, for example from a gasification pressure of 0.1 MPa at 650°C to 0.9 MPa at 750°C. Due to this reason atmospheric gasification and capture is only suitable for biomass, whereas the utilisation of brown coal requires pressurised conditions. The potential to further increase the pressure is limited, because increasing the pressure promotes methane formation and reduces carbon capture. The exothermic CO_2 absorption reaction provides heat for the endothermic gasification.
- In a second reactor, CO_2 is released by calcination of $CaCO_3$, producing a high-purity CO_2 stream for storage. The regenerated CaO is recycled to the gasifier. At atmospheric pressure, the temperature required for regeneration is about 900°C. Higher pressures require higher temperatures for regeneration, which results in an energy penalty. The heat required for regeneration can be provided by the combustion of residual char from the gasifier. If a concentrated CO_2 stream is required, oxygen is needed for combustion.

7.6.6 Components and Integration

7.6.6.1 Gas Turbines

Gas turbines are designed for natural gas and liquid fuels, but are also commercially available for operation using syngas. The two coal IGCC demonstration plants in the USA (Tampa and Wabash) each use a GE 7FA turbine, while the two European plants at Buggenum and Puertollano use the Siemens SGT5-2000E (previously called V94.2) and the SGT5-4000F (previously called V94.3) (Maurstad 2005).

Syngas Combustion

Synthesis gas from high-temperature gasification essentially consists of carbon monoxide and hydrogen. Hydrogen is very reactive and increases the flame velocity, as opposed to fuels containing hydrocarbons, which decrease it. The calorific value of syngas from oxygen-blown gasifiers lies between about 8 and 11 MJ/kg, while it lies between 4 and 6 MJ/kg for air-blown gasifiers, which dilute the gas with nitrogen. These latter values are roughly 4–10 times lower than the calorific values of common gas turbine fuels. Given the lower air demand of H_2 and CO, the adiabatic combustion temperatures – under standard conditions for the combustion air (1 bar, 15°C) – are around 2,040°C and thus 100°C higher than in the combustion of natural gas. By admixing the nitrogen fraction removed in air separation, the adiabatic combustion temperatures can be lowered to about 1,580°C. These flame temperatures are sufficiently low to limit the formation of thermal nitrogen in diffusion flames (Schetter et al. 1991).

Gas turbines which run on natural gas normally use the so-called dry low-NO_x (DLN) combustors. In these combustion chambers, the fuel is pre-mixed with the total air to lower the flame temperatures and reduce the formation of thermal NO_x. They are designated "dry" because no injection of water or steam is used to reduce flame temperatures. However, DLN combustors are currently not used with syngas or hydrogen mixtures as fuels because of the danger of flashback caused by the high flame propagation speed of hydrogen. The current practice is to use traditional diffusion combustors, which require a diluent to reduce flame temperatures for NO_x control. The existing IGCC demonstration plants use nitrogen from the air separation unit, saturation of syngas with water or a combination of both to dilute the fuel before combustion with air. By these methods it is possible to reach a NO_x concentration of around 10 ppmv (at 15% O_2) in the exhaust gas.

For coal gas burners, emissions have to be low both for coal gas and for natural gas firing. Figure 7.82 shows a standard hybrid burner and a modified hybrid burner, the latter of which is used in Buggenum and in Puertollano and operated as a simple diffusion burner. The syngas injection is via an annular swirl nozzle with a large cross-section around the centrally mounted natural gas or crude oil nozzle. The syngas flames in Buggenum are stable up to at least a calorific value of 4 MJ/kg and a 40% output and potentially more (Huth et al. 1998; Hannemann et al. 2003).

When a gas turbine is run on syngas, which has a higher share of hydrogen compared to natural gas, the exhaust gas may consist of more than the usual 8% by

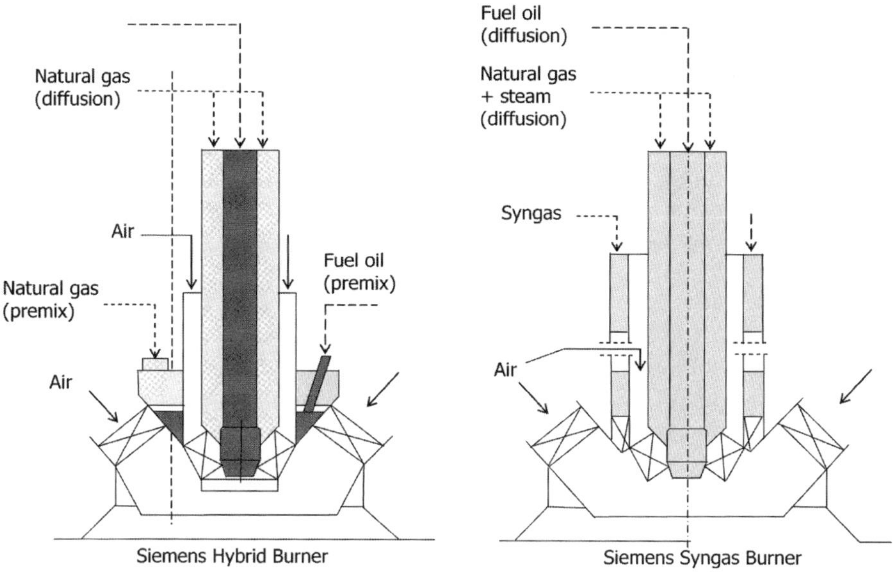

Fig. 7.82 A burner for syngas applications (Huth et al. 1998)

volume of water vapour. The water vapour concentration can vary between 5% (for the Puertollano fuel gas) and 14% (for pure hydrogen). The significance of a higher volume percentage of steam in the exhaust is an increased heat transfer, which raises the metal temperatures, thus shortening the lifetime of the turbine materials. Additionally, the increased mass flow through the turbine also results in an enhanced heat transfer.

While there are several gas turbines running on syngas, there are no turbines running solely on hydrogen. For pure hydrogen to be the fuel, the turbine inlet temperature (TIT) of the gas turbine would probably have to be reduced to avoid shorter lifetimes of the turbine blades (both the base materials and the coatings). This reduction in the TIT would reduce the efficiency of the combined cycle. In IGCC applications with CO_2 removal, hydrogen can be diluted by water saturation or by nitrogen.

DLN burners for syngas or hydrogen are not state of the art, but their use would be beneficial, because they do not require such a massive dilution for NO_x control. Dilution with steam and nitrogen down to hydrogen concentrations of 50–70% is the countermeasure to control the high flame propagation speed. DLN burners are currently under development.

Gas Turbine Power and Compressor Surge

As explained above, syngas use features a higher mass flow in comparison to natural gas. The increased mass flow of fuel and therefore the higher mass flowrate through the turbine will increase the power output from the turbine. If the fuel is diluted with nitrogen or water, the potential for increased GT power output is even higher. Depending on the turbine, there may be several limitations to the realisation of the increased power output, such as the risk of compressor surge and limitations on the gas turbine torque and turbine inlet temperatures.

A higher mass flowrate through the turbine stage will increase the pressure at the turbine inlet and thus also at the compressor outlet, so that the compressor can run into surge, with the air flow no longer maintained. The amount of pressure increase the compressor can tolerate is defined as the compressor surge margin, which depends on the design of a given compressor. There are several other possible strategies to resolve the surge limitation problem:

- Modify the gas turbine of the GT: The turbine itself can be modified to have an increased cross-sectional area to allow a higher flowrate.
- Modify the compressor of the GT: With an additional compressor stage the gas turbine can operate at a higher overall pressure ratio without surge problems.
- Bleeding off air from the compressor: This solution is possible for plants with air integration. Bleeding off more air mass flow than the mass flow of the nitrogen brought back from the ASU to the turbine reduces the mass flow through the turbine. Air integration therefore provides the potential to use gas turbines which would otherwise need to be redesigned to work with syngas (Maurstad 2005).

7.6.6.2 Air Separation Unit (ASU)

The oxygen supply to a gasifier is one of the most expensive single parts of any gasification process. The capital cost of the ASU, with its associated compressors, is about 10–15% of the total plant cost. The commercial technology used for oxygen production in IGCC plants is cryogenic air separation, which may be defined as the separation of air into component gases by distillation at low temperatures. Cryogenic air separation has single-train O_2 production capacities of 3,200 t/day and is recognised for its high reliability.

The major energy requirement of the process is for the air compression, which is in the order of magnitude of 5–7% of the gross generator output. Typically, the air to the ASU is compressed to around 5 bar, and the oxygen (typically 95% O_2, 3.5% Ar and 1.5% N_2 by volume) and nitrogen product streams become available at around 1 bar. The process may also operate at elevated pressure so that part or all of the ASU air is supplied from the gas turbine compressor. In this case, the ASU product streams are at around 5 bar, so this reduces the recompression work.

Alternative processes for air separation are limited in capacity and achievable oxygen purity. Pressure swing absorption units are available up to a capacity of about 140 t/day, but they can only reach a purity of about 95%. The product quality obtainable with polymer membrane technology is about 40% oxygen, with capacities of up to 20 t/day. Oxygen purity is a limiting factor for chemical applications. Most probable gasification applications for polymer membrane technology and pressure swing absorption are biomass power applications, where the sizes are also at the lower end of the scale. Both technologies have the advantage of a quick start-up compared to cryogenic units (Higman and van der Burgt 2008; Maurstad 2005).

7.6.6.3 Integration

In a natural gas fired gas and steam turbine power plant, the gas turbine and steam processes are linked to each other only by the flue gas feed to the heat recovery steam generator. In contrast, in an IGCC power plant, there are several possibilities for coupling the air separation unit, the gasifier, the gas turbine and the steam generator. The purpose of integration – hence the term "integrated gasification process" – is to maximise the efficiency. A high degree of integration, however, can imply disadvantages affecting the operating performance at different loads and diminish the availability of the process as a whole.

Figure 7.83 shows the different possibilities for coupling:

Steamside integration: In an IGCC power plant, steam is generated in the heat recovery steam generator (HRSG), the gasifier and the raw gas cooler. In a natural gas fired HRSG, the flue gas temperature determines the live steam temperature via the pinch point of the heat exchanger. This problem is diminished in an IGCC plant by the large evaporative surfaces of the syngas cooler, which can superheat more saturated steam in the waste heat boiler. By doing so, the efficiency of the steam production rises from about 40 to 42%. Steamside integration is applied in all demonstration plants.

Airside integration: About 15–25% of the total air flow has to be fed to the air separation unit (ASU) for the generation of oxygen; the rest serves to oxidise the syngas in the gas turbine combustor. It is deemed full integration if the total air flow needed for the ASU is drawn off after the gas turbine compressor (GT compressor). The better compression efficiency of the gas turbine helps to reduce the energy demand for the compression as a whole. The air separation unit is then operated under pressure; the pressure rise of the oxygen up to gasification pressure is

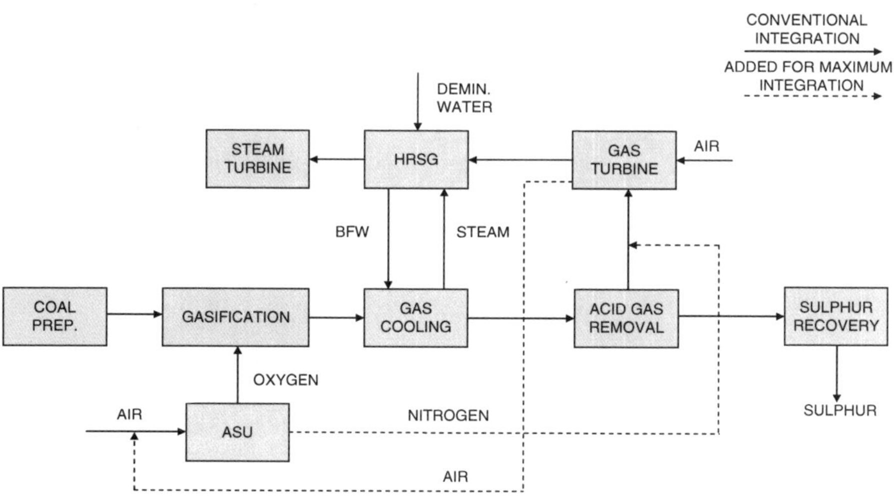

Fig. 7.83 Integrated IGCC power Plants – level of integration (from Higman and van der Burgt 2008, © 2008, with permission from Elsevier)

therefore smaller. The degree of (air) integration is usually defined as the percentage of the total ASU air required coming from the GT compressor. The two existing US IGCC demonstration plants started with 0% integration, while the two European plants had 100% integration.

The possible benefits of integration are an increased efficiency, increased power output and reduced capital cost (e.g. savings on the ASU air compressor). The drawbacks of high integration are a possible reduced availability of the process, less operational flexibility, lengthy start-up times and the fact that the ASU cannot start without the GT running. An integration of 100% will always yield the maximum efficiency, but not necessarily the maximum power.

Because of the lower process availability and operational flexibility during start-up and shutdown, a partial integration scheme is implemented in newer plants as a compromise between availability and operational flexibility on the one hand and efficiency on the other. In such schemes, only part of the air flow for the ASU is taken from the gas turbine, and the rest of the air is fed by way of a separate compressor.

Integration on the nitrogen side: In order to reduce the flame temperatures in the gas turbine and the nitrogen oxide emissions, nitrogen is fed before or into the gas turbine combustor. Similarly to air integration, prior pressurisation of the nitrogen reduces the necessary pressure increase up to the gasification pressure. Even if the air integration is 0%, it may still be beneficial to use nitrogen from the ASU for NO_x reduction (Geosits and Schmoe 2005; Higman and van der Burgt 2008; Eurlings and Ploeg 1999; Maurstad 2005).

7.6.7 State of the Art and Perspectives

7.6.7.1 IGCC Plants in Operation

The worldwide installed gasification capacity of about 70 GW_{th} predominantly serves to make chemical products and fuels; only about 20% of the capacity is used to generate electrical power. For power generation, residues from refinery processes are mostly used – the generation of electrical power by coal in IGCC power plants worldwide is limited to a few plants with a total fuel capacity of 3 GW (NETL 2007; Minchener 2005). IGCC technology for coal was demonstrated, using public funding, at several locations in the 1990s in Europe and the USA. The following plants were part of those demonstrations and are being operated commercially at the moment:

- In 1994, the 253 MW_{el} (semi-)commercial IGCC power plant Demkolec was commissioned at Buggenum, the Netherlands. The net generating efficiency of this power plant, which uses Shell gasification technology, is 43.2%.
- In 1995, the 252 MW_{el} Wabash River IGCC went into operation in Terre Haute, Indiana (USA), based on the E-Gasification process. Wabash River was a repowering of a small steam power plant and not a greenfield project. The IGCC has a net efficiency of 39.9%.

- In 1996, a 250 MW_{el} IGCC was commissioned in Polk County, Florida (USA). This IGCC power plant, based on Texaco gasification technology, has a net generating efficiency of 38.0% (Tampa_Electric 2002; Tampa_Electric 2004).
- In 1996, a 318 MW_{el} IGCC went into service at Puertollano (Spain). Until 1998, the power plant was operated using natural gas as a (secondary) fuel (Hannemann et al. 2003; Mendez-Vigo et al. 1998; Cortes 1999). When the gasifier was ready for operation in 1998, hard coal and petroleum cokes (50/50%) became the main fuels. Puertollano makes use of Prenflo gasification technology and has a net efficiency of 45%.

The plant data is compiled in Table 7.18. The two European plants, due to their gas turbines, dry feeding, dry quenching and high degree of integration of the air separation unit, have a higher efficiency.

7.6.7.2 Description of the Puertollano Plant

The IGCC plant in Puertollano is the biggest coal-based IGCC power plant in the world, having a net capacity of 305 MW_{el} and a design efficiency of 45% at a condenser pressure of 0.0715 bar. The fuel used is a mixture, consisting half of a Spanish coal with a very high ash content (47% ash) and half of a high-sulphur petrol coke. The efficiency is highly dependent on the fuel; running with the petrol coke alone, an efficiency of only 42% is achieved. Despite the high-sulphur fuel, the emissions of SO_x remains below 20 mg/Nm^3 (at 6% O_2), which is below the EU limit by one order of magnitude.

The entrained-flow gasifier is fed with oxygen of 85% purity, and at high temperatures (above 1,600°C) the finely milled fuel gets converted. The fuel is blown in pneumatically with a nitrogen stream which is tapped after the air separation unit. The pressure in the reactor is 25 bar; the temperature at the outlet of the gasifier is about 1,600°C. The employed gasification system is a Prenflo entrained-flow gasifier, which is roughly similar in construction to the Shell gasifier. About 30–50% of the cold syngas is recycled to quench the hot syngas from the gasifier to about 900°C. A convective heat exchanger further reduces the temperature to about 200°C. The particles are separated from the cooled raw gas in a ceramic cartridge filter, and there is the potential to return the fly ash to the gasifier by way of a lock-hopper system, thus removing almost the entire ash load from the process as unleachable slag. Subsequently, the raw gas is subjected to wet gas cleaning, consisting of an MDEA scrubber, an upstream COS hydrolysis unit and a Claus tail-gas cleaning process for sulphur production. The residual sulphur content of the coal gas is then less than 25 mg/m^3. After the cleaning process, the syngas is diluted with nitrogen from the air separation unit and burned in the gas turbine combustor. The flue gases, with a gas turbine entry temperature of 1,150°C, expand in the gas turbine. The remaining heat of 500°C as it exits the turbine is used in the HRSG (Coca 2003).

The air separation unit has a high degree of integration and is fed with a partial flow from the air compressors. Part of the nitrogen stream produced during air

Table 7.18 Data for IGCC power plants in operation (Hannemann et al. 2003; Lako 2004; Tampa_Electric 2002; Tampa_Electric 2004; Holt 2003; Coca 2003)

IGCC power plant	Wabash river	Buggenum	Puertollano	Tampa
Year (operational)	1995	1994	1998	1996
State/Country	Indiana/USA	The Netherlands	Spain	Florida/USA
Feedstock	Hard coal	Hard coal	Lignite and petr. coke	Hard coal
Secondary fuel	–	Natural gas	Natural gas	
Gasification process	E-Gas	Shell	Prenflo	GE
Level of integration	Low	High	High	Low
Raw gas (after the gasifier)				
Fuel gas temperature [°C]	300	300	302	330
H_2 [%vol.]	34.4	25.5	22.1	36.4
CO [%vol.]	45.3	62.7	60.5	42.8
CO_2 [%vol.]	15.8	2.2	3.9	14.4
N_2 [%vol.]	2.9	8.8	12.5	3.3
Ar [%vol.]	0.6	0.8	1.0	0.9
H_2/CO ratio (vol.)	0.76	0.4	0.36	0.85
HHV [MJ/m^3]	10.3	10.4	10.5	9.9
Fuel gas (fired in the gas turbine)				
N_2 [%vol.]		42.0	53.1	
H_2O [%vol.]		19.1	4.2	
LHV [MJ/m^3]		4.3	4.3	
Emissions				
SO_2 emission [mg/m^3]	40	35	25	40
NO_x emission [mg/m^3]	100	25	150	100
Integration				
Air extracted from GT related to ASU [%]	0	100	100	0
Related to compressor [%]	0	16	18	0
Nitrogen integration [%]	0	100	100	100
Output				
Gas turbine [MW]	192	156	179	192
Steam turbine [MW]	96	128	137	120–135
Net power output [MW]	252	253	276	250
Net efficiency [LHV]				
High-quality hard coal [%]	39.9	43.2	45.0	38.0
Lignite and petcoke [%]	–	–	42.0	
Petroleum cokes [%]		–	–	
Natural gas [%]	–	52	52.4	

separation is used to transport the dry and milled fuel mixture into the gasifier, while another part of it cleans the raw gas dedusting cartridge filters. The remaining nitrogen stream is mixed in with the cleaned coal gas before entering the gas turbine combustor to lower the combustion temperature and comply with the NO_x emission limit without flue gas cleaning.

In Fig. 7.86, an optimised IGCC configuration based on the configuration of the plant in Puertollano is shown.

7.6.7.3 Process Availability and Costs of IGCC Plants

In all the IGCC power plants in operation, a great number of problems arose in the first years of operation in relation to the gasifier, the gas turbine and the gas cleaning. The high integration degree of the ASU, too, gave rise to difficulties and resulted in the use of new control techniques. Today, the process availabilities of the power plants range between about 70 and 80% and are thus markedly lower than the availabilities of steam power plants. Figure 7.84 shows this graphically, making it clear that the main problem of this technology is its lower availability. The ability to compete with other power plant technologies requires availabilities higher than 90%. IGCC plants in refineries show higher availabilities when tried-and-tested technologies and lower degrees of integration are used (Folke 2006; Higman 2005). New data from Buggenum with availabilities of 90% underlines that from an engineering point of view it should be possible to achieve similar availabilities than in conventional power plants.

Another problem is the considerably higher cost of IGCC power plants. Figure 7.85 pits the capital costs of the installed IGCC power plants against the capital costs of conventional pulverised coal fired steam power plants, using Euros in 2004 as a basis. The capital costs of the erected plants were between 30 and 100% above the costs of conventional steam power plants; currently, it is assumed that costs are 40% higher. The expectation is, however, that this cost difference will decrease as advancement along the learning curve for IGCC technology occurs (Lako 2004). In addition, if CO_2 capture is considered, it is expected that IGCC power plants with CO_2 capture and steam power plants with CO_2 capture will have roughly the same level of capital costs.

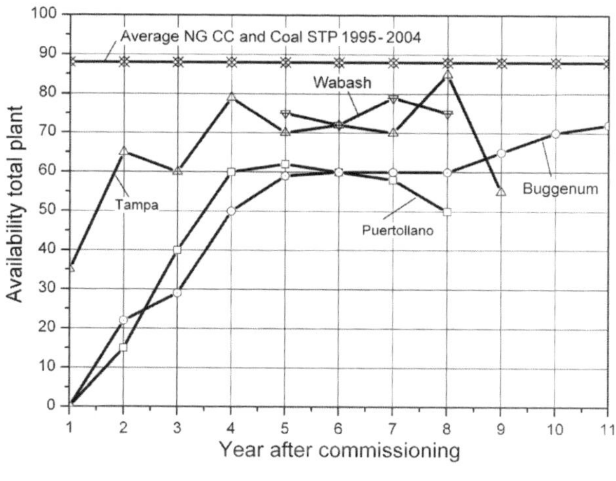

Fig. 7.84 Process availability of existing IGCC plants (Folke 2006)

Fig. 7.85 Cost of IGCC plants in comparison to conventional steam power plants (Lako 2004)

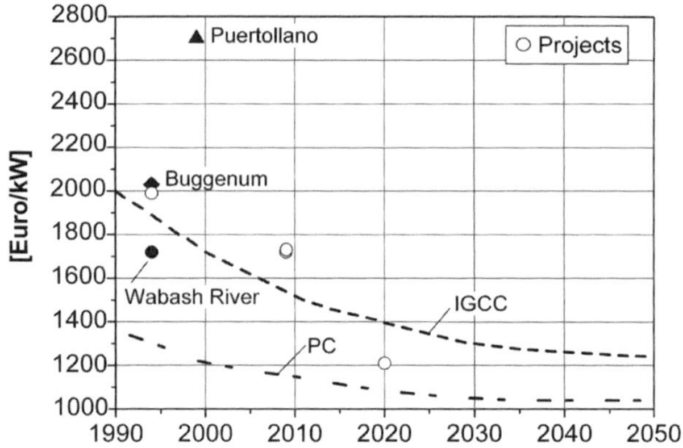

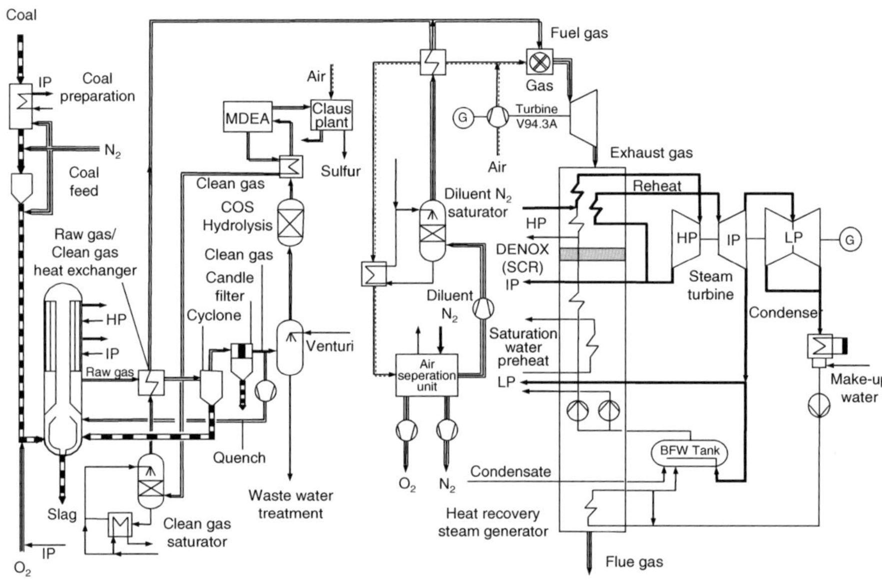

Fig. 7.86 Process flow diagram of IGCC 98 (Pruschek 2002)

7.6.7.4 Efficiency Potential

IGCC power plants in operation today achieve electrical efficiencies of up to about 45%. Using today's state-of-the-art technology, without CO_2 removal, IGCC power plants could be designed and built with an electrical efficiency of over 50%. If the classical single components and the overall IGCC concept were to be further optimised, electrical efficiencies of 55% could be achieved within 15 years. Estimates of efficiency are based both on achieved efficiencies at operating plants (and the potential to optimise them) and on the expected future advancements in technology, particularly for the gas turbine.

Extensive studies on the potential of IGCC power plants were carried out by Pruschek, who, within the framework of European projects, and comparing to the IGCC plant in Puertollano, investigated the design and the efficiency of an IGCC power plant feasible in 1998 (Pruschek 1998). Its configuration, shown in Fig. 7.86, is similar to the configuration of the Puertollano IGCC power plant. The entrained-flow gasification is operated at a pressure of 29 bar with 95% oxygen as the oxidiser. The gas exiting the gasifier first gets cooled to about 900°C with recirculated raw gas, then the remaining flue gas heat is used to generate steam and to reheat the cleaned fuel gas. The air separation unit is fully integrated, so that the fresh air is also compressed in the gas turbine compressor and part of the waste nitrogen is mixed back into the fuel gas before entering the gas turbine combustor. The waste nitrogen and the cleaned fuel gas are humidified with the necessary water fraction by saturators before the mixing is performed, at a temperature as low as possible. At the given ambient conditions and at a gas turbine entry temperature of 1,250°C (ISO), the IGCC achieves an efficiency of 51.5%. The efficiency owes its increase, in essence, to the raising of the gas turbine entry temperature to 1,250°C, as well as further measures such as the increase of the steam conditions and more intensive fuel gas preheating (Haupt et al. 1998; Pruschek 1998; Pruschek et al. 1997).

Over and above the 1998 IGCC concept design, there are considerable potentials for increasing the efficiency of IGCC power plants. Applying the measures shown in Fig. 7.87, efficiencies up to 58% could be achieved. It becomes clear that development of gas turbines would yield the largest efficiency increases. Hot dry gas cleaning would raise the efficiency by about 0.7% (Pruschek 1998).

7.6.7.5 IGCC Concept Designs with CO_2 Removal

IGCC power plants can separate CO_2 with few additional components and can be regarded as the most advanced power plant technology for CO_2 separation, because the necessary additional components (CO shift and CO_2 scrubbing units)

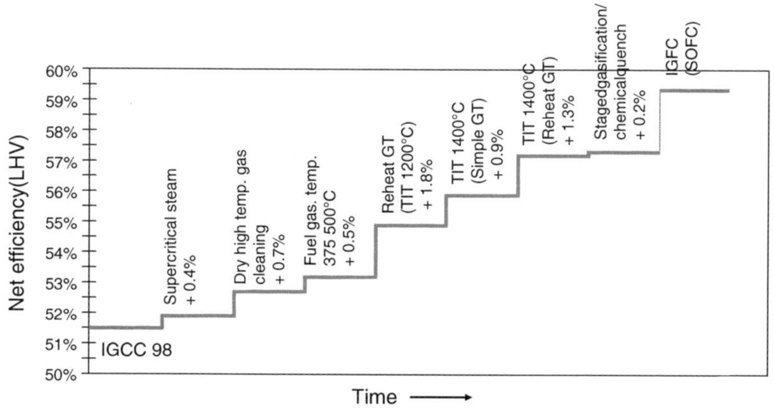

Fig. 7.87 Potential future development of IGCC power plants (Pruschek 1998)

are employed already for the production of hydrogen. An IGCC design with CO_2 capture is shown in Fig. 8.22. Expectations are that the efficiency will drop by about 8–10% compared to a process without CO_2 capture – starting out from the IGCC 98 concept, having 52% efficiency, the efficiency decreases to 42%. In Chap. 8, the IGCC technology with CO_2 removal is compared to competing methods for CO_2 separation during power production.

Various projects examining the design and construction of a CO_2-free power plant using IGCC technology are currently in progress, such as the following:

- RWE: Construction of a 360 MW_{el} lignite IGCC-CCS power plant by 2014 (Lambertz and Ewers 2006). The current design results in low efficiencies in the range of about 35% because first priority has been given to the plant availability. For this reason proven technologies and a low level of integration have been chosen.
- FutureGen (USA): Construction of a 275 MW_{el} IGCC-CCS power plant, a project lasting until 2012 (FutureGen 2007)
- GreenGen (China): Development of its own coal gasification technology by 2009 followed by construction of a 400 MW_{el} demonstration plant for hydrogen production with CO_2 capture (Folke 2006).

7.6.7.6 Long-Term Perspectives

Further development must aim to eliminate the disadvantages of IGCC power plants, namely the low availability and the high costs. If these problems can be solved, IGCC could provide a substantial increase in the efficiency of power generation from coal.

In gasification, the major part of the fuel energy gets converted into chemically bound energy and a smaller part into heat. A level of conversion into chemically bound energy as high as possible and a heat exploitation as efficient as possible should be aimed at.

The quality of the conversion into chemically bound energy is described by the cold gas efficiency. The maximum conversion of fuel energy into chemical energy can be achieved using chemical quenches or by internal heat utilisation. Gasification methods should be sought that combine the benefit of fluidised bed gasification (high cold gas efficiency) with the benefit of entrained-flow gasification (ash removal, fuel flexibility, compactness).

The utilisation of the heat of hot syngases would be optimal if the gases could be fed directly into the gas turbine. However, for gas cleaning, these gases have to be cooled down to very low temperatures. The heat removed when cooling to these temperatures can be partly used, being converted into mechanical energy with the lower efficiency of the steam cycle. On top of this, the requirements of gas cleaning involve cooling and heating processes that cause additional exergy losses. In order to avoid them, gas cleaning processes need to be developed that run at higher temperatures and in dry conditions to the greatest possible extent. The aim is not to cool the gas below the temperature of the following step in the process, so that a continuous extraction of useful heat is possible.

A further significant increase in the energy efficiency can be expected by using hydrogen membranes. With H_2 membranes, only a stoichiometric steam-to-CO ratio would be necessary. This could decrease the demand of steam significantly, while also making it possible to avoid the cooling to the thermodynamically necessary low temperatures for the shift reaction, thus providing a hotter gas for the burner of the gas turbine. In this respect, high-temperature membrane shift reactors are desired, which, in one unit, perform both the gas conversion (shift) and the gas separation.

The long-term development objective for a CO_2-free IGCC power plant with the highest possible efficiency is therefore a high-pressure, high-temperature gasifier with integrated hot gas cleaning, H_2 separation by catalytic high-temperature membranes at the highest possible temperatures and an H_2 gas turbine being fed with hot fuel gas.

In the long term, it will also be reasonable to combine gasification and solid oxide fuel cells (SOFCs). The technology suited in particular for this process is allothermal fluidised bed gasification, which utilises the waste heat of an SOFC to supply gasification heat.

References

Aho, M., Hernberg, R. and Häyrinen, V. (1995). Fuel reactivity and release of pollutants and alkali vapours in pressurized combustion for combined cycle power generation. Final Report, Contract JOU2-CT92-0037.

Almhem, P. (1996). PFBC-Heizkraftwerk Cottbus, Wirbelschicht-Kombi-Anlage für Trockenbraunkohle. *VDI-GET Tagung "Entwicklungslinien der Energie- und Kraftwerkstechnik"*. Siegen.

Aquaro, D. and Pieve, M. (2007). High temperature heat exchangers for power plants: performance of advanced metallic recuperators. *Applied Thermal Engineering* **27**(2–3): 389–400.

Aravind, P. V. (2007). Studies on High Efficiency Energy Systems Based on Biomass Gasifiers and Solid Oxide Fuel Cells with Ni/GDC Anodes. Delft, TU Delft. Ph.D.

Asai, A., Izaki, K., Egami, Y., Tsuji, K., Kumagai, S. and Nishijima, Y. (2004). System outline and operational status of Karita Power Station New Unit 1 (PFBC). *JSME International Journal Series B-Fluids and Thermal Engineering* **47**(2): 193–199.

Atimay, A. T. (2001). Cleaner energy production with integrated gasification combined cycle systems and use of metal oxide sorbents for H2S cleanup from coal gas. *Clean Products and Processes* **2**: 197–208.

Bammert, K. (1986). Operating experience and measurements on turbo sets of CCGT-cogeneration plants in Germany. International Gas Turbine Conference, Düsseldorf, ASME-Paper 86-GT-101.

Bannister, R. L., Newby, R. A. and Diehl, R. C. (1992). Development of a direct coal fired combined cycle. *Mechanical Engineering* **114**(12): 64–70.

Bannister, R. L., Pillsbury, P. W., Diehl, R. C. and Loftus, P. J. (1990). Recent test results in the direct coal fired 80 MW combustion turbine program. International Gas Turbine and Aeroengine Congress, Brussels, ASME-Paper 90-GT-58.

Bauer, D. A. and Marocco, M. (1995). Tidd PFBC demonstration plant, summary of operations. Book 3, Part C, Vol. 2, pp. 109–121, Anaheim, USA, Power Gen 95 Americas.

Bauer, F., Meier, H. J., Smuda, J. V. and Stuhlmüller, F. (1994). PFBC Development for Lignite Fired CC Power Plants. Power-Gen Europe, Köln, Band III, pp. 101–120.
Baum, J. (2001). Untersuchungen zum extern befeuerten Gasturbinenprozeß mit Hochtemperaturwärmeübertrager. VDI Fortschritt-Berichte Reihe 6 Energietechnik, Nr. 456. Düsseldorf, VDI.
Benson, S. (2000). Ceramics for advanced power generation. London, IEA Coal Research, CCC/37.
Bliem, C. et al. (1985). Ceramic heat exchanger concepts and materials technology. Park Ridge, NJ, Noyes.
Böhm, H. (1994). Fossilbefeuerte Kraftwerke. Stand, Entwicklungstendenzen. *VGB Kraftwerkstechnik* **74**(3): 173–186.
Bohn, D. (2005). Technologien für Gasturbinen der übernächsten Generation. *VGB PowerTech* **85**(7): 65–71.
Bohn, D. (2007). Improved cooling concept for turbine blades of high-temperature gas turbines. *VGB PowerTech* **87**(10): 96–103.
Bonn, B. (1996). On-line Messung von Alkalidampf-Konzentrationen in heißen Rauchgasen. *BMBF-Statusseminar Heißgasreinigung*. Karlsruhe.
Bunthoff, D. and Meier, H. J. (1987). Umweltfreundliches Kraftwerk mit Druckwirbelschichtfeuerung, Planungskriterien für die Dampferzeuger. *VGB Kraftwerkstechnik* **67**(8): 751–757.
Bunthoff, D., Rehwinkel, H. and von Wedel, G. (1990). Druckwirbelschichtfeuerung: Fortschrittliche Konzepte und aktueller Entwicklungsstand. *Technische Mitteilungen* **83**(1): 25–32.
Chalupnik, R. W., Krautz, H. J., Wirtz, M. and Stuhlmüller, F. (2001). Applied research for a new generation of lignite-fired combined cycle power plants using circulating pressurized fluidized bed combustion. Proceedings of the 16th International Conference on Fluidized Bed Combustion. May 13–16, Reno, NV.
Coca, M. (2003). Integrated gasification combined cycle technology – Its actual application in Spain: ELCOGAS- Puertollano. From http://www.elcogas.es/en_home_set.htm
Coombs, M., Kotchick, D. and Waren, H (1979). High-temperature ceramic heat exchanger. Final report EPRI-EP-1127, July 1979. Torrance, CA, AiResearch Mfg. Co.
COORETEC (2003). Forschungs– und Entwicklungskonzept für emissionsarme fossil befeuerte Kraftwerke – Bericht der COORETEC-Arbeitsgruppen. Bundesministerium für Wirtschaft und Arbeit, December 2003, ISSN 0342-9288 (BMWA-Dokumentation), From http://www.bmwa.bund.de.
Cortes, J. C. (1999). Puertollano IGCC; operation status. Gasification Technologies Conference. October 12–20, San Francisco CA.
Cowell, L. H. and LeCren, R. T. (1992). Experimental evaluation of a two stage slagging combustor design for a coal-fueled industrial gas turbine. International Gas Turbine and Aeroengine Congress. Cologne, ASME-Paper 92-GT-259.
Cowell, L. H., Hasen, A. M., LeCren, R. T. and Stephenson, M. D. (1992a). Coal fueled two-stage slagging combustion island and cleanup system for gas turbine application. International Gas Turbine and Aeroengine Congress. Brussels, ASME-Paper 90-GT-59.
Cowell, L. H., LeCren, R. T. and Tenbrook, C. E. (1992b). Two-Stage Slagging Combustor Design for a Coal-Fueled Industrial Gas Turbine. *Journal of Engineering for Gas Turbines and Power* **114**(4): 359–366.
Daijou, Y., Suzuki, K., Ishinomori, T., Shigeta, J., Yanagisawa, T. and Tsumita, Y. (1997). Alkali-vapor measurements in the Wakamatsu PFBC plant. Proceedings of the 14th International Conference on Fluidized Bed Combustion. Vancouver, Canada.
Dibelius, G. and Pitt, R. (1989). Gasturbinenkonzepte für Druckwirbelschichtanlagen – Konzepte, Auslegung, Betriebsverhalten. *VDI-Berichte* **715**: 213–236.
DOE (1997). Clean coal technology demonstration program. Washington, DC, October.
DOE (2003a). Clean coal technology program, McIntosh Unit 4A PCFB Demonstration Project.
DOE (2003b). Clean coal technology program, McIntosh Unit 4B Topped PCFB Demonstration Project.

Edelmann, H. and Stuhlmüller, F. (1997). EFCC – Ein zukünftiges Konzept für Kohle-Kombikraftwerke. *VGB Kraftwerkstechnik* **77**(7): 537–543.

Emsperger, W. and Brückner, H. (1986). Kombiprozesse mit atmosphärischen und druckbetriebenen Wirbelschichtsystemen. *VDI-Berichte Nr* **601**: 257–272.

Escobar, I. and Muller, M. (2007). Alkali removal at about 1400 degrees C for the pressurized pulverized coal combustion combined cycle. 2. Sorbents and sorption mechanisms. *Energy & Fuels* **21**(2): 735–743.

Escobar, I., Oleschko, H., Wolf, K.-J. and Muller, M. (2008). Alkali removal from hot flue gas by solid sorbents in pressurized pulverized coal combustion. *Powder Technology* **180**(1–2): 51–56.

Eurlings, J. and Ploeg, J. (1999). Process performance of the SCGP at Buggenum IGCC. Gasification Technologies Conference. October 18–20, San Francisco, CA.

Fichtner (1992). Der Kombiprozeß mit indirekt kohlegefeuerter Gasturbine. Abschlußbericht, BMFT-Förderkennzeichen 0326746, September.

Florin, N. H. and Harris, A. T. (2008). Enhanced hydrogen production from biomass with in situ carbon dioxide capture using calcium oxide sorbents. *Chemical Engineering Science* **63**(2): 287–316.

Folke, C. (2006). Kohlvergasung – Option zur CO2-Abtrennung. Kraftwerkstechnisches Kolloquium Dresden 2006. Vortrag V7.10.

Förster, M., Hannes, K. and Telöken, R. (2001). Combined cycle power plant with Pressurized Coal Combustion (PPCC): State of development and perspectives. *VGB PowerTech* **81**(9): 30–35.

Förster, M., Neumann, F., Telöken, R. and Pavone, D. (2005). Druckkohlenstaubfeuerung: Aktueller Fortschritt des kohlebasierten GuD-Prozesses in Richtung Gasturbinenverträglichkeit und Stand des Entwicklungsprojektes. *VGB PowerTech* **85**(6): 38–41.

FutureGen (2007). Initial conceptual design report. 2009, from http://www.futuregenalliance.org/publications.stm.

Geosits, R. F. and Schmoe, L. A. (2005). IGCC – The challenges of integration. ASME Turbo Expo 2005: Power for Land, Sea and Air, June 6–9, 2005, Reno-Tahoe, NV. Paper GT2005-68997.

Gockel, B. (1994). Ein mathematisches Modell zur Berechnung der Kohlenstaubverbrennung unter Berücksichtigung höherer Drücke. Dissertation Universität Bochum.

Goto, H. (1995). Operation experience from the 71 MW Wakamatsu PFBC demonstration plant. Proceedings of the 13th International Conference on Fluidized Bed Combustion. Orlando, FL. **2**: 911–918.

Gottwald, U., Monkhouse, P. and Bonn, B. (2001). Dependence of alkali emissions in PFB combustion on coal composition. *Fuel* **80**(13): 1893–1899.

Groll, M. (1980). Grundlagen und Anwendungen des Wärmerohres. *Naturwissenschaften* **67**: 72–79.

Hannemann, F. et al. (2003). Pushing forward IGCC technology at Siemens. Gasification Technologies 2003. 13 October 2003, San Francisco, CA.

Hannes, K. (1986). Kohlenstaub-Druckverbrennung. Entwicklungsstand und Anforderungen des Prozesses an die Verbrennungsführung *VGB-Kraftwerkstechnik* **77**(5): 393–400.

Hannes, K. (1996). Entwicklungsprogramm Druckkohlenstaubfeuerung. *BMBF-Statusseminar Heißgasreinigung*. Karlsruhe.

Hannes, K. W. (2002). Kohle-Kombi-Kraftwerke mit Druckkohlenstaubfeuerung: Das Druckflamm-Forschungsprogramm. *VGB PowerTech* **82**(8): 56–60.

Hannes, K. W., Reichert, G. and Weber, E. (1989). Pressurized pulverized coal combustion with molten ash removal. Conference on Technologies for Producing Electricity in the 21st Century. San Francisco, USA.

Harkins, B. D. and Ward, M. E. (1989). Preliminary design of a high pressure, high temperature ceramic air heat exchanger. In: Waste heat utilization, the winter annual meeting of the ASME. San Francisco, CA, December 10–15, Band HTD-118, pp. 51–61.

Haupt, G., Zimmermann, G., Baumann, H.-R., Ullrich, N., Pruschek, R. and Oeljeklaus, G. (1998). New design of IGGCC for competitive power generation. Gasification Technologies Conference. October 4–7, San Francisco, CA.

Higman, C. (2005). The reliability of integrated gasification combined cycle (IGCC) power generation units. Gasification Technologies Conference. San Francisco, CA.

Higman, C. and van der Burgt, M. (2008). Gasification. Amsterdam [u.a.], Elsevier/GPP, Gulf Professional Publishing.

Hippinen, I., Jahkola, A. and Lu, Y. (1991). Pressurized fluidized bed combustion of different type of solid fuels. Proceedings of the 11th International Conference on Fluidized Bed Combustion. Montreal, ASME.

Hoberg, H. and Gudenau, H. W. (1998). Heißgasreinigung für Kombiprozesse mit Druckkohlenstaubfeuerung. VGB-Konferenz Forschung in der Kraftwerkstechnik. Essen.

Hokari, N., Miyamoto, T., Ito, O., Mizumoto, M., Tomuro, J., Iwase, T., Yoshioka, S. and Ootani, Y. (2001). Sulfur dioxide removal performance of 250 MW PFBC at Osaki. Proceedings of the 16th International Conference on Fluidized Bed Combustion. May 13–16, Reno, NV.

Holt, N. A. H. (2003). Operating experience and improvement opportunities for coal-based IGCC plants. *Materials at High Temperatures* **20**(1): 1–6.

Hübner, K., Miao, J. and Görner, K. (2003). Untersuchungen zur Alkalisorption mit Unterstützung elektrischer Felder bei extremen Temperaturen oberhalb von 1300°C. 21. Deutscher Flammentag, Cottbus, VDI.

Hübner, K., Pavone, D., Schmidt, D. and Weber, E. (1988). Möglichkeiten der Gasreinigung im Temperaturbereich oberhalb von 1000°C. *VGB Kraftwerkstechnik* **68**(9): 931–935.

Hurley, J. P., Seery, D. J. and Robson, F. L. (2003). Experience with an ODS high-temperature heat exchanger in a pilot-scale HiPPS plant. *Materials at High Temperatures* **20**(1): 39–44.

Huth, M., Vortmeyer, N., Schetter, B. and Karg, J. (1998). Verbrennung von Synthesegas in Gasturbinen. *BWK* **50**(9): 35–39.

Jahraus, B. and Dieckmann, R. (1989). Der Kombiprozeß mit indirekt gefeuerter Gasturbine. *VDI-Berichte* **1182**: 287–300.

Jansson, S.A. (1995a). ABB's PFBC technology operation experience and standard products. EPRI Conference on New Power Generation Technology. San Francisco, CA.

Jansson, S. A. (1995b). *PFBC – clean coal technology – status and experience*. Power-Gen Europe '95. Amsterdam, The Netherlands. **4**: 97–109.

Jansson, S. A. (1996). *Gas turbine tolerance to particulates and chemical constituents*. BMBF Statusseminar "Heißgasreinigung in der Kraftwerkstechnik". Germany, Karlsruhe.

Jansson, S. A. and Svensson, J.-A. (1997). Experiences from the IEA hotgas filter testing project. Proceedings of the 14th International Conference on Fluidized Bed Combustion. Vancouver, Canada.

Jansson, S. and Anderson, J. (1999). Progress of ABB's PFBC projects. Proceedings of the 15th International Conference on Fluidized Bed Combustion. May 16–19, Savannah, GA.

Jansson, S., Kemmer, L. and Flodin, E. (1996). Pressurized fluidized bed power plants. Energy Conference. Haifa.

JBDT (1992). Jahrbuch der Dampferzeugertechnik, 7. Ausgabe, Essen Vulkan.

Jopp, K. (2005). Neue Maßstäbe im GuD-Prozess. *BWK* **67**(6): 52–54.

Jüntgen, H. and van Heek, K. H. (1981). Kohlevergasung – Grundlagen und Technische Anwendung. München, Verlag Karl Thiemig.

Kail, C. and Rukes, B. (1995). Fortschrittliche Gas- und Dampfturbinenprozesse zur Wirkungsgrad- und Leistungssteigerung bei GuD-Kraftwerken. *VDI-Berichte* **1182**: 71–87.

Kainer, H. (1988). Keramische Wärmetauscher. In: Technische Keramik. pp. 325–333 Essen, Vulkan.

Kainer, H. and Willmann, G. (1987). Komponenten für den Bau keramischer Rekuperatoren – Möglichkeiten und Beschränkungen. *Gas Wärme International* **36**(6): 332–340.

Käser, J. (1996). *Alkali Release under PFBC Conditions*. Diplomarbeit, Institut für Verfahrenstechnik und Dampfkesselwesen, Universität Stuttgart und ABB Carbon.

Keller, C. (1946). The Escher Wyss-AK closed-cycle turbine, its actual development and future prospects. *Transactions of the ASME* **68**: 791–822.

Keller, C. and Gaehler, W. (1961). The coal burning closed-cycle gas turbine. Gas Turbine Power Conference. ASME-Paper 61-GTP-2, Washington, DC

Keppel, W. (1995). Die Gasturbine als Schlüsselkomponente in der modernen Kraftwerkstechnik. *VDI-Berichte* **1182**: pp. 1–25.

Klara, J. M. (1994a). High performance power systems for the near-term and beyond. International Gas Turbine and Aeroengine Congress and Exposition. Netherlands, ASME-Paper 94-GT-315.

Klara, J. M. (1994b). HIPPS can compete with conventional PC systems. *Part II. Power Engineering* **98**: pp. 33–36.

Kloster, R. (1999). Thermodynamische Analyse und Optimierung von Gas-/Dampfturbinen-Kombi-Kraftwerke mit integrierter Kohlevergasung, VDI-Fortschritt-Berichte Reihe 6, Nr. 409 Dissertation Universität Essen.

Koike, J., Nakamura, S., Watanabe, H. and Imaizumi, T. (2003). Manufacturing and construction, operation of Karita 360 MW unit. Proceedings of the 17th International Fluidized Bed Combustion Conference. May 18–21, 2003, Jacksonville, FL.

Korens, N., Simbeck, D. R. and Wilhelm, D. J. (2002). Process screening analysis of alternative gas treating and sulfur removal for gasification. Pittsburgh, CA USDOE.

Kuhn, S. (2007). Funktions- und Betriebssicherheit eines druckaufgeladenen Wirbelschichtvergasers. Doctoral thesis, Technische Universität München, München.

Kuhnle, J., Wolf, H., Maile, K., Kußmaul, K., Baum, J., Spliethoff, H. and Hein, K. R. G. (1997). Untersuchungen zum Einsatz und zur Entwicklung eines Hochtemperaturwärmetauschers in Kohlenstaubfeuerungen. Abschlussbericht, Stiftung Energieforschung Baden-Württemberg, Förderkennzeichen A 000 4793/4893.

LaHaye, P. G. and Bary, M. R. (1994). Effective means of rejuvenation for older coal-fired stations. International Gas Turbine and Aeroengine Congress and Exposition. The Hague/The Netherlands, ASME-Paper 94-GT-483.

LaHaye, P. G. and Feldmann, B. E. (1986). Externally fired gas turbine/combined cycle. International Coal Utilization Conference and Exhibition. Pittsburgh, CA.

LaHaye, P. G., Briggs, G. F., Mazer, J. M. and Seger, J. L. (1990). Results of preliminary coal-fired tests on high pressure ceramic air heater for indirectly-fired gas turbine applications. Proceedings of the 7th Annual Coal-fueled Heat Engines and Gas Stream Center (METC).

LaHaye, P. G. and Zabolotny, E. (1989). Externally-fired combined cycle. ASME Cogen-Turbo Meeting. Nice, France.

Lako, P. (2004). '*Coal-fired power technologies*.' Coal-fired power options on the brink of climate policies. ECN-C–04-076, October 2004.

Lambertz, J. and Ewers, J. (2006). Clean coal power – Die Antwort der Kraftwerkstechnik auf die Herausforderungen der Klimavorsorge. *VGB PowerTech* **86**(5): 71–77.

Landfermann, H. and Hausner, H. (1988). Keramische Werkstoffe – Übersicht, Herstellung – Eigenschaften – Anwendungen. In: Technische Keramik. Essen, Vulkan, pp. 11–15.

Lee, S. H. D. and Carls, K. M. (1990). Measurement of sodium and potassium vapours in pressurized-bed combustion of Beluah lignite. *Journal of the Institute of Energy* **66**(12): 203–210.

Lee, S. H. D. and Swift, W. M. (1991). A fixed granular-bed sorber for measurement and control of alkali vapors in PFBC. 11th International Conference on Fluidized Bed Combustion. pp. 1095–1103 Montreal, ASME.

Leufert, J. (1993). Abschlußbericht Verbundprojekt Druckkohlenstaubfeuerung, Phase I, Forschungsteilbereich Kohleumwandlungssystem. BMBF-Förderkennzeichen FKZ 03266.

Mann, M. D. and Ludlow, D. K. (1997). Capture of alkali during pressurized fluidized-bed combustion using in-bed sorbents. Proceedings of the 14th International Conference on Fluidized Bed Combustion. Vancouver, Canada, ASME.

Martinez Crespo, P. and Menedez Perez, J. A. E. (1995). Operating experiences at the escatron pressurized fluid bed combustion demonstration plant. Power-Europe, Amsterdam, The Netherlands, Vol. 4, pp. 113–130.

Maurstad, O. (2005). An overview of coal based integrated gasification combined cycle (IGCC) technology, from http://lfee.mit.edu/publications.
McMullan, J. (2004). Fossil fuel power generation, State of the art. Power Clean Network.
Mendez-Vigo, I., Schellberg, W. and Karg, J. (1998). The Puertollano IGCC plant: status update. Gasification Technologies Conference. October 4–7, San Francisco, CA.
Minchener, A. J. (2005). Coal gasification for advanced power generation. *Fuel* **84**: 2222–2235.
Mitchell, S. (1997). Hot gas particulate filtration. London, IEA Clean Coal Centre, IEACR/95.
Moersch, O., Nagel, H., Spliethoff, H. and Hein K. R. G. (1999). Influence of various operating conditions on advanced PFBC with staged combustion. Proceedings of the 15th International Conference on Fluidized Bed Combustion. May 16–19, Savannah, GA.
Mojtahedi, W. and Backman, R. (1989). The fate of sodium and potassium in the pressurized fluidized-bed combustion and gasification of peat. *Journal of the Institute of Energy*, December, 189–196, from http://www.ingentaconnect.com/content/maney/eni.
Monkhouse, P. (2002). On-line diagnostic methods for metal species in industrial process gas. *Progress in Energy and Combustion Science* **28**(4): 331–381.
Monkhouse, P. B., Gottwald, U. A., Davidsson, K. O., Lonn, B., Engvall, K. and Pettersson, J. B. C. (2003). Phase discrimination of alkali species in PCFB combustion flue gas using simultaneous monitoring by surface ionisation and photofragmentation fluorescence*. *Fuel* **82**(4): 365–371.
Müller, M. (2008). Freisetzung und Einbindung von Alkalimetallverbindungen in zukünftigen, kohlebefeuerten Kombikraftwerken, RWTH Aachen. Habilitation.
Müller, M., Förster, M. and Pavone, D. (2008). Enforced binding of alkali metals into liquid slags by weak electrical fields. Proceedings of the 7th International Symposium on Gas Cleaning at High Temperatures. Shoal Bay, Australia.
Müller, M., Pavone, D. and Abraham, R. (2009). Hot fuel gas cleaning in IGCC at gasification temperature. Proceedings of the 4th International Conference on Clean Coal Technologies for our Future (CCT). Dresden.
Mustonen, J. P., Bossart, S. J. and Durner, M. W. (1991). Technical and economic analysis of advanced particle filters for PFBC applications. Proceedings of the 11th International Conference on Fluidized Bed Combustion. Montreal, ASME.
Nagel, H. (2002). Untersuchungen zum Emissionsverhalten und Wirkungsgradpotential von Druckwirbelschichtfeuerungen der ersten und zweiten Generation. Düsseldorf, VDI.
Nagel, H., Spliethoff, H. and Hein, K. R. G. (1998). Untersuchungen zum Einfluß des Hybridkonzeptes auf den Betrieb einer Druckwirbelschichtfeuerung. Proceedings of the 10th International VGB-Konferenz Forschung in der Kraftwerkstechnik. Essen.
NETL (2007). Gasification database, from http://www.netl.doe.gov/technologies/coalpower/gasification/database/database.html.
Neumann, F., Hannes, K., Wirtz, S. and Kremer, H. (1996). Untersuchungen zur Auslegung eines Hochtemperaturwärmetauschers für die Integration in einem kombinierten Gas- und Dampfturbinenprozeß mit druckaufgeladener Kohlenstaubfeuerung. *VDI-Berichte Nr* **1280**: 473–498.
Newby, R. A., Lippert, T., Alvin, M. A., Bruck, G. J., Sanjana, Z. N. and Smeltzer, E. E. (2001). Hot gas filter status and innovations for PFBC. Proceedings of the 16th International Conference on Fluidized Bed Combustion. May 13–16, Reno, NV.
Newby, R. A., Lippert, T., Bruck, G. J., Alvin, M. A. and Smeltzer, E. E. (1999). Development of advanced hot gas ceramic filter systems. Proceedings of the 15th International Conference on Fluidized Bed Combustion. May 16–19, Savannah, GA.
Ogriseck, K. and Meyer, B. (2005). Erzeugung wasserstoffreicher Brennstoffe in IGCC-Kraftwerksprozessen für eine CO2-arme Braunkohlenutzung. *VGB Powertech* **85**(11): 54–59.
Oleschko, H. and Muller, M. (2007). Influence of coal composition and operating conditions on the release of alkali species during combustion of hard coal. *Energy & Fuels* **21**: 3240–3248.
Parsons, E. L. and Byam, J. W. (1989). Overview of gas turbine coal-fired combustor concepts. Gas Turbine and Aeroengine Congress. Toronto, ASME-Paper 89-GT-252.
Pietsch, A. (1978). Coal fired prototype high temperature continuous flow heat exchanger AF-684 research project 545-1. Final report, Feb 1978. Phoenix, AZ, AiResearch Mfg. Co. of Arizona.

Pillsbury, P. W., Bannister, R. L., Diehl, R. C. and Loftus, P. J. (1989). Direct coal firing for large combustion turbines: what do economic projections and subscale combustor tests show? Joint ASME/IEEE Power Generation Conference, Dallas, ASME-Paper 89-JPGC/GT-4.

Preußer, P. and Spindler, K. (1988). Druckkohlenstaubfeuerung: Stand, Wirkungsgrad und Entwicklungsziele. *VGB Kraftwerkstechnik* **68**(9): 917–921.

Pruschek, R. (1998). Improvement of IGCC power plants starting from the state of art, JOF3-CT95-0004, Final Report.

Pruschek, R. (2002). Elektrizitätserzeugung aus fossilen Brennstoffen in Kraftwerken. In: E. Rebhan (ed.) Energiehandbuch: Gewinnung, Wandlung und Nutzung von Energie. Berlin, Springer.

Pruschek, R., Oeljeklaus, G., Haupt, G. and Zimmermann, G. (1997). Kohlekraftwerke der Zukunft. Teil 2: Stand und Entwicklungspotentiale von GUD-Prozessen mit integrierter Kohlevergasung. *BWK* **49**(11/12): 48–52.

Pruschek, R., Renz, U. and Weber, E. (1990). Kohlekraftwerke der Zukunft – Stand der Entwicklung, Erprobung und Planung neuer Kohlekraftwerkstechnologien, Ministerium für Wirtschaft, Mittelstand und Technologie des Landes Nordrhein-Westfalen.

Punjak, W. A., Uberoi, M. and Shadman, F. (1989). High-temperature Adsorption of Alkali Vapors on Solid Sorbents. *AIChE Journal* **35**(7): 1186–1194.

Radhakrishnan, P., Zakkay, V. and Agnone, A. (1986). Alkali and gas emissions from PFB combustion of lignite. *Combustion Science and Technology* **50**: 271–281.

Radtke, K. R., Henritz-Adrian, M. and Marsico, C. (2006). New wave of coal-to liquids. *VGB PowerTech* **86**(5): 78–84.

Radtke, K. R., Battensby, D. and Marscio, C. (2005). Renaissance of gasification based on cutting edge technologies. *VGB PowerTech* **85**(9): 72–77.

Ratcliff, P., Garbett, P. and Fischer, W. (2007). The new Siemens gas turbine SGT5-8000H for more customer benefit. *VGB PowerTech* **87**(9): 128–132.

Rehwinkel, H. (1989). Konzepte, Entwicklungsziele und Einsatzschwerpunkte für druckaufgeladene Wirbelschichtfeuerungen. *VDI-Berichte* **715**: 237–258.

Rehwinkel, H., Meier, H.-J. and von Wedel, G. (1993). Stationäre und zirkulierende Druckwirbelschichtfeuerung: Versuchsergebnisse und Anlagenkonzepte. *VGB Kraftwerkstechnik* **73**(6): 529–535.

Rehwinkel, H., Meier, H.-J. and von Wedel, G. (1992). Stationäre und zirkulierende Druckwirbelschichtfeuerung: Versuchsergebnisse und Anlagenkonzepte. VGB-Konferenz Wirbelschichtsysteme. Essen, VGB-TB 214.

Reichelt, T. (2001). *Freisetzung gasförmiger Alkaliverbindungen bei atmosphärischer und druckaufgeladenener Verbrennung*. Düsseldorf, Fortschritt-Berichte VDI, Reihe 3, Nr. 687, VDI.

Reichert, G., Thelen, F., Schmidt, D. and Weber, E. (1988). Untersuchung an Druckkohlenstaubfeuerungen. *VGB Kraftwerkstechnik* **68**(10): 1017–1021.

Renz, U. (1993). Heißgasfiltration bei Wirbelschichtfeuerungen. VDI-Berichte 1081.

Renz, U. (1994). An assessment of PFBC technology, present and future. International Clean Coal Technology Symposium on PFBC. Japan, Kitakyushu.

Riedle, K., Rukes, B. and Wittchow, E. (1990). Die Erhöhung des Kraftwerkswirkungsgrades in der Vergangenheit und in der Zukunft. VGB Konferenz "Kraftwerkstechnik 2000. Essen, VGB TB 120.

Robertson, A., Fan, Z., Goldstein, H., Horazak, D., Newby, R. A. and Bose, A. C. (2005). 2nd generation PFB plant with super critical pressure steam turbine. Proceedings of the 22nd International Pittsburgh Coal Conference. September 12–15.

Robertson, A., Goldstein, H., Horazak, D. and Newby, R. A. (2001). Second-generation PFB plant performance with W501G gas turbine. Proceedings of the 16th International Conference on Fluidized Bed Combustion. May 13–16, Reno, NV.

Rogbeck, J. (1996). Utilisation of residues from PFBC. International Coal Technology Symposium on Coal Ash Utilisation. Tokyo.

Rukes, B. (1993). Kraftwerkskonzepte für fossile Brennstoffe VDI-Berichte 1023, pp. 3–40.

Ruth, L. A. (1997). The US department of energy's combustion 2000 program: clean, efficient electricity from coal. *Energy Conversion and Management* **38**(10–13): 1249–1257.

Ruth, L. A. (2001). Advanced coal-fired power plants. *Journal of Energy Resources Technology-Transactions of the ASME* **123**(1): 4–9.

Sasatu, H., Tazawa, K., Goto, H., Misawa, N. and Sakuno, S. (2001). CTF development at Wakamatsu 71 MWe PFBC combined cycle power plant. Proceedings of the 16th International Conference on Fluidized Bed Combustion. May 13–16, Reno, NV, ASME.

Scandrett, L. A. and Clift, R. (1984). The thermodynamics of alkali removal from coal-derived gases. *Journal of the Institute of Energy*, from http://www.ingentaconnect.com/content/maney/eni.(Dec): 391–397.

Schemenau, W. (1993). Druckwirbelschichtfeuerung – fortschrittliche Kohleverstromung mit Betriebserfahrung. *BWK* **45**(1/2): 23–26.

Schemenau, W. and Anderson, J. (1992). Druckwirbelschichtfeuerung – betriebsbewährte Kohleverstromung mit geringer Emission und gutem Wirkungsgrad. VGB-Konferenz Wirbelschichtsysteme, VGB-TB 214. Essen.

Schemenau, W. and van den Bergh, C. (1993). Betriebserfahrungen bei druckaufgeladenen Wirbelschichtfeuerungen und neue Konzepte von Druckwirbelschichtfeuerungsanlagen großer Leistung. *VDI-Bericht 1081*, VDI: 19–44.

Schetter, B., Becker, B. and Egener, E. G. (1991). Schadstoffarme Kohlegasverbrennung in modernen Gasturbinen. VGB-Konferenz Kohlevergasung 1991. Mai 16–17, Dortmund, VGB.

Schiffer, H. P. (1989). Heißgasentstaubung. VDI-Berichte 715.

Schöler, D. (2007). NGF/SGT5-8000H – Gasturbinentechnologie. 39. Kraftwerkstechnisches Kolloquium Dresden.

Schuknecht, M. (2003). Entwicklungspotential eines Kombikraftwerks mit Druckkohlenstaubfeuerung, Universität Essen. Dissertation.

Schürmann, H., Unterberger, S., Hein, K. R. G., Monkhouse, P. B. and Gottwald, U. (2001). The influence of fuel additives on the behaviour of gaseous alkali compounds during pulverised coal combustion. *Faraday Discussions* **119**: 433–444.

Shimuzu, M. and Itoh, M. (2001). Development and operation results of Osaki 250 MW commercial PFBC plant (Osaki 1-1 PFBC Plant). 16th International Conference on Fluidized Bed Combustion. May 13–16, Reno, NV.

Singer, J. G. (1991). Combustion fossil power. Windsor, VT, Combustion Engineering.

SolarTurbines (1980). Development of a ceramic tube heat exchanger with relaxing joint. Final report, DOE contract No. EF-77-C-01-2556, FE-2556-30.

Spliethoff, H. (2000). *Verbrennung fester Brennstoffe zur Strom- und Wärmeerzeugung: Verfahren und Stand der Technik – Wirkungsgrad, Betrieb, Emissionen und Reststoffe. VDI Fortschritt-Berichte, Reihe 6 Energietechnik, Nr. 443*. Düsseldorf, VDI.

Spliethoff, H. and Baum, J. (2002a). Externally fired combined cycle – Potential and state of the art (Part 1): *BWK* **54**(5): 70–75.

Spliethoff, H. and Baum, J. (2002b). Externally fired combined cycle-Potential and state of the art (Part 2): *BWK* **54**(6): 52–56.

Spliethoff, H., Meyer, B., Müller, M. and Hack, K. (2009). Investigations on high temperature gasification and gas cleaning – The research project HotVeGas. Proceedings of the 4th International Conference on Clean Coal Technologies for our Future (CCT), Dresden.

Stoll, K. E. and Bleif, F. (1986). Entwicklung, Errichtung und Erprobung eines umweltfreundlichen Steinkohlekraftwerkes – Modellkraftwerk Völklingen. BMFT-Forschungsbericht T 86-025.

Strauß, K. (2006). *Kraftwerkstechnik: zur Nutzung fossiler, nuklearer und regenerativer Energiequellen*. Berlin [u.a.], Springer.

Stringer, J. (1989). Erosion and corrosion in PFBC systems. *VDI-Berichte* **715**: 57–82.

Stuhlmüller, F. and Schauenburg, G. (2001). The gas turbine in a power plant with pressurized circulating fluidized bed combustion. Proceedings of the 16th International Conference on Fluidized Bed Combustion. May 13–16, Reno, NV.

Stuhlmüller, F., Schauenburg, G. and Waldinger, D. (1995). Die Standard-Gasturbine im Druckwirbelschicht-Einsatz. *VGB Kraftwerkstechnik* **75**(12): 1037–1042.

Takahashi, M., Nakabayashi, Y. and Kimura, N. (1995). Aktueller Stand der 350 MW-Wirbelschichtfeuerung Takehara und der 71-MW-Druckwirbelschichtfeuerung Wakamatsu der EPDC sowie der fortschrittlichen Stromerzeugung in Japan. *VGB Kraftwerkstechnik* **75**(5): 427–432.

Tampa_Electric (1996). The Tampa electric integrated gasification combined-cycle project, From http://www.fossil.energy.gov/programs/powersystems/publications/Clean_Coal_Topical_Reports/topical19.pdf.

Tampa_Electric (2002). Tampa electric polk power station IGCC project – Final technical report, August 2002, From http://www.netl.doe.gov/technologies/coalpower/cctc/cctdp/bibliography/demonstration/aepg/baepgig_tampaig.html.

Tampa_Electric (2004). Project performance summary, clean coal technology demonstration program, June 2004, from http://www.netl.doe.gov/technologies/coalpower/cctc/cctdp/bibliography/demonstration/aepg/baepgig_tampaig.html.

Terhaag, U., Renz, U., Dibelius, G., Bohn, D., Reinartz, A. and Steven, H. (1995). Ergebnisse aus dem Betrieb des druckaufgeladenenen Wirbelschicht-Dampferzeugers im Heizkraftwerk der RWTH Aachen. *VGB Kraftwerkstechnik* **75**(3): 243–246.

Thambimuthu, K. V. (1993). Gas cleaning for advanced coal-based power generation. London, IEA Coal Research. ieacr/53.

Thelen, F. (1993). Gas-/Dampfturbinenprozesse für feste Brennstoffe mit Druckwirbelschicht-feuerung. *VGB Kraftwerkstechnik* **73**(8): 671–677.

Toriyama, A., Higashi, K., Maeno, H., Saito, T., Mori, M. and Hori, J. (1999). Operating results of the advanced ceramic tube filter (ACTF) at Wakamatsu 71 MW PFBC demonstration plant. Proceedings of the 15th International Conference on Fluidized Bed Combustion. May 16–19, Savannah, GA.

Turn, S. Q., Kinoshita, C. M., Ishimura, D. M., Zhou, J., Hiraki, T. T. and Masutani, S. M. (1998). A review of sorbent materials for fixed bed alkali getter systems in biomass gasifier combined cycle applications. *Journal of the Institute of Energy* **71**: 163–177.

Uberoi, M. (1990). The kinetics and mechanism of alkali removal from flue gases by solid sorbents. *Progress in Energy Combustion Science* **16**: 205–211.

Uhde (2008). PRENFLO gasification. Dortmund, Uhde.

Vandervort, C. L. (1991). High pressure ceramic air heater for indirectly fired gas turbine applications. Proceedings of the 8th Coal-Fueled Heat Engines and Gas Stream Cleanup Systems Contractors Review Meeting. Morgantown, WV.

Vandervort, C. L. and Orozco, N. J. (1992). Development status of a utility-scale externally fired combined cycle. Proceedings of the 9th Annual International Pittsburgh Coal Conference, Pittsburgh.

Walter, E., Krautz, H. J. and Almhem, P. (1997). Lignite fired combined cycle heat and power plant using pressurized fluidized bed combustion. Proceedings of the 14th International Conference on Fluidized Bed Combustion. ASME, Vancouver, Canada.

Ward, M. E., Metcalfe, A. G. and Dapkunas, S. J. (1983). Ceramic tube heat exchanger technology development for indirect-fired gas turbine cycle. *Transactions of the ASME* **105**: 310–316.

Warner, J. and Nielsen, H. (1993). Umfassendes Auswahlverfahren ermöglicht optimale Auslegung von Kombikraftwerken. ABB Technik, Nr. 8.

Wauschkuhn, A. (1994). Kostenvergleich von Kohlekraftwerken. Studienarbeit, Institut für Verfahrenstechnik und Dampfkesselwesen der Universität Stuttgart.

Weber, E. and Pavone, D. (1990). Entwicklungslinien der Gasreinigung bei höchsten Gastemperaturen. Symposium Emissionsminderung mit dem Schwerpunkt Hochtemperatur- und Hochdruck-Gasreinigung, Essen.

Weber, E., Hübner, K., Pavone, D., Schulz, R. and Wiggers, H. (1993). Entwicklungsergebnisse bei der Druckkohlenstaubfeuerung. *VGB Kraftwerkstechnik* **73**(7): 602–607.

Weimer, T., Berger, R., Hawthorne, C. and Abanades, J. C. (2008). Lime enhanced gasification of solid fuels: examination of a process for simultaneous hydrogen production and CO2 capture. *Fuel* **87**(8–9): 1678–1686.

Weitzel, P. S. and McDonald, D. K. (1999). PFBC design and arrangement improvements due to the application of ceramic tube filters. Proceedings of the 15th International Conference on Fluidized Bed Combustion. ASME, May 16–19 Savannah, GA.

Weitzel, P. S., McDonald, D. K., Whitney, S. A. and Oda, N. (1996). Directions and trends for commercial PFBC and hot gas clean up. Pittsburgh, PA, Pittsburgh Coal Conference.

Wen, C. S., Cowell, L. H., Smit, F. J., Boyd, J. D. and LeCren, R. T. (1992). Coal alkali retention in a slagging combustor. *Fuel* **71**(2): 219–224.

Wheeldon, J. M., Bonsu, A. K., Foote, J. P., Mortan, F. C., Romans, D. E. and Zoldak, F. D. (2001). Commissioning of the circulating PFBC in the Foster Wheeler advanced PFBC train at the PDSF. 16th International Conference on Fluidized Bed Combustion. May 13–16, Reno, NV.

Willenborg, W., Muller, M. and Hilpert, K. (2006). Alkali removal at about 1400 degrees C for the pressurized pulverized coal combustion combined cycle. 1. Thermodynamics and concept. *Energy & Fuels* **20**(6): 2593–2598.

Wilson, D. G. (1993a). Low-leakage and high flow regenerators for gas turbine engines. *Proceedings Institution of Mechanical Engineers* **207**: 195–202.

Wilson, D. G. (1993b). The supplementary-fired exhaust-heated cycle for coal, wood and refuse-derived fuel. *Proceedings Institution of Mechanical Engineers* **207**: pp. 203–208.

Wilson, D. G., Frenkel, R. G., Kowalick, D. J., Nahatis, H. M., Silverstein, S. M. and Tampe, L. A. (1991). Coal-burning exhaust-heated-cycle gas turbine with a regenerative heat exchanger. international gas turbine and aeroengine. Congress and Exposition, ASME-Paper 91-GT-342 Orlando, FL.

Wittchow, E. and Müller, R. (1993). Fortgeschrittene Kraftwerkskonzepte mit druckaufgeladenen Kohleumwandlungsverfahren. VDI-Berichte 715, 1989, pp. 21–44.

Wright, I. G., Stringer, J. and Wheeldon, J. M. (2003). Material issues in bubbling PFBC systems. *Materials At High Temperatures* **20**(2): 219–232.

Wu, Z. (2006). Developments in fluidised bed combustion technology. London, IEA Clean Coal Centre.

Zakkay, V., Gbordzoe, E., Radhakrishnan, R., Sellakumar, K. M., Patel, J., Kasinathan, R., Haas, W. J. and Eckels, D. E. (1989). Particulate and alkali capture from PFBC flue gas utilizing granular bed filter (GBF). *Combustion Science and Technology* **68**: 113–130.

Chapter 8
Carbon Capture and Storage (CCS)

8.1 Potential for Carbon Capture and Storage

Worldwide anthropogenic CO_2 emissions were around 26 Gt CO_2/year in the year 2005. This quantity can be attributed to the use of crude oil, coal and natural gas, contributing 40, 40 and 20%, respectively. Around 60% of the total emissions can be put down to roughly 8,000 big emitters, each with annual CO_2 emissions larger than 0.1 Mt CO_2/year. Electrical power production, with roughly 5,000 large power plants having emissions of more than 0.1 Mt CO_2/year, has a share of around 45% of the emissions worldwide.

Energy agencies typically assume that the contribution of the various energy carriers and the share of the primary energy used in electric power production (presented in Chap. 1) will not change substantially in the medium term. In consequence, with the predicted rates of increase of energy consumption and electric power production, CO_2 emissions will drastically rise in the next 20 years. Predictions based on various sources forecast an increase in CO_2 emissions to about 38 Gt CO_2/year up to the years 2025–2030 (IPCC 2005; IEA 2002, 2006; DoE 2005) – see also Chap. 1.

It is obvious that, in the long term, humanity will not succeed in restricting the release of CO_2 emissions by more efficient power plant technologies alone. Fossil fuels can ensure the supply of electric power in an energy mix with renewable energy sources only when there is successful separation of carbon dioxide from the power production process and storage of it in a safe way.

Carbon capture and storage (CCS) is seen today as a way to satisfy the global hunger for energy from fossil fuels on one hand and to limit the impacts on the Earth's climate on the other. The projected potential of CO_2 capture has been estimated at an annual 2.6–4.9 Gt CO_2 by 2020 (0.7–1.3 GtC) and 4.7–37.5 Gt CO_2 by 2050 (1.3–10 GtC) (DoE 2005). This will only remain an option if suitable methods are developed to separate the carbon dioxide from power production processes and to store it underground.

The following technical tasks arise as a consequence:

- Separation of carbon dioxide in the power production process
- Conditioning and transport of carbon dioxide

H. Spliethoff, *Power Generation from Solid Fuels*, Power Systems,
DOI 10.1007/978-3-642-02856-4_8,

– Safe and long-term storage of carbon dioxide (VGB 2002; IPCC 2005; Linßen et al. 2006; Seifritz 1989; Smith and Thambimuthu 1991; Pruschek and Oeljeklaus 1992; Radgen et al. 2005)

8.2 Properties and Transport of CO_2

The physical state of carbon dioxide varies with temperature and pressure as shown in Fig. 8.1. At normal temperatures and pressure, carbon dioxide is a gas. At atmospheric pressure and temperatures far below ambient temperatures, carbon dioxide is a solid. With increasing temperature, the solid will sublime directly into the vapour phase – at atmospheric pressure, carbon dioxide sublimes at a temperature of −78.5 °C. At intermediate temperatures between the temperature of the triple point (−56.5 °C, 5.2 bar) and the temperature of the critical point (31.1 °C, 73.9 bar), a pressure increase results in a gradual state change as a two-phase gas – liquid mixture. At ambient temperature and pressures above 60 bar, carbon dioxide is a liquid. At temperatures higher than 31.1 °C and pressures above 73.9 bar, carbon dioxide is in a supercritical state, where it behaves as a gas. The density of carbon dioxide is given in Fig. 8.2 as a function of pressure and temperature.

The transport of the carbon dioxide is the tie between the emitter and the CO_2 storage location. Depending on the transport and storage mechanisms, CO_2 has to have a certain condition. To give an example, for the injection of CO_2 into depleted oil fields or gas reservoirs, the carbon dioxide ought to be provided pressurised in

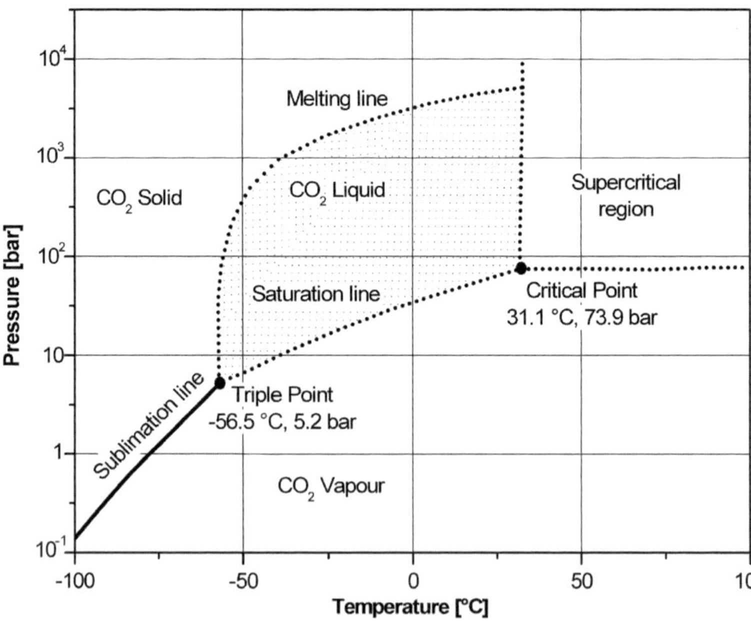

Fig. 8.1 Phase diagram of CO_2 (Ritter et al. 2007)

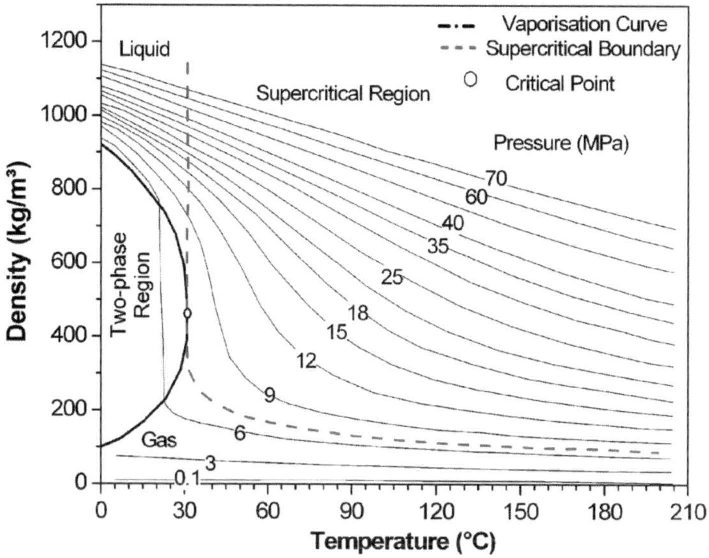

Fig. 8.2 CO_2 density as a function of temperature and pressure (IPCC 2005)

a gaseous/liquid state. Transport via pipeline ought to be carried out at ambient temperatures with the carbon dioxide in the gas or liquid phase; via ship, it is also possible to carry it supercooled or as dry ice (carbon dioxide ice).

An economic balance needs to be struck between a density as high as possible for transport and an expenditure of energy as low as possible on compression, liquefaction or solidification.

In principle, the following possibilities present themselves:

- Compression to above the critical pressure, typically around or above 100 bar, then transport at ambient temperatures. At ambient temperatures and pressures above the saturation pressure, carbon dioxide is in a liquid or supercritical state.
- Liquefaction by refrigeration at 25 bar and −15 °C, then cooled transport.
- Solidification by refrigeration at atmospheric pressure to produce CO_2 ice, then cooled transport.

The achievable densities and the energy expenditures involved are listed in Table 8.1. Liquefaction by compression is the preferred option because it involves the lowest expenditures of energy.

Table 8.1 Energy requirements for liquefaction and freezing (Göttlicher 1999)

	Electrical energy requirement	Energy required related to coal
Compression (11 MPa)	0.11–0.13 kWh/kg CO_2	~ 3.5% of LHV
Liquefaction by refrigeration	0.16 kWh/kg CO_2 (25 bar, −15 °C)	~ 5.2% of LHV
Freezing	0.26–0.42 kWh/kg CO_2	~ 8.4–13.5% of LHV

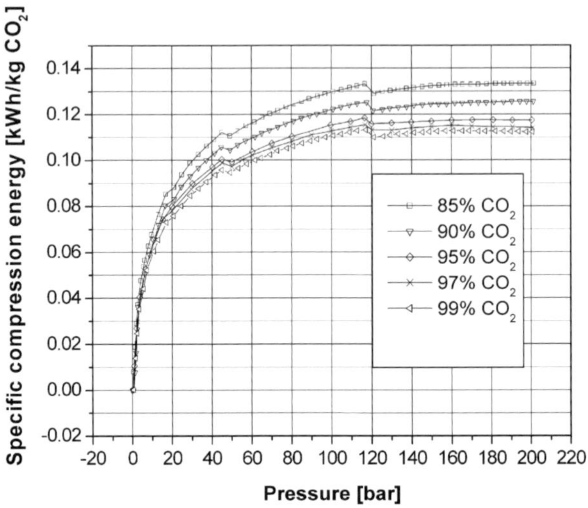

Fig. 8.3 Specific compression energy as a function of pressure and CO_2 purity (Göttlicher 1999)

Today, pipelines are the main method considered as a reasonable solution for transporting compressed supercritical carbon dioxide and are considered state of the art. In this process, carbon dioxide is compressed to pressures higher than 100 bar. An alternative is the transport of gaseous, liquid or solid carbon dioxide by ship, but this is cost-effective only for distances of more than 1,000 km.

For transport in pipelines, it is necessary to avoid impurities in the carbon dioxide, such as water vapour or sulphur compounds, as this prevents damage of the transport vessels or pipes through condensation of water and corrosion. When there are water vapour fractions in the carbon dioxide, undesirable hydrate crystals may form under high pressure, which can result in blockages in pumps or pipes. The fraction of noncondensable gases, like nitrogen or oxygen for instance, increases the energy required for compression and liquefaction. Figure 8.3 shows the specific compression work as a function of the pressure and the CO_2 purity (Göttlicher 1999).

8.3 CO_2 Storage

The safe and long-term sequestration of carbon dioxide is the prerequisite for the applicability of CCS technology. What follows is a discussion of the principal possibilities for CO_2 storage (IPCC 2005; Radgen et al. 2005; Seifritz 1989).

8.3.1 Industrial Use

Industrial utilisation of carbon dioxide can, in principle, contribute to keeping carbon dioxide out of the atmosphere by storing it in manufactured products. However, as a measure for mitigating climate change, this option is not of great importance because the quantities and duration of carbon dioxide storage are not significant.

The utilisation as a material in industrial processes worldwide amounts to approximately 120 Mt/CO_2 (excluding the use for EOR (enhanced oil recovery)) and thus lies below 1% of global anthropogenic CO_2 emissions. Carbon dioxide is a reactant in urea and methanol production, and it is directly used in various applications in the horticulture industry, refrigeration, food packaging, welding, beverages and fire extinguishers. Most (two thirds of the total) is used to produce urea, which in turn is used in manufacturing fertilisers and other products.

Most of the carbon dioxide currently used by industrial processes has storage times of only days to months; after this period, the stored carbon dioxide is emitted to the atmosphere. Such short timescales do not contribute significantly to climate change mitigation. Some industrial processes store roughly 20 Mt CO_2/year for up to several decades. The amount of long-term storage on a century scale is only in the order of 1 Mt CO_2/year or less, with no prospects for major increases.

8.3.2 Geological Storage

Geological storage is the injection of carbon dioxide in a dense form into a rock formation below the Earth's surface. The different geological storage options, in depleted oil and gas reservoirs, deep saline formations and unmineable coal seams, are shown in Fig. 8.4. Porous rock formations that hold or have previously held fluids, such as natural gas, oil or brines, are potential sites for CO_2 storage. Coal beds can also be used for the storage of carbon dioxide wherever it is unlikely that the coal will later be mined and provided that the permeability is sufficient. The injection of carbon dioxide into deep geological formations involves the same technologies that are applied in the oil and gas exploration and production industry.

Long-term CO_2 storage in hydrocarbon reservoirs or deep saline formations has to take place at depths below 800 m, where, due to the high pressures, carbon dioxide

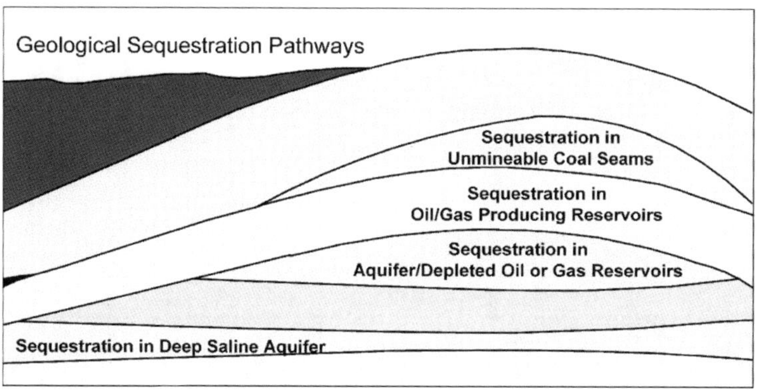

Fig. 8.4 Options for geological storage

is in a liquid or supercritical state. Under these conditions, the density of carbon dioxide ranges from 50 to 80% of the density of water, resulting in buoyant forces that drive carbon dioxide upwards. Consequently, a well-sealed cap rock over the storage reservoir has to be in place to ensure that carbon dioxide remains trapped underground. When injected underground, the carbon dioxide compresses and fills the pore space by partially displacing the fluids that are already present.

Once injected, the storage depends on a combination of physical and geochemical trapping mechanisms. Physical trapping occurs by blocking upward migration of carbon dioxide by the cap rock, typically an impermeable layer of shale and clay rock. Additional physical trapping can be provided by capillary forces that hold carbon dioxide in the pore spaces. Geochemical trapping occurs as the carbon dioxide reacts with the fluids and host rock. First, carbon dioxide dissolves in the in situ water. Over timescales of hundreds to thousands of years, the CO_2-laden water becomes denser and sinks lower into the rock formation. In a timeframe of millions of years, a fraction of the injected carbon dioxide will be converted to solid carbonate minerals. Another type of trapping mechanism is the adsorption of carbon dioxide onto coal or organic-rich shales, replacing gases such as methane. In these cases, carbon dioxide will remain trapped as long as pressures and temperatures remain stable.

8.3.2.1 Existing CO_2 Storage Projects

Geological storage of carbon dioxide is ongoing in industrial-scale projects: the Sleipner project in the North Sea, the Weyburn project in Canada and the In Salah project in Algeria. About 3–4 Mt CO_2 that would otherwise be released to the atmosphere is captured and stored annually in geological formations.

In addition to the CCS projects currently in place, 30 Mt CO_2 is injected annually for EOR, mostly in Texas, USA, where EOR commenced in the early 1970s. The option of storing carbon dioxide in coal beds and enhancing methane production is still in the demonstration phase.

8.3.2.2 Capacity of Storage Sites

Regions with sedimentary basins that are potentially suitable for CO_2 storage exist around the world, both onshore and offshore. In comparison to the estimated potential of CCS in 2020 of about 2–5 Gt CO_2/year and in 2020 of 5–40 Gt/year, this

Table 8.2 Technical potential of geological storage options (IPCC 2005)

Reservoir type	Lower estimate of storage capacity (Gt CO_2)	Upper estimate of storage capacity (Gt CO_2)
Oil and gas fields	675	900
Unmineable coal seams (ECBM)	3–15	200
Deep saline formations	1,000	Uncertain, but possibly 10^4

would be sufficient to store the sequestable carbon dioxide produced over at least one century.

8.3.2.3 Risks and Open Questions

Open questions over CCS technology mainly concern the safety of the storage, the impact on the environment and legal aspects. There are two types of leakage scenarios: abrupt leakage, through injection well failure or leakage up through an abandoned well, and gradual leakage, through undetected faults, fractures or wells. The impacts of CO_2 leakage of the reservoir and release into the atmosphere could include lethal effects on plants and subsoil animals and the contamination of groundwater. High fluxes in conjunction with stable atmospheric conditions could lead to local high CO_2 concentrations in the air. A concentration of carbon dioxide greater than 7–10% in air would cause immediate dangers to human life and health.

It is expected that local health, safety and environmental risks from geological storage would be comparable to those from the existing activities of natural gas storage and EOR. CO_2 leakages have to be controlled by appropriate site selection, a monitoring programme for CO_2 flows in the storage system and the use of remediation methods. Features of storage sites with a low probability of leakage include highly impermeable cap rocks, geological stability, absence of leakage paths and effective trapping mechanisms. If leakage occurs at a storage site, measures to stop the leakage could involve standard well repair techniques or the interception and extraction of the carbon dioxide before it leaks into any shallow groundwater aquifer. Monitoring tools are available from gas and oil exploration and are being developed in current research activities. The effectiveness of risk management methods still needs to be demonstrated.

No legal and regulatory frameworks for long-term CO_2 storage exist today (2009) in any country. Long-term liability issues associated with the leakage of carbon dioxide to the atmosphere and local environmental impacts have to be solved. Finally, public acceptance for CO_2 storage is required (IPCC 2005).

8.3.2.4 Ocean Storage

A potential CO_2 storage option is to inject captured carbon dioxide directly into the deep ocean at depths greater than 1,000 m. Because mixing between deep ocean water and surface water is limited, it would be isolated from the atmosphere for at least several hundreds of years, with the fraction retained tending to be higher with deeper injection. In contrast, the residence time of carbon dioxide in surface waters would be only about 100 years. For deep-ocean injection, carbon dioxide is transported as a gas or liquid via pipelines or ships to an ocean storage site where it is injected into the water column of the ocean or at the sea floor. Below a depth of approximately 2,600 m, and with a water temperature of 2 °C, the density of liquid carbon dioxide is greater than that of the seawater and the injected carbon dioxide sinks to the seafloor.

Natural exchanges of carbon dioxide occur between the atmosphere and water at the ocean surface until equilibrium is reached. This means that as the atmospheric concentration of carbon dioxide increases, the ocean gradually takes up additional CO_2.

In this way, the oceans have taken up about 500 Gt of the total 1,300 Gt of anthropogenic CO_2 emissions released to the atmosphere over the past 200 years. As a result of the increased atmospheric CO_2 concentrations from human activities, the oceans are currently taking up carbon dioxide at a rate of about 7 Gt CO_2/year. Most of this carbon dioxide is retained in the upper ocean and has resulted in an acidity increase of about 0.1 pH at the ocean surface because of the acidic nature of carbon dioxide in water. To date, however, there has been virtually no change in acidity in the deep ocean. Models predict that over several centuries the oceans can take up most of the carbon dioxide released to the atmosphere as carbon dioxide is dissolved at the ocean surface. An acidity change of more than 0.25 pH at the ocean surface is estimated due to equilibration when the atmospheric concentration of CO_2 reaches 550 ppm.

Carbon dioxide injected into the deep sea is released again on a millennial timescale. The amount of carbon dioxide then stored in the ocean will depend on the equilibrium with the atmosphere.

Adding carbon dioxide to the ocean or forming pools of liquid carbon dioxide on the ocean floor will change the local chemical environment. Experiments have shown that sustained high CO_2 concentrations would cause mortality of ocean organisms. The effects of direct CO_2 injection into the ocean on ecosystems over large ocean areas and long timescales have not yet been studied.

Ocean storage of carbon dioxide is currently not considered as a promising option (IPCC 2005).

8.3.2.5 Mineral Carbonation

Mineral carbonation refers to the chemical fixation of carbon dioxide using alkaline and alkaline earth oxides, such as magnesium oxide (MgO) and calcium oxide (CaO), which are abundant in naturally occurring silicate minerals such as serpentine and olivine. Reactions between these materials and carbon dioxide produce stable compounds such as magnesium carbonate, $MgCO_3$, and calcium carbonate, $CaCO_3$, known as limestone.

The process of mineral carbonation occurs naturally and is termed "weathering". The natural reaction is very slow and has to be enhanced for technical applications by pre-treatment of the minerals, which at present is very energy intensive. The carbonation process is as such: mined material is ground and may require thermal pre-treatment prior to carbonation with carbon dioxide at 187 bar/155 °C. Approximately 80% of the serpentine (a Mg silicate) is converted to $MgCO_3$ within half an hour under stoichiometric conditions. The mineral carbonation process would require 1.6–3.7 t of silicates to be mined per tonne of carbon dioxide and produce 2.6–4.7 t of materials to be disposed per tonne of carbon dioxide stored as carbonates. This would therefore be a large operation, with an environmental impact similar to that of current large-scale surface mining operations. The carbonation process energy required would be 30–50% of the output from the plant where the CO_2 is captured.

Mineral carbonation technology using natural silicates is in the research phase, but some processes using industrial wastes are in the demonstration phase. Assessments required include the technical feasibility and the corresponding energy requirements at large scales and also the fraction of silicate reserves that can be technically and economically exploited for CO_2 storage (IPCC 2005).

8.4 Overview of Capture Technologies

8.4.1 Technology Overview

The purpose of CO_2 capture is to produce a concentrated stream of carbon dioxide at high pressure that can be transported to a storage site. Although, in principle, the entire flue gas stream containing low concentrations of carbon dioxide could be transported and injected underground, energy costs and other associated costs generally make this approach impractical. It is therefore necessary to produce a nearly pure CO_2 stream for transport and storage.

There are three main approaches to capture CO_2:

Post-combustion systems separate carbon dioxide from the flue gases produced by the combustion of the carbonaceous fuel with air. These systems normally use a liquid solvent to capture the small fraction of carbon dioxide (typically 3–15% by volume) present in a flue gas stream in which the main constituent is nitrogen.

Pre-combustion systems remove the carbon dioxide prior to combustion. They can be used in power plants that employ integrated gasification combined cycle (IGCC) technology. The primary fuel is converted in a gasifier to a synthesis gas consisting mainly of carbon monoxide and hydrogen. The carbon monoxide reacts in a shift reactor with steam to produce additional hydrogen, together with CO_2. The mixture of hydrogen and carbon dioxide can then be divided into a CO_2 gas stream and a stream of hydrogen. The hydrogen is burned to generate power and/or heat. The high concentrations of carbon dioxide produced by the shift reactor (typically 15–60% by volume on a dry basis) and the high pressures are favourable for CO_2 separation.

Oxy-fuel combustion systems use oxygen instead of air for the combustion of the primary fuel to produce a flue gas that is mainly water vapour and CO_2. The water vapour is then removed by cooling and compressing the gas stream. Oxy-fuel combustion requires upstream oxygen separation from air, with a purity of 95–99% oxygen required in most current designs. Further treatment of the flue gas may be needed to remove pollutants before the carbon dioxide is sent to storage.

Figure 8.5 is a schematic diagram of the main capture processes. All require a step involving the separation of CO_2, H_2 or O_2 from a bulk gas stream such as flue gas, synthesis gas or air. Separation can be accomplished by means of physical or chemical solvents, membranes, solid sorbents, or by cryogenic separation. The choice of a specific capture technology is determined by the process conditions under which it must operate. Typically, post-combustion technology uses chemical solvents to remove the carbon dioxide from the atmospheric flue gas; pre-combustion employs a physical solvent to separate carbon dioxide from the high-pressure synthesis gas; and for oxy-fuel combustion, cryogenic air separation is the standard technology. Membrane technologies are still in the development stage but would offer a much lower energy requirement for CO_2 removal.

The different technologies can capture 85–95% of the carbon dioxide that is produced from coal-fired power plants. Higher capture efficiencies are possible, but separation devices then become considerably larger, more energy intensive and more costly. Capture and compression needs roughly 10–40% more energy than an equivalent plant without capture, depending on the type of system. Due

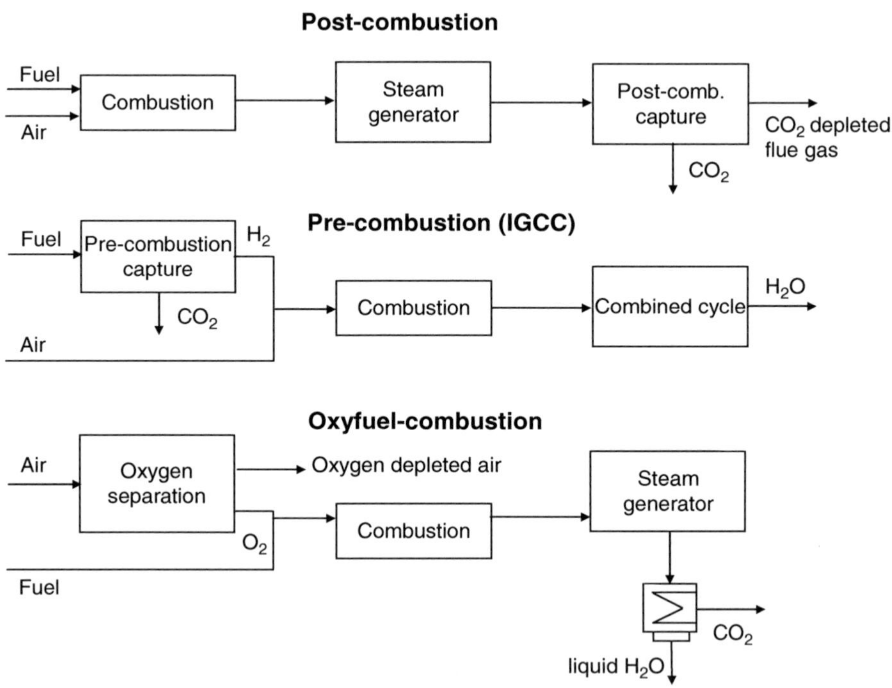

Fig. 8.5 Classification of CO_2 sequestration technologies

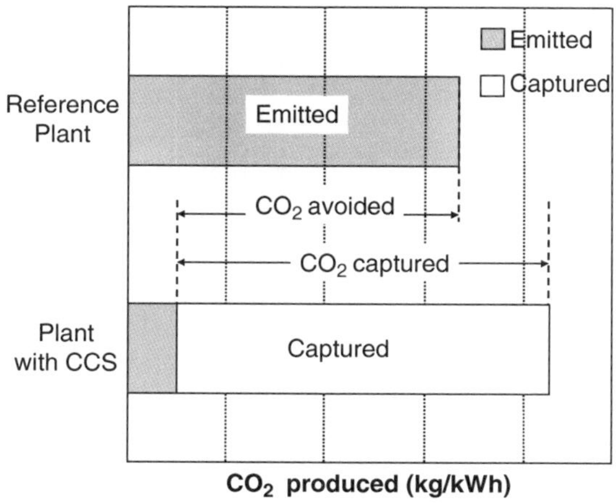

Fig. 8.6 CO_2 emissions from power plants with CO_2 capture and storage (IPCC 2005)

to the associated CO_2 emissions, the net amount of carbon dioxide captured is approximately 80–90% – see Fig. 8.6 (IPCC 2005).

8.4.2 Separation Technologies

Different gas separation technologies are applied and integrated in the three CO_2 capture systems (post-combustion, pre-combustion and oxy-fuel combustion). The principle separation technologies are given below (Göttlicher 1999; Radgen et al. 2005; IPCC 2005).

8.4.2.1 Separation with Sorbents or Solvents

Figure 8.7a shows a general schematic of separation by chemical absorption and physical absorption and adsorption. The separation process, which uses liquid absorbents or solid sorbents, consists of two steps – the absorption step and the regeneration step. In the absorber vessel, the CO_2-containing gas is brought into contact with the sorbent that captures the CO_2. The sorbent, loaded with carbon dioxide, is then transported to a second reactor, where carbon dioxide is released (regeneration) after being heated. The regenerated sorbent is sent back to capture more carbon dioxide in a cyclic process. A make-up flow of fresh sorbent is always required to compensate for deactivation or loss of the sorbent. In some variants of this scheme, the sorbent does not circulate between vessels because sorption and regeneration are achieved by cyclic changes in pressure or temperature in the vessel where the sorbent is contained.

The separation process is based on the reversibility of the absorption and desorption reaction. In the absorption step, heat is released (exothermic), while in the regeneration process the same amount of heat has to be added (endothermic).

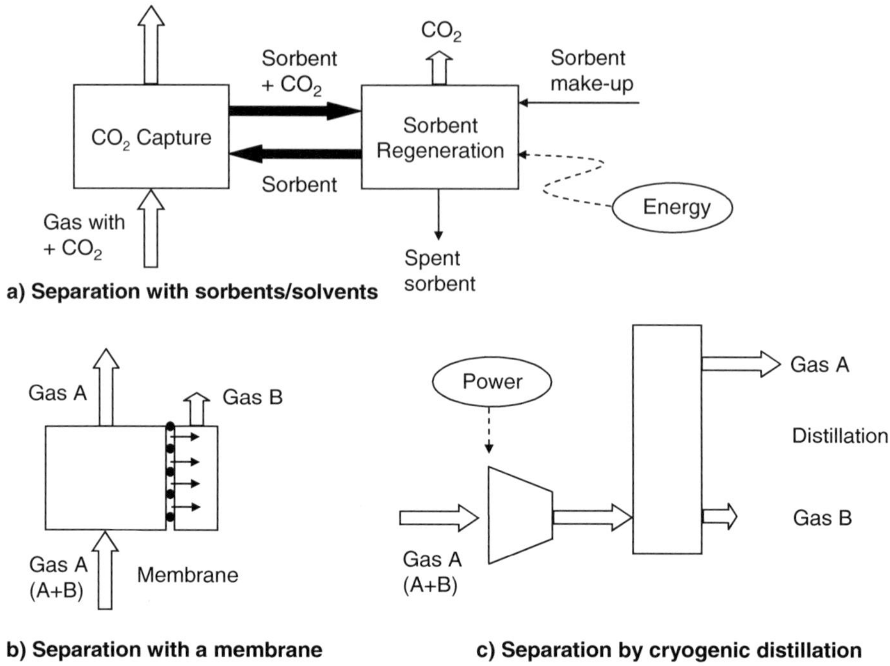

Fig. 8.7 Schematic diagram of separation processes (IPCC 2005)

Because the heat release occurs at a low temperature, the heat cannot be utilised. The consequence is that the heat added at a higher temperature in the regenerator is lost and causes an energy penalty. The lower the temperature of regeneration, the lower the energy penalty.

8.4.2.2 Separation with Membranes

Membranes (Fig. 8.7b) are specially manufactured materials that allow the selective permeation of a gas. The selectivity of the membrane to different gases is related to the material, and the gas flow through the membrane is driven by the pressure difference across it. Therefore, a high pressure is preferred for membrane separation. There are different types of membrane materials (polymeric, metallic, ceramic) that may find application in CO_2 capture systems to preferentially separate H_2 from a fuel gas stream, carbon dioxide from a flue gas stream or O_2 from air. Reliable and low-cost membrane technologies are not yet available for the large-scale and demanding conditions required by CO_2 capture systems. A large worldwide R&D effort is in progress aimed at developing suitable membrane materials for CO_2 capture.

8.4.2.3 Distillation of a Liquefied Gas Stream and Refrigerated Separation

A gas can be liquefied by a series of compression, cooling and expansion steps. In the liquid state, the components of the gas can be separated in a distillation column. In the case of air, this process is commercially available on a large scale. Oxygen separation, as shown in Fig. 8.7c, is used in a range of CO_2 capture systems, such as oxy-fuel combustion and pre-combustion capture. The key issue for these systems is the large flow of oxygen required. Refrigerated separation can also be used to separate carbon dioxide from other gases and to separate impurities from relatively high-purity CO_2 streams, for example from oxy-fuel combustion.

8.4.2.4 Separation Work

The minimum energy required for the separation is termed the reversible separation energy. The reversible molar separation energy w_r is the compression work which is necessary to raise the component with the mass fraction x_i from the partial pressure p_i of the mixture to the total pressure p_tot:

$$w_\mathrm{r} = x_i R_i T_0 \ln\left(\frac{p_\mathrm{tot}}{p_i}\right) \tag{8.1}$$

The reversible separation energy is hence a function of the carbon dioxide concentration (see Fig. 8.8).

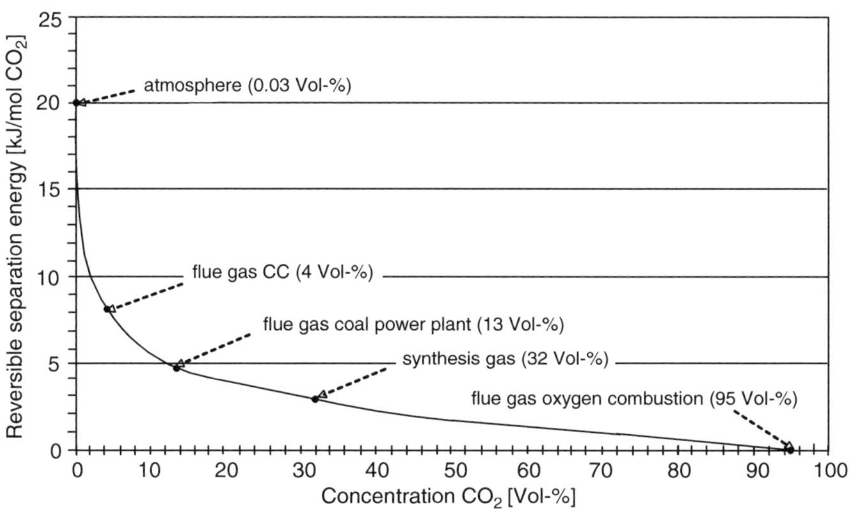

Fig. 8.8 Reversible separation energy (Göttlicher 1999)

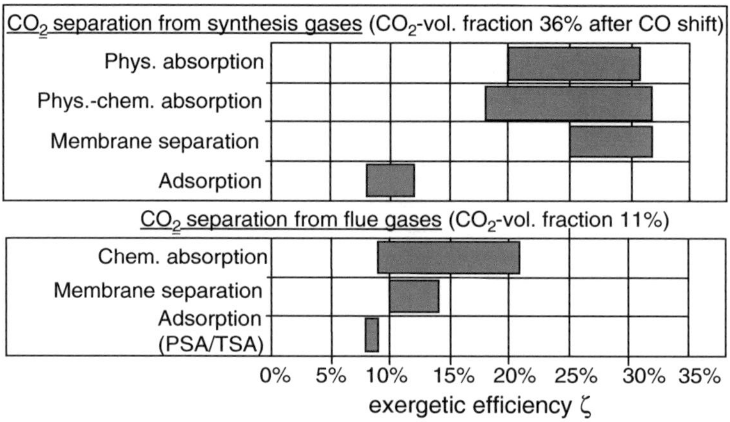

Fig. 8.9 Exergetic efficiency of CO_2 separation processes (Göttlicher 1999). *Bars* indicate range of efficiency

The energy actually required in modern separation processes is many times higher than the reversible separation energy. The ratio is given by the exergetic efficiency $\zeta_{\text{separation}}$ of the scrubbing process:

$$\zeta_{\text{separation}} = \frac{w_r}{w_G + x_i R_i T_0 \ln\left(\frac{p_{\text{tot}}}{p_{\text{desorber}}}\right)} \tag{8.2}$$

with w_G as the real separation work. The term $x_i R_i T_0 \ln(p_{\text{tot}}/p_{\text{desorber}})$ takes the work to compress the component from the desorption pressure p_{desorber} to the total pressure p_{tot} into account. Figure 8.9 indicates typical exergetic efficiencies in flue gases and synthesis gases. Absorption of CO_2 in the synthesis gas yields the highest exergetic efficiency. Taking into account the low reversible separation energy due to the high CO_2 concentration (see Fig. 8.8), the separation energy in the synthesis gas will yield the lowest value of the required separation energy (separation energy = reversible separation energy/exergetic efficiency).

8.5 Post-combustion Technologies

8.5.1 Chemical Absorption

Chemical absorption processes based on organic solvents such as amines are currently the preferred option for post-combustion CO_2 capture. Absorption processes using amines are mainly employed in the chemical industry and are commercially available for post-combustion CO_2 capture systems, but not on the scale required for power plant flue gases. As the experience of chemical absorption for coal combustion is limited, the interaction between solvents and flue gas constituents is an open issue, one that needs to be addressed, as it might result in solvent degradation, increased corrosiveness of the solvent or plugging of the absorber or regenerator.

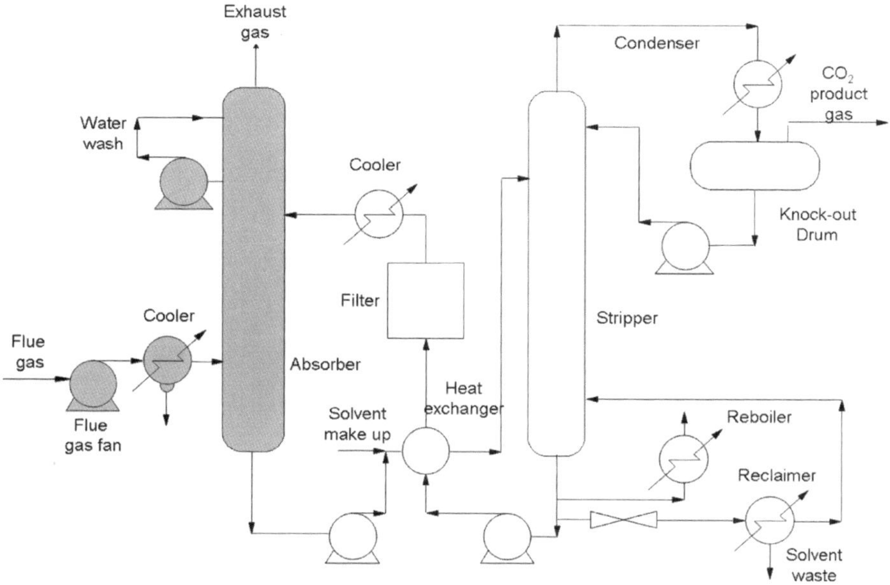

Fig. 8.10 CO_2 recovery by chemical absorption (IPCC 2005)

Figure 8.10 shows a scrubbing system used to separate carbon dioxide from flue gas by chemical absorption. The system consists of two main elements – an absorber in which the carbon dioxide is removed and a regenerator (stripper) in which the carbon dioxide is released in a concentrated form and the solvent is recovered. Prior to CO_2 removal, the flue gas is typically cooled to temperatures between 40 and 60 °C and then treated to reduce particulates (which cause operational problems) and other impurities (for example SO_x, NO_x, HCl, Hg) which would otherwise cause costly losses of the solvent. The amine solvent absorbs the carbon dioxide (together with traces of SO_x and NO_x) by chemical reaction to form a loosely bound compound. The regeneration of the chemical solvent is carried out in the stripper at elevated temperatures between 100 and 140 °C and pressures not very much higher than atmospheric pressure. Regeneration requires a large amount of heat, which is typically extracted from the steam cycle, reducing the net efficiency of the power plant significantly. As for all other separation technologies for CO_2, electrical energy is also needed to compress the captured carbon dioxide for transportation to the storage site (IPCC 2005).

The key parameters determining the technical and economic operation of a CO_2 absorption system are the

- *Flue gas flowrate*: The flue gas flowrate will determine the size of the absorber, which represents a sizeable contribution to the overall cost.
- *CO_2 content in the flue gas*: Since flue gas is usually at atmospheric pressure, the partial pressure of carbon dioxide is low, around 3–15 kPa. With these low partial pressures, aqueous amines (chemical solvents) are the most suitable absorption solvents.

- *CO_2 removal*: Amine absorption can be designed to capture up to 85–95% of the carbon dioxide in the flue gas and produce carbon dioxide with a purity of above 99.95%. Both the level of recovery and the CO_2 purity require economic optimisation, however. A higher recovery will lead to a taller absorption column and higher energy penalties and hence increased costs.
- *Solvent flowrate*: The solvent flowrate will determine the size of most equipment apart from the absorber. For a given solvent, the flowrate will be fixed by the parameters above and also the chosen CO_2 concentrations within the lean and the rich solutions.
- *Energy requirement*: The energy consumption of the process is the sum of the thermal energy needed to regenerate the solvents and the electrical energy required to operate the liquid pumps and the flue gas blower or fan. Energy is also required to compress the recovered carbon dioxide to the final pressure required for transport and storage.
- *Cooling requirement*: Cooling is needed to bring the flue gas and solvent temperatures down to the temperatures required for efficient absorption of CO_2.

8.5.1.1 Solvents (Amines)

The choice of solvent used for absorption is an issue of optimisation. Important considerations include the CO_2 loading (mol CO_2/mol amine), the solvent concentration in the aqueous solution, the heat of reaction, the heat of vaporisation, the reaction rate and the temperature required for regeneration. Additionally, the stability of the solvent can be a problem. Solvents can degrade thermally or by interactions with the flue gas components, of which oxygen is the most important. Degradation is often linked with corrosion because degradation products are associated with the corrosiveness of the solvent. All of these parameters are obviously not optimal simultaneously for any one solvent; for example high absorption rates generally cause high reaction heat rates.

The commercially available absorbents active enough for the recovery of dilute carbon dioxide at atmospheric pressure are aqueous solutions of alkanolamines such as

- primary amines: monoethanolamine (MEA),
- secondary amines: diethanolamine (DEA),
- tertiary amines: methyldiethanolamine (MDEA) and
- hindered amines.

Alternative solvents are discussed later.

8.5.1.2 Energy Requirements

The energy requirement when using absorption is a key consideration. A large amount of heat is required, mainly to regenerate the amine, and electricity is consumed by the necessary fans and pumps. The absorption solvents active at low

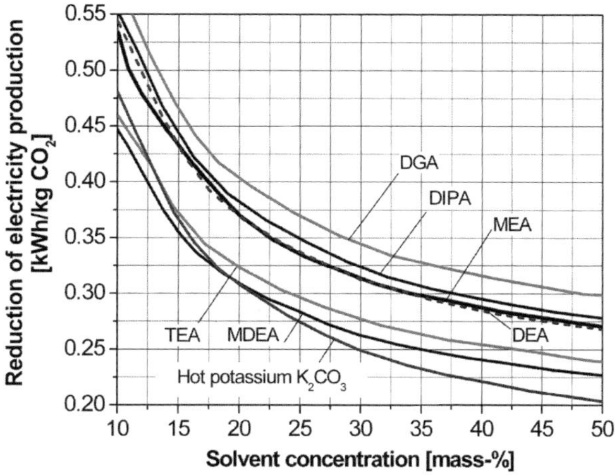

Fig. 8.11 Energy demand for chemical absorption of CO_2 from flue gases (Göttlicher 1999)

partial pressures are those with higher reaction energies and which require more energy for regeneration. The design challenges are to minimise the regeneration energy by selecting a solvent or mixture of solvents with a low reaction energy and to use a low-value heat source to provide this energy. The lowest values for the heat requirement for regeneration are between 2.7 and 3.3 GJ/t CO_2 (0.75–0.91 kWh/kg CO_2), depending on the solvent process. The goal in ongoing research projects is to reach 2 GJ/t. Steam with a pressure of about 3–4.5 bar is used to regenerate the solvent. This steam has to be extracted from the steam turbine and thereby reduces the mass flow through the turbine and therefore its power output.

The total heat requirements of various solvents, 80% of which is heat for regeneration, are shown in Fig. 8.11, calculated as $KWh_{el}/kg\ CO_2$. The calculated electricity consumption corresponds to the power reduction by steam extraction, assuming a conversion of heat to power of 19% at the temperature of extraction (Göttlicher 1999). Typical values for the electricity requirement are between 0.02 and 0.03 kWh/kg CO_2 for post-combustion capture in coal-fired power plants. Compression of the carbon dioxide to 110 bar will require around 0.13 kWh/kg CO_2. Consequently, the total power requirement for CO_2 separation and compression is in the range of 0.35–0.5 kWh/kg CO_2. For a coal-fired power station with a baseline efficiency of 45% this means an efficiency reduction between 10 and 14%.

8.5.1.3 Flue Gas Pre-treatment

Because most of the experience of chemical absorption processes has been gained in the chemical industry, attention has to be paid to the interactions between the flue gas components and the solvent when it comes to applying such processes in coal-fired power stations. These interactions can result in the decomposition of the solvent by irreversible reactions, an increase in the metallic corrosive attack by solvents and an increase in the plugging of equipment by corrosion and decomposition products. Interactions are dependent on the characteristics of the solvent and the composition of the flue gas. Measures to prevent side-effects have to be taken either by advanced flue gas cleaning or by modifying the properties of the solvent.

NO_x, SO_x: Acid gas components such as NO_x and SO_x will, similarly to CO_2, react with the solvent. This interaction leads to the formation of heat-stable salts and hence a loss in the absorption capacity of the solvent and the risk of solids formation in the solution. Therefore, the reduction of NO_x and SO_x to very low concentrations before CO_2 recovery becomes essential. Depending on the cost of the solvent, SO_2 concentrations of around 10 ppm may be required to keep solvent consumption and make-up costs at reasonable values – which often means that additional flue gas desulphurisation is needed. For NO_x, it is the NO_2 which leads to the formation of heat-stable salts. Because the level of NO_2 is usually less than 10% of the overall NO_x content in a flue gas, state-of-the-art $DeNO_x$ systems are sufficient to achieve the recommended levels of less than 20 ppm.

Fly ash: Careful attention must also be paid to the fly ash and soot present in the flue gas, as they might plug the absorber and increase corrosion and solvent loss by chemical degradation if contaminants levels are too high.

Oxygen: The presence of oxygen in the flue gas can increase corrosion and solvent degradation in the absorption system. Uninhibited alkanolamines such as MEA and DEA can be oxidised to produce carboxylic acids and heat-stable amine salts. A solution to this problem is to apply an inhibitor to both passivate the metal and inhibit amine degradation (IPCC 2005; VGB 2002).

Further Development

Various novel solvents are being investigated with the objective of achieving a lower solvent regeneration energy consumption. Research is focussed on aqueous solutions of

- alternative alkanolamines,
- amino acid salts,
- alkali or earth alkali carbonate solutions and
- ammonia.

For a process with chilled ammonia, a 50% lower energy penalty in comparison to an MEA solvent is indicated. The absorber has to operate at a temperature between 2 and 16 °C to minimise ammonia losses.

Research is also being carried out to improve existing process methods and packing types, for example to replace the absorption columns by spray washers in order to reduce pressure losses in the flue gas path (Davidson 2007).

8.5.2 Solid Sorbents

Post-combustion systems are being proposed that make use of regenerable solid sorbents to remove carbon dioxide at relatively high temperatures. The application of high temperatures in the CO_2 separation step has the potential to have higher

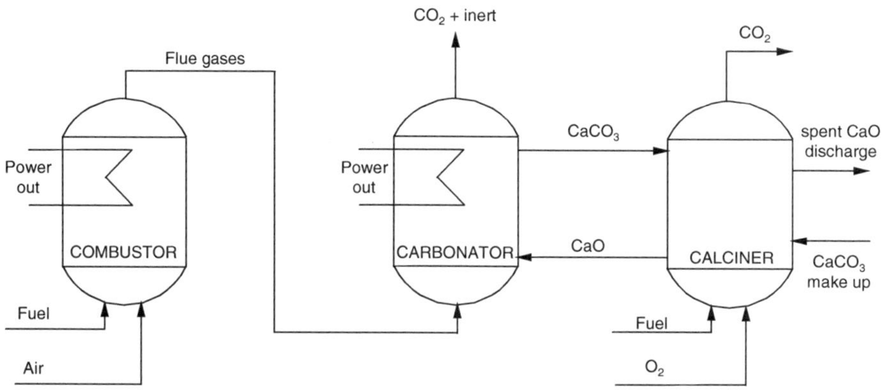

Fig. 8.12 CO_2 recovery with a $CaCO_3$ sorbent

efficiencies in comparison to wet-absorption methods, because the absorption heat is released at a temperature which can be used for power production (IPCC 2005).

The solid sorbents being investigated for large-scale CO_2 capture purposes are sodium and potassium oxides and carbonates (to produce bicarbonate). Also, high-temperature Li-based and CaO-based sorbents have shown potential.

The use of CaO as a regenerable CO_2 sorbent has been proposed in several processes dating back to the 19th century. The carbonation reaction of CaO to separate carbon dioxide from hot gases ($T > 600\,^{\circ}\mathrm{C}$) is very fast and the regeneration of the sorbent by calcining the $CaCO_3$ into CaO and pure CO_2 is favoured at $T > 900\,^{\circ}\mathrm{C}$ (at a partial pressure of carbon dioxide of 0.1 MPa). Figure 8.12 shows a process with CaO as the sorbent. For both carbonation and calcination, fluidised bed reactors are used, operating at 650 and 900 °C. Due to the release of absorption heat at a high temperature, power can be produced in the carbonator. The temperature increase to 900 °C in the calciner is achieved by combustion of a fossil fuel which has to be burned with oxygen to produce a pure CO_2 stream. About one third of the total fuel input is required for the calciner.

A key issue for these systems is the sorbent itself, which has to have a good CO_2 absorption capacity and chemical and mechanical stability for long periods of operation in repeated cycles. Natural sorbents like limestone and dolomite deactivate rapidly, and a large make-up flow of sorbent, in the order of the mass flow of fuel entering the plant, is required to maintain the activity in the capture-regeneration loop (Shimizu et al. 1999; Abanades et al. 2004, 2005; Sivalingam et al. 2009).

The absorption of CO_2 can also be integrated into a gasification reactor, which is discussed in Sect. 7.6.5.7.

8.6 Oxy-fuel Combustion

Oxy-fuel firing involves burning a carbon-containing fuel in either pure oxygen or a mixture of pure oxygen and a CO_2-rich recycled flue gas. The oxygen is provided by an air separation plant. Because this method eliminates nitrogen from the comburent, a flue gas mixture is obtained with carbon dioxide and water vapour as the essential components. The flue gas, after cooling to condense the water vapour, contains about 80–98% carbon dioxide, depending on the oxygen purity, the fuel in use and the particular oxy-fuel combustion process. Impurities in the carbon dioxide are gas components deriving from the fuel, such as SO_x, NO_x, HCl and Hg, and gas components, such as nitrogen, argon and oxygen, contained in the oxygen fed into the system or from air leakage. This concentrated CO_2 stream can be compressed, dried or further purified before delivery into a pipeline for storage (IPCC 2005).

The key separation step in oxy-fuel combustion capture systems is air separation. The current methods of oxygen production by air separation are cryogenic distillation, adsorption using multi-bed pressure swing units and polymeric membranes. For oxy-fuel firing requiring less than 200 t O_2/day, the adsorption system will be economic. For all larger applications, which include power station boilers, the most economic solution is cryogenic air separation in an air fractionation unit. This method in particular causes a significant efficiency loss over the entire process because of its considerable energy demand.

Figure 8.13 provides an indication of the electrical energy needed for oxygen production in an air separation unit with two columns (Göttlicher 1999). The higher the purity of the oxygen, the higher will be the expenditure of energy per mass of oxygen. For combustion using oxygen, the purities required range between 95 and 99%, where the purity chosen is the result obtained by optimisation of the energy demand. Roughly assuming a net efficiency of 45% of a power plant without CO_2 separation, about 0.6 kg of oxygen per produced kWh_{el} (excess oxygen 10%) is needed for oxy-fuel combustion. With an oxygen purity of 99.5%, an energy demand of 0.29 kWh/kg O_2 for the separation of the oxygen is required. This results in an energy expenditure of 0.175 kWh per kWh of electrical power produced, which diminishes the efficiency by almost 8% through the necessary air

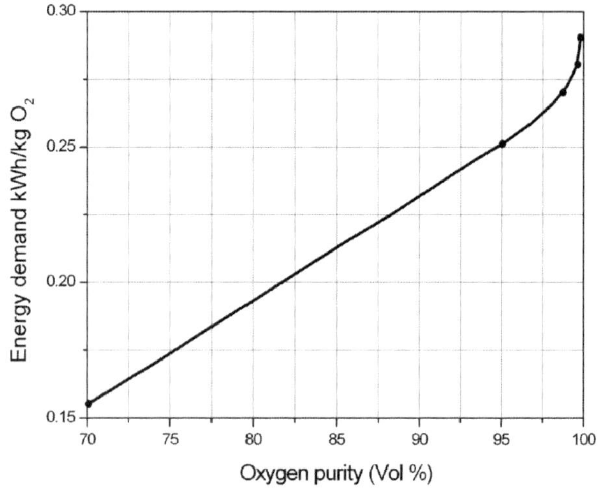

Fig. 8.13 Energy requirement for cryogenic air separation (Göttlicher 1999)

fractionation alone. Current oxy-fuel designs assume an oxygen purity of 95% with an energy requirement of 0.23–0.25 kWh/kg O_2 (corresponding to a 6.2–6.8% efficiency loss). Optimisation of the air fractionation process, for example by introducing a three-column process, could reduce the energy requirements to 0.2 kWh/kg O_2 (a 5.4% efficiency loss); a further reduction to 0.16 kWh/kg O_2 (a 4.3% efficiency loss) is expected by integrating the air fractionation process into the power plant. However, a further loss of about 3.5% is incurred through the necessary work for the compression of the CO_2 to 110 bar for transport. The total loss in efficiency is then about 11–12%, with a potential reduction down to 8% (Kather et al. 2007b).

Although elements of oxy-fuel combustion technologies are in use in the aluminium, iron and steel and glass melting industries today, oxy-fuel technologies for CO_2 capture have yet to be deployed on a commercial scale. The so-called oxy-fuel process gathered attention in the early 1990s with the increasing interest in carbon capture and sequestration. At that time one of the first investigations into oxy-coal combustion was carried out by the IFRF (International Flame Research Foundation) (Tan et al. 2005). In recent years, several research initiatives have been started to study the effect of oxy-fuel combustion atmospheres on combustion behaviour, heat transfer, emissions and operational behaviour. Vattenfall has constructed an oxy-fuel power demonstration plant (30 MW pilot plant) at *Schwarze Pumpe* which went into service in 2008 (Burchardt and Radunsky 2007; Kluger et al. 2006; Burchhardt and Jacoby 2008).

One of the big advantages of the oxy-fuel process is the simple way a conventional coal-fired power-generating design can be adapted to CO_2 separation. With the existing components being largely kept in place, it is possible in general to perform a successful retrofit on power plants.

8.6.1 Oxy-fuel Steam Generator Concepts

The combustion of coal in a pure oxygen atmosphere can lead to flame temperatures rising above 3,000 °C, far too high for typical power plant materials. What is more, temperatures so high would volatilise a substantial part of the ash and result in heavy fouling of the convective heating surfaces. The actual combustion temperature should not exceed 1,500–1,800 °C.

Figure 8.14 shows calculated flame temperatures as a function of the oxygen/fuel stoichiometry for different ratios of flue gas recirculation. In principle, there are two possible methods to limit the combustion temperature. The first is to control the temperature by the suitable admixture of a thermal ballast, for instance in the form of recirculated flue gas, solids or water vapour. The second is to carry out the combustion process at lower or higher values than the stoichiometric ratio of oxygen to fuel. Both oxygen and oxygen-deficient combustion result in a reduction of the adiabatic combustion temperature until the stability limit of the flame is reached.

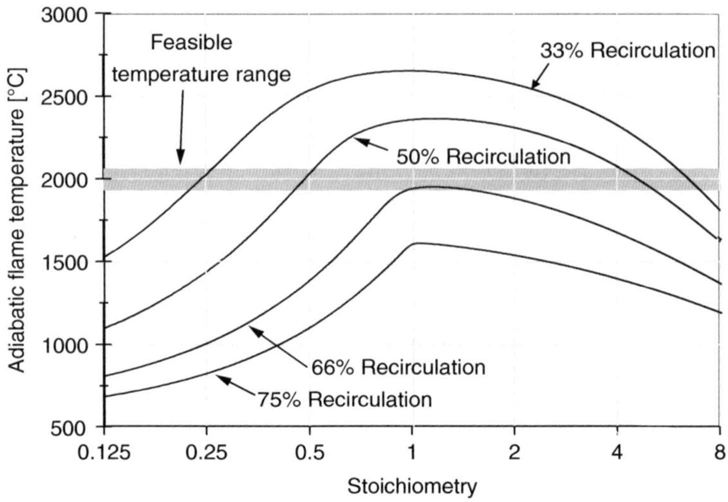

Fig. 8.14 Adiabatic flame temperatures as a function of stoichiometry for different flue gas recirculation ratios, calculated by Factsage (Bale et al. 2002)

8.6.1.1 Flue Gas Recirculation

Nearly all oxy-fuel combustion concepts considered at present rely on external flue gas recirculation to control the combustion temperature. The goal is to limit combustion temperatures to values similar to those in combustion with air. Cool flue gas is extracted either before (wet recirculation, flue gas temperature about 150 °C) or after vapour condensation (dry recirculation, temperature 50 °C) and recirculated to the furnace. Depending on the fuel and the temperature of the recirculated flue gas, differing but high volumetric flowrates are needed to reduce the combustion temperature. Assuming a flue gas recirculation temperature between 200 and 300 °C, about two thirds of the flue gas produced in the steam generator has to be recirculated to achieve temperatures similar to air combustion. The high volumetric flow that needs to be transported involves an increase in the dimensions of the plant and the auxiliary power requirement for the recirculation fan. External recirculation is an established technology, but one that entails a number of problems (e.g. the construction size, wear and tear, corrosion when temperatures fall below the dew point, distribution of the individual flows). Owing to the possibility of retrofitting existing power stations with it, recirculation is a solution at the logical beginning of the realisation of CO_2-emission-free power plants.

Another method is to recirculate the flue gas in the boiler internally. Every conventional jet and swirl burner has a well-developed recirculation zone to homogenise the combustion and to avoid hot spots. Technologies with extensive internal flue gas recirculation such as flameless oxidation burners have been successfully applied in the steel industry. For the duty range of power plants, the use of flameless oxidation burners is unknown so far. It remains an open question whether it will be possible by this technique exclusively to recirculate sufficiently large flue gas quantities to achieve adequate cooling. It is certain, though, that internal recirculation can contribute to obtaining local uniformity of the flue gas temperatures in combustion with oxygen.

8.6.1.2 Water/Steam Spraying

The intention of this method is to cool the flame by injecting water and/or steam. For gas turbines in particular, water-cooled burners are a cost-effective and well-tried option. In steel furnaces, these burners belong to the first generation of oxy-fuel burners in practical application. However, the heat loss in the steam generator increases because the vaporisation enthalpy cannot be used for power generation.

8.6.1.3 Controlled Fuel/Oxygen Staging with Rich/Lean Burners

The central problem in realising a firing system with reduced flue gas recirculation lies in the need to control the flame temperatures and to distribute the heat release over a greater part of the furnace. Staged combustion technologies such as air or fuel staging, which are employed in conventional steam generators to reduce nitrogen oxide emissions, are possible methods to delay and therefore distribute the heat release in the oxy-fuel combustion process. In addition, the application of fuel-rich/lean burners offers the potential to reduce the adiabatic temperature and to control the peak temperature in the burner.

The so-called "controlled fuel/oxygen staging with rich/lean burners" concept applies both methods to control the temperature in the flame and furnace while using reduced flue gas recirculation rates than would otherwise be necessary. The concept is illustrated in Fig. 8.15 for controlled fuel staging and oxygen staging. Staged addition of oxygen in the case of oxygen staging, or fuel in the case of fuel staging, result in incremental oxidation and heat release. Between the stages, heat is transferred by radiation to the furnace walls. The lower adiabatic temperature of fuel-rich/lean burners reduces the peak temperature. The advantage of the two schemes is the spatial distribution of the heat release. Both concepts avoid the mixing problem of conventional fuel or oxygen staging by means of the higher impulse of the burners operating with oxygen deficiency or excess oxygen, respectively (Becher et al. 2007; Spliethoff 2006).

8.6.2 *Impact of Oxy-fuel Combustion*

8.6.2.1 Flue Gas Composition

The nitrogen which is present in the flue gas from combustion with air is removed to a great extent in the air separation unit of oxy-coal processes prior to combustion, so only minor amounts are in the flue gas. In consequence the flue gas is mainly composed of carbon dioxide and water vapour. If the circulation of the flue gas is increased, its composition does not change because the recirculated flue gas has the

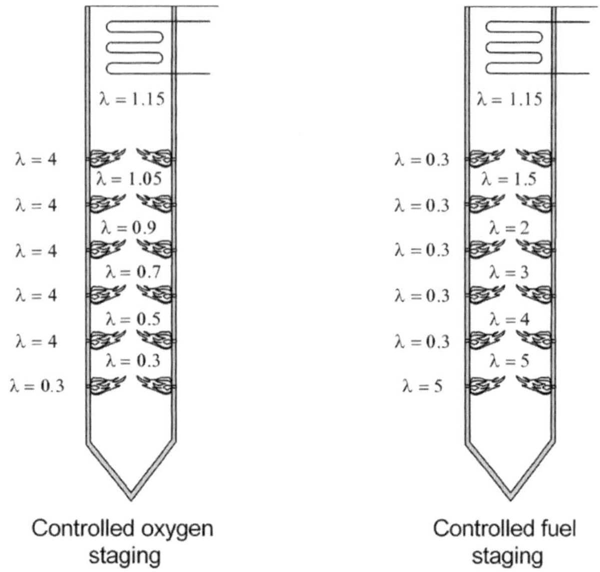

Fig. 8.15 Controlled fuel/oxygen staging in the furnace. λ is the ratio of the supplied comburent to the stoichiometric comburent requirement

same composition as the products of combustion and is not involved in the combustion process. An exception is the recirculation of dry flue gas, where water vapour is condensed and extracted.

8.6.2.2 Thermodynamic Properties

The thermodynamic properties of the flue gas from oxy-fuel combustion are different from those of the flue gas from air combustion, due to the differing composition. The density of the flue gas from oxy-coal processes is greater, because carbon dioxide (with 44 kg/kmol) has a larger molecular weight than nitrogen (28 kg/kmol). In wet flue gas recirculation, the lower molecular weight of water (18 kg/kmol) can partly compensate the effect of CO_2. The molar thermal capacity of the flue gases C_p increases due to the higher concentrations of CO_2 and H_2O. The triatomic molecules have more degrees of vibrational freedom and can store more heat energy. The specific (mass-related) thermal capacity c_p increases as well; due to the high specific thermal capacity of water vapour, the effect is more pronounced for wet flue gas (see Table 8.3).

The higher concentrations of CO_2 and H_2O, which are band emitters, in the flue gas intensify the heat transfer by radiation (Gupta et al. 2006). In pulverised coal fired furnaces, though, the emissivity of the hot flue gases in the burner zone is dominated by the solid-state radiation of the char and ash particles. The emissivity given for the combustion zone using air as the comburent is between 0.8 and 0.9 (Blokh and Viskanta 1988), so only a slight increase is possible using oxygen as the comburent. With the completion of char burnout above the burner zone, the emissivity drops and a more pronounced effect can be expected.

Table 8.3 Composition of the flue gases of firing systems with air and with oxygen (fuel: hard coal, $\lambda = 1.15$; gas properties from Kretzschmar et al. 2005)

	Air combustion	Oxy-coal combustion dry recycling	Oxy-coal combustion wet recycling
N_2 [vol.%]	77%	0%	0%
CO_2 [vol.%]	16%	87%	74%
H_2O [vol.%]	5%	8%	22%
O_2 [vol.%]	3%	5%	4%
$\rho(300\,°C)$ [kg/m^3]	0.64	0.88	0.81
$\overline{C}_{p,300\,°C-1,200\,°C}$ [kJ/kmol K]	27.4	39.0	37.8
$\overline{c}_{p,300\,°C-1,200\,°C}$ [kJ/kg K]	0.91	0.94	1.00
M [kg/kmol]	30.46	41.3	37.8

8.6.2.3 Heat Transfer

For the design of furnaces, the heat quantities to be transferred in the furnace and in the convective heat exchangers are the decisive factors. Given that the furnace outlet temperature is defined by the ash deformation behaviour of the fuel, the division of heat transfer to the furnace walls and to the convective heat exchangers is a function of the adiabatic combustion temperature and hence substantially depends on the flue gas recirculation rate. This correlation is plotted in Fig. 8.16 and compared with a furnace using air. For the calculations, a furnace outlet temperature of 1,200 °C and an air ratio of 1.15 were assumed. This makes it clear that for oxy-coal power plants with a low flue gas recirculation rate, more heat has to be transferred to the furnace.

It is common practice to choose the recirculation rate so that the furnace outlet temperature and the corresponding heat transfer characteristic are similar to firing systems using air. In contrast to combustion with air, lower mass flows are produced due to the higher specific heat capacity, and lower volume flows due to the higher molar heat capacity. If the same volumetric flow as in a furnace firing with air was required, the resulting recirculation rate for a bituminous coal would be 78%. The oxygen content of the oxidising agent, i.e. the mixture of oxygen and the recirculated flue gas, would then be 21 vol.%. If setting the same temperatures and heat transfer characteristic for a bituminous coal, the recirculation rate is 67% (wet recirculation). The absolute volumetric flows are plotted in Fig. 8.17. It shows clearly that for an oxy-coal firing system designed to use the same temperatures as air combustion, the volumetric flows in the furnace are about 35% lower. The corresponding mass flow is 18% lower and the oxygen content of the oxidising agent is about 32 vol.%.

Due to the lower volumetric flowrate, the cross-section of an oxy-fuel steam generator needs to be reduced from air-firing sizes in order to induce velocities similar to those in air combustion, which produce a good heat transfer coefficient in the convective heat transfer region. Assuming heat transfer coefficients similar to air firing, the height of the oxy-fuel boiler has to be increased to install the required heat exchanger surface. However, both radiative and convective heat transfer improve under oxy-fuel conditions, partly reducing the necessary increase of height. As previously mentioned, the increased concentrations of CO_2 and H_2O will

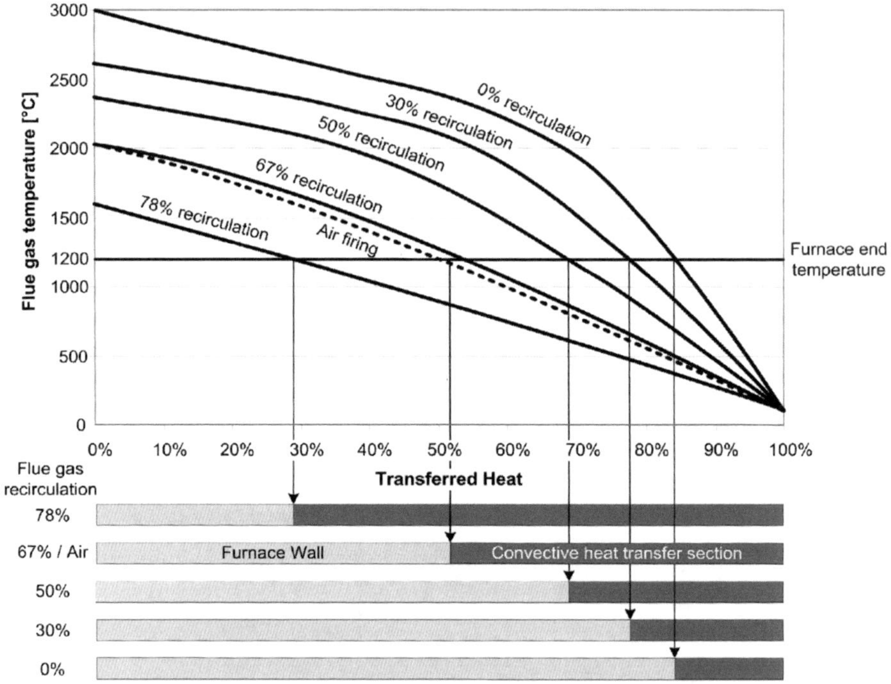

Fig. 8.16 Temperature-heat diagram for different recirculation ratios (wet flue gas recirculation, recirculation temperature 300 °C, bituminous coal)

improve the radiative heat transfer in the furnace, but only slightly. In the convective heat exchangers the improvement of the heat transfer coefficient by both radiation (due to the emissivity of CO_2 and H_2O) and convection (due to the higher thermal conductivity of CO_2 and H_2O) is more pronounced. This results in a more compact convective heat exchanger (Hellfritsch et al. 2007; Kakaras et al. 2007).

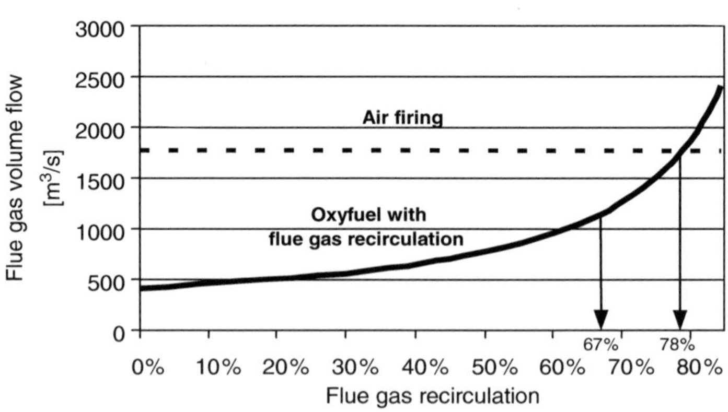

Fig. 8.17 Flue gas volume as a function of the recirculation ratio for a bituminous coal (1,000 MW_{Fuel})

An oxy-fuel retrofit of an air-fired boiler is a compromise to meet contrasting requirements. On one hand, the given cross-section requires higher velocities and recirculation ratios to ensure particle entrainment in the furnace and sufficient heat transfer coefficients in the convective heat exchangers. Higher recirculation ratios, however, shift the heat transfer to the convective heat exchangers, as can be seen from Fig. 8.16. For a retrofit design, recirculation ratios are 8–10% higher than for a greenfield oxy-fuel plant. Another difficulty for retrofits are the air leakages from the (existing) boilers, which limit the achievable CO_2 captures (see Sect. 8.6.3).

8.6.2.4 Emissions

When considering the emissions from oxy-fuel firing, the first issue to take into account is the increased concentration of all pollutants in the flue gas. In combustion with air, roughly 10 kg of air is required for 1 kg of bituminous coal, whereas in oxy-fuel combustion, only about 2 kg of oxygen is fed. The lower flue gas mass flow leads to an increase of the mass concentration of the combustion products by a factor of around 3.6. Two other effects cause an additional increase in the measured concentrations (mostly volume fractions in the dry flue gas). The flue gas of an oxy-fuel process has a density roughly 1.25 times higher than that from air combustion, which leads to a further increase of the volumetric concentrations by the same factor. Since the measurements are carried out in the dry flue gas, removing the high water fraction of about 25% by volume from the oxy-fuel gas means additionally higher pollutant concentrations. The sum of these effects are increased concentration factors of 4.5 in the case of humid and 5.7 in the case of dry flue gas. This is reflected, using the example of NO_x, in Fig. 8.18, which assumes the same pollutant loads as in the process with air. In order to make the emissions of different processes comparable, the measurements ought to be translated into specific conversion rates, as shown in Fig. 8.18, or be referred to the energy content of the fuel (Kather et al. 2007a).

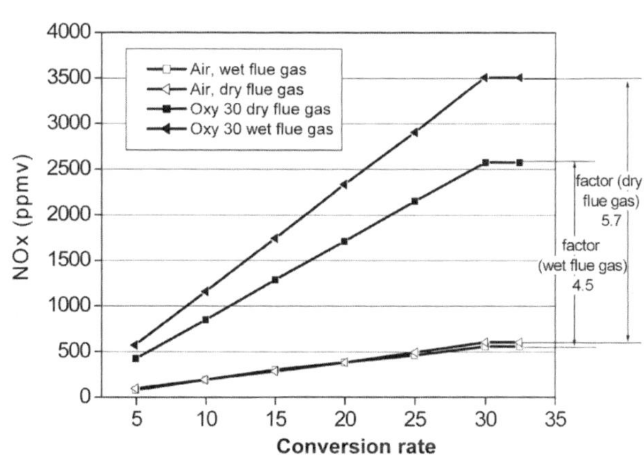

Fig. 8.18 Relation between pollution conversion rate and concentration (Kather et al. 2007a)

The increased concentration of pollutants is counteracted by lower conversion rates of the pollutants from the fuel (and/or comburent) into the flue gas. The conversion rate or the NO_x emissions generated per unit of energy can be reduced by up to 70% in oxy-fuel combustion, depending on the burner design and operation. The NO_x reduction is thought to be the result of several mechanisms: a decrease of thermal NO_x due to the very low concentration of N_2 in the comburent, the reduction of recycled NO_x as it is reburned in the volatile matter release region of the flame and the reaction between recycled NO_x and char. Both homogeneous and heterogeneous reduction reactions are favoured by the higher NO concentrations.

The SO_2 emissions per energy from the burned fuel may be lowered through sulphur retention in both the ash and deposits and are typically lower by 20%, depending on the ash composition. The absolute SO_2 concentrations are typically 2.5–3 times higher than in air firing, and it has been observed that the conversion ratio of SO_2 to SO_3 is higher under oxy-fuel conditions. This higher SO_3 concentration in the flue gas in combination with the higher moisture content increases the dew point temperature and limits the waste heat utilisation. Whereas for an air-fired boiler, the typical dew point temperature for a hard coal is in the range of 100–110 °C, it increases to about 140–150 °C under oxy-fuel conditions (Scheffknecht and Maier 2008).

The effect of oxy-fuel combustion on trace element emissions and mineral matter transformation is uncertain, but it can be expected that the behaviour of certain minerals (in particular carbonates) will be affected by the change in the flue gas environment (Wall 2007; Maier et al. 2007).

8.6.3 *Oxy-fuel Configurations*

Figure 8.19 shows the typical configuration of oxy-coal combustion processes. For heat transfer purposes, two thirds of the flue gas is extracted after the steam generator and conducted back to the furnace to cool the newly produced flue gases. The un-recirculated portion of the flue gas, volumetrically two thirds smaller than from combustion plants using air, gets dehumidified after flue gas cleaning and is then either compressed or liquefied.

8.6.3.1 CO_2 Purity

The purity of the carbon dioxide has a decisive influence on the energy input necessary for the compression process. The fraction of other non condensable gases in the CO_2 after dehumidification is shown as 11% in Fig. 8.19, an amount which is repeated in various sources. This value can be optimised to a limited extent by the design. The calculations in Fig. 8.19 assume an oxygen purity of 99.5%, whereas most designs are based on 95%.

A large fraction of the inert gases consists of argon and nitrogen, the resulting percentage depending on the purity of the produced oxygen. An increasing purity

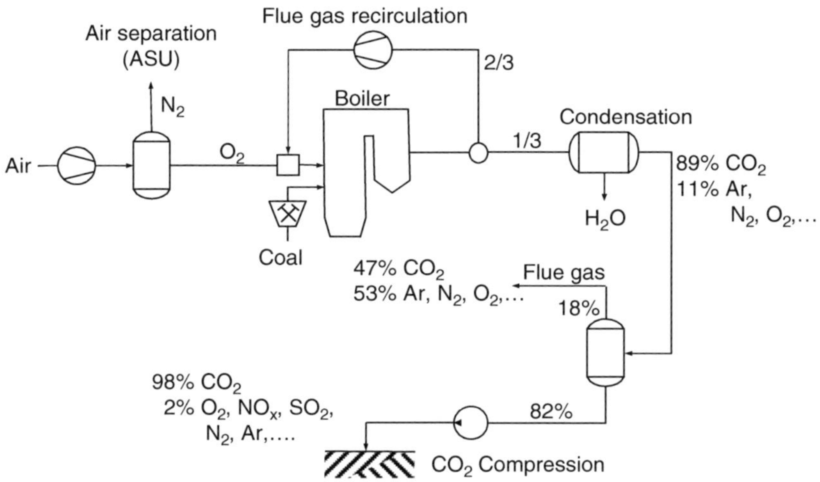

Fig. 8.19 An oxy-fuel process diagram (air leakage 1%, oxygen purity 99.5%, excess air 15%) (Kather et al. 2007a)

of the oxygen will increase the CO_2 concentration in the flue gas. However, more energy has to be expended for air separation if this is to be achieved (see Fig. 8.13).

Furthermore, there are fractions of oxygen in the carbon dioxide which originate from the set excess oxygen. For achieving a complete burnout and preventing corrosion in the furnace, about 15% excess oxygen is chosen in modern coal-fired power plants using air, which corresponds to a content of oxygen in the flue gas of 4.5–5% in an oxy-fuel firing system. If the pulverised coal is distributed more evenly, it seems possible to reduce the amount of excess air to about 10%, corresponding to an oxygen content of 3–3.5% by volume in oxy-fuel firing.

In industrial-scale combustors in service today, the entire flue gas path, from the burner throat to the induced-draught fan, is operated at a slight negative pressure. Since the flue gas path does not have a completely air-tight design for economic reasons, certain amounts of leakage air enter the flue gas. Today, it is assumed for power plant furnaces in service that about 4% of the combustion air leaks in, an amount increasing to about 10% as the plant ages. A large proportion of this leakage occurs in the regenerative air preheater, where, because of the way it is constructed, leaked air enters the flue gas flow. While the consequences of leakage air are moderate in boilers operated with air, leakage air in oxy-coal boilers increases the proportion of undesirable inert gases. In order to achieve a purity of 90% CO_2, the leakage air fraction has to be limited to 1% if the excess air fraction in the combustion is 15% and the oxygen purity is 99.5%. This necessitates expenditure to seal up the entire flue gas duct and excludes the use of a regenerative air preheater (Kather et al. 2007a).

The achievable CO_2 purity in an oxy-fuel process is considered to be about 90%; if higher CO_2 concentrations are required, an additional separation step is required as shown in Fig. 8.19. The costs of an oxy-fuel process are therefore dependent on the CO_2 purity requirements.

8.6.3.2 Waste Heat Recovery

In a conventional plant operated with air, the flue gases can only be cooled to about 350–380 °C by using feed water as the working fluid, because nowadays, for reasons of efficiency, the feed water is preheated to temperatures of around 300 °C. The flue gas heat is used to preheat the cold combustion air in a regenerative air preheater.

Due to the significantly lower mass flow after the recirculation branch (where the recirculation gas is taken out of the post-furnace flue gas flow), the sensible heat of the flue gas, at 300–350 °C from oxy-fuel firing, is comparable to a conventional air-fired boiler at a boiler exit temperature of 130 °C. Therefore the sensible heat contained in the flue gas could in theory be neglected without having a lower steam generator efficiency in comparison to an air fuel fired power plant. However, with every lowering of the flue gas temperature through flue gas heat utilisation after the branching into recirculation, the flue gas losses decrease and the efficiency of the total process increases. Since the flue gas has to be cooled significantly before CO_2 liquefaction, not least because of the necessary dehumidification, the potential for waste heat recovery is relatively high. The increased concentration of pollutant gases, though, restricts the recovery of waste heat. For example, oxygen preheating and heat transfer in the high-pressure feed water preheater may be restricted because the sulphuric acid dew point, depending on the coal type, may rise to 160 °C.

In principle, the heat transfer to the oxygen and to the recirculating flue gas offers itself as an effective means to utilise the flue gas heat, because the heat, as in conventional processes, is returned directly to the furnace. Preheating the oxygen to temperatures above 200 °C requires the use of high-quality heat exchanger materials. At the dimensions required in power plants, this is not yet the state of the art (Kather et al. 2007a; Hellfritsch et al. 2004).

8.6.3.3 Flue Gas Recirculation

Various methods for the recirculation of flue gas, differing according to the temperature and pollutant load of the recirculated flue gases, are available. The possible temperatures lie between 100 and 350 °C. The lower the temperature of the recirculated flue gases, the smaller the recirculation flows necessary for furnace cooling and the higher the density of the flue gas. So, owing to the lower volumetric flows, the flue gas ducts can be built with smaller cross-sections. As another consequence, however, the ducts have to be longer.

When considering the pollutant load, one has to distinguish between the recirculation of untreated and treated flue gas (see Fig. 8.20). Feeding back treated and cold flue gas is the better variant for operation with a minimum of maintenance because flue gas desulphurisation, dust removal and predrying substantially decrease the risk of corrosion and material wear in the entire flue gas and coal-handling system. The drawbacks of using treated flue gas are the construction sizes of the cleaning

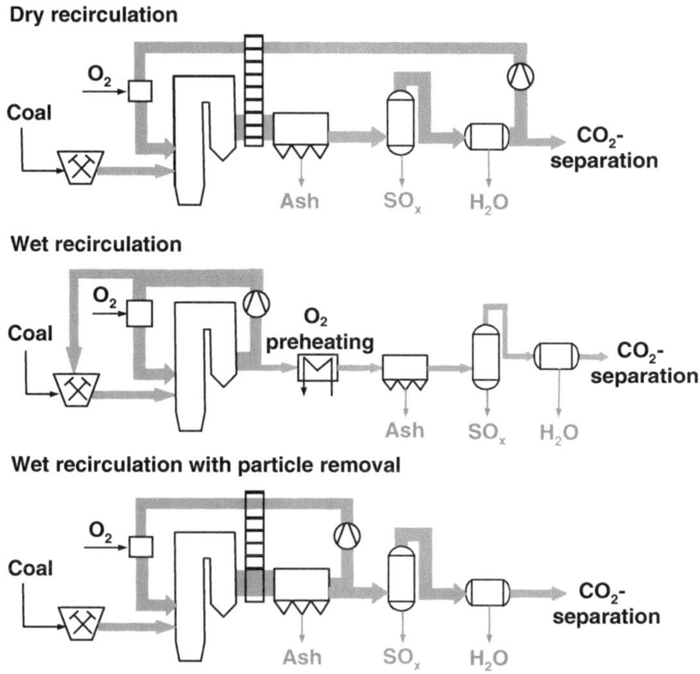

Fig. 8.20 Flue gas recirculation concepts for oxy-fuel combustion (Kather et al. 2007a) and amendments

facilities, which have to be designed for the triply high volumetric flow in the recirculation piping, and the energy losses through the condensation of the water vapour.

If untreated hot flue gas is recirculated, the length of the flue gas recirculation pipes can be kept short and a heat transfer system for reheating the cooled flue gas is not necessary. If the flue gas temperature is sufficiently high, the recirculating gas can be used to dry the bituminous coal in the combined drying and pulverising section of the plant. The large dust loads and the high temperatures restrict the choice of efficient recirculating fans.

A reasonable process variant is therefore the recirculation of the flue gases after dust removal and cooling. Since the collection efficiency of ESPs diminishes with increasing temperatures, the flue gases are first cooled and de-dusted and then reheated again by means of a heat transfer system.

8.6.4 Chemical-Looping Combustion

In the chemical-looping combustion (CLC) process, the oxygen needed for combustion is provided by means of an oxygen carrier, Fig. 8.21. mostly a metal oxide, which circulates between two separate reactors – see Fig. 8.21 (Lyngfelt et al. 2001; Tan and Santos 2006; Ryden et al. 2008; Anthony 2008). In the reduction or fuel reactor, the metal oxide is deoxidised while oxidising the fuel. For a fuel with the

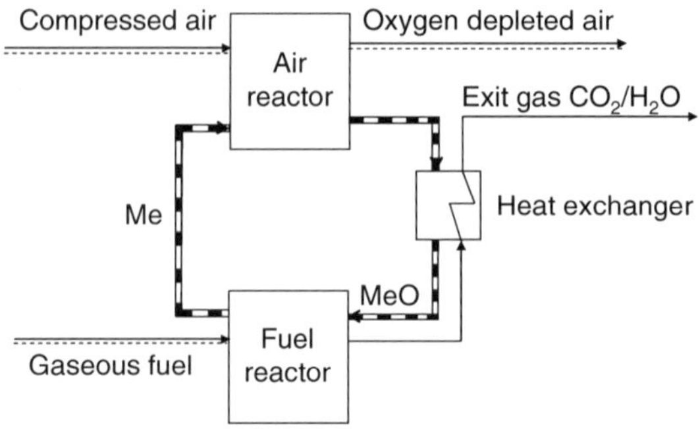

Fig. 8.21 Chemical looping process diagram

composition C_nH_{2m} for instance, the following reaction takes place:

$$C_nH_{2m} + (2n+m)\text{MeO} \rightarrow n\text{CO}_2 + m\text{H}_2\text{O} + (2n+m)\text{Me} \qquad (8.3)$$

The reduced metal is then fed to an oxidation reactor in order to reform a metal oxide using the oxygen in the air:

$$\text{Me} + \frac{1}{2}\text{O}_2 \rightarrow \text{MeO} \qquad (8.4)$$

Since the oxidation reactor effects the separation of the oxygen from the air, no energy-consuming air separation unit is necessary. The oxygen carrier is reduced again afterwards by the fuel.

The energy released by these two reactions corresponds to the reaction enthalpy of conventional combustion. The advantage of having combustion in two reactors compared to conventional combustion in a single stage is that the carbon dioxide is not diluted with nitrogen gas but is almost pure after the separation from water, without requiring any extra energy and costly external equipment for CO_2 separation (IPCC 2005).

The recycle rate of the solid material between the two reactors and the average solids residence time in each reactor control the heat balance and the temperatures in both reactors. The temperatures in the reactors are within the range of 800–1,200 °C. Possible metals for oxidation are those such as iron, nickel, copper and manganese. Particles with diameters from 100 to 500 μm move between the two reactors, being fluidised in each reactor. This method also ensures efficient heat and mass transfer between the gases and the particles. One of the beds gets fluidised with air, the other one with fuel. For this reason, this method is most suitable for gaseous fuels. The chemical-looping method using natural gas as the fuel is not yet technically mature, although the fundamental idea was known as early as the late 1960s. Work on chemical-looping combustion is currently in the pilot-plant and materials research stage. A critical issue is the long-term mechanical and chemical stability of the particles that have to undergo repeated cycles of oxidation and reduction. A minimum of material make-up flow must be achieved for the process to be economic. In order to avoid deposits of carbon in the reduction reactor, it is necessary to incorporate some steam into the fuel flow.

The interest in using CLC for solid fuels is huge, because the substantial efficiency drop associated with other CO_2 removal technologies is considerably reduced. There are different design concepts for using CLC for solid fuels. By an additional gasification step the solid fuel can be converted into a gas, which allows for gas–solid reactions in the fuel reactor. The gasification steps can be external; development is ongoing to integrate a gasification step (by H_2O or CO_2) into the fuel reactor. A disadvantage of gasification in the fuel reactor is the slow gasification kinetics, which requires a longer residence time to reach a sufficiently complete state. Another problem may be the negative impact of fuel ash on the lifetime of the oxygen carrier (Berguerand and Lyngfelt 2008; Leion et al. 2008; Cao et al. 2006). A novel concept proposes to uncouple the oxygen release and fuel reaction in the fuel reactor. The oxygen is released in an intermediate step after the air reactor to directly react with the solid fuel (Mattisson et al. 2009).

8.7 Integrated Gasification Combined Cycles with Carbon Capture and Storage

Integrated coal gasification technology with CO_2 removal has been presented in detail in Sect. 7.5. In this section, the efficiency losses will be covered. Figure 8.22 shows schematically the design of an IGCC-CCS power plant. Compared to an IGCC power plant without CO_2 removal (see Fig. 7.2), it requires two additional components – a CO shift reactor and a CO_2 scrubber.

The liquefaction process included, a decrease of the efficiency by about 8–10%, at a CO_2 removal rate of 90%, is the result of adding the extra process steps, taking as a reference the IGCC power plant 1998 with an efficiency of 52% (discussed in Sect. 7.5). The efficiency loss can be put down to the following causes:

- Shift conversion: Due to the exothermic conversion reaction, part of the syngas heating value is converted to heat. In the case of a typical synthesis gas from a high-temperature gasifier with about 60% carbon monoxide and 30% hydrogen, the heating value of the fuel gas is reduced by about 10%. The arising reaction heat can be utilised at the temperature of the shift reaction. The maximum efficiency loss of about 5% can be limited to 2.5% or so by heat exploitation.
- The separation of the CO_2 from the syngas requires energy to regenerate the scrubbing agent. Owing to the high partial pressure of the CO_2, this input is significantly lower than for CO_2 scrubbing of atmospheric flue gases. The energy demand ranges around 1–2%.
- The separated volumetric CO_2 flow does not get expanded in the gas turbine. The consequent diminished gas turbine output results in an efficiency loss of about 1.2%.
- By compression to 110 bar, the CO_2 is turned into a liquid state. The efficiency loss through the necessary compression energy amounts to about 3–3.5%.

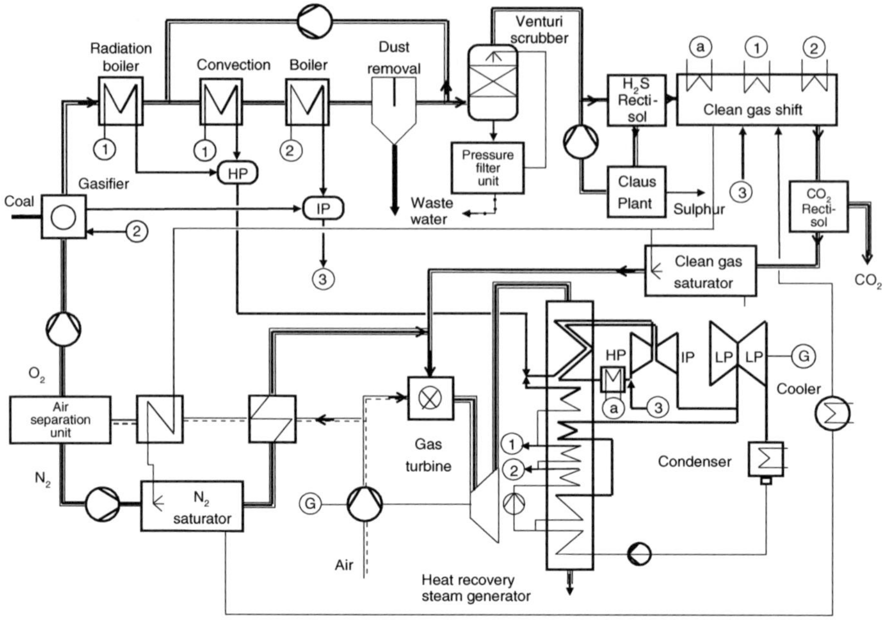

Fig. 8.22 Schematic diagram of IGCC with CO_2 capture (Pruschek 2002)

Figure 8.23 shows the efficiency losses at a CO_2 removal rate of 90% calculated during a comparative study of the IGCC 98 study. Figure 8.24 shows the efficiency loss as a function of the CO_2 separation rate (Göttlicher 1999). Expectations are that the efficiency will drop by about 8–10% compared to a process without CO_2 capture. Starting out from the IGCC 98 concept (having an efficiency of 52%), the efficiency decreases to 42%. The values have to be seen as target values, as designs of actual plants feature much lower efficiencies because of the selection of proven technology and a low level of integration in order to achieve a high process availability.

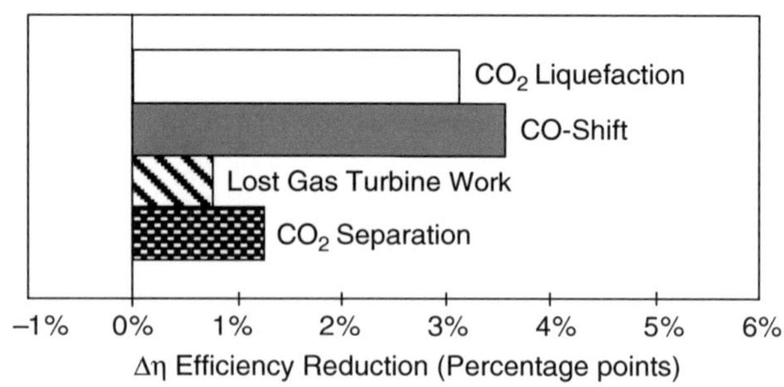

Fig. 8.23 Energy losses due to CO_2 capture from IGCC syngas (Göttlicher 1999)

Fig. 8.24 Effect of the CO_2 capture ratio on the efficiency loss and the specific energy requirement (Göttlicher 1999)

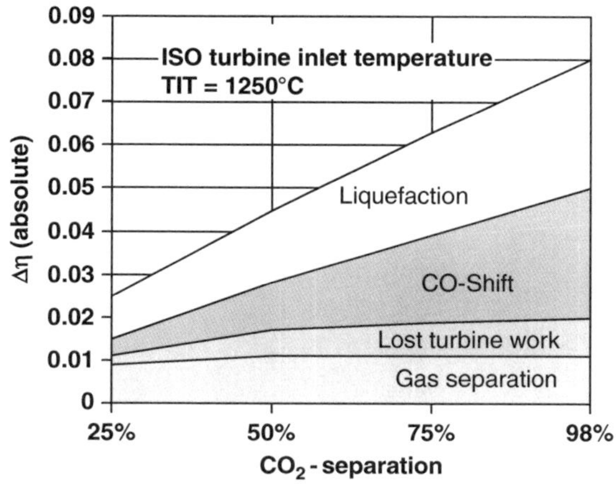

8.8 Comparison of CCS Technologies

A summarising evaluation and a comparison of the different technologies for the separation of carbon dioxide from coal-fuelled power plants is shown in Table 8.4. The IGCC-CCS power plant has a number of advantages in comparison with other technologies, including that it is already available and that it has the highest potential efficiency in the long term. In addition, IGCC-CCS has the flexibility of being able to produce products other than electricity, such as synthesis gas or liquid fuels. However, the main problem of the low process availability of IGCC power plants in operation today still has to be solved, and the costs still have to be proven. Designing IGCC plants for a high reliability will reduce their efficiency at first. The essential advantage of downstream CO_2 scrubbing is the potential of retrofitting it to existing power plants, but this comes at the expense of efficiency. The development of the technology of oxy-fuel combustion is at the stage of pilot-plant demonstration. It could offer a cost-effective method for CCS combined with the reliability of steam power plants; however, costs of an oxy-fuel process are dependent on the CO_2 purity requirements.

In Fig. 8.25, the different technologies are compared with each other in terms of efficiency, capital and operating costs, using brown or hard coal as the fuel

Table 8.4 Comparison of CCS technologies

	Flue gas scrubbing	IGCC	Oxy-fuel
State of the art	Pilot scale	Large scale (with exception of H_2 turbine)	Pilot scale
Potential efficiency costs	−	+	0
	−	+	+
Possibility of retrofit of conventional power plants	+	−	0

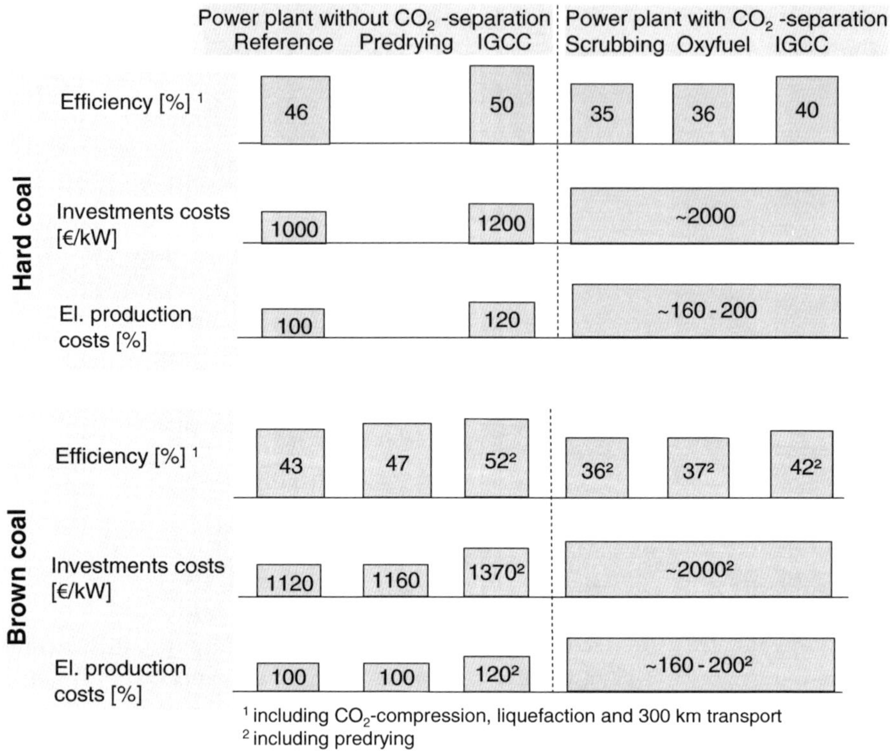

Fig. 8.25 Comparison of costs and efficiencies of CCS technologies

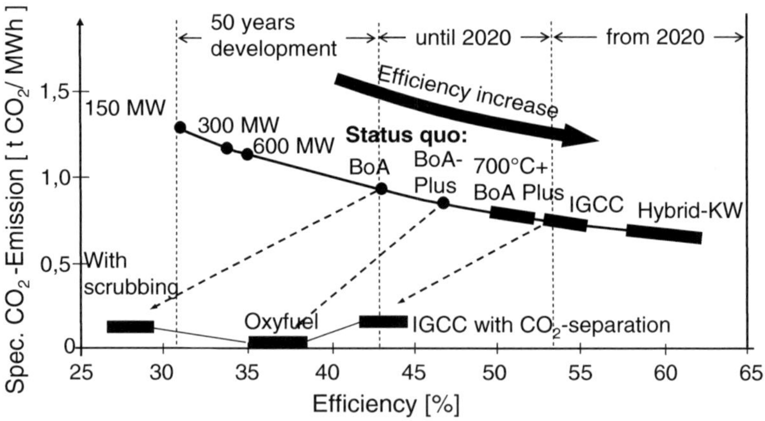

Fig. 8.26 Future improvement in efficiency of various technologies with CO_2 separation using lignite (Ewers and Renzenbrink 2005)

(Ewers and Renzenbrink 2005; Linßen et al. 2006; ENCAP 2009). It is evident that CO_2 removal and liquefaction presently result in efficiency losses of at least 10%. It has to be pointed out that the comparison in Fig. 8.25 can only be indicative, and that any concluding assessment recommending a particular technology as best suited for CO_2 removal is not possible today because of the comprehensive research and development work still required. Large-scale plants have not yet been built. Figure 8.26 illustrates possible developments of power generation by lignite (Ewers and Renzenbrink 2005).

References

Abanades, J. C., Anthony, E. J., Lu, D. Y., Salvador, C. and Alvarez, D. (2004). Capture of CO2 from combustion gases in a fluidized bed of CaO. *AIChE Journal* **50**(7): 1614–1622.

Abanades, J. C., Anthony, E. J., Wang, J. S. and Oakey, J. E. (2005). Fluidized bed combustion systems integrating CO2 capture with CaO. *Environmental Science & Technology* **39**(8): 2861–2866.

Anthony, E. J. (2008). Solid looping cycles: a new technology for coal conversion. *Industrial & Engineering Chemistry Research* **47**(6): 1747–1754.

Bale, C. W., Chartrand, P., Degterov, S. A., Eriksson, G., Hack, K., Ben Mahfoud, R., Melançon, J., Pelton, A. D. and Petersen, S. (2002). FactSage thermochemical software and databases. *Calphad* **26**(2): 189–228.

Becher, V., Goanta, A., Gleis, S. and Spliethoff, H. (2007). Controlled staging with non-stoichiometric burners for oxyfuel processes. Proceedings of the 32nd International Technical Conference on Coal Utilization & Fuel Systems. 10–15 July, Clearwater, FL.

Berguerand, N. and Lyngfelt, A. (2008). Design and operation of a 10 kW(th) chemical-looping combustor for solid fuels – testing with South African coal. *Fuel* **87**(12): 2713–2726.

Blokh, A. G. and Viskanta, R. (1988). Heat transfer in steam boiler furnaces. Washington, DC, Hemisphere Publishing Corporation.

Burchardt, U. and Radunsky, D. (2007). Erfahrungen aus der Planung und Genehmigung der Oxyfuel-Forschungsanlage von Vattenfall. 39. Kraftwerkstechnisches Kolloquium, 11–12 Oktober 2007, Vortrag V 18, Dresden.

Burchhardt, U. and Jacoby, J. (2008). Erfahrungen aus der Inbetriebnahme und erste Eregbnisse der Oxyfuel-Forschungsanlage von Vattenfall. 40. Kraftwerkstechnisches Kolloquium, Dresden.

Cao, Y., Casenas, B. and Pan, W. P. (2006). Investigation of chemical looping combustion by solid fuels. 2. Redox reaction kinetics and product characterization with coal, biomass, and solid waste as solid fuels and CuO as an oxygen carrier. *Energy & Fuels* **20**(5): 1845–1854.

Davidson, R. M. (2007). Post-combustion carbon capture from coal fired plants – solvent scrubbing. London, IEA Coal Research, CCC/125.

DoE (2005). International energy outlook, energy information administration, Department of energy, from www.eia.doe.gov/oiaf/ieo/index.html.

ENCAP (2009). Public summary reports. Retrieved 20th May 2009, from http:// www.encapco2.org/publicsumreports.htm.

Ewers, J. and Renzenbrink, W. (2005). Bestandsaufnahme und Einordnung der verschiedenen Technologien zur CO2-Minderung. *VGB PowerTech* **85**(4): 46–51.

Göttlicher, G. (1999). Energetik der Kohlendioxidrückhaltung in Kraftwerken. VDI-Fortschrittberichte, Reihe 6 Energietechnik Nr. 421. Düsseldorf, VDI Verlag.

Gupta, R., Khare, S., Wall, T. F., Eriksson, K., Lundstrom, D. and Eriksson, J. (2006). Adaptation of gas emissivity models for CFD based radiative transfer in large air-fired and oxy-fired furnaces. 31. International Technical Conference on Coal Utilization and Fuel Systems. Clearwater, FL Coal Technology Association.

Hellfritsch, S., Gilli, P. G. and Jentsch, N. (2004). Concept for a lignite-fired power plant based on the optimised oyxfuel process with CO2 recovery. *VGB Powertech* **84**(8): 76–82.

Hellfritsch, S., Kluger, F. and Bergins, C. (2007). Studie eines braunkohlegefeuerten Oxyfuel-Dampferzeugers. 39. Kraftwerkstechnisches Kolloquium, 11–12 Oktober 2007, Vortrag V 21, Dresden.

IEA (2002). World energy outlook 2002. Paris, IEA.

IEA (2006). World energy outlook 2006. Paris, IEA.

IPCC (2005). IPCC special report on carbon dioxide capture and storage. Working Group Technical Unit, Intergovernmental Panel on Climate Change. Cambridge [u.a.], Cambridge University Press.

Kakaras, E., Koumanakos, A., Doukelis, A., Giannakopoulos, D. and Vorrias, I. (2007). Oxyfuel boiler design in a lignite-fired power plant. *Fuel* **86**: 2144–2150.

Kather, A., Klostermann, M. and Hermsdorf, C. (2007a). Steinkohlekraftwerke mit CO2-Abtrennung auf Basis des Oxyfuel-Prozesses. 39. Kraftwerkstechnisches Kolloquium. 11–12 Oktober 2007, Dresden.

Kather, A., Hermsdorf, C. and Klostermann, M. (2007b). Der kohlebefeuerte Oxyfuel-Prozess. *VGB PowerTech* **87**: 84–91.

Kluger, F., Lysk, S., Altmann, H. and Krohmer, B. (2006). 30 MWth Oxyfuel-Pilotanlage – Untersuchungsschwerpunkte und Auslegung des Dampferzeugers. 38. Kraftwerkstechnisches Kolloquium. 24–25 Oktober 2006, Vortrag V 7.5, Dresden.

Kretzschmar, H.-J., Stöcker, I., Jähne, I., Knobloch, K., Kleemann, L. and Seibt, D. (2005). Software for the calculation of the properties of humid gas mixtures. Zittau, Germany, University of Applied Sciences of Zittau and Görlitz.

Leion, H., Mattisson, T. and Lyngfelt, A. (2008). Solid fuels in chemical-looping combustion. *International Journal of Greenhouse Gas Control* **2**(2): 180–193.

Linßen, J., Markewitz, P., Martinsen, D. and Walbeck, M. (2006). Zukünftige Energieversorgung unter den Rahmenbedingungen einer großtechnischen CO2-Abscheidung und Speicherung. Abschlussbericht des FKZ 0326889.

Lyngfelt, A., Leckner, B. and Mattisson, T. (2001). A fluidised-bed combustion process with inherent CO2 separation: Application of chemical looping combustion. *Chemical Engineering Science* **56**: 3101–3113.

Maier, J., Dhungel, B., Mönckert, P. and Scheffknecht, G. (2007). Combustion and emission behaviour under oxyfuel conditions. 39. Kraftwerkstechnisches Kolloquium. 11–12 Oktober 2007, Vortrag V 20, Dresden.

Mattisson, T., Lyngfelt, A. and Leion, H. (2009). Chemical-looping with oxygen uncoupling for combustion of solid fuels. *International Journal of Greenhouse Gas Control* **3**(1): 11–19.

Pruschek, R. (2002). Elektrizitätserzeugung aus fossilen Brennstoffen in Kraftwerken. In: E. Rebhan (ed.) Energiehandbuch: Gewinnung, Wandlung und Nutzung von Energie. Berlin, Springer.

Pruschek, R. and Oeljeklaus, G: (1992). CO2-Rückhaltung und CO2-Entsorgung Düsseldorf, VDI-Berichte Nr. 1016, pp. 103–124.

Radgen, P., Cremer, C., Warkentin, S., Gerling, P., May, F. and Knopf, S. (2005). Bewertung von Verfahren zur CO2-Abscheidung und -Deponierung Fraunhofer Institut für Systemtechnik und Innovationsforschung (ISI), Karlsruhe und Bundesanstalt für Geowissenschaften und Rohstoffe, Hannover, März 2005, Abschlußbericht F+E-Vorhaben Nr. 203 41 110 des Umweltbundesamtes.

Ritter, R., Holling, B., Altmann, H. and Biele, M. (2007). Konzepte und Ausblick für eine CO2-Anlage eines Oxyfuel-Kraftwerkes am Beispiel Schwarze Pumpe. 39. Kraftwerkstechnisches Kolloquium, Dresden.

Ryden, M., Lyngfelt, A. and Mattisson, T. (2008). Chemical-looping combustion and chemical-looping reforming in a circulating fluidized-bed reactor using Ni-based oxygen carriers. *Energy & Fuels* **22**(4): 2585–2597.

Scheffknecht, G. and Maier, J. (2008). Firing issues related to the oxyfuel process. *VGB Power Tech* **88**(11): 91–97.

Seifritz, W. (1989). Über die Möglichkeiten einer CO2-Entsorgung. VDI-Berichte Nr. 809, pp. 223–245. Düsseldorf.

Sivalingam, S., Gleis, S., Hartmut, S., Yrjas, P. and Hupa, M. (2009). Cyclic carbonation calcination studies of limestone and dolomite for CO2 separation from combustion flue gases. *Journal of Engineering for Gas Turbines and Power* **131** (1): 011801–011808.

Shimizu, T., Hirama, T., Hosoda, H., Kitano, K., Inagaki, M. and Tejima, K. (1999). A twin fluid-bed reactor for removal of CO2 from combustion processes. *Chemical Engineering Research & Design* **77**(A1): 62–68.

Smith, I. M. and Thambimuthu, K. V. (1991). Greenhouse Gases, Abatement and Control: the role of coal. *IEACR/39*. London, IEA Coal Research.

Spliethoff, H. (2006). Advanced steam generator concepts for oxy-fuel processes. Advances in New and Sustainable Energy Conversion and Storage Technologies, IUC Conference. pp. 23–25 September, Grand Hotel Park, Dubrovnik, Croatia.

Tan, R., Corragio, G. and Santos, S. (2005). Technology review – Oxy-coal combustion with flue gas recycle for the power generation industry. International Flame Research Foundation. Report-No. 23/y/1. Velsen, NL.

Tan, R. and Santos, S. (2006). Technology review – Chemical looping combustion for fossil fuel utilisation with carbon sequestration. Report-No. G 23/y/2. Velsen, NL, IFRF.

VGB (2002). CO2 Capture and Storage. VGB report on the state of the art. Essen, VGB Powertech.

Wall, T. F. (2007). Combustion processes for carbon capture. *Proceedings of the Combustion Institute* **31**(1): 31–47.

Index

H. Spliethoff, *Power Generation from Solid Fuels,* Power Systems,
DOI 10.1007/978-3-642-02856-4,